Ecological Studies

Analysis and Synthesis

Volume 240

Ecological Studies is Springer's premier book series treating all aspects of ecology. These volumes, either authored or edited collections, appear several times each year. They are intended to analyze and synthesize our understanding of natural and managed ecosystems and their constituent organisms and resources at different scales from the biosphere to communities, populations, individual organisms and molecular interactions. Many volumes constitute case studies illustrating and synthesizing ecological principles for an intended audience of scientists, students, environmental managers and policy experts. Recent volumes address biodiversity, global change, landscape ecology, air pollution, ecosystem analysis, microbial ecology, ecophysiology and molecular ecology.

More information about this series at http://www.springer.com/series/86

Delphis F. Levia
Editor

Darryl E. Carlyle-Moses • Shin'ichi Iida
Beate Michalzik • Kazuki Nanko
Alexander Tischer
Co-Editors

Forest-Water Interactions

Springer

EXTRAS ONLINE

Editor
Delphis F. Levia
Departments of Geography and
Plant & Soil Sciences
University of Delaware
Newark, Delaware, USA

Co-Editors
Darryl E. Carlyle-Moses
Department of Geography and
Environmental Studies
Thompson Rivers University
Kamloops, British Columbia,
Canada

Beate Michalzik
Institute of Geography
Friedrich Schiller University Jena
Jena, Germany

Alexander Tischer
Institute of Geography
Friedrich Schiller University Jena
Jena, Germany

Shin'ichi Iida
Department of Disaster Prevention, Meteorology
and Hydrology
Forestry and Forest Products Research Institute
Tsukuba, Ibaraki, Japan

Kazuki Nanko
Department of Disaster Prevention, Meteorology
and Hydrology
Forestry and Forest Products Research Institute
Tsukuba, Ibaraki, Japan

ISSN 0070-8356 ISSN 2196-971X (electronic)
Ecological Studies
ISBN 978-3-030-26085-9 ISBN 978-3-030-26086-6 (eBook)
https://doi.org/10.1007/978-3-030-26086-6

This Springer imprint is published by the registered company Springer Nature Switzerland AG.
The registered company address is: Gewerbestrasse 11, 6330 Cham, Switzerland

Foreword

There are now several books that synthesize research on forest and water interactions. These volumes largely consider past research and provide insight for future advances in the field. Notable examples include *Forest Influences* by Kittredge (1948) where the concept of forest hydrology was introduced, and the seminal proceedings volume from the *International Symposium on Forest Hydrology* held at the Pennsylvania State University in 1965 (Sopper and Lull 1967; IAHS 1966) at the beginning of the first International Hydrological Decade (Nace 1964). These volumes and others defined and addressed the perplexing questions in forest hydrology, which set the stage for research over the next several decades. However, by 2010, much had changed in terms of tools, techniques, theory, and approaches that were being used to study forest and water interactions. A major shift also occurred over this period where hydrology and biogeochemistry studies in forested ecosystems became more integrated. Recognizing this, Levia, Carlyle-Moses, and Tanaka published an edited volume called *Forest Hydrology and Biogeochemistry: Synthesis of Past Research and Future Directions*, which established a new synthesis baseline at the interface between forest hydrology and biogeochemistry. They included updated approaches and research themes that were being used to investigate forest and water interactions in the decades since the 1965 symposium. Inclusion of biogeochemistry was a signature of this volume. The book was well-received and is now a go-to resource for graduate students and researchers in forest hydrology and biogeochemistry who are looking to understand what has been accomplished to date and what new research directions and opportunities lie ahead.

Forest-Water Interactions builds on this previous synthesis with a new group of international coeditors, namely, Michalzik, Tischer, Iida, and Nanko, who joined Levia and Carlyle-Moses. This new volume is unique with high-level detail on new methods and statistical analyses that will guide students and seasoned researchers alike and also provides the same robust synthesis as *Forest Hydrology and Biogeochemistry: Synthesis of Past Research and Future Directions*.

Forest-Water Interactions is divided into sections that focus on novel methods, statistical analyses, innovative process-based studies, urban forests, and long-term studies. Much of the book is devoted to new forward-looking approaches rather than a synthesis of past work. It is clear that developments in forest hydrology and ecohydrology have kept pace with technological advances such as the use of unpiloted aerial vehicles, LiDAR, wireless sensors, and numerical analysis and modeling, which provide the focus of the first half of *Forest-Water Interactions*. The book brilliantly includes code written in R language that follows several of the chapters on data analysis and statistical modeling so that readers can use, understand, and adapt advanced techniques easily. Furthermore, the opening chapter of the book emphasizes the importance of open science (data, code, technology, etc.) and its role in advancing hydrology. Process studies are also included, with attention given more to unresolved or poorly understood processes such as throughfall partitioning, leaf water repellency, and stemflow contributions to preferential infiltration. Also, compared to other texts and syntheses in forest hydrology, a unique perspective on urban forest ecosystems is introduced where the editors have selected chapters that outline various influences urban trees and canopies have on hydrological processes. This is a welcomed section given the growing interest in urban ecosystem management and integration of concepts from forest ecosystems into built environments and green infrastructure. Urban planning, management, and site design are rapidly expanding utilization of ecohydrological services provided by urban trees as a solution to urban water problems. Chapters in this section not only address water quantity issues such as stormwater mitigation but also discuss how urban tree canopies affect water quality. The last section of the book builds on analyses from long-term experimental watersheds, which have been a hallmark of forest and water interaction research, addressing the value of what can be learned from intensive, long-term, place-based science in hydrology. The book concludes with a rich synthesis on forest and water interactions under global change that was adapted from an International Union of Forest Research Organizations (IUFRO) report from scientists who were part of the Global Forest Expert Panels on Forests and Water (Creed and van Noordwijk 2018). The chapter offers insightful views on the effects of global change on forest-water interactions and water availability to ecosystems and people. The global perspective, human dimension, and future outlook make this the perfect culmination to the book. The authors of this chapter and throughout the book demonstrate diverse contributions from around the world in the research of forest and water interactions that provides a blueprint for future research direction.

Similar to the early synthesis volumes in forest hydrology by Kittredge (1948), Sopper and Lull (1967), and Swank and Crossley (1988), *Forest-Water Interactions* presents contemporary research and synthesis with a focus on opportunities for advancing forest hydrological research and application. By including topics such as novel environmental sensing methods, statistical analyses with supporting computer code, and a section on urban forest-water interactions, this book is distinctive and will be an important resource for researchers moving forward.

Department of Forest Resources and
Environmental Conservation and
Virginia Water Resources Research Center
Virginia Tech, Blacksburg, VA, USA
May 2019

Kevin J. McGuire

References

Creed IF, van Noordwijk M (eds) (2018) Forest and water on a changing planet: vulnerability, adaptation and governance opportunities. A global assessment report. IUFRO World Series, vol 38. Vienna

IAHS (1966) International symposium on forest hydrology: final report. Bull Int Assoc Sci Hydrol 11:161–170

Kittredge J (1948) Forest influences. McGraw-Hill, New York

Levia DF, Carlyle-Moses D, Tanaka, T (eds) (2011) Forest hydrology and biogeochemistry: synthesis of past research and future directions. Ecological studies series, vol 216. Springer Science & Business Media, Dordrecht

Nace RL (1964) The international hydrological decade. Eos, Trans AGU 45(3):413–421

Sopper WE, Lull HW (eds) (1967) International symposium on forest hydrology. In: Proceedings of a National Science Foundation advanced science seminar held at the Pennsylvania State University, University Park, Pennsylvania, Aug 29–Sept 10 1965

Swank WT, Crossley Jr DA (eds) (1988) Forest hydrology and ecology at Coweeta. Ecological studies series, vol 66. Springer Verlag, New York

Preface

It is an important time for hydrology. Some parts of the world are suffering from extreme floods, others severe drought, and yet others from a lack of potable water that exact a terrible toll on humankind. These water problems are exacerbated by changes in human populations and human systems and the vagaries and complexities of global environmental change. Accordingly, the International Association of Hydrological Sciences has designated 2013–2022 as the scientific decade of "Panta Rhei-Everything Flows." Recognizing the inextricable linkages between water and society, the overarching aim of Panta Rhei is to encourage hydrological research that leads to a better understanding of the role of changes in the processes constituting the hydrologic cycle in relation to the human domain (see https://iahs.info/Commissions--W-Groups/Working-Groups/Panta-Rhei.do). In addition, a declaration from the United Nations has selected 2018–2028 as the International Decade for Action on Water for Sustainable Development (see https://www.un.org/en/events/waterdecade/). As the central underpinning of our book, it is our contention that the hydrologic community can conduct better science and make a more meaningful impact to the world's water crisis if hydrologists are (1) better equipped to utilize new methods and harness big data from either or both high-frequency sensors and long-term research watersheds and (2) aware of new developments in our process-based understanding of the hydrological cycle in both natural and urban settings.

The hydrologic sciences have entered a new era. This era will be dominated by automated high-frequency sensing, UAVs, open source code, and big data. It is important to note these fundamental wholesale shifts in the very nature of hydrologic inquiry are changing and will change the way hydrologists and those of allied disciplines work. The work of CUAHSI (see https://www.cuahsi.org/) and the Critical Zone Observatories (CZOs) (e.g., for the USA, see http://criticalzone.org/national/; for Europe, see https://esdac.jrc.ec.europa.eu/projects/critical-zone-observatories) will likely play ever more important roles in the hydrologic sciences. Of course, field data and modeling work will still be important. In fact, with well-designed experiments that leverage the strength of new sensing technologies to gain unprecedented insights into hydrological processes and advances in data science that make the data explosion meaningful, field work and modeling work will continue in our data-rich environment in a synergistic manner. It is to the harnessing of this technological and data science revolution in a manner that leads to consequential science for the improvement of the human condition that this book is devoted.

To best serve the scientific community and provide a base from which to leverage fruitfully all of the aforementioned changes and developments in and around the hydrologic sciences, *Forest-Water Interactions* is divided into six parts. Part I, entitled "Novel Methods: Environmental Sensing Opportunities for Forest-Water Interactions," opens with a chapter on open-source ecohydrology to avail readers of the wide array of open-source options for hydrologic research and its potential ability to transform our understanding of hydrological processes. Other chapters in Part I focus on the importance of sensor calibration and the use of unpiloted aerial vehicle (UAV) technology, LiDAR, and isotopes to improve our knowledge of forest-water interactions. The last chapter in Part I highlights the important roles of instrument development and operational environmental sensing to the practice of hydrologic science that can make a positive impact on humankind. Part II, "Analyzing and Synthesizing Ecohydrological Patterns and Functional Relationships with Statistics," is intended to serve as a "statistical cookbook" so readers can better harness data transformations, spatial and temporal statistics, Bayesian statistics, and machine learning to help advance hydrologic science. R code is supplied to accompany the data sets for chapters in Part II to aid readers in employing these techniques in their own research. Part III, entitled "Synthesizing Process-Based Understanding of Forest-Water Interactions: Some Recent Advances," is intended to provide readers with an up-to-date insight and areas fertile for new exploration in some areas of forest-water interactions research where there has been a substantial increase in knowledge. Some specific chapters are devoted to leaf water repellency, root-water interactions, and radiocesium cycling. Part IV, "Forest-Water Interactions in Urban Ecosystems," summarizes the burgeoning growth of forest-water interactions research in cities and identifies avenues for further research. Individual chapters focus on hydrological, biogeochemical, and modeling work. The last chapter in the section examines the application of planning principles to the management of stormwater runoff and water quality in urban areas. With research from Hubbard Brook Experimental Forest and long-term research sites in Japan and Ecuador, Part V, "Forest-Water Interactions over the Long Term: Analysis and Synthesis from Several Experimental Watersheds," illustrates the utility and importance of long-term data in analyzing the role of human activities or disturbances (natural and anthropogenic) on temporal changes in hydrologic regimes at the watershed scale. Finally, Part VI, "Final Synthesis: Future Forest Hydrology" (consisting of the last chapter), serves as a conclusion to the book and focuses on future forest-water interactions in the Anthropocene, thereby reminding us of the interconnected nature of water and humankind in the spirit of Panta Rhei and the Earth's future.

We gratefully acknowledge the enthusiasm, dedication, and Herculean work ethic of all the chapter authors who made this book possible. The quality of the chapters is excellent and we owe them a debt of gratitude. The authors who served as peer reviewers for chapters are thanked as well since the careful and judicious comments of the peer reviewers surely improved the quality of the chapters and book as a whole. We further acknowledge the work of external scientists who reviewed chapters within the book (listed alphabetically): Martin De Kauwe, Michael Dietze, Florian Hartig, Jürgen Knauer, Ken'ichirou Kosugi, and Takahisa Mizuyama. Kevin McGuire is thanked for writing the book's Foreword. Rob Jackson, the Ecological Studies Series Editor handling this volume, is gratefully acknowledged for his comments that have improved the quality of the book as well. We also thank Andrea Schlitzberger of Springer's Ecological Studies Series for her dedication to our book and the Ecological Studies Series. Production Editor James Alexander is thanked for careful oversight of the book's production. Finally, we wish to recognize and thank Robert Doe, Springer Publishing Editor, for his moral support and enthusiasm for our book. Editor Levia fondly remembers initial discussions about this book with Andrea Schlitzberger and Robert Doe in Heidelberg; the ideas generated in that meeting led to the successful launch of this book project.

We live in a highly interconnected world endowed with a tremendous amount of intellectual capital. Leveraging the global knowledge hub of renowned hydrologists and rising stars in forest-water interactions, this book represents the ideas and talents of many individuals who have managed to operate effectively as a whole. It is our genuine collective conviction that this book will be valued by researchers for years to come and serve as a base from which more thoughtful, innovative, impactful, and consequential research on forest-water interactions is realized.

Newark, DE, USA	Delphis F. Levia
Kamloops, British Columbia, Canada	Darryl E. Carlyle-Moses
Tsukuba, Ibaraki, Japan	Shin'ichi Iida
Jena, Thuringia, Germany	Beate Michalzik
Tsukuba, Ibaraki, Japan	Kazuki Nanko
Jena, Thuringia, Germany	Alexander Tischer
May 2019	

Contents

Part I
Novel Methods: Environmental Sensing Opportunities for Forest-Water Interactions

Chapter 1
Cracking "Open" Technology in Ecohydrology

B. Turner, D. J. Hill, and K. Caton

1.1 Introduction

In recent years, the adjective "open" has been applied to many aspects of scientific knowledge discovery and dissemination, including open-source software and hardware, open access journal articles, massive open online courses, and open data. Applying the term open to these entities emphasizes an intention for them to be accessible—a quality increasingly emphasized as desirable in science. But there is nothing new about the idea of "open science." Science shed its character of secrecy in sixteenth-century Europe, when the wealthy players in the patronage system that was increasingly supporting scientific inquiry at that time found it in their interest for the scientists they supported to go public with their discoveries, as a form of ornamental display to enhance the patrons' own reputations and power (David 2004). Around this time, scientific discovery also became so complex and mathematically grounded that the patrons could no longer assess quality on their own; fearing charlatanism, patrons began encouraging systems of peer vetting, in which scientists would open their work to other scientists for the purpose of review and confirmation (David 2004). Clearly, the historically instituted practices of peer review and experiment replication are still followed in the current era, so what is

The original version of this chapter was published non-open access. It has now been made open access at the Editor's request. The correction to this chapter is available at https://doi.org/10.1007/978-3-030-26086-6_25

B. Turner · D. J. Hill (✉)
Department of Geography and Environmental Studies, Thompson Rivers University, Kamloops, BC, Canada
e-mail: dhill@tru.ca

K. Caton
Department of Tourism Management, Thompson Rivers University, Kamloops, BC, Canada

D. F. Levia et al. (eds.), *Forest-Water Interactions*, Ecological Studies 240,
https://doi.org/10.1007/978-3-030-26086-6_1

different about science today that has led to the emergence of "open" as a buzzword for so many initiatives?

Whereas in the Renaissance, the new and esoteric language of mathematics drove the requirement of peer review, today, the turn toward computational and data intensive science in ecohydrology is driving new mechanisms for enabling replication, validation, and extension of research results. At the same time, governments are increasingly demanding that the results from publicly funded research be released into the public domain (e.g., the United States Fair Access to Public Research Act) to return knowledge to the taxpayers who funded its creation. Furthermore, the specter of cataclysmic consequences of climate change is increasing calls for more efficient collaboration around this shared global problem. Just as climate change recognizes no private ownership boundaries, so too, it is argued, must environmental knowledge come to be understood as a shared, open resource to combat this and other threats to the global common good.

This chapter explores what it means for technology to be open and how open technology is transforming the field of ecohydrology today. Starting from the concept of open science, the next section explores the open science ideals of open source, open method, open data, and open hardware and develops a consistent definition of open technology. Based on this definition, a review of open technology applications and development within the field of hydrology is presented that categorizes technology into truly open and quasi-open and discusses how this technology is enabling hydrologic research. The chapter then concludes with a discussion of the potential of open technology to advance the field of ecohydrology.

1.2 Contextualizing Open Technology

1.2.1 The Philosophy of Open Movements

Open science is driven by at least two distinct philosophical agendas: one centered on scientific efficiency and the other on humanistic principles. In the case of the first, advocates argue that scientific processes can be optimized through openness (Fecher and Friesike 2014). Open data makes way for reanalysis of one's work by other researchers and permits comparisons between different researchers' methods for prediction and analysis. Open hardware can be leveraged for new projects, rather than scholars having to reinvent the wheel. Open access to scientific literature allows a larger pool of scientists to correctly spot the research frontiers of a given discipline and maximizes the amount of researcher skill and perspective that can be brought to bear on particular problems. By increasing the number of expert eyes on a research problem and reducing redundancy of effort, the overall scientific enterprise advances further.

In the case of the humanistic agenda driving open science, advocates assert that knowledge is fundamental to human development. Some working within this perspective focus on the importance of facilitating direct citizen participation in research efforts (i.e., "citizen science"), arguing that such involvement can demystify the process of scientific knowledge creation, empower citizens at problem-solving, promote engagement with the natural world, and create appreciation for the scientific endeavor (Bonney et al. 2014; Cohn 2008). Others focus on the importance of citizen access to research results—either in traditional academic form or in adapted formats designed for a nontechnical audience. Arguments for open access to academic outputs include the ethical stance that, because much research is funded by taxpayers, it rightfully belongs to them; when public libraries must purchase subscriptions (to journals and so forth) to make research available to readers, taxpayers effectively "pay twice for research," once to support its production and once to support public purchase of access to its results (Phelps et al. 2012). They also include the moral stance that knowledge should not be used to exacerbate a world of haves and have-nots. Scientific knowledge is integral to human development and well-being in the twenty-first century, and its power must not be taken lightly (Fecher and Friesike 2014). When it comes to scientific knowledge about the environment—such as in the study of water resources—an additional case can be made based on the object of study's status as a common good. In economic theory, a common good is one that is of a shared, public nature, such that individual ownership rights cannot be assigned, but also of a nature that one person's use of that good can reduce another person's ability to use it. Environmental resources are common goods *par excellence*, with all needing to partake of them in order to survive, but all (and some more than others) having the ability to affect them in ways that make them less usable to others.

In this chapter, we take the position that, while scientific efficiency is certainly valuable, the moral imperative posed by the reality of water resources as a global common good, and the power of knowledge to enable their better management, is sufficient to champion openness as a goal for the future of hydrologic science. In taking this position, however, we are aware that complete openness in science faces many implementation obstacles. Thus, we invoke openness as a value statement—a principle to work toward—while recognizing that pragmatic compromises will often make sense in pursuit of more open ends.

1.2.2 The Open-Source-Software Movement

The open-source movement, originating in the 1990s, is arguably the most widely known open movement. The name of the open-source movement was selected to distinguish it from the earlier free software movement, which originated in the 1980s. The free software movement seeks to protect the freedom of computer users to run, study, change, and redistribute copies or derivative products of software (Stallman n.d.). The free software movement espouses a strong moral responsibility to end the creation and use of proprietary software through the creation and adoption of alternative free software applications. Free software, as defined by the free software movement, is software that gives the user freedom to distribute copies of the software, access the source code of the software, change the software, or use parts of the software in new free software applications. Free software does not, however, necessitate that the software have no cost (GPL v3). To protect these freedoms, free software is distributed under a free software license, such as the GNU[1] Public License (GPL, https://www.gnu.org/licenses/gpl.html).

The open-source software movement emerged out of the free software movement due to fundamental differences in philosophy. The underlying philosophy behind free software is that non-free software is a social ill, whereas the philosophy behind open-source software development is that it generates a better solution by involving a broader developer base than non-open-source-software. In practice, what this means is that software developed as open-source software may include components or dependencies, which constitute proprietary software. Thus, while free software is accessible to anyone, open-source software is accessible only to those who possess any proprietary software necessary to run the open-source software. For example, a plugin developed for a proprietary software application, such as a geographic information system (GIS), could be developed as an open-source software project, whereas such a project could not be developed as a free software project due to the requirement of the proprietary GIS.

The goals of the open-source software movement are to motivate and support a community of like-minded software developers to collaboratively create and improve openly available source code to create better software solutions to computational problems. To accomplish these goals, the open-source software movement has established a set of commonly accepted practices surrounding the sharing, use, and distribution of software source code that is defined by the Open Source Definition (OSD) and maintained by the Open Source Initiative (OSI). The fundamental tenets of open-source software are that software developed under this paradigm is released in a manner which allows for it to be redistributed without discrimination, without charge or royalty, and that derivative works may be produced (Open Source Initiative 2018). To protect the freedoms of open-source software users and the rights of open-source software developers, open-source software is distributed under an open-source license such as the Apache License (http://www.apache.org/licenses/) or the Mozilla Public License (MPL, https://www.mozilla.org/en-US/MPL/) or under a free software license such as the GPL.

Despite the philosophical differences between the free and open-source software communities, which revolve around the motivation for sharing source code and derivative software products, the software products developed as part of the open-source software and free software movements are very similar in terms of their ability to be run, studied, modified, and redistributed, as well as in the terms of user freedoms and developer rights that are protected by the various open or free software licenses. In fact, all free software can be considered open-source software, while most, but not all, open-source software qualifies as free. To describe software that has the qualities of both free and open source, regardless of the underlying philosophy of the developer, the acronym FOSS (free and open-source software) is used.

---

[1]GNU is a recursive acronym meaning "GNU's not Unix."

Because these software are free, they must adhere to the more restrictive philosophy of free software, which eschews non-free software dependencies. Thus, FOSS provides access to the software to run, study, modify, and distribute copies or derivative products in accordance with FOSS licenses. Within FOSS communities, there is a wide array of licenses designed to protect authors and enshrine legal frameworks to enforce specific principles. Table 1.1 provides a listing of several commonly used licenses, their abbreviations, and a brief description of their unique attributes. Some FOSS licenses, such as GPL (https://www.gnu.org/licenses/gpl.html), are designed such that all subsequent work built upon the original licensed work must also inherit the same licensing term (sometimes derogatorily termed a "viral" license). This style of licensing differs from the more permissive licenses, such as the MPL mentioned previously, which permit derivative works to implement restrictions (including patents). Fundamentally, it is a matter of the developer's preference as to whether an open-source software project is developed as FOSS or simply open source. It is necessary to be aware of the differences between the two when leveraging previously published work.

Notable FOSS projects relevant to ecohydrology include Linux,[2] R, QGIS, GRASS, MapServer, OpenLayers, GDAL/OGR, PostgreSQL, and Hadoop. The Linux kernel is a UNIX-like environment that is maintained by the Linux Foundation and distributed under the GPL. R is a statistical computing environment that is maintained by the R Core Team and distributed under the GPL. QGIS[3]and GRASS[4] are desktop GISs that are maintained by the Open Source Geospatial Foundation (OSGeo) and released under the GPL. MapServer[5] is a server-side GIS that is maintained by OSGeo and distributed under the X/MIT license. OpenLayers[6] is a JavaScript library for displaying maps in Web browsers. OpenLayers is maintained by OSGeo and distributed under the Berkley Software Distribution (BSD) license. The geospatial data abstraction library (GDAL) and OGR simple features library[7] are software libraries for the manipulation of vector and raster spatial data features that are maintained by OSGeo and distributed under the X/MIT license. PostgreSQL[8] is a database management system released under the PostgreSQL License. Hadoop[9] is a suite of utilities that enable the creation of distributed computing architectures. Hadoop is maintained by the Apache Software Foundation and released under the Apache License.

---

[2]https://www.kernel.org/

[3]https://www.qgis.org/

[4]https://www.grass.osgeo.org/

[5]https://www.mapserver.org/

[6]https://www.openlayers.org

[7]https://www.gdal.org/

[8]https://www.postgresql.org/

[9]https://www.hadoop.apache.org/

Table 1.1 Commonly used free and open licenses for software, hardware, and data

General		
License name	Abbreviation	Summary
Mozilla public license	MPL	The MPL is a license that straddles a middle ground between the GNU public license and the more permissive free licenses such as MIT. It allows for selected portions of software code licensed under it to be proprietary
GNU public license	GPL	The GPL is a copyleft license which requires all works which build upon it to inherit its licensing terms. It enshrines the rights to freedom to use, modify, and redistribute technology and creative works which use it
Massachusetts Institute of Technology license	MIT	The MIT license is among the most popular of all free and open source licenses. This popularity is due to its permissive nature, which makes software licensed under it compatible with numerous other FOSS licenses. It offers an allowance for the integration of proprietary and copyrighted material in the scope of a project
Creative commons license	CC CC0 CC BY CC BY SA	The creative commons license is an umbrella under which a number of variant licenses exist. It is designed to allow creators the ability to control how their work is shared. Only three sublicenses within the creative commons family are recognized as truly open: Creative commons zero (which is a public domain license), creative commons attribution, and creative commons attribution share alike
Hardware specific		
License name	Abbreviation	Summary
Tucson amateur packet Radio corporation (TAPR) open hardware license	TAPR OHL	Both these licenses are copyleft and specifically tailored to licensing hardware. They do not prohibit the creation of derivative, patented works but do protect original creators from legal action
European Organization of Nuclear Research (CERN) open hardware license	CERN OHL	
Data specific		
License name	Abbreviation	Summary
Open data commons public domain Dedication and license	PDDL	The PDDL license is designed for use with databases and enshrines the contents of the database as public domain, waving all rights
Open data commons open database license	ODbL	The open database license enshrines freedom to share contents, create new works, and adapt information within a database, on the condition that all subsequent works share the same license, that they remain open, and that the original author is attributed

1.2.3 Open-Source Hardware

The open-source hardware movement builds upon the principles of open-source software, extending them to the material world. It leverages developmental approaches similar to those utilized in the open-source software community to create, share, modify, and reuse designs and processes related to objects ranging from electronics to industrial machinery. The movement has roots in the amateur radio and homebrew computing communities common in the 1970s and 1980s. These groups of enthusiasts shared designs and expertise to facilitate development of hardware and propagate understanding. The definition of open-source hardware is curated by the Open Source Hardware Association (OSHWA) and is based upon the OSD.

The open-source hardware movement is built upon sharing the knowledge necessary for the design and construction of hardware. The movement's motivations mirror those of the open-source software movement, wherein it both facilitates and encourages the creation, use, distribution, and/or modification of community-developed resources. Differences between the two arise from the material nature of hardware. With open-source hardware, the expectation is that all involved software as well as original design files, instructive documentation, and component lists necessary for reproduction are included with the project's public release and that each possesses a compliant open license (Ackermann 2009; Open Source Hardware Association n.d.). Open-source hardware is also expected (though not necessarily required) to be created from readily available, standardized components (i.e., common fasteners, mass-produced integrated circuits), which ensures that open-source hardware users can replicate or adapt designs in an affordable manner. While comparable to open-source software in many ways, open-source hardware is incompatible with the FOSS movement due to the use of non-free and proprietary components and designs. Open-source software has the luxury of reinventing necessary proprietary software, thanks to the inherent portability and accessibility of computer code. The legal, logistical, and financial barriers of reinventing necessary technologies in the physical world afford the open-source hardware movement much less flexibility when creating 100% open solutions.

The goal of the open-source hardware movement is to develop products, platforms, tools, and devices that facilitate freedom to control technology, share knowledge, and openly exchange designs (Open Source Hardware Association, n.d.). The open-source hardware movement seeks to enable communities of non-expert end users to create and adapt technology which they may not otherwise feel capable of interfacing with. An intended byproduct of the movement is to propagate understanding of how common hardware functions. The long-term objective of the open-source hardware movement is to improve quality of life and the environment globally, by means of the innovation and dissemination enabled by the open-source hardware philosophy. This philosophy enables the creation and recreation of universally accessible devices that are integral to modern society.

Given the hobbyist roots of the open-source hardware movement, as well as its relatively recent mainstream emergence, there are few widely known projects at this time. Among the most recognizable are the Arduino[10] and the RepRap[11] projects. Arduino is a brand associated with a range of microcontroller-powered electronic boards. Initially envisioned as a tool for designers with no prior electronics experience, it has since become a mainstay of science, technology, engineering, and math (STEM) education, Internet of Things (IoT) project development, low-cost environmental sensing platforms (Fisher and Gould 2012; Prescott et al. 2016), and do-it-yourself lab equipment (Koenka et al. 2014; Pearce 2012). The RepRap project is a community developed around the creation of affordable and accessible 3D printers capable of recreating themselves.

1.2.4 Open Data

The open data movement originated in the mid-1990s, growing from the tradition of knowledge sharing in science. The term "open data" was first applied to represent the notion of sharing data in the geosciences to facilitate a holistic understanding of the biosphere which "transcends borders" (National Research Council 1995). The modern understanding of open data emerged in 2007, when leaders from both the free software and open-source software movements established the principles of open access and public ownership, which permitted open data to expand beyond the scientific community (Chignard 2013). Their work and advocacy represented the first push toward open public data (Chignard 2013). Building upon the base principles set forth in 2007, several organizations have asserted competing definitions of open data. For this discussion, the Open Definition (www.opendefinition.org), which is maintained by the Open Knowledge Initiative (OKI), will be adopted to define open data. This definition was selected to guide this discussion due to its emphasis on the freedoms associated with open data, which, in some ways, parallel FOSS, and due to its adoption by the Government of Canada (the jurisdiction in which the authors of this chapter conduct research) for its open government initiative (Government of Canada 2017).

The core philosophy of the modern open data movement can be traced back to Mertonian philosophies of communalism (and, therefore, communal knowledge sharing) as an inherent character of the scientific pursuit (Merton 1973). The movement seeks to establish and maintain a commons in which data can be freely accessed and utilized. Unlike the usage of "free" when referring to free software, free in the context of open data implies both freedom of use, a trait of free software, and freedom from cost, not necessarily a trait of free software. The open data movement is motivated by two primary concerns. The first is the need to release data,

[10]https://www.arduino.cc/

[11]https://reprap.org/wiki/RepRap

particularly government data, as a means of democratizing knowledge and enabling stakeholder participation. Several governments and governing bodies have made commitments to open data in recent years, including the United States, Canada, and the European Union. The second is the realization that, for further progress to be made in science and for it to benefit the greatest number of people, free access to usable data is critical. With these motivations in mind, open data is meant to be an invitation for any individual, academic or otherwise, to investigate and contribute to understanding the massive volumes of information accumulated in the Internet age.

The goal of the open data movement is to have all relevant information stored in a manner which allows for it to be easily and freely retrieved in a nonproprietary, open, machine-readable format. The open data movement expects that, given time, global access to truly open data will enable better science (Heidorn 2008; Molloy 2011). The objective of open data, however, cannot yet be fully realized due to costs associated with data storage and transmission. Thus, the open definition implements a pragmatic stance by stating that a reasonable one-time reproduction fee may be charged. For example, a data consumer may be asked to pay for the services of an archivist when requested data must be retrieved from a physical, non-networked archive.

Creators of open data are encouraged to release their work under an established open data license such as Creative Commons[12] or Open Data Commons,[13] when sharing it with the public (Open Data Handbook n.d.). Often, data creators are inclined to release their work into the public domain unlicensed believing that it will be accessible to all. Being unlicensed in the public domain is anathema to open data practices, however, as it is likely to cause confusion over who owns the data, especially once time has passed since publication. An analogy would be to place a stack of money in a public space with the expectation that everyone will be willing and able to take from it. Without indication that it is free for everyone, many will avoid it fearing that they are stealing, while others may take it and claim that it was theirs all along. Placing a sign with the cash stating that it is free for all to take and that it is a public good removes the confusion that the unexplained stack of money would cause. This illustrates the value of an open license: it does not alter the spirit of the data shared, but it does enshrine its availability for the greatest number of users.

Within ecohydrology research, leveraging and/or creating open data is not uncommon. Notable examples of organizations working to curate open datasets for the discipline include the Consortium of Universities for the Advancement of Hydrologic Sciences, Inc. (CUAHSI) and Data Observation Network for Earth (DataONE), as well as government-sponsored data providers such as Natural Resources Canada. CUAHSI encourages the publication of data through its Web portal, making it searchable for other interested users. DataONE hosts a series of coordinating server nodes, which index several distributed data servers, making them searchable and allowing the data to reach end users more easily. Natural

[12]https://creativecommons.org

[13]https://opendatacommons.org/

Resources Canada offers data related to geomorphology, geology, and hydrology which has been gathered through publicly funded research and institutions. Outside of ecohydrology, exemplars of open data in action include the Human Genome Project and Statistics Canada.

Two important ecohydrology data portals that do not meet our definition of open also deserve mention, the FLUXNET and the long-term ecological research network (LTER) portals. Both of these portals offer a two-tier licensing system. The tier 1 FLUXNET data are distributed under a custom fair use license, which requires that the authors of the data be notified of who is using the data and for what purposes the data will be used before the data are accessed, and that an appropriate citation or acknowledgment be made in all published work that uses the data. The LTER tier 1 data, on the other hand, are distributed under a Creative Commons BY ATTRIBUTION license. Giving credit to data owners through appropriate citations to datasets is an important tenant of professional research ethics that is recognized by the open technology movement and enshrined in open data licenses, such as those applied to the FLUXNET and LTER datasets. However, there is an important difference in how these two datasets are served to potential users. The FLUXNET dataset license requires that the potential users notify data owners before the data can be downloaded and implements this rule by requiring all potential users to create a FLUXNET account which will be used to connect the potential user to the data owner and communicate the intended use of the data. On the other hand, tier 1 LTER data can be downloaded anonymously by clicking a weblink to the datasets. It could be argued that the requirement of notifying the data owners before using the data could be considered to fetter access to the data, and thus the FLUXNET dataset would not meet our definition of open data. Furthermore, tier 2 data in both portals are restricted for use only by select individuals, and tier 2 status supersedes tier 1 status, so data requests that include both tier 1 and 2 data (even if they are from the same instrument at the same location) are considered tier 2. For these reasons, these portals do not meet our definition of open data portals.

1.2.5 What Is Open Technology

In this chapter, technology is defined broadly as a knowledge-based output that can be put to use toward a specific purpose. This definition encompasses such things as computing hardware, software, electronic data repositories (particularly those that are Web-enabled), and Web services. Thus, open technology can be seen as a superset of the type of products developed through the free or open-source software, open hardware, and open data movements.

1.2.6 Trends in Open Technology in Ecohydrology

Figure 1.1 illustrates the growth of "open technology" in the scientific literature since the year 2000. The figure shows two distinct periods when open technology research accelerated. The first period is between 2002 and 2009, followed by a brief plateau ending in 2012. The second phase of growth in open technology research happened between 2012 and 2016.

Data illustrated in Fig. 1.1 were gathered using a keyword search on the Web of Science index, which includes the Science Citation Index (SCI-EXPANDED), Conference Proceedings Citation Index–Science (CPCI-S), and the Emerging Sources Citation Index (ESCI) from 2000 to 2018. Searches were constrained by a wildcard involving "∗hydrolog∗" plus one of four key phrases: "open source," "open hardware," "open source hardware," and "open data." The four key phrases collectively contribute to the total open technology results gathered. The wildcard was used to capture results from the maximum number of hydrology-related topics. A secondary search was completed, which was not constrained to the hydrologic sciences, using the same four key phrases and date range. This secondary search determined the trend of open technology across the entirety of the selected Web of Science indices. The results from each search were grouped as "Hydrologic" and "All," respectively, for the purpose of analysis. Each data point represents the

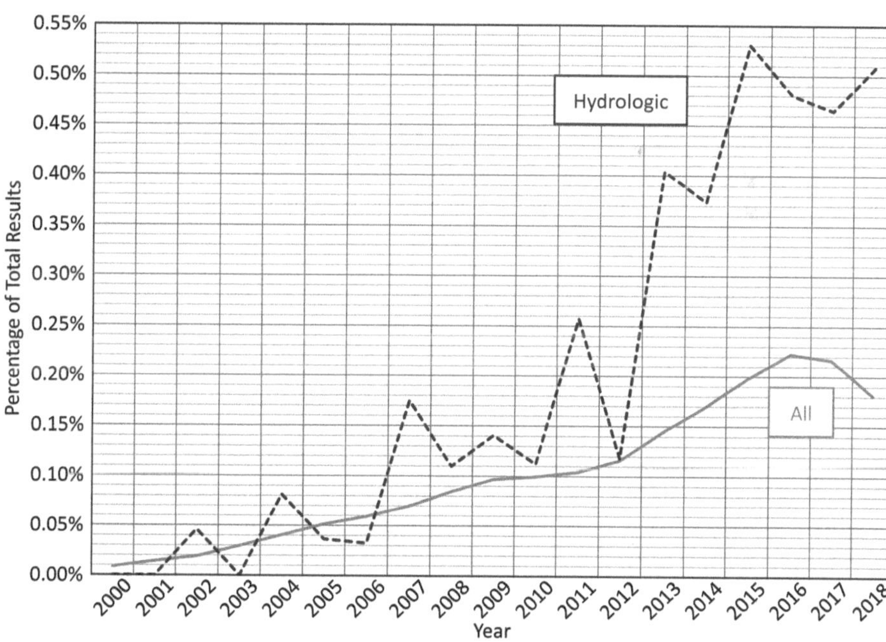

Fig. 1.1 Open related sources as a percentage of total returned results

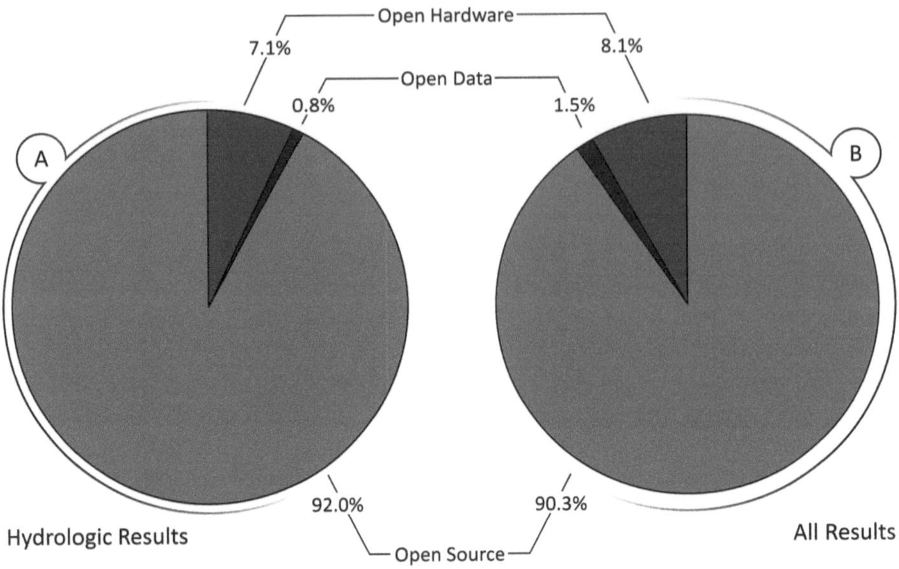

Fig. 1.2 Displays (**a**) the share of each aspect of open technology within the open results from the hydrology-relevant literature and (**b**) the share of each aspect of open technology within the open results from all open literature

percentage of the total returned results that were open technology related against the sum of results over the period unconstrained by "open" keywords (total papers indexed).

Figure 1.1 indicates a clear overall upward trend in the adoption of open technology in the literature. As a percentage of the returned results, hydrology appears to be slightly ahead of the curve when compared to the entirety of the searched indices. However, due to the comparatively insignificant number of results matching open technology terms in the body of hydrology research, as well as a lack of additional constraints on the broad search of Web of Science, it is not possible to state that it is indeed more common within the discipline.

Figure 1.2 illustrates the distribution of the various components of open technology over the period of 2000–2018 for both groups of search results. Figure 1.2a, b demonstrates a similar distribution, with open source being the most dominant paradigm present in the open technology literature. This dominance is both a product of its relative maturity, when compared to all other open movements, as well as the increasing adoption of open-source software packages as de facto tools across a variety of disciplines. Much of the open data and open hardware research illustrated in these figures has emerged in the last 5 years.

The outlook for open technology over the next decade is uncertain and will depend on adoption of open ideals by developers, adoption of open technology by users, and advances in computing and network technology. In the case of open-source software, the largest component of open technology, the advances that researchers make are being drawn into, and developed within, large existing open-source projects, to increase their reach and capability. Examples of this include the development of plugins which interface common modeling software (such as SWAT[14]) with open-source geographic information systems. This will inevitably lead to a slow reduction in the number of open-source articles published over time, as there is a finite number of plugins that need creation. Open data is likely to increase, as the need for large, well-documented, and publicly available data becomes more pressing. Its growth will also be aided by the requirements of public institutions, such as the National Science Foundation of the United States, that require all research funded with public money to be publicly accessible. Open hardware is slowly gaining momentum in mainstream research. In the near future, we expect it will be a very minor contributor to open technology within ecohydrology. Many of the applications of open hardware which are relevant to the discipline are not demonstrated within the literature, but instead are found in spaces of amateur or citizen science or in grey publications; however, this may change as hydrology journals offer increasing opportunities to publish the results of open technology development, for example, as "methods papers."

1.3 Open Technology in Review

Open technology has been applied to a variety of topics of interest in the field of hydrology. A literature review was completed to highlight many of the mainstream applications exhibiting the ideals and promise of open technology. The results of this literature review have been divided into two categories based on the intended target of the research: (1) water resource management and (2) hydrologic system dynamics and interactions work. The first category captures research concerned with the quantification and/or conservation of water resources, as well as systems and standards for data and data sharing (a key component of management). The second category includes research concerned with topics ranging from rainfall interception to runoff mechanics; it is intended to capture all aspects of how water moves through and interacts with the built and/or natural environment. Each category is further divided to outline projects by their focus: software and computational modeling, hardware, data, and services or platforms.

All articles were reviewed from the perspective of ecohydrology. Despite casting a wide net, each of the selected papers is pertinent to the discipline. A broad search was completed to highlight as many projects within the discipline as possible that leverage open technology. Where relevant, notes are added regarding the current state of projects (whether they are still accessible/supported). Comprehensive assessments of the level of openness of individual projects have not been included here. For more information on the difference between a truly open and quasi-open project,

[14]Soil and Water Assessment Tool.

please refer to the discussion section. It is worth noting that the literature review captured very few research projects that leveraged open governmental data, such as that which is provided by Natural Resources Canada.

1.3.1 Water Resource Management

1.3.1.1 Software and Computational Modeling

Many projects concerned with water resource management have grown around the need to enhance the capabilities of GISs through extensions and integration. Researchers have approached via different avenues for enhancing the analytical abilities of these software packages. The most common is the adaptation of open hydrologic models (Pontes et al. 2017; George and Leon 2008; van Griensven et al. 2006) to a particular GIS suite, such as open-source software like QGIS (Pontes et al. 2017; Dile et al. 2016) and Map window (Rahman et al. 2017; George and Leon 2008), or proprietary software like ArcGIS (Holmes et al. 2017) and ArcView (van Griensven et al. 2006). In the case of the latter two projects, which are based on proprietary software, their openness is questionable, as they rely upon software that is not readily available and modifiable. Another adaptation involving GIS software is to better integrate it into workflows, either by demonstrating the robustness of a suite's capabilities (Rudiyanto et al. 2018) or by leveraging it as a tool for simulation and prediction (Holmes et al. 2017; Rahman et al. 2017; Sanhouse-Garcia et al. 2017; Thorp and Bronson 2013).

As the breadth and depth of water resource management research grow and the capability of both data collection methods and computing hardware increases, there is a need for open software, which can adapt to meet new demands. To this end, researchers have undertaken projects focused on simulation and model-building, which seek to optimize and support processing and decision-making. For example, Hofierka et al. (2017) utilized the open-source GRASS GIS software to build an open, fully parallelized processing pipeline for large-scale geodatasets. The parallelization method they demonstrated allowed for a near threefold increase in the processing speed of large-scale datasets. In another example, Kneis (2015) created a software solution that serves as a generic basis upon which a variety of catchment models may be constructed.

1.3.1.2 Hardware

Open hardware is the least leveraged portion of open technology, particularly in the segment of water resource management. That said, there are still stand-out examples of what is possible when open hardware is applied to research. Bartos et al. (2018) used open hardware to create a system of storm water monitors. The goal of the project was to create a network of sensors which would be capable of informing flood control and stormwater management infrastructure in real time, to maintain a balance of optimal retention and discharge. Prescott et al. (2016) created a system called HydroSense, which was built on open software, hardware, and standards and served as a universal hub/datalogger for sensors. Their system employed a standardized interface, which allowed it to be used with a range of proprietary and open sensors. This allows the system to be flexible and adaptable to many different applications. In both projects, however, the researchers created customized circuit boards to accommodate the designs of their projects, potentially limiting the portability of their core design to new research, as these circuit boards are not commercially available.

1.3.1.3 Data

Much of the concern surrounding data in research involves finding methods to deliver and maintain vital datasets of varying sizes (Flint et al. 2017). In most cases, vital data is scattered and poorly documented, making research arduous (Heidorn 2008). As a result, many initiatives now exist to take openly available datasets and collate them into singular, accessible, and well-documented databases (Flint et al. 2017; Horsburgh et al. 2016; Soranno et al. 2015). These databases are designed to serve multiple stakeholders across disciplines. In the case of Flint et al. (2017), their iUtah database included hydrologic data as well as social data regarding citizen attitudes toward water resources. Related to the iUtah project is the work of Jones et al. (2015), which discusses the necessary workflows and procedures that allow for the creation of a robust and well-documented database.

Beyond databases, there is also the issue of data standards. To ensure that information is presented in a manner which is widely adopted, machine readability and platform independence are central to open data sharing. To this end, Swain et al. (2015) undertook a review of the various data standards maintained by the open geospatial consortium (OGC). Their work specifically addressed the use of customized versions of OGC standards across a number of popular Web applications. The authors noted that there was more value in sticking with the general implementation of data standards, as it increased portability. Relevant to this is the discussion surrounding the use of the WaterML standard for hydrologic data (Challco et al. 2017; Yu et al. 2015). Work in this area is iterative and serves as a check-and-balance between what the standards offer and what they necessarily sacrifice in the interest of universality.

1.3.1.4 Platforms and Services

There is a need within the water resource management community for a means by which stakeholders can interface with and understand the outcomes and results of data collection, modeling, and research. To this end, a number of open technology projects have been undertaken, which seek to streamline the interaction between end-users and knowledge producers. Web applications built upon open-source software, open data standards, and open technology serve as data portals (Horsburgh et al. 2016; Soranno et al. 2015; Steiner et al. 2009) and decision support systems (Casadei et al. 2018; Sommerlot et al. 2016). Services may also include Web interfaces, which serve as a GIS system capable of outputting information specific to a topic (Siles et al. 2018; Hill et al. 2011). Swain et al. (2015) created a tool known as Tehys suite, which was designed to be a simple-to-use package for the creation and distribution of geographic Web applications.

1.3.2 Hydrologic Systems Dynamics and Interactions

1.3.2.1 Software and Computational Modeling

Modeling is the primary application of open-source code in the area of hydrologic systems dynamics and interactions. Research in this area is largely concerned with modeling of watershed-scale inputs and outputs (Brown et al. 2018; Tesfatsion et al. 2017; Hartanto et al. 2017; Sommerlot et al. 2016; Lampert and Wu 2015; López-Vicente et al. 2014; Zhang et al. 2011). Noteworthy among these projects is a reduced emphasis on building upon or implementing existing modeling software. Tesfatsion et al. (2017) took an open-source modeling tool from the OpenDanubia project and stripped it of a number of elements, including its graphical user interface, in order to expose model functions as an application programming interface (API), and thus make it more generally applicable. Also of interest within this area of research are the effects of hydrologic processes on terrestrial systems, such as shorelines (McCall et al. 2014), and their impacts on geomorphological processes, such as landslides (Strauch et al. 2017).

1.3.2.2 Hardware

Within the space of hydrologic interactions, open hardware has a wide range of applications. Examples range from soil temperature monitoring (Bitella et al. 2014) and stemflow measurement (Turner et al. 2019) to open satellite remote sensing via CubeSat (McCabe et al. 2017), which offers a lower cost, highly customizable satellite remote sensing platform. Lee et al. (2016) offer a more common application of open hardware in their research. They leveraged open-source hardware to log data and control aspects of an experiment on brackish water wetlands.

1.3.2.3 Data

Open data for the study of hydrologic systems dynamics and interactions is vital. In order for deeper linkages to be made between the multitudes of physical processes which relate to water, it is necessary to pull data from many different possible sources (Hill et al. 2014). To this end, many researchers have leveraged open datasets, as well as preserved their data for others to use. As an example of the latter point, Strauch et al. (2017), in their project on landslide probability, ensured that the data they produced was appropriately documented and shared by first entering it into CUAHSI's HydroShare. They also devoted a portion of the final publication to explaining how and where to find datasets and software used in their work. Other applications of open data have seen researchers attempt to enhance ground measurements using small datasets gathered from nonprofit organizations and citizen science groups, in addition to public and open government data (Flint et al. 2017; Niemi et al. 2017; Soranno et al. 2015). These fused data resources, combined with openly available earth observation resources, offer an exciting frontier for understanding the impact of hydrologic processes on the environment (McCabe et al. 2017).

1.3.2.4 Platforms and Services

Projects are developed with the intent that they are to be utilized by specific and insular research or professional communities. Software such as OpenGeoSys (Kolditz et al. 2012), which is a tool for developing simulations of water flow through porous media, exemplifies the focused scope and technical demand of tools in this area of research. Tesfatsion et al. (2017) similarly developed a platform known as WACCShed, which was targeted at the development of watershed-scale models and follows the same pattern of technicality and precision in one target area of inquiry. Orfeo, a toolbox program designed as an accompaniment to the commercial European Pleiades Earth-observing satellites, is an open-source remote sensing platform, which is capable of handling data from many different sensors (Tinel et al. 2012). Orfeo is targeted at a professional and expert community to serve as an intermediary tool for working with satellite imagery.

1.4 Discussion

Open has become a widely used descriptor of research and technology development in recent years. Some projects fully adhere to the definition of openness described in this chapter, while others, for reasons of pragmatism, do not; the latter can thus better be described as quasi-open. As open technology continues to grow, it will be necessary for new projects to integrate and construct upon the existing body of open projects. If the project a researcher is building upon is quasi-open, then there may be unforeseen barriers to development and dissemination. Such barriers could be as basic as a lack of metadata for a critical dataset or as serious as a copyright infringement.

Water is a basic human right (United Nations General Assembly 2010) and a common good. In the face of a changing climate that is predicted to increase water stress globally (Vörösmarty et al. 2000), water researchers have a moral obligation to adopt open principles and practices. Mainstream media outlets have started to raise the issue of morality in the fight against climate change (CBC Radio 2018). Ecohydrology considers the interaction between hydrologic and ecologic systems and is thus in a prime position to benefit from openness in terms of research efficiency while at the same time enhancing social good though the democratization of research outputs.

1.4.1 Truly Open Versus Quasi-Open Projects

Truly open projects are those which are released and maintained according to the principles of a widely accepted open organization's open definition (e.g., OSI, OSHW, or OKI). These projects are released with as few barriers to their advancement as possible. An example of truly open projects is the R project for statistical computing. Created as a GNU-compliant version of the S programming language, R has been widely adopted. The Central R Archive Network (CRAN), at the time of writing, includes over 12,000 hosted packages, each of which is FOSS and each of which extends the core capabilities of the language, allowing it to adapt to and evolve with changing research needs. R stands as an example of a platform that fosters innovation, due to its mature nature, open community, and transparent source.

Quasi-open projects are those which are released to the public domain but are not maintained, are not readily available, or rely upon some form of closed or constrained resource. Quasi-open projects do not offer a simple means of open collaboration or widespread adoption. Examples of quasi-open projects include open-source extensions to proprietary software packages. While there is value in these extensions, they cannot be easily tested or implemented by the global scientific community (including academics, governments, and citizens) without a researcher first acquiring the proprietary software that they are seeking to extend. As an example, Holmes et al. (2017) created a package for ArcGIS that was capable of identifying natural water storage basins. While the code for the package is open, it relies upon an expensive proprietary software package, which is inaccessible to many researchers. Quasi-open projects may also have an appearance of openness as an afterthought. In this case, the creators of the project may have made a decision late in the project development that the data, code, or designs should be open and then decided to release them in an "as is, where is" state that lacks appropriate documentation, accessibility, and/or licensing.

True openness should be the ultimate goal of all research projects, but the reality of the current research culture means that it rarely is. The result is a large number of quasi-open projects within the open technology arena that fail to achieve true openness for one or more of several reasons. First, in cases where the decision to make a project open comes late in the development process, documentation of the project may be insufficient or absent. Second, researchers will sometimes opt to not make their project open at all, due to the time-cost of this documentation, in light of productivity metrics to that do not value such contributions (Heidorn 2008). Third, researchers may choose to make their project only selectively open, by sharing the project information and products only in instances in which they expect reciprocal benefit (Campbell et al. 2002; Haeussler 2011). Fourth, journals that require data submission as supplementary information may hold copyrights to these data, thus preventing the original data collector from sharing the data openly (Fecher and Friesike 2014). Finally, researchers may choose to withhold their data from the public domain because data collection without accompanying analysis has no currency in academic career advancement, and thus they may fear that they would have insufficient advantage over others to publish outcomes from their data collection and receive credit, if others were to have unfettered access to their data.

1.4.2 Successes

There are a number of individual projects which stand out as successes within open technology when viewed from the perspective of ecohydrology. Modeling tools such as the Soil and Water Assessment Tool (SWAT) and the Hydrological Simulation Program (HSP) have been widely adopted and heavily integrated into the research of the community at large. Academic communities such as CUAHSI have demonstrated how open technology could potentially be integrated to form a feature-rich and robust data repository.

As shown in Fig. 1.2 and discussed previously, open-source software comprises the most significant portion of the broader open technology movement. The open-source software movement has become a powerful force in the digital age. Since the mid-2000s, the FOSS community has gained mainstream recognition, and the projects created there have begun to rival the closed source, proprietary, and expensive software on the market (Haefliger et al. 2008). The FOSS movement has now redeveloped, implemented, and optimized a number of core computing technologies, placing them firmly in the public sphere. The open exposure of these software packages allows for researchers to delve into the source code to understand how each portion of the software operates and potentially improve upon it. As an example, it is possible to modify the source code of an open-source GIS, such as GRASS, and optimize the way it processes data to attain better performance (Hofierka et al. 2017).

The same principles which have led to the success of the open-software movement are now carrying over into other open movements. Open-source hardware has seen a number of proponents leveraging simple and commonly available technology (some of which may be proprietary but still highly accessible) in order to reimagine the basic tools of laboratory science (Fisher and Gould 2012) and environmental monitoring technology (Bartos et al. 2018; McCabe et al. 2017; Bitella et al. 2014). Developing open hardware, however, is not just about constructing new hardware from open components. It also requires a commitment to openly share all information necessary to replicate the hardware project. For example, Oregon State's OPEnS Lab[15] demonstrates this philosophy by providing all relevant instruction files, software code, and metadata necessary to replicate the environmental monitoring solutions they have created via an open license.

Open data appears to be at a point not unlike where open-source software was in the mid-2000s. We are starting to see serious efforts toward optimization and standardization in how we collect and store data. Exemplifying this is the check-and-balance we are seeing with open standards such as WaterML (Yu et al. 2015). This type of review of a common standard is similar to the optimization and adoption process of open-source software. An additional success for open data is that it has become mainstream and has been adopted as a governmental mandate by a number of governing bodies around the world (including Canada, much of the European Union, and Mexico).

1.4.3 Challenges

In our experience with open technology development in the hydrologic sciences, we have identified four key challenges to the growth of open projects and adoption of open principles within hydrology and forest-water interactions. First, current research culture hinders or discourages researchers from embracing open practices in project design and in the publication of research outputs, including data, software, and hardware (Molloy 2011; Soranno et al. 2015; Flint et al. 2017). Development of open projects requires a commitment to the creation of metadata necessary for others to leverage the project outputs, a time-consuming process, which is not rewarded in typical metrics for career progression (Heidorn 2008). The withholding of research outputs is not unique to the hydrologic sciences, but rather endemic to academic research practices in general. For example, in a survey of academic geneticists, Campbell et al. (2002) found that only around 25% of researchers share their data, even upon request. This study found that the most common reason stated for choosing to not to share data was the time required to properly format and document the data—time that could be used to perform activities for which one would be rewarded by career advancement. When researchers do decide to share data, this decision is not always altruistically driven, but rather based on the perception that the act of sharing data will increase one's social capital (Haeussler 2011); thus, even shared data may not be shared in a way that is truly democratic. There are also few avenues for publishing open data, software, or hardware designs, and so these publications often end up in the gray literature, which is generally held in less regard. Important research products, including data, software, and hardware designs, thus tend to get lost, becoming "dark" research products (Heidorn 2008), not visible to others who could use them fruitfully.

[15] www.open-sensing.org

Second, the movement by academic journals in the fields of hydrology and forest-water interactions to encourage or require data, software, or hardware designs to accompany manuscripts as supplementary information helps to get data into the public sphere. Although the intention of publishing data, software, and hardware produced in the course of a research project is to increase access, when this supplementary information is held in copyright by the publisher, it limits access and renders the data, software, or hardware designs closed (Fecher and Friesike 2014). Third, open source, open data, and open hardware all have multiple user groups, communities, or organizations with subtly differing definitions of what constitutes open. Without agreeing to a common standard, these groups are challenged in working together to advance open technology. Finally, the open hardware movement is facing challenges in gaining mainstream acceptance. We suspect that this is due to the jack-of-all-trades nature of many open hardware designs. Researchers may choose to design their own more limited capability hardware, rather than use the original open hardware project (e.g., Kerkez et al. 2012), or use of the hardware may come with additional regulatory oversight (as discussed in this volume by Hill et al., with respect to open flight controllers for unpiloted aerial vehicles).

1.4.4 Opportunities

As open technology projects increase within the discipline of ecohydrology, there are many opportunities to improve knowledge production in hydrology and forest-water interactions by addressing some of the challenges noted above. First, projects working to integrate open technology to produce higher-level analysis tools can prevent important research outcomes from being lost, becoming dark technology (Heidorn 2008). For example, Sorrano et al. (2015) addressed this problem by integrating numerous small and distributed datasets to create a unified data portal for analyzing lake ecosystems at the macroscale. In our opinion, the rise of system-oriented analysis as the dominant paradigm for understanding and managing complex phenomena, such as forest-water interactions (Liu et al. 2007), could be greatly enhanced by the availability of open technology that can be leveraged for analyzing systems-level interactions.

Second, mechanisms for publishing open datasets, created by individual researchers, are becoming increasingly available in hydrology and forest-water interactions research. In addition to Web repositories hosted by universities and research collaboratives (e.g., CUAHSI), academic journals are beginning to publish data not only as supplementary information but also as "data papers," which provide metadata describing the data and make the data easier for other researchers to access. Similarly, mechanisms to publish open hardware are also becoming more available. In addition to methods-oriented journals (e.g., *Sensors*, MDPI and *MethodsX*, Elsevier), "methods papers" are increasingly published in hydrologic journals. These journal-based opportunities for publishing open technology development permit researchers to get credit for the substantial amount of time required to create the technology, potentially increasing the appeal of creating open technology. However, for technology published as a data or methods paper to be considered open, not only must the journal provide open access to these papers, but also the technology itself must be designed in accordance with open principles and be distributed with an open license.

Third, joining forces with movements such as slow science can help to reform metric-based academic cultures that promote speed of discovery at the expense of openness. The slow science movement advocates for research impact over volume, recognizing that impactful research is that which can be easily replicated, validated, and expanded upon (Stengers 2018)—goals facilitated by open principles. The slow science movement recognizes the additional, currently unrewarded, time required to properly document and share one's research and advocates for such invisible costs of true impact to be brought into account.

Finally, there exist a number of opportunities resulting from advancements in electronics and computing hardware that could enhance research in hydrology and forest-water interactions. With cellular and satellite technology continuing to become smaller, more capable, and more affordable, there exist opportunities to create larger-scale, more robust sensor networks and remote sensing devices that improve our understanding of Earth systems processes. We are also presented with more powerful computing technology, which can handle increasingly demanding simulation tasks. Changes in research practices and improvements in infrastructure supporting research, such as data storage and archiving, as well as tools for collaboration, will support the further emergence of open technology and open science as a whole (Fecher and Friesike 2014).

Acknowledgments This work is supported in part by the Natural Science and Engineering Research Council of Canada through grant number RGPIN 2014-06114 (Hill). Any use of trade, firm, or product names is for descriptive purposes only and does not imply endorsement by the authors.

References

Ackermann JR (2009) Toward open source hardware. U Dayton L Rev 34:41

Bartos M, Wong B, Kerkez B (2018) Open storm: a complete framework for sensing and control of urban watersheds. Environ Sci Water Res Technol 4:346–358. https://doi.org/10.1039/C7EW00374A

Bitella G, Rossi R, Bochicchio R, Perniola M, Amato M (2014) A novel low-cost open-hardware platform for monitoring soil water content and multiple soil-air-vegetation parameters. Sensors-Basel 14:19639–19659. https://doi.org/10.3390/s141019639

Bonney R, Shirk J, Phillips T, Wiggins A, Ballard H, Miller-Rushing A et al (2014) Next steps for citizen science. Science 343:1436–1437. https://doi.org/10.1126/science.1251554

Brown H, Carrick S, Müller K, Thomas S, Sharp J, Cichota R et al (2018) Modelling soil-water dynamics in the rootzone of structured and water-repellent soils. Comput Geosci 113:33–42. https://doi.org/10.1016/j.cageo.2018.01.014

Campbell EG, Clarridge BR, Gokhale M, Birenbaum L, Hilgartner S, Holtzman NA et al (2002) Data withholding in academic genetics: evidence from a national survey. JAMA-J Am Med Assoc 287:473. https://doi.org/10.1001/jama.287.4.473

Casadei S, Pierleoni A, Bellezza M (2018) Sustainability of water withdrawals in the Tiber River basin (Central Italy). Sustainability-Basel 10:485. https://doi.org/10.3390/su10020485

Challco BMS, Masó JP, Pesquer LM (2017) Estudio del estándar WATERML. Geofocus Revista Internacional de Ciencia y Tecnología de la información Geográfica 19:129–150. https://doi.org/10.21138/GF.485

Chignard S (2013, March 29) A brief history of open data. Paris Innovation Review Retrieved June 27, 2018, from http://parisinnovationreview.com/articles-en/a-brief-history-of-open-data

Cohn J (2008) Citizen science: can volunteers do real research? Bioscience 58:192–197. https://doi.org/10.1641/B580303

David PA (2004) Understanding the emergence of “open science” institutions: functionalist economics in historical context. Ind Corp Chang 13:571–589. https://doi.org/10.1093/icc/dth023

Dile YT, Daggupati P, George C, Srinivasan R, Arnold J (2016) Introducing a new open source GIS user interface for the SWAT model. Environ Model Softw 85:129–138. https://doi.org/10.1016/j.envsoft.2016.08.004

Fecher B, Friesike S (2014) Open science: one term, five schools of thought. In: Bartling S, Friesike S (eds) Opening science: the evolving guide on how the internet is changing research, collaboration and scholarly publishing. Springer, Cham, pp 17–47. https://doi.org/10.1007/978-3-319-00026-8_2

Fisher DK, Gould P (2012) Open-source hardware is a low-cost alternative for scientific instrumentation and research. Modern Instrum 01:8–20. https://doi.org/10.4236/mi.2012.12002

Flint CG, Jones AS, Horsburgh JS (2017) Data management dimensions of social water science: the iUtah experience. J Am Water Resour As 53:988–996. https://doi.org/10.1111/1752-1688.12568

George C, Leon LF (2008) Waterbase: SWAT in an open source GIS. Open Hydrol J 2:1–6. https://doi.org/10.2174/1874378100802010001

Government of Canada (2017) Open data 101. Retrieved October 1, 2018, from https://open.canada.ca/en/open-data-principles

Haefliger S, von Krogh G, Spaeth S (2008) Code reuse in open source software. Manag Sci 54:180–193. https://doi.org/10.1287/mnsc.1070.0748

Haeussler C (2011) Information-sharing in academia and the industry: a comparative study. Res Policy 40:105–122. https://doi.org/10.1016/j.respol.2010.08.007

Hartanto IM, van der Kwast J, Alexandridis TK, Almeida W, Song Y, van Andel SJ et al (2017) Data assimilation of satellite-based actual evapotranspiration in a distributed hydrological model of a controlled water system. Int J Appl Earth Obs 57:123–135. https://doi.org/10.1016/j.jag.2016.12.015

Heidorn PB (2008) Shedding light on the dark data in the long tail of science. Libr Trends 57:280–299. https://doi.org/10.1353/lib.0.0036

Hill DJ, Liu Y, Marini L, Kooper R, Rodriguez A, Futrelle J et al (2011) A virtual sensor system for user-generated, real-time environmental data products. Environ Model Softw 26:1710–1724. https://doi.org/10.1016/j.envsoft.2011.09.001

Hill DJ, Kerkez B, Rasekh A, Ostfeld A, Minsker B, Banks MK (2014) Sensing and cyberinfrastructure for smarter water management: the promise and challenge of ubiquity. J Water Resour Plan Manag 140:01814002. https://doi.org/10.1061/(ASCE)WR.1943-5452.0000449

Hofierka J, Lacko M, Zubal S (2017) Parallelization of interpolation, solar radiation and water flow simulation modules in GRASS GIS using OpenMP. Comput Geosci 107:20–27. https://doi.org/10.1016/j.cageo.2017.07.007

Holmes D, McEvoy J, Dixon J, Payne S (2017) A geospatial approach for identifying and exploring potential natural water storage sites. Water-Base 9:585. https://doi.org/10.3390/w9080585

Horsburgh JS, Morsy MM, Castronova AM, Goodall JL, Gan T, Yi H et al (2016) HydroShare: sharing diverse environmental data types and models as social objects with application to the hydrology domain. J Am Water Res As 52:873–889. https://doi.org/10.1111/1752-1688.12363

Jones AS, Horsburgh JS, Reeder SL, Ramírez M, Caraballo J (2015) A data management and publication workflow for a large-scale, heterogeneous sensor network. Environ Monit Assess 187:348. https://doi.org/10.1007/s10661-015-4594-3

Kerkez B, Glasser SD, Bales RC, Meadows W (2012) Design and performance of a wireless sensor network for catchment-scale snow and soil moisture measurements. Water Resour Res 48: W09515. https://doi.org/10.1029/2011WR011214

Kneis D (2015) A lightweight framework for rapid development of object-based hydrological model engines. Environ Model Softw 68:110–121. https://doi.org/10.1016/j.envsoft.2015.02.009

Koenka IJ, Sáiz J, Hauser PC (2014) Instrumentino: an open-source modular Python framework for controlling Arduino based experimental instruments. Comput Phys Commun 185:2724–2729. https://doi.org/10.1016/j.cpc.2014.06.007

Kolditz O, Bauer S, Bilke L, Böttcher N, Delfs JO, Fischer T et al (2012) OpenGeoSys: an open-source initiative for numerical simulation of thermo-hydro-mechanical/chemical (THM/C) processes in porous media. Environ Earth Sci 67:589–599. https://doi.org/10.1007/s12665-012-1546-x

Lampert DJ, Wu M (2015) Development of an open-source software package for watershed modeling with the hydrological simulation program in Fortran. Environ Model Softw 68:166–174. https://doi.org/10.1016/j.envsoft.2015.02.018

Lee DY, De Meo OA, Thomas RB, Tillett AL, Neubauer SC (2016) Design and construction of an automated irrigation system for simulating saltwater intrusion in a tidal freshwater wetland. Wetlands 36:889–898. https://doi.org/10.1007/s13157-016-0801-4

Liu J, Dietz T, Carpenter SR, Alberti M, Folke C, Moran E et al (2007) Complexity of coupled human and natural systems. Science 317:1513–1516. https://doi.org/10.1126/science.1144004

López-Vicente M, Pérez-Bielsa C, López-Montero T, Lambán LJ, Navas A (2014) Runoff simulation with eight different flow accumulation algorithms: recommendations using a spatially distributed and open-source model. Environ Model Softw 62:11–21. https://doi.org/10.1016/j.envsoft.2014.08.025

McCabe MF, Rodell M, Alsdorf DE, Miralles DG, Uijlenhoet R, Wagner W et al (2017) The future of Earth observation in hydrology. Hydrol Earth Syst Sci 21:3879–3914. https://doi.org/10.5194/hess-21-3879-2017

McCall RT, Masselink G, Poate TG, Roelvink JA, Almeida LP, Davidson M, Russell PE (2014) Modelling storm hydrodynamics on gravel beaches with XBeach-G. Coast Eng 91:231–250. https://doi.org/10.1016/j.coastaleng.2014.06.007

Merton RK (1973) The sociology of science: theoretical and empirical investigations. University of Chicago Press, Chicago

Molloy JC (2011) The Open Knowledge Foundation: open data means better science. PLoS Biol 9: e1001195. https://doi.org/10.1371/journal.pbio.1001195

National Research Council (1995) On the full and open exchange of scientific data. National Academies Press, Washington, DC. https://doi.org/10.17226/18769

Niemi TJ, Warsta L, Taka M, Hickman B, Pulkkinen S, Krebs G et al (2017) Applicability of open rainfall data to event-scale urban rainfall-runoff modelling. J Hydrol 547:143–155. https://doi.org/10.1016/j.jhydrol.2017.01.056

Open Source Hardware Association (n.d.). Definition. Retrieved June 27, 2018, from https://www.oshwa.org/definition/

Pearce JM (2012) Building research equipment with free, open-source hardware. Science 337 (6100):1303–1304. https://doi.org/10.1126/science.1228183

Phelps L, Fox B, Marincola F (2012) Supporting the advancement of science: open access publishing and the role of mandates. J Transl Med 10:13. https://doi.org/10.1186/1479-5876-10-13

Pontes PRM, Fan FM, Fleischmann AS, de Paiva RCD, Buarque DC, Siqueira VA et al (2017) MGB-IPH model for hydrological and hydraulic simulation of large floodplain river systems coupled with open source GIS. Environ Model Softw 94:1–20. https://doi.org/10.1016/j.envsoft.2017.03.029

Prescott E, Rome C, Heinzmann C, Hawkins M, Marchiori A, Hayes BR (2016) Hydrosense: an open platform for hydroclimatic monitoring. In: Proceedings of the 2016 IEEE international conference on smart computing (SMARTCOMP), pp 1–5. https://https://doi.org/10.1109/SMARTCOMP.2016.7501695

Radio CBC (2018, August 10) "losing earth": do we have a collective moral responsibility to fight climate change? CBC Radio. Canadian Broadcasting Corporation. Retrieved from https://www.cbc.ca/radio/day6/episode-402-saudi-trolls-vs-canada-alex-jones-s-precarious-empire-losing-earth-pampered-poultry-and-more-1.4777781/losing-earth-do-we-have-a-collective-moral-responsibility-to-fight-climate-change-1.4777798

Rahman K, Ray N, Giuliani G, Maringanti C, George C, Lehmann A (2017) Breaking walls towards fully open source hydrological modeling. Water Res 44:23–30. https://doi.org/10.1134/S0097807817010067

Rudiyanto MB, Setiawan BI, Saptomo SK, McBratney AB (2018) Open digital mapping as a cost-effective method for mapping peat thickness and assessing the carbon stock of tropical peatlands. Geoderma 313:25–40. https://doi.org/10.1016/j.geoderma.2017.10.018

Sanhouse-Garcia AJ, Bustos-Terrones Y, Rangel-Peraza JG, Quevedo-Castro A, Pacheco C (2017) Multi-temporal analysis for land use and land cover changes in an agricultural region using open source tools. Remote Sens Appl Soc Environ 8:278–290. https://doi.org/10.1016/j.rsase.2016.11.002

Siles G, Voirin Y, Bénié GB (2018) Open-source based geo-platform to support management of wetlands and biodiversity in Quebec. Ecol Inform 43:84–95. https://doi.org/10.1016/j.ecoinf.2017.11.005

Sommerlot AR, Wagena MB, Fuka DR, Easton ZM (2016) Coupling the short-term global forecast system weather data with a variable source area hydrologic model. Environ Model Softw 86:68–80. https://doi.org/10.1016/j.envsoft.2016.09.008

Soranno PA, Bissell EG, Cheruvelil KS, Christel ST, Collins SM, Fergus CE et al (2015) Building a multi-scaled geospatial temporal ecology database from disparate data sources: fostering open science and data reuse. GigaScience 4:28. https://doi.org/10.1186/s13742-015-0067-4

Stallman R (n.d.) Why open source misses the point of free software. Retrieved October 1, 2018, from https://www.gnu.org/philosophy/open-source-misses-the-point.html

Steiner JL, Sadler EJ, Hatfield JL, Wilson G, James D, Vandenberg B et al (2009) Data management to enhance long-term watershed research capacity: context and STEWARDS case study. Ecohydrology 2:391–398. https://doi.org/10.1002/eco.89

Stengers I (2018) Another science is possible: a manifesto for slow science. Polity, Cambridge

Strauch R, Istanbulluoglu E, Nudurupati SS, Bandaragoda C, Gasparini NM, Tucker GE (2017) A hydro-climatological approach to predicting regional landslide probability using Landlab. Earth Surf Dynam 6:49–75. https://doi.org/10.5194/esurf-2017-39

Swain NR, Latu K, Christensen SD, Jones NL, Nelson EJ, Ames DP, Williams GP (2015) A review of open source software solutions for developing water resources web applications. Environ Model Softw 67:108–117. https://doi.org/10.1016/j.envsoft.2015.01.014

Tesfatsion L, Rehmann CR, Cardoso DS, Jie Y, Gutowski WJ (2017) An agent-based platform for the study of watersheds as coupled natural and human systems. Environ Model Softw 89:40–60. https://doi.org/10.1016/j.envsoft.2016.11.021

Thorp KR, Bronson KF (2013) A model-independent open-source geospatial tool for managing point-based environmental model simulations at multiple spatial locations. Environ Model Softw 50:25–36. https://doi.org/10.1016/j.envsoft.2013.09.002

Tinel C, Grizonnet M, Fontannaz D, de Boissezon H, Giros A (2012) Orfeo, the pleiades accompaniment program and its users thematic commissioning. Int Arch Photogramm XXXIX-B3:569–572. https://doi.org/10.5194/isprsarchives-XXXIX-B3-569-2012

Turner B, Hill DJ, Carlyle-Moses DE, Rahman M (2019) Low-cost, high-resolution stemflow sensing. J Hydrol 570:62–68. https://doi.org/10.1016/j.hydrol.2018.12.072

United Nations General Assembly (2010) Resolution 64-292 adopted by the General Assembly on 28 July 2010. United Nations

van Griensven A, Breuer L, Di Luzio M, Vandenberghe V, Goethals P, Meixner T et al (2006) Environmental and ecological hydroinformatics to support the implementation of the European Water Framework Directive for river basin management. J Hydroinf 8:239–252. https://doi.org/10.2166/hydro.2006.010

Vörösmarty CJ, Green P, Salisbury J, Lammers RB (2000) Global water resources: vulnerability from climate change and population growth. Science 289:284–288. https://doi.org/10.1126/science.289.5477.284

Yu J, Taylor P, Cox SJD, Walker G (2015) Validating observation data in WaterML 2.0. Comput Geosci 82:98–110. https://doi.org/10.1016/j.cageo.2015.06.001

Zhang Y, Zhang Z, Reed S, Koren V (2011) An enhanced and automated approach for deriving a priori SAC-SMA parameters from the soil survey geographic database. Comput Geosci 37:219–231. https://doi.org/10.1016/j.cageo.2010.05.016

Chapter 2
The Necessity of Sensor Calibration for the Precise Measurement of Water Fluxes in Forest Ecosystems

Shin'ichi Iida, Takanori Shimizu, Yoshinori Shinohara, Shin'ichi Takeuchi, and Tomo'omi Kumagai

2.1 Introduction

The amount of runoff from a forested watershed changes depending on the degree of tree removal (e.g., Bosch and Hewlett 1982). The forest ecosystem can greatly influence the water balance in a watershed, which, at the annual time scale, may be expressed as:

$$P = E + Q_R \tag{2.1}$$

where P is precipitation, E is evapotranspiration, and Q_R represents runoff. Evapotranspiration may be further subdivided into three components:

$$E = E_T + E_I + E_F \tag{2.2}$$

where E_T is transpiration from forest ecosystem, E_I is interception loss, and E_F is evaporation from forest floor.

As Oki and Kanae (2006) state, water resources engineers consider $P - E$ as a measure of the maximum renewable freshwater resources for a watershed (blue water). Thus, the accurate measurement of P and E, with the latter being mostly

S. Iida (✉)
Department of Disaster Prevention, Meteorology and Hydrology, Forestry and Forest Products Research Institute, Tsukuba, Ibaraki, Japan
e-mail: iishin@ffpri.affrc.go.jp

T. Shimizu
Forestry and Forest Products Research Institute, Tsukuba, Ibaraki, Japan

Y. Shinohara
University of Miyazaki, Miyazaki, Miyazaki, Japan

S. Takeuchi
Tokai University, Shizuoka, Shizuoka, Japan

T. Kumagai
University of Tokyo, Bunkyoku, Tokyo, Japan

D. F. Levia et al. (eds.), *Forest-Water Interactions*, Ecological Studies 240,
https://doi.org/10.1007/978-3-030-26086-6_2

comprised of E_T and E_I for closed-canopy forests (e.g., Wilson et al. 2001), is critical for estimating runoff generation in forest ecosystems.

Interception loss, E_I, is usually estimated as the difference between rain falling on the forest canopy and the proportion of that rain delivered to the forest floor (Helvey and Partic 1965; Carlyle-Moses and Gash 2011). Rain depth is usually measured using a tipping-bucket rain gauge or some type of tipping-bucket flow meter (e.g., Reid and Lewis 2009). However, tipping-bucket rain gauges and flow meters commonly underestimate inflow rates, especially for higher intensity rains and flow inputs (e.g., Edwards et al. 1974; Iida et al. 2018; Shimizu et al. 2018), and, as a consequence, a minimum 10% uncertainty in E_I estimates should be expected when uncorrected gauge and flow meter data are used (Iida et al. 2012). Transpiration, E_T, is evaluated using sap flow measurements with sensors inserted into the boles of trees (e.g., Kumagai et al. 2014); however, this technique may underestimate sap flux densities. For example, Steppe et al. (2010) determined that laboratory-derived sap flux densities and thus, by extension, transpiration rates of freshly cut stem segments of *Fagus grandifolia* trees were underestimated by 60%. In order to evaluate E_I and E_T correctly, the measurement uncertainties associated with tipping-bucket gauges and flow meters, as well as sap flow sensors, must be defined and taken into consideration when performing instrument calibration.

In this chapter, we describe in detail how to calibrate tipping-bucket rain gauges, tipping-bucket flow meters, and sap flow sensors (Fig. 2.1). Additionally, based on laboratory calibrations, we evaluate the effect of applying calibration on E_I estimates. Tree-to-tree and site-specific differences in calibration results of sap flow sensors and the degree of uncertainty that can be expected in estimating E_T are also evaluated and discussed.

2.2 Correction of Tipping-Bucket Rain Gauges and Flow Meters for Interception Loss Estimates

When rain falls on a forest, a proportion of the rain is stored on the tree canopies and boles that comprise some of this storage being evaporated back to the atmosphere. This interception loss (E_I) can be appreciable, accounting for 10–50% of precipitation (Roth et al. 2007). Although some researchers evaluate E_I based on the estimation of rainwater stored on the tree by detecting stem compression (e.g., Friesen et al. 2008) or a much more promising method of detecting shifts of tree sway frequency using accelerometers (e.g., van Emmerik et al. 2017; see Chap. 6 of this volume), E_I in forests is more as commonly estimated as (e.g., Carlyle-Moses et al. 2018):

$$E_I = P - (T_f + S_f) \tag{2.3}$$

where T_f is throughfall and S_f is stemflow.

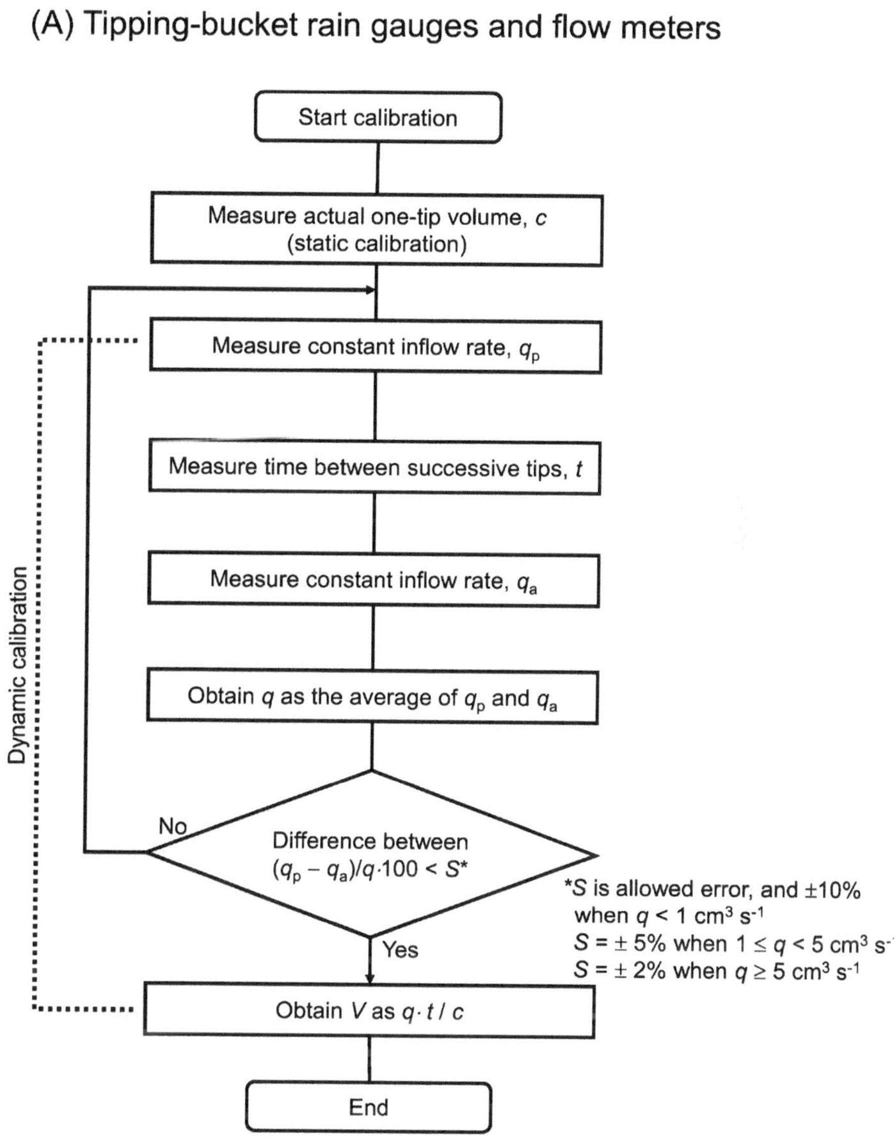

Fig. 2.1 Calibration procedures for (**a**) tipping-bucket rain gauges and flow meters and (**b**, continued on the following page) three sap flow techniques of thermal dissipation (TD) method, heat field deformation (HFD) method, and heat ratio (HR) method

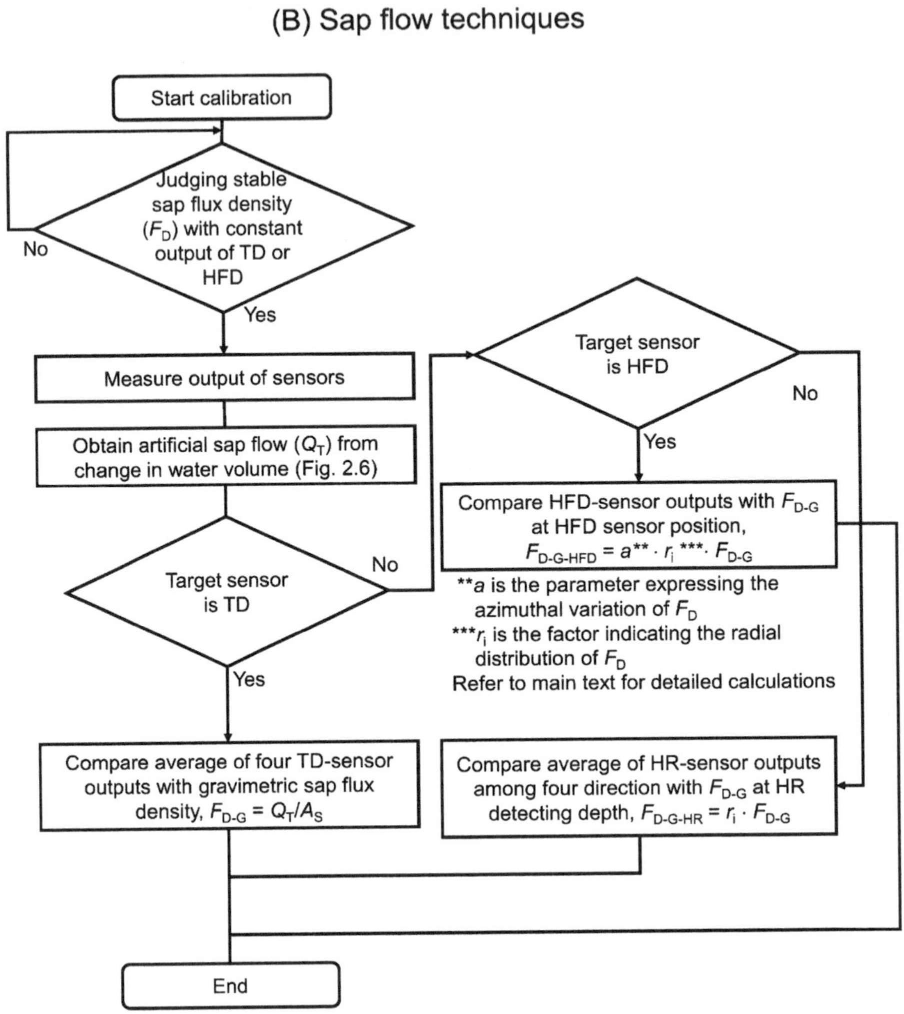

Fig. 2.1 (continued)

It is well known that T_f commonly exhibits high spatial heterogeneity with some areas of the forest floor receiving T_f inputs greater than P (e.g., Lloyd and Marques 1988; Carlyle-Moses and Lishman 2015). In order to obtain spatially representative T_f measurements, large numbers of gauges and/or gauges designed for integrating the spatial variability of T_f (e.g., trough-style gauge) are required (Carlyle-Moses et al. 2014; Su et al. 2016). Iida (2009), for example, found that for precise estimates of T_f (error $\leq 5\%$) more than 1 m^2 of gauge collection area must be distributed beneath the forest canopy of interest. Thus, the heterogeneity associated with forest overstory and the associated water routing through the canopy mean that meeting the statistical objectives of a T_f measurement campaign will often require many more gauges than that required to estimate precipitation (e.g., Kimmins 1973; Carlyle-Moses et al. 2004; Ziegler et al. 2009). The minimum number of gauges required to estimate T_f for a desired error (ε) and confidence level (t) may be estimated using the following (Kimmins 1973; Puckett 1991):

$$n' = \frac{t^2 \cdot CV^2}{\varepsilon^2} \tag{2.4}$$

where n' is the estimated number of gauges required to sample T_f and CV is the coefficient of variation (%) of T_f measurements.

Stemflow represents water that is routed by the tree canopy to the bole where it then flows to the forest floor. Like T_f, S_f can be highly spatially variable at the plot scale since S_f yields can vary greatly among and within tree species due to factors such as tree age, canopy structure, and canopy rain shadows (Levia and Germer 2015). The quantitative importance of S_f at the individual tree scale may be expressed using the funneling ratio (Herwitz 1985) and at the forest-scale using the stand-scale funneling ratio (Levia and Gemer 2015; Carlyle-Moses et al. 2018). The funneling ratio represents the ratio of the S_f volume generated by a tree or trees to the volume of rain that would have been captured by a rain gauge having a diameter equal to that of the tree bole(s) at breast height (DBH). Funneling ratios at both the tree and forest stand-scale are often much greater than unity (e.g., Levia and Germer 2015; Carlyle-Moses et al. 2018), indicating that S_f volume is often much more than P at basal area scale. Thus, in order to obtain reasonable estimates of T_f and S_f, and therefore E_I, highly variable and often voluminous water inputs need to be measured with a high degree of accuracy.

Rainfall event scale E_I is commonly estimated based on event-scale measurements of P, T_f, and S_f (e.g., van Dijk et al. 2015), employing storage-type gauges and large-capacity collection reservoirs. However, for a better understanding of the interception process, including the impact of meteorological factors and tree characteristics, more work examining the dynamics of E_I is necessary at the intra-event time scale using high temporal resolution T_f and S_f measurements by tipping-bucket rain gauges and flow meters (e.g., Reid and Lewis 2009; Iida et al. 2017) as well as instrument systems utilizing ultrasonic rangefinders to monitor changes in collected flow depths (Turner et al. 2019). In such cases, the well-known systematic biases of tipping-bucket rain gauges and flow meters must be considered. When the bucket mechanism of a tipping-bucket gauge or flow meter is filled with water, it tips the other bucket into position to continue receiving water input. The water flux is then measured as the number of tips over a duration with this information being stored on a datalogger. If, however, water flows continuously during the period between one bucket tipping and the other bucket being brought into position, a certain amount of water does not flow into either bucket and is not registered with the degree of water input underestimation increasing with increasing intensity of inflow. This systematic underestimation has been reported and dynamically calibrated (e.g., Edwards et al. 1974). Moreover, the static amount of a tip (c), which is the volume

Table 2.1 A list of calibration curves for various tipping-bucket rain gauges and flow meters

Type	One tip volume (c_m, mL)	Calibration curve		Source
Tipping-bucket rain gauge (TBRG)				
OC[a]	3.73	$V = -0.0352Q^2 + 0.418Q + 1$	$R^2 = 0.99$	Iida et al. (2012)
OT[b]	15.7	$V = -0.119Q^2 + 0.454Q + 1$	$R^2 = 0.97$	Iida et al. (2012)
DI[c]	4.28	$V = -0.2005Q^2 + 0.702Q + 1$	$R^2 = 0.95$	Iida et al. (2018)
Tipping-bucket flow meter (TBFM)				
U50[d]	50	$V = -0.189Q^2 + 0.531Q + 1$	$R^2 = 0.99$	This study
U100[e]	100	$V = -0.906Q^2 + 0.971Q + 1$	$R^2 = 0.96$	This study
U200[f]	200	$V = -0.492Q^2 + 0.719Q + 1$	$R^2 = 0.98$	Iida et al. (2012)
U200[f]	200	$V = -0.489Q^2 + 0.672Q + 1$	$R^2 = 0.85$	Shimizu et al. (2018)
U500[g]	500	$V = -0.632Q^2 + 0.814Q + 1$	$R^2 = 0.98$	This study
I200[h]	200	$V = -0.161Q^2 + 0.554Q + 1$	$R^2 = 0.98$	Iida et al. (2012)
I400[i]	400	$V = -0.756Q^2 + 0.719Q + 1$	$R^2 = 0.95$	Iida et al. (2012)
Y500[j]	500	$V = -0.258Q^2 + 0.907Q + 1$	$R^2 = 0.97$	Shimizu et al. (2018)

[a]RG-3 M, Onset Computer Corp., Massachusetts
[b]OW-34-BP, Ota keiki seisakusho Co., Ltd., Tokyo, Japan
[c]Rain Collector II, Davis Instruments, California
[d]UIZ-TB-50, Uizin Co., Ltd, Tokyo., Japan
[e]UIZ-TB-100, Uizin Co., Ltd, Tokyo., Japan
[f]UIZ-TB-200, Uizin Co., Ltd, Tokyo., Japan
[g]UIZ-TB-500, Uizin Co., Ltd, Tokyo., Japan
[h]TXQ-200, Ikeda keiki seisakusho Co., Ltd., Tokyo, Japan
[i]TXQ-400, Ikeda keiki seisakusho Co., Ltd., Tokyo, Japan
[j]500 mL model Yokogawa Electric Corp., Tokyo, Japan

associated with one tip of the tipping-bucket mechanism under very low intensity inflows, is frequently found to be different from that stipulated by the gauge or flow meter manufacturer (c_m) (Shedekar et al. 2016; Iida et al. 2018; Shimizu et al. 2018). Thus, static calibration is critical for measuring water fluxes correctly with tipping-bucket rain gauges and flow meters. The necessary calibrations for several tipping-bucket and flow meter models have been reported (Iida et al. 2012, 2018; Shimizu et al. 2018; see Table 2.1). Based on these results and the established need to correct for the under-catch of these gauges and flow meters, we explain and detail the procedure necessary to properly calibrate these instruments.

2.2.1 *Static Calibration Methods*

The calibration results of three tipping-bucket rain gauge models and eight tipping-bucket flow meter models amalgamated from the existing literature and this study are presented in Table 2.1. At first, static calibration, which quantifies the volume of one tip (c) directly, should be conducted (Fig. 2.1). Once c has been established, the relationship between the inflow rate and the actual amount one tip is determined by dynamic calibration.

Since manufacture stated tip-inducing volumes (c_m) for the tipping-bucket rain gauges are all <16 mL (Table 2.1), we used injectors to fill the tipping-bucket mechanism with water on a drop by drop basis in order to obtain the volume of water required to cause a single tip (c). More specifically, c was found as the difference in the mass, using an electronic scale, of the injector plus the water within the injector at the start of the calibration procedure and the mass of the injector plus the water remaining in the injector once a tip had occurred. However, tipping-bucket flow meters have stated c_m values > 50 mL and, in order to avoid injection errors while determining c for these instruments, a beaker was used to help facilitate the calibration procedure. When calibrating the U50 flow meter, for example, 40-mL of water was initially added to the bucket with the injector being used for subsequent additions of water until c was realized.

2.2.2 Dynamic Calibration Methods

For dynamic calibration, it is imperative that a constant inflow rate (q, mL s^{-1}) be obtained. The key to maintain a stable q during the calibration procedure is to keep a constant difference in head between the inlet and outlet of water flow. A simple apparatus, depicted in Fig. 2.2, is used to generate q. The constant difference in head can be attained by maintaining an overflow of water (Fig. 2.2). The value of q is given as the mass of stored water, measured by an electronic balance, over a certain duration determined by a stopwatch. For each intensity of q, at least 10 tips are

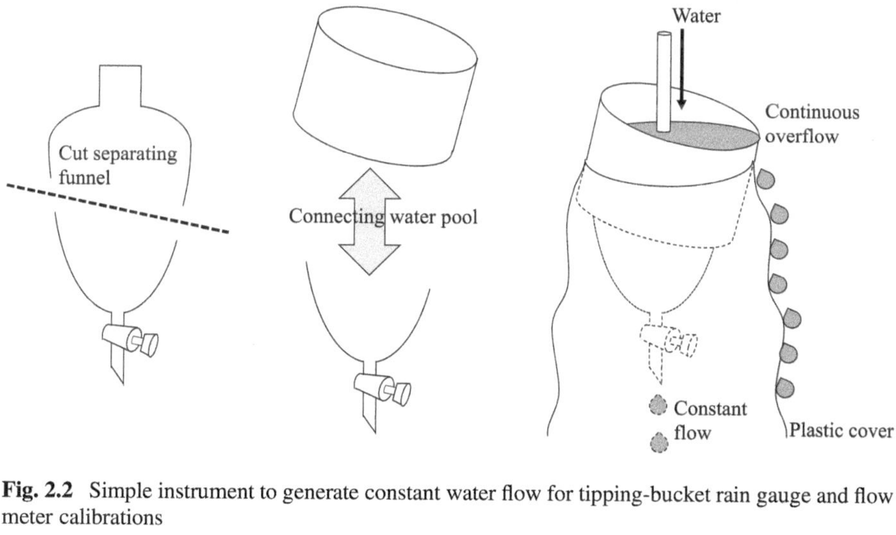

Fig. 2.2 Simple instrument to generate constant water flow for tipping-bucket rain gauge and flow meter calibrations

recorded as the times of each tip with 10 Hz with a datalogger (type CR1000, Campbell Scientific Inc., Logan, Utah). The average time between tips (t, s) is derived, while the volume associated with a tip (v, mL) is found as:

$$v = q \cdot t. \tag{2.5}$$

Constant inflow-volume relationships may vary even for the same type of tipping-bucket rain gauge or flow meter model. Shiraki and Yamato (2004) proposed that a common relationship for the same model of gauge or flow meter can be obtained by scaling q and v with c, with $Q = q/c$ and $V = v/c$. The best fit of the $Q - V$ relationships for all tipping-bucket rain gauges and flow meters evaluated is a quadratic curve (Table 2.1).

$$V = d \cdot Q^2 + e \cdot Q + 1 \tag{2.6}$$

where d and e are the fitting parameters.

As aforementioned, the underlying cause of the underestimation in water input rates is the failure of capturing q during the time of the tip (Δt). If Δt is constant to Q, the $Q - V$ relationship must be fitted with a linear equation ($V = \Delta t \cdot Q + 1$). However, since a quadratic curve has been found to be a better fit of the Q-V relationship and not the linear line, this implies that Δt decreases with increasing Q (Iida et al. 2012). Shiraki et al. (2018) measured Δt by high-speed-digital video, and confirmed the decreasing trend in Δt with increasing Q. Shiraki et al. (2018) also noted that the movement of stored water resulting from the high kinetic energy of Q, decreases the value of V compared with the predicted value from a linear relationship derived using a smaller range of Q.

Constant inflow, Q, can be obtained from Eq. 2.6 and that V can be equated with $Q \cdot t$:

$$Q = \frac{-(e - t) - \sqrt{(e - t)^2 - 4d}}{2d} \tag{2.7}$$

For the application of the calibration curves to data obtained in the field, Q is first calculated by substituting t into Eq.2.7, while the volume associated with a single tip is obtained as the product of c, Q, and t (Eq. 2.5), that is, $v = q \cdot t = c \cdot Q \cdot t$.

2.2.3 Dynamic Calibration of Tipping-Bucket Rain Gauges and Flow Meters

Of all the tipping-bucket rain gauges and flow meters evaluated, the U100 tipping-bucket flow meter shows the most significant underestimation of V for smaller values of Q (Table 2.1, Fig. 2.3). For this model, a 10% underestimation of V (i.e., $V = v/c = 1.1$) occurs when $Q > 0.12\ \mathrm{s}^{-1}$. For tipping-bucket rain gauges,

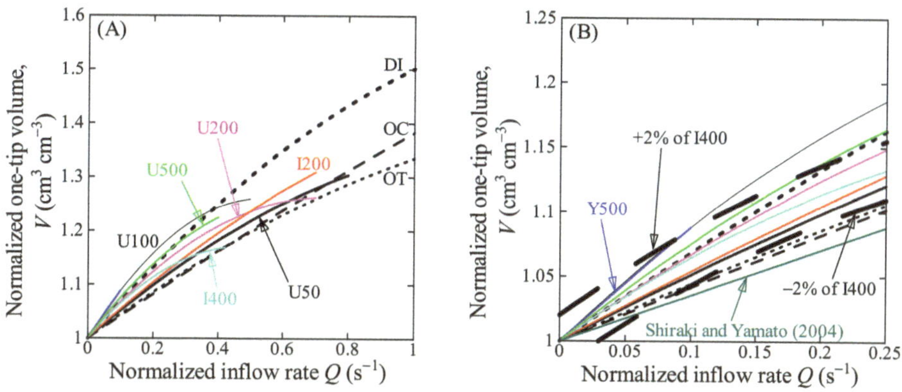

Fig. 2.3 Relationships between normalized inflow rate (Q) and one-tip volume (V) for a range of Q from (**a**) 0 to 1 and (**b**) 0 to 0.25 s^{-1}. Refer to Table 2.1 for details of target tipping-bucket rain gauges and flow meters and calibration curves

the OT and OC models have a relatively small amount of underestimation, and 10% errors are not detected for Q values less than 0.24 s^{-1} (Fig. 2.3). If U100 has $c = 100$ mL and $Q = 0.12\ s^{-1}$ then q ($q = c \cdot Q$) is 12 mL s^{-1}. Additionally, if the c of the OT and OC rain gauges are 0.5 and 0.2 mm, respectively, then $Q = 0.24\ s^{-1}$ and thus $q = 0.12$ mm s^{-1} and 0.048 mm s^{-1}, respectively.

For almost all types of large tipping-bucket flow meters ($c_m \geq$ 150–200 mL), Shimizu et al. (2018) suggest that the $Q - V$ relationships obtained through laboratory tests can be plotted within the range of ±2% of that for the I400 model (Table 2.1) when Q is less than 0.1–0.2 s^{-1} (corresponding to 0.1–0.2 Hz tipping rate). This indicates the calibration equation for I400 can be applicable for most types of large tipping-bucket flow meters. However, our calibration results for the U100 model found that the $Q - V$ relationship exceeds the +2% range of the I400 gauge when $Q \geq 0.1\ s^{-1}$ (Fig. 2.3b). Considering that a high frequency tip rate occurs more readily on a tipping-bucket flow meter with a relatively small c_m, precaution must be taken when applying the equation for the I400 flow meter to flow meters with $c_m \leq 100$ mL.

Similar to the U100 gauge, the $Q - V$ relationships for the OT and OC tipping-bucket rain gauges lie outside of the −2% range of the I400 flow meter even when $Q \approx 0.1\ s^{-1}$ or less (Fig. 2.3b). The DI rain gauge (Table 2.1), however, was found to have a $Q - V$ relationship that was within the +2% range when $Q < 0.22$ (Fig. 2.3b). These findings suggest the difficulty in establishing a representative dynamic calibration equation which would cover most tipping-bucket rain gauges. Shiraki and Yamato (2004) proposed a common calibration eq. ($V = 0.353Q + 1$) for tipping-bucket rain gauges; however, it should be noted that the equation lies within the lowermost of the plots in Fig. 2.3b. As such, the application of the Shiraki and Yamato (2004) equation underestimates the water input for all tipping-bucket rain gauges. Determining if a common calibration equation that could be applied to most types of tipping-bucket rain gauges can be derived should be revisited once laboratory and field tests for larger samples of several of tipping-bucket rain gauges and flow meters has been accomplished.

2.2.4 Cautions for Applying the Calibration to Measured Data

Applying static calibration to tipping-bucket rain gauges and flow meters is critical to measure q correctly. Shimizu et al. (2018) reported a maximum difference between c and c_m of +2.5% for larger tipping-bucket flow meters. Maximum differences between c and c_m among the tipping-bucket rain gauges and 200-mL flow meters calibrated by Iida et al. (2012) were −15% and +3.2%, respectively. Additionally, Iida et al. (2018) found that the difference between c and c_m was −16% for the DI-type tipping-bucket rain gauge. These examples clearly illustrate the need for the static calibration of all types of tipping-bucket rain gauges and flow meters.

High-frequency data (e.g., 10 Hz) were used for the dynamic calibrations of the tipping-bucket rain gauges and flow meters. However, field measurements are usually recorded as the accumulated number of tips, n_t, during a given duration, D. In these circumstances, instead of actual t, the average time between tipping T, found as $T = D/n_t$, is substituted into (Eq. 2.7) and the corrected v is derived. However, if D is longer than a suitable range, T can be overestimated and the degree of correction to v would be smaller than that based using actual t. Iida et al. (2012) investigated the suitable range of D for a temperate forest in Japan and a tropical forest in Cambodia. The D values obtained were up to 60 min and 10 min for P and T_f measurements in the temperate and the tropical forest, respectively. Stemflow measurements required $D < 150$ s for a tree having a DBH of 25.5 cm and funneling ratios > 10 in the tropical forest, while in the temperate forest $D = 600$ s was suitable for a tree with a DBH of 27.1 cm and funneling ratios < 20 (Iida et al. 2012). However, these examples of D may not be applicable to measurements of P, T_f and S_f for different forest ecosystems. As such, we recommend that D be set to the shortest time interval permissible based on the measurements of t for interception studies.

2.2.5 Effect of Dynamic Calibration on Interception Loss Estimates

Interception loss (E_I) is calculated as the difference between P and the sum of T_f and S_f (see Eq. 2.3). P is the largest component of (Eq. 2.3), and therefore, in most cases, the amount of correction by dynamic calibration is largest for P. Thus, if the dynamic calibration is not applied, E_I would be underestimated. However, there are some cases that E_I is overestimated for combinations of different types of tipping-bucket rain gauges (Fig. 2.4, Iida et al. 2018). We investigated the nine combinations of P and T_f measurements with three tipping-bucket rain gauge models (type OT, OC, and DI; Table 2.1). When the same model of tipping-bucket rain gauge is used to measure both P and T_f, the effects on E_I are relatively small (Fig. 2.4a, b, and c). The

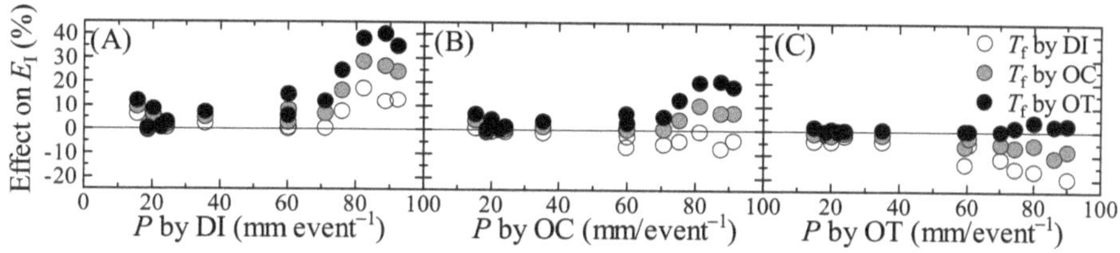

Fig. 2.4 Effect of dynamic calibrated gross rainfall (P) and throughfall (T_f) on interception loss (E_I). (**a**) P is measured by Rain Collector II, Davis Instruments (DI), (**b**) RG-3M, Onset Computer Corp. (OC), and (**c**) OW-34-BP, Ota keiki seisakusho Co., Ltd. (OT). 10% effect means 10% increase in E_I after applying the correction, that is, 10% underestimation. Data are cited from Iida et al. (2018)

best scenario is the combination of both P and T_f measured with the OT gauge, in which a maximum effect on E_I of +4% is detected (Fig. 2.4c). A +4% error, however, is not negligible, and thus, even under optimum conditions, tipping-bucket gauges require correction using the dynamic calibration curve. The fact that evaluation of spatially representative T_f needs many tipping-bucket rain gauges, it can be expected that lower-cost tipping-bucket rain gauges, such as the DI type, would be used for T_f measurements. However, a −20% error in E_I (20% overestimation) is found when T_f is measured with the type DI gauge (Fig. 2.4c), having the largest systematic bias among the three tipping-bucket rain gauges evaluated (Fig. 2.3a). If the DI-type tipping-bucket rain gauge is also used to measure P, E_I is underestimated by as much as 40% (Fig. 2.4a).

This section discussed the error associated with tipping-bucket rain gauges and flow meters at the rainfall-event scale. If, however, the dynamics of the interception process during the rainfall event are to be investigated (Reid and Lewis 2009, Iida et al. 2017), then dynamic calibrations at the hourly time scale are highly recommended.

2.3 Calibration of Sap Flow Sensors Measuring Forest Transpiration

Sap flux density, F_D, measurements are widely used to evaluate forest transpiration, E_T (Wilson et al. 2001; Kumagai et al. 2008; Oishi et al. 2008). Transpiration from a tree, using sap flow (Q_T) as a surrogate, can be estimated as the product of sapwood area (A_S) and F_D averaged over sapwood area, $\overline{F_D}$:

$$Q_T = A_S \cdot \overline{F_D}. \tag{2.8}$$

Transpiration can be estimated by scaling Q_T up from the tree to the stand level (e.g., Kumagai et al. 2008). Both A_S and $\overline{F_D}$ are required in order to estimate E_T, and a detailed analysis is necessary to derive the number of samples required to estimate the representative values of A_S and $\overline{F_D}$ (Kumagai et al. 2005a, b).

The focus of this section is on the required calibration associated with three different sap flow techniques, namely, (i) thermal dissipation (TD) (Granier 1985), (ii) heat ratio (HR) (Burgess et al. 2001), and (iii) heat field deformation (HFD) (Nadezhdina et al. 2012) (Figs. 2.1 and 2.5a). All three techniques insert probes and a heater into a stem and detect the F_D by using heat as a tracer. Recently, some studies report that TD underestimates F_D (e.g., Steppe et al. 2010; Peters et al. 2018) and uncertainties of measuring F_D with HFD (e.g., Steppe et al. 2010; Fuchs et al. 2017). We sampled stem segments of Japanese cedar (*Cryptomeria japonica*) planted in the nursery of the Forestry and Forest Products Research Institute (FFPRI), Tsukuba, approximately 50 km northeast of Tokyo, and in the Nagasaka Experimental Watershed, northern Honshu island (Table 2.2). An artificial Q_T is

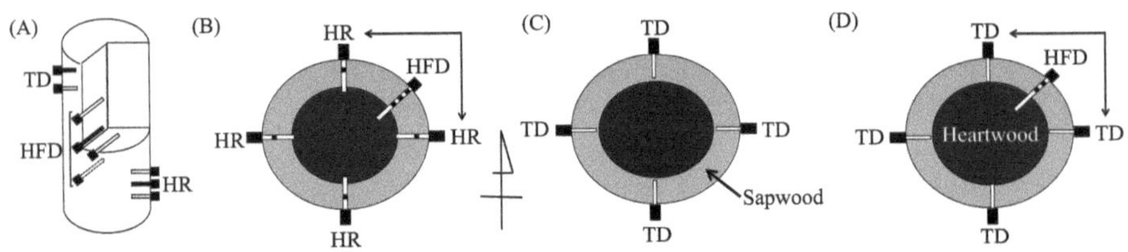

Fig. 2.5 Schematic diagram showing (**a**) sap flow sensors of three methods of thermal dissipation (TD), heat field deformation (HFD), and heat ratio (HR) and (**b**, **c**, **d**) arrangements of sensors used in calibration. Black colored probe includes heater. Arrangements of sensors for tree segments of (**b**) F1 to F4, (**c**) F5 to F8, and (**d**) N1 to N5 are indicated (refer to Table 2.2). Note that the measurement points are marked as black squares for sensors of HFD and HR methods (**b** and **d**). Sensors indicated by arrows are used to calculate the coefficient of *a* (**b** and **d**)

Table 2.2 Characteristics of the Japanese cedar test trees

Segment	DBH (cm)	TH (m)	*SW* (cm)	A_S (cm^2)	Calibration
FFPRI nursery					
F1	11.1	9.7	3.6	78.6	HR, HFD
F2	10.9	9.0	2.8	68.1	HR, HFD
F3	12.2	10.0	3.0	81.1	HR, HFD
F4	12.5	9.1	3.1	86.2	HR, HFD
F5	11.6	9.5	2.8	73.6	TD
F6	11.5	9.8	2.5	69.2	TD
F7	11.2	9.3	2.4	60.2	TD
F8	11.5	9.2	2.4	65.0	TD
Nagasaka					
N1	13.4	16.5	2.1	68.9	TD, HFD
N2	12.5	14.1	1.9	59.1	TD, HFD
N3	13.9	14.7	1.7	61.2	TD, HFD
N4	12.0	14.7	1.9	55.5	TD, HFD
N5	11.8	9.9	1.9	56.1	TD, HFD

generated by applying negative pressure at the upper surface of the segment with a vacuum pump and calibration of the three techniques was conducted in the laboratory.

2.3.1 Characteristics of the Three Sap Flow Techniques

The thermal dissipation, TD, method proposed by Granier (1985) inserts a pair of thermocouples into a plant stem (Fig. 2.5a). The upper probe measures the temperature of an included heater to generate constant heat of 0.2 W, while the lower probe measures the xylem temperature as a reference. The space between probes should be more than 10 cm (Iida and Tanaka 2010), while the length of the probes is 2.0 cm. The temperature difference between the probes (ΔT) increases during low F_D and decreases with increasing F_D. Granier (1985) obtained the following calibration curve between the parameter K_{TD} and F_D:

$$F_{D-TD} = 42.84 \cdot {K_{TD}}^{1.231} = 42.84 \cdot \left(\frac{\Delta T_0 - \Delta T}{\Delta T}\right)^{1.231} \tag{2.9}$$

where $F_{D\text{-}TD}$ is the F_D measured by the TD method (cm^3 cm^{-2} h^{-1}), and ΔT_0 is ΔT when $F_D = 0$. We employed handmade sensors described by Kumagai et al. (2005a). The TD sensor detected the average F_D across the range of probe.

The heat field deformation (HFD) method also uses heat to measure F_D. A continuously powered heater is inserted at the center of the upper and lower temperature probes with an additional temperature probe installed tangentially next to the heater (Fig. 2.5a). F_D is calculated by the following Nadezhdina et al. (2012):

$$F_{D-HFD} = \frac{D(K_{HFD} + dT_{s-a})Z_{ax}}{dT_{as} \cdot Z_{tg} \cdot SW} \cdot 3600. \tag{2.10}$$

where $F_{D\text{-}HFD}$ is F_D measured by HFD (cm^3 cm^{-2} h^{-1}), $dT_{s\text{-}a}$ is the temperature difference between the upper and tangential probes (°C), or can be equated with the difference between dT_{sym} and dT_{as}, which represent the temperature differences (°C) between the upper and lower probes and between the tangential and lower probes, respectively, Z_{ax} is the difference between the heater and upper or lower temperature probe (1.5 cm), Z_{tg} is the distance between the heater and the tangential probe (0.5 cm), and SW is the width of sapwood (cm), while K_{HFD} is the $dT_{s\text{-}a}$ when F_D is equal to zero.

Nadezhdina et al. (2012) found that the ratio of dT_{sym} to dT_{as} (dT_{sym}/dT_{as}) has a linear relationship with F_D and that K_{HFD} may be equated with the y-intercept value of the linear regression between dT_{sym}/dT_{as} (dependent variable) and dT_{as} (independent variable). Since the absolute value of $dT_{s\text{-}a}$ is equal to that of dT_{as} when F_D is zero, the absolute value of y-intercept derived from the linear regression between dT_{sym}/dT_{as} and $dT_{s\text{-}a}$ must be equal to that between dT_{sym}/dT_{as} and dT_{as}. The

HFD method determines K_{HFD} objectively based on two linear regressions which meet the condition of having the same absolute values of y-intercepts, representing an advantage of the HFD method (Nadezhdina et al. 2012). We used a commercial HFD sensor (HFD8, ICT international Pty. Ltd.). The length of the heater and temperature probes are 11.7 and 9.7 cm, respectively, and eight thermistors are positioned in a 1-cm span from the probes. We inserted temperature probes into tree stem segments to locate the first thermistor at the depth of 0.5 cm in sapwood.

The heat ratio, HR, method, which is an alternate version of heat pulse method, calculates heat pulse velocity (HPV_r, cm h^{-1}) based on the temperature changes of the xylem caused by induced heat by a heater and measured with two temperature probes located at the same distance from the heater (Fig. 2.5a). HPV_r is then found as (Marshall 1958; Burguess et al. 2001):

$$HPV_r = \frac{D}{x} \ln\left(\frac{v_1}{v_2}\right) \cdot 3600. \tag{2.11}$$

where D is thermal diffusivity of fresh wood (2.5×10^{-3} cm^2 s^{-1}; Marshall 1958), x is the distance between the heater and the temperature probe (0.5 cm), v_1 and v_2 are the increase of xylem temperature (°C) by heat detected by the upper and lower temperature probes, respectively.

HPV_r is affected by the wounding effect caused by the nonconducting wood around the probes, which results from the mechanical damage to the xylem tissue during sensor installation (Swanson and Whitfield 1981; Burgess et al. 2001). Burgess et al. (2001) proposed the following equation to correct for the wounding effect:

$$HPV_c = B \cdot HPV_r. \tag{2.12}$$

where HPV_c is corrected HPV_r by considering the wounding effect, B is the correction coefficient. In this study, we assume the wound width of 0.17 cm and applied a B value of 1.7283 (Burgess et al. 2001). Sap flux density, F_D, measured with the HR method ($F_{D\text{-}HR}$) is then derived as:

$$F_{D-HR} = \frac{\rho_b}{\rho_s}\left(m_c + \frac{c_{dw}}{c_s}\right) HPV_c. \tag{2.13}$$

where ρ_b is dry wood density of Japanese cedar (0.314 g cm^{-3}; Fujiwara et al. 2004), ρ_b is density of sap assumed to be equal to the density of water (1.0 g cm^{-3}), m_c is the mass of the water content of sapwood relative to dry weight of sapwood (1.78), c_{dw} is the specific heat capacity of oven-dried wood (J g^{-1} °C^{-1}), and c_s is the specific heat capacity of sap, assumed to be equal to that of water (4.186 J g^{-1} °C^{-1}).

The ratio of c_{dw} to c_s is often assumed to be constant at 0.33 (=1.380/4.186; Dunlap 1912; Edwards and Warwick 1984; Steppe et al. 2010). We used a commercial HR sensor (type SFM1, ICT international Pty. Ltd.), whose sensor length is

3.5 cm including two thermistors to detect v at different depths. We installed sensors into stem segments to detect $F_{D\text{-}HR}$ at the depth of 1 cm of sapwood.

2.3.2 *Artificial Sap Flow Generated by a Vacuum Pump*

In order to calibrate the sap flow techniques, an artificial and known Q_T is required. Granier (1985) generated Q_T by applying positive water pressure on the tree segment. Similar methods of generating Q_T using positive pressures have also been proposed (Herbst et al. 2007; Steppe et al. 2010; Hubbard et al. 2010; Bosch et al. 2014; Fuchs et al. 2017; Ouyang et al. 2018; Peters et al. 2018). In some instances, a negative pressure has been used to control Q_T (Taneda and Sperry 2008; Bush et al. 2010; Schmidt-Walter et al. 2014). Additionally, the amount of transpiration from a tree, which equates with Q_T, has been found by monitoring the change in mass of trees planted in a lysimeter (Lu and Chako 1998; Mcculloh et al. 2007). Takeuchi et al. (2017) excavated a tree, including the root-ball—a common practice for transplanting trees—with Q_T equated to the changes in the total mass of the tree and root-ball, a technique the authors termed the "weighing root-ball" method. Sun et al. (2012) estimated Q_T using potometer experiments, in which cut foliated tree branches were submerged into water and the amount of absorbed water was measured, while Lopez et al. (In press) cut *Eucalyptus grandis* trees ranging from 3 to 6 cm in diameter and submerged them into a fixed-volume reservoir with Q_T found as the change in the volume of water held in the reservoir.

In this study, we apply negative pressure to generate Q_T within a cut stem segment of Japanese cedar (Fig. 2.6) (Shinohara et al. 2016). An attachment designed to clean the inside of the PVC pipe (Fig. 2.6) was fixed to the upper surface of the cut segment and connected to a vacuum pump. Then, the segment was

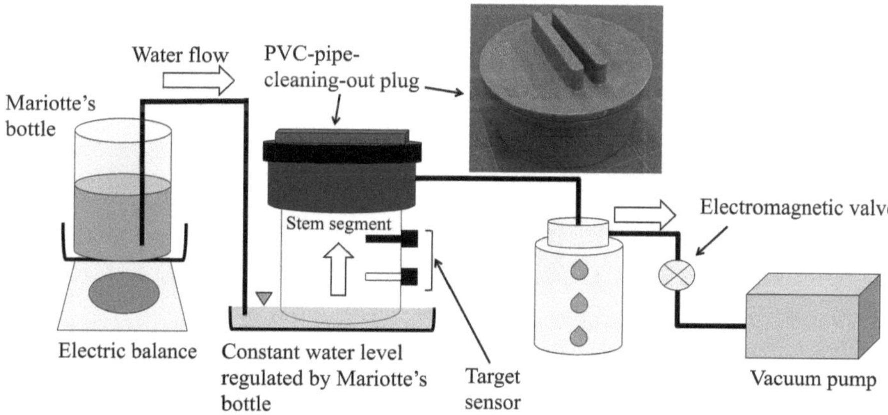

Fig. 2.6 A vacuum pump system to generate artificial sap flow within a stem segment, originally devised by Shinohara et al. (2016)

suspended by the cable ties, and its lower surface was submerged into free water surface. Depending on the pressure gradient, which was typically between 0.03 and 0.14 MPa/m, the lower surface absorbs KCl solution. This water surface was kept at a constant level with a Mariotte's bottle. The water flow, Q_T, was then found as the decrease in weight of the Mariotte's bottle with an electric balance with a 10-sec temporal resolution.

Precise calibration of the sap flow techniques is accomplished by comparing the actual F_D at the sensor position to the F_D measured by the sensor. When cut stem segments are used, the actual value of Q_T is obtained and the gravimetric sap flux density, $F_{D\text{-}G}$, is derived within a segment as Q_T/A_S (Fig. 2.1, Eq. 2.8). For each of the segments calibrated, a clear radial trend in $F_{D\text{-}HFD}$ was found (Fig. 2.7a, Table 2.2), and different $F_{D\text{-}TD}$ were measured among the four cardinal directions (i.e., north, east, south, and west) (Fig. 2.7b, Table 2.2). Similar results of radial and azimuthal distribution of the F_D within stem segments have been reported by Steppe et al. (2010). Thus, it is likely that the actual F_D derived at sensor position is not equal to $F_{D\text{-}G}$ and that the sensor output at the single position is not sufficient to perform suitable calibrations. In this study, measurements of F_D variations over sapwood area for all cut stem segments were made (Fig. 2.5b, c, d).

2.3.3 Calibration of the TD Method

Tree stem segments with SW being approximately 2.0 cm (equal to the senor length) were selected for the TD calibrations (Table 2.2). The four TD sensors were inserted into the stem segments so that the azimuthal variation of F_D (north, east, south, and

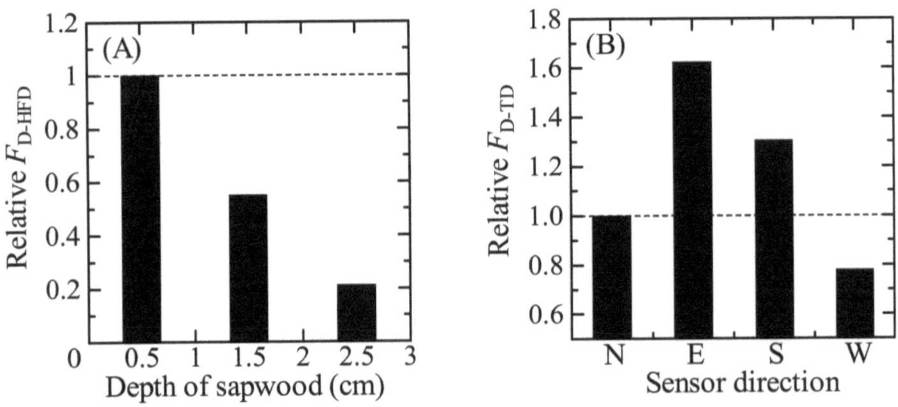

Fig. 2.7 Panel (**a**) shows radial distribution of sap flux density detected by HFD sensor ($F_{D\text{-}HFD}$) in cut segment of F4 (see Table 2.2). $F_{D\text{-}HFD}$ was shown as relative to $F_{D\text{-}HFD}$ measured at 0.5 cm depth (see Fig. 2.5B). Panel (**b**) indicates azimuthal variation in sap flux density measured by four TD sensors ($F_{D\text{-}TD}$) installed in a cut segment of N2 (see Table 2.2). North, east, south, and west $F_{D\text{-}TD}$ are shown as relative to north $F_{D\text{-}TD}$ (see Fig. 2.5d)

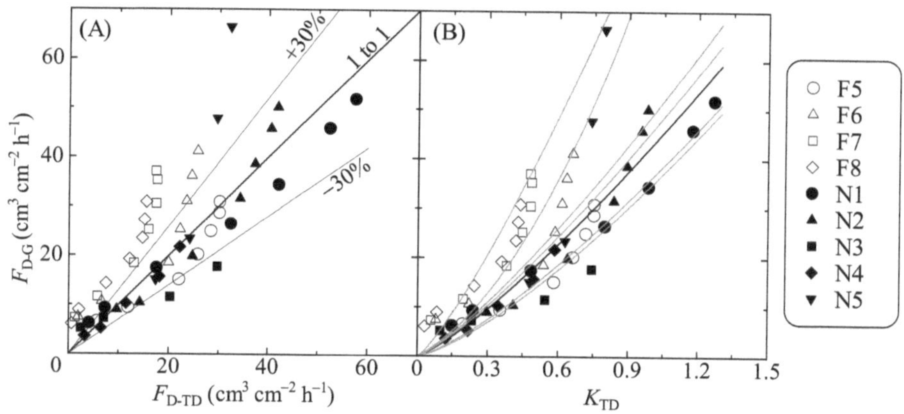

Fig. 2.8 Comparisons of gravimetric sap flux density ($F_{D\text{-}G}$) and (**a**) measured sap flux density with TD method ($F_{D\text{-}TD}$), and (**b**) parameter K_{TD}. Open and black symbols show the data measured at the segments of Japanese cedar sampled in FFPRI nursery and Nagasaka site, respectively. Black curve is the original calibration by Granier (1985), and the gray curves are previously published calibrations for softwood species (Sun et al. 2012; Bosch et al. 2014; Ouyang et al. 2018; Peters et al. 2018)

west directions) was accounted for and the averaged $F_{D\text{-}TD}$ is compared with $F_{D\text{-}G}$ (Fig. 2.1). Note that when *SW* is less than 2.0 cm, the correction proposed by Clearwater et al. (1999) should be applied. The $F_{D\text{-}TD}$ values were found to be within ±30% of the $F_{D\text{-}G}$ of stem segments sampled at Nagasaka site (Fig. 2.8a). Thus, for the Nagasaka site stem segments, clear underestimations of $F_{D\text{-}TD}$ were not found. However, for stem segments at the FFPRI nursery, most $F_{D\text{-}TD}$ values were found to be less than $F_{D\text{-}G}$ and underestimations of $F_{D\text{-}TD}$ were detected. Relatively high tree-specific variations among the $F_{D\text{-}TD}$ data were found for the four FFPRI stem segments, illustrating the difficulty of establishing a calibration equation. Almost all the variations in the relationship between the parameter K_{TD} and $F_{D\text{-}G}$ for the nine stem segments (87% of data plotted in Fig. 2.8b) are in the range of the reported calibration curves.

2.3.4 *Calibration of the HFD Method*

It is important to recognize that the heat field deformation, HFD, method measures point F_D at the detection depths in rather than measuring the average F_D along the TD sensor. We derived the gravimetric F_D values at the HFD sensor measurement points, $F_{D\text{-}G\text{-}HFD}$, as follows: Four FFPRI stem segments and five Nagasaka stem segments (Table 2.2) were used with F_D measured in each of the four cardinal directions with a HR sensor (Fig. 2.5b) for the FFPRI stem segments, while for the Nagasaka stem segments, four TD sensors were installed in each direction

(Fig. 2.5d). The ratio of the average of two-direction values measured by the HR or TD method to the average of all four direction values (a) was calculated (2.5b, d). The radial distribution of F_D is obtained from $F_{D\text{-}HFD}$ measurements at the different depths within the sapwood (Fig. 2.5b, d). We calculate the ratio of $F_{D\text{-}HFD}$ at a certain depth (i) to the average $F_{D\text{-}HFD}$ for all depths within the sapwood (r_i), and gravimetric sap flux density at the detection point of the HFD sensor, $F_{D\text{-}G\text{-}HFD}$, at the depth of i can be obtained as $a \cdot r_i \cdot F_{D\text{-}G}$ (Fig. 2.1).

The plots of $F_{D\text{-}HFD}$ were found to be distributed around the 1:1 relationship with $F_{D\text{-}G\text{-}HFD}$, however, relatively high dispersion of $F_{D\text{-}HFD}$ with values overlying +30% or underlying −30% was also found (Fig. 2.9a), indicating large tree-to-tree differences. Focusing on FFPRI stem segments, the HFD method clearly underestimated $F_{D\text{-}G\text{-}HFD}$ with the linear relationship of $F_{D\text{-}G\text{-}HFD} = 1.33\ F_{D\text{-}HFD}$, $R^2 = 0.72$, being derived (Fig. 2.9b). However, when the Nagasaka stem segments are considered, the values show relatively high variability and trends of overestimation or underestimation were not detected ($F_{D\text{-}G\text{-}HFD} = 0.94\ F_{D\text{-}HFD}$, $R^2 = 0.69$).

2.3.5 *Calibration of the HR Method*

The heat ratio, HR, method was calibrated using the four FFPRI stem segments (Table 2.2, Fig. 2.5b). Similar to the HFD method, the HR sensor measures point F_D. The gravimetric F_D at the detecting point of the HR sensor ($F_{D\text{-}G\text{-}HR}$) is calculated using the similar methodology as for the HFD calibrations. In this case, four-direction values of $F_{D\text{-}HR}$ were measured, and we compared the averaged $F_{D\text{-}HR}$

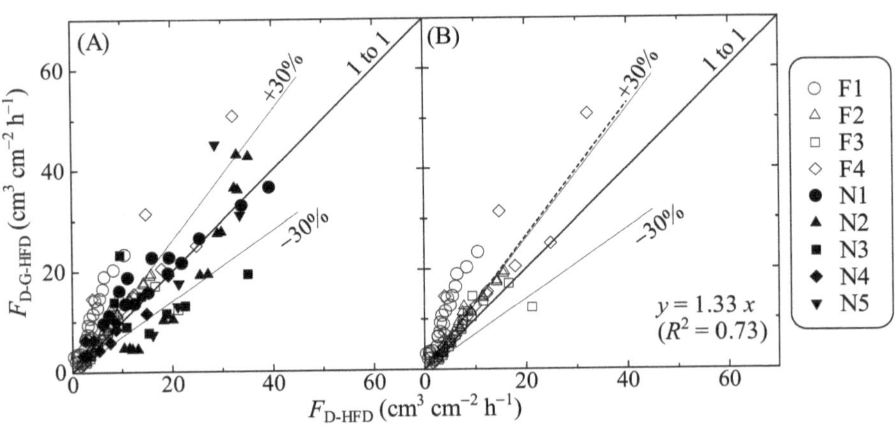

Fig. 2.9 Relationship between gravimetric sap flux density at sensor position of heat field deformation (HFD) method ($F_{D\text{-}G\text{-}HFD}$) and measured sap flux density with HFD ($F_{D\text{-}HFD}$) for (**a**) all cut segments and (**b**) FFPRI nursery segments (F1 to F4 in Table 2.2). The dashed line shows the linear regression for FFPRI stem segments ($y = 1.33x$, $R^2 = 0.72$)

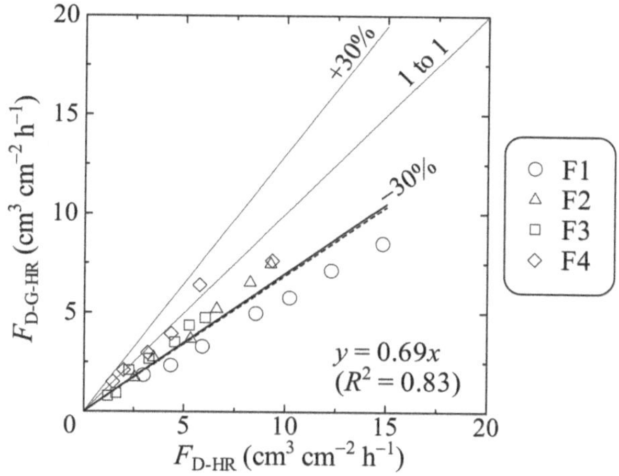

Fig. 2.10 Relationship between gravimetric sap flux density at the detecting depth of heat ratio (HR) method ($F_{D\text{-}G\text{-}HR}$) and measured sap flux density with HR ($F_{D\text{-}HR}$). The dashed line indicates the regression line ($y = 0.69x$, $R^2 = 0.83$)

values among four directions with the gravimetric sap flux density at the measuring depth of HR, $F_{D\text{-}G\text{-}HR}$, obtained as $r_i \cdot F_{D\text{-}G}$ (Fig. 2.1). A very high correlation between $F_{D\text{-}HR}$ and $F_{D\text{-}G\text{-}HR}$ was found, with almost all $F_{D\text{-}HR}$ values being larger than the $F_{D\text{-}G\text{-}HR}$ values, thereby indicating the tendency of the HR method to overestimate F_D (Fig. 2.10). When the HR method was applied to the Japanese cedar trees in the FFPRI nursery, reasonable estimations of F_D were obtained by the linear regression: $F_{D\text{-}G\text{-}HR} = 0.69\ F_{D\text{-}HR}$, $R^2 = 0.83$. In this study, the wounding width of 0.17 cm was assumed resulting in $B = 1.7283$. $F_{D\text{-}HR}$ can vary depending on B (see Eq. 2.12 and 2.13). There are some possibilities that the actual width of the wounding is smaller than 0.17 cm and the overestimation of the wounding effect on $F_{D\text{-}HR}$ would occur. For future calibrations, directly measurements of the wounding width are required.

2.3.6 Summary of Sap Flow Calibration for the Three Techniques

We strongly suggest that the radial and azimuthal variations in F_D occur in the cross-section of cut segments (Fig. 2.7). As a consequence, these variations must be measured in order to compare sensor output with the gravimetric values of F_D. Although the Nagasaka stem segments showed relatively low coefficients of determination associated with the calibration curves of the TD and HFD methods, any tendencies, that is overestimations or underestimations, were not found (Fig. 2.8a, 2.9a). These results indicate that stand-scale measurements of F_D with the TD and HFD methods would be accurate and representative and that no calibrations were required at the Nagasaka site. However, for FFPRI stem segments, the TD and HFD methods underestimated F_D by 30% or more (Fig. 2.8a and 2.9b), while for the HR method at this site, an overestimation of approximately 30% was found (Fig. 2.10). Thus, if no corrections were applied to these stem segments the TD and HFD methods would have underestimated E_T by at least 30% and often more, while the HR method would have resulted in an overestimate of E_T of approximately 30%.

Compared with the TD and HFD techniques, the HR method shows very small tree-to-tree differences in calibration results (Fig. 2.10). HR is the only technique among three methods that is based on the physics of heat dissipation (e.g., Fuchs et al. 2017). Sun et al. (2012) showed that the accuracy of the TD method is not dependent on differences in xylem anatomy of the target tree species (i.e., diffuse-porous, ring-porous, and tracheid). However, Fan et al. (2018) suggest that the point or "discrete" F_D measurements derived using the HR and HFD methods are affected by the differences in both hydraulic conductivity and thermal diffusivity between the earlywood and latewood of pine trees. The tree-to-tree and site-specific differences found in this study are probably caused by the heterogeneity of the xylem anatomy including the spatial distribution of the earlywood and latewood. The precise explanation of the uncertainty of sap flow techniques is still unknown (e.g., Steppe et al. 2010), although site- and species-specific calibrations are likely suitable to obtain the most accurate values of F_D and E_T (e.g., Peters et al. 2018; Flo et al. 2019).

2.4 Future Directions in Calibration Studies for Tipping-Bucket Rain Gauges, Tipping-Bucket Flow Meters, and Sap Flow Techniques

The underestimation of rainfall and/or water inflow by tipping-bucket rain gauges and flow meters, respectively, creates a systematic bias dependent on the intensities of inflow. Fortunately, the establishment of correction curves is relatively straightforward, and the equipment required to generate constant inflow (Fig. 2.2), which is a key part of the calibration system, is relatively easy to construct. It is highly recommended to calibrate tipping-bucket rain gauges and flow meters when used for E_I studies. With the exception of Takahashi et al. (2011), Iida et al. (2012), Iida et al. (2018) and Shimizu et al. (2018), among a few others, relatively few studies have evaluated the effects of applying calibration curves on P, T_f, S_f, and/or E_I. Given the findings of work reviewed in this chapter, the degree of systematic underestimations on T_f and S_f, and uncertainties in E_I caused by tipping-bucket rain gauges and flow meters should be investigated for many types of forest ecosystems. We should note that the biases are larger for higher inflow rates (i.e., heavy rainfall) and over shorter time scales (i.e., hourly scale rather than event scale). Studies examining the intrastorm dynamics of interception processes with tipping-bucket rain gauges and flow meters are especially encouraged to utilize calibration curves (e.g., Iida et al. 2017). If tipping-bucket rain gauges and flow meters are already installed in the field without any established calibration curves, in situ static calibrations are strongly recommended. If practicable, the tipping-bucket gauges could be returned to the laboratory for dynamic calibrations, or, following Shimizu et al. (2018), a generalized correction equation could be employed for larger flow meters with $c_{\mathrm{m}} \geq 150$–200 mL. If the target equipment is out of the application range of the generalized equation, we argue choosing similar types of gauges from Table 2.1 and estimating the potential degree of underestimation. We believe that researchers should recognize the substantial effects of systematic biases of tipping-bucket rain gauges and flow meters on P, T_{f}, S_{f}, and E_{I}, and hope that all manufacturers will provide more detailed calibration results for users in near future.

The IPCC's Fifth Assessment Report (AR5) predicts that extremely large rainfalls will occur more frequently in mid-latitude and tropical areas (Stocker et al. 2014). In Japan, as elsewhere, disasters can be triggered by the localization of heavy rainfall (Tokyo Climate Center, Japan Meteorological Agency 2018). Valid measurements of rainfall are a fundamental requirement for a more precise understanding of the phenomena of localized heavy rainfall. The total degree of underestimation of a rainfall event is affected by not only the inflow intensity but also total number of tips. As such, parallel measurements of rainfall with a 0.5 mm or less tipping-bucket rain gauge and a 1.0 mm or more gauge at the same location are reasonable path forward, provided that data from the lager tipping-bucket rain gauge is selected for events with extremely high intensity. Tipping-bucket rain gauges which regulate high inflow rates using a siphon system are commercially available (e.g., type TB3, HyQuest Solutions Pty. Ltd., Australia). Shedekar et al. (2016) dynamically calibrated the TB3 and showed that the measurement error decreases and stabilizes with increasing inflow rates. Electronic weighing rain gauges, which store rainwater and weigh its amount with a load cell, are available (e.g., Seibert and Morén 1999). Turner et al. (2019) developed an instrument system to measure the change in the depth of stemflow by ultrasonic rangefinders. Theoretically, weighing and depth-measurement approaches have some advantage over tipping-bucket gauges, especially at higher rainfall intensities due to the lack of a tipping-bucket mechanism. However, very few comparisons of rainfall measurements have been made between gauges with inflow regulated and those that have nonregulated flow. Accordingly, further work could compare the differences to interception loss measurements among various tipping-bucket, weighing, and/or depth-measurement gauge combinations under various climate settings. Regardless, as regions of the world experience higher intensity rainfall events, the importance of gauge-corrected measurements from sound calibration procedures will only increase in the future, especially in water-stressed areas where rainfall regimes are changing.

Spurred by Steppe et al. (2010), a large number of TD calibration studies (e.g., Peters et al. 2018) have reported the underestimation of F_{D}. Such findings raise the importance of TD sensor calibration to accurately estimate E_{T}. Accordingly, we highlight the necessity of using stem segments in calibration procedures. We would note that the artificial Q_{T} can be measured directly for stem segments, but artificial F_{D} cannot be obtained from the single sensor measurement due to the azimuthal and radial trends in F_{D}. For TD sensors detecting F_{D} along a sensor length of 2.0 cm, we recommend using stem segments with a sapwood width of ~2.0 cm and applying enough TD sensors to take the azimuthal variation into account. On the other hand, HFD and HR methods detect F_D at the certain point of sensors. Thus, careful calculations of F_D at each detecting point are required for HFD and HR calibrations. As adopted in this chapter, the HFD method is relatively advantageous to evaluate the radial trend of F_D over the cross section of segments. To increase the number and quality of calibrations for HFD and HR methods, we make a call for future studies to pinpoint the actual F_D at any given sensor position. In addition, although there are numerous calibration studies, the mechanism causing the uncertainties in sap flow techniques is still unknown and should be clarified.

Despite the existence of numerous laboratory-based TD calibrations existed, the application of calibration equations to living trees in forest settings is very limited, with the notable exception of Steppe et al. (2010). Thus, we make a call for both sap flow sensor calibration studies and their applications to field data in order to improve E_T estimates. Wilson et al. (2001) estimated E using both the watershed water balance and eddy covariance methods and evaluated its three components: E_I, E_T, and E_F (Eq. 2.2). Based on the comparison between E and the sum of E_T, derived using the TD method, computed E_I and measured E_F, they suggest the possibility of E_T being underestimated using TD method. Shimizu et al. (2015), however, reported the sum of the three components correspond well with E obtained from both the water balance and eddy covariance methods. Similar findings were reported by Oishi et al. (2008). Compared with the number of studies whose focus is on calibration for the TD method, only a few studies evaluate how well E corresponds to the sum of its components, including E_T. In particular, no studies have reported the application of TD calibration to improve imbalances between E and the sum of E_I, E_T, and E_F. Based on the laboratory calibrations of the TD method, its future application to actual forests is highly recommended to evaluate E_T correctly and to derive a suitable strategy for the precise understanding of the hydrologic cycle in forested watersheds.

Acknowledgments We wish to thank Akita Forestry Research and Training Center for supporting our studies in the Nagasaka Experimental Watershed. This study was partially supported by the project "Research on adaptation to climate change for forestry and fisheries" founded by the Agriculture, Forestry and Fisheries Research Council, Japan, the Global Environmental Research Coordination System from Ministry of the Environment of Japan, and JSPS KAKENHI Grant Numbers JP21710021, JP26450495, JP18K05714 and JP19K06135. We also wish to recognize the anonymous reviewer as well as editors of this volume, especially DE Carlyle-Moses and DF Levia whose comments greatly improved this chapter.

References

Bosch JM, Hewlett JD (1982) A review of catchment experiments to determine the effect of vegetation changes on water yield and evapotranspiration. J Hydrol 55:3–23. https://doi.org/10.1016/0022-1694(82)90117-2

Bosch DD, Marshall LK, Teskey R (2014) Forest transpiration from sap flux density measurements in a southeastern coastal plain riparian buffer system. Agric For Meteorol 187:72–82. https://doi.org/10.1016/j.agrformet.2013.12.002

Burgess SSO, Adams MA, Turner NC, Beverly CR, Ong CK, Khan AA et al (2001) An improved heat pulse method to measure low and reverse rates of sap flow in woody plants. Tree Physiol 21:589–598. https://doi.org/10.1093/treephys/21.9.589

Bush SE, Hultine KR, Sperry JS, Ehleringer JR (2010) Calibration of thermal dissipation sap flow probes for ring-and diffuse-porous trees. Tree Physiol 30:1545–1554. https://doi.org/10.1093/treephys/tpq096

Carlyle-Moses DE, Gash JHC (2011) Rainfall interception loss by forest canopies. In: Levia DF, Carlyle-Moses DE, Tanaka T (eds) Forest ecology and biogeochemistry: synthesis of past research and future directions, ecol studies 216. Springer, Dordrecht, pp 407–423. https://doi.org/10.1007/978-94-007-1363-5_20

Carlyle-Moses DE, Lishman CE (2015) Temporal persistence of throughfall heterogeneity below and between the canopies of juvenile lodgepole pine (Pinus contorta). Hydrol Process 29:4051–4067. https://doi.org/10.1002/hyp.10494

Carlyle-Moses DE, Laureano JF, Price AG (2004) Throughfall and throughfall spatial variability in Madrean oak forest communities of northeastern Mexico. J Hydrol 297:124–135. https://doi.org/10.1016/j.jhydrol.2004.04.007

Carlyle-Moses DE, Lishman CE, McKee AJ (2014) A preliminary evaluation of throughfall sampling techniques in a mature coniferous forest. J For Res 25:407–413. https://doi.org/10.1007/s11676-014-0468-8

Carlyle-Moses DE, Iida S, Germer S, Llorens P, Michalzik B, Nanko K et al (2018) Expressing stemflow commensurate with its ecohydrological importance. Adv Water Resour 121:472–479. https://doi.org/10.1016/j.advwatres.2018.08.015

Clearwater MJ, Meinzer FC, Andrade JL, Goldstein G, Holbrook NM (1999) Potential errors in measurement of nonuniform sap flow using heat dissipation probes. Tree Physiol 19:681–687. https://doi.org/10.1093/treephys/19.10.681

Dunlap F (1912) The specific heat of wood. USDA Forest Service Bulletin No. 110, p 28

Edwards WRN, Warwick NWM (1984) Transpiration from a kiwifruit vine as estimated by the heat pulse technique and the penman-Monteith equation. New Zealand J Agric Res 27:537–543. https://doi.org/10.1080/00288233.1984.10418016

Edwards IJ, Jackson WD, Fleming PM (1974) Tipping bucket gauges for measuring runoff from experimental plots. Agric Meteorol 13:189–201. https://doi.org/10.1016/0002-1571(74)90046-6

Fan J, Guyot A, Ostergaard KT, Lockington DA (2018) Effects of earlywood and latewood on sap flux density-based transpiration estimates in conifers. Agric Meteorol 249:264–274. https://doi.org/10.1016/j.agrformet.2017.11.006

Flo V, Martinez-Vilalta J, Steppe K, Schuldt B, Poyatos R (2019) A synthesis of bias and uncertainty in sap flow methods. Agric For Meteorol 271:362–374. https://doi.org/10.1016/j.agrformet.2019.03.012

Friesen J, Van Beek C, Selker J, Savenije HHG, Van de Giesen N (2008) Tree rainfall interception measured by stem compression. Water Resour Res 44:W00D15. https://doi.org/10.1029/2008WR007074

Fuchs S, Leuschner C, Link R, Coners H, Schuldt B (2017) Calibration and comparison of thermal dissipation, heat ratio and heat field deformation sap flow probes for diffuse-porous trees. Agric Meteorol 244:151–161. https://doi.org/10.1016/j.agrformet.2017.04.003

Fujiwara K, Yamashita K, Hirakawa Y (2004) Mean basic density and density variation within individual trees in major plantation species. Bull FFPRI 3:341–348. (in Japanese with English abstract)

Granier A (1985) Une nouvelle méthod pour la mesure du flux de sève brute dans le tronc des arbres. Ann Sci For 42:193–200. https://doi.org/10.1051/forest:19850204. [Granier A (1985) A new method of sap flow measurement in tree trunks. English translation by Gash JHC, Granier A (2007) In Evaporation, Benchmark Papers in Hydrology 2. Gash JHC, Shuttleworth, WJ (eds). IAHS Press: Oxfordshire; 61–63.]

Helvey JD, Patric JH (1965) Canopy and litter interception of rainfall by hardwoods of eastern United States. Water Resour Res 1:193–206. https://doi.org/10.1029/WR001i002p00193

Herbst M, Roberts JM, Rosier PT, Gowing DJ (2007) Seasonal and interannual variability of canopy transpiration of a hedgerow in southern England. Tree Physiol 27:321–333. https://doi.org/10.1093/treephys/27.3.321

Herwitz SR (1985) Interception storage capacities of tropical rainforest canopy trees. J Hydrol 77:237–252. https://doi.org/10.1016/0022-1694(85)90209-4

Hubbard RM, Stape J, Ryan MG, Almeida AC, Rojas J (2010) Effects of irrigation on water use and water use efficiency in two fast growing Eucalyptus plantations. For Ecol Manag 259:1714–1721. https://doi.org/10.1016/j.foreco.2009.10.028

Iida S (2009) Rainfall redistribution by vegetation. In: Sugita M, Tanaka T (eds) Hydrologic laboratory of the university of Tsukuba, Japan hydrologic science. Kyoritsushuppan, Tokyo, pp 103–117. (in Japanese)

Iida S, Tanaka T (2010) Effect of the span length of Granier-type thermal dissipation probes on sap flux density measurements. Ann For Sci 67:408. https://doi.org/10.1051/forest/2009128

Iida S, Shimizu T, Kabeya N, Nobuhiro T, Tamai K, Shimizu A et al (2012) Calibration of tipping-bucket flow meters and rain gauges to measure gross rainfall, throughfall, and stemflow applied to data from a Japanese temperate coniferous forest and a Cambodian tropical deciduous forest. Hydrol Process 26:2445–2454. https://doi.org/10.1002/hyp.9462

Iida S, Levia DF, Shimizu A, Shimizu T, Tamai K, Nobuhiro T et al (2017) Intrastorm scale rainfall interception dynamics in a mature coniferous forest stand. J Hydrol 548:770–783. https://doi.org/10.1016/j.jhydrol.2017.03.009

Iida S, Levia DF, Nanko K, Sun X, Shimizu T, Tamai K et al (2018) Correction of canopy interception loss measurements in temperate forests: a comparison of necessary adjustments among three different rain gauges based on a dynamic calibration procedure. J Hydrometeorol 19:547–553. https://doi.org/10.1175/JHM-D-17-0124.1

Kimmins JP (1973) Some statistical aspects of sampling throughfall precipitation in nutrient cycling studies in British Columbian coastal forests. Ecology 54:1008–1019. https://doi.org/10.2307/1935567

Kumagai T, Aoki S, Nagasawa H, Mabuchi T, Kubota K, Inoue S et al (2005a) Effects of tree-to-tree and radial variations on sap flow estimates of transpiration in Japanese cedar. Agric Forest Meteorol 135:110–116. https://doi.org/10.1016/j.agrformet.2005.11.007

Kumagai T, Nagasawa H, Mabuchi T, Ohsaki S, Kubota K, Kogi K et al (2005b) Sources of error in estimating stand transpiration using allometric relationships between stem diameter and sapwood area for *Cryptomeria japonica* and *Chamaecyparis obtusa*. For Ecol Manag 206:191–195. https://doi.org/10.1016/j.foreco.2004.10.066

Kumagai T, Tateishi M, Shimizu T, Otsuki K (2008) Transpiration and canopy conductance at two slope positions in a Japanese cedar forest watershed. Agric Forest Meteorol 148:1444–1455. https://doi.org/10.1016/j.agrformet.2008.04.010

Kumagai T, Tateishi M, Miyazawa Y, Kobayashi M, Yoshifuji N, Komatsu H et al (2014) Estimation of annual forest evapotranspiration from a coniferous plantation watershed in Japan (1): water use components in Japanese cedar stands. J Hydrol 508:66–76. https://doi.org/10.1016/j.jhydrol.2013.10.047

Levia DF, Germer S (2015) A review of stemflow generation dynamics and stemflow-environment interactions in forests and shrublands. Rev Geophys 53:673–714. https://doi.org/10.1002/2015RG000479

Lloyd CR, Marques ADO (1988) Spatial variability of throughfall and stemflow measurements in Amazonian rainforest. Agric Forest Meteorol 42:63–73. https://doi.org/10.1016/0168-1923(88)90067-6

Lopez JG, Licata J, Pypker T, Asbjornsen H (In press) Effects of heater wattage on sap flux density estimates using an improved tree-cut experiment. Tree Physiol. https://doi.org/10.1093/treephys/tpy137

Lu P, Chacko E (1998) Evaluation of Granier's sap flux sensor in young mango trees. Agronomie 18:461–471. https://doi.org/10.1051/agro:19980703

Marshall DC (1958) Measurement of sap flow in conifers by heat transport. Plant Physiol 33:385–396. https://doi.org/10.1104/pp.33.6.385

McCulloh KA, Winter K, Meinzer FC, Garcia M, Aranda J, Lachenbruch B (2007) A comparison of daily water use estimates derived from constant-heat sap-flow probe values and gravimetric measurements in pot-grown saplings. Tree Physiol 27:1355–1360. https://doi.org/10.1093/treephys/27.9.1355

Nadezhdina N, Vandegehuchte MW, Steppe K (2012) Sap flux density measurements based on the heat field deformation method. Trees 26:1439–1448. https://doi.org/10.1007/s00468-012-0718-3

Oishi AC, Oren R, Stoy PC (2008) Estimating components of forest evapotranspiration: a footprint approach for scaling sap flux measurements. Agric Forest Meteorol 148:1719–1732. https://doi.org/10.1016/j.agrformet.2008.06.013

Oki T, Kanae S (2006) Global hydrological cycles and world water resources. Science 313:1068–1072. https://doi.org/10.1126/science.1128845

Ouyang S, Xiao K, Zhao Z, Xiang W, Xu C, Lei P et al (2018) Stand transpiration estimates from recalibrated parameters for the granier equation in a Chinese fir (*Cunninghamia lanceolata*) plantation in southern China. Forests 9:162. https://doi.org/10.3390/f9040162

Peters RL, Fonti P, Frank DC, Poyatos R, Pappas C, Kahmen A et al (2018) Quantification of uncertainties in conifer sap flow measured with the thermal dissipation method. New Phytol 219:1283–1299. https://doi.org/10.1111/nph.15241

Puckett LJ (1991) Spatial variability and collector requirements for sampling throughfall volume and chemistry under a mixed-hardwood canopy. Can J For Res 21:1581–1588. https://doi.org/10.1139/x91-220

Reid LM, Lewis J (2009) Rates, timing, and mechanisms of rainfall interception loss in a coastal redwood forest. J Hydrol 375:459–470. https://doi.org/10.1016/j.jhydrol.2009.06.048

Roth BE, Slatton KC, Cohen MJ (2007) On the potential for high-resolution lidar to improve rainfall interception estimates in forest ecosystems. Front Ecol Environ 5:421–428. https://doi.org/10.1890/060119.1

Schmidt-Walter P, Richter F, Herbst M, Schuldt B, Lamersdorf NP (2014) Transpiration and water use strategies of a young and a full-grown short rotation coppice differing in canopy cover and leaf area. Agric For Meteorol 195:165–178. https://doi.org/10.1016/j.agrformet.2014.05.006

Seibert J, Morén AS (1999) Reducing systematic errors in rainfall measurements using a new type of gauge. Agric For Meteorol 98:341–348. https://doi.org/10.1016/S0168-1923(99)00107-0

Shedekar VS, King KW, Fausey NR, Soboyejo AB, Harmel RD, Brown LC (2016) Assessment of measurement errors and dynamic calibration methods for three different tipping bucket rain gauges. Atmos Res 178:445–458. https://doi.org/10.1016/j.atmosres.2016.04.016

Shimizu T, Kumagai T, Kobayashi M, Tamai K, Iida S, Kabeya N et al (2015) Estimation of annual forest evapotranspiration from a coniferous plantation watershed in Japan (2): comparison of eddy covariance, water budget and sap-flow plus interception loss. J Hydrol 522:250–264. https://doi.org/10.1016/j.jhydrol.2014.12.021

Shimizu T, Kobayashi M, Iida S, Levia DF (2018) A generalized correction equation for large tipping-bucket flow meters for use in hydrological applications. J Hydrol 563:1051–1056. https://doi.org/10.1016/j.jhydrol.2018.06.036

Shinohara Y, Oda T, Kume T, Iida S, Chiu CW, Katayama A et al (2016) Calibration of granier method for Japanese larch and oak. In: 63rd annual meeting of ecological society of Japan, P2–069 (in Japanese)

Shiraki K, Yamato T (2004) Compensation of tipping-bucket flow meters. J Japan Soc Hydrol Water Resour 17:159–162. https://doi.org/10.3178/jjshwr.17.159. (in Japanese with English abstract)

Shiraki K, Sun J, Kagami S, Nagai K, Yokoyama Y, Koyama Y et al (2018) Accracy of low-cost flow meters for stemflow observation including handmade tipping buckets flow meter. J Japan Soc Hydrol Water Resour 31:380–392. https://doi.org/10.3178/jjshwr.31.380. (in Japanese with English abstract)

Steppe K, DJW DP, Doody TM, Teskey RO (2010) A comparison of sap flux density using thermal dissipation, heat pulse velocity and heat field deformation methods. Agric Forest Meteorol 150:1046–1056. https://doi.org/10.1016/j.agrformet.2010.04.004

Stocker TF, Qin D, Plattner G-K, Tignor M, Allen SK, Boschung J et al (eds) (2014) Climate change 2013: the physical science basis: working group I contribution to the fifth assessment report of the intergovernmental panel on climate change. Cambridge University Press, Cambridge/New York, p 1535

Su L, Zhao C, Xu W, Xie Z (2016) Modelling interception loss using the revised gash model: a case study in a mixed evergreen and deciduous broadleaved forest in China. Ecohydrology 9:1580–1589. https://doi.org/10.1002/eco.1749

Sun H, Aubrey DP, Teskey RO (2012) A simple calibration improved the accuracy of the thermal dissipation technique for sap flow measurements in juvenile trees of six species. Trees 26:631–640. https://doi.org/10.1007/s00468-011-0631-1

Swanson RH, Whitfield DWA (1981) A numerical analysis of heat pulse velocity theory and practice. J Exp Bot 32:221–239. https://doi.org/10.1093/jxb/32.1.221

Takahashi M, Giambelluca TW, Mudd RG, DeLay JK, Nullet MA, Asner GP (2011) Rainfall partitioning and cloud water interception in native forest and invaded forest in Hawai'i volcanoes National Park. Hydrol Process 25:448–464. https://doi.org/10.1002/hyp.7797

Takeuchi S, Sugio Y, Shinozaki K, Matsushima D, Iida S (2017) Calibration of heat ratio method by direct measurements of transpiration with weighing root-ball method: a study with Acer palmatum Thunb. J Japan Soc Reveg Tech 43:109–114. https://doi.org/10.7211/jjsrt.43.109. (in Japanese with English abstract)

Taneda H, Sperry JS (2008) A case-study of water transport in co-occurring ring-versus diffuse-porous trees: contrasts in water-status, conducting capacity, cavitation and vessel refilling. Tree Physiol 28:1641–1651. https://doi.org/10.1093/treephys/28.11.1641

Tokyo Climate Center, Japan Meteorological Agency (2018) Primary factors behind the heavy rain event of July 2018 and the subsequent heatwave in Japan from Mid-July Onward. https://ds.data.jma.go.jp/tcc/tcc/news/press_20180822.pdf

Turner B, Hill DJ, Carlyle-Moses DE, Rahman M (2019) Low-cost, high-resolution stemflow sensing. J Hydrol 570:62–68. https://doi.org/10.1016/j.jhydrol.2018.12.072

van Dijk AI, Gash JH, van Gorsel E, Blanken PD, Cescatti A, Emmel C et al (2015) Rainfall interception and the coupled surface water and energy balance. Agric Forest Meteorol 214:402–415. https://doi.org/10.1016/j.agrformet.2015.09.006

van Emmerik T, Steele-Dunne S, Hut R, Gentine P, Guerin M, Oliveira RS et al (2017) Measuring tree properties and responses using low-cost accelerometers. Sensors 17:1098. https://doi.org/10.3390/s17051098

Wilson KB, Hanson PJ, Mulholland PJ, Baldocchi DD, Wullschleger SD (2001) A comparison of methods for determining forest evapotranspiration and its components: sap-flow, soil water budget, eddy covariance and catchment water balance. Agric Forest Meteorol 106:153–168. https://doi.org/10.1016/S0168-1923(00)00199-4

Ziegler AD, Giambelluca TW, Nullet MA, Sutherland RA, Tantasarin C, Vogler JB et al (2009) Throughfall in an evergreen-dominated forest stand in northern Thailand: comparison of mobile and stationary methods. Agric Forest Meteorol 149:373–384. https://doi.org/10.1016/j.agrformet.2008.09.002

Chapter 3
Applications of Unpiloted Aerial Vehicles (UAVs) in Forest Hydrology

D. J. Hill, T. G. Pypker, and J. Church

3.1 Introduction

The field of ecohydrology has benefited greatly from the availability of remotely sensed data, particularly when considering large spatial scales (e.g., a forest watershed). Since the 1970s, satellite technology has provided spatial measurement of landscape characteristics across broad areas. Optical sensors were the first, and are still the most common, type of satellite-borne remote sensors. Optical sensors measure the intensity of solar energy that is reflected by the Earth's surface or by particulates or aerosols within the Earth's atmosphere. Due to the spectrum of electromagnetic radiation (EMR) emitted by the sun, and the atmospheric characteristics of the Earth, optical sensors are limited to measurement within the visible through infrared regime (~0.4–2.5 μm) of the electromagnetic spectrum (ES). Multispectral remote sensors capture variability in the spectral dimension of reflected energy by quantifying the reflected energy within many discrete spectral bands, whereas hyperspectral sensors divide reflected solar energy into a greater number of narrower spectral bands. Satellite-borne thermal infrared remote (TIR) sensors measure the intensity of EMR emitted from the Earth in the TIR regime of the ES (~3–100 μm). Often TIR measurements are used to estimate skin temperatures, such as the land surface temperature or cloud top temperature. Satellite-borne microwave sensors measure the intensity of the reflection of a microwave signal generated by the satellite off the Earth's surface or atmospheric particles, aerosols, or hydrometeors. Finally, satellite-borne light distancing and ranging (LiDAR) estimates the distance between the sensor and the land surface based on the time

D. J. Hill (✉)
Department of Geography and Environmental Studies, Thompson Rivers University, Kamloops, BC, Canada
e-mail: dhill@tru.ca

T. G. Pypker · J. Church
Department of Natural Resource Science, Thompson Rivers University, Kamloops, BC, Canada

D. F. Levia et al. (eds.), *Forest-Water Interactions*, Ecological Studies 240,
https://doi.org/10.1007/978-3-030-26086-6_3

required for laser light, emitted by the sensor, to reflect off the land surface and return to the sensor. Data collected by satellite-based remote sensing have transformed our understanding and ability to model the interactions of water within ecosystems that occur at the plot, regional and continental scales. Notably, satellite remote sensing has been instrumental in modelling evapotranspiration (Meijerink 2002), mapping forest floor vegetation (Tuanmu et al. 2010; Becker et al. 2013), quantifying forest structure (Lefsky 2010; Hansen et al. 2013; Persson and Perko 2016; Babcock et al. 2018; Hawrylo and Wężyk 2018) and health (Coops et al. 2010; Ortiz et al. 2013; Wulder et al. 2006) and estimating forest biomass (Lefsky et al. 2005; Boudreau et al. 2008; Nelson et al. 2009).

More recently, conventional aircraft have been employed as remote sensing platforms. Though remote sensing performed by conventional aircraft is restricted to regional-scale sensing missions, the relative ease of mission scheduling and the lower operating altitudes of conventional aircraft gives this approach several benefits over satellite remote sensing. While the time of a satellite overpass at a field site is predetermined by the orbital characteristics of the satellite, aerial remote sensing missions can be flown at nearly any desired time and repeat frequency. Thus, conventional aircraft-borne sensing can be scheduled during favourable weather (e.g., low cloud cover) or during critical times for the land surface characteristic being measured (e.g., during flowering of vegetation). Furthermore, conventional aircraft can be scheduled to fly missions in rapid succession, enabling high temporal resolution of remotely sensed data. The lower altitude flight of conventional aircraft compared to satellites reduces the distance between the sensor and the surface reflecting/emitting the measured EMR, resulting in higher spatial resolution multi-/hyperspectral, TIR and LiDAR measurements than are possible using satellite-based remote sensing. The shorter distance the EMR has to travel also reduces atmospheric scattering and absorption of the EMR before it reaches the sensor, reducing measurement errors.

Today, unpiloted aerial vehicles (UAVs) are transforming the field of ecohydrology as remote sensing platforms. UAV flight characteristics bring them closer to the land surface, further increasing the achievable spatial resolution of multi-/hyperspectral, TIR and LiDAR measurements and enabling measurement of the land surface at times when clouds obscure the view of satellite- and conventional aircraft-borne sensors. However, it is the low cost of UAVs and the lower operator risk associated with their use that truly define their role as more than a novelty. This chapter will explore the impact of UAV technology on advancing the field of ecohydrology. In particular, the chapter will focus on consumer-grade UAVs as platforms for remote sensing, because this category of UAV is much more accessible in terms of cost and regulation to use by the ecohydrologist than military- or aerospace-industry-derived UAVs. We define consumer-grade technology as technology that appeals to a wide consumer base and has relatively few barriers to its operation; generating an economy of scale that increases its availability in the marketplace and reduces its cost. The next section will review the state of development of consumer-grade UAVs and the remote sensors that can be integrated into these platforms. The next three sections will then discuss the role of UAVs in advancing methods to map forest stand composition, evapotranspiration and snowpack hydrology, respectively. Finally, the chapter will conclude with a discussion of future needs and potential research directions.

3.2 UAVs as a Sensing Platform

UAV technology traces its origins to military hardware (Newcome 2004), and early concerns regarding their safe use for civilian applications slowed their emergence as a tool for environmental monitoring (Newcome 2004). By the early 2000s, however, modified military UAVs were already showing promise for advancing the geosciences and natural resource management. For example, the First Response Experiment (FiRE) demonstrated the use of a General Atomics MQ-9 Reaper, which had originally been designed for the United States Air Force, for forest fire monitoring (Ambrosia et al. 2011, 2003). Because of the military origins of the UAV, the FiRE project was conducted with the assistance of pilots from the United States (US) National Aeronautics and Space Administration (NASA) (Merlin 2009). These early successes fostered a nascent civilian UAV market for geoscience applications. At first, however, civilian UAVs were expensive, and regulations on UAV operation remained a significant barrier to adoption of UAV technology. For example, Rango et al. (2009) reviewed five commercially available UAV sensing systems, with prices ranging from 50,000 to 350,000 USD. As regulations evolved to accommodate the use of UAVs, particularly for aerial photography, the demand for capable lower-cost UAVs continued to grow. This section reviews the variety of consumer-grade UAV platforms available today, as well as the sensing hardware and data processing software that together enable affordable, timely, high-resolution remote sensing of the environment.

3.2.1 *UAV Platforms*

UAV platforms can be roughly divided into two categories: fixed-wing and rotary-wing UAVs. Rotary-wing UAVs can be further divided into standard and multicopter. Muticopters have more than two rotors and are by far the most prevalent low-cost UAV platform today. As a remote sensing platform, rotary- and fixed-wing UAVs have different strengths. First, fixed-wing aircraft can generally carry heavier sensor payloads than copter UAVs (Jayathunga et al. 2018). However, these payloads must be integrated into the body of the fixed-wing aircraft to avoid disturbing the aerodynamics, and thus, the controllability, of the aircraft (González-Jorge et al. 2017). In contrast, the authors have found multicopter UAVs to be tolerant of externally mounted sensors, so long as the centre of balance of the aircraft is not disrupted too much. For this reason, it is generally easier to add stand-alone sensor payloads to a multicopter UAV without requiring a high degree of technical knowledge or fabrication. Second, fixed-wing UAVs can generally fly faster, at higher altitude, and for longer than multicopter UAVs (González-Jorge et al. 2017; Jayathunga et al. 2018). Thus, fixed-wing UAVs can travel farther from the pilot during operation and survey larger areas. However, fixed-wing UAVs require greater space to manoeuvre, take off and land (González-Jorge et al. 2017), and thus, it is difficult to operate fixed-wing UAVs in moderate to dense vegetation. Additionally, due to the space required to manoeuvre, it can be difficult to operate a fixed-wing UAV at low elevations above a vegetative canopy. This latter point is important, because of the relationship between elevation of a multi-/hyperspectral or TIR sensor over a measured surface and the on-the-ground pixel size, or ground-resolved distance (GRD). As the distance between the sensor and measured surface increases, the GRD increases, so lower flight elevations result in increased spatial resolution of the measurements. Finally, multicopters are more stable and exhibit fewer vibrations than fixed-wing platforms, which makes them more suited for high-resolution imaging missions (Wallace et al. 2012; Anderson and Gaston 2013; Fernández-Guisuraga et al. 2018).

Some of the earliest work applying UAV technology in ecohydrology employed large fixed-wing UAVs that were derived from military hardware, such as the Ikhana (Ambrosia et al. 2003; Ambrosia et al. 2011) and BAT-3 (Rango et al. 2009; Laliberte and Rango 2017). The success of this work drove the availability of small commercial-grade fixed-wing UAVs by companies including Quest UAV (Mlambo et al. 2017; Seier et al. 2017), senseFly (Chianucci et al. 2015; Leitão et al. 2016), and Trimble (Jayathunga et al. 2018), which are designed for landscape imaging. Due to the importance of aerodynamics to stable flight of these platforms, the sensors carried by these UAVs tend to be tightly integrated with the UAV's systems and contained within the UAV's airframe. This limits the ability for these UAVs to be modified to carry custom sensor payloads. However, the long flight time, around 60 min, enables fixed-wing UAVs to image large spatial areas during a single flight (González-Jorge et al. 2017).

Multicopter-style UAVs are more prevalent in the ecohydrology literature, likely because they are considerably less expensive than fixed-wing alternatives. The reason for this cost differential is because of the considerably larger market for multicopters than for fixed-wing UAVs, which is likely due to the ability of multicopters to take off and land vertically and to hover. These flight characteristics make multicopters popular aerial photographic platforms for a wide range of recreational, commercial, and civil government uses. According to a recent market survey conducted by Goldman Sachs (2016), the recreational market for UAVs is the largest non-military market segment; however, the growth in this market is outpaced by growth in the commercial and civil government sector. Together, these non-military market segments create an economy of scale that serves to drive down the cost of multicopter UAVs.

There are a wide variety of multicopter UAVs that have been applied in the field of ecohydrology; however, the Phantom series quadcopters from Dà-Jiāng Innovations Science and Technology Co. Ltd. (DJI) appear to be the most popular (e.g., Detert and Weitbrecht 2015; David and Ballado 2016; Perks et al. 2016,

Cardil et al. 2017; Hill et al. 2017; Mlambo et al. 2017; Thumser et al. 2017; Alonzo et al. 2018; Baron et al. 2018; Iizuka et al. 2018; Leduc and Knudby 2018). The Phantom series UAVs are inexpensive (~1000 USD for the current model at the time of writing), and, based on the authors' experience, feature reliable flight platforms, which likely leads to their popularity. Additionally, because they weigh less than 2 kg, they may be operated with few regulatory impediments in many jurisdictions (González-Jorge et al. 2017). Other UAVs that have been mentioned in the literature include the DJI Matrice series (Ishida et al. 2018; Zakaria et al. 2018), the 3D Robotics Iris (Chen et al. 2017), and custom-built UAVs (Dandois and Ellis 2013; Näsi et al. 2015; Nevalainen et al. 2017; Avanzi 2018; Saarinen et al. 2018).

Popular multicopter designs have four (quadcopter), six (hexacopter), or eight (octocopter) rotors, each powered by a separate motor. These designs provide varying levels of resilience to a motor failure, with both hexa- and octocopters being capable of landing in the event of a motor failure (González-Jorge et al. 2017). Despite differences in size, commonly available UAV multicopters have flight times of 30 min or less (González-Jorge et al. 2017). Thus, the difference between UAV platforms generally consists mainly of size, lifting capacity and the flight control system. The last of which is important when integrating remote sensors into a UAV system. There are both proprietary, closed flight control systems and open-source flight control systems. Closed systems, such as the one offered by DJI, protect access to sensor data collected by the flight control system, such as GNSS location, and access to services central to the operation of the UAV, such as battery management. Although an applications programming interface (API) is often available, it may be challenging to deeply embed a remote sensor within a UAV operating with a closed flight control system. However, independent sensor payloads, which contain their own global navigation satellite system (GNSS), processor and power source, can easily be loosely integrated with a UAV with sufficient lifting capacity. Open-source flight controllers, such as the one used by 3D Robotics and most custom-built UAVs, enable users to deeply embed sensors within the UAV system, by permitting access to power and data available from the batteries, inertial measurement unit (IMU), GNSS and flight plan. However, it is important that the sensing overhead does not overwhelm the available resources on the flight control system or an accident could occur. Because of this, and because of the ability to program a UAV controlled by an open-source controller to adapt the flight plan based on sensing inputs, operation of a UAV controlled by an open-source controller may be subject to more regulatory hurdles than the operation of a UAV controlled by a well-known closed flight controller.

Finally, it is important to note the importance of flight planning software, and the capability of UAVs for autonomous flight. Mapping with multi-/hyperspectral, TIR or LiDAR requires measurements at regularly spaced intervals over the land surface, which is generally accomplished by tasking the UAV to autonomously follow a predefined pattern over the landscape while acquiring measurements at intervals that have been coordinated with the flight speed to provide the necessary spacing along the ground, a task that is accomplished using flight planning software. For multicopter UAVs, there is a wide range of available flight planning software applications, each of which will work with a select group of UAV and ground-station hardware. For example, autonomous flights for DJI UAVs that the authors have experience with (Phantom 3 and 4 and Matrice 100 and 600) can be planned with applications offered by the UAV manufacturer (e.g., DJI), companies offering UAV-acquired data processing software (e.g., Pix4D, DroneDeploy), or third parties (e.g., QGroundControl). Because the DJI UAVs utilize an IOS or Android tablet or smartphone to provide the human-UAV interface, these flight planning software packages are offered as Android or IOS apps. These apps enable a user to design a flight plan that meets the needs of the remote sensing mission and falls within the performance envelope of the UAV. They also provide UAV systems data in real time during the flight, so that the pilot can monitor the location of the UAV vis-a-vis the flight plan and battery performance, among other things. Data telemetry is usually accomplished through a Wi-Fi link established between the UAV and tablet/smartphone, and the data are presented to the pilot on the smartphone/tablet screen. The authors have faced two challenges regarding the use of the available flight planning software. First, many flight planning software restricts the UAV to a constant altitude and cannot control the UAV to follow the terrain. This means that when operating a UAV on a hillside or in an area with complex terrain the pilot must carefully select the operating height to avoid the flight plan intersecting the land surface, resulting in a controlled flight into the terrain. This also means that the UAV elevation over the land surface is not constant, which will lead to a variable GRD in the acquired imagery from multi-/ hyperspectral and TIR sensors (also noted by Alonzo et al. 2018). Second, due to the open-source nature of the Android operating system, each hardware manufacturer utilizes a slightly different flavour of Android, and the authors have found that this can cause intermittent errors to arise when using an Android device as the human-UAV interface.

3.2.2 Sensors

Remote sensing payloads for UAVs are becoming increasingly powerful, and currently the commercial market offers several multispectral, hyperspectral, and TIR imagers, as well as LiDAR units that are suitable for UAV applications. Consumer-grade UAVs come equipped with an integrated digital camera to aid the pilot. In most cases, this camera is mounted on a gimbal, a motorized, gyroscopically stabilized bracket, which helps to maintain sensor stability against changes in the UAV's attitude (roll, pitch and yaw) while the sensor is positioned in a downward facing (nadir) perspective (González-Jorge et al. 2017). The images acquired by the digital camera provide data on reflection of red, green and blue (RGB) light from the land surface; however, in many cases these data cannot be considered radiometrically accurate because they are colour- and contrast-enhanced by proprietary methods of the camera manufacturer (Richardson et al. 2009; Klosterman and

Richardson 2017). Despite this limitation, several researchers have investigated the use of images acquired by UAV digital cameras for landscape mapping (Cardil et al. 2017; Chen et al. 2017; Hill et al. 2017; Klosterman and Richardson 2017; Avanzi et al. 2018; Baron et al. 2018; Leduc and Knudby 2018) and forest structure estimation (Dandois and Ellis 2013; Mlambo et al. 2017; Alonzo et al. 2018; Iizuka et al. 2018).

Several researchers have investigated modifying the integrated digital camera by removing the infrared filter from the lens to attempt to estimate a vegetation index similar to the normalized difference vegetation index (Puliti et al. 2015; Zhou et al. 2016). The authors (Hill and Church, unpublished data) have compared the spatial patterns of vegetation index computed from data collected by a modified digital camera to the NDVI calculated from measurements made by a four-band multispectral imager (Parrot Sequoia), which measures reflectance in the green, red, red-edge and NIR spectral bands. To account for differences in the range of the index values and the difference in the radiometric resolution between the modified RGB camera and the Sequoia multispectral imager, the index values were placed into five categories (very low, low, moderate, high and very high), where the category breaks were defined by the 20th, 40th, 60th and 80th percentiles of the values of each index measured across the same scene. A pairwise Mann-Whitney test was then used to evaluate the null hypothesis that for the same location within the scene, the vegetation index computed from the modified RGB camera fell into the same category as the NDVI calculated from the multispectral imager. The results of this test indicate a significant difference between the categorized vegetation indices calculated at co-located pixels within the scene ($p = 0.01$), suggesting that the spatial pattern exhibited by these two indices is different.

Multispectral, and more recently, hyperspectral, imagers are available in sufficiently small and lightweight packages to be integrated with consumer-grade UAVs. Due to the success of NDVI for precision agriculture and vegetation mapping, many multispectral sensors are focused on capturing the red-edge transition from red to NIR (such as the Sequoia sensor mentioned above), which is characteristic of the spectral signature of vegetation, although a few manufacturers will custom build imagers with user-specified spectral bands. Due to their lightweight and small size, multispectral imagers are popular payloads for low-cost multirotor UAVs (Hill et al. 2017; Fernández-Guisuraga et al. 2018; Baron et al. 2018). Hyperspectral imagers capable of being lifted by copter-style UAVs are being exploited by the research community (Calderón et al. 2013; Näsi et al. 2015; Nevalainen et al. 2017; Sankey et al. 2017; Cao et al. 2018a, b; Ishida et al. 2018; Saarinen et al. 2018); however, these sensors are costly, and due to their weight, they generally require a much larger UAV platform than a multispectral imager.

Thermal infrared imagers are also available in sufficiently small and lightweight packages to be carried by consumer-grade UAVs. However, due to weight constraints, these sensors are generally uncooled, which means that they are subject to measurement noise due to the temperature of the sensor. To overcome this thermally induced noise, it is necessary to increase the integration time or the area of the measurement. This means that UAV-borne TIR sensors generally have a slower response time and larger GRD than would be possible with a cooled TIR sensor. Additionally, these uncooled sensors are insufficiently accurate to measure small changes in temperature of the scene being viewed. Despite these limitations, UAV-acquired TIR imagery has been investigated for supporting forest fire management (Casbeer et al. 2006), forest canopy temperature (Webster et al. 2018) and tree health (Calderón et al. 2013).

Finally, LiDAR systems are also beginning to be integrated with UAV platforms. UAV-borne LiDAR requires a stable flight platform as well as an IMU to provide attitude data that is crucial for LiDAR data processing. This means that LiDAR payloads generally require a large copter UAV. Furthermore, laser emitting payloads are more heavily regulated than multi-hyperspectral and TIR imagers. However, the value of LiDAR data for ecosystem mapping and analysis is driving research leveraging this technology (Ganthaler et al. 2018).

The cost of RGB cameras, multi-/hyperspectral imagers, TIR imagers and LiDAR systems suitable for UAV use continues to decrease. However, in today's market, most multi-/hyperspectral, TIR and LiDAR solutions remain too expensive to be described as low-cost (Table 3.1). For this reason, there has been great interest in advancements in analysis methods applicable to high-resolution RGB imagery and colour infrared (consisting of three spectral bands measuring green, red and NIR, respectively) that increase the information that can be extracted from these types of images.

3.2.3 Structure from Motion/Multi-View Stereo

UAV-acquired remote sensing images represent small segments of the landscape and exhibit significant distortions (e.g., relief displacement). For these reasons, these images are of little utility on their own and require processing to be integrated into an orthomosaic of the region of interest. The emergence of software packages implementing and combining structure from motion (SfM) and multi-view stereo (MVS) algorithms, and affordable computer hardware that can run these packages, have been integral to the success of UAV-based remote sensing in the field of ecohydrology.

SfM is a computational method that simultaneously solves for intrinsic (e.g., focal length) and extrinsic (e.g., location and orientation) imager parameters to generate a sparse three-dimensional point cloud representing the location of the imaged surface. MVS employs principles of stereoscopy to enhance the sparse point cloud generated by SfM, increasing the point cloud density by an order of magnitude or more (Carrivick et al. 2016). The point cloud can then be used to orthorectify and mosaic the images used to generate it. Foundational to the SfM/MVS procedure is the presence of many easily identifiable features in the landscape being imaged, which can be automatically detected and matched across multiple images. Clearly, not all of the images will include all of these features; however, the more of such features in each image, the higher the quality of the resulting point cloud. Thus, SfM/MVS can

Table 3.1 Sensor hardware, example applications reported in the literature, and relative cost

Manufacturer	Model	Type	Applications	Cost
DJI	Phantom	Digital camera	Aerial photography, canopy height estimation, tree health monitoring, phenology	$
Sony	Alpha 600	Digital camera		$
Canon	Powershot a3300	Digital camera		$
Sony	Nex-5T	Digital camera		$
GoPro	Hero	Digital camera		$
Parrot	Sequoia	Multispectral (4-band)	Vegetative species mapping, vegetation health monitoring	$$
MicaSense	Red edge	Multispectral (5-band)		$$
Tetracam	ADC snap	Multispectral (4-band)		$$
Tetracam	Micro MCA-6	Multispectral (6-band)	Vegetation health monitoring	$$$
Tetracam	Micro MCA-12	Multispectral (12-band)		$$$
FLIR	Tau 2	Thermal infrared	Canopy temperature monitoring	$$$
Headwall photonics	Micro-hyperspec	Hyperspectral	Tree species identification, tree health monitoring, biodiversity estimation	$$$$
Velodyne	VLP-16	LiDAR	Digital elevation model generation, canopy height model generation, Forest structure estimation	$$$

Sensor cost is reported in USD, where "$" indicates hundreds of USD, "$$" indicates thousands of USD, "$$$" indicates tens of thousands of USD and "$$$$" indicates hundreds of thousands of USD

have difficulty constructing a point cloud, and consequently an orthomosaic, from images of a homogeneous landscape, such as flat water or snow (Mlambo et al. 2017; Avanzi et al. 2018; Webster et al. 2018). Similarly, SfM will be challenged by images representing very small regions within the landscape. For this reason, wider-angle lenses are preferable for low-elevation imaging flights performed by copter UAVs (Mlambo et al. 2017; Webster et al. 2018). With wider-angle lenses, it is easier to achieve a significant proportion of image overlap (~80% or more), which is necessary to ensure that individual features are identifiable in many of the acquired images (Avanzi et al. 2018). Finally, the more spectral information (i.e., bands) present in the images, the easier it is to identify landscape features to use in the SfM/MVS process. Because TIR imagery constitutes only a single band, the authors and others (e.g., Webster et al. 2018) have noted that TIR imagery can challenge the process of constructing an orthomosaic by SfM/MVS.

Because SfM infers positional information of image features and the imaging platform, the position of the imager does not need to be recorded for each acquired image; however, this information can improve the SfM processing speed and accuracy. For this reason, imaging sensor payloads intended for UAVs do not require integration with a GNSS or IMU, which significantly reduces their cost and weight. Geocoordinates, such as those provided by a GNSS, however, are required to georeference the output of SfM/MVS. Geocoordinates of the image acquisition location, provided by a GNSS embedded in the imaging sensor can be leveraged by some SfM/MVS software packages (e.g., Pix4D) to georeference the resulting point cloud and orthomosaic. However, while this georeferencing can be used as a first-order estimate, it often has considerable error and must be corrected before attempting to integrate these data with other geodata. Ground control points, locations within the field site marked with easily identifiable objects where the geographic coordinates are known to a high degree of accuracy (e.g., through a use of a real-time kinematic GNSS, RTK-GNSS), are often used to georeference data products generated by SfM/MVS. Ground control points can be based on features naturally occurring within the scene (e.g., large trees, Klosterman et al. 2018) or specially made targets are placed within the field site to serve this function. While human-made targets have well-defined centre points which make them easy to use, they are often impractical in densely forested environments where placing and recovering them may take considerable effort or may be infeasible. Once the SfM/MVS procedure is complete, the known geographic coordinates of the ground control points can be associated with their location within the orthomosaic, allowing the software to refine its positioning calculations to minimize the geopositioning error across these known locations.

Not only can the SfM/MVS-derived point cloud be used to produce an orthomosaic from a collection of images acquired over a region of interest, but it can also create a digital surface model (DSM) of the landscape. As has been reported elsewhere (Näsi et al. 2015; Chen et al. 2017; Mlambo et al. 2017; Saarinen et al. 2018), the authors (Hill, unpublished data) have found that SfM/MVS-derived DSMs are reasonably accurate, although the vertical datum may exhibit significant error (Dandois and Ellis 2013; Näsi et al. 2015; Ruzgienė et al. 2015; Fernández-Guisuraga et al. 2018; Nevalainen et al. 2017; Mlambo et al. 2017). The authors attribute this vertical positioning error to the reliance on GNSS-derived altitude, a measure that is known to have more significant errors than GNSS-derived horizontal position (Shaw et al. 2000). While this error in the datum does not impact analyses regarding surface heights relative to the DSM (e.g., Näsi et al. 2015; Mlambo et al. 2017; Cao et al. 2018a, b), it will negatively impact analyses that require the comparison of a SfM-derived DSM with a digital elevation model (DEM) or DSM generated by a different method or from data acquired at a different time.

3.2.4 Regulation

The rapid adoption of UAV technology by commercial and recreational users has left many jurisdictions struggling to regulate the safe use of this technology. This has resulted in a period of rapidly changing regulations. For example, in Canada, where the authors' research is based, there have been three significant changes in federal UAV regulations between 2014 and 2019. In 2014, Transport Canada (TC) exempted the use of a UAV of no more than 2 kg in weight from more stringent regulation required by Canada's Aviation Regulations if the operator applied for an exemption; met certain age, training and insurance criteria; and observed specific geographical constraints during flight (e.g., observe exclusion zones around airports). In 2016, TC amended this exemption to distinguish between the use of UAVs weighing no more than 1 kg and UAVs weighing more than 1 kg but no less than 25 kg. Clearly, this revision increased the weight of the UAVs that qualified for exempt flights; however, the exempt operation of the heavier class of UAVs required that operators demonstrate evidence of appropriate training. The 2016 regulations also introduced a mechanism, the "no drone zone", for landowners to restrict drone use on their property. Most recently, in 2019 TC revised the Aviation Regulations to permit the use of UAVs weighing less than 25 kg, operated by a pilot who has completed written knowledge exam, and observing specific geographic restrictions. As in previous years, operation of UAVs outside these parameters requires a special flight operations certificate (SFOC) as per the Air Regulations. This pattern of rapidly changing regulation is likely to have been echoed elsewhere around the world (e.g., Newcome 2000). In addition to age, insurance, training and daytime-flight requirements for UAV pilots, many jurisdictions also regulate the maximum flight elevation above ground level and the distance a UAV may travel from the pilot. Low-elevation, short-distance flights are generally minimally regulated, whereas high-elevation, long-range flights are more strictly controlled. Regulations also restrict the use of gasoline or other combustible fuels, limiting the power storage onboard consumer-grade UAV to batteries. The weight of current battery technology and the limited power density, relative to combustible fuels, further limits the endurance of consumer-grade UAV technology. Such regulations limit the capacity for UAVs to perform remote sensing of large regions, while being relatively permissive of field-scale use. Furthermore, flights in populated areas are more heavily regulated than flights in areas where few people are expected to be, which makes UAV technology much more appropriate for use in rural settings than in urban areas. Finally, although regulations are generally permissive of consumer-grade copter UAVs capable of carrying a wide array of remote sensors, special permission is required for UAVs carrying LiDAR payloads due to concerns over air traffic safety risks associated with the laser (e.g., Aeronautics Act 2002).

3.2.5 UAV Technology Summary

The combined demand for consumer-grade UAV technology by recreational and commercial users has significantly driven down prices on a wide variety of multicopter-style UAVs capable of lifting sensor payloads in excess of 5 kg. At the same time, miniaturization in sensor hardware has led to multi-/hyperspectral, TIR and LiDAR payloads suitable for use on these platforms, and inexpensive computing power has enabled SfM/MVS software to be run on consumer-grade desktop computers. Additionally, changes to aviation regulations have lowered barriers to use of these "small" UAVs in many jurisdictions, under certain conditions that are generally favourable for multi-/hyperspectral or TIR remote sensing in unpopulated areas. Together, these events have enabled low-cost UAV remote sensing of forested ecosystems, which has shown promise for their study and management.

Consumer-grade multicopter UAV technology provides remote sensing platforms that can provide more dense spatial data than field methods (Anderson and Gaston 2013; Näsi et al. 2015; Chen et al. 2017; Fernández-Guisuraga et al. 2018; Nevalainen et al. 2017; Avanzi et al. 2018; Cao et al. 2018a, b; Ganthaler et al. 2018; Klosterman et al. 2018) and can operate in difficult-to-access and/or dangerous environments (Cardil et al. 2017; Chen et al. 2017; Hill et al. 2017; Nevalainen et al. 2017; Alonzo et al. 2018; Avanzi et al. 2018; Cao et al. 2018a, b; Webster et al. 2018). Multicopter-style UAVs are particularly well-suited for low-level remote sensing, because they are more manoeuvrable and able to fly more slowly than fixed-wing alternatives (González-Jorge et al. 2017). This enables UAV-borne remote sensors to acquire images with exceptionally high spatial resolution. However, it is the low cost of these systems that lowers barriers to access to high-resolution remote sensing data (Stafford 2007; Anderson and Gaston 2013; Näsi et al. 2015; Cardil et al. 2017; Chen et al. 2017; Fernández-Guisuraga et al. 2018; Mlambo et al. 2017; Nevalainen et al. 2017; Sankey et al. 2017; Alonzo et al. 2018; Avanzi et al. 2018; Cao et al. 2018a, b; Demir 2018; Ganthaler et al. 2018; Ishida et al. 2018; Leduc and Knudby 2018; Saarinen et al. 2018; Webster et al. 2018), and that makes them transformative to the practice of ecohydrology. Today, acquiring high-resolution remote sensing data for a specific field site is easier than ever, and data can be acquired on-demand repeatedly throughout the year. Furthermore, unlike satellite- and conventional aircraft-based remote sensing, UAV image acquisition occurs below the clouds, such that the landscape can continue to be imaged during cloudy weather (Wing et al. 2013; Klosterman et al. 2018). These characteristics enable the acquisition of data during/after critical events, such as fire (Casbeer et al. 2006) or changes in the phenological stage of vegetation (Klosterman and Richardson 2017; Klosterman et al. 2018). It also enables multi-temporal analysis of landscape dynamics.

Figure 3.1 illustrates the improvement in spatial resolution achievable with UAV-acquired imagery, compared with imagery acquired by conventional aircraft or satellite-borne sensors, by showing image orthomosaics with spatial resolutions

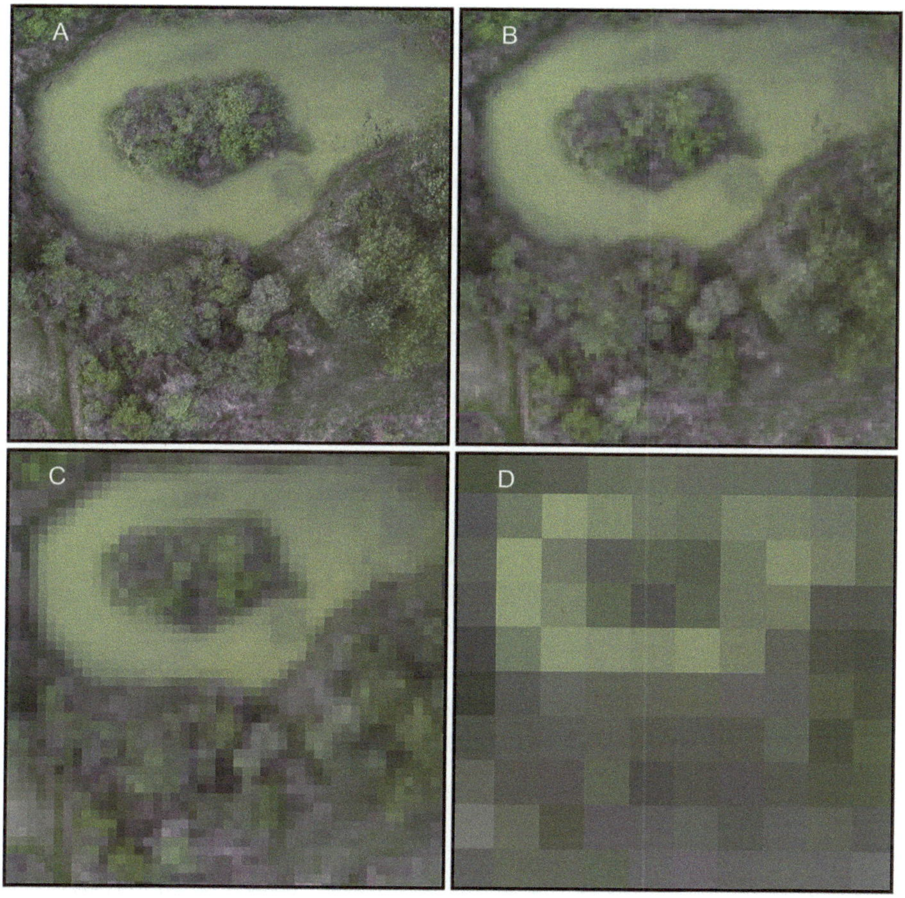

Fig. 3.1 Comparison of spatial resolutions typical of UAV, conventional aircraft and satellite imaging technology. Image (**a**) has a spatial resolution of 1-cm and was generated by orthomosaicking images acquired by a DJI Phantom-3 UAV. Images (**b**), (**c**) and (**d**) were created by reaggregating this image to 50-cm (characteristic of conventional aircraft orthoimagery), 1.2-m (characteristic of WorldView-3 satellite imagery) and 6-m (characteristic of SPOT-6 satellite imagery), respectively

typical of these three types of sensing platforms. The area represented in Fig. 3.1 is a 1.44 ha region of a forested wetland near Creston, British Columbia that we imaged in April 2017 as part of a larger (39 ha) aerial imaging survey. The 1-cm resolution image was created by orthomosaicking the images acquired by the digital camera embedded within the DJI Phantom 3 UAV used for the imaging flights. The lower-resolution images were created by reaggregating these data to resolutions of 50-cm, the resolution of the conventional-aircraft acquired orthoimagery available from the Province of British Columbia; 1.2-m, the resolution of imagery acquired by the WorldView-3 satellite; and 6-m, the resolution of imagery acquired by the SPOT-6 satellite. The two satellites were selected to represent satellite-acquired imagery, because WorldView-3 imagery is arguably the highest resolution satellite imagery that can be acquired, albeit at significant cost, for non-military applications, whereas SPOT-6 imagery is arguably the highest resolution satellite imagery that is freely available. As can be seen in Fig. 3.1, individual tree crowns are clearly visible in the UAV-acquired imagery, whereas at the lower resolutions, these image details are increasingly obscured. In fact, the difference between trees and other vegetative cover can no longer be resolved in the 1.2-cm resolution image, whereas, even the lake in the centre of the image cannot be discriminated from vegetative land cover in the 600-cm resolution image.

Low-cost multicopter UAV technology, however, does not replace the need for conventional aircraft-based remote sensing, or even for expensive UAV systems. Because of their small size and multicopter design, low-cost UAVs have limited endurance and lifting capacity. This limits the area that can be mapped by a UAV-borne remote sensor in a single flight and the sensor payloads that can be carried during flight. Furthermore, current aviation regulations strictly control night-time operation (González-Jorge et al. 2017) when TIR mapping is most useful, as well as the use of laser emitting sensors (e.g., LiDAR or other type of laser range finder) (Aeronautics Act 2002). However, the ubiquity of low-cost UAV copters renders them important as a sensing platform in study and practice of forest ecohydrology.

3.3 Utility of UAVs in Estimating Forest Stand Composition

Several studies have explored the utility of UAV-borne remote sensing for estimating biophysical and biochemical parameters at the scale of individual trees. When combined with spectral data, these parameters can be used to infer tree species, leaf area index (LAI), phenophase, or health, which can be analysed at the stand scale to compute measures of stand-level biodiversity. Deriving tree-scale parameters from remotely sensed data covering an entire forest requires that the remotely sensed data can be assigned to specific individual trees within the forest. This is usually accomplished by first generating a canopy height model (CHM) and applying a segmentation algorithm, such as watershed segmentation (Näsi et al. 2015; Alonzo et al. 2018; Saarinen et al. 2018), to delineate individual tree crowns within the forest. CHMs are created by subtracting a DEM, which describes the topography of the land surface, from a DSM, which describes the topography of the forest canopy. Due to the dense spatial sampling and narrow laser beam width, LiDAR can reliably produce both DEM and DSM of a forest stand to generate a CHM (e.g., Cardil et al. 2017; Sankey et al. 2017). However, due to the cost of LiDAR data acquisition, many researchers (Chen et al. 2017; Nevalainen et al. 2017; Alonzo et al. 2018; Avanzi et al. 2018; Iizuka et al. 2018; Webster et al. 2018) have explored the use of UAV-acquired imagery and SfM/MVS to derive the DEM and DSM used to compute the CHM and LAI (e.g., Matthews and Jensen 2013). Not only does using SfM/MVS to generate the DEM and DSM eliminate the need for a costly LiDAR survey, but it also permits the data for both the tree crown delineation and spectral analysis to be acquired in a single UAV mission. Although there have been many successes of developing accurate CHMs from SfM/MVS-derived DEMs and DSMs (Alonzo et al. 2018; Avanzi et al. 2018; Webster et al. 2018), it has been shown that SfM/MVS-derived point clouds have fewer points representing the land surface than are present in LiDAR-derived point clouds, a problem that gets worse as the canopy density increases (Dandois and Ellis 2013; Chen et al. 2017; Jayathunga et al. 2018; Klosterman et al. 2018). Thus, as the canopy becomes increasingly dense, SfM/MVS-based methods must interpolate between increasingly sparse measurements to generate a DEM, resulting in inaccuracies, especially in complex terrain (Chen et al. 2017; Iizuka et al. 2018; Jayathunga et al. 2018; Klosterman et al. 2018). Because land surface topography is unlikely to change quickly, other researchers have justified combining a historical LiDAR-derived DEM with an SfM/MVS-derived DSM to compute the CHM (Klosterman et al. 2018; Saarinen et al. 2018). In these cases, many control points with highly accurate three-dimensional positioning, such as would be acquired by a survey-grade GNSS, were required to georeference the DSM with the DEM. Because GNSS positioning accuracy decreases with increasing canopy closure, this requirement reduces the practicality of these methods in dense forest environments. These results suggest that, while SfM/MVS may be able to generate reasonable CHMs in sparse canopy conditions, LiDAR provides a more robust solution that will provide reasonable results even under dense canopy and complex terrain conditions.

Once tree crown delineation has been performed, SfM/MVS point cloud statistics and spectral data acquired by multi-/hyperspectral and TIR sensors can be attributed to individual tree crowns. These data can be combined with the CHM for each tree crown and can be used to estimate several biophysical and biochemical parameters at the tree scale. LAI can be estimated using SfM/MVS-derived point cloud statistics with accuracies similar to LiDAR (e.g., Matthews and Jensen 2013). Crown metrics including area, diameter, and perimeter can be directly calculated from the crown delineation (Shrestha and Wynne 2012), while tree height can be reliably estimated as the largest value of the CHM within the crown boundary (Shrestha and Wynne 2012; Chen et al. 2017; Mlambo et al. 2017; Demir 2018; Iizuka et al. 2018; Jayathunga et al. 2018; Saarinen et al. 2018). Crown temperature can be estimated using the average TIR response from the delineated crown area transformed into temperature using the Stefan–Boltzmann equation (e.g., Calderón et al. 2013; Webster et al. 2018). Derived tree height and crown-metrics have been used to estimate diameter at breast height (DBH) (Shrestha and Wynne 2012; Iizuka et al. 2018) and above-ground biomass (Shrestha and Wynne 2012; Dandois and Ellis 2013; Alonzo et al. 2018) using regression techniques, and recently, biochemical properties, namely chlorophyll and carotenoid content, of individual tree crowns have been estimated from hyperspectral measurements of individual tree crowns (Ferreira et al. 2018).

In addition to tree biophysical and biochemical parameters, UAV-acquired remote sensing imagery has been valuable in determining tree species, phenophase and health. The high-spatial resolution of UAV-acquired multi- or hyperspectral imagery enables the use of object-based image analysis (OBIA, Blaschke 2010; Blaschke et al. 2014), where the objects (e.g., individual tree crowns) are delineated using an image segmentation based on the CHM. Statistics of the spectral measurements from within each image object can be computed to generate spectral and textural features that can be used to classify the objects. While many studies investigating the classification of tree species within a forest stand leverage the measurement density of hyperspectral imagery (Nevalainen et al. 2017; Cao et al. 2018a, b; Saarinen et al. 2018), RGB imagery acquired by low-cost UAVs has also been demonstrated to be successful in classifying tree species in a boreal forest stand (Alonzo et al. 2018). UAV-acquired RGB imagery has also been demonstrated to be useful in inferring tree phenophase by analysing changes in the level of greenness apparent in a sequence of imagery acquired throughout the growing season (Klosterman and Richardson 2017; Klosterman et al. 2018). Although this analysis was applied at the pixel level (due to lower-resolution, 10 m imagery), higher-resolution imagery may enable the method to be applied at the level of independent tree crowns. Similarly, pest damage to trees has also been analysed from UAV-acquired spectral imagery, using both pixel-based (i.e., pixel-by-pixel) analyses (Näsi et al. 2015; Cardil et al. 2017) and OBIA (Ganthaler et al. 2018). Pixel-based analysis enabled researchers to quantify the proportion of damage to a tree crown, whereas, OBIA enabled the identification of individual pest-damaged trees. In both pixel-based analysis and OBIA, a machine learning classifier is used to classify the pixel or object based on the spectral features associated with the pixel/object, where the classification is based on ground reference data collected in the field. There are many types of classifiers and classifier selection is often application dependent (Lu and Weng 2005).

Measures of forest stand productivity and biodiversity can also be estimated from UAV-acquired remote sensing data. Productivity can be estimated by phenophase transitions (Klosterman and Richardson 2017; Jayathunga et al. 2018), while biodiversity metrics, such as species diversity, evenness and composition, can be computed from the outcomes of tree species classification performed by OBIA (Saarinen et al. 2018). Other measures of biodiversity, including amount of deadwood, tree density and successional stage, have also been predicted by regression using both hyperspectral (Saarinen et al. 2018) and RGB (Alonzo et al. 2018) imagery. At the same time, new indicators of biodiversity computed from UAV-acquired remote sensing data, such as forest gap structure (Getzin et al. 2012), the presence of key understory vegetation (Getzin et al. 2012; Leduc and Knudby 2018), or the variation in hyperspectral reflectance measured across the forest stand (Wang et al. 2018), are being reported in the literature.

Modelling the relationship between remote sensing measurements (e.g., spectral data) and derived tree/canopy metrics (e.g., DBH), tree classification (e.g., species), or forest-stand characteristics (e.g., biodiversity) depends heavily on the availability of a large number of ground-reference data, often collected by field surveys. These ground reference data represent the characteristic the model is designed to predict given the remote sensing data. Spatial alignment of the ground-reference data with the remote sensing data is extremely important to ensure that the modelling is based on the remote sensing data that was measured at the same location as the ground reference data. Inaccurate georeferencing increase the noise in the dataset and result in reduced accuracy of the model.

3.4 Utility of UAVs in Measuring Evapotranspiration

Of the water entering a forested ecosystem as precipitation, frequently more than half is returned to the atmosphere as evaporation or transpiration (ET). Understanding the mechanisms controlling the loss of water via ET requires an understanding of the physical structure of the canopy, the microclimate and the plant physiology (e.g., Jarvis and McNaughton 1986; Monteith and Unsworth 2007). Typically, researchers are unable to directly measure all the components of ET, and therefore rely on indirect techniques. Common methods for quantifying ET include eddy covariance and sapflux techniques (e.g., Granier 1985; Baldocchi et al. 1988; Goulden and Field 1994). Eddy covariance estimates ET over areas of approximately 1–100 ha, but the footprint of the measurement varies with time (e.g., Finnigan 2004; Lee 2004; Baldocchi 1997). Sapflux techniques can estimate fluxes from individual trees, but the measurements often have low sample sizes because of instrumentation constraints (e.g., Köstner et al. 1998; Wright et al. 2018). UAV-acquired TIR imaging can provide data critical to understanding the processes controlling the flux of water to and from the vegetation at scales between individual trees (sapflux) and landscape scale (eddy covariance) (Costa et al. 2013). TIR imagery collected by a tripod-mounted imaging system with 75-cm resolution has been used to map the temperature of the vegetation canopy in alpine landscapes and quantify vegetation microhabitats (Scherrer and Körner 2010, 2011). Using methods that have been developed for satellite-acquired (Anderson et al. 2008) and conventional aircraft-acquired (Chávez et al. 2008) TIR imagery, these measurements could also be analysed to estimate water fluxes and map ET at the plot scale. Use of a UAV-borne TIR imager, as opposed to a ground-based tripod mount, would permit wall-to-wall imaging of the entire field site and potentially increase resolution, as the resolution in the work by Scherrer and Körner (2010, 2011) was influenced by the siting requirements of the tripod mount. In this way, UAV-acquired TIR imagery could provide thermal maps of forest canopies that could be used to estimate water fluxes and ET at the forest stand scale.

3.4.1 Utility of UAVs to Improve Estimates of Transpiration

By enhancing measurements of canopy structure and biophysical characteristics, UAV-based remote sensing has the potential to improve models of transpiration. Canopy structure directly influences the exchange of energy between the atmosphere and the canopy, which drives ET. The Penman–Monteith equation is a common model that is used to estimate ET:

$$LE = \frac{\Delta(R_n - G) + \rho c_p(e_s - e_a)\frac{1}{r_a}}{\Delta + \gamma\left(1 + \frac{r_s}{r_a}\right)} \quad (3.1)$$

where LE is the water vapour flux (W m^{-2}), R_n is the canopy net radiation budget (W m^{-2}), G is the ground heat flux (W m^{-2}), ρ is the density of dry air (kg m^{-3}), c_p is the specific heat capacity of air (J kg^{-1} K^{-1}), e_s is the saturation vapour pressure deficit (Pa) at a given temperature (T_a), e_a is the actual vapour pressure (Pa), r_a is the atmospheric resistance (s m^{-1}), r_s is the surface resistance (s m^{-1}), γ is the psychrometric constant (Pa K^{-1}) and Δ is the rate of change of saturation vapour pressure with air temperature (Pa K^{-1}).

Canopy structure and biophysical traits directly influence many of the parameters of the Penman-Monteith equation. First, the albedo of the canopy surface will directly influence the net radiation budget described in Eq. (3.1) by altering the absorption of shortwave radiation. However, albedo is not uniform in time and space; rather, it varies with species, canopy height, season and canopy wetness (Moore et al. 1996; McPherson 2007; Arora and Montenegro 2011). These changes will directly influence the energy budget of the forest at scales ranging from the leaf to the watershed. UAV-borne pyranometers can be used to provide denser spatial albedo measurements than would be feasible with field-based pyranometer towers that also exhibit higher accuracy than satellite-derived albedo estimates (Levy et al. 2018). Alternatively, multi-spectral imaging data could also be used to estimate canopy albedo at high spatial resolution using methods similar to those used to derive satellite-image-based albedo estimates. For example, Cao et al. (2018a, b) developed a workflow to estimate albedo from an orthomosaic of images acquired by a UAV-borne uncalibrated RGB camera. This workflow first required the calibration of the UAV-acquired imagery using ground-based spectrometer measurements. The calibrated UAV-acquired RGB data were then processed using a version of the Landsat 8 albedo algorithm (Liang 2000) that was modified to include only red, green, and blue spectral reflectances. While the method presented by Cao et al. (2018a, b) produces reasonable results, it could be improved by using a radiometric multi-spectral imager that measures in both the visible and near infrared, such as a MicaSense or Tetracam.

Canopy roughness directly influences r_a (Monteith 1965). Under stable atmospheric conditions, r_a is defined by

$$r_a = \frac{\left[ln\left(\frac{z-d}{z_0}\right)\right]^2}{k^2\, u(z)} \quad (3.2)$$

where z is canopy height (m), d is the zone of zero displacement (m), z_0 is the roughness length (m), k is von Karman's constant and $u(z)$ is the wind speed of the height z (m s^{-1}). Taller canopies and canopies with a rougher surface can exchange water vapour more easily because they have a lower r_a. Using techniques outlined in Sect. 3.3, UAV-acquired data can be used to improve the accuracy of canopy height estimates, thus improving estimates of r_a.

The ground heat flux (G) is frequently assumed to be 10–20% of R_n, yet the actual value ranges from 1% to 50% of R_n (Sauer and Horton 2005). For example, in forests with large gap fractions, the importance of G generally increases (e.g., Sauer and Horton 2005; Chan et al. 2018). Therefore, accurate quantification of the amount of bare ground and its subsequent impact on LE is important. Ground-based fisheye lens photos can be used to estimate the gap fraction. However, orthoimagery constructed from UAV-acquired imagery can provide estimates of the gap fraction over larger areas and provide a better understanding of the spatial distribution of canopy gaps.

The ability of the canopy to remove energy via *LE*, sensible heat (H) and longwave radiation directly influences $e_s - e_a$ because e_s increases exponentially with air temperature. Researchers often assume that air temperature is equal to leaf temperature because leaf temperatures are difficult to measure. The atmospheric vapour pressure deficit ($e_S - e_a$) is then used to calculate the vapour pressure gradient between the leaf and the atmosphere ($e_L - e_a$), a quantity that is central to determining the flux of water from leaves (e.g., Jarvis and McNaugton 1986).

However, air temperature is a poor estimator of leaf temperature. The majority of transpiration comes from the sunlit portions of large canopy trees (e.g., Meinzer et al. 2001; Cristiano et al. 2015; Kunert et al. 2017; Brum et al. in press). On sunny days, sunlit leaves are intercepting more direct beam solar radiation than other portions of the canopy. The ability of the leaf to dissipate the absorbed solar radiation will affect the leaf temperature and subsequently the vapour pressure gradient between the leaf and the atmosphere. Thus, the vapour pressure gradient will vary depending on the leaf structure. Under the same conditions, large flat leaves are likely to experience higher leaf temperatures relative to thin, cylindrical leaves (Monteith and Unsworth 2007). The difference is directly related to the change in boundary layer resistance, with flat leaves having a greater resistance (Monteith and Unsworth 2007). In instances when the canopy is dry, the temperature of the leaf will often be greater than the air temperature. The difference will increase based upon the structure of the leaves and the canopy (Monteith and Unsworth 2007). For example, conifer needles typically dissipate absorbed energy quicker than deciduous leaves, and rougher canopies will have a lower boundary layer resistance, and therefore, lower leaf temperatures (Monteith and Unsworth 2007).

In addition to estimating canopy characteristics (see Sect. 3.3), UAV-borne TIR imagers can provide detailed information on leaf temperature (e.g., Maes and Steppe 2012; Costa et al. 2013; Webster et al. 2018). The ability to quantify leaf temperature will improve spatial estimates of forest canopy transpiration, enabling researchers to identify regions of the canopy transpiring the most/least water. For example, Page et al. (2018) recently demonstrated that combining physiological models with UAV-acquired TIR imagery allowed for accurate estimates of ET by soybean leaves without either a wet or dry reference measurement. With additional work, research conducted on crops, such as this study, may be transferred to forested systems.

3.4.2 *Utility of UAVs for Estimating Evaporation*

UAVs show promise for improving estimates of evaporation from forested systems by providing critical measurements of phenology and crown architecture that may serve to improve models of within-canopy fluxes of water and nutrients. Hydrologic fluxes within forest canopies and soil have long been treated as a black box (Pypker et al. 2011). Researchers typically monitor rainfall above and below the forest canopy to estimate the proportion of precipitation that is intercepted by the canopy (and returned to the atmosphere by evaporation) and the proportion that transits the canopy (and either infiltrates into the soil or becomes runoff). Recently, however, interest has risen in modelling the behaviour of precipitation within forest canopies, driving research into the mechanisms and spatial patterns of throughfall (*TF*), stemflow (*SF*) and interception losses.

Past research on the spatial patterning of *TF* under forest canopies has documented that persistent patterns of *TF* can occur (e.g., Keim et al. 2005) and that these patterns create "hot spots" of *TF* and, presumably, nutrients; but, insufficient research exists on the latter (Pypker et al. 2011). The pattern of these hot spots varies and likely depends on meteorological conditions, plant phenology and canopy structure. However, past work, relying on measurements taken from below the canopy, has had limited success linking canopy elements to *TF* patterns (e.g., Pypker et al. 2011). Estimates of phenology, canopy structure and cover, LAI and tree density generated from UAV-acquired measurements may provide the necessary data to derive new spatially explicit models of *TF* that can be used to calculate the spatial distribution of *TF* as well as the direct *TF* fraction at the stand scale (e.g., Flerchinger and Saxton 1989; Jetten 1996; Pypker et al. 2011). Notably, gap fraction is frequently used to estimate direct *TF* fraction (e.g., Link et al. 2004). UAVs mounted with LiDAR or high-definition cameras can provide estimates of gap fraction as well as its spatial distribution, as described in Sect. 3.3.

SF generated by individual trees is influenced by phenology and canopy structure and concentrates precipitation at the tree base, resulting in a significant point input of nutrients and water to the forest floor (e.g., Weathers et al. 2001; Pypker et al. 2011). At the stand scale, the net input of nutrients and water due to *SF* is further driven by vegetation cover and tree density (e.g., Levia and Frost 2003). Estimates of phenology, canopy structure, vegetation cover and tree density derived from UAV-acquired data (as described in Sect. 3.3) can enable research into the relationships between these forest characteristics and *SF*, resulting in new spatially explicit models of *SF* inputs of water and nutrients into the surface and subsurface environments.

Finally, canopy rainfall interception models rely on estimates of canopy characteristics to calculate interception loss (Muzylo et al. 2009). Common parameters needed to calculate rainfall interception loss include direct *TF* fraction, canopy water storage and the ratio of evaporation to rainfall intensity (*E/R*). As discussed above, estimates of direct *TF* fraction may serve to improve estimates of direct *TF* fraction. Similarly, due to the strong correlation between LAI and canopy water storage (e.g., Deguchi et al. 2006), spatial estimates of LAI generated from UAV-acquired data could provide both the stand average LAI and the spatial and temporal distribution of LAI in forests. More work would be needed for canopies where the correlation between LAI and canopy water storage is weak (e.g., canopies with extensive epiphytes—Van Stan and Pypker 2015). Finally, E/R has been associated with canopy structure (e.g., Flerchinger and Saxton 1989; Jetten 1996; Pypker et al. 2011), including canopy characteristics such as canopy height, roughness and leaf area distribution (Pypker et al. 2011). Canopies with leaf area distributions closer to the forest floor may have a greater r_a and a subsequently reduced E/R (e.g., Pypker et al. 2005). Given the capacity for UAV-acquired measurements to improve accuracy and spatial/temporal resolution of forest composition and structure (Sect. 3.3), UAV-borne remote sensing may play a pivotal role in the development of new spatially explicit models of canopy interception.

3.4.3 Use of UAVs in Eddy Covariance Studies

Measurements of eddy covariance (EC), from EC towers, allows for continuous estimates of ET from forested watersheds. EC towers measure the spatially averaged flux of atmospheric gasses from an upwind area called the footprint of the tower. The footprint of a tower is time variable and depends on microclimate conditions (e.g., Goulden and Field 1994). Typically, models are used to estimate the tower footprint (Baldocchi 1997; Finnigan 2004). Recently, Vivoni et al. (2014) demonstrated that UAVs are useful for characterizing the area around EC towers though analysis of UAV-acquired imagery to create a high-resolution land cover map. This map can be used to predict the spatial distribution of soil moisture and temperature between point measurements collected by an on-the-ground network of soil moisture and soil temperature sensors. This spatial characterization of the area around the flux tower could be used to infer the footprint of the tower by comparing the EC tower measurements to spatial aggregation of expected fluxes from different soil conditions and land cover types (Vivoni et al. 2014). UAV-acquired imagery could also be used to quantify phenological changes and thereby allow models to better estimate the footprint size throughout the year (Vivoni et al. 2014), or to upscale fluxes to satellite observations for application over broad areas (Anderson and Gaston 2013).

Although the work reported by Vivoni et al. (2014) utilized a single-rotor, gasoline-powered UAV (Rotomotion SR30), which does not meet our definition of consumer-grade technology due to its use of a gasoline combustion engine and concomitant high acquisition cost (>$50 k USD), the RGB imagery (5-cm resolution) collected by this platform could have been collected by a consumer-grade UAV (e.g., Matrice 600) with no loss of value to the study.

3.5 Utility of UAVs for Snow Hydrology

In the past, snow depth was assessed using point measurements from snow pillows, ultrasonic sensors, or manual measurements. Snow pillows and ultrasonic sensors provide accurate depth estimates, but their measurements are representative of a very small area. Manual measurements are labour intensive yet still represent a limited spatial area. Terrestrial laser scanning (i.e., ground-based LiDAR) can provide spatial estimates of snow depth (Deems et al. 2013, Hedrik et al. 2015), but this technology is expensive and, therefore, is less feasible than other methods.

DSMs generated by SfM/MVS from images acquired by UAV-borne digital cameras provide an opportunity to monitor snow depth over large areas at significantly less cost than terrestrial laser scanning. DSMs generated from UAV-acquired imagery have been widely used in a variety of fields, including archaeology, geomorphology and ecology (e.g., Eisenbeiss and Zhang 2006; Jaakkola et al. 2010; Lucieers et al. 2013; Colomina and Molina 2014), and past work indicates that snow depth can be accurately estimated on flat terrain using SfM/MVS (De Michele et al. 2015), suggesting that UAVs can enable frequent spatial estimates of snow depth at the field scale, requiring substantially less labour than a field survey and substantially less cost than an airborne LiDAR survey to accomplish the same goal. However, UAV-enabled snow depth monitoring in forested environments is challenged in two ways. First, regulations which require the operator to maintain visual line-of-sight with the UAV throughout the aerial imaging mission may limit the applicability of UAV-enabled snowpack mapping to regions with low tree density or areas where management activities have removed the vegetation. Second, technical challenges relating to complex terrain and blowing snow must be surmounted. The authors (Pypker, unpublished data) recently concluded that the accuracy of snow depth measurements derived from SfM/MVS analysis of images acquired by a low-cost UAV (DJI Phantom 4) was limited by blowing snow and inaccurate vertical positioning (see Sect. 2.3). Future work that addresses these two technical challenges may allow for improved wall-to-wall estimates of snowpack depth, thereby removing the need for labour-intensive snow courses.

3.6 Future Needs and Direction

Presently, UAVs provide a cost-effective tool to assist in ecohydrological research. They provide greater resolution than coarser-scale satellite products (e.g., Babcock et al. 2018; Wang et al. 2018). UAVs can provide information on physical attributes of the canopy that directly relate to ecohydrology from the scale of the leaf to the watershed. However, to reach the full potential of UAVs in ecohydrology, future work needs to focus on regulatory change, improved capacity of UAVs to fly longer distances and in inclement weather, and improved sensor technology and methodology.

In many regions, regulations require the operator fly the UAV within their line-of-sight. When mapping vegetation in larger watersheds or within tall canopies, maintaining line-of-sight can be challenging (e.g., Leduc and Knudby 2018). If we are to effectively use UAVs for research purposes in forested watersheds, working with government regulators to modify current regulations to allow researchers to effectively use these tools over larger areas will be beneficial.

The current suite of small, cost-effective UAVs has physical limitations that limit their use. Presently, consumer-grade UAVs can only be used for small regions (e.g., Leduc and Knudby 2018). Part of this limitation is a direct result of the aforementioned line-of-sight regulations. The second limitation is a short battery life that restricts flight times for most consumer-grade UAVs to approximately 30 minutes. The flight time is further reduced in cold weather and windy conditions. In addition, at the time of writing, there were no commercially available UAVs designed to fly in rain, thereby greatly limiting the use of UAVs in ecohydrology. The inability to fly in the rain is a significant drawback, because it limits measurement of the canopy during precipitation events. In addition, ice rimming on the rotors, which results in decreased lift, in subzero temperatures limits their ability to fly in cold, damp conditions. Future work to improve the durability and endurance of low-cost UAVs will greatly advance their use in ecohydrological studies.

Testing of additional sensors would expand the suite of measurements possible with UAVs. Recently, active longwave infrared (LWIR) spectroscopy has been used to identify leaves from different tree species (e.g., Harrison et al. 2018). With additional research, LWIR imagers mounted on UAVs may be able to map tree species over large areas. Martin et al. (2018) were able to quantify canopy water content of leaves using VIS-SWIR hyperspectral mounted on airplanes. Using a combination of LiDAR data and VIS-SWIR, Marin et al. could quantify drought stress in giant sequoia at a resolution of $2m^2$. As noted in Table 3.1, hyperspectral imagers designed for use with UAVs are commercially available, but, at the time of writing, are prohibitively expensive. When carried aloft by UAVs, these imagers could produce data of such resolution as to enable better understanding of the spatial distribution of drought stress in the canopy of individual trees and possibility water storage on leaves.

Additionally, refinement of current methodology is needed to address ecohydrology issues. Assessment of large areas of forested regions by LiDAR is cost prohibitive. Recent work by Babcock et al. (2018) suggests that incomplete datasets from LiDAR can be combined with models to extract accurate estimates of forest cover over large areas. Similarly, Fernández-Guisuraga et al. (2018) used UAV-borne multispectral imagers to assess a large burned area. However, further research is needed, as issues with image quality and long data processing times limit the applicability of the large datasets acquired by UAV-borne imaging sensors. This problem will be further exacerbated as multispectral imagers with a larger spectral resolution and hyperspectral imagers become more affordable. Development of new data fusion and spatial modelling methods (e.g., Babcock et al. 2018) may enable UAV-based remote sensing over increasingly large areas.

Instrumentation mounted on UAVs may be able to directly measure transpiration, evaporation and snowpack depth, but further work is needed to improve accuracy. Transpiration has been measured in crops (Page et al. 2018); however, transferring the technique to structurally complex forest canopies requires additional research. If a new technique is developed, the use of UAVs could provide a better understanding of the spatial distribution of transpiration by different tree species over large areas. Additionally, estimates of evaporation using rainfall interception models (e.g., Muzylo et al. 2009) could be improved if these models were combined with canopy structural characteristics provided by measurements enabled by UAVs. The structural data could provide better canopy structural inputs into the models and, with additional research, provide insights into the spatial distribution of rainfall beneath forest canopies. Similarly, snowpack depth has been accurately measured on flat terrain using SfM/MVS (De Michele et al. 2015). However, the use of SfM/MVS to generate DSMs is challenged by canopy density and complex terrain, a challenge that may be partially addressed as real-time kinetic (RTK) GNSS receivers become more affordable for UAV applications. Furthermore, errors in the SfM/MVS-derived point cloud, attributed to blowing snow, must also be addressed before this method may provide an economical method of snow depth mapping. While the availability of more affordable LiDAR systems for UAV applications may improve snow depth mapping, blowing snow may still confound the LiDAR-derived point clouds, thus rendering methods to spatially smooth the point cloud necessary to improve snow depth mapping accuracy.

Acknowledgements This work is supported in part by the Natural Science and Engineering Research Council of Canada through the following research grants: RGPIN 2014-06114 (Hill), EGP 502265-16 (Hill), RGPIN-2018-06766 (Pypker) and EGP 505371-16 (Pypker). Any use of trade, firm, or product names is for descriptive purposes only and does not imply endorsement by the authors.

References

Aeronautics Act: Regulations Amending the Canadian Aviation Regulations (Part VI) (2002) Canada Gazette Part II, 136(11). Retrieved January 16, 2019, from the Canada Gazette website: http://www.gazette.gc.ca/rp-pr/publications-eng.html#wb-cont

Alonzo MH, Andersen E, Morton DC, Cook BD (2018) Quantifying boreal forest structure and composition using UAV structure from motion. Forests 9:119. https://doi.org/10.3390/f9030119

Ambrosia VG, Wenger SS, Sullivan DV, Buechel SW, Dunagan SE, Brass JA et al (2003) Demonstrating UAV-acquired real-time thermal data over fires. Photogramm Eng Remote Sens 69:391–402. https://doi.org/10.14358/PERS.69.4.391

Ambrosia VG, Wegener SS, Zajkowski T, Sullivan DV, Buechel S, Enomoto F et al (2011) The Ikhana UAS western states fire imaging missions: from concept to reality (2006–2011). Geocarto Int 26:85–101. https://doi.org/10.1080/10106049.2010.539302

Anderson K, Gaston KJ (2013) Lightweight unmanned aerial vehicles will revolutionize spatial ecology. Front Ecol Environ 11:138–146. https://doi.org/10.1890/120150

Anderson MC, Norman JM, Kustas WP, Houborg R, Starks PJ, Agam N (2008) A thermal-based remote sensing technique for routine mapping of land-surface carbon, water and energy fluxes from field to regional scales. Remote Sens Environ 112:4227–4241. https://doi.org/10.1016/j.rse.2008.07.009

Arora VK, Montenegro A (2011) Small temperature benefits provided by realistic afforestation efforts. Nat Geosci 1:514. https://doi.org/10.1038/ngeo1182

Avanzi F, Bianchi A, Cina A, De Michele C, Maschio P, Pagliari D et al (2018) Centimetric accuracy in snow depth using unmanned aerial system photogrammetry and a multistation. Remote Sens 10:765. https://doi.org/10.3390/rs10050765

Babcock CAO, Finley H-E, Andersen R, Pattison R, Cook BD, Morton M et al (2018) Geostatistical estimation of forest biomass in interior Alaska combining Landsat-derived tree cover, sampled airborne lidar and field observations. Remote Sens Environ 212:212–230. https://doi.org/10.1016/j.rse.2018.04.044

Baldocchi DD (1997) Flux footprints within and over forest canopies. Boundary-Layer Meteorol 85:273–292. https://doi.org/10.1023/A:1000472717236

Baldocchi DD, Hicks BB, Meyers TP (1988) Measuring biosphere-atmosphere exchanges of biologically related gases with micrometeorological methods. Ecology 69:1331–1340. https://doi.org/10.2307/1941631

Baron J, Hill DJ, Elmigili H (2018) Combining image processing and machine learning to identify invasive plants in high resolution images. Int J Remote Sens 39:5099–5118. https://doi.org/10.1080/01431161.2017.1420940

Becker RH, Zmijewski KA, Crail T (2013) Seeing the forest for the invasives: mapping buckthorn in the oak openings. Biol Invasions 15:315–326. https://doi.org/10.1007/s10530-012-0288-8

Blaschke T (2010) Object based image analysis for remote sensing. ISPRS J Photogramm 65:2–16. https://doi.org/10.1016/j.isprsjprs.2009.06.004

Blaschke T, Hay GJ, Kelly M, Lang S, Hofmann P, Addink E et al (2014) Geographic object-based image analysis – towards a new paradigm. ISPRS J Photogramm 87:180–191. https://doi.org/10.1016/j.isprsjprs.2013.09.014

Boudreau J, Nelson RF, Margolis HA, Beaudoin A, Guindon L, Kimes DS (2008) Regional aboveground forest biomass using airborne and spaceborne LiDAR in Québec. Remote Sens Environ 112:3876–3890. https://doi.org/10.1016/j.rse.2008.06.003

Calderón R, Navas-Cortés JA, Lucena C, Zarco-Tejada PJ (2013) High-resolution airborne hyperspectral and thermal imagery for early detection of Verticillium wilt of olive using fluorescence, temperature and narrow-band spectral indices. Remote Sens Environ 139:231–245. https://doi.org/10.1016/j.rse.2013.07.031

Cao C, Lee X, Muhlhausen J, Bonneau L, Xu J (2018a) Measuring landscape albedo using unmanned aerial vehicles. Remote Sens 10:1812. https://doi.org/10.3390/rs10111812

Cao J, Leng W, Liu K, Liu L, He Z, Zhu Y (2018b) Object-based mangrove species classification using unmanned aerial vehicle hyperspectral images and digital surface models. Remote Sens 10:89. https://doi.org/10.3390/rs10010089

Cardil A, Vepakomma U, Brotons L (2017) Assessing pine processionary moth defoliation using unmanned aerial systems. Forests 8:402. https://doi.org/10.3390/f8100402

Carrivick JL, Smith MW, Quincey DJ (2016) Structure from motion in the geosciences. Wiley, Hoboken

Casbeer DW, Kingston DB, Beard RW, McLain TW (2006) Cooperative forest fire surveillance using a team of small unmanned air vehicles. Int J Syst Sci 37:351–360. https://doi.org/10.1080/00207720500438480

Chan FCC, Arain MA, Khomik M, Brodeur JJ, Peichl M, Restrepo-Coupe N et al (2018) Carbon, water and energy exchange dynamics of a young pine plantation forest during the initial fourteen years of growth. For Ecol Manag 410:12–26. https://doi.org/10.1016/j.foreco.2017.12.024

Chávez JL, Neale CMU, Prueger JH, Kustas WP (2008) Daily evapotranspiration estimates from extrapolating instantaneous airborne remote sensing ET values. Irrig Sci 27:67–81. https://doi.org/10.1007/s00271-008-0122-3

Chen S, McDermid GJ, Castilla G (2017) Measuring vegetation height in linear disturbances in the boreal forest with UAV photogrammetry. Remote Sens 9:1257. https://doi.org/10.3390/rs9121257

Chianucci F, Disperati L, Guzzi D (2015) Estimation of canopy attributes in beech forests using true colour digital images from a small fixed-wing UAV. Int J App Earth Obs 47:60–68. https://doi.org/10.1016/j.jag.2015.12.005

Colomina I, Molina P (2014) Unmanned aerial systems for photogrammetry and remote sensing: a review. ISPRS J Photogramm 92:79–97. https://doi.org/10.1016/j.isprsjprs.2014.02.013

Coops NC, Gillanders SN, Wulder MA (2010) Assessing changes in forest fragmentation following infestation using time series Landsat imagery. For Ecol Manag 259:2355–2365. https://doi.org/10.1016/j.foreco.2010.03.008

Costa JM, Grant OM, Chaves MM (2013) Thermography to explore plant–environment interactions. J Exp Bot 64:3937–3949. https://doi.org/10.1093/jxb/ert029

Cristiano PM, Campanello PI, Bucci SJ, Rodriguez SA, Lezcano OA, Scholz FG et al (2015) Evapotranspiration of subtropical forests and tree plantations: a comparative analysis at different temporal and spatial scales. Agric For Meteorol 203:96–106. https://doi.org/10.1016/j.agrformet.2015.01.007

Dandois JP, Ellis EC (2013) High spatial resolution three-dimensional mapping of vegetation spectral dynamics using computer vision. Remote Sens Environ 136:259–276. https://doi.org/10.1016/j.rse.2013.04.005

David LCG, Ballado AH (2016) Vegetation indices and textures in object-based detection from UAV imagery. In: 6th IEEE international conference on control system, computing and engineering (ICCSCE), 25–27 Nov. 2016, Batu Ferringhi, Malaysia. https://doi.org/10.1109/ICCSCE.2016.7893584

De Michele C, Avanzi F, Passoni D, Della Vedova G (2015) Microscale variability of snow depth using UAS technology. Cryosphere Discuss 9:1047–1075. https://doi.org/10.5194/tcd-9-1047-2015

Deems JS, Painter TH, Finnegan DC (2013) Lidar measurement of snow depth: a review. J Glaciol 59:467–479. https://doi.org/10.3189/2013JoG12J154

Deguchi A, Hattori S, Park HT (2006) The influence of seasonal changes in canopy structure on interception loss: application of the revised Gash model. J Hydrol 318:80–102. https://doi.org/10.1016/j.jhydrol.2005.06.005

Demir N (2018) Using UAVs for detection of trees from digital surface models. J For Res 29:813–821. https://doi.org/10.1007/s11676-017-0473-9

Detert M, Weitbrecht V (2015) A low-cost airborne velocimetry system: proof of concept. J Hydraul Res 53:532–539. https://doi.org/10.1080/00221686.2015.1054322

Eisenbeiss H, Zhang L (2006) Comparison of DSMs generated from mini UAS imagery and terrestrial laser scanner in a cultural heritage application. Int Arch Photogramm 36:90–96

Fernández-Guisuraga JM, Sanz-Ablanedo E, Suárez-Seoane CL (2018) Using unmanned aerial vehicles in postfire vegetation survey campaigns through large and heterogeneous areas: opportunities and challenges. Sensors 18:586. https://doi.org/10.3390/s18020586

Ferreira MP, Féret JB, Grau E, Gastellu-Etchegorry J-P, do Amaral CH, Shimabukuro YE et al (2018) Retrieving structural and chemical properties of individual tree crowns in a highly diverse tropical forest with 3D radiative transfer modeling and imaging spectroscopy. Remote Sens Environ 211:276–291. https://doi.org/10.1016/j.rse.2018.04.023

Finnigan JJ (2004) A re-evaluation of long-term flux measurement techniques part II: coordinate systems. Bound-Layer Meteorol 113(1). https://doi.org/10.1023/B:BOUN.0000037348.64252.45

Flerchinger GN, Saxton KE (1989) Simultaneous heat and water model of a freezing snow-residue-soil system. I. Theory and development. Trans ASAE 32:565–571. https://doi.org/10.13031/2013.31040

Ganthaler A, Losso A, Mayer S (2018) Using image analysis for quantitative assessment of needle bladder rust disease of Norway spruce. Plant Pathol 67:1122–1130. https://doi.org/10.1111/ppa.12842

Getzin S, Wiegand K, Schöning I (2012) Assessing biodiversity in forests using very high resolution images and unmanned aerial vehicles. Methods Ecol Evol 3:397–404. https://doi.org/10.1111/j.2041-210X.2011.00158.x

Goldmean Sachs (2016) Drones reporting for work. Accessed online June 27, 2018: http://www.goldmansachs.com/our-thinking/technology-driving-innovation/drones/

González-Jorge H, Martínez-Sánchez J, Bueno M, Arias P (2017) Unmanned aerial systems for civil applications: a review. Drones 1:2. https://doi.org/10.3390/drones1010002

Goulden ML, Field CB (1994) Methods for monitoring the gas-exchange of individual tree canopies – ventilated-chamber, sap-flow and Penman-Monteith measurements on evergreen oaks. Funct Ecol 8:125–135. https://doi.org/10.2307/2390121

Granier A (1985) Une nouvelle méthode pour la mesure du flux de sève brute dans le tronc des arbres. Annals For Sci 42:193–200. https://doi.org/10.1051/forest:19850204

Hansen MC, Potapov PV, Moore R, Hancher M, Turubanova SA, Tyukavina A et al (2013) High resolution global maps of 21st-century forest cover change. Science 342:850–853. https://doi.org/10.1126/science.1244693

Harrison D, Rivard B, Sànchez-Azofeifa A (2018) Classification of tree species based on long wave hyperspectral data from leaves, a case study for a tropical dry forest. Int J Appl Earth Obs Geoinf 66:93–105. https://doi.org/10.1016/j.jag.2017.11.009

Hawrylo P, Wężyk P (2018) Predicting growing stock volume of Scots Pine stands using Sentinel-2 satellite imagery and airborne image-derived point clouds. Forests 9:274. https://doi.org/10.3390/f9050274

Hedrick A, Marshall H-P, Winstral A (2015) Independent evaluation of the SNODAS snow depth product using regional-scale lidar-derived measurements. Cryosphere 9:13–23. https://doi.org/10.5194/tc-9-13-2015

Hill DJ, Tarasoff C, Whitworth GE, Baron J, Bradshaw JL, Church J (2017) Utility of unmanned aerial vehicles for mapping invasive plant species: a case study on yellow flag iris (*Iris pseudacorus* L.). Int J Remote Sens 38:2083–2105. https://doi.org/10.1080/01431161.2016.1264030

Iizuka K, Yonehara T, Itoh M, Kosugi Y (2018) Estimating tree height and diameter at breast height (DBH) from digital surface models and orthophotos obtained with an unmanned aerial system for a Japanese Cypress (*Chamaecyparis obtusa*) forest. Remote Sens 10:13. https://doi.org/10.3390/rs10010013

Ishida T, Kurihara J, Vivray FA, Namuco SB, Paringit EC, Perez Y et al (2018) A novel approach for vegetation classification using UAV-based hyperspectral imaging. Comput Electron Agric 144:80–85. https://doi.org/10.11016/j.compag.2017.11.027

Jaakkola A, Hyyppä J, Kukko A, Yu X, Kaartinen H, Lehtomäki M et al (2010) A low-cost multi-sensoral mobile mapping system and its feasibility for tree measurements. ISPRS J Photogramm 65:514–522. https://doi.org/10.1016/j.isprsjprs.2010.08.002

Jarvis PG, McNaughton KG (1986) Stomatal control of transpiration: scaling up from leaf to region. Adv Ecol Res 15:1–49. https://doi.org/10.1016/S0065-2504(08)60119-1

Jayathunga S, Owari T, Tsuyuki S (2018) Evaluating the performance of photogrammetric products using fixed-wing UAV imagery over mixed conifer-broadleaf forest: comparison with airborne laser scanning. Remote Sens 10:187. https://doi.org/10.3390/rs10020187

Jetten VG (1996) Interception of tropical rainforest: performance of a canopy water balance model. Hydrol Process 10:671–685. https://doi.org/10.1002/(SICI)1099-1085(199605)10:5<671::AID-HYP310>3.0.CO;2-A

Keim RF, Skaugset AE, Weiler M (2005) Temporal persistence of spatial patterns in throughfall. J Hydrol 314:263–274. https://doi.org/10.1016/j.jhydrol.2005.03.021

Klosterman S, Richardson A (2017) Observing spring and fall phenology in a deciduous forest with aerial drone imagery. Sensors 17:2852. https://doi.org/10.3390/s17122852

Klosterman S, Melaas E, Wang JA, Martinez A, Frederick S, O'Keefe J (2018) Fine-scale perspectives on landscape phenology from unmanned aerial vehicle (UAV) photography. Agric For Meteorol 248:397–407. https://doi.org/10.1016/j.agformet.2017.10.015

Köstner B, Granier A, Cermák J (1998) Sapflow measurements in forest stands: methods and uncertainties. Ann Sci For 55:13–27. https://doi.org/10.1051/forest:19980102
Kunert N, Aparecido LMT, Wolff S, Higuchi N, dos Santos J, de Araujo C et al (2017) A revised hydrological model for the Central Amazon: the importance of emergent canopy trees in the forest water budget. Agric For Meteorol 239:47–57. https://doi.org/10.1016/j.agrformet.2017.03.002s
Laliberte AS, Rango A (2017) Image processing and classification procedures for analysis of sub-decimeter imagery acquired with an unmanned aircraft over arid rangelands. GISci Remote Sens 48(1):4–23. https://doi.org/10.2747/1548-1603.48.1.4
Leduc M-B, Knudby AJ (2018) Mapping wild leak though the forest canopy using a UAV. Remote Sens 10:70. https://doi.org/10.3390/rs10010070
Lee X (2004) A model for scalar advection inside canopies and application to footprint investigation. Agric For Meteorol 127:131–141. https://doi.org/10.1016/j.agrformet.2004.07.009
Lefsky MAA (2010) A global forest canopy height map from the moderate resolution imaging spectroradiometer and the geoscience laser altimeter system. Geophys Res Lett 37. https://doi.org/10.1029/2010GL043622
Lefsky MA, Harding DJ, Keller M, Cohen WB, Carabajal CC, Del Bom E-SF et al (2005) Estimates of forest canopy height and aboveground biomass using ICESat. Geophys Res Lett 32:L22S02. https://doi.org/10.1029/2005GL023971
Leitão JP, Moy de Vitry M, Scheidegger A (2016) Assessing the quality of digital elevation models obtained from mini unmanned aerial vehicles for overland flow modelling in urban areas. Hydrol Earth Syst Sci 20:1637–1653. https://doi.org/10.5194/hess-20-1637-2016
Levia DF Jr, Frost EE (2003) A review and evaluation of stemflow literature in the hydrologic and biogeochemical cycles of forested and agricultural ecosystems. J Hydrol 274:1–29. https://doi.org/10.1016/S0022-1694(02)00399-2
Levy CR, Burakowski E, Richardson AD (2018) Novel measurements of fine-scale albedo: using a commercial quadcopter to measure radiation fluxes. Remote Sens 10:1303. https://doi.org/10.3390/rs10081303
Liang S (2000) Narrowband to broadband conversions of land surface albedo I: algorithms. Remote Sens Environ 76:213–238. https://doi.org/10.1016/S0034-4257(00)00205-4
Link TE, Unsworth M, Marks D (2004) The dynamics of rainfall by a seasonal temperate rainforest. Agric For Met 124(3–4):171–191. https://doi.org/10.1016/j.agrformet.2004.01.010
Lu D, Weng Q (2005) A survey of image classification method and techniques for improved classification performance. Int J Remote Sens 28:823–870. https://doi.org/10.1080/01431160600746456
Lucieers A, de Jong SM, Turner D (2013) Mapping landslide displacements using structure from motion (SfM) and image correlation of multi-temporal UAS photography. Prog Phys Geogr 38:97–116. https://doi.org/10.1177/0309133313515293
Maes WH, Steppe K (2012) Estimating evapotranspiration and drought stress with ground-based thermal remote sensing in agriculture: a review. J Exp Bot 63:4671–4712. https://doi.org/10.1093/jxb/ers165
Martin RE, Asner GP, Francis E, Ambrose A, Baxter DAJ et al (2018) Remote measurement of canopy water content in giant sequoias (*Sequoiadendron giganteum*) during drought. For Ecol Manag 419–420:279–290. https://doi.org/10.1016/j.foreco.2017.12.002
Matthews AJ, Jensen LJR (2013) Visualizing and quantifying vineyard canopy LAI using an unmanned aerial vehicle (UAV) collected high density structure from motion point cloud. Remote Sens 5:2164–2183. https://doi.org/10.3390/rs5052164
McPherson RA (2007) A review of vegetation-atmosphere interactions and their influences on mesoscale phenomena. Prog Phys Geogr 31:261–285. https://doi.org/10.1177/0309133307079055
Meijerink AMJ (2002) Satellite eco-hydrology – a review. Trop Ecol 43:91–106

Meinzer FC, Goldstein G, Andrade JL (2001) Regulation of water flux through tropical forest canopy trees: do universal rules apply? Tree Physiol 21:19–26. https://doi.org/10.1093/treephys/21.1.19

Merlin PW (2009) Ikhana: unmanned aircraft system western states fire missions. Monographs in aerospace history no. 44 (SP-2009-4544). National Aeronautics and Space Administration (NASA) History Office, Washington, DC

Mlambo RI, Woodhouse H, Gerard F, Anderson K (2017) Structure from motion (SfM) photogrammetry with drone data: a low cost method for monitoring greenhouse gas emissions from forests in developing countries. Forests 8:68. https://doi.org/10.3390/f8030068

Monteith JL (1965) Evaporation and environment. Symo Soc Exp Biol 19:205–234

Monteith JL, Unsworth MH (2007) Principles of environmental physics. Elsevier, New York, 418p

Moore KE, Fitzjarrald DR, Sakai RK, Goulden ML, Munger JW, Wofsy SC (1996) Seasonal variation in radiative and turbulent exchange at a deciduous forest in Central Massachusetts. J Appl Meterol 40:1297–1309. https://doi.org/10.1175/1520-0450(1996)035<0122:SVIRAT>2.0.CO;2

Muzylo A, Llorens P, Valente F, Keizer JJ, Domingo F, Gash JHC (2009) A review of rainfall interception modelling. J Hydrol 370:1–4. https://doi.org/10.1016/j.jhydrol.2009.02.058

Näsi R, Honkavaara E, Lyytikäinen-Saarenmaa P, Blomqvist M, Litkey P, Hakala T et al (2015) Using UAV-based photogrammetry and hyperspectral imaging for mapping bark beetle damage at tree-level. Remote Sens 7:15467–15493. https://doi.org/10.3390/rs71115467

Nelson R, Ranson KJ, Sun G, Kimes DS, Kharuk V, Montesano P (2009) Estimating Siberian timber volume using MODIS and ICESat/GLAS. Remote Sens Environ 113:691–701. https://doi.org/10.1016/j.rse.2008.11.010

Nevalainen O, Honkavaara E, Tuominen S, Viljanen N, Hakala T, Yu X et al (2017) Individual tree detection and classification with UAV-based photogrammetric point clouds and hyperspectral imaging. Remote Sens 7:9185. https://doi.org/10.3390/rs9030185

Newcome L (2000) Commercial UAV operations in civil airspace. In: Proceedings of SPIE 4127 airborne reconnaissance XXIV. https://doi.org/10.1117/12.408698

Newcome LR (2004) Unmanned aviation: a brief history of unmanned aerial vehicles. American Institute of Aeronautics and Astronautics, Reston, Va

Ortiz S, Breidenbach J, Kändler G (2013) Early detection of bark beetle green attack using TerraSAR-X and RapidEye data. Remote Sens 5:1912–1931. https://doi.org/10.3390/rs5041912

Page GFM, Liénard JF, Pruett MJ, Moffett KB (2018) Spatiotemporal dynamics of leaf transpiration quantified with time-series thermal imaging. Agric For Met 256–257(15):304–314. https://doi.org/10.1016/j.agrformet.2018.02.023

Perks MT, Russell AJ, Large ARG (2016) Advances in flash flood monitoring using unmanned aerial vehicles (UAVs). Hydrol Earth Syst Sci 20:4005–4015. https://doi.org/10.5194/hess-20-4005-2016

Persson HJ, Perko R (2016) Assessment of boreal forest height from WorldView-2 satellite stereo images. Remote Sens Lett 7:1150–1159. https://doi.org/10.1080/2150704X.2016.1219424

Puliti S, Ørka HO, Gobakken T, Næsset E (2015) Inventory of small forest areas using an unmanned aerial system. Remote Sens 7:9632–9654. https://doi.org/10.3390/rs70809632

Pypker TG, Bond BJ, Link TE, Marks D, Unsworth MH (2005) The importance of canopy structure in controlling the interception loss: examples from a young and old-growth Douglas-fir forests. Agric For Meteorol 130:113–129. https://doi.org/10.1016/j.agrformet.2005.03.003

Pypker TG, Levia DF Jr, Staelens J, Van Stan IIJT (2011) Chapter 18: canopy structure in relation to hydrological and biogeochemical fluxes. In: Levia DF Jr, Carlyle-Moses DE, Tanaka T (eds) Forest hydrology and biogeochemistry: synthesis of past research and future directions, Ecological studies series, vol 216. Springer, Berlin. https://doi.org/10.1007/978-94-007-1363-5_18

Rango A, Laliberte AS, Herrick JE, Havstad KM (2009) Unmanned aerial vehicle-based remote sensing for rangeland assessment, monitoring, and management. J Remote Sens 3:033542. https://doi.org/10.1117/1.3216822

Richardson AD, Braswell BH, Hollinger DY, Jenkins JP, Ollinger SV (2009) Near-surface remote sensing of spatial and temporal variation in canopy phenology. Ecol Appl 19:1417–1428. https://doi.org/10.1890/08-2022.1

Ruzgienė B, Berteška T, Gečyte S, Jakubauskienė E, Aksamitauskas VČ (2015) The surface modelling based on UAV photogrammetry and qualitative estimation. Measurement 73:619–627. https://doi.org/10.1016/j.measurement.2015.04.018

Saarinen N, Vastaranta M, Näsi R, Rosnell T, Hakala T, Honkavaara E et al (2018) Assessing biodiversity in boreal forests with UAV-based photogrammetric point clouds and hyperspectral imaging. Remote Sens 10:338. https://doi.org/10.3390/rs10020338

Sankey TT, McVay J, Swetnam TL, McClaran MP, Heilman P, Nichols M (2017) UAV hyperspectral and lidar data and their fusion for arid and semi-arid land vegetation monitoring. Remote Sen Ecol Conserv 4:20–33. https://doi.org/10.1002/rse2.44

Sauer TJ, Horton R (2005) Soil heat flux. In: Hatfield JL, Baker JM (eds) Micrometerology in agricultural systems, ASA monograph 47. American Society of Agronomy, Madison, WI, pp 131–154. https://doi.org/10.2134/agronmonogr47.c7

Scherrer D, Körner C (2010) Infra-red thermometry of alpine landscapes challenges climatic warming projections. Glob Chang Biol 16:2602–2613. https://doi.org/10.1111/j.1365-2486.2009.02122.x

Scherrer D, Körner C (2011) Topographically controlled thermal-habitat differentiation buffers alpine plant diversity against climate warming. J Biogeogr 38:406–416. https://doi.org/10.1111/j.1365-2699.2010.02407.x

Seier G, Stangl J, Schöttl S (2017) UAV and TLS for monitoring a creek in an alpine environment, Styria, Austria. Int J Remote Sens 38(8–10):2903–2920. https://doi.org/10.1080/01431161.2016.1277045

Shaw M, Sandhoo K, Turner T (2000) Modernization of the global positioning system. GPS World 11:36–40

Shrestha R, Wynne RH (2012) Estimating biophysical parameters of individual trees in an urban environment using small footprint discrete-return imaging lidar. Remote Sens 4:484–508. https://doi.org/10.3390/rs4020484

Stafford N (2007) Spy in the sky. Nature 445:808–809. https://doi.org/10.1038/445808a

Thumser P, Haas C, Tuhtan JA, Fuentes-Pérez JF, Toming G (2017) RAPTOR-UAV: real-time particle tracking in rivers using unmanned aerial vehicle. Earth Surf Process Landf 42:2439–2446. https://doi.org/10.1002/esp.4199

Tuanmu M-N, Viña A, Bearer S, Xu W, Ouyang Z, Zhang H, Jianguo L (2010) Mapping understory vegetation using phenological characteristics derived from remotely sensed data. Remote Sens Environ 114:1833–1844. https://doi.org/10.1016/j.rse.2010.03.008

Van Stan II JT, Pypker TG (2015) A review and evaluation of forest canopy epiphyte roles in the partitioning and chemical alteration of precipitation. Sci Total Environ 236:813–824. https://doi.org/10.1016/j.scitotenv.2015.07.134

Vivoni ER, Rango A, Anderson CA, Pierini NA, Schreiner-McGraw AP, Saripalli S et al (2014) Ecohydrology with unmanned aerial vehicles. Ecosphere 5:130. https://doi.org/10.1890/ES14-00217.1

Wallace L, Lucier A, Watson C, Turner D (2012) Development of a UAV-LiDAR system with application to forest inventory. Remote Sens 4:1519–1543. https://doi.org/10.3390/rs4061519

Wang R, Gamon JA, Schweiger AK, Cavender-Bares J, Townsend PA, Zygielbaum AI et al (2018) Influence of species richness, evenness, and composition on optical diversity: a simulation study. Remote Sens Environ 211:218–228. https://doi.org/10.1016/j.rse.2018.04.010

Weathers KC, Cadenasso ML, Pickett STA (2001) Forest edges as nutrient and pollutant concentrators: potential synergisms between fragmentation, forest canopies, and the atmosphere. Conserv Biol 15:1506–1514. https://doi.org/10.1046/j.1523-1739.2001.01090.x

Webster C, Westoby M, Rutter N, Jonas T (2018) Three-dimensional thermal characterization of forest canopies using UAV photogrammetry. Remote Sens Environ 209:835–8476. https://doi.org/10.1016/j.rse.2017.09.033

Wing MG, Burnett J, Sessions J, Brungardt J, Cordell V, Dobler D et al (2013) Eyes in the sky: remote sensing technology development using small unmanned aircraft systems. J For 111:341–347. https://doi.org/10.5849/jof.12-117

Wright C, Kagawa-Viviani A, Gerlein-Safdi C, Mosquera GM, Poca M, Tseng H et al (2018) Advancing ecohydrology in the changing tropics perspectives from early career scientists. Ecohydrology 11:e1918. https://doi.org/10.1002/eco.1918

Wulder MA, Dymond CC, White JC, Leckie DG, Carroll AL (2006) Surveying mountain pine beetle damage of forests: a review of remote sensing opportunities. For Ecol Manag 221:27–41. https://doi.org/10.1016/j.foreco.2005.09.021

Zakaria S, Hahadi MR, Abdullah AF et al (2018) Aerial platform reliability for flood monitoring under various weather conditions: a review. Int Arch Photogramm Remote Sens Spat Inf Sci XLII-3/W4:591–602. https://doi.org/10.5194/isprs-archives-XLII-3-W4-591-2018

Zhou JM, Pavek J, Shelton SC, Holden ZJ, Sankaran S (2016) Aerial multispectral imaging for crop hail damage assessment in potato. Comput Electron Agric 127:406–412. https://doi.org/10.1016/j.compag.2016.06.019

Chapter 4
LiDAR Applications to Forest-Water Interactions

Johannes Schumacher and Jesper Riis Christiansen

4.1 Introduction

Light detection and ranging (LiDAR), also known as laser altimetry or laser scanning, is a remote sensing technology, which provides 3D information of surfaces. LiDAR data can be used to create terrain and surface models, even below dense forest canopies, and to capture the vertical structure of the vegetation cover. This information can be used for a broad range of applications, such as assessment of topography and structural information about the vegetation cover, including canopy density and canopy architecture, which are relevant to hydrological studies.

A common task in hydrological studies is to assess the magnitude of water fluxes and stores within a particular forest stand or watershed. The main factors influencing these water fluxes and stores are terrain, vegetation, and stream properties. Soil properties are also important factors; however, they cannot directly be assessed with LiDAR. Terrain properties, such as slope and aspect of the surface, determine the speed and direction of water flow, and thereby infiltration into the ground. Vegetation cover, such as forests, influences canopy interception, throughfall, stemflow, and evapotranspiration, which, in combination, shape the water retention, infiltration, and runoff properties of a catchment. Streams and rivers transport surface water and contribute to the export of water from the forest system. To better understand and quantify water fluxes and stores within forest ecosystems, manual, plot-based measurements of the aforementioned terrain, vegetation, and stream properties, are aimed to gain a deeper understanding of the mechanisms influencing the hydrology

---

J. Schumacher (✉)
Forest Research Institute Baden-Württemberg (FVA), Freiburg, Germany

Present Address: Norwegian Institute of Bioeconomy Research (NIBIO), Ås, Norway
e-mail: johannes.schumacher@nibio.no

J. R. Christiansen
Department of Geosciences and Natural Resource Management, University of Copenhagen, Frederiksberg, Denmark

D. F. Levia et al. (eds.), *Forest-Water Interactions*, Ecological Studies 240,
https://doi.org/10.1007/978-3-030-26086-6_4

of these systems. However, to assess and quantify water fluxes and stores throughout an entire watershed, area-wide information describing these properties is needed. With LiDAR data, it is possible to obtain area-wide estimates of these properties. Given the explosive growth in the use of LiDAR in the geosciences and ecology, it is important to review the current use of LiDAR data in hydrological research, to address how these new types of data can help to improve the basic understanding of vegetation-hydrology interactions and how these data can be applied to develop hydrological models for better predictions.

The generation of digital terrain models (DTMs, also called digital elevation models, DEMs) provides a good example of the possible applications of LiDAR in hydrology. LiDAR technology enables terrain properties at unprecedented scales, ranging from m^2 to km^2 and beyond. With higher precision DTMs, delineation of watersheds and estimation of topographic features, such as slopes, become more precise and, in turn, can improve hydrological modelling of watersheds when applied in a proper physical context (Creed and Sass 2011). Even though DTMs, per se, are not related to forest-water interactions, they comprise an important LiDAR product, which, in many cases, are critical for scaling forest water fluxes and stores to the watershed scale.

In studies of water balances at larger scales (i.e., at scales beyond the plot scale), it has become apparent that integrating detailed physical descriptions of the forest vegetation composition is instrumental for the correct estimation of evapotranspiration and runoff (Carlyle-Moses and Gash 2011). Forest canopy structure affects interception of precipitation by the canopy and strongly modulates the amount of water transferred to the forest floor (via throughfall and stemflow) that may eventually become part of the runoff. Vegetation properties relevant to these important water fluxes in forests, such as canopy architecture and canopy density, can be assessed by analyzing LiDAR data.

Finally, riparian forest zones are transitional ecosystems in the watershed at the border between terrestrial to aqueous ecosystems. These riparian forests play a major role for water quality as they retain and release nutrients and pollutants to streams and affect biodiversity in both the forest and the stream. Delineating riparian forests in watersheds is important for closing water and nutrient balances and can be performed using DTMs, as well as structural vegetation properties, including tree height, crown diameter, canopy closure, and vegetation density, which also can be derived from LiDAR data.

In this chapter we provide a brief overview of LiDAR remote sensing (Sect. 4.2) and then present applications and case studies (Sect. 4.3) covering the topics "DTMs in hydrological modelling of watersheds" (Sect. 4.3.1), "LiDAR-based assessment of vegetation structure relevant to hydrological modelling" (Sect. 4.3.2), and "LiDAR remote sensing for assessment of riparian zones" (Sect. 4.3.3). We primarily focus on airborne LiDAR applications since there have been studies using these data directly for the assessment of forest-water interactions. Terrestrial LiDAR data have been used in several studies for the assessment of forest structural properties (Hopkinson et al. 2004; Danson et al. 2007; Côté et al. 2011; Wilkes et al. 2017), but we will address this topic only briefly. We conclude with future research directions (Sect. 4.4).

4.2 Fundamentals of LiDAR Remote Sensing

In LiDAR remote sensing, a laser generates pulses of light, which are directed to a surface, and a sensor detects the energy reflected from the surface. Therefore, LiDAR is an active remote sensing technology. The principle in laser scanning is the measurement of distances by calculating the time the light travels from the laser emitter, mounted on a platform with known position, to a surface and back to the sensor, usually mounted on the same platform (see Fig. 4.1). As the laser travels from the laser generator towards the surface of an object it diverges, so that the diameter of the laser (called a footprint when hitting a target) increases with increasing travel distance. Beam divergence depends on the laser frequency; i.e., at higher laser frequencies, beams generally have lower divergence resulting in a smaller footprint, resulting in better focus of the laser on the surface. If the footprint

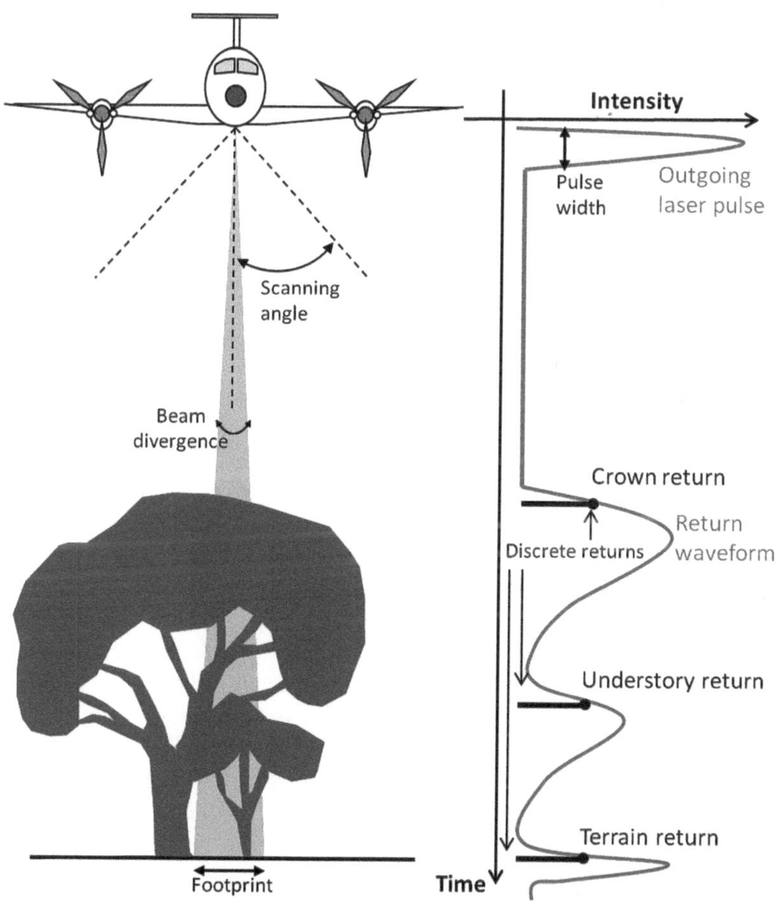

Fig. 4.1 Principles of airborne laser scanning

is larger than the object it hits (imagine a small branch), parts of the laser continue traveling until the next object is reached (e.g., the ground). The sensor can then detect reflected energy from that pulse both from the branch and from the ground. The intensity of the energy that is reflected at the surface and detected by the sensor needs to exceed a certain threshold in order to be recorded, usually called a "return", while the collective measurements from all returns are referred to as a "point cloud". In its very basic form, these returns are represented by points with x, y, z coordinates (northing, easting, and elevation). Further information can be included such as the intensity of each return, return number, and number of returns – in case of multiple returns from one pulse, scan angle, GPS time, and some more (see https://www.asprs.org/ for laser file format). Another recording mode of the sensor is the recording of the full waveform. In this case, the reflected energy does not have to exceed a threshold to be recorded as a return; instead the entire return waveform is recorded capturing much more detail and information but also making the dataset much larger and its analysis more complex.

Laser scanning is a measurement method that combines LiDAR with a moving platform that is rotated in a specific pattern (e.g., oscillating or rotating perpendicular to the direction of movement). We usually speak of airborne laser scanning (ALS) when the scanning LiDAR is mounted on a moving airborne platform. This results in coverage of a larger area compared to a profiling LiDAR, which is pointed in only one direction. Common wavelengths of airborne LiDAR applications are shortwave infrared (1064 nm or 1540 nm) and near- to shortwave infrared (900–1100 nm) used for terrain and vegetation assessments, or green (523 nm), which is used in bathymetry due to the ability of energy at this wavelength to penetrate water. For a comprehensive reference on ALS, interested readers are referred to Renslow (2012). Analogously, in terrestrial laser scanning (TLS), the LiDAR unit is mounted on a stationary terrestrial platform (e.g., a tripod) or a mobile platform (e.g., handheld, fitted into a backpack, or moving vehicle) and rotated in a specific pattern. Currently, there is no operating space-borne LiDAR system, since GLAS (Geoscience Laser Altimeter System) on ICESat (Ice, Cloud and Land Elevation Satellite) retired in 2010. However, NASA's GEDI (Global Ecosystem Dynamics Investigation LiDAR) was launched on December 5, 2018, on board the SpaceX CRS-16 mission to be attached to the ISS (International Space Station) to measure nearly all tropical and temperate forests. GEDI will advance applications for a number of domains, namely, forest management, topography, water resource management, and weather prediction. Data collected by GEDI will be used, for example, for studying forest height and vertical structure, habitat quality and biodiversity, forest carbon sinks and sources, parameterisation of ecosystem models, and creation of accurate bare earth and under canopy topographic elevations (https://gedi.umd.edu). Unpiloted aerial vehicles (UAVs) are another platform which can carry remote sensing sensors including laser scanners (see Chap. 3). Due to the low-level flight characteristics of UAV-LiDAR, the laser beam width is decreased and, due to the slow flying speed, the spatial density of the point cloud is increased.

Some uncertainties and errors are inherent to LiDAR data collection and processing. For example, objects such as walls or tree trunks can reflect LiDAR beams multiple times before the light is reflected back to the sensor, thereby increasing the travelling time of the light and falsely indicating an object further away from the sensor. Extreme outliers can be filtered by LiDAR processing software, while others can cause inaccuracies in the data. LiDAR-derived DTMs are created by software and algorithms interpolating between LiDAR ground returns. The quality of the point cloud (density, and horizontal and vertical accuracy) determines the accuracy of the resulting DTM.

4.3 Applications: Preview to Case Studies

The concept of using airborne LiDAR data for distance measurements for assessment of the Earth's surface can be traced to oceanographic applications in the 1960s and 1970s. First demonstrated for bathymetric applications, LiDAR data was used to calculate water depth by calculating the distance between a first peak in returned energy (water surface) and a second (presumably ocean ground) based on the speed of light (Hickman and Hogg 1969). The properties which make airborne LiDAR so interesting and useful for water depth measurements (namely, its ability to pass through different layers of the surface cover and thereby producing multiple returns) are relevant for resolving problems of measurement of the terrestrial realm as well. Initial interest focused on terrain mapping, and once geolocation accuracy was greatly improved by the emergence of GPS technology in the 1980s, applications for topographic mapping arose. A useful property of LiDAR data was its ability to penetrate vegetation and reach the ground. LiDAR returns from vegetation were considered noise which needed to be removed to obtain measurements of the ground surface below. However, dense vegetation prevents LiDAR beams from reaching the ground and thereby influences the accuracy of the obtained terrain surface; however, it was realised that this effect of the vegetation canopy allowed for the estimation of the vegetation height (Schreier et al. 1984). Since the 1980s, the possibilities of assessing terrain, vegetation, and water bodies with LiDAR data have been explored in many experiments and studies (e.g., Nelson et al. 1988; Næsset 2002; Zimble et al. 2003; Popescu and Zhao 2008; Smeeckaert et al. 2013; Crasto et al. 2015; Nilsson et al. 2017; Valbuena et al. 2017). It is increasingly appreciated that new insight into the hydrological functioning of forests can be obtained using LiDAR, assisting laborious manual field assessments of vegetation properties while, at the same time, increasing the detail and accuracy of how vegetation is represented in hydrological models. However, much work still needs to be done in reconciling the highly dynamic hydrological cycle at various temporal scales (hourly to annual) and the more static vegetation properties derived with LiDAR.

Inherent to all approaches, where remote sensing data are combined with field data for model development, calibration, or estimation, is the challenge of accurate geolocation of the field plots. Even though today's global navigation satellite systems (GNSS) can obtain sub-meter accuracies under optimal conditions and with differential correction methods, accurate positioning under forest canopy is often hampered due to limited satellite view and signal reflections. This can cause a positional deviation between the actual and the assumed plot location of several meters. Nevertheless, LiDAR remote sensing approaches offer the great potential to obtain information on large scales which is not feasible from field work campaigns.

The following subsections present examples of recent applications of LiDAR technologies to the study of forest-water interactions and discuss the current state and future directions for research.

4.3.1 Use of DTMs in Hydrological Modelling of Watersheds

LiDAR-derived DTMs depict the terrain surface in great detail and enable us to assess surface topography and delineate watersheds. They are, thus, an integral part of distributed hydrological models (e.g., for the prediction of hillslope flow paths, or to calculate depressional storage, which is needed to adjust peak flow rates during model calibration) and thereby enable us to assess entire catchment areas (Quinn et al. 1991; Amatya et al. 2013). Furthermore, they can be used to detect historic land use features below the forest canopy, such as dikes or drainage ditches, to detect stream networks (e.g., meandering channels), and to assess the effects of these features (e.g., diversion ditches interrupting surface runoff) on watershed hydrology (Amatya et al. 2013).

Terrain properties (such as slope and aspect) control the watershed extent, sub-basin locations, and stream network topology, and therefore outputs from hydrological models such as water yield, flow rate, hydrograph timing, and water quality. LiDAR-derived DTMs have been adopted for hydrological models due to their high spatial resolution and accuracy compared to DTMs obtained from other data sources (e.g., Remmel et al. 2008; Goulden et al. 2016). In a case study presented by Goulden et al. (2016), DTM uncertainty generally ranged between 0.025 and 0.060 m, slope uncertainty ranged between 0.6 and 1.5°, and aspect uncertainty ranged between 2.7 and 24.1°. Compared to other available DTMs (e.g., the Shuttle Radar Topography Mission (SRTM), DTMs calculated from contour layers, or the National Elevation Dataset (NED) in the United States), LiDAR-derived DTMs improve the delineation of stream segments and watershed boundaries and the computation of slope, aspect, and other DTM-derived data layers due to their high-spatial resolution (Remmel et al. 2008). Zhang et al. (2009) assessed LiDAR-based DTMs (at 4, 10, and 30 m ground resolution), DTMs obtained from the NED (at 10 and 30 m ground resolution), and the SRTM (at 30 m ground resolution) for hydrological and erosion simulation using the Water Erosion Prediction Project (WEPP) model for forest watersheds in northern Idaho, USA. Runoff simulations using the WEPP model configured with these different DTMs did not differ substantially. However, the WEPP model configured with the 10 m LiDAR DTM produced predictions of total and seasonal patterns of watershed discharge that were closer to field observations than the predictions using the WEPP model configured with all the other DTMs. Similar observations regarding the influence of DTM scale for streamflow simulation were made by Yang et al. (2014), where LiDAR-based DTMs with finer resolution (compared to LiDAR-based DTMs with coarser resolution) resulted in more accurate representation of terrain features; however, they did not necessarily improve watershed scale streamflow simulation modelling. Best fits between reference and modelled streams were observed at an intermediate cell size (10 m cell size DTMs) instead of finest resolution (1 m cell size DTMs). This can be expected, since LiDAR does not account for geology which is just as important as topography for determining watershed dynamics. However, higher topographic resolution DTMs from LiDAR may work better on smaller catchments in high-order streams (high in the catchment) where streamflow is more connected to surface infiltration and surface runoff.

A typical hydrologic property in forested watersheds is that subsurface lateral flow is predominant (Zhang et al. 2009). This is because the infiltration capacity of forest soils is generally higher relative to the intensity of the throughfall below the canopy, effectively reducing the surface runoff compared to non-vegetated surfaces. Therefore, decreasing the cell size of DTMs will be less effective in improving runoff prediction in forested watersheds, as surface runoff in these systems will be less sensitive to fine-scale topography compared to non-vegetated areas. Rather, in forested catchments, vegetation composition and structure are significant drivers of hydrologic processes moderating the amount of rain reaching the ground, partitioning the precipitation into interception loss, throughfall, and stemflow, as well as dampening the intensity of precipitation reaching the forest floor (Varhola and Coops 2013).

Even though LiDAR-derived DTMs are probably the most accurate terrain models that can be obtained for larger regions, a general challenge is that LiDAR beams might not always penetrate all the way through dense vegetation and therefore not represent the ground. DTM accuracy degrades as canopies become denser and fewer LiDAR beams reach the ground since DTMs are created by interpolating between LiDAR ground returns.

4.3.2 LiDAR-Based Estimation of Vegetation Structure Relevant to Hydrological Modelling

In forest ecosystems, precipitation is either intercepted by tree crowns and evaporated back to the atmosphere (a process known as interception loss) or transferred to the ground directly by falling through canopy gaps or indirectly by dripping or splashing off foliage and branches as throughfall or running on the outside of tree boles as stemflow. Canopy interception loss has been found to range from 10% to 50% of gross rainfall (Roth et al. 2007) and the amount of throughfall is a function of both abiotic and biotic factors, such as weather conditions and canopy density (Rutter et al. 1971; Levia et al. 2017). Assessments of the global interception loss from forests put this estimate at 15% for tropical areas, dominated by short-duration convective rainfall events and large annual precipitation depths, and at up to 30% for climatic zones in which rainfall is dominated by long-duration synoptic events (Miralles et al. 2010; Miralles et al. 2016). Thus, interception loss reduces the absolute amount of water reaching the soil and directly impacts availability of water in the root zone in the short term and groundwater regeneration in the long term.

In relation to interception and throughfall in forest ecosystems, key canopy features include the vertical distribution and density of foliage cover or leaf surfaces that intercept water from the canopy surface to the ground, and the spatial distribution of individual trees within larger forest ecosystems. Thus, in forests there is a spatial heterogeneity of canopy cover properties that need to be characterised in order to evaluate the impact of vegetation on the water balance at scales relevant to ecological processes or forest management practices. LiDAR data can be used to assess vegetation properties at these spatial scales and can be related to measures of interception, throughfall, and stemflow, thereby making these data highly relevant for hydrological modelling. However, so far, only a few studies have used forest structure-related LiDAR variables to estimate hydrologically relevant variables (e.g., Chasmer et al. 2011; Mitchell et al. 2012; Varhola et al. 2012; Schumacher and Christiansen 2015; Sutherland et al. 2017).

The most commonly used hydrological models use specific sets of variables that usually represent conditions at the tree or forest plot level. Muzylo et al. (2009) reviewed rainfall interception models and separated them into two groups. The first group models the redistribution of rainfall volume using a mass balance equation and is further divided into Rutter- and Gash-type models (Rutter 1971; Gash 1979). The second group models the interception of rainfall drops using a probability distribution and consists of the Calder one- (Calder 1986) and two-layer (Calder 1996) models (using the Poisson and the Poisson and Marshall-Palmer (Marshall and Palmer 1948) distributions, respectively). These models and further improved models include forest canopy variables such as forest tree height, canopy cover fraction, leaf area index (LAI), and number of vegetation layers. The Rutter model represents the interception process by a running water balance of rainfall input, storage, and output in the form of drainage and evaporation. The Gash model was the first analytical model of canopy interception loss and, according to Muzylo et al. (2009), is used most commonly in the literature. The popularity of this type of model is probably due to its ease of use in terms of workload and costs, parameter requirements, data input, and low conceptual and programming complexity.

Besides interception, transpiration of trees and direct soil evaporation also need to be assessed in order to account for all the evapotranspiration from a forest. Tree transpiration depends on sap flow velocity and is therefore related to sapwood area which, in turn, is correlated with basal area that can be modelled using LiDAR data (Mitchell et al. 2012), while soil evaporation is mostly dependent on the available solar energy reaching the forest floor.

Whereas some of the aforementioned structural related variables can directly be obtained from LiDAR data (e.g., tree height, forest cover fraction), others need to be modelled (e.g., basal area, LAI). Tree height is an important conceptual component of hydrological models and is related to surface boundary conditions including wind speed reduction, surface roughness, and canopy resistance that all are involved in calculating evapotranspiration (Pomeroy et al. 2007). Forest cover modifies solar radiation transmission to the ground, throughfall and interception of precipitation, and snow load (Ellis et al. 2010). Basal area and sapwood area are closely related to each other and determine crown dimensions and evapotranspiration (Haydon et al. 1997; Benyon et al. 2015). LAI is an important vegetation index characterising vegetation structure and density and remains the primary conceptual vegetation structure parameter used in hydrological models (Varhola et al. 2012) to calculate both interception loss and transpiration. Interception storage capacity is conceptually considered directly proportional to LAI making it a very convenient variable to parameterise models across forest types and tree species, but measurements of LAI mainly include the leaves and not branches and stems. However, using LAI comes with a number of limitations: different tree species may have the same LAI, but different leaf surface properties (such as surface wax and hydrophobicity) that impact interception capacities (Holder 2012; Klamerus-Iwan and Błońska 2018). This can partly be compensated by multiplying the LAI with an arbitrary tree species-specific parameter. However, this may be insufficient as the vertical structure of the canopy also dictates interception capacity and this differs between forests. For example, in a multilayered canopy, existing methods for measuring LAI will be challenged because of the undergrowth canopy beneath the primary canopy. It has also been shown that LAI of tree canopies change on an annual basis in response to environmental factors or management. For example, drought decreased maximal LAI in stands with high LAI the year following drought (Le Dantec et al. 2000) and in the same study it was shown that soil fertility (humus quality) and thinning had a significant impact on the spatial variation in maximal LAI. Thus, LAI is by no means a static parameter in space and time calling for repeated measurements of the forest structure for the whole stand or forest in question. Capturing the spatiotemporal variations in the canopy density is important to estimate the distribution of interception evaporation and hence the below canopy water fluxes to the soil that impact seepage, biodiversity, and nutrient cycling.

The areal coverage and type of bark is an important factor in determining the total interception capacity of forests and is closely related to tree species (Herwitz 1985; Levia and Herwitz 2005). The water-holding capacity of bark regulates the influx of water to the soil through stemflow. Thus, interception capacity may be underestimated if only LAI is used for model simulations of interception loss. High-resolution LiDAR offers an opportunity to improve LAI estimation methods because it can measure forest structure in greater detail over larger areas. However, LiDAR captures all vegetation elements such as leaves, branches, and stems. This is also the case for common field-based LAI measuring methods such as hemispherical photographs or light radiation sensors (e.g., Li-COR Plant Canopy Analyzer LAI-2000). As opposed to LAI, plant area index PAI (also called effective leaf area index, eLAI) is mathematically related to gap fraction (GF) and includes the areas of branches and stems (Cescatti 2007) and is used as an input parameter in physically based hydrologic models (Ellis et al. 2010). Thus, use of LiDAR-derived surface estimates may improve interception estimates as all intercepting plant surfaces (leaves, branches, and trunks) from the surface to canopy top in principle can be estimated.

The methods inherent to the LAI-2000 and hemispherical photographs calculate PAI from measured radiation transmittance (gap fraction) and assume randomly distributed canopy elements. While the assumption of randomly distributed foliage holds true for leaves, needles are not randomly distributed in the canopy but are rather clumped within shoots (Stenberg 1996), on branches, whorls, and crowns (Chen et al. 1997). Therefore, PAI obtained from such methods differs from the true LAI and underestimates LAI where the foliage is clumped (Chen et al. 1997); however, this can be mended by applying a correction factor (Gower and Norman 1991). Here, we refer to the commonly measured and used version of LAI, namely PAI or eLAI. LAI might be derived for a given forest area from the penetration rate of airborne LiDAR pulses through the canopy layer (e.g., Riano et al. 2003; Lefsky et al. 2005; Solberg et al. 2006; Morsdorf et al. 2006; Sasaki et al. 2008; Solberg et al. 2009; Zhao and Popescu 2009). Other approaches for obtaining LAI have been proposed using airborne full waveform LiDAR data (Ma et al. 2015), terrestrial discrete LiDAR (Danson et al. 2007; Moorthy et al. 2008), terrestrial full waveform LiDAR (Strahler et al. 2008; Jupp et al. 2009), and space-borne LiDAR data (Tang et al. 2014). van Leeuwen and Nieuwenhuis (2010) give a review on the use of terrestrial, airborne, and space-borne LiDAR techniques for the estimation of forest inventory parameters and structural characteristics including tree height, forest cover fraction, basal area, and LAI mentioned above.

When a relationship between a variable of interest (i.e., a reference variable, e.g., LAI or basal area obtained from a field survey) and LiDAR-based variables is established in the form of a statistical model (e.g., a regression model), this model can then be applied to the entire area for which the LiDAR data are available and which is represented by the ground reference data (see scheme depicting the sequence of steps in Fig. 4.2).

We now present and discuss three studies in which GF, LAI, sky-view factor (SVN, defined as the proportion of visible sky above a certain observation point), and solar radiation transmission metrics (Varhola et al. 2012), evapotranspiration (Mitchell et al. 2012), and throughfall (Schumacher and Christiansen 2015) were estimated using LiDAR data.

Varhola et al. (2012) obtained area-wide forest canopy GF, LAI, SVF, and solar radiation transmission metrics from synthetic hemispherical images from LiDAR data. Collectively, the GF and SVF factors describe the spatial variability in forest canopy density which is important to consider when parameterising distributed hydrological models as this controls the distribution and partitioning of interception and throughfall in the model domain. The solar radiation transmission metrics are used to derive realistic estimates of the available energy at the soil surface which, in turn, drives soil evaporation.

The authors converted the Cartesian coordinates of airborne LiDAR data to a polar projection and generated synthetic hemispherical images, which are directly comparable to those from optical photography. Fig. 4.3 (from Varhola et al. (2012),

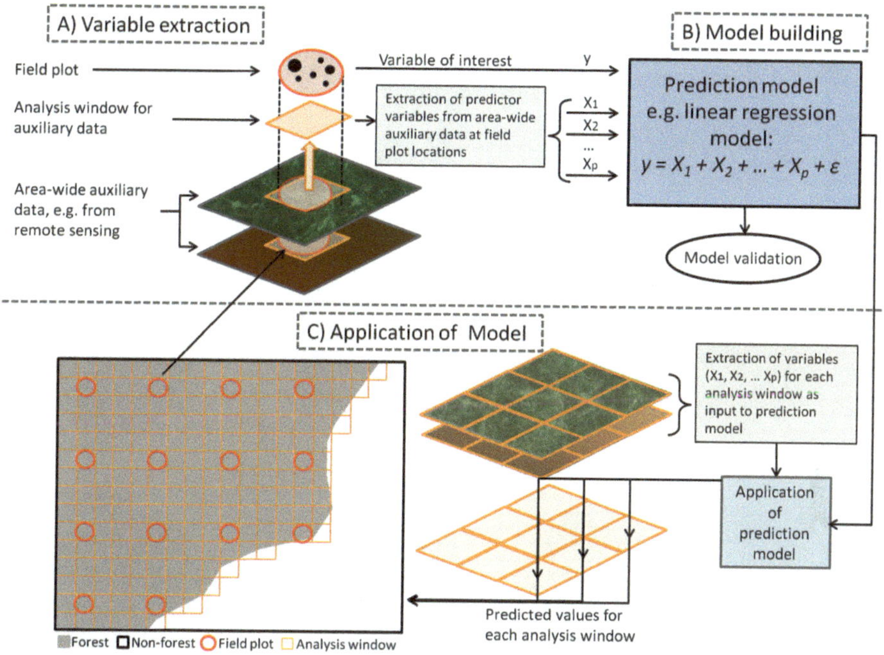

Fig. 4.2 Schematic depiction of a possible design and processing steps for remote sensing-assisted assessment of forest properties. Red circles represent field plots; orange rectangles represent analyses windows for auxiliary data. Area-wide auxiliary data can be data from, for example, remote sensing (e.g., digital surface models, digital terrain models) or climatic data (e.g., air temperature, precipitation). The spatial distribution of subplots depends on the specific study. It can be a random, systematic (as depicted here), or stratified design. The number of field plots depends on the subject and the required accuracy. For model validation, it is possible to reserve a set of field plots only for validation, or, if available, use separate independent field dataset for validation

reused here with permission) shows examples of LiDAR point clouds for sample plots with corresponding synthetic and field-based hemispherical images. The advantage of generating hemispherical photographs from LiDAR is that the variables of GF and LAI can be estimated for a much larger number of plots than through manual measurements and can be readily used in the appropriate hydrological models. The authors calibrated airborne LiDAR-based synthetic hemispherical images with field measured optical hemispherical photographs to obtain forest canopy GF, LAI, SVF, and solar radiation transmission metrics. LAI in this study corresponded to PAI, since in hemispherical photography stems and branches are also captured.

In general, synthetic hemispherical photograph (HP) images were similar to real optical HP images. However, individual trees were difficult to identify, and LiDAR returns sometimes appeared where there were no canopy elements in the HP. Possible explanations for these differences are that the density of the LiDAR point cloud was not high enough to capture low crown elements which are apparent

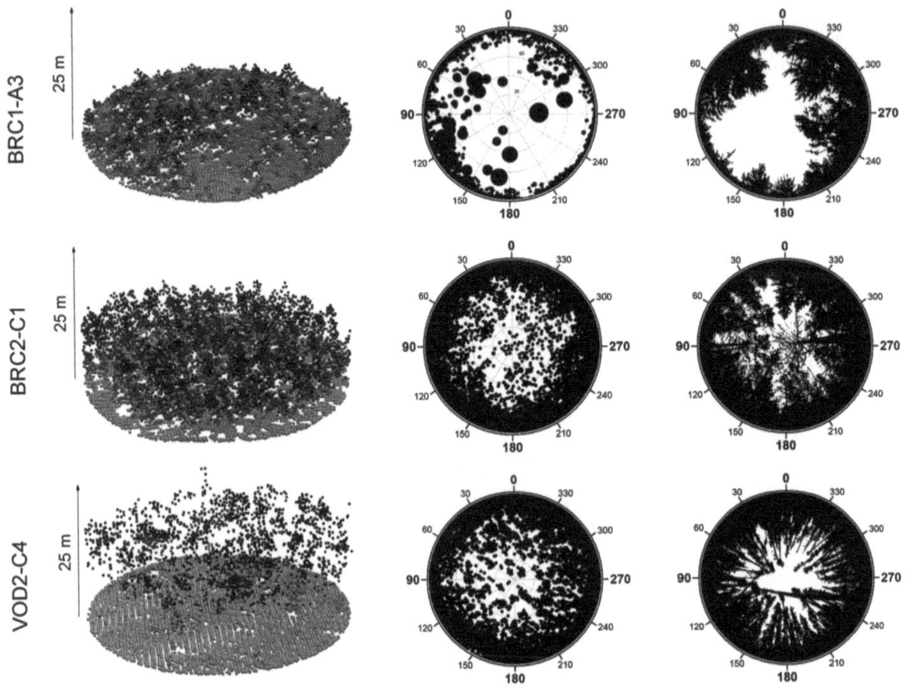

Fig. 4.3 Representative examples of ALS point clouds (left), ALS synthetic hemispherical images (centre), and real optical hemispherical photographs (HP) (right) for each stand; azimuths (°) are shown on the hemispherical illustrations. (Adapted from Varhola et al. (2012) under creative common license: https://creativecommons.org/licenses/by/3.0/)

in the HP. Furthermore, HPs were collected approximately half a year after the LiDAR data and changes in stand structure (e.g., crown damage, tree fall, etc.) might have occurred. Another possibility is a common problem when relating remotely sensed data with field data – namely, GNSS positional errors in plot locations under the canopy are large and cause differences in the features that are captured and compared.

Varhola et al. (2012) estimated GF, LAI, SVF, and radiation transmittance obtained from HPs using linear regression. First, simple linear regression was applied where each of the metrics (GF, LAI, SVF, and radiation transmittance) was estimated with the corresponding metric obtained from the synthetic HP images. Second, multiple linear regression was applied where each of the metrics was estimated with the corresponding metric obtained from the synthetic HP and with the LiDAR point cloud-based variables "ratio of ground returns and total returns" and "spatial density of LiDAR returns". Only with the second approach (multiple linear regression) could the metrics GF, LAI, SVF, and radiation transmittance obtained from HPs be successfully estimated with LiDAR metrics, since both forest structure (described by GF, LAI, SVF, or radiation transmittance) and data configuration properties (described by "Ratio of ground returns and total returns" and "spatial density of LiDAR returns") were accounted for. The models' R^2 values were above 0.80 and normalised RMSE was below 10%. This indicates that there is more information in the LiDAR data (e.g., percentage of returns above a specific height threshold indicating the spatial vertical distribution of vegetation elements) than the information that can be directly extracted from it (i.e., the gap fraction) and that the combination of LiDAR-derived metrics can increase its explanatory power so long as they are uncorrelated among each other.

The advantage of the approach of transforming LiDAR coordinates to generate synthetic HP images is (1) its simplicity, because it is based on the LiDAR point cloud, and therefore no other canopy models need to be processed; (2) pointwise local information of the hydrological relevant variables are obtained at the same spatial scale as field measurements, instead of, for example, averages of larger spatial units from other remote sensing techniques with coarser spatial resolution; and (3) it is compatible with software specialised to estimate GF, LAI, SVF, and solar radiation based on synthetic HPs. This makes the information applicable for the entire area in which LiDAR data are available for and, therefore, allows for the creation of maps of these variables at a cell size corresponding to the spatial representation of field measurements. Even for areas where no LiDAR data is available, this approach can be applied by extrapolating the LiDAR-derived variables to a larger area by correlating the LiDAR metrics with satellite images, for example, Landsat spectral indices (Varhola and Coops 2013; Babcock et al. 2018) or MODIS data (Chasmer et al. 2011). However, a study in which this approach is directly tested in hydrological models to clarify whether they can be inputted to fully distributed hydrologic models has yet to be undertaken.

Mitchell et al. (2012) used LiDAR-based variables to predict the following forest structural attributes for a eucalypt-dominated catchment in Victoria, Australia: basal area, sapwood area, LAI, and canopy cover. These variables were subsequently used in empirical models to quantify forest transpiration (E_{sap}) and in numerical models to quantify interception loss (E_i) and forest floor evapotranspiration (E_{floor}) on the catchment scale. In 18 plots measuring 20 m × 20 m, field measurements of diameter at breast height (dbh, at 1.3 m aboveground) were conducted, basal area was calculated, and sapwood area was modelled by a power function using basal area as the predictor. Cover photography (i.e., a camera was placed in such a way that the lens pointed directly upwards) was used to model LAI and canopy cover. For the same plots, the following LiDAR-based variables were computed: 95th, 80th, and 50th height percentile, mean height, standard deviation of return heights, proportion of returns higher than 2 m above ground to all returns (compared to variable "IR" in Schumacher and Christiansen (2015), see below), the number of all returns, and interquartile range. Additionally, to obtain vertical forest structure as distinct vegetation layers, continuous distribution functions were fitted to the vertical forest profiles depicted by the LiDAR returns using the method described by Jaskierniak et al. (2011). These variables were also calculated for 20 m × 20 m grid cells over the entire catchment area. Based on the 18 sample plots, sapwood area, canopy cover, and LAI were regressed with each LiDAR variable (Table 4.1). Furthermore, on four

Table 4.1 Forest structure metrics, LiDAR variables used in regression models of forest structure metrics, and corresponding $\boldsymbol{R}^2$ and P values (data from Mitchell et al. 2012)

Forest structure metrics	LiDAR-based predictor variable	R^2	P value
ln(sapwood area)	H_{95} (95th percentile of LiDAR return heights)	0.70	<0.01
ln(crown cover)	H_{med} (median of LiDAR return heights)	0.80	<0.01
ln(LAI)	H_{med} (median of LiDAR return heights)	0.76	<0.01

plots of 20 m × 20 m, measurements of transpiration (E_{sap}, based on sap velocity and sapwood area), interception loss (E_i, total precipitation minus throughfall minus stemflow), and forest floor evapotranspiration (E_{floor}, measured with a portable evaporation dome) were conducted. Relative humidity, air temperature, net radiation, rainfall, and wind speeds were measured by a meteorological station in a clearing at the bottom of the catchment.

To obtain predictions of E_{sap} for the entire catchment, a relationship between E_{sap} and the forest structure variable sapwood area (SA) obtained from field measurements was established (not more information on that was given). Subsequently, the LiDAR-SA model was applied to each grid cell in the catchment to predict SA. The relationship between E_{sap} and SA was used to predict E_{sap} values for the entire catchment.

To predict E_i for each 20 m × 20 m grid cell in the entire catchment, the revised Gash interception model (Gash et al. 1995) was applied, in which canopy cover predictions, based on the LiDAR regressions, were used. Similarly, to make catchment-wide predictions of E_{floor}, a two-stage bucket model was applied, in which the variable Rn (energy reaching the forest floor) was estimated from the LiDAR-derived LAI. Rn was calculated from the Rn at the weather station as 1 − [exp(k LAI)], where k is the extinction coefficient, assumed to be 0.5 for the eucalypt canopy.

The Mitchell et al. (2012) study is one of the first to scale forest water use at the catchment level by employing LiDAR-based forest structure variables. The study shows that it is possible to relate LiDAR-based structural metrics to hydrology for spatial scales that are relevant to watershed management practices. This represents a large potential in other settings for improved management of water resources within an environmental context. However, there are also several shortcomings that need to be highlighted which can guide future development. For example, there were only few plots used for establishing empirical relationships, which were central for upscaling, which inadvertently adds uncertainty in the catchment scale estimates.

Surprisingly and contradictory to the study by Schumacher and Christiansen (2015; see next example below), the LiDAR variable "proportion of returns higher than 2 m above ground to all returns" performed poorly in the regression models estimating forest structure. This might be due to particular conditions in eucalypt forests, the limited number of plots used (resulting in a limited number of forest structure types being captured as the authors describe these forests as being dominated by the overstory strata), or the precipitation regime. This emphasises the need for multiple plots covering the spatial variability of the forest stand or catchment in question. Naturally, there is an upper limit for the number of plots that can be studied calling for future coordination and collaboration between traditional catchment hydrologists and LiDAR experts. The flow chart in Fig. 4.2 conceptualises the design of such a study where ground-based measurements are combined with airborne LiDAR data for the purpose of fitting a statistical model, which subsequently can be applied to the entire area for which the LiDAR data are available and which is represented by the field reference data.

Schumacher and Christiansen (2015) estimated in situ throughfall measurements for a range of Danish forests across a precipitation gradient with an ALS-based canopy structure metric termed "interception ration" (IR, calculated as the percentage of all returns above 1 m) (Fig. 4.4) and the precipitation and air temperature using mixed linear models. Throughfall was measured as monthly sums in monoculture forest stands of approximately 0.25 ha representing five deciduous and one coniferous tree species (Fig. 4.4, the fifth deciduous tree species in this study (Ash, *Fraxinus excelsior*) is not present in this plot). The duration of throughfall measurement differed among locations with a minimum of 2 years to more than

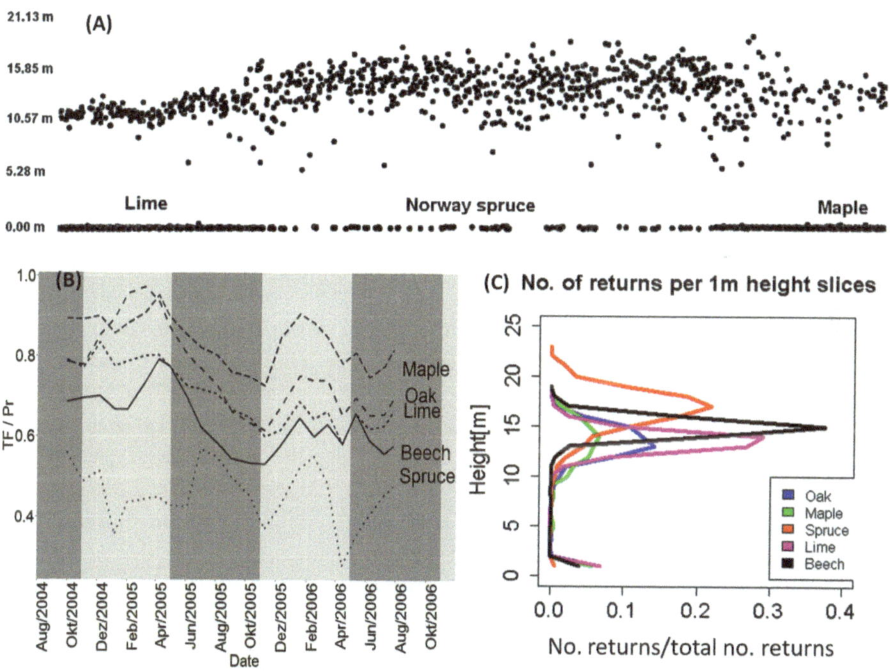

Fig. 4.4 (**a**) Three plots stocked with lime trees (*Tilia cordata*), Norway spruce (*Picea abies* (L.) Karst.), and maple (*Acer pseudoplatanus*) viewed from the side as scanned with LiDAR data (from Schumacher and Christiansen (2015)); (**b**) variation of ratios of throughfall (TF) and total precipitation (Pr) for five tree species, maple, oak (*Quercus robur*), lime, beech (*Fagus sylvatica* L.), and Norway spruce over time; (**c**) ratio of number of LiDAR returns in 1 m height slices and total number of LiDAR returns for five tree species maple, oak, lime, beech, and Norway spruce in the same plots as shown in (**b**)

10 years of measurements. Precipitation was also measured as monthly sums outside the forest close to the throughfall monitoring plots. Precipitation, throughfall, and air temperature were aggregated for seasons (summer (April–September) and winter (October–March)) coinciding roughly with the periods of leaf-on and leaf-off, respectively, and years to explore if the IR variable explained throughfall differently across various hydrological time scales. As expected, in all models monthly, seasonal, and annual, bulk precipitation was found to be the dominant factor explaining the throughfall variability. However, the proportion of explained variance of the LiDAR variable increased from 1.7% for the monthly period to 12.2% and 19.5% for seasonal and annual data sets, respectively. Surprisingly, there was no difference in model results for the two contrasting seasonal models.

The relationship between interception and canopy density can be exemplified by comparing the throughfall/precipitation ratio with the vertical density of five different tree species (Fig. 4.4). The small tree height and less dense canopy of Maple trees is associated with the highest throughfall/precipitation ratio whereas the taller and denser Norway spruce trees have the lowest throughfall/precipitation ratio in the example. A surprising conclusion by Schumacher and Christiansen (2015) was that the relationship between observed and predicted throughfall was similar for conifer and broadleaved species. Given that the relationship for the broadleaf forest was derived from point clouds in the leaf-off season, this could suggest that other factors related to canopy architecture, and not just the interception capacity of leaf surfaces (where surface wax and hydrophobicity play important roles (see Holder 2012)), determine interception capacity; however, these factors have not been identified. The explanation could also be that precipitation is driving most of the variation (80–90%) in throughfall obscuring any differences in IR between species. Furthermore, the rather long time between throughfall collection at the sites does not permit the study of the potential interaction between single rain events and tree species specific crown architecture where the effect of structure on throughfall may be more evident between contrasting tree species. Although the study of Schumacher and Christiansen (2015) is the first to estimate throughfall using ALS-derived structural metrics, any deeper understanding of the relationship between LiDAR-derived metrics and hydrology should rely on investigating the influence of tree crown architecture and structure on throughfall at a finer temporal scale (e.g., single rain events), and with higher-resolution LiDAR data, which could be derived with terrestrial laser scanning techniques.

An important lesson to be learned from these three case studies is the power of LiDAR data analyses for predicting and mapping forest structure on large spatial scales. Due to horizontal and vertical variability in forest structure, forest-water interactions change on a small spatial scale within forests, which can be detected by LiDAR applications (Fig. 4.5). Comparing the forest area in the left and right parts of Fig. 4.5, conifers generally have lower throughfall values (darker regions in the left and right parts of Fig. 4.5) than broadleaves (brighter regions in left and right parts of Fig. 4.5). More important is that even within conifer and broadleaf forest, predicted

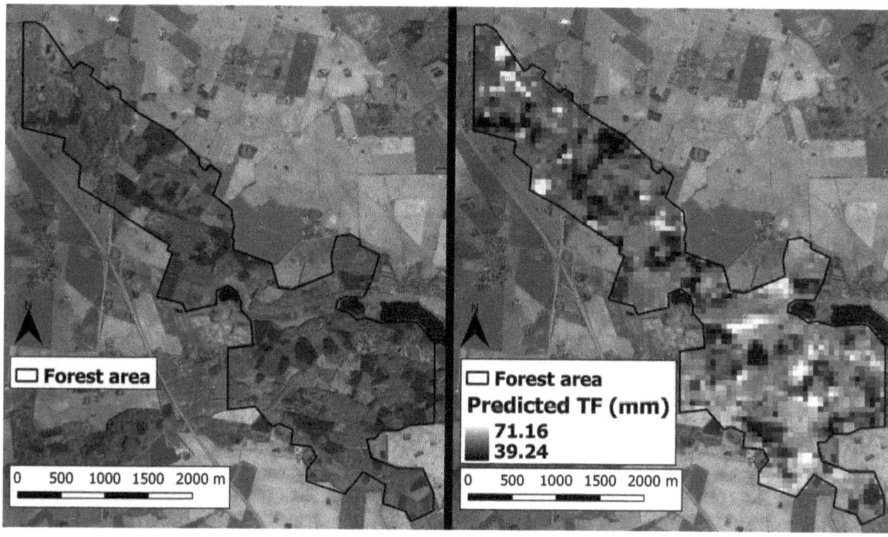

Fig. 4.5 Left, orthophoto of the "Hov" forest located in Zealand, Denmark; darker forest patches are conifer forest and lighter patches are broadleaf forest; right, map of LiDAR-predicted throughfall (TF) for "Hov" forest. (Figure adapted and modified from Schumacher and Christiansen 2015)

throughfall values vary substantially on a small spatial scale. These differences over a larger area cannot be detected with field survey plots, which only represent small areas. This makes LiDAR applications highly relevant to use for parameterisation of spatially explicit hydrological models that are able to include vegetation characteristics to calculate interception, throughfall, and evaporation at the watershed level (compare modelling concept in Fig. 4.2).

The emerging research theme of coupling LiDAR to forest hydrology is confined to relatively few studies and an important task ahead is to establish whether there are global relationships between canopy structure and associated water fluxes. The main question is whether there are uniform relationships between canopy structure and interception and throughfall across forest types, or whether this relationship is context-dependent, for example, strongly dependent on functional traits of tree species with different leaf and needle surface properties. A main hypothesis to test in this regard is to establish if interception loss is proportional to the structural complexity of the forest. That is, to determine if interception per unit rainfall depth is greater in complex canopies with multiple layers compared to more homogeneous canopies, such as in plantations, monocultures, or boreal forests. Tree species-specific properties of leaves and needles in terms of interception capacity and transpiration will interact with this. The examples presented in this chapter show the breadth of possible canopy structure metrics that can be derived and linked to forest hydrology providing future researchers in this field a foundation from which to start; however, it also presents a challenge as there is currently no consensus on what defines structural complexity of a canopy relevant for hydrological applications.

If longer drought periods and fewer but larger magnitude rain events are to become more common in the future, the need to manage water resources more carefully will arise. This may lead to water-oriented forest management, in which the influence of forest structure and tree species composition on interception, throughfall, and evapotranspiration is taken into account when planning forest management activities. The relevance of such a concept is apparent in Denmark, for example, where afforestation of former agricultural soils is done to protect groundwater; however, tree species traits in relation to hydrological management are not yet considered.

4.3.3 Assessment of Riparian Zones

Riparian zones are the interface between the land and a river or a stream and play an important role in ecology, environmental management, and civil engineering due to their function for mitigating stream flow, preventing erosion, soil conservation, and their specific habitat for flora and fauna. Riparian forest zones contribute to forest ecosystem functions including nutrient cycling, vegetative communities, and water quality (Naiman et al. 1993). Their narrow and fragmented nature makes them prone to a number of disturbances that can easily degrade their ecological function. LiDAR applications to forest water interactions in riparian zones present a special case in this chapter, where the components of terrain, river bed and stream cross-section, and vegetation assessment are combined to describe and quantify the status of the riparian biosphere.

Structural characteristics of vegetation, including tree species composition (broadleaved and conifer) and canopy closure, were found to be different between riparian zones and forest stands within the same watershed (Congalton et al. 2002). In this study, these differences could only poorly be detected using Landsat TM satellite imagery and a combined supervised/unsupervised classification approach. The extreme diversity and linear arrangement of the riparian vegetation created classification problems and resulted in the Landsat TM imagery being inadequate due to its coarse pixel size of 30 × 30 m. Also, analysis of other satellite images with finer ground resolution (SPOT-5 with 10 m and QuickBird with 2.4 m pixel size) performed poorer in mapping riparian zone attributes compared to LiDAR data (Johansen et al. 2010). In this study, object-based image and regression analyses were applied.

There are several studies that have addressed the assessment and mapping of aspects of riparian zones using LiDAR remote sensing: the aspects of riparian zones that can be characterised by LiDAR data are stream order and magnitude (James et al. 2007), gradient (Cavalli et al. 2008; Vianello et al. 2009), stream bed and riparian zone width (Johansen et al. 2010, 2011; Michez et al. 2013), sinuosity (the ratio between channel length and a straight distance between channel beginning and end) (McKean et al. 2009; Tompalski et al. 2017), channel morphology (Cavalli et al. 2008), flow paths (Wallace et al. 2018), shading influencing water temperature

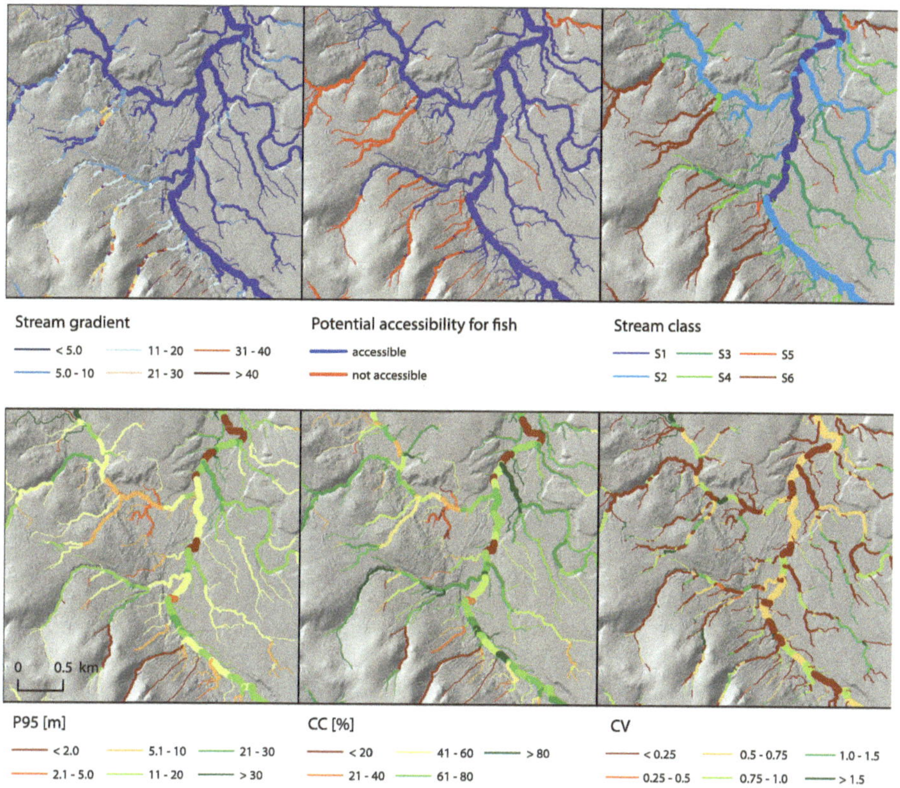

Fig. 4.6 Stream gradient, potential accessibility, class, stand height (P95), canopy cover (CC), and vertical variability (CV) derived for each stream segment demonstrated on a portion of study area. Notice that for some streams, segments of higher and lower gradient are alternating. Accessibility for fish for such streams is defined by the lowest segment exceeding 20% gradient. Height and canopy cover are calculated for stands in direct vicinity of streams (20 m buffer). Values are averaged for each stream segment. (Source: Reprinted from *Remote Sensing of Environment,* Vol. 192, Tompalski et al., Characterizing streams and riparian areas with airborne laser scanning data, 73–86, 2017, with permission from Elsevier)

(Greenberg et al. 2012), and vegetation properties including canopy cover, height, connectivity, structure, and tree types (broadleaved and conifer trees) (Goetz 2006; Johansen et al. 2010; Michez et al. 2013; Wasser et al. 2013). Using bathymetric LiDAR, stream water properties such as water level can be mapped (McKean et al. 2009). Tompalski et al. (2017) characterised streams and riparian areas in a 52,000 ha study area and identified and described streams with a total length of 6422 km. In this area they delineated the stream networks and assessed riparian attributes such as stream sinuosity (based on stream segments), stream order, gradient and width (based on a LiDAR DTM), and vegetation properties such as height, canopy cover, vertical structure (Fig. 4.6, Fig. 11 in Tompalski et al. 2017), and hours of shading of the stream (based on the LiDAR point cloud). Therefore, with LiDAR data analyses it is possible to obtain a holistic characterisation of the riparian biosphere and acquire useful information on both stream and vegetation properties influencing the hydrology of a watershed. Similarly, riparian buffer zones associated with the entire river network of Wallonia, southern Belgium, were characterised and mapped on local (>10 km) to regional scale (>12,000 km) by Michez et al. (2017). The focus of the exercise was to provide accurate information on the physical parameters of channel width and depth, the forest properties of height, longitudinal connectivity, and water accessibility, and to explore the factors controlling these channel and vegetation characteristics, as well as to propose riparian zone management indicators.

Characterisation of riparian buffer zones and their conditions is an important part for ecologically sound resource management and the maintenance of aquatic ecosystem integrity. LiDAR data analyses can help to quantify riparian buffer zone conditions at spatial extents and resolutions not possible using conventional manual methods. However, a holistic example where the combined LiDAR data analyses for the characterisation of riparian zones, watershed topography, and forest vegetation are integrated into the assessment of all aspects of an entire watershed has yet to be conducted and validated.

4.4 Conclusions and Future Research Directions

In this chapter, we highlighted three LiDAR applications in the study of forest-water interactions: the use of DTMs in watershed modelling, quantification of vegetation structure based on LiDAR point clouds and its interaction with water fluxes, and assessment of riparian buffer zones with a combination of LiDAR-based DTMs and LiDAR point clouds. For all three applications, LiDAR data enabled analyses at small spatial scales that would not be feasible using manual field techniques and provided a level of detail not obtainable with classification of spectral satellite imagery. It should be acknowledged, however, that the data types produced by LiDAR and spectral satellite sensors are inherently different and cannot be compared directly.

LiDAR-based DTMs accurately depict the land surface and allow for the modelling of hydrological processes such as water runoff and discharge, flow paths, and streamflow at the watershed scale. Comparisons of DTM resolutions suggest that a moderate spatial resolution (i.e., 10 m) produces the most accurate outputs from hydrologic models.

Considering the effect of vegetation on forest water fluxes and stores, additional information for watershed modelling can be provided by LiDAR data analyses. Forest canopy interception maps or other attributes of the vegetation cover relevant for hydrology can be derived based on LiDAR data. Interception and throughfall can be partitioned in layered canopies, and structural differences in the top and intermediate canopy layers can be identified and, therefore, differences in interception storage capacity can be found. Furthermore, the effect of undergrowth, such as shrubs and regeneration, on net throughfall to the soil surface can be better assessed.

Characterisation of the riparian forest zone is an important first step in ecologically sound resource management for maintaining or restoring such ecosystems. Several authors have studied the assessment of riparian zones using LiDAR data, and large-scale characterisations of thousands of kilometres of rivers have been conducted in two studies.

For the development of prediction models and their area-wide application, the design of data acquisition and processing is important. Several studies described this approach in a forest inventory context in the past (e.g., Næsset 2002; McRoberts et al. 2013; Mandallaz et al. 2013; Magnussen et al. 2014). In Fig. 4.2, processing steps for remote sensing-assisted assessment of forest properties are presented.

One obvious application of LiDAR-derived metrics is their integration into hydrological models; however, this application is currently limited by the need to transform LiDAR or other remote sensing variables specific to the model being used. For example, to include vegetation properties based on LiDAR data in hydrological models, LiDAR data are usually adapted or transformed to represent indices required by these models (e.g., gap fraction, LAI, etc.). In doing so, information on the complexity of the vegetation is inadvertently lost or confounded. Therefore, the question arises whether we should use LiDAR data in hydrological models by converting them to well-known variables, or if it is possible to use them untransformed? This is an important question as we know the shortcomings of current metrics that are used to characterise forest systems, such as LAI, and thus exploring new ways to develop hydrological models based on direct measurements of canopy structure may be justified.

Spatial scale is an important topic when addressing LiDAR applications. In most of the aforementioned applications, airborne LiDAR data was used to obtain information for larger areas. However, water fluxes and stores in forest canopy research are measured at much smaller scales, and interception loss is often estimated indirectly using throughfall measurements. The use of TLS may offer a solution to reconcile this disparity. For example, Baptista et al. (2018) calculated canopy metrics from TLS data of individual trees under lab conditions and related these to interception loss parameters. TLS data, including data from mobile platforms such as a handheld LiDAR system, may be used to study forest-water interactions at the plot scale with a higher resolution than what airborne LiDAR can offer. The relationship between ground truth hydrologic data and TLS data could then be upscaled with airborne LiDAR data to achieve area-wide predictions. This represents an unexplored field of research. Furthermore, the development of unpiloted aerial vehicles (UAVs) is progressing rapidly. UAVs for the collection of LiDAR data may provide a very high level of detail at local to regional spatial scales (see Chap. 3).

Finally, to connect plot-level information to even larger scales where no other data are available, a correlation between, for example, ALS- or UAV-borne data and satellite imagery is worth exploring. This requires that the link between the radiative properties of the surface and canopy structure can be established, as shown by Varhola and Coops (2013) and Babcock et al. (2018). The space-borne LiDAR GEDI was launched at the end of 2018 and will allow for scaling up LiDAR applications from local to regional scales and from the regional scale to the global scale.

References

Amatya D, Trettin C, Panda S, Ssegane H (2013) Application of LiDAR Data for hydrologic assessments of low-gradient coastal watershed drainage characteristics. J Geogr Inf Syst 5:175–191. https://doi.org/10.4236/jgis.2013.52017

Babcock C, Finley AO, Andersen H-E, Pattison R, Cook BD, Morton DC et al (2018) Geostatistical estimation of forest biomass in interior Alaska combining Landsat-derived tree cover, sampled airborne lidar and field observations. Remote Sens Environ 212:212–230. https://doi.org/10.1016/j.rse.2018.04.044

Baptista MD, Livesley SJ, Parmehr EG, Neave M, Amati M (2018) Terrestrial laser scanning to predict canopy area metrics, water storage capacity, and throughfall redistribution in small trees. Remote Sens 10:1958. https://doi.org/10.3390/rs10121958

Benyon RG, Lane PNJ, Jaskierniak D, Kuczera G, Haydon SR (2015) Use of a forest sapwood area index to explain long-term variability in mean annual evapotranspiration and streamflow in moist eucalypt forests. Water Resour Res 51:5318–5331. https://doi.org/10.1002/2015WR017321

Calder IR (1986) A stochastic model of rainfall interception. J Hydrol 89:65–71. https://doi.org/10.1016/0022-1694(86)90143-5

Calder IR (1996) Dependence of rainfall interception on drop size: 1. Development of the two-layer stochastic model. J Hydrol 185:363–378. https://doi.org/10.1016/0022-1694(95)02998-2

Carlyle-Moses DE, Gash JHC (2011) Rainfall interception loss by forest canopies. In: Levia DF, Carlyle-Moses DE, Tanaka T (eds) Forest hydrology and biogeochemistry, ecological studies, vol 216. Springer, Dordrecht, pp 407–423. https://doi.org/10.1007/978-94-007-1363-5_20

Cavalli M, Tarolli P, Marchi L, Fontana GD (2008) The effectiveness of airborne lidar data in the recognition of channel-bed morphology. Catena 73:249–260. https://doi.org/10.1016/j.catena.2007.11.001

Cescatti A (2007) Indirect estimates of canopy gap fraction based on the linear conversion of hemispherical photographs: methodology and comparison with standard thresholding techniques. Agric For Meteorol 143:1–12. https://doi.org/10.1016/j.agrformet.2006.04.009

Chasmer L, Kljun N, Hopkinson C, Brown S, Milne T, Giroux K, Barr A, Devito K, Creed I, Petrone R (2011) Characterizing vegetation structural and topographic characteristics sampled by eddy covariance within two mature aspen stands using lidar and a flux footprint model: scaling to MODIS. J Geophys Res 116:G02026. https://doi.org/10.1029/2010JG001567

Chen JM, Rich PM, Gower ST, Norman JM, Plummer S (1997) Leaf area index of boreal forests: theory, techniques, and measurements. J Geophys Res-Atmos 102:29429–29443. https://doi.org/10.1029/97JD01107

Congalton RG, Birch K, Jones R, Schriever J (2002) Evaluating remotely sensed techniques for mapping riparian vegetation. Comput Electron Agric 37:113–126. https://doi.org/10.1016/S0168-1699(02)00108-4

Côté J-F, Fournier RA, Egli R (2011) An architectural model of trees to estimate forest structural attributes using terrestrial lidar. Environ Model Softw 26:761–777. https://doi.org/10.1016/j.envsoft.2010.12.008

Crasto N, Hopkinson C, Forbes D, Lesack L, Marsh P, Spooner I et al (2015) A lidar-based decision-tree classification of open water surfaces in an arctic delta. Remote Sens Environ 164:90–102. https://doi.org/10.1016/j.rse.2015.04.011

Creed IF, Sass GZ (2011) Digital terrain analysis approaches for tracking hydrological and biogeochemical pathways and processes in forested landscapes. In: Levia DF, Carlyle-Moses DE, Tanaka T (eds) Forest hydrology and biogeochemistry, ecological studies, vol 216. Springer, Dordrecht, pp 69–100. https://doi.org/10.1007/978-94-007-1363-5_4

Danson FM, Hetherington D, Morsdorf F, Koetz B, Allgower B (2007) Forest canopy gap fraction from terrestrial laser scanning. IEEE Geosci Remote Sens Lett 4:157–160. https://doi.org/10.1109/LGRS.2006.887064

Ellis CR, Pomeroy JW, Brown T, MacDonald J (2010) Simulation of snow accumulation and melt in needleleaf forest environments. Hydrol Earth Syst Sci 14:925–940. https://doi.org/10.5194/hess-14-925-2010

Gash JHC (1979) An analytical model of rainfall interception by forests. Q J R Meteorol Soc 105(443):43–55. https://doi.org/10.1002/qj.49710544304

Gash JHC, Lloyd CR, Lachaud G (1995) Estimating sparse forest rainfall interception with an analytical model. J Hydrol 170:79–86. https://doi.org/10.1016/0022-1694(95)02697-N

Goetz SJ (2006) Remote sensing of riparian buffers: past progress and future prospects. J Am Water Resour Assoc 42:133–143. https://doi.org/10.1111/j.1752-1688.2006.tb03829.x

Goulden T, Hopkinson C, Jamieson R, Sterling S (2016) Sensitivity of dem, slope, aspect and watershed attributes to lidar measurement uncertainty. Remote Sens Environ 179:23–35. https://doi.org/10.1016/j.rse.2016.03.005

Gower S, Norman J (1991) Rapid estimation of leaf area index in conifer and broad-leaf plantations. Ecology 72:1896–1900. https://doi.org/10.2307/1940988

Greenberg JA, Hestir EL, Riano D, Scheer GJ, Ustin SL (2012) Using LiDAR data analysis to estimate changes in insolation under large-scale riparian deforestation. J Am Water Resour Assoc 48:939–948. https://doi.org/10.1111/j.1752-1688.2012.00664.x

Haydon SR, Benyon RG, Lewis R (1997) Variation in sapwood area and throughfall with forest age in mountain ash (*Eucalyptus regnans* F. Muell.). J Hydrol 187:351–366. https://doi.org/10.1016/S0022-1694(96)03016-8

Herwitz SR (1985) Interception storage capacities of tropical rainforest canopy trees. J Hydrol 77:237–252. https://doi.org/10.1016/0022-1694(85)90209-4

Hickman GD, Hogg JE (1969) Application of an airborne pulsed laser for near shore bathymetric measurements. Remote Sens Environ 1:47–58. https://doi.org/10.1016/S0034-4257(69)90088-1

Holder CD (2012) The relationship between leaf hydrophobicity, water droplet retention, and leaf angle of common species in a semi-arid region of the western United States. Agric For Meteorol 152:11–16. https://doi.org/10.1016/j.agrformet.2011.08.005

Hopkinson C, Chasmer L, Young-Pow C, Treitz P (2004) Assessing forest metrics with a ground-based scanning lidar. Canada J For Res 34:573–583. https://doi.org/10.1139/x03-225

James LA, Watson DG, Hansen WF (2007) Using lidar data to map gullies and headwater streams under forest canopy: South Carolina, USA. Catena 71:132–144. https://doi.org/10.1016/j.catena.2006.10.010

Jaskierniak D, Lane PN, Robinson A, Lucieer A (2011) Extracting lidar indices to characterise multilayered forest structure using mixture distribution functions. Remote Sens Environ 115:573–585. https://doi.org/10.1016/j.rse.2010.10.003

Johansen K, Phinn S, Witte C (2010) Mapping of riparian zone attributes using discrete return LiDAR, QuickBird and SPOT-5 imagery: assessing accuracy and costs. Remote Sens Environ 114:2679–2691. https://doi.org/10.1016/j.rse.2010.06.004

Johansen K, Tiede D, Blaschke T, Arroyo LA, Phinn S (2011) Automatic geographic object based mapping of streambed and riparian zone extent from LiDAR data in a temperate rural urban environment, Australia. Remote Sens 3:1139–1156. https://doi.org/10.3390/rs3061139

Jupp DL, Culvenor D, Lovell J, Newnham G, Strahler A, Woodcock C (2009) Estimating forest LAI profiles and structural parameters using a groundbased laser called 'Echidna®. Tree Physiol 29:171–181. https://doi.org/10.1093/treephys/tpn022

Klamerus-Iwan A, Błońska E (2018) Canopy storage capacity and wettability of leaves and needles: the effect of water temperature changes. J Hydrol 559:534–540. https://doi.org/10.1016/j.jhydrol.2018.02.032

Le Dantec V, Dufrêne E, Saugier B (2000) Interannual and spatial variation in maximum leaf area index of temperate deciduous stands. For Ecol Manag 134:71–81. https://doi.org/10.1016/S0378-1127(99)00246-7

Lefsky MA, Hudak AT, Cohen WB, Acker S (2005) Geographic variability in lidar predictions of forest stand structure in the Pacific Northwest. Remote Sens Environ 95:532–548. https://doi.org/10.1016/j.rse.2005.01.010

Levia DF, Herwitz SR (2005) Interspecific variation of bark water storage capacity of three deciduous tree species in relation to stemflow yield and solute flux to forest soils. Catena 64:117–137. https://doi.org/10.1016/j.catena.2005.08.001

Levia DF, Hudson SA, Llorens P, Nanko K (2017) Throughfall drop size distributions: a review and prospectus for future research. WIRES Water 4:e1225. https://doi.org/10.1002/wat2.1225

Ma H, Song J, Wang J (2015) Forest canopy LAI and vertical FAVD profile inversion from airborne full-waveform LiDAR data based on a radiative transfer model. Remote Sens 7:1897–1914. https://doi.org/10.3390/rs70201897

Magnussen S, Mandallaz D, Breidenbach J, Lanz A, Ginzler C (2014) National forest inventories in the service of small area estimation of stem volume. Can J For Res 44:1079–1090. https://doi.org/10.1139/cjfr-2013-0448

Mandallaz D, Breschan J, Hill A (2013) New regression estimators in forest inventories with two-phase sampling and partially exhaustive information: a design-based monte carlo approach with applications to small-area estimation. Can J For Res 43:1023–1031. https://doi.org/10.1139/cjfr-2013-0181

Marshall JS, Palmer WMK (1948) The distribution of raindrops with size. J Meteorol 5:165–166. https://doi.org/10.1175/1520-0469(1948)005<0165:TDORWS>2.0.CO;2

McKean J, Nagel D, Tonina D, Bailey P, Wright CW, Bohn C et al (2009) Remote sensing of channels and riparian zones with a narrow-beam aquatic-terrestrial lidar. Remote Sens 1:1065–1096. https://doi.org/10.3390/rs1041065

McRoberts RE, Næsset E, Gobakken T (2013) Inference for lidar-assisted estimation of forest growing stock volume. Remote Sens Environ 128:268–275. https://doi.org/10.1016/j.rse.2012.10.007

Michez A, Piégay H, Toromanoff F, Brogna D, Bonnet S, Lejeune P (2013) LiDAR derived ecological integrity indicators for riparian zones: application to the Houille river in southern Belgium/northern France. Ecol Indic 34:627–640. https://doi.org/10.1016/j.ecolind.2013.06.024

Michez A, Piégay H, Lejeune P, Claessens H (2017) Multi-temporal monitoring of a regional riparian buffer network (>12,000 km) with lidar and photogrammetric point clouds. J Environ Manag 202:424–436. https://doi.org/10.1016/j.jenvman.2017.02.034

Miralles DG, Gash JH, Holmes TRH, de Jeu RAM, Dolman AJ (2010) Global canopy interception from satellite observations. J Geophys Res Atmos 115:D16122. https://doi.org/10.1029/2009JD013530

Miralles DG, Jiménez C, Jung M, Michel D, Ershadi A, McCabe MF et al (2016) The WACMOS-ET project – part 2: evaluation of global terrestrial evaporation data sets. Hydrol Earth Syst Sci 20:823–842. https://doi.org/10.5194/hess-20-823-2016

Mitchell PJ, Lane PNJ, Benyon RG (2012) Capturing within catchment variation in evapotranspiration from montane forests using LiDAR canopy profiles with measured and modelled fluxes of water. Ecohydrology 5:708–720. https://doi.org/10.1002/eco.255

Moorthy I, Miller JR, Hu B, Chen J, Li Q (2008) Retrieving crown leaf area index from an individual tree using ground-based lidar data. Can J Remote Sens 34:320–332. https://doi.org/10.5589/m08-027

Morsdorf F, Kötz B, Meier E, Itten KI, Allgöwer B (2006) Estimation of LAI and fractional cover from small footprint airborne laser scanning data based on gap fraction. Remote Sens Environ 104:50–61. https://doi.org/10.1016/j.rse.2006.04.019

Muzylo A, Llorens P, Valente F, Keizer J, Domingo F, Gash J (2009) A review of rainfall interception modelling. J Hydrol 370:191–206. https://doi.org/10.1016/j.jhydrol.2009.02.058

Næsset E (2002) Predicting forest stand characteristics with airborne scanning laser using a practical two-stage procedure and field data. Remote Sens Environ 80:88–99. https://doi.org/10.1016/S0034-4257(01)00290-5

Naiman RJ, Decamps H, Pollock M (1993) The role of riparian corridors in maintaining regional biodiversity. Ecol Appl 3:209–212. https://doi.org/10.2307/1941822

Nelson R, Krabill W, Tonelli J (1988) Estimating forest biomass and volume using airborne laser data. Remote Sens Environ 24:247–267. https://doi.org/10.1016/0034-4257(88)90028-4

Nilsson M, Nordkvist K, Jonzén J, Lindgren N, Axensten P, Wallerman J et al (2017) A nationwide forest attribute map of Sweden predicted using airborne laser scanning data and field data from the National Forest Inventory. Remote Sens Environ 194:447–454. https://doi.org/10.1016/j.rse.2016.10.022

Pomeroy JW, Gray DM, Brown T, Hedstrom NR, Quinton WL, Granger RJ et al (2007) The cold regions hydrological model: a platform for basing process representation and model structure on physical evidence. Hydrol Process 21:2650–2667. https://doi.org/10.1002/hyp.6787

Popescu SC, Zhao K (2008) A voxel based lidar method for estimating crown base height for deciduous and pine trees. Remote Sens Environ 112:767–781. https://doi.org/10.1016/j.rse.2007.06.011

Quinn P, Beven K, Chevallier P, Planchon O (1991) The prediction of hillslope flow paths for distributed hydrological modelling using digital terrain models. Hydrol Process 5:59–79. https://doi.org/10.1002/hyp.3360050106

Remmel TK, Todd KW, Buttle J (2008) A comparison of existing surficial hydrological data layers in a low-relief forested Ontario landscape with those derived from a LiDAR DEM. For Chron 84:850–865. https://doi.org/10.5558/tfc84850-6

Renslow MS (2012) Manual of airborne topographic lidar. American Society for photogrammetry remote sensing.

Riaño D, Meier E, Allgöwer B, Chuvieco E, Ustin SL (2003) Modeling airborne laser scanning data for the spatial generation of critical forest parameters in fire behavior modeling. Remote Sens Environ 86:177–186. https://doi.org/10.1016/S0034-4257(03)00098-1

Roth BE, Slatton KC, Cohen MJ (2007) On the potential for high- resolution lidar to improve rainfall interception estimates in forest ecosystems. Front Ecol Environ 5:421–428. https://doi.org/10.1890/060119.1

Rutter AJ, Kershaw KA, Robins PC, Morton AJ (1971) A predictive model of rainfall interception in forests, 1. derivation of the model from observations in a plantation of Corsican pine. Agric Meteorol 9:367–384. https://doi.org/10.1016/0002-1571(71)90034-3

Sasaki T, Imanishi J, Ioki K, Morimoto Y, Kitada K (2008) Estimation of leaf area index and canopy openness in broad-leaved forest using an airborne laser scanner in comparison with high-resolution near-infrared digital photography. Landsc Ecol Eng 4:47–55. https://doi.org/10.1007/s11355-008-0041-8

Schreier H, Lougheed J, Gibson JR, Russell J (1984) Calibrating an airborne laser profiling system. Photogramm Eng Remote Sens 50:1591–1598

Schumacher J, Christiansen JR (2015) Forest canopy water fluxes can be estimated using canopy structure metrics derived from airborne light detection and ranging (LiDAR). Agric For Meteorol 203:131–141. https://doi.org/10.1016/j.agrformet.2014.12.007

Smeeckaert J, Mallet C, David N, Chehata N, Ferraz A (2013) Largescale classification of water areas using airborne topographic lidar data. Remote Sens Environ 138:134–148. https://doi.org/10.1016/j.rse.2013.07.004

Solberg S, Næsset E, Hanssen KH, Christiansen E (2006) Mapping defoliation during a severe insect attack on Scots pine using airborne laser scanning. Remote Sens Environ 102:364–376. https://doi.org/10.1016/j.rse.2006.03.001

Solberg S, Brunner A, Hanssen KH, Lange H, Næsset E, Rautiainen M et al (2009) Mapping LAI in a Norway spruce forest using airborne laser scanning. Remote Sens Environ 113:2317–2327. https://doi.org/10.1016/j.rse.2009.06.010

Stenberg P (1996) Correcting LAI-2000 estimates for the clumping of needles in shoots of conifers. Agric For Meteorol 79:1–8. https://doi.org/10.1016/0168-1923(95)02274-0

Strahler AH, Jupp DL, Woodcock CE, Schaaf CB, Yao T, Zhao F et al (2008) Retrieval of forest structural parameters using a ground-based lidar instrument (Echidna®). Can J Remote Sens 34:426–440. https://doi.org/10.5589/m08-046

Sutherland G, Chasmer LE, Kljun N, Devito KJ, Petrone RM (2017) Using high resolution LiDAR data and a flux footprint parameterization to scale evapotranspiration estimates to lower pixel resolutions. Can J Remote Sens 43:215–229. https://doi.org/10.1080/07038992.2017.1291338

Tang H, Brolly M, Zhao F, Strahler AH, Schaaf CL, Ganguly S et al (2014) Deriving and validating leaf area index (LAI) at multiple spatial scales through lidar remote sensing: a case study in Sierra National Forest, CA. Remote Sens Environ 143:131–141. https://doi.org/10.1016/j.rse.2013.12.007

Tompalski P, Coops NC, White JC, Wulder MA, Yuill A (2017) Characterizing streams and riparian areas with airborne laser scanning data. Remote Sens Environ 192:73–86. https://doi.org/10.1016/j.rse.2017.01.038

Valbuena R, Maltamo M, Mehtätalo L, Packalen P (2017) Key structural features of boreal forests may be detected directly using L-moments from airborne lidar data. Remote Sens Environ 194:437–446. https://doi.org/10.1016/j.rse.2016.10.024

van Leeuwen M, Nieuwenhuis M (2010) Retrieval of forest structural parameters using lidar remote sensing. Eur J For Res 129:749–770. https://doi.org/10.1007/s10342-010-0381-4

Varhola A, Coops NC (2013) Estimation of watershed-level distributed forest structure metrics relevant to hydrologic modeling using LiDAR and Landsat. J Hydrol 487:70–86. https://doi.org/10.1016/j.jhydrol.2013.02.032

Varhola A, Frazer GW, Teti P, Coops NC (2012) Estimation of forest structure metrics relevant to hydrologic modelling using coordinate transformation of airborne laser scanning data. Hydrol Earth Syst Sci 16:3749–3766. https://doi.org/10.5194/hess-16-3749-2012

Vianello A, Cavalli M, Tarolli P (2009) LiDAR-derived slopes for headwater channel network analysis. Catena 76:97–106. https://doi.org/10.1016/j.catena.2008.09.012

Wallace CW, McCarty G, Lee S, Brooks RP, Veith TL, Kleinman PJA et al (2018) Evaluating concentrated flowpaths in riparian forest buffer contributing areas using LiDAR imagery and topographic metrics. Remote Sens 10:614. https://doi.org/10.3390/rs10040614

Wasser L, Day R, Chasmer L, Taylor A (2013) Influence of vegetation structure on Lidar-derived canopy height and fractional cover in forested riparian buffers during leaf-off and leaf-On conditions. PLoS One 8:1–13. https://doi.org/10.1371/journal.pone.0054776

Wilkes P, Lau A, Disney M, Calders K, Burt A, de Tanago JG et al (2017) Data acquisition considerations for terrestrial laser scanning of forest plots. Remote Sens Environ 196:140–153. https://doi.org/10.1016/j.rse.2017.04.030

Yang P, Ames DP, Fonseca A, Anderson D, Shrestha R, Glenn NF et al (2014) What is the effect of lidar-derived dem resolution on large-scale watershed model results? Environ Model Softw 58:48–57. https://doi.org/10.1016/j.envsoft.2014.04.005

Zhang JX, Wu JQ, Chang K, Elliot WJ, Dun S (2009) Effects of DEM source and resolution on WEPP hydrologic and erosion simulation: a case study of two forest watersheds in northern Idaho. Trans ASABE 52:447–457. https://doi.org/10.13031/2013.26838

Zhao K, Popescu S (2009) Lidar-based mapping of leaf area index and its use for validating GLOBCARBON satellite LAI product in a temperate forest of the southern USA. Remote Sens Environ 113(8):1628–1645. https://doi.org/10.1016/j.rse.2009.03.006

Zimble DA, Evans DL, Carlson GC, Parker RC, Grado SC, Gerard PD (2003) Characterizing vertical forest structure using small-footprint airborne LiDAR. Remote Sens Environ 87:171–182. https://doi.org/10.1016/S0034-4257(03)00139-1

Chapter 5
On Complementing the Tracer Toolbox for Quantifying Hydrological Connectivity: Insights Gained from Terrestrial Diatom Tracer Experiments

L. Pfister, S. T. Allen, C. E. Wetzel, and N. Martínez-Carreras

5.1 Introduction

Characterising where and when water travels is important for understanding how water shapes the landscapes and their ecosystems' functions. The flow paths of water ultimately control how much and how quickly water will reach stream channels (or return to the atmosphere) and what will be transported from the land to streams. To understand catchment flow routing, it is necessary not only to monitor the magnitude of different fluxes (e.g., streamflow, aquifer recharge, evapotranspiration) but also to know water sources, flow paths and transit times.

Perceived storm response mechanisms of catchments have gradually evolved since the early 'Hortonian' concept of dominating surface flow (Horton 1933) towards the variable source area concept (VSA) (Hewlett and Hibbert 1967; Dunne and Black 1970). However, a persistent fundamental question is what fraction of streamflow is composed of recent precipitation. Tracers have played a key role in understanding water travel pathways, and they have been pivotal in

L. Pfister (✉)
Environmental Research and Innovation Department, Catchment and Eco-hydrology Research Group, Luxembourg Institute of Science and Technology, Belvaux, Luxembourg

Faculty of Science, Technology and Communication, University of Luxembourg, Esch-sur-Alzette, Luxembourg
e-mail: laurent.pfister@list.lu

S. T. Allen
ETHZ, Institute of Terrestrial Ecosystems, Zürich, Switzerland

C. E. Wetzel
Environmental Research and Innovation Department, Environmental Microbiology and Biotechnology, Luxembourg Institute of Science and Technology, Belvaux, Luxembourg

N. Martínez-Carreras
Environmental Research and Innovation Department, Catchment and Eco-hydrology Research Group, Luxembourg Institute of Science and Technology, Belvaux, Luxembourg

D. F. Levia et al. (eds.), *Forest-Water Interactions*, Ecological Studies 240,
https://doi.org/10.1007/978-3-030-26086-6_5

improving our basic and applied understanding of catchment hydrology. The advent of geochemical and isotope fingerprinting of water sources in the late 1960s has unveiled the baffling diversity of streamflow generating processes across multiple spatial and temporal scales (e.g., Hooper et al. 1990; Sklash and Farvolden 1979)—preferably in combination with accompanying hydrometric information in order to avoid misleading interpretations (Uhlenbrook and Hoeg 2003).

Prior to the widespread use of tracers, early forest hydrology studies relied mostly on hydrometric measurements (soil moisture content, groundwater levels, discharge, precipitation inputs). A dominant pre-tracer paradigm was that high precipitation intensities resulted in rapid runoff responses when infiltration rates were exceeded by precipitation rates, as deduced from hydrometric data (Horton 1933). Subsequent studies formulated more nuanced descriptions, accounting for varying causes of rapid flow to streams (infiltration excess/saturation excess overland flow versus subsurface stormflow (Hewlett and Hibbert 1967) and exfiltrating return flow (Dunne and Black 1970)) and their differential occurrence and contribution to streamflow from different landscape units (e.g., hillslopes versus riparian areas (Dunne et al. 1975)).

Early hydrologic tracer applications included performing hydrograph separations to understand the fractions of event and pre-event water in streams (Sklash and Farvolden 1979). Tracers revealed that stream water is often minimally composed of recent precipitation, so surface runoff was not likely to explain observed hydrographic response (as reviewed in Buttle and Peters (1997) and Klaus and McDonnell (2013)), and thus new concepts were needed. For example, McGlynn and McDonnell (2003) combined hydrometric measurements with isotopic and solute dynamics and were able to quantify contributions from riparian zones and hillslopes. Using additional tracers, in the Panola catchment, Burns et al. (2001) applied an end-member mixing analysis model (EMMA) and quantified contributions of three geographic stormflow sources. With longer series of tracer measurements, more sophisticated inferences became possible; for example, the temporal origins of water fluxes can also be distinguished in transit time distributions (McGuire et al. 2005) and young water fractions (Kirchner 2016).

In this chapter, we provide a primer on tracer applications in studying the hydrological connections in catchments and forests. After providing an overview of general basics in tracer applications, we focus on a new type of tracer—terrestrial diatoms—and give some insights into their use for tracing the onset/cessation of surface hydrological connectivity. Given that this is a newly developing tracer method, we demonstrate the process of exploring and testing tracer applications. Furthermore, we discuss the differences among and complementarity between diatoms and more traditional tracers (e.g., solutes).

5.2 Overview of Tracers

5.2.1 Definition

Hydrological tracers are any detectable constituent of water, substance or property that is transported or transmitted in water, and is useful for understanding how that water moves. For a classic in-stream tracer example, an orange (with specific gravity ≈ 1) could be dropped into a river to measure velocity, just by measuring the distance travelled over an elapsed time. Alternatively, a bucket of oranges could be dumped into a river—together they provide directly observable variations in flow paths, a transit time distribution, and also some indicator of how long some oranges (and thus water) may reside in certain parts of the channel. However, such simple tracers are not useful in forest hydrology for understanding terrestrial aquatic linkages.

Following Leibundgut et al. (2009), we distinguish between 'artificial tracers', which are injected and then tracked, and 'environmental tracers', which are naturally present substances or properties. The environmental tracers that are most used in hydrology are the isotopes of elements of the water molecule (e.g., deuterium, oxygen-18 and tritium), naturally occurring solutes (from precipitation, geologic materials or biological processes), physicochemical properties of the water (e.g., temperature, electrical conductivity) and particulates (e.g., microorganisms, DNA); we refer readers to a more complete discussion of these specific tracers in Kendall and McDonnell (1998) and Leibundgut et al. (2009). Note that although pollutants (e.g., nitrate) or other anthropogenic substances released to the atmosphere (e.g., so-called bomb tritium) are artificial, we characterise them as environmental tracers because they are used similarly to natural tracers—relying on spatial and temporal variations in sources to identify processes (Leibundgut et al. 2009).

In this chapter, we focus on environmental tracers, which tend to be useful over longer time scales and at larger spatial scales. They are naturally abundant and naturally varying, potentially constantly mobilizing towards water.

5.2.2 Necessary Conditions for Tracer Applications

Natural tracer signals have to involve distinct spatial or temporal variations to be useful. For example, precipitation solute concentrations are usually very distinct from those in soil waters or groundwaters, potentially allowing the relative contributions from those sources to be distinguished. With water stable isotopes, precipitation varies spatiotemporally within events (Coplen et al. 2015), across events (Fischer et al. 2017) and across seasons; these signals support inferring mixing and transit processes at various resolutions.

A disadvantage of natural tracers is that their input function is often difficult to define. Another limitation inherent to natural tracers may stem from weak input and/or output signals. By determining which and how many end-members will eventually be identified, tracer set size and composition will ultimately also impact the conceptual model inferred from an end-member mixing analysis (EMMA; Hooper et al. 1990). From their analysis of various tracer set sizes and compositions, Barthold et al. (2011) concluded that (i) major elements are not always the most suitable tracers and (ii) large tracer sets may help to avoid false conclusions on catchment functioning. Furthermore, it is only after establishing whether a tracer is conservative (or, at least understanding what physical or biological processes affect the signal) and how various relevant source signatures vary (in space and time), that processes can be inferred.

5.2.3 Examples of the Potential and Limitations of Tracer-Based Analytical Approaches

5.2.3.1 Hydrograph Separation

A major contribution of (environmental and artificial) tracers to hydrological sciences is the advent of hydrograph separation studies. The goal of pre-event versus event hydrograph separation is to identify the proportion of new water traveling into streams. Isotope hydrograph separations have been carried out across a wide range of climatic and physiographic contexts (Klaus and McDonnell 2013). In catchments with abundant precipitation that are response-dominated, large volumes of pre-event water commonly depend on precipitation intensity that controls saturation overland flow (Bonell et al. 1999). In other areas, with shallow soils and steep slopes—such as the Maimai catchment in New Zealand—macropore flow delivers considerable volumes of pre-event water to storm hydrographs (McDonnell 1990). This is in strong contrast to monsoonal flash flood events in low elevation and low relief semiarid catchments, where flood hydrographs are solely composed of event water (Ingraham et al. 1999). However, Desilets et al. (2008) have shown significant contributions of pre-event water in the flood peak for catchments located in semiarid terrain using isotope-based hydrograph separation.

Hydrograph separation studies also infer geographic sources of stormflow. Combining hydrometric measurements with isotopic and solute dynamics to quantify various contributions from interacting dynamic riparian zones and hillslopes over storm hydrographs, McGlynn and McDonnell (2003) found that riparian water dominated between the events, throughout a small runoff event, and in early portions of a large event. In the large event, proportions of riparian and hillslope runoff were similar. In the Panola catchment, Burns et al. (2001) applied an end-member mixing analysis model for quantifying contributions of three different stormflow sources. For their analysis, they relied on five solutes determined from stream water, runoff from an outcrop and hillslope subsurface stormflow, and concluded that on a whole-storm basis, outcrop runoff and riparian groundwater runoff were the dominant sources.

While their contribution to a better understanding of runoff generation processes is undeniable, it is problematic that some of the basic assumptions in hydrograph separation are only rarely met. Eventually, the uncertainties in geochemistry and isotope-based stormflow-hydrograph separations need to be considered (Buttle and Peters 1997; Richey et al. 1998)—a fact that had gradually led to a decrease in the 'initial thrill' that had surrounded the geochemistry and isotope based stormflow-hydrograph separations (Burns 2002).

5.2.3.2 Transit Time Distribution of Water as a Fundamental Descriptor of a Catchment's Hydrological Functions

Streamflow transit time distribution (TTD)—commonly inferred from isotopes (McGuire and McDonnell 2006) or conservative chemicals (Kirchner et al. 2010)—is a fundamental descriptor of hydrological functions (Soulsby et al. 2009; McDonnell et al. 2010). TTD characterises the proportions of water moving via rapid flow paths (e.g., surface runoff, throughflow) versus slow pathways (e.g., the displacement of aquifer storage). The mean transit time (MTT) of water is also a useful metric for comparing hydrological behaviours of different catchments with contrasted physiographical characteristics (McDonnell et al. 2010). Isotope transit time analyses have revealed that older groundwater contributions to stream water can be large and were often otherwise neglected.

Different TTD may result from tritium on the one hand and $\delta^{18}O$, $\delta^{2}H$ and conservative solutes on the other hand (Stewart et al. 2010). When taken individually, tracers provide information on transit time only over limited parts of the complete TTD (McDonnell et al. 2010). Until recently, most TTD studies were based on stable isotopes and solute variation analysis (McGuire and McDonnell 2006). However, with preliminary data from New Zealand—where bomb tritium in groundwater is close to cosmogenic tritium levels—Morgenstern et al. (2010) have shown that unique MTT values can be derived from only a few stream water tritium measurements.

Another evolving area of catchment hydrology is based on understanding how stream MTT vary under different flow conditions, presumably due to the activation of different storages (Stewart and Thomas 2008; Morgenstern et al. 2010). Constant, or time invariant, storage is approached by steady-state assumptions (e.g., Maloszewski and Zuber 1993), whereas in catchments with time variant storage non-steady-state assumptions are applied (e.g., Foussereau et al. 2001). Given that precipitation and evapo-transpiration are highly variable in space and time, most catchments exhibit non-steady-state conditions (flow path lengths, flow velocities, etc.). This in turn leads the shape of catchment TTD to be time variant (McDonnell et al. 2010) and controlled by storage (e.g., Benettin et al. 2013; Harman 2015; Klaus et al. 2015a; Rodriguez et al. 2018). Moreover, in catchments dominated by more permeable bedrock and more seasonality, transit time bimodality (i.e., streamflow being constituted of two or more major age components) may be the norm (e.g., Maloszewski et al. 1983; Uhlenbrook et al. 2002; Stewart et al. 2007; Stewart and Thomas 2008; Stewart and Fahey 2010).

5.3 Towards the Use of Diatoms as Biological Tracers for Quantifying Land-Stream Connections?

5.3.1 The Challenge: Understanding the Role of Surface Connectivity in Rainfall-Runoff Response

One key need in forest hydrology is to develop a better understanding of how surface waters connect and thus facilitate transport. The importance of differentiating between variable 'active' areas and 'contributing', i.e., temporarily connected areas, has been emphasized by Ambroise (2004). Connectivity and threshold measures have been proposed for moving beyond the status quo of the simplistic variable source area concept (Ambroise 2004; Bracken and Croke 2007; McDonnell et al. 2007). Headwater streams sustain hydrological connectivity between uplands, riparian areas and streams (Pringle 2003; Freeman et al. 2007)—thereby structuring both ecosystems and hydrologic behaviour across a wide range of scales (McDonnell et al. 2007). Tetzlaff et al. (2007) have proposed the connectivity between landscapes and riverscapes as a unifying theme—ultimately integrating hydrology and ecology in catchment science. Surface water connectivity is especially important because it allows for rapid transport, which can have either positive effects (species dispersal) or negative effects (pollution transport).

Stable isotopes and geochemical tracers do not specifically show surface connectivity, allowing for ambiguity in our understanding of how 'event water' actually reaches streams. Furthermore, there are challenges in aggregating these tracer measurements across diverse landscapes to understand larger scale connectivity patterns. Thus, there was need to find new tracers to specifically quantify surface water connectivity across scales.

5.3.2 Diatoms: Widespread Organisms and Indicators Used in Environmental Monitoring and Research

Diatoms are unicellular eukaryotic algae—with some species forming filament, or ribbon-shaped colonies. They are contained within a frustule that has its cell wall made of silica and that exhibits a sheer endless variety in shapes and forms. Diatoms are ubiquitous in the terrestrial and aquatic environments and their highly diversified species distributions are largely controlled by the environmental conditions (including anthropogenic controls, such as organic pollution sources and eutrophication) and physiographic characteristics that prevail in their habitat (e.g., Ector and Rimet 2005). The very pronounced sensitivity of diatoms to multiple environmental variables—including light, moisture conditions, temperature, current velocity, salinity, pH, oxygen, inorganic nutrients (carbon, phosphorous, nitrogen, silica), organic carbon and organic nitrogen—has them commonly applied to geological, archaeological and water quality research (Van Dam et al. 1994; Stoermer and Smol 2010).

The range of their application across various fields in Earth Sciences is very large, with diatoms serving as bioindicators of water quality (Lobo et al. 2016), biostratigraphical markers in marine deposits (Barron and Baldauf 1995), stratigraphic indicators for mineral and petroleum exploration (Shukla and Mohan 2012) or indicators of climate change (Fritz et al. 1991). Boulêtreau et al. (2006) suggested hydrological applications of diatoms, demonstrating hydro-geochemical controls on freshwater phytoplankton (drift) and benthic diatoms in rivers.

5.3.3 *Some Initial Work That Suggested Diatoms Could be Useful Hydrological Tracers*

The extraordinary variety of diatom species (estimates of 200,000 species; ≈10–200 μm) and their strong dependence on environmental conditions prevailing in their habitats preconfigured the original working hypothesis ideated by Pfister et al. (2009). They hypothesized that diatoms living predominantly in terrestrial habitats could be mobilized by overland flow during precipitation events. In case of the onset of overland flow connectivity within the hillslope-riparian zone-stream (HRS) continuum, aerial diatom frustules would have to be found in stream water samples collected during the rising and falling limbs of the storm hydrograph.

Pfister et al. (2009) carried out proof-of-concept work in the Attert River basin, where three locations had been selected for their distinct physiogeographic characteristics: (i) the forested Weierbach catchment (0.45 km^2), where bedrock geology is dominated by Devonian schist and phyllades and quartzite; (ii) the mainly forested Huewelerbach catchment (2.7 km^2), dominated by alternating layers of sandstone and marls; and (iii) the Attert River basin in Useldange (250 km^2), where bedrock is composed by schists, marls and sandstone. The experimental protocol essentially consisted of collecting water samples in the three locations before the rainfall event, during the rising limb of the hydrograph and at peak flow. Sample processing consisted in preparing microscopic slides for identifying and counting diatom species with a light microscope (Leica DM-RX). The Van Dam et al. (1994) diatom ecological classification system was used for distributing the identified diatom frustules among five classes of occurrence—each class corresponding to distinct habitat wetness conditions. Those five diatom classes aligned with five hydrological functional units (see Sect. 3.4), suggesting diatom's usefulness in tracing hydrological connectivity across the HRS continuum.

When used in the Attert River basin by Pfister et al. (2009), these sampling and analytical protocols revealed substantial increases in diatom frustules in stream water samples during the rising limb of two storm hydrographs. While aquatic species appeared to compose most of the sampled drift material, terrestrial species gradually increased in relative proportions along the rising limb of the storm hydrographs. Consequently, this exploratory work confirmed that terrestrial diatoms were flushed from their terrestrial habitats to the stream—thereby validating that diatoms are responsive to flow conditions and that they vary with the onset/cessation of surface hydrological connectivity.

5.3.4 Diatom Assemblages Across the Landscape

Prior to proposing terrestrial diatoms as a new tracer for hydrological process investigations, two major assumptions had to be tested. One must first verify that aquatic and terrestrial diatom communities can be discretized across the HRS continuum and between catchments with contrasted physiographic characteristics. The second feature that must be documented is that terrestrial diatoms are indeed flushed to the stream during precipitation events. Here these two fundamental assumptions are briefly discussed:

(i) Can aquatic and terrestrial diatom communities be discretized across the hillslope-riparian zone-stream continuum and between catchments with contrasted physiographic characteristics?

With a view to characterise terrestrial diatom habitats, Martínez-Carreras et al. (2015) investigated terrestrial diatom communities in the experimental forested Weierbach catchment from November 2010 to December 2011. They sampled various habitats, including epilithon, epipelon and stream water samples—eventually revealing 230 taxa. The number of diatom valves found in overland flow, or on hillslope litter, was low in comparison to sites with bryophytes. The largest diatom reservoirs were found in the riparian zones of the catchment. The sampling campaigns had revealed no substantial seasonal changes in the relative abundances of areal diatom species across the range of investigated habitats.

Through a statistical ordination (non-metric multi-dimensional scaling, NMDS) of several hundreds of diatom slides sampled in various experimental catchments located in the Attert River basin (Luxembourg), Pfister et al. (2017) were able to discern distinct spatial patterns in the organisation of diatom communities across hillslopes (mosses), riparian zones (mosses, soil, leaves) and streams (epilithon, epipelon/drift at baseflow/drift at storm events).

Pfister et al. (2017) equally found a large dissimilarity among diatom communities sampled on the riparian zone and hillslope areas of catchments with contrasted bedrock geologies (e.g., Weierbach (schists) and Huewelerbach (sandstone) catchments; Fig. 5.1). Of the approximately 400 identified diatom species, none were common to both geological bedrock types of the Weierbach and Huewelerbach catchments. Eventually, these findings supported the existence of highly contrasted communities—both within and between catchments.

(ii) Are terrestrial diatoms flushed to the stream during precipitation events?

For documenting a systematic flushing of terrestrial diatoms to the stream during precipitation events, stream water was sampled across all four seasons at baseflow,

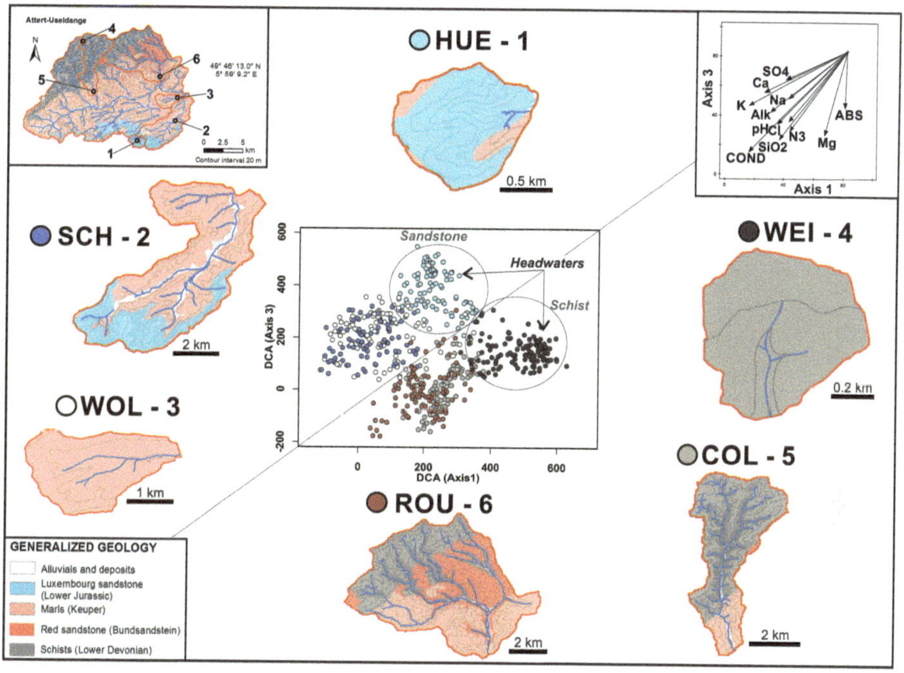

Fig. 5.1 Detrended correspondence analysis (DCA) ordination using diatom species data from six distinct geological units collected with ISCO automatic samplers (Teledyne Isco, Inc., USA) during rainfall events in the Attert River basin, Luxembourg. Each point represents a diatom species. Groups were defined using an indicator species analysis based on the subcatchments. Catchment abbreviations: *HUE* Upper Huewelerbach, *SCH* Schwebich, *WOL* Wollefsbach, *WEI* Weierbach, *COL* Colpach, *ROU* Roudbach. (Reprinted from Pfister et al. (2017) under creative common license: https://creativecommons.org/licenses/by-nc/4.0/)

as well as during the rising and falling limbs of storm hydrographs (Pfister et al. 2009; Pfister et al. 2017). The percentage of terrestrial diatoms in stream water samples was substantially higher during summer storm events (up to 30%) than during winter storm events (around 15%).

In the schistous and forested Weierbach catchment (0.47 km^2) hydrograph response is characterised by a threshold behaviour. Single peak responses to precipitation occur as long as catchment storage remains below a certain threshold. The nearly immediate response to precipitation is mainly generated by water quickly reaching the stream from nearby riparian zones. In autumn and winter, the gradual filling of catchment storage leads to a substantial increase in baseflow, alongside a so-called double-peak response to precipitation. While the nearly immediate response to precipitation is mainly generated by water quickly reaching the stream, the delayed secondary peak is triggered by contributions from both hillslopes and plateaus (Martínez-Carreras et al. 2016).

Sampling of terrestrial diatom frustules during these double-peak events have clearly shown increases in the percentage of terrestrial diatoms in stream water

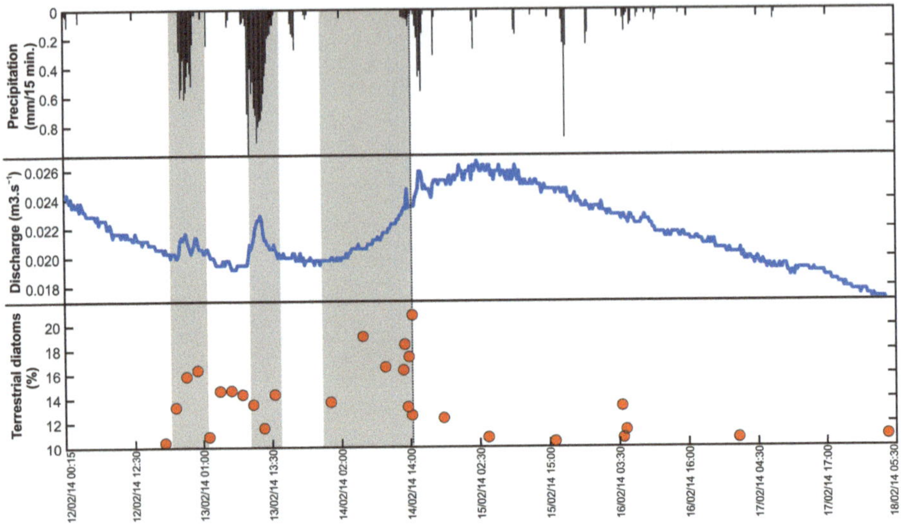

Fig. 5.2 Hydrograph response (blue line) to precipitation (black bars) in the schistous Weierbach catchment. Terrestrial diatom concentration [in %] in drift samples (red dots). Shaded areas: periods with rising hydrograph and increase in terrestrial diatom concentrations

mirroring the nearly instantaneous response to incident precipitation (Fig. 5.2). As shown by hydro-chemical and isotope tracers, these increases in discharge mainly correspond to contributions from the riparian zone of the Weierbach (Martínez-Carreras et al. 2015; Wrede et al. 2015). Dominated by rapid surface runoff, these responses mobilize substantial amounts of terrestrial diatoms that are flushed to the creek during precipitation events. Interestingly, during the delayed discharge peaks the percentages in terrestrial diatoms tend to gradually decrease. This decrease in terrestrial diatom percentages reflects the relative decrease in rapid surface runoff contributions against rising contributions triggered by sub-surface processes.

5.4 Discussion and Outlook

5.4.1 On Fundamental Assumptions of Terrestrial Diatom Reservoir Size, Distribution and Dynamics

While proof-of-concept work had demonstrated the potential for terrestrial diatoms to serve as a tracer of hydrological connectivity within the HRS continuum, a fundamental assumption still needed verification. If terrestrial diatoms were to be routinely used as a hydrological tracer, the perennial character of their reservoir would ultimately have to be verified. Likewise, their potential for tracing surface hydrological connectivity over larger scales (e.g., up to mesoscale) also requires a distribution of terrestrial diatom species across a wide range of terrestrial habitats.

The proof-of-concept work carried out in the Attert River basin also stressed the need for additional research into fundamental questions related to potential aerial diatom reservoir depletion, as well as vertical percolation and lateral subsurface transport of diatoms. In this context, Coles et al. (2016) tested the hypothesis that the supply of terrestrial diatoms is infinite during rainfall events. They carried out artificial rainfall experiments in selected riparian soil plots located in the Weierbach experimental catchment, simulating a 1- in 10-year rainfall event. At regular intervals they collected soil and overland flow samples in order to determine diatom population size and assemblages. As a result of the sprinkling experiment, the population size decreased from 96,100 to 27,200 diatoms per cm^2 (i.e., a decrease of 72%). The main finding of their study was that while the aerial diatoms were not of infinite supply, they nonetheless continued to be flushed throughout the entire simulated rainfall event. Hence, the artificial rainfall experiment supported the initial working hypothesis, identifying terrestrial diatoms as biological indicators of the onset/cessation of overland flow across the HRS continuum. In order to assess the potential for terrestrial diatoms to trace hydrological connectivity over larger scales (e.g., up to mesoscale), additional research into the spatial and temporal variability of diatom assemblages through heterogeneous landscapes was needed.

As a potential tracer for detecting hydrological surface connectivity, another major assumption stipulates that terrestrial diatoms are not percolating through soils. This assumption was verified by Tauro et al. (2016) via laboratory experiments with labelled frustules of purified diatomite and *Conticribra weissflogii* (Grunow) Stachura-Suchoples and D.M. Williams cultures with rhodamine 123 and 2-(4-pyridyl)-5[4-dimethylaminoethyl-aminocarbamoyl)-methoxy]phenyloxazole (PDMPO). During the soil column experiments, 3600 ml of water was applied for a maximum duration of 10 hours over soils covered with labelled diatoms. No fluorescent diatom frustules were eventually found in the percolated material collected at the bottom of the soil columns.

De Graaf et al. (2015) carried out another series of laboratory experiments, targeting diatom percolation (speed of percolation, speed of percolation over time and species distribution) through undisturbed soil columns. The three soil samples had been collected in three catchments (Wollefsbach, Huewelerbach, Weierbach)—with distinct geological bedrock settings (marls, sandstone and schists)—within the Attert River basin. Unlike for the previous laboratory experiment with dyed diatom frustules, artificial rainfall sprinkling caused *Pseudostaurosira* sp. and *Melosira* sp. (used as percolation proxies) to be rapidly transported through all three soil columns. Eventually, they followed breakthrough curves previously determined for rhodamine dye experiments. Macroporosity within the samples soil columns were suspected to have controlled the observed percolation of diatom frustules.

5.4.2 *Terrestrial Diatom Tracing of Hydrological Connectivity Across Scales*

Following up on promising proof-of-concept work related to terrestrial diatom reservoir extension, percolation and flushing during precipitation events, Klaus et al. (2015b) tested the potential for terrestrial diatoms to trace hydrological connectivity across scales. In a nested set-up of seven catchments (0.45–247 km^2) with clean and mixed physiographic characteristics, they assessed terrestrial diatom transport between their habitats and the stream network.

Focusing at a single sequence of rainfall events recorded in December 2011, they collected and analysed 149 samples from hillslope, riparian and in-stream habitats. They identified 421 different diatom species, among which 85 were terrestrial species. Across the range of catchments, the number of species was highly variable (e.g., 93 species found in the Weierbach catchment, 221 species found in the Colpach catchment). Whereas catchment size was not a controlling factor for species richness, the proportion of terrestrial diatoms in stream water clearly decreased with increasing catchment size. Throughout their investigated sequence of precipitation events, Klaus et al. (2015b) did not find any evidence for terrestrial diatom depletion—confirming the findings of Coles et al. (2016).

Ultimately, Klaus et al. (2015b) found terrestrial diatom species proper to distinct physiographic source areas to map across the entire set of seven nested catchments—documenting their potential for tracing hydrological connectivity across a wide range of catchment scales (up to 250 km^2).

5.4.3 *On the Need for a Better Understanding of Terrestrial Diatom Assemblage Composition Dynamics*

In contrast to species-environment relations of benthic diatom assemblages that are well documented for aquatic ecosystems, the composition dynamics of diatom assemblages in the water column of headwater lotic ecosystems remain poorly understood. We lack understanding on how diatom drift is interacting with autochthonous aquatic assemblages and communities from surrounding subaerial and terrestrial habitats (such as riparian zones and uplands). Diatoms are proven highly sensitive to alterations in the extent of impervious cover (IC) in catchments (surfaces impermeable to water, such as roads and buildings). Likewise, the percentages of riparian area and forest cover are likely to affect hydrologic, nutrient, and sediment regimes, which then impact diatom community physiognomy and taxa sensitive to nutrients and conductivity (Smucker et al. 2013). Thus, forest alteration has a direct impact on such freshwater communities.

Consequently, better understanding of diatom ecology (e.g., variables controlling drift composition and habitat interactions) and sampling protocols (Wu et al. 2016) is needed to advance tracer applications based on terrestrial diatoms. Given that hydrological systems largely control the spatial distribution of diatom species assemblages, the latter equally offer potential for identifying coherent and reproducible flow patterns—ultimately pointing towards underlying catchment behaviour and organising principles (Pfister et al. 2009; Wu et al. 2018). Just as new technology to efficiently analyse stable isotopes revolutionized tracer applications in hydrology, we expect that advancements in terrestrial diatom assessment will facilitate new discoveries of processes and applications.

5.5 Conclusion

Great progress has been made with the conventional tracer toolbox on water source, flow paths and transit times. But, these approaches are based on assumptions that are only rarely met. In order to partially overcome this issue, proof-of-concept work has documented the potential for terrestrial diatoms to serve as a tool for tracing the onset/cessation of hydrological connectivity within the HRS continuum across a wide range of scales and physiographic settings. Fundamental assumptions needed for further application of terrestrial diatoms in tracer hydrology have been verified—including terrestrial diatom reservoir sustainability and limited percolation of diatom frustules through soils.

Meanwhile, disentangling the origin and pathways of terrestrial diatoms remains challenging at larger scales, and thereby across more diverse substrates and flow conditions. Further unanswered questions relate to the potential impact of seasonal variations in meteorological conditions on terrestrial diatom assemblages, terrestrial diatom reservoir dynamics in space and across climate zones or terrestrial diatom reservoir dynamics (across different habitats and precipitation events of contrasting intensity, duration and magnitude).

Ongoing work is focusing on species assemblage and distribution modelling of terrestrial diatoms—as a prerequisite for testing the potential for individual species to serve as source area indicators in hydrological connectivity investigations. For this purpose, new terrestrial diatom classification schemes are urgently needed.

Acknowledgements Research on the potential for terrestrial diatoms to serve as hydrological tracers has been funded through the National Research Fund of Luxembourg (Grants C09/SR/14—BIGSTREAM & C12/SR/4018854—ECSTREAM). The authors would like to acknowledge Jeffrey J. McDonnell for providing comments on an earlier version of this chapter.

References

Ambroise B (2004) Variable 'active' versus 'contributing' areas or periods: a necessary distinction. Hydrol Process 18:1149–1155. https://doi.org/10.1002/hyp.5536

Barron JA, Baldauf JG (1995) Cenozoic marine diatom biostratigraphy and applications to paleoclimatology and paleoceanograhy. In: Blome CD et al (eds) (Convenors) Siliceous microfossils, paleontological society short courses in paleontology, vol 8, pp 107–118

Barthold FK, Tyralla C, Schneider K, Vaché KB, Frede HG, Breuer L (2011) How many tracers do we need for end member mixing analysis (EMMA)? A sensitivity analysis. Water Resour Res 47:W08519. https://doi.org/10.1029/2011WR010604

Benettin P, Velde Y, van der Zee S, Rinaldo A, Botter G (2013) Chloride circulation in a lowland catchment and the formulation of transport by travel time distributions. Water Resour Res 49:4619–4632. https://doi.org/10.1002/wrcr.20309

Bonell M, Barnes CJ, Grant CR, Howard A, Burns J (1999) High rainfall, response-dominated catchments: a comparative study of experiments in tropical northeast Queensland with temperate New Zealand. In: Kendall C, JJ MD (eds) Isotope tracers in catchment hydrology. Elsevier, Amsterdam, pp 347–390. https://doi.org/10.1016/B978-0-444-81546-0.50018-5

Boulêtreau S, Garabétian F, Sauvage S, Sánchez-Pérez JM (2006) Assessing the importance of a self-generated detachment process in river biofilm models. Freshw Biol 51:901–912. https://doi.org/10.1111/j.1365-2427.2006.01541.x

Bracken L, Croke J (2007) The concept of hydrological connectivity and its contribution to understanding runoff-dominated geomorphic systems. Hydrol Process 21:1749–1763. https://doi.org/10.1002/hyp.6313

Burns DA, McDonnell JJ, Hooper RP, Peters NE, Freer JE, Kendall C et al (2001) Quantifying contributions to storm runoff through end-member mixing analyses and hydrologic measurements at the Panola Mountain Research Watershed (Georgia, USA). Hydrol Process 15: 1903–1924. https://doi.org/10.1002/hyp.246

Burns DA (2002) Stormflow hydrograph separation based on isotopes: the thrill is gone – what's next? Hydrol Process 16:1515–1517. https://doi.org/10.1002/hyp.5008

Buttle JM, Peters DL (1997) Inferring hydrological processes in a temperate basin using isotopic a geochemical hydrograph separation: a re-evaluation. Hydrol Process 11:557–573. https://doi.org/10.1002/(SICI)1099-1085(199705)11:6<557::AID-HYP477>3.0.CO;2-Y

Coles AE, Wetzel CE, Martínez-Carreras N, Ector L, McDonnell JJ, Frentress J et al (2016) Diatom as a tracer of hydrological connectivity: are they supply limited? Ecohydrology 9:631–645. https://doi.org/10.1002/eco.1662

Coplen TB, Neiman PJ, White AB, Ralph FM (2015) Categorisation of northern California rainfall for periods with and without a radar brightband using stable isotopes and a novel automated precipitation collector. Tellus Ser B Chem Phys Meteorol 67:28574. https://doi.org/10.3402/tellusb.v67.28574

De Graaf L, Cammeraat E, Pfister L, Wetzel C, Klaus J, Hissler C (2015) Do diatoms percolate through soil and can they be used for tracing the origin of runoff? EGU General Assembly Conference Abstracts 2015 17:EGU2015-8961

Desilets SLE, Ferré TPA, Ekwurzel B (2008) Flash flood dynamics and composition in a semiarid mountain watershed. Water Resour Res 44:W12436. https://doi.org/10.1029/2007WR006159

Dunne T, Black RD (1970) Partial area contributions to storm runoff in a small New England watershed. Water Resour Res 6:1296–1311. https://doi.org/10.1029/WR006i005p01296

Dunne T, Moore TR, Taylor CH (1975) Recognition and prediction of runoff-producing areas in humid regions. Hydrol Sci Bull 20:305–327

Ector L, Rimet F (2005) Using bioindicators to assess rivers in Europe: an overview. In: Lek S, Scardi M, PFM V, Descy JP, Park YS (eds) Modelling community structure in freshwater ecosystems. Springer, Berlin/Heidelberg, pp 7–19. https://doi.org/10.1007/3-540-26894-4_2

Fischer BM, van Meerveld HI, Seibert J (2017) Spatial variability in the isotopic composition of rainfall in a small headwater catchment and its effect on hydrograph separation. J Hydrol 547:755–769. https://doi.org/10.1016/j.jhydrol.2017.01.045

Foussereau X, Graham WD, Akpoji GA, Destouni G, Rao PSC (2001) Solute transport through a heterogeneous coupled vadose-saturated zone system with temporally random rainfall. Water Resour Res 37:1577–1588. https://doi.org/10.1029/2000WR900389

Freeman MC, Pringle CM, Jackson CR (2007) Hydrologic connectivity and the contribution of stream headwaters to ecological integrity at regional scales. J Am Water Resour Assoc 43:5–14

Fritz SC, Juggins S, Battarbee RW, Engstrom DR (1991) Reconstruction of past changes in salinity and climate using a diatom-based transfer function. Nature 352:706–708
Harman CJ (2015) Time-variable transit time distributions and transport: theory and application to storage-dependent transport of chloride in a watershed. Water Resour Res 51:1–30. https://doi.org/10.1002/2014WR015707
Hewlett JD, Hibbert AR (1967) Factors affecting the response of small watersheds to precipitation in humid areas. In: Sopper WE, Lull HW (eds) International symposium on forest hydrology. Pergamon, Oxford, pp 275–290
Hooper RP, Christophersen N, Peters NE (1990) Modelling streamwater chemistry as a mixture of soil water end-members – an application to the Panola Mountain catchment, Georgia, USA. J Hydrol 116:321–343. https://doi.org/10.1016/0022-1694(90)90131-G
Horton RE (1933) The role of infiltration in the hydrologic cycle. Trans Am Geophys Union 14:446–460. https://doi.org/10.1029/TR014i001p00446
Ingraham NL, Caldwell EA, Verhagen BT (1999) Arid catchments. In: Kendall C, JJ MD (eds) Isotope tracers in catchment hydrology. Elsevier, Amsterdam, pp 435–465. https://doi.org/10.1016/B978-0-444-81546-0.50020-3
Kendall C, McDonnell JJ (1998) Isotope tracers in catchment hydrology. Elsevier Science Publishers BV, Amsterdam
Kirchner JW, Tetzlaff D, Soulsby C (2010) Comparing chloride and water isotopes as hydrological tracers in two Scottish catchments. Hydrol Process 24:1631–1645. https://doi.org/10.1002/hyp.7676
Kirchner JW (2016) Aggregation in environmental systems – part 1: seasonal tracer cycles quantify young water fractions, but not mean transit times, in spatially heterogeneous catchments. Hydrol Earth Syst Sci 20:279–297. https://doi.org/10.5194/hess-20-279-2016
Klaus J, McDonnell JJ (2013) Hydrograph separation using stable isotopes: review and evaluation. J Hydrol 505:47–64. https://doi.org/10.1016/j.jhydrol.2013.09.006
Klaus J, Chun KP, McGuire KJ, McDonnell JJ (2015a) Temporal dynamics of catchment transit times from stable isotope data. Water Resour Res 51:4208–4223. https://doi.org/10.1002/2014WR016247
Klaus J, Wetzel CE, Martínez-Carreras N, Ector L, Pfister L (2015b) A tracer to bridge the scales: on the value of diatoms for tracing fast flow path connectivity from headwaters to meso-scale catchments. Hydrol Process 29:5275–5289. https://doi.org/10.1002/hyp.10628
Leibundgut C, Malozewski P, Külls C (2009) Tracers in hydrology. Wiley, Chichester
Lobo EA, Heinrich CG, Schuch M, Wetzel CE, Ector L (2016) Diatoms as bioindicators in rivers. In: Neccho O (ed) River algae. Springer, Heidelberg, pp 245–271. https://doi.org/10.1007/978-3-319-31984-1_11
Maloszewski P, Rauert W, Stichler W, Herrmann A (1983) Application of flow models in an alpine catchment area using tritium and deuterium data. J Hydrol 66:319–330. https://doi.org/10.1016/0022-1694(83)90193-2
Maloszewski P, Zuber A (1993) Principles and practice of calibration and validation of mathematical models for the interpretation of environmental tracer data in aquifers. Adv Water Resour 16:173–190. https://doi.org/10.1016/0309-1708(93)90036-F
Martínez-Carreras N, Wetzel CE, Frentress J, Ector L, McDonnell JJ, Hoffmann L et al (2015) Hydrological connectivity inferred from diatom transport through the riparian-stream system. Hydrol Earth Syst Sci 19: 3133–3151. https://doi.org/10.5194/hess-19-3133-2015
Martínez-Carreras N, Hissler C, Gourdol L, Klaus J, Juilleret J, Iffly JF et al (2016) Storage controls on the generation of double peak hydrographs in a forested headwater catchment. J Hydrol 543:255–269. https://doi.org/10.1016/j.jhydrol.2016.10.004
McDonnell JJ (1990) A rationale for old water discharge through macropores in a steep, humid catchment. Water Resour Res 26:2821–2832. https://doi.org/10.1029/WR026i011p02821
McDonnell JJ, Sivapalan M, Vaché K, Dunn S, Grant G, Haggerty R et al (2007) Moving beyond heterogeneity and process complexity: a new vision for watershed hydrology. Water Resour Res 43:W07301. https://doi.org/10.1029/2006WR005467

McDonnell JJ, McGuire K, Aggarwal P, Beven KJ, Biondi D, Destouni G et al (2010) How old is streamwater? Open questions in catchment transit time conceptualization, modelling and analysis. Hydrol Process 24:1745–1754. https://doi.org/10.1002/hyp.7796

McGlynn BL, McDonnell JJ (2003) Quantifying the relative contributions of riparian and hillslope zones to catchment runoff. Water Resour Res 39:1310. https://doi.org/10.1029/2003WR002091

McGuire KJ, McDonnell JJ, Weiler M, Kendall C, McGlynn BL, Welker JM et al (2005) The role of topography on catchment-scale water residence time. Water Resour Res 41:W05002. https://doi.org/10.1029/2004WR003657

McGuire KJ, McDonnell JJ (2006) A review and evaluation of catchment transit time modeling. J Hydrol 330: 543–563. https://doi.org/10.1016/j.jhydrol.2006.04.020

Morgenstern U, Stewart MK, Stenger R (2010) Dating of streamwater using tritium in a post nuclear bomb pulse world: continuous variation of mean transit time with streamflow. Hydrol Earth Syst Sci 14:2289–2301. https://doi.org/10.5194/hess-14-2289-2010

Pfister L, McDonnell JJ, Wrede S, Hlúbiková D, Matgen P, Fenicia F et al (2009) The rivers are alive: on the potential for diatoms as a tracer of water source and hydrological connectivity. Hydrol Process 23:2841–2845. https://doi.org/10.1002/hyp.7426

Pfister L, Wetzel CE, Klaus J, Martínez-Carreras N, Antonelli M, Teuling A et al (2017) Terrestrial diatoms as tracers in catchment hydrology: a review. WIREs Water 4:e1241. https://doi.org/10.1002/wat2.1241

Pringle C (2003) The need for a more predictive understanding of hydrologic connectivity. Aquat Conserv 13:467–471. https://doi.org/10.1002/aqc.603

Richey DG, McDonnell JJ, Erbe MW, Hurd TM (1998) Hydrograph separations based on chemical and isotopic concentrations: a critical appraisal of published studies from New Zealand, North America and Europe. J Hydrol (NZ) 37:95–111

Rodriguez NB, McGuire KJ, Klaus J (2018) Time-varying storage–water age relationships in a catchment with a Mediterranean climate. Water Resour Res 54:3988–4008. https://doi.org/10.1029/2017WR021964

Shukla SK, Mohan R (2012) The contribution of diatoms to worldwide crude oil deposits. In: Gordon R, Seckbach J (eds) The science of algal fuels. Springer, Dordrecht, pp 355–382. https://doi.org/10.1007/978-94-007-5110-1_20

Sklash MG, Farvolden RN (1979) The role of groundwater in storm runoff. J Hydrol 43:45–65. https://doi.org/10.1016/0022-1694(79)90164-1

Smucker NJ, Detenbeck NE, Morrison AC (2013) Diatom responses to watershed development and potential moderating effects of near-stream forest and wetland cover. Freshwater Sci 32:230–249. https://doi.org/10.1899/11-171.1

Soulsby C, Tetzlaff D, Hrachowitz M (2009) Tracers and transit times: windows for viewing catchment scale storage? Hydrol Process 23:3503–3507. https://doi.org/10.1002/hyp.7501

Stewart MK, Mehlhorn J, Elliott S (2007) Hydrometric and natural tracer (oxygen-18, silica, tritium and sulphur hexafluoride) evidence for a dominant groundwater contribution to Pukemanga Stream, New Zealand. Hydrol Process 21:3340–3356. https://doi.org/10.1002/hyp.6557

Stewart MK, Thomas JT (2008) A conceptual model of flow to the Waikoropupu Springs, NW Nelson, New Zealand, based on hydrometric and tracer (^{18}O, Cl, ^{3}H and CFC) evidence. Hydrol Earth Syst Sci 12:1–19. https://doi.org/10.5194/hess-12-1-2008

Stewart MK, Morgenstern U, McDonnell JJ (2010) Truncation of stream residence time: how the use of stable isotopes has skewed our concept of streamwater age and origin. Hydrol Process 24:1646–1659. https://doi.org/10.1002/hyp.7576

Stewart MK, Fahey BD (2010) Runoff generating processes in adjacent tussock grassland and pine plantation catchments as indicated by mean transit time estimation using tritium. Hydrol Earth Syst Sci 14:1021–1032. https://doi.org/10.5194/hess-14-1021-2010

Stoermer EF, Smol JP (2010) The diatoms: applications for the environmental and earth sciences. Cambridge University Press, Cambridge

Tauro F, Martínez-Carreras N, Barnich F, Juilleret J, Wetzel CE, Ector L et al (2016) Diatom percolation through soils: a proof of concept laboratory experiment. Ecohydrology 9: 753–764. https://doi.org/10.1002/eco.1671

Tetzlaff D, Soulsby C, Bacon PJ, Youngson AF, Gibbins C, Malcolm IA (2007) Connectivity between landscapes and riverscapes – a unifying theme in integrating hydrology and ecology in catchment science. Hydrol Process 21:1385–1389. https://doi.org/10.1002/hyp.6701

Uhlenbrook S, Frey M, Leibundgut C, Maloszewski P (2002) Hydrograph separations in a mesoscale mountainous basin at event and seasonal timescales. Water Resour Res 38. https://doi.org/10.1029/2001WR000938

Uhlenbrook S, Hoeg S (2003) Quantifying uncertainties in tracer-based hydrograph separations: a case study for two-, three- and five-component hydrograph separations in a mountainous catchment. Hydrol Process 17:431–453. https://doi.org/10.1002/hyp.1134

Van Dam H, Mertens A, Sinkeldam J (1994) A coded checklist and ecological indicator values of freshwater diatoms from The Netherlands. Neth J Aquat Ecol 28:117–133. https://doi.org/10.1007/BF02334251

Wrede S, Fenicia F, Martínez-Carreras N, Juilleret J, Hissler C, Krein A et al (2015) Towards more systematic perceptual model development: a case study using 3 Luxembourgish catchments. Hydrol Process 29:2731–2750. https://doi.org/10.1002/hyp.10393

Wu N, Faber C, Sun X, Qu Y, Wang C, Ivetic S et al (2016) Importance of sampling frequency when collecting diatoms. Sci Rep 6:36950. https://doi.org/10.1038/srep36950

Wu N, Faber C, Ulrich U, Fohrer N (2018) Diatoms as an indicator for tile drainage flow in a German lowland catchment. Environ Sci Eur 30. https://doi.org/10.1186/s12302-018-0133-5

Chapter 6
Lessons in New Measurement Technologies: From Instrumenting Trees to the Trans-African Hydrometeorological Observatory

J. S. Selker, F. Selker, R. Llamas, A. Kruger, J. Niemeier, M. R. Abou Najm, N. van de Giesen, R. Hut, T. van Emmerik, J. W. Lane, D. E. Rupp, H. Lintz, R. D. Stewart, and K. McCulloh

6.1 Introduction and Context

Despite the increasing demands on global natural resources, many observation systems are in significant decline (Fekete et al. 2012). This seems to be at odds with the emergence of less-expensive sensors, more ubiquitous access to wireless data transfer, and improved solar energy generation and storage. We consider two case studies here: one involving instrument development and the other involving operational environmental observation, illustrating both the opportunity and the challenges of making ecological observations leveraging these emerging opportunities.

Ecological processes are characterized by complexity, heterogeneity, and rapid change. Further, ecological systems aggressively fill niches, including orifices of measurement instruments. These features indicate that monitoring of ecological processes can benefit from techniques that are inexpensive enough to replicate in many locations, which take informative data at high frequency, and are robust to environmental insults and weathering.

J. S. Selker (✉)
Department of Biological and Ecological Engineering, Oregon State University, Corvallis, OR, USA
e-mail: John.Selker@oregonstate.edu

F. Selker
SelkerMetrics, Portland, OR, USA

R. Llamas · A. Kruger
Department of Electrical and Computer Engineering and IIHR-Hydroscience & Engineering, University of Iowa, Iowa City, IA, USA

D. F. Levia et al. (eds.), *Forest-Water Interactions*, Ecological Studies 240,
https://doi.org/10.1007/978-3-030-26086-6_6

While satellites and other remote sensing technologies (e.g., Lidar, hyperspectral, thermal infrared, optical structure from motion) are critical to contemporary observation, on-the-ground observations are still fundamental to many environmental measurement goals and will be the focus of this chapter. One of the costliest aspects of on-the-ground environmental sensing is putting people into the field to install, maintain, download, and retrieve sensing systems. Thus, when evaluating the budgetary aspects of a sensing system we must consider not only the purchase, but rather the lifetime cost of each measurement. Estimation of these costs can be challenging in the case of emerging methods as they intrinsically lack a track record by which expected maintenance can be determined.

To make this concrete we recount a recent experience in Africa of the Trans-African Hydrometeorological Observatory (TAHMO). A wireless firmware update to TAHMO stations scattered across 20 countries resulted in installation of a faulty system that drained many of the station batteries, rendering them impossible to repair without visiting each station. This unanticipated event alone increased the 10-year lifetime cost of these stations by approximately 10%, and more importantly, doubled the rate of expenditure on station operations from the budgeted value at a time when the system was most vulnerable to financial shocks. As it turns out, this event followed on the heels of a need to replace an interior three-dimensional printed part that failed in most stations; a sonic anemometer which had a code error that required in-person repair; and other "early generation" defects. Taken together, these situations fundamentally altered the economic framework for the system, increasing our costs by over 50%. While the details of these issues may not have been possible to anticipate, we do think that adoption of emerging technologies should be anticipated to require several in-person visits to resolve emerging issues, and a factor of two multiplier on anticipated maintenance costs would not be unreasonable. One

J. Niemeier
IIHR-Hydroscience & Engineering, The University of Iowa, Iowa City, IA, USA

M. R. A. Najm
University of California Davis, Davis, CA, USA

N. van de Giesen · R. Hut · T. van Emmerik
Faculty of Civil Engineering and Geosciences, Delft University of Technology, Delft, Holland

J. W. Lane
USGS Office of Geophysics, Storrs, CT, USA

D. E. Rupp · H. Lintz
Oregon Climate Change Research Institute, College of Earth, Ocean, and Atmospheric Sciences, Oregon State University, Corvallis, OR, USA

R. D. Stewart
Department of Crop & Soil Environmental Sciences, Virginia Tech, Blacksburg, VA, USA

K. McCulloh
Department of Botany, University of Wisconsin-Madison, Madison, WI, USA

should also think carefully about the launch of new instruments, being sure to also leave time to debug the system prior to expecting full performance. Being in a hurry tends to be expensive. Laboratory and close-by deployments have many advantages in preparing for the installation of novel instrumentation.

While terms such as "transformational" and "revolutionary" are touted as laudable goals, incremental advancement is recommended as the most likely to produce successful observations. One should view environmental sensing systems as chains of linked technologies, including the sensing element, the power system, the enclosure system, the data recording system, the communication system, and the data management system. Beyond the core technology, each of these elements has complexity with respect to the environmental challenges, embedded code, weather-related factors, etc. Thus, each novel aspect has elevated potential for failure due to lack of experience in the field, and lack of previous integration into the observation chain. From this perspective, one should be wary of "stacking" novelty, introducing multiple unproven elements in this chain. This also reflects the fact that introduction of one new element can reveal previously unseen limitations and defects in systems that were previously found to be reliable. Clearly all systems must be validated in their complete functional form prior to deployment as part of a measurement campaign.

We therefore stress the difference between instrument development and operational environmental observation. Both activities are central to the advancement of ecological knowledge, but we argue that they are best differentiated as independent tasks. In this chapter, we will review examples of each of these activities. Specifically, we will describe advances in instrumentation to monitor aboveground tree biomass to illustrate the tremendous opportunities of new sensing methods, and we will review the progress of the TAHMO program as it leverages the combination of newly available instrumentation with the global growth of GPRS wireless communication.

6.2 Aboveground Biomass

Measurement of change in aboveground tree biomass or weight through time is important to the management of ecosystems, agriculture, and the biosphere. Tree mass measurements can estimate dry matter, nutrients, rain and snow interception, water content, transpiration, productivity, ecosystem products, and services such as carbon sequestration, timber yield and quality, and agricultural capital. Short-term change in tree mass can also offer insights into hydraulic strategy, growth trajectories, and the timing of biological events like vegetative bud break, fruit production, and flowering. However, it is a challenge to nondestructively measure tree mass through time.

Aside from putting a tree in a weighing lysimeter, the only direct nondestructive method presented in the literature is the calibration of mass to stem compression, which was developed to measure rainfall interception on tree leaves (Friesen et al.

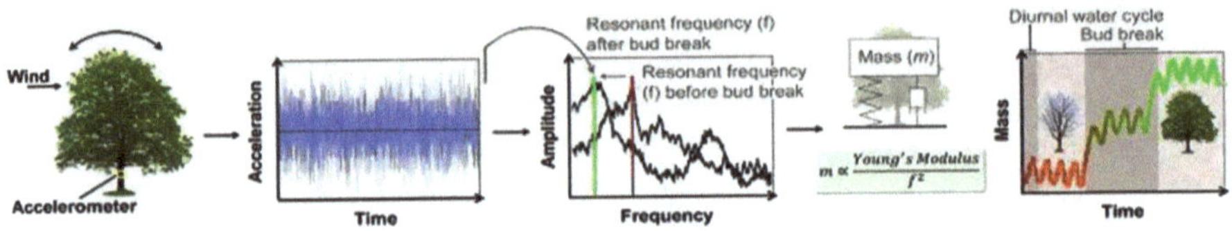

Fig. 6.1 Conceptual diagram of method using tree acceleration to estimate biomass change. A data-logging accelerometer is placed on a tree to collect data on tree acceleration in the wind through time (left panel). The resonant frequency of the tree is derived from acceleration data using signal processing (second panel), and the resonant frequency forms the basis for deriving tree mass, assuming the tree is a simple damped oscillator (middle panels). Time series of mass estimates for a tree can reveal biological traits and adaptive events like the timing of bud break caused by changes in mass (right panel)

2015). Mass may be indirectly estimated by measuring key dimensions of the trees such as diameter and height and then applying a biomass equation (Kloeppel et al. 2017). However, employment of biomass equations does not facilitate measurement of transient changes such as water loss due to transpiration. Tree harvest and weighing are required for the validation of biomass equations. While harvest plays an important role in validating biomass equations, alternative methods of estimating and measuring tree mass are needed to assess dynamic change and improve certainty of estimates.

The resonant frequency is an intrinsic property of a vibrating system largely independent of extrinsic applied forces, so long as displacements are small enough to stay in the linear Young's modulus regime. The resonant frequency depends on the physical properties of the system including mass, material properties, and geometry. For example, the concept that the resonant frequency measured from wind-induced oscillations can estimate mass has been implemented with rain gauges to capture accurately minor changes in mass due to precipitation and evaporation to and from the gauge (Stewart et al. 2012). Here, we show that accelerometers can effectively measure changes in tree mass including changes from diurnal water flux (see Fig. 6.1 for conceptual diagram of approach). We also demonstrate the detection of leaf-out using a lower sampling rate over a longer time interval.

We report on two experiments to demonstrate proof of concept that accelerometers can be used to measure biomass changes for individual trees. We placed an accelerometer on the trunk of an Oregon white oak (*Quercus garryana* Dougl.) before and after leaf-out to determine mass change from leaf out and rainfall interception (Fig. 6.2). We also placed an accelerometer on, and true Time Domain Reflectometer

Fig. 6.2 (continued) the number of points in the smoothing window; (**b**) change in resonance suggesting 4 mm interception before and after a 6.6 mm rainfall event June 13, 2011. Total power is a function of wind strength during data collection, which varied by day and time, so overall relative heights of curves are not directly comparable; (**c**) shift in resonance due to addition of 71 kg calibration weight

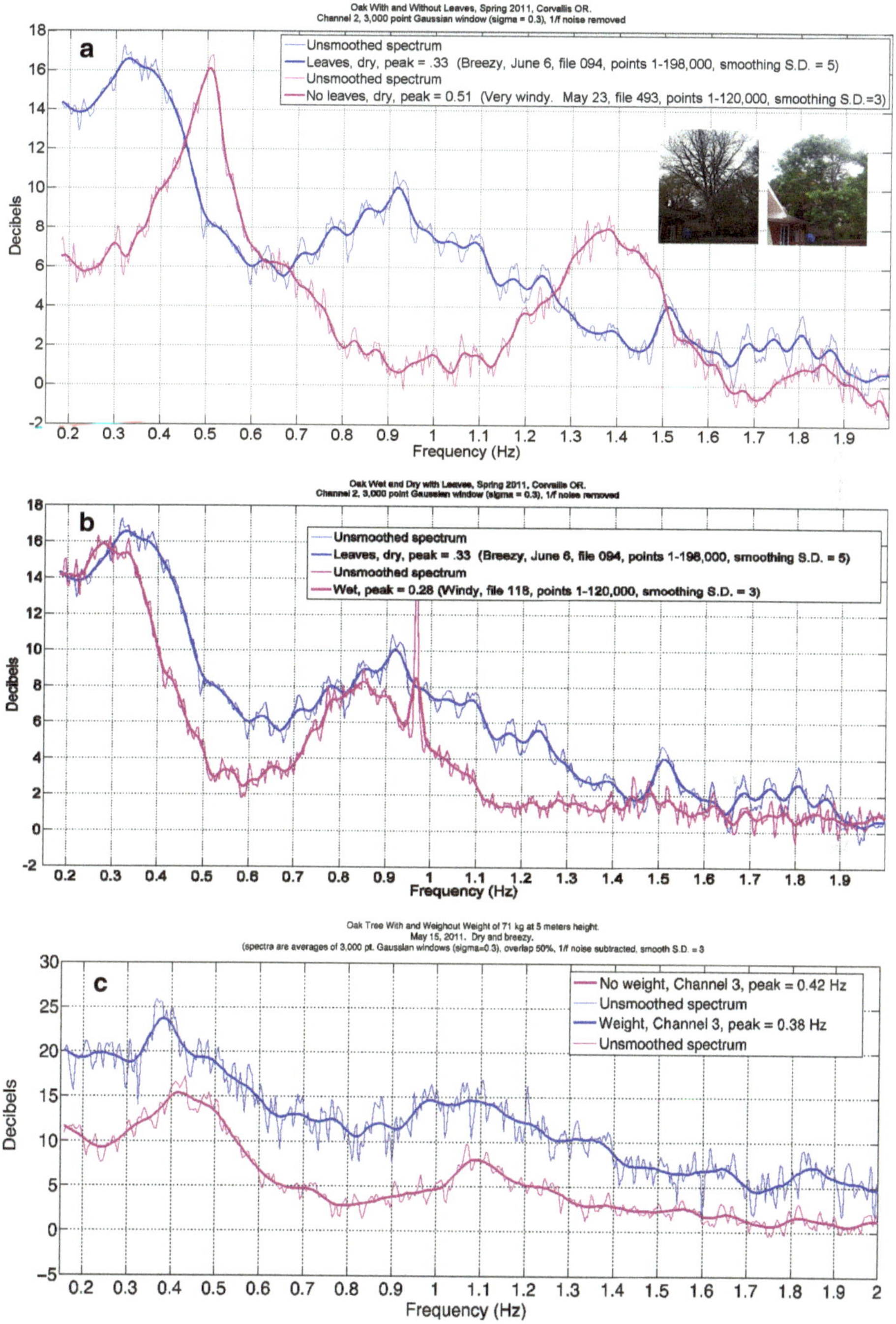

Fig. 6.2 (**a**) Power spectra derived for the white oak tree before (magenta) and after (blue) leaf-out on May 23 and June 6, 2011, respectively. The dominant resonant frequency is indicated with a vertical line for each date. Photos show the tree before and after leaf-out. "Smoothing S.D." refers to

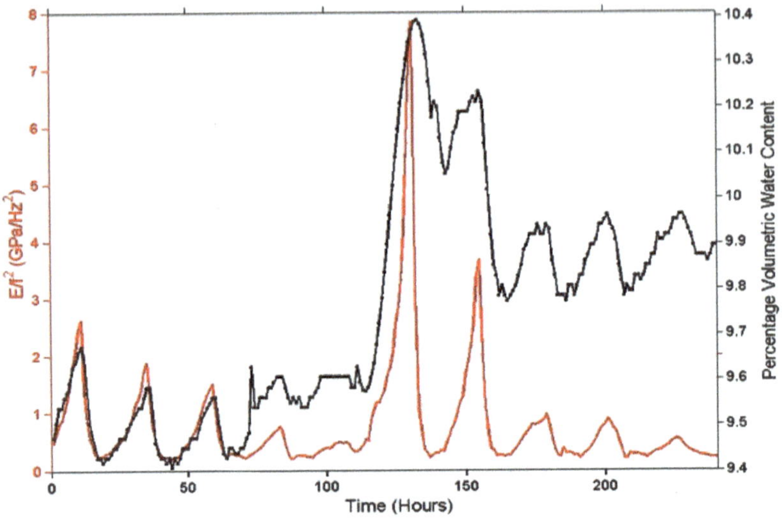

Fig. 6.3 (Left-axis, red) Time series of estimates of the Young's modulus (*E*) divided by the square of the resonant frequency (f^2) (the metric which theory would suggest to be proportional to mass) of a sugar maple. The resonant frequency *f* was derived from hourly measurements of tree acceleration (see Appendix for details on derivation of *f* and *E*). (Right-axis, black) Percentage volumetric water content measured hourly in the main trunk of the same tree from a TDR probe. The time series spans a 10-day period with rain falling around 120 h. The estimates of mass made from resonant frequency vary in synchrony with diurnal volumetric water content, and both peak during a rainstorm suggesting rainfall interception, while the tree stem appeared to retain moisture in great amount than the overall tree mass, as one might expect due to rapid loss of wetness on the leaves via evaporation

(TDR) probes in, the trunk of a sugar maple (*Acer saccharum* Marsh.) to determine if the accelerometer can measure diurnal change in tree water content (Fig. 6.3).

Assuming a whole tree functions as a standard damped oscillator, (Sellier and Fourcand 2005; Bruchert and Gardiner 2006), we estimated the hourly change in mass as a function of resonant frequency (see Appendix, Fig. 6.6). Standard damped oscillator behavior indicates that system mass is proportional to Young's modulus and inversely proportional to the square of the resonant frequency. In other words, Young's modulus (*E*) divided by the square of the resonant frequency (f^2) is a direct metric of (changes in) mass. In the first experiment, we sought to measure both the mass change of bud break (Fig. 6.2a) and rainfall interception (Fig. 6.2b). As predicted, the dominant resonant frequency of the oak tree decreased in each case, suggesting a 4 mm rainfall interception and more than doubling of aboveground mass after leaf-out. In the same tree, a 71 kg mass was placed in the tree to calibrate the mass-resonance relationship (Fig. 6.2c). A second experiment showed that the metric, E/f^2, changed in synchrony with tree water content following a diurnal cycle for a 10-day measurement period (Fig. 6.3). The metric also responded quickly to a rain event, showing a rapid, transient response to the rain. The resonant frequency did not increase in exact synchrony with changes in volumetric water content but did follow similar patterns. This illustrates that, for example, the mass change due to canopy interception immediately following a rainfall (a short-lived condition to which the accelerometer is sensitive) was distinct from the longer duration but small change in stored water in the trunk revealed by the TDR.

In each experiment, resonant frequency fell with events expected to increase tree mass. We did not directly measure tree mass, so we cannot rule out changes in tree modulus and damping as sources of changed resonance, but we verified that data acquired by accelerometers can measure changes in tree resonance, and results are suggestive that this reflects tree biomass. The placement of a calibration mass in the oak study did allow direct computation of relative changes in resonant frequency, and thus estimated mass change. Further study of the optimal placement of calibration mass (e.g., high in the canopy or on the main stem) would be required to have known levels of confidence in mass quantification.

Our results encourage application of tree accelerometry to measure change in mass. For example, this method can contribute to observation of phenology by providing a method that can be implemented broadly to investigate processes such as hydraulic strategy and the timing in biological events that elicit mass (Houle et al. 2010; Neale and Kremer 2011).

The use of tree accelerometers for applications that require quantification of absolute mass merits further research. Additional work is needed to improve our physical understanding of trees as oscillating systems to confirm the relationship between resonant frequency and mass (Bruchert et al. 2003; Spatz et al. 2007; Rodriguez et al. 2008). For example, the effect of damping on observed frequencies is an important consideration; trees have efficient damping properties of self-similar systems (Spatz et al. 2007; Rodriguez et al. 2008). Tree branches, when excised from the main tree, have similar resonant frequencies to the tree itself (Spatz et al. 2007). Additionally, physical and theoretical works on fractal resonance reveal that objects with fractal perimeters exhibit local modes of resonance, and this may have relevance to trees (Sapoval 1989; Sapoval and Gobron 1993; Hobiki et al. 1996; Rodriguez et al. 2008). Here, we assumed the effect of damping to be small given our theoretical model. Changes in resonance and mass were consistent with what may be expected if trees operate as simple damped oscillators using an instrument costing about US $100.

The use of accelerometers on trees can yield insight in fields including ecology, hydrology, and climate change mitigation. In hydrology, there is a paucity of practicable methods to measure dynamic mass interception of precipitation (rain and snow), as well as the loss of this mass through sublimation and evaporation. Given that interception can account for up to 20% of the incident precipitation, this gap in understanding is critical to advancement of hydrologic processes in many catchments (Xiao 2000). Considerable work has also supported the importance that interception can have on the water budget of plants because of direct absorption through leaves and/or bark (Stone 1957; Simonin, et al. 2009), yet the current methods to estimate this input are time-consuming and often require keeping plants in a controlled environment such as a greenhouse. Additionally, cost-effective biomass measurement can reduce uncertainty in carbon balance calculations. Current methods of estimation of live carbon content of trees from biomass equations can introduce 20–40% of uncertainty in biomass estimates (Melson et al. 2011). The magnitude of this uncertainty challenges our capacity to meet accuracy and precision standards required for carbon accounting protocols (Melson et al. 2011). Forests can function as carbon sinks (Luyssaert et al. 2008), and international programs dedicated to harnessing this potential to mitigate climate change lack cost-effective methods in carbon accounting (Fry 2011).

Having shown that resonant frequency can be revealing of changes in aboveground canopy mass, we note that accelerometers are not the only way to measure tree oscillation to derive resonance. For example, the use of computer analysis of video images might be used to estimate resonant frequencies. This could be more cost-effective to implement than small accelerometers if seeking to analyze many trees simultaneously.

The uncertainty in the dynamic change in aboveground mass of growing trees is great and impedes accurate prediction of this important element of the global carbon cycle (Clough et al. 2016). Our results suggest that measurement of tree resonance could provide a calibrated method to estimate this term, with potential utility to hydrologists, plant physiologists, and environmental modelers.

6.3 Climate Observations in Under-Resourced Areas

As earth's resources are stressed to provide for the ever-expanding population, anticipating and strategically responding to climatic dynamics is of increasing value. While satellites have vastly increased the extent of detailed observation, these data are often highly uncertain. For example, daily rainfall estimates from satellite data over Africa have proven to be biased by over 100% (Fig. 6.4). Thus, on-the-ground observations of key parameters remains essential. In Africa, the density of climate observations has fallen by more than tenfold in the last 60 years (Fig. 6.5).

In this context we observe that African cellular phone (GPRS) coverage is now comprehensive in villages, meaning that local coverage is available within 20 km of most locations across the continent. With GPRS, an automatic weather station can be in continuous bidirectional contact with the World Wide Web at all times. This provides for technician-free operation of stations so long as local hosts provide basic cleaning and protection.

Cost models are also much reduced due to the development of reliable and inexpensive all-in-one weather stations. The station employed by TAHMO integrates observation of rainfall, temperature, humidity, wind speed and direction, barometric pressure, solar radiation, lightning, GPS location, and XYZ acceleration. Additionally, 5-min resolution data is both logged locally (with 3-month data capacity) and sent via the GPRS hourly. The station is self-powered via a small solar panel. The station has no moving parts, using a sonic anemometer and a

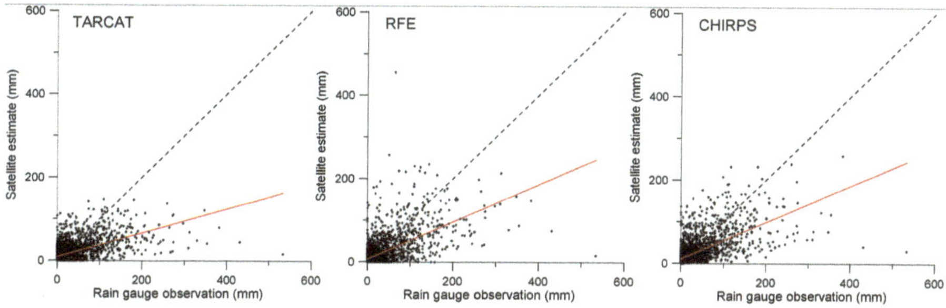

Fig. 6.4 Plots of satellite estimated daily rainfall compared to on-the-ground observations from Mozambique for currently leading models of satellite estimation of rainfall. The estimates were on average off by a factor of two in absolute event magnitude. (Reprinted from Toté et al. (2015) under creative common license: https://creativecommons.org/licenses/by/4.0/)

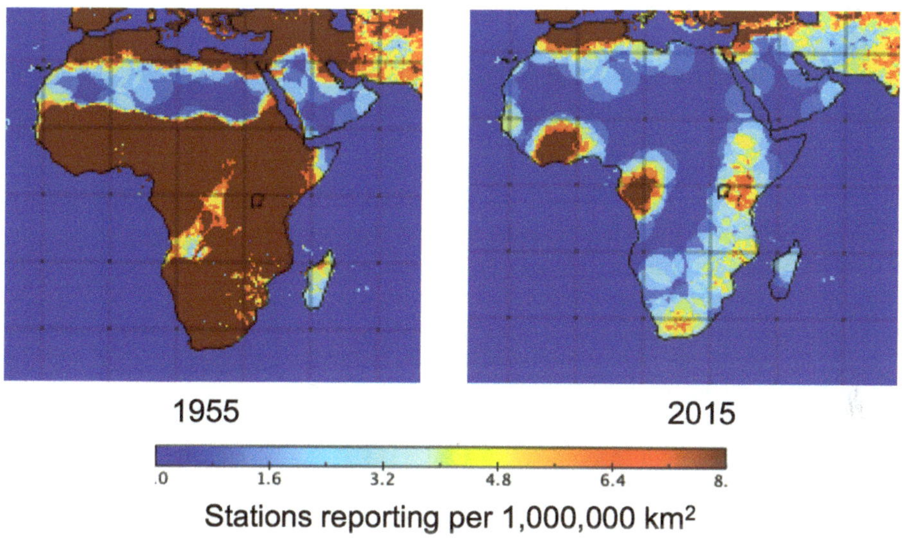

Fig. 6.5 The density of weather stations reporting from Africa from 1955 to 2015 dropped from over one station per 100,000 km^{-2} to below one station per million km^{-2} over this period

drip-counting rain gauge. This further improves station performance (for instance, the sonic anemometer is far more sensitive than would be a mechanical device, and has an expanded range of accurate reading, while the drip-counting approach has approximately 10 times the volumetric resolution compared to the standard tipping bucket mechanism) and, moreover, dramatically increases station robustness. Consistent instrumentation at all sites, mass-produced all-in-one sensor organization and bi-directional GPRS allows for dramatically reduced-cost operational models compared to traditional methods of climate and weather observation.

Today the TAHMO network has over 500 operating stations that span 21 African countries, as well as stations in demonstration sites in Costa Rica, Guatemala, the USA, and Europe. The goal of the organization is to install 20,000 stations that span the African continent. The challenges to completing the TAHMO network have been both predictable and surprising. It was understood that working across the massive and diverse continent would be challenging. From extreme desert conditions to rainforests with over 10 m of rain per year, the climatic conditions test the extremes of heat, water, and dust. For TAHMO to be successful, the hardware needed to be the most robust ever created. At the same time, we need to work with every government, and gain official permission to place observational hardware in their countries. This requires written agreements with empowered officials. Finally, we needed to have top-flight technical team able to service equipment and social aspects of the program in every corner of the continent.

TAHMO has a zero tolerance for participation in corruption, which to our great relief, has to this point presented no obstacle to our work. Still, getting meteorology offices to collaborate with TAHMO has often taken several years per country. While from the outside it would seem that the offer of no-cost data streams would be immediately compelling, in fact meteorology offices often are protective of the rights to make meteorological observations, seeming to fear that they might become redundant if other organizations take on this task. Due to the combined effects of dramatic reductions in on-the-ground observations, and the inaccessible complexity of data-assimilating numerical climate models, it is difficult for individual country teams in Africa to produce predictions that are as reliable or long-range as those available on people's smartphones. That said, the vast majority of citizens of the African continent do not have access to extended weather predictions, so for the foreseeable future the state climate offices are needed to make actionable weather predictions available to the largely agricultural communities.

Most surprisingly was the difficulty of developing the needed hardware and software. Working with leading teams in the USA and Europe we found that the specified levels of performance and reliability in the one-piece station took over 5 years to achieve in a mass-produced design. At the same time, the database system for TAHMO has taken over 2 years to develop. Our data quality control and assurance software has also gone into production data processing much more slowly than we had envisioned.

6.4 Conclusions

Improved accuracy and spatial resolution in environmental observation have emerged in the past 10 years that leverage the parallel developments in sensor technology, wireless communications, and three-dimensional fabrication. These tools offer two lines of advancement: sufficient detail in observations to test new theory and ubiquity of observation so the entire globe could be continuously monitored. These advances allow for scientifically based management of the earth's ecosystem. Advanced knowledge of droughts, flooding, and disease-breeding conditions will allow humanity to improve the consistency of harvests as well as assuring that critical habitat conditions are maintained. Our hope is that the presented models and technologies will be adopted by the environmental sensing community.

Acknowledgements The authors wish to thank the Plant Genome Research Program and the National Science Foundation for funding H.L., A.K., J.N., and R.L. on Award # 1238246, and J.S. on Awards # EAR 0930061 and #EAR 1551483 that contributed to this work.

Appendix: Additional Details on Methods

Accelerometer, Power Supply, Memory We used the X16-1C 3-axis USB accelerometer with 15-bit resolution made by Gulf Coast Data Concepts, LLC. This is a low cost and low power device. The USB-Accelerometer acts as a USB mass storage device with a removable Micro SD memory card (here used with a 2 GB card).

Experiments We performed the experiment relating tree water content to resonance on a sugar maple tree in a residential area in Cedar Rapids, Iowa. The tree dimensions are approximately 12 m high and 5 m wide. The accelerometer was glued in a waterproof enclosure constructed from PVC pipe fittings. The accelerometer was tightly strapped to the tree at 4.6 m high on the main stem. It was located immediately below the point where the main stem divided into multiple branches. We performed the oak tree experiment on a white oak tree in Corvallis, Oregon, with the same type of accelerometer mounted 5.5 m above the ground. The tree is approximately 8.2 m high and 5.3 m wide at the widest part of the tree. The accelerometer was encased in a waterproof bottle stuffed tightly with cotton and strapped to the tree at approximately one half the height of the tree. Acceleration was measured for windy periods before and after leaf-out.

Accelerometer Configuration and Data Collection *For the sugar maple experiment*, we configured the accelerometer with a sampling rate of 1.6 kHz. While the commercial specifications for that accelerometer recommend usage at several hundred Hz, we contacted the manufacturer and they said it can be used up to 1.6 kHz. We measured three-axis acceleration set to the most sensitive acceleration range of 2 g recorded the first 10 min of each hour. All three axes were combined into a single acceleration vector for the analysis. A 10-minute sample window was chosen to conserve battery life and sensor memory; 528 h of acceleration data were collected, though not all data were usable. Data files created when the temperature dropped below freezing were discarded. Additionally, some data files exceeded size limits of the device or were missing lines of data due to write-errors of the accelerometer. These issues were largely resolved by the end of the experimental effort. *For the oak tree experiment*, the accelerometer sampling rate was set to 10 Hz. This sampling rate differed from the above-described experiment because the objective differed. In this case, we were looking for the resonant frequency before and after leaf-out rather than the diurnal change in resonant frequency, which otherwise would require more battery power and a higher sampling rate for shorter periods (on the hour). Data were collected during windy weather before leaf-out (or 178,200 data points; about 5 h) and after leaf-out (or 12,000 points; about 20 min). One of the three acceleration axes was found to yield the highest magnitude data both before and after leaf-out, and those data were used. While we used data from windy periods for the oak tree experiment, all data were used for the maple experiment regardless of the presence of wind.

TDR Probe To verify that the accelerometer was able to capture the diurnal cycle in water content of the maple, we installed a Time Domain Reflectometer (model CS616, Campbell Scientific, Inc.) adjacent to the accelerometer. The volumetric water content of the tree was recorded every 15 min. The probe was inserted into holes that were drilled 30.5 cm into the main trunk on the tree. These measurements were averaged by the hour.

Data processing and spectral analysis to derive resonant frequency *For the maple tree experiment*, in order to find the dominant resonant frequency induced by the wind, we used the Welch periodogram (50% overlap) and the Multiple Signal Classification (MUSIC) (Schmidt 1986) spectral estimation methods, as implemented in the pmusic function of the software MATLAB R2012a. The spectra showed similar peaks and overall congruence with spectral estimation provided by Fourier analysis that employs smooth windowing. *For the oak tree experiment*, we used a Fourier spectral analysis to generate the spectra for the leaf-out experiment in Fig. 6.2. The results provided by the Fourier analysis were corroborated when checked against MUSIC.

Derivation of mass from frequency We assumed that the tree acts as a simple mass-spring oscillator and the relationship between mass and frequency is:

$$m \propto \frac{E}{f^2}$$

where m is an estimate of mass of the tree, f is dominant resonant frequency of tree, and E is Young's modulus. We assumed that we may neglect the effect of damping given the following. Consider the damping ratio as $\zeta = c/(2m\omega_n)$ and the damped natural frequency as $\omega_d = \omega_n\ (1\text{-}\zeta^2)^{1/2}$, where ω_n is the resonant frequency without damping in rad per second (OMEGA $= 2 * \text{PI} * f$), and c is the damping coefficient. As c increases, ζ increases. As ζ increases, damping will affect the damped resonant frequency $\omega_n\ (1\text{-}\zeta^2)^{1/2}$. Thus, the damping value, c, must be very large to substantially affect the resonant frequency. Current estimates of c for trees are around 0.5 (Kooreman 2013), which would yield a very small effect on the resonant frequency compared to the effect of mass.

We derived E individually for each time point by employing eq. 4-3 on page 4–34 of Green et al. (1999), where E is a function of: volumetric water content (measured), the modulus of elasticity of green wood of sugar maple from Table 4-3a in Green

Fig. 6.6 Accelerometer and waterproof enclosure installed on sugar maple tree (top left), X-16C accelerometer with U.S. quarter shown for scale (bottom left), and full-sized maple tree employed in these experiments (right). The TDR probe is adjacent to the accelerometer and above it approximately 0.35 m, although not pictured

et al. (1999), the modulus of elasticity of sugar maple at 12% water content from Table 4-3a in Green et al. (1999), and the point at which the moisture-strength line of green wood meets that of dry wood assumed to be 25 following Green et al. (1999), page 4–34. We discovered that accounting for variation in volumetric water content through time mattered little in the estimate of E and E/f^2 (Fig. 6.6).

References

Bruchert F, Gardiner B (2006) The effect of wind exposure on the tree aerial architecture and biomechanics of Sitka spruce (Picea sitchensis, Pinaceae). Am J Bot 93:1512–1521. https://doi.org/10.3732/ajb.93.10.1512

Bruchert F, Speck O, Spatz H (2003) Oscillations of plants' stems and their damping: theory and experimentation. Philos Trans R Soc London, Ser B 358:1487–1492. https://doi.org/10.1098/rstb.2003.1348

Clough BJ, Matthew RB, Domke GM (2016) Quantifying allometric model uncertainty for plot-level live tree biomass stocks with a data-driven, hierarchical framework. For Ecol Manag 372:175–188. https://doi.org/10.1016/j.foreco.2016.04.001

Fekete BM, Looser U, Pietroniro A et al (2012) Rationale for monitoring discharge on the ground. J Hydrometeorol 13:1977–1986. https://doi.org/10.1175/JHM-D-11-0126.1

Friesen JJ, Lundquist D, Van Stan JT (2015) Evolution of forest precipitation water storage measurement methods. Hydrol Process 29:2504–2520. https://doi.org/10.1002/hyp.10376

Fry BP (2011) Community forest monitoring in REDD+: the 'M' in MRV? Environ Sci Policy 14:181–187. https://doi.org/10.1016/j.envsci.2010.12.004

Green DW, Winandy JE, Kretschmann DE (1999) Chapter 4: Mechanical properties of wood. In: Wood handbook – wood as an engineering material, Gen Tech Rep FPL-GTR-113. USDA, US Forest Service, Madison, pp 4-1–4-45

Hobiki Y, Yakubo K, Nakayama T (1996) Spectral characteristics in resonators with fractal boundaries. Phys Rev E 54:1997–2004. https://doi.org/10.1103/PhysRevE.54.1997

Houle D, Govindaraju DR, Omholt S (2010) Phenomics: the next challenge. Nat Rev Genet 11:855–866. https://doi.org/10.1038/nrg2897

Kloeppel B, Harmon ME, Fahey TJ (2017) Estimating aboveground net primary productivity in forest-dominated ecosystems. In: Fahey T, Eds KA (eds) Principles and standards for measuring net primary production. Oxford University Press, New York, pp 63–81

Kooreman BD (2013) Measuring weight fluctuations in trees based on natural frequency. MSc thesis, TU Delft, 56 p

Luyssaert S, Schulze ED, Börner A et al (2008) Old-growth forests as global carbon sink. Nature 455.7210:213–215. https://doi.org/10.1038/nature07276

Melson SL, Harmon ME, Fried JS et al (2011) Estimates of live-tree carbon stores in the Pacific Northwest are sen-sitive to model selection. Carbon Balance Manag 6:1–16. https://doi.org/10.1186/1750-0680-6-2

Neale D, Kremer A (2011) Forest tree genomics: growing resources and applications. Nat Rev Genet 12:111–122. https://doi.org/10.1038/nrg2931

Rodriguez M, de Langre E, Moulia B (2008) A scaling law for the effects of architecture and allometry on tree vibration modes suggests a biological tuning to modal compartmentalization. Am J Bot 95:1523–1537. https://doi.org/10.3732/ajb.0800161

Sapoval B (1989) Experimental observation of local modes in fractal drums. Physica D 38:296–298. https://doi.org/10.1016/0167-2789(89)90209-1

Sapoval B, Gobron T (1993) Vibrations of strongly irregular or fractal resonators. Phys Rev E 47:3013. https://doi.org/10.1103/PhysRevE.47.3013

Schmidt RO (1986) Multiple emitter location and signal parameter estimation. IEEE Trans Antennas Propag AP34:276–280. https://doi.org/10.1109/TAP.1986.1143830

Sellier D, Fourcand T (2005) A mechanical analysis of the relationship between free oscillations of Pinus pinaster Ait. saplings and their aerial architecture. J Exp Bot 56:1563–1573. https://doi.org/10.1093/jxb/eril151

Simonin KA, Santiago LS, Dawson TE (2009) Fog interception by Sequoia sempervirens (D. Don) crowns decouples physiology from soil water deficit. Plant Cell Environ 32:882–892. https://doi.org/10.1111/j.1365-3040.2009.01967.x

Spatz H, Bruchert F, Pfisterer J (2007) Multiple resonance damping or how do trees escape dangerously large oscillations? Am J Bot 94:1603–1611. https://doi.org/10.3732/ajb.94.10.1603

Stewart RD, Hut R, Rupp DE et al (2012) A resonating rainfall and evaporation recorder. Water Resour Res 48:W08601. https://doi.org/10.1029/2011WR011529

Stone EC (1957) Dew as an ecological factor: II. The effect of artificial dew on the survival of Pinus Ponderosa and associated species. Ecology 38:414–422. https://doi.org/10.2307/1929884

Toté C, Patricio D, Boogaard H et al (2015) Evaluation of satellite rainfall estimates for drought and flood monitoring in Mozambique. Remote Sens 7:1758–1776. https://doi.org/10.3390/rs70201758

Xiao Q, McPherson EG, Ustin SL et al (2000) Winter rainfall interception by two mature open-grown trees in Davis, California. Hydrol Process 14:763–784. https://doi.org/10.1002/(SICI)1099-1085(200003)14:4<763::AID-HYP971>3.0.CO;2-7

Part II
Analyzing and Synthesizing Ecohydrological Patterns and Functional Relationships with Statistics

Chapter 7
Primary Steps in Analyzing Data: Tasks and Tools for a Systematic Data Exploration

Martin Zwanzig, Robert Schlicht, Nico Frischbier, and Uta Berger

7.1 Introduction

The question of the appropriateness of a particular statistical procedure for a given dataset is fundamental. This applies to all forms of statistical analysis, including, for example, the classical frequentist's approach of 'null hypothesis significance testing' (NHST), conventional regression analysis using ordinary least squares and information theoretic approaches such as model selection based on Akaike's information criterion (Freckleton 2011). A systematic exploratory data analysis (EDA) can help to answer this question. On the one hand, any analysis requires that certain assumptions hold and manifold aspects such as independence of observations, the shape of their distribution, the influence of outliers, homogeneity of residual variances and the correlation of explanatory variables have been clearly addressed (Quinn and Keough 2002; Zuur et al. 2010). If this is ignored, the conclusions of an analysis could likely be distorted. On the other hand, if violations have been detected, data transformation can help to overcome consequential restrictions in the application of a statistical method such as ANOVA or linear mixed effects model, although this should be treated with caution as explained in Sect. 7.8. The sequence of steps for a systematic data exploration is not fixed and depends on the statistical techniques and the specific dataset.

Electronic Supplementary Material The online version of this chapter (https://doi.org/10.1007/978-3-030-26086-6_7) contains supplementary material, which is available to authorized users.

M. Zwanzig (✉) · R. Schlicht · U. Berger
Institute of Forest Growth and Forest Computer Sciences, Technische Universität Dresden, Tharandt, Germany
e-mail: martin.zwanzig@tu-dresden.de

N. Frischbier
Forestry Research and Competence Centre, ThüringenForst, Gotha, Germany

D. F. Levia et al. (eds.), *Forest-Water Interactions*, Ecological Studies 240,
https://doi.org/10.1007/978-3-030-26086-6_7

Any scientific study usually begins with hypotheses grounded on the understanding of the system. Data is then collected in a way that permits to test these hypotheses by experimental manipulations or observations in natural systems. This means data are collected with regard to the hypotheses and not just randomly. The same principle applies to data exploration. Specifically, hypotheses generated by data exploration and not determined a priori cannot be statistically tested using the same data, but require new independent data providing information not already included in the formulation of the hypotheses. Note that this problem does not apply to hypotheses constructed by theoretical reasoning such as those based on fundamental ecological principles or on the understanding of the system behaviour according to previous empirical results.

Moreover, it is highly recommended to think about the data and the way it can be analyzed to answer a certain research question before it is collected. According to Dytham (2006), for a successful data analysis, it can be helpful to analyze some artificial data beforehand. Such dummy data should be generated according to the design of the experiment and its expected outcome. If the analysis fails, the hypotheses and the underlying design of experiment can be adjusted. The relevance of such an exercise is to see if the sampling design and the roughly estimated effects fit to each other and if data structured in the way it is planned to be sampled can be analyzed appropriately with the intended statistical method. It is of course not easy to estimate effect sizes and the associated variability before real data are collected. For that purpose, conducting preliminary experiments can be appropriate. It additionally demonstrates the effort that is required to take samples, which can help to estimate realizable sampling frequencies (e.g., Frischbier 2012).

This chapter presents a guideline for data exploration and data transformation, illustrated by some data from a case study by Frischbier (2012), who performed spatially explicit, single-rain-event-based, replicated measurements of gross precipitation and throughfall during and outside the vegetation period within a mixed spruce-beech forest stand. We will explain and demonstrate how to

- ensure independence of observations (or consider dependence structures),
- detect outliers (and, for example, remove them by transformations),
- check distributions (and transform them to normality),
- examine variances (and ensure homoscedasticity),
- explore relationships (and linearize them),
- find interactions and collinearity among explanatory variables (and account for them).

For interested readers, the example data and the full R code that was used to create the figures presented here are provided as online electronic supplementary material to this chapter (see Extras Springer online). This material offers a hands-on introduction to basic *R* functions as well as advanced data visualization techniques, using recent packages ggplot2 (Wickham 2016) and plotly (Sievert 2018). It demonstrates, for instance, how to create plots that are interactive, a feature that significantly enhances data exploration, since you can access all the information about a single data point in the figure that you created, just by moving your mouse to the respective point in the plot. We hope that this commented code can serve as an outline and cheat sheet for own surveys.

7.1.1 Example: Throughfall Dataset

Let us now describe the data, referred to as 'throughfall dataset' below, that our exploratory data analysis is based on. It represents a subset of data from Frischbier (2012), parts of which have also been used for an analysis presented in Frischbier and Wagner (2015). They tested the hypothesis that lateral water translocation within the crown can be determined from simultaneous records of precipitation at defined measurement points below and above the canopy by taking single-tree characteristics such as species and crown width into account. Consequently, the sampling design was optimized with regard to the observation of throughfall at varying distance classes (balanced sampling of each distance class for each tree species).

The throughfall dataset contains temporally repeated observations of the amount of precipitation water collected by funnels that are either placed in forest gap positions or under the canopy of Norway spruce (*Picea abies* (L.) Karst.) or European beech (*Fagus sylvatica* L.) trees, which were further characterized by tree characteristics.

Specifically, throughfall was measured with 175 funnels placed along transect lines between single trees or gaps. Funnels were either placed in canopy gap positions ($n = 15$) or under beech ($n = 76$), spruce ($n = 75$) or mixed canopy ($n = 9$). Each funnel is additionally characterized by its canopy position, that is, the distance to the stem of its dominating tree in units of the crown edge's distance. These trees, 17 of which European beech and 19 Norway spruce trees, were further characterized by tree height, diameter at breast height, crown radius and crown length. In total 50 rain events were recorded during and outside the vegetation period, 33 of which fulfilled a list of requirements of the idea of ideal rain events: (i) They originated from a single precipitation event, here defined as a period of precipitation preceded by at least 4 h of no rain; (ii) wind speeds did not exceed 6 m s^{-1}; (iii) synchronized readings of gross precipitation from five reference funnels on an adjacent meadow were almost identical, i.e., with a coefficient of variation less than 6%; and (iv) gross precipitation (0–49.0 mm) was consistent at canopy gap positions inside the stand and in the adjacent meadow. This aimed at allowing to record crown dripping and simultaneously minimizing evaporation losses. More technical details on the data collection can be found in Frischbier and Wagner (2015) and in Frischbier (2012).

7.1.2 *Outline of Questions to Address by Data Exploration*

A proper data exploration should provide an answer to the following ten questions, which will also be used as a roadmap for the sections that follow in this chapter:

Understanding the structure of the dataset	*1. How is the dataset structured* (e.g., *grouping of observations)?*
	2. Are there missing values (variables without data for single observations)?
	3. Is the dataset balanced with respect to its structure (e.g., *same number of observations per group)?*
Locating measurement points	*4. Where in the study area are the measurement points? What might this mean for spatial analysis?*
Detecting dependence among observations	*5. Is there an inherent dependency due to spatial or temporal correlations* (e.g., *repeated measures)?*
Identifying outliers	*6. Are there outliers in the response or explanatory variables?*
Checking for normality and homogeneity	*7. Can we assume normality and homogeneity for the response variable (and of its residuals)?*
Exploring the relationships of variables	*8. Are the relationships to the response variable linear or non-linear?*
	9. Are there interactions and a sufficient number of observations to allow for them (e.g., *group-specific outcomes of a statistical analysis)?*
	10. How strong are the explanatory variables correlated with each other (collinearity)?

Finally, the section 'transformations as potential actions' describes approaches to deal with outliers, non-normality, heterogeneity of variances and non-linear relations.

7.2 Understanding the Structure of the Dataset

Datasets are usually based on the design of an experiment aimed at testing one or more hypotheses, as mentioned in the introduction. In our example, it is hypothesized that the amount of throughfall is influenced by gross precipitation and single-tree characteristics of the canopy tree such as species and crown width.

The first step, after measurements have been performed and data was stored in databases[1], is to get an overview of the given data. The description in Sect. 7.1.1 already provided some details, but this is not yet a complete solution to our question *1. How is the dataset structured?* At the beginning we have to clarify which variables are given, which scale they have and which of these is the response and which are the explanatory variables. One way to address these points is to create a tabular

[1] *See the DataONE primer on data management for an overview on how to effectively manage data: https://www.dataone.org/sites/all/documents/DataONE_BP_Primer_020212.pdf*

Table 7.1 Dataset variables and their scale

Object	Variable	Scale	Explanation
Funnels	Throughfall	Metric	Precipitation amount measured by funnel in mm or equivalently as l m^{-2} (response variable)
	Gross precipitation	Metric	Mean precipitation in adjacent meadow (= total precipitation of a single rainfall event) in mm
	Observation date	Interval	Time point of measurement (date of single precipitation event)
	Period/state of canopy tree	Nominal	2 categories: Leafed or non-leafed or NA•
	Funnel ID	Nominal	Identity number of funnels
	Funnel coordinates	Interval	x- and y-coordinates of funnel position according to the sample grid in m
	Canopy type	Nominal	4 categories: Beech, spruce, mixed or none (canopy gap) as a measure of forest structure
	Absolute distance to canopy tree stem	Metric	Funnel distance to the canopy tree stem in m or NA•
	Relative position under canopy tree crown	Ratio	Funnel distance to the canopy tree stem in units of the crown edge's distance or NA•
Trees	Tree ID	Nominal	Identity number of canopy tree
	Tree coordinates	Interval	*x*- and *y*-coordinates of tree stem position according to the sample grid in m
	Tree species	Nominal	European beech or Norway spruce
	Tree height	Metric	Absolute height of canopy tree in m
	Tree diameter at breast height	Metric	Stem width of canopy tree in cm
	Crown radius	Metric	Crown radius of canopy tree in m
	Crown beginning	Metric	Lowest height of the canopy tree's crown in m

•Funnels with mixed canopy or in gap position are marked with NA (not available), since they do not have a dominating tree that this variable refers to

overview listing this information and a brief explanation, as shown in Table 7.1. Here we see that the throughfall dataset contains various variables of different scale, representing properties either at the funnel or the tree level. A number of quantitative variables (i.e., on a metric scale) represent the results of field measurements. We consider throughfall as the response variable, since we aim to understand the variability of throughfall as affected by tree species, crown section and forest structure. In fact, we will focus on the proportion of throughfall on gross precipitation, as this provides some advantages that will be addressed later. According to the particular research questions, it is also possible to consider more than one variable as a response. All other variables were recorded to examine their influence on throughfall, i.e., they serve as explanatory variables.

The next step is to understand how observations can be grouped and if this involves some nesting. In general, nested and crossed data structures are distinguished, since their structure can influence the interpretation of statistical models

(Schielzeth and Nakagawa 2013). A nested structure is given when each level of a categorical explanatory variable is uniquely associated with only one level at the higher-level categorical variable. Observing, for example, birch and poplar trees in the mountains and beech and spruce trees in the lowland, we have a nested structure. It is claimed that the interaction variance inflates the variances of the main effects in such cases (Schielzeth and Nakagawa 2013). In a crossed design, for example, when all tree species are observed in all landscapes, both components (tree species and landscapes) can be distinguished.

The qualitative variables (nominal scale) distinguish between crossed and nested designs, since they determine the categories the observations belong to, such as 'beech canopy' or 'leafed period'. Such categorical variables are also called 'factors' and their categories 'levels' or 'groups'. Ecological data often contain more than one factor, as the throughfall dataset illustrates: Observations were made in different periods at various precipitation events at funnels placed in a certain canopy position, whereas some funnels are associated with the same specific tree. Since each canopy level was observed in each period level, the throughfall dataset has a crossed design. Considering that the funnels are associated with a single tree that belongs either to beech or to spruce, this also involves some nesting. It often helps to illustrate the data structure graphically (e.g., by a sketch) as shown in Fig.7.1, instead of explaining the sampling design in words.

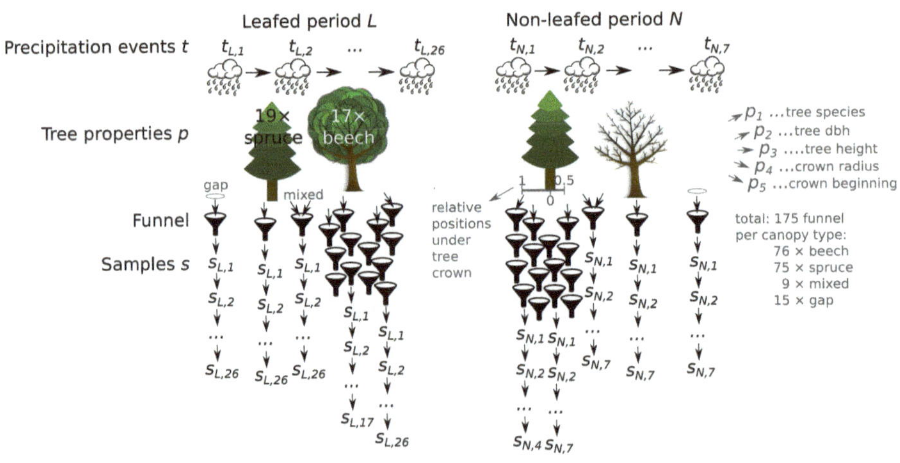

Fig. 7.1 Sketch of the data structure. The throughfall dataset consists in total of 33 ideal rainfall events (out of 50 measured events) t, 26 in the leafed period (L) and 7 in the non-leafed period (N). Samples of each rainfall event (s) were taken at funnels that were either placed in a gap position ($n = 15$) or under beech ($n = 76$), spruce ($n = 75$) or mixed canopy ($n = 9$). Single spruce trees dominated 1–15 funnels and single beech trees 1–13 funnels. The individual positions under beech or spruce canopy tree crowns were optimized with respect to the frequency of funnels belonging to single relative distance classes. A few funnels recorded only 17 out of 26 rain events in the leafed period or only 4 out of 7 events in the non-leafed period. Each tree was characterized by measuring five different properties p onetime. In addition, the coordinates of the funnels and trees have been recorded, although not specified here

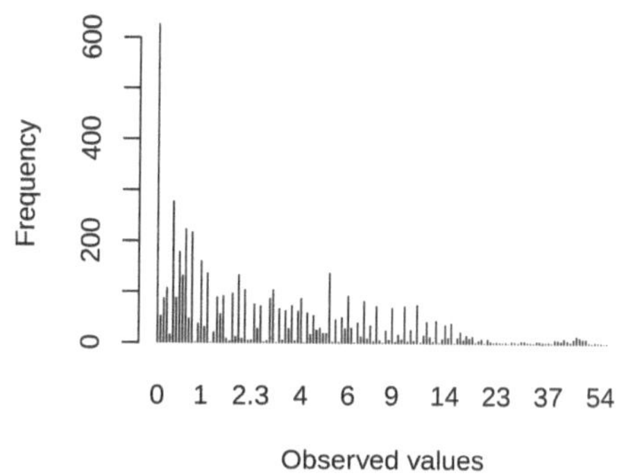

Fig. 7.2 Frequency plot of throughfall [mm] observations. This figure summarizes the number of observations with the same throughfall, e.g., more than 600 zero values. Note that distances at the *x*-axis are not equidistant (as in a histogram)

A proper understanding of the structure of the data is an important step, since we have to know what the data actually represents. If certain features of the structure of the data are not considered, we likely build the wrong model and end up with wrong conclusions. For example, if the period is omitted as a factor in the throughfall dataset, we might not be able to detect or explain that the periodical increase in throughfall proportion of beech trees is due to defoliation.

After a conceptual representation of the data structure has been set up, the raw data can be checked for consistency regarding our question 2. *Are there missing values?* This means, the data table might be incomplete at certain observations, since the data for single variables is missing (as represented by 'NA', '9999', etc.). In this case, the knowledge acquired about the data structure can help to understand the causes, as it often reflects variables that could not be measured for each object (e.g., tree species or tree age) or at any time (e.g., only in winter). Besides, missing data might reflect real erroneous measurements, for example, caused by data loggers that failed. From a practical point of view, it is often easiest to treat missing data as if it had not been measured at all. This is permissible if the process leading to the missing data is stochastically independent of the data itself. Otherwise, the absence of data can be treated as an additional value that can be investigated, for example, by including the processes that cause the absence of data in the model. The throughfall dataset contains, for instance, more than 600 zeros for the response variable throughfall (Fig. 7.2). This represents highly valuable information, since it originates from rainfall events with precipitation amounts lower than the canopy storage capacity (at some funnel positions; see Chap. 8).

Another part of data exploration at this stage is to check question 3. *Is the dataset balanced with respect to its structure?* Clearly, this can only be appropriately addressed when the (nesting) structure of categorical variables has been resolved as we did before. Figure 7.1 already mentioned the numbers of observations that belong to the same groups. This was determined with some convenient functions in *R*, namely, 'table' (incl. 'with') and 'tapply', which can be used to count the number of observations that belong to a certain group (as defined by a combination of categories for the various categorical variables, e.g., beech canopy in leafed period). The dataset is unbalanced according to the different number of observations at single funnels (e.g., in leafed period, $n = 17$ to $n = 26$), but the number of funnels that belong either to spruce ($n = 75$) or beech trees ($n = 76$) is nearly identical. Nonetheless, if it comes to the individual trees, we see that some are only measured by one and other with up to 15 funnels. Such an imbalance is not present regarding to the relative position under the canopy tree crown, because Frischbier and Wagner (2015), who collected the data, focused on small-scale differences under tree canopies and optimized the sampling design with respect to a balanced frequency of observations at different distance classes.

Another important point is that the number of precipitation events differs greatly between the 'leafed' ($n = 26$) and 'non-leafed' period ($n = 7$). Although most statistical methods can theoretically cope with such imbalance, it can lead to less efficient numerics and various computational difficulties. It is therefore important to know whether a particular type of analysis can (or has to) be chosen in order to answer research questions that focus on differences between such groups of observations. The analysis presented by Frischbier and Wagner (2015) was, for instance, restricted to data from the leafed period.

7.3 Locating Measurement Points

An illustration of the position of measurement points is an important part of documentation, often required for reports or publications. It reveals spatial aspects of the data structure that might otherwise be less obvious or just buried and leads to question *4. Where in the study area are the measurement points?* The location of funnels (Fig. 7.3) demonstrates that measurements are not randomly placed. Instead, they form transects, which sometimes intersect. In our case study, this means that changes in throughfall due to canopy type may be confounding with transects, since transects often represent only a single canopy type. Similar problems are given for many studies. Of course, the situation would be worse if our data would represent two study sites, one in a beech and another in a spruce stand since then it would not even be possible to distinguish the effects of site and tree species from one another. This again demonstrates the importance of keeping in mind the focus of the study. The design of the throughfall dataset is, for instance, ideal to focus on the effect of canopy type with respect to distance classes, as Frischbier and Wagner (2015) did.

7.4 Detecting Dependence Among Observations

Although spatial dependence structures can be included in the model (see Sect. 7.4.3), a crucial assumption of most statistical tests is independence of observations (Hurlbert 1984). This topic should already be carefully addressed at the very first beginning, ideally before a study or experiment is conducted. Indeed, as dependence can arise in an uncountable number of ways and statistical procedures can only detect specific forms of dependence, independence is preferably justified

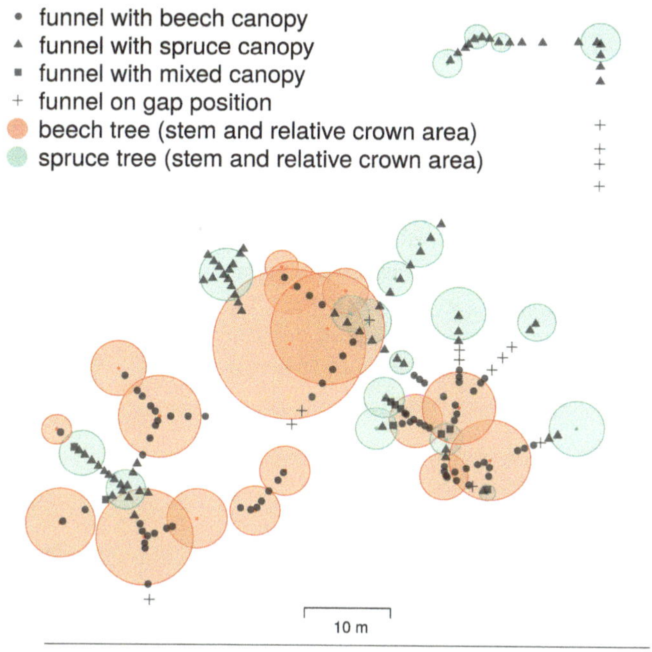

Fig. 7.3 Location of measurement points. Tree crown areas of characterized trees in relation to their crown dimensions and position of funnels. Note that crowns of higher trees overlap those of smaller ones, but crown areas are simplified and in reality neither symmetric nor centred at the stem

based on theoretical knowledge of the structure of the experiment. A dependence structure in the data arises from the design of the study or experiment. When samples are collected from the same tree, they are likely not independent. Strictly speaking, this might only represent an issue as long as there are substantial differences between the observations at single trees that cannot be sufficiently explained by other explanatory variables. For instance, it might be argued that independence of observations is given, if one has considered canopy type (tree species) as well as the relative position of funnels under a tree's canopy, since the remaining effects of individual tree variability might be too small to be important. Nevertheless, any suspicion of dependency should be repeatedly monitored, which means that also the residuals of a model are checked for dependencies, e.g., by plotting them against tree identity. Mixed effects models are seen as the superior method in such cases (Zuur and Ieno 2015), since they can include the individual variability as random effects, and they can account for temporal dependence structures. Nevertheless, the setup of such models requires some caution, particularly regarding the role of the random effects, which may in some situations appear as random slopes instead of random intercepts (Schielzeth and Forstmeier 2009). This topic is addressed more in detail in Chap. 8 of this volume.

Apart from dependencies related to the nested structure of the data, observations should always be checked with respect to question 5. *Is there an inherent dependency due to spatial or temporal correlations?* Clearly, this is only an issue if measurement points have been repeatedly sampled or if they represent spatially explicit observations. Sometimes observations are carried out at different locations at any time, for instance using the ‘roving gauge (= funnel) method’ for throughfall measurements (Carlyle-Moses et al. 2014). This can be more advantageous for some research questions. However, our exemplary throughfall dataset was collected with the ‘stationary gauge method’, referring to repeated observations at the same measurement point (= funnel). Therefore it has to be tested if observations at funnels very close to each other share the same characteristics. The same holds for observations at time points very close to each other. A common approach is to plot the response variable versus time and spatial coordinates, whereas any clear pattern indicates a dependency (Zuur and Ieno 2015).

7.4.1 Temporal Dependence

Temporal patterns can be revealed through time series analysis. An introduction to this huge topic is, for example, given by Wilks (2006). Variables representing repeated temporal observations are called autocorrelated if they correlate with their own past and future. This is usually computed as Pearson’s r for lagged data pairs, representing a shift of the series by one or more units of time. Such statistical dependence of time series is also called persistence.

Atmospheric data often shows a positive serial dependence, describing the fact that large values tend to be followed by relatively large values, and small values by relatively small. The typical course of temperature over the day is autocorrelated, since the values gradually increase from night to day and vice versa. For precipitation this is much more complicated, since precipitation is not continuously present, but represents a series of discrete events. So firstly rain falls or not. It is possible to estimate the conditional probability for the number of days with precipitation that follow days with precipitation and to compare it to the conditional probability for the number of days with precipitation that follow days without precipitation. If both probabilities differ, a serial dependency is given, which typically means that there is a tendency of precipitation to occur in runs (followed by a series of dry days).

A more formal way is provided by the autocorrelation function, which can be used when observations are regularly spaced in time (Zuur et al. 2010). However, the temporal distance of observations is often irregular. Our throughfall dataset represents, for instance, observations that belong to single rainfall events, which are not evenly spaced in time (Fig. 7.4a). In order to be able to apply the autocorrelation function to such data, it is possible to define rainfall events as a time unit. This makes it possible to assess whether the throughfall observations are correlated to those measured at the previous and the following rainfall event at the same place (refers to a lag of 1 ‘rainfall event’) or whether the observations correlate within an increased number of rainfall events (refers to lags >1). Figure 7.4b presents a typical plot of the autocorrelation function for the time series of a chosen single funnel (ID 105). A lag

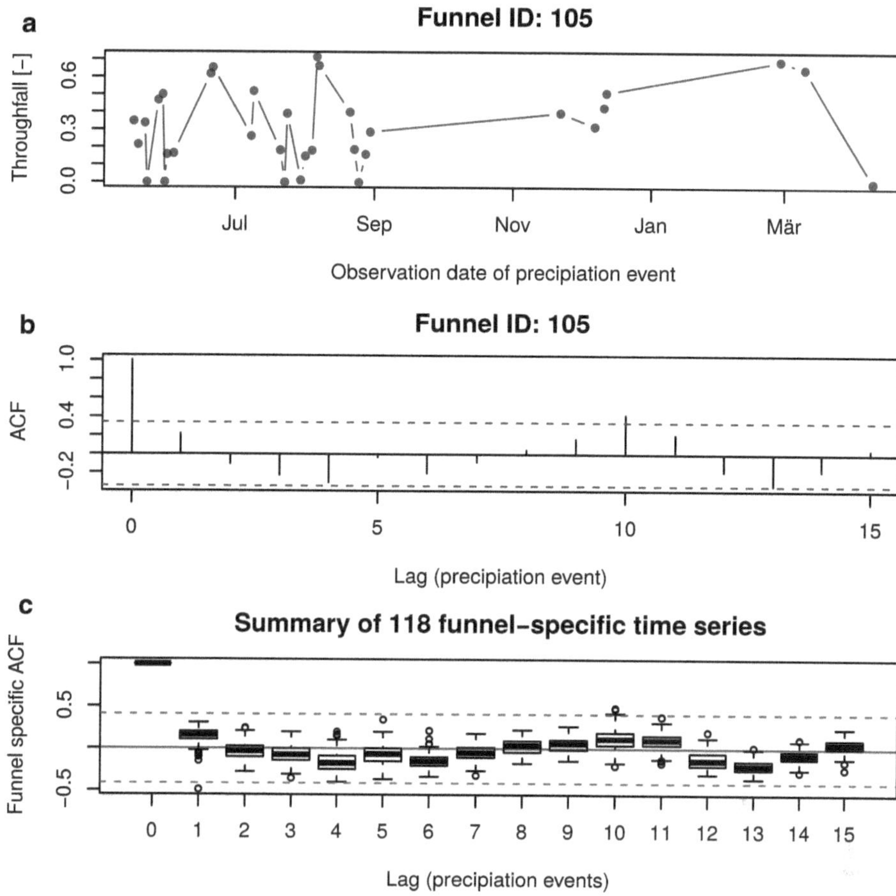

Fig. 7.4 Temporal dependence structures of observations. Plotting the response variable (here the proportion of throughfall on gross precipitation) vs. time reveals temporal dependencies among observations. Any pattern is a sign of dependence, but has to be assessed with respect to nesting factors. (**a**) single time series of throughfall at funnel 105; (**b**) its autocorrelation for varying lags as computed with the R function 'acf' (autocorrelation function (ACF)). For independent observations, we would mostly expect autocorrelations between the blue dashed lines; (**c**) boxplots summarizing funnel specific acf for all those funnels that registered any precipitation event (= complete time series; blue dashed lines, 95% CI)

time of 0 represents the correlation of the time series to the unshifted version of the time series, meaning to itself, which by definition is always 1 (Sun et al. 2018). The results for a lag >0 show that the autocorrelation is indistinguishable from 0, as most values are within the limits of the confidence interval. Since our data represents in total 175 time series (one for each funnel), a summary statistic has to be computed. This is given in Fig. 7.4c for all those 118 funnels representing consistent time series, i.e., containing data for all observation dates. Each boxplot in this figure therefore represents the distribution of all 118 estimates of Pearson's $\boldsymbol{r}$ that result from the autocorrelation that was individually determined for each funnel specific time series. Even if some estimated correlations exceed the limits of the confidence interval, this is not worrying because the boxes are all close to 0 and their whiskers all range from negative to positive correlations (for an explanation of boxplots see Sect. 7.5.1). Nevertheless, the series of boxplots reproduces the wave-like pattern that was already found for the autocorrelation of Funnel 105, with peaks of negative correlation around a lag of 4 and 13 and a peak of positive correlation around a lag of 10 precipitation events. When inference statistics are applied, it is recommended to keep an eye on such patterns, even if they are not statistically significant.

Some care has to be taken, when interpreting the results presented in Fig. 7.4, since the approach presented in the paragraph above is only applicable when all rainfall events within the respective period of time are considered. This is not the case for this throughfall dataset, since some rainfall events were excluded due to high wind speed or because of other factors listed in Sect. 7.2 ('Understanding the structure of the dataset'). A shift of the time series by a unit of time therefore generates data pairs that do not represent the same lag time. Let us, for instance, assume there is another rainfall event that was not considered, but occurred between the ninth and tenth rainfall events in the list of the given data. In this case the autocorrelation with a lag of one rainfall event is computed correctly for all observations but the ninth and tenth rainfall event, because they are compared with each other despite the fact that their real temporal distance is two rainfall events. In spite of such problems with our dataset, we have used the autocorrelation method here to demonstrate its application and the interpretation of its results. If you are not sure about patterns in your data, a purely random time series can serve as a reference for an independent dataset (Sun et al. 2018). This can, for example, be generated by randomly reordering the given sequence of rainfall events (see the electronic supplemental material R code for further details).

7.4.2 *Spatial Dependence*

Let us further inspect the spatial correlation of the throughfall dataset. Again, this addresses a broad topic and readers interested in spatial data analysis with R are referred to Bivand et al. (2008). Since we know that our data contains multiple observations at the same measurement point (funnel), we actually have to investigate spatio-temporal patterns. The computation of variograms represents a formal way to assess such patterns (Keim et al. 2005; see Figs. 8.3 and 8.11 in Chap. 8). Another less formal approach is to plot the response (throughfall) against spatial coordinates (Zuur et al. 2010). First, we can leave the raw data untouched and create single plots for each observation date and inspect them individually (Fig. 7.5a). This reveals if spatial dependence structures are given for single rainfall events.

In addition, a form of aggregated information on spatial dependence for a period of time is required, since this shows if some small differences in single observation dates may accumulate to a real pattern. It is therefore appropriate to calculate a single

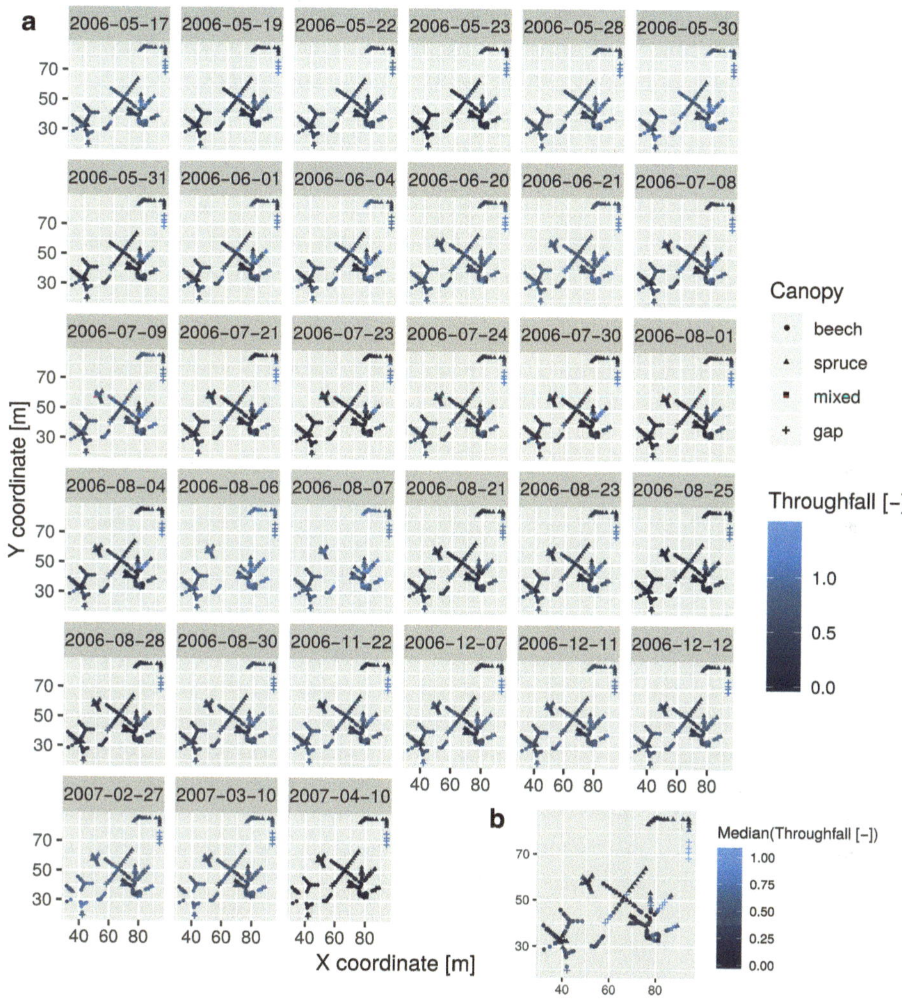

Fig. 7.5 Spatial dependence structures of observations. Plotting the response variable (here the proportion of throughfall on gross precipitation) vs. space reveals spatial dependencies among observations. Any pattern is a sign of dependence, but has to be assessed with respect to nesting factors. (**a**) spatial correlation structure for single precipitation events; (**b**) spatial correlation structure for the median estimated by all observations registered at a funnel in the leafed period

estimate for each funnel over all observation dates of the leafed or the non-leafed period and to plot the results in a single map (Fig. 7.5b). Any pattern is a sign of dependence, but has to be assessed with respect to the (remaining/not aggregated) factor levels (e.g., canopy types and period). The selection of an estimate that is used for such an aggregation depends on the type of data. For metric data the median serves well, since it is less sensitive to outliers than the mean. Precipitation could also be represented by cumulative sums.

Instead of estimating quantities, it is further interesting to investigate the spatial variability of throughfall variability by a calculation of the coefficient of variation for the observations given at any location. Whatever estimate is chosen, care has to be taken when the data is unbalanced. Throughfall was, for instance, not recorded at all funnels at each observation date. Either incomplete observation dates can be excluded from the calculation of these estimates (as done for Fig. 7.5b) or the data structure indeed prevents a satisfying aggregation. Ignoring this, you may see patterns that merely arise from the influence of events that were only measured at some locations and affect their overall mean or, if the number of observations differs greatly, even the median.

Some studies intend to measure dependencies to address specific hypotheses. For example, Frischbier and Wagner (2015) placed the funnels in a way to determine the spatial dependency of throughfall. Data exploration can therefore also serve as a first analytic framework to investigate such desired relationships by various visualizations.

7.4.3 General Remarks

Solutions to a violation of the independence assumption are to explicitly model the spatial or temporal relationship, e.g., using time points (observation date) or a two-dimensional smoother (spatial coordinates) as an explanatory variable (Zuur et al. 2010) or the previously mentioned mixed effects models that can include a temporal or spatial dependence structure in the model (Pinheiro and Bates 2010; see Chap. 8 of Tischer et al. for a demonstration of various techniques). Note that any such method requires justification, which means that the (standardized) model residuals have to be tested for a dependency to any given explanatory variable, those that are included in the model and those that are not. If there is a dependence, the model might be improved by including the respective variables. This directly points to a problem that can be unsolvable, namely, the correlation structure of the model residuals can be caused by variables that were not measured at all.

7.5 Identifying Outliers

Data often contains observations that appear unusual in comparison to the rest of observations. A proper data exploration should identify such 'outliers' and thus address question *6. Are there outliers in the response or explanatory variables?* In detail, an outlier can represent an extraordinary deviation from the common pattern observed in one-, two-, or even higher-dimensional space. While the precise interpretation depends on the intended application, this essentially means an observation represents a rather large or small value compared to the majority of observations, also with respect to a combination of variables. It is important to note that any such deviation might be associated with the measurement of a rare event or component that is an inherent part of the nature of the object under study. Accordingly, you certainly do not want to remove outliers. Otherwise, outliers can also arise from measurement errors, which means that they should be dropped. Most care has to be taken regarding the fact that outliers can seriously influence the outcome of some statistical methods as regression models, principal component analysis or discriminant analysis (Daszykowski et al. 2007; Zuur and Ieno 2015).

First of all, observations that do not match the regular pattern have to be identified and will be determined as 'potential outliers', owing to the fact that they potentially affect the results of an analysis. Actions such as data transformation or outlier removal can help to cope with such issues, but do not necessarily follow these first steps of outlier detection. We will discuss this later on. In the following, we will present various graphical methods to detect outliers and a tool applying various formal outlier tests.

7.5.1 Simple Plots

In one-dimensional space, simple boxplots as well as Cleveland dotplots can be applied to the observations given for the response variable(s) and explanatory variable(s). Cleveland dotplots illustrate the value of each observation (x-axis) versus its row number in the data frame (y-axis) and are as simple as powerful, since any point sticking out from the usual pattern can be regarded as a potential outlier (Zuur et al. 2010). In contrary to this, a boxplot delivers information in a more aggregated form. It usually visualizes (i) a horizontal line, representing the median as the centre of the data; (ii) a box that represents the data within the first and third quartile (25% and 75% percentiles); (iii) some outgoing lines called 'whisker', which denote the distance of the last data point that does not exceed a distance from the box of 1.5 times the length of the box (= quartile range); and, if at all, (iv) single points that have a larger distance from the box, representing outliers. The maximal whisker length can be adjusted in *R* and might not be the same in other statistics software. This should be checked, before boxplots are interpreted. The described default definition for whisker length makes sense considering a normal distribution, since in this case around 95% of the data are represented by the box-whisker plot and roughly the 2.5% most extreme points at both ends are shown as potential outliers.

7.5.2 Conditional Plots

The insights provided by Cleveland dotplots and boxplots can be readily enhanced by taking the structure of the data into account. This is achieved by including the group membership of observations as an attribute to the plot. Throughfall can, for

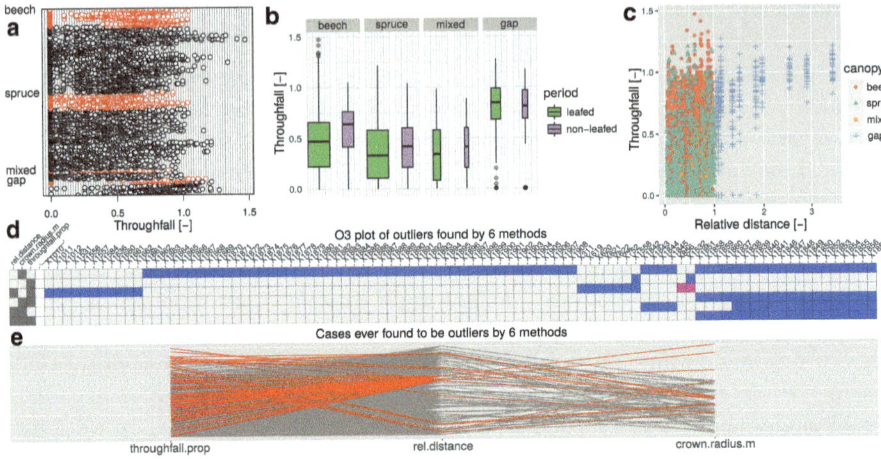

Fig. 7.6 Outliers in observations. (**a**) Cleveland dotplot of proportional throughfall conditional on canopy type (labelled) and period ('non-leafed' in red, 'leafed' in black). This indicates that the majority of observations (proportional throughfall) is between 0 and 1, but two observations under beech slightly stick out on the right (also identified as outliers by formal tests as shown in panel **d**). (**b**) Conditional boxplot on proportional throughfall indicates most pronounced group differences for gap compared to all other canopy types in both periods. Potential outliers are those observations in single groups that are illustrated as points (can be identified using *plotly* routines in *R*, see electronic supplement). (**c**) Scatterplot allowing to identify observations that deviate from the regular pattern found in the *x*-*y*-space, which may not be outliers considering group memberships (all funnels in gap position have relative distances larger than 1). (**d**) Outliers found by various formal tests in one- and multidimensional space for the three tested variables (proportional throughfall, relative distance and crown radius) using the functionality of the *OutliersO3* package (Unwin 2018) in *R*; grey boxes on the left indicate the (combination of) variables for which the outliers to the right have been detected by one (in blue) or two methods (in pink); e.g., first row shows outliers for 'crown.radius.m' with a high number of outliers detected by one detection method; 4 of 6 methods detected no outliers at all; other variables were not tested in the illustrated case. (**e**) Parallel coordinate plot with outlier cases in red; shows the combination of values that each (outlier) observation represents in each of the three tested variables, e.g., an outlier might represent the combination of a high throughfall proportion, a low relative distance and a moderate crown radius. The number of lines to 'throughfall.prop' is higher than those to 'crown.radius.m', since some outlying throughfall observations have been detected at particular funnels, each of them representing only one combination of 'rel.distance' and 'crown.radius.m'

example, be plotted according to canopy type and period of the respective observation, which already represent a nesting by more than one factor (Fig. 7.6a and b). In general, conditional boxplots and conditional dotplots allow to check if the distribution or the pattern that is associated with the data is different between the given groups. This makes it possible to identify observations that represent potential outliers with respect to the general pattern of the other observations in the same group. Such potential outliers can be important for group associated comparisons, but may not be identified when the nested structure of the data is ignored, meaning only simple plots for each variable are generated for outlier detection. Another advantage of conditional boxplots is that they can be used to compare group median values and to check visually if group associated residuals in regression type models are likely to have similar variances. This refers to homoscedasticity, which will be further addressed later on.

7.5.3 Scatterplots

Plotting every variable against every other variable shows if some observations deviate from the regular pattern that can be seen in two-dimensional space (Fig. 7.6c gives an example). It enables to identify so called *x*-*y*-outliers, which cannot be detected when the *x*- and *y*-variables are plotted alone. This should in principle be examined for every variable combination and for more than two dimensions, when multiple explanatory variables are given. It is, for example, worth to plot total throughfall against gross precipitation in order to identify potential outliers in this space. But, when this is applied, the factors canopy and period should not be ignored. Instead, a multi-panel plot should be used, which means that every canopy-type to period-type combination should have its own plot in order to identify outliers in the respective two-dimensional space (Fig. 7.7e). Such explorations can represent a heavy workload and might be complemented by the application of formal tests and tools such as those introduced in the following.

7.5.4 Formal Outlier Tests

Several formal tests for outlier detection exist. Results of various outlier detection methods can be compared with the R-package OutliersO3 (Unwin 2018). This package draws a plot-like overview of outliers (Fig. 7.6d), denoting all observations (data frame rows) that represent univariate outliers (considering each variable alone) or multivariate outliers (considering all combinations of variables) according to one or several identification methods and classified by one or several outlier tolerance levels. Although a controversy about formal outlier tests exists in the literature (Zuur and Ieno 2015), this method has the advantage of providing a multi-objective framework for outlier detection. This enables to combine the advantages of each single identification method to detect outliers that other methods might ignore (possibly for good reasons) in order to create a 'full outlier collection' (Fig. 7.6e). In contrast to this, it is also possible to create an 'intersecting outlier collection', consisting of all those potential outliers that were identified as such by each of the identification methods in use. These collections represent a set of observations the analyst should keep an eye on. If it comes to hypothesis testing with statistical models, these observations might influence the results. This can be tested by a comparison of the test results obtained using the original data to those that exclude the 'full outlier collection' or the 'intersecting outlier collection'. It is possible that none of the outlier collections is influential but some observations that were not detected by this approach are because the influence of an observation depends on the statistical method that might rearrange the data in a way that cannot be explored prior to its application.

7.5.5 *Influential Observations*

A dataset may contain multiple observations that are identified as potential outliers, but it is not clear how they affect, for instance, a certain type of regression model. In addition to the classification of observations as 'potential outliers' by graphical methods or the application of formal tests, the influence of each observation on the results of a statistical test can be computed. This approach is most target-aimed, since it evaluates the role of an observation with respect to the results. Effects of explanatory variables that are used to generate these results are consequently considered in this evaluation. Some observations may have an influence on the model structured in one way and others on a model structured another way. It might not be wise to exclude them all, since sample sizes are often small and it is not necessary to remove observations that do not really influence the results that are obtained by an otherwise valid model.

A common method applied to linear models and generalised linear models is Cook's distance plot (Fig. 7.7d). It refers to Cook's distance statistic D_i that represents a measure of the influence of each observation on the regression parameters (Quinn and Keough 2002). This is computed for each observation by comparing the model results when the observation is included to those when the observation is dropped. Observations associated with a D_i value greater than 1 are considered influential. A transformation or restructuring of the model can help to reduce the influence of such outliers. Otherwise, the observation may be dropped. The leverage that is shown on the *x*-axis in Fig. 7.7d indicates how strong an observations explanatory variable(s) deviate(s) from the majority of observations. For the calculation of D_i both residuals and leverage are used, whereby the Cook's distance (dashed line) represents the critical threshold isolines for a D_i of 0.5 (severe) and 1 (even more severe) in the residual-leverage-space.

Ordination techniques can serve as a method for the detection of outliers in higher dimensions. In principal component analysis (PCA), the biplot can show observations that stick out from the rest of clustered points in the centre. This can reveal potential outliers that have not been detected with other univariate methods before (Zuur and Ieno 2015). Furthermore, these are likely 'influential observations' with respect to the PCA results, since the importance of the first few axes can be inflated due to outliers (Joliffe and Cadima 2016). Applying the PCA with and without these observations enables to identify their influence.

Although 'influential observations' can be considered as a more severe type of outlier than 'potential outliers', and they do not necessarily have to be removed. If it can be argued that they represent genuine variation, it is recommended to determine and report their specific weights on the model results (Zuur and Ieno 2015).

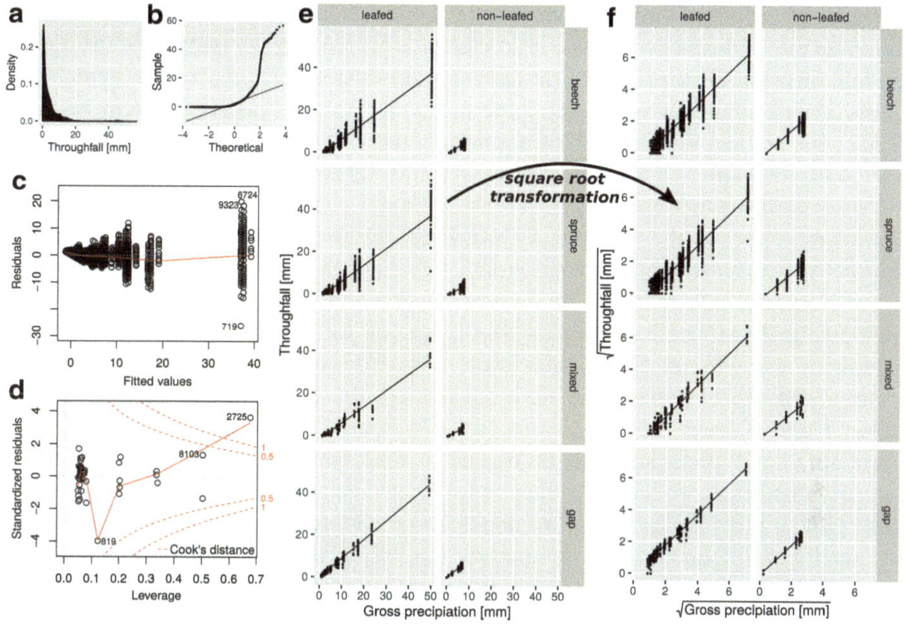

Fig. 7.7 Normality, homoscedasticity, influential observations and the effect of transformations. (**a**) A density plot of throughfall shows that this variable is right-skewed. (**b**) The Q-Q-plot reveals substantial deviations of throughfall quantiles (Sample) to those of a theoretical normal distribution (the solid line connects its first and third quartile). (**c**) A residual plot of a linear model on throughfall vs. gross precipitation, canopy type and period indicates that residuals might be normally distributed around any fitted value but that the associated variance is heterogeneous. (**d**) A Cook's distance plot of an extremely short arbitrary subset of the data illustrates the relation of residual to leverage size of any given observation and that this severely affects the model parameters for observation '2725' and less intense for two others that are labelled as well. (**e**) Multi-panel scatterplot showing the relation of throughfall and gross precipitation and its potential interaction with canopy type (beech, spruce, ...) and period (leafed, non-leafed). (**f**) Demonstrating that a square root transformation of throughfall and gross precipitation can substantially reduce the heterogeneity of variance over gross precipitation as observed in subfigure (**e**)

7.6 Checking Normality and Homoscedasticity

In the introduction to this chapter, we raised question *7. Can we assume normality and homogeneity for the response variable (of the residuals)?* Indeed, we do not have to be as strict with both principles as this question may suggest. Here, we will present some graphical diagnosis tools and guidance that help to address this topic appropriately.

Normality refers to a Gaussian distribution (also called normal distribution), which is often used for modelling continuous data. However, it is not needed for every statistical technique. The specific assumptions on normality also depend on the statistical test. In simple linear regression, normality is, for instance, assumed for the response variable at any value of the explanatory variable. This actually means that there can be a non-Gaussian distribution for the response variable, but when the effect of the explanatory variable is considered, the remaining residuals of the model should be normally distributed. In many circumstances linear regression models are fairly robust against a violation of the normality assumption (Quinn and Keough 2002), but not to a violation of the assumption on variance homogeneity (Zuur et al. 2010). If you explore your data and see that it is not normally distributed, you don't necessarily have to transform it. Set up a model and check the residuals, which reflect the effect of the explanatory variables on the response. If the residuals are normally distributed and, more important, show similar variances along the explanatory variable, this is often an indication that the model assumptions are not fundamentally violated. If not, transformations of the response variable, the application of generalized linear models or, when the variance follows a consistent pattern, the use of generalized least squares for the fit of linear models can help (Quinn and Keough 2002). For example, a square root transformation of total throughfall and gross precipitation helped to overcome the problem with heterogeneous variances as observed for the untransformed data (Fig. 7.7e and f). However, if the residual plots are examined in detail, some patterns might still be found and a stronger type of transformation such as the cubic root transformation should be considered. For the throughfall dataset, new problems arise with this procedure, since zeros in the original data are not accordingly transformed. To achieve this, we have to add a constant (e.g., 1) to the variable prior to transformation. But as we do so, the resulting variable becomes more complicated to interpret than after a simple square root transformation. In general, model results will be less intuitive to interpret when transformations are applied (see Sect. 7.8 for a detailed discussion on transformations). The selection of appropriate statistical tools and non-normal distributions should therefore be prioritized.

Theoretical considerations should come first when trying to find out which distribution the observations come from. Thus it is necessary to ask questions about the nature of the data. What processes generated the data? Is it, for instance, like a collection of raindrops, which refers to count data with a corresponding distribution? Is it continuous (e.g., tree height) or discrete (e.g., the number of trees in a stand)? Is it restricted in its range (e.g., precipitation cannot take on negative values)? All these questions can help to identify the most important properties that determine the real distribution of some observations.

In addition, there are various graphical techniques that can be used to explore the distribution of your empirical data prior to analysis. Common approaches are to draw a histogram (showing the frequency of classes), a density curve (estimating the relative frequency on a continuous scale) or a Q-Q-plot (comparing the observed quantiles with those of a theoretical normal distribution). Plotting, for example, throughfall amount shows if this variable is normally distributed or if it might be skewed to the left or right (Fig. 7.7a and b). We can probably answer this question in advance by considering the underlying processes that generate this kind of data. This means knowing that precipitation refers to single events that cannot be associated with negative values and likely consist of more small than large values, since every rainfall refers to a certain amount of raindrops (rain intensity) that fall within a certain period of time (rain duration). Decreasing the measurement interval will reduce the number of counted raindrops. Other climatic factors such as, for instance, temperature would not be affected by the measurement interval, since the underlying processes are different. In the case of precipitation and throughfall, it refers to count data that is often right-skewed, including some extreme events with high precipitation, but none smaller than 0.

A transformation of the data to approach normality is, for instance, required for *t*-tests (Townend 2008). This test splits the data into two groups and does not account for the effects of explanatory variables, which could be associated with the non-normal distribution of the response. Nonetheless, *t*-tests are also quite robust against a violation of the normality assumption (Zuur et al. 2010).

In ANOVA and other types of regression techniques, the variance should be checked while accounting for the group structure associated with some variables or their interaction (Zuur et al. 2010). For example, considering the throughfall dataset, variation under beech canopy should be similar to variation under spruce canopy, mixed canopy and in gap position or variation between these canopy types should be similar within leafed period compared to their variation within the non-leafed period.

Unequal variance is a common phenomenon, since large values commonly have more spread in their value than small ones. Simple formal tests to compare variances are the F-test (for two groups), Bartlett's test (for more than two groups) or Levene's test (less sensitive to departures from normality) or the Brown and Forsythe test (a modified version of Levene's test). Nonetheless, an examination of graphs is often more reliable, since these tests are insensitive for small sample sizes and are too sensitive for large sample sizes (Townend 2008). In simple linear regression, it is most useful to check the pattern of the residuals for the different values of the explanatory variable. Figure 7.7c shows, for example, that the residuals of total throughfall increase with the explanatory variable gross precipitation, which indicates a violation of a core assumption (similar residual variances along the values/levels of explanatory variables) of the underlying linear model. If such patterns are revealed by data exploration, it can be tested if the pattern is removed either by modifying the model using more or other explanatory variables and or interactions terms or by a transformation according to the recommendations given in the subsection 'transformations as potential actions'. In our example, the square root transformation of both response and the continuous explanatory variable gross precipitation removed the pattern in the residuals of the linear model (not shown), whereas a change in the model structure was less helpful. In Chap. 8 of this volume, Tischer et al. (2019) are using the same dataset and provide more details on how to appropriately address such issues related to temporal and spatial patterns.

7.7 Exploring the Relationships of Variables

7.7.1 Interactions

This part of data exploration focuses on the association of variables. Several graphical tools can be used for this purpose. Some of them, such as the scatterplot, have already been introduced. Instead of identifying potential outliers in the *x*-*y*-space, we can use this type of figure to study the relation of variables to each other, thus addressing question *8. Are the relationships (to the response) linear or non-linear?*—and question *9. Are there interactions and a sufficient number of observations to allow for them?* Both questions can be addressed with respect to the relation of throughfall, gross precipitation, canopy type and period when we have a look on the multi-panel scatterplots presented in Fig. 7.7e and f. First, we can check the pattern of points in each scatterplot, which demonstrates that the relation of throughfall and gross precipitation is likely linear. Moreover, the regression lines in each panel can be compared to see if the slope varies between groups of the data, which is only marginally pronounced for the combinations of the canopy and period categories. In general, such interactions are present when the outcome of a statistical analysis differs between groups that are underlying the data structure. In a strong form, the slope can be positive in one and negative in another group.

When no continuous explanatory variable is given, interaction plots can be created to detect interactions. Figure 7.8e shows the mean values of proportional throughfall for each canopy type in each period, whereas estimates belonging to the same period are connected by a line. If lines cross multiple times, there is a strong interaction. If this happens only once, it is a moderate form of interaction and when the lines are almost parallel, as for canopy type and period, it indicates no interaction.

In linear regression it is recommended to centre and standardize the input variables, especially when interactions are considered (Schielzeth 2010). This avoids a misinterpretation of main effects and makes the resulting model parameters and effects sizes comparable in magnitude within a model and between studies. Moreover, a common misconception about linear regression is that it is limited to linear relationships, which is not true. For example, quadratic terms or interactions make it possible to fit nonlinear patterns between response and explanatory variables (Zuur et al. 2010).

7.7.2 Collinearity

Multiple regression is a popular statistical method to identify the relevant environmental drivers (Zuur et al. 2010). It relates the response to a set of explanatory variables. Let us imagine we model the proportion of throughfall as a function of tree species, height, diameter at breast height and crown radius. We imagine that changes in data or the structure of the model, e.g., adding or deleting an explanatory variable,

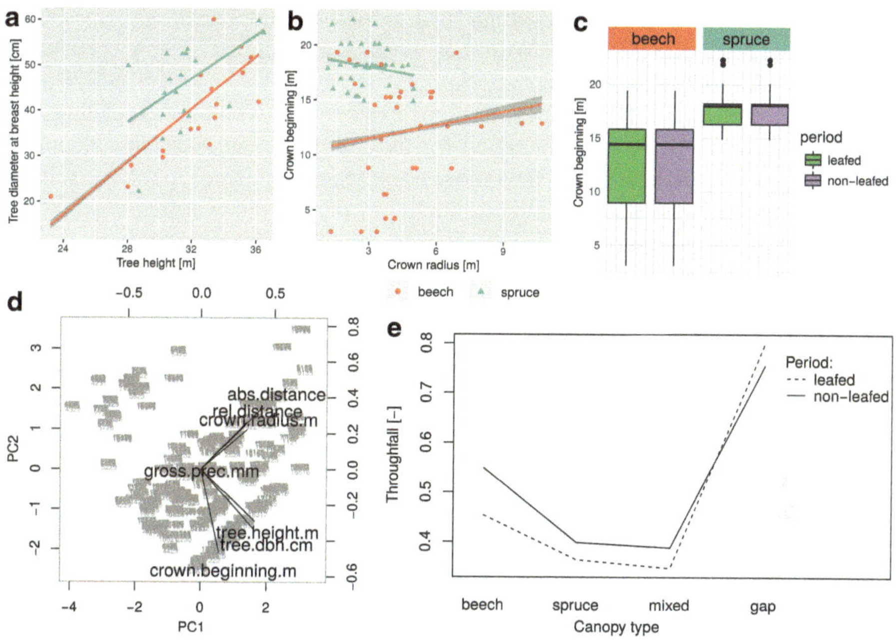

Fig. 7.8 Interactions, collinearity and confounding. (**a**) Scatterplot showing the collinearity between tree height and tree diameter at breast height ($R^2_{beech} = 0.77$, $R^2_{spruce} = 0.72$). (**b**) Scatterplot showing that data for beech and spruce is located in different clusters in the *x*-*y*-space of crown radius and lowest height of the tree crown (crown beginning). (**c**) Boxplots of a continuous (crown beginning) and two categorical explanatory variables (tree species and period); Note the confounding problem due to the range differences of species, which makes it almost impossible to distinguish between the effect of species and crown beginning. (**d**) PCA correlation biplot indicating collinearity for the continuous explanatory variables grouped together; Variables were standardized prior to PCA analysis. (**e**) Interaction plot showing the interaction of canopy type and period on proportional throughfall. As lines are almost parallel, there is no considerable interaction present

can have a high impact on the regression parameters, particularly the slope. Should we be alarmed? Indeed, since such a case describes the effect of collinearity that refers to the correlation of explanatory variables (Zuur and Ieno 2015), which is likely given considering the selection of explanatory variables in our model. Although such models can be significant, the model parameters are associated with highly inflated standard errors (Quinn and Keough 2002). This is because no real optimum can be found if more than one factor can be equally used to explain similar parts of the existing variance. Using another sample we likely end up with other model parameters, which makes any inference based on these model results not much convincing.

Biological data likely shows some collinearity, which leads us to our final question *10. How strong are the explanatory variables correlated with each other?* The correlation of explanatory variables can be calculated using Pearson's *r* (identifying linear correlations) and visualized using scatterplots (also identifying nonlinear relations) or a biplot resulting from principal component analysis (PCA) (Fig. 7.8d) (Zuur and Ieno 2015). It has been demonstrated that a correlation coefficient |r| > 0.7 represents an appropriate threshold for a severe collinearity effect (Dormann et al. 2013). Thus, pairs of explanatory variables exceeding this value should not be used in conjunction for the multiple regression. For example, tree height and diameter at breast height represent redundant, since highly correlated information (Fig. 7.8a) and one of them should be omitted. It is worth noting that the correlation of these variables is also clear by biological reasoning (i.e., due to allometric relationships). Otherwise, a strong collinearity can even arise (incidentally) from outliers, which might be removed or transformed with the affected variable prior to analysis (Zuur and Ieno 2015) or down-weighted using a collinearity-weighted regression (CWR) approach (Dormann et al. 2013). However, some collinearity is acceptable, especially when the pattern between response and explanatory variables is strong (Zuur and Ieno 2015). But it is recommended to interpret results with caution and to avoid predictions to new collinearity structures (Dormann et al. 2013).

When the dataset represents a mix of continuous and categorical explanatory variables, conditional boxplots (Fig. 7.8c) or cluster analysis can visualize and help to investigate collinearity (Zuur and Ieno 2015). A problem with collinearity is that the real driver of some observation cannot be traced back to a single environmental effect when this is only one of a set of collinear explanatory variables that was selected for an analysis. Imagine the cluster of variables found by PCA (Fig. 7.8d). The real driver could even be one of the other variables not included in the model or another variable altogether unknown. Collinearity is also ‘implemented’ in a model when it contains an interaction of two continuous explanatory variables. This is because the interaction term simply represents the product of both variables, which likely correlates with both variables (Zuur and Ieno 2015).

An estimate for the collinearity of an explanatory variable in an ordinary least squares regression type model is the variance inflation factor (VIF). It can be computed for each explanatory variable using it as a response and all other explanatory variables as explanatory variables. The higher the fraction of variance that can be explained by the other explanatory variables, the more redundant the information that can be provided by the explanatory variable that was used as a response. VIF values greater than 10 suggest strong collinearity (Quinn and Keough 2002). In *R*, the package *car* provides a function to compute the VIF for linear, generalized linear and other models (Fox and Weisberg 2011).

Splitting the data into subsets for training and testing can help to evaluate a model’s performance, also with respect to collinearity (Dormann et al. 2013). Furthermore, this enables to optimize the variance-bias-trade-off, which refers to the problem that the more explanatory variables are used, the more variance is explained (better model fit), but the less generalizable are the model results (Burnham and Anderson 2002). This can be achieved by evaluating the performance of competing candidate models using model quality criteria such as AIC and BIC (accounting for model performance and ‘costs’ of each variable) on the training data and the prediction quality, for example, in form of Pearsons R^2 on the test dataset.

7.7.3 *Confounding*

Many observations in biology are the result of interactions. But even controlled experiments can become difficult to interpret regarding the effect of a particular factor. On the one hand, this is because statistical methods can analyze relationships in the data, which do not have to arise from direct natural causality. On the other hand, the objects under study rarely differ in a single characteristic only. Beech and spruce trees, for example, differ in their general characteristic of belonging to deciduous and coniferous trees, respectively. In order to assess the effect of this characteristic on the proportion of throughfall, we can place funnels under the canopy of both types. But, as we begin to relate throughfall to the trees single characteristics, we may not be able to differentiate between the tree type effect and the effect of the other characteristics, since those are by themselves too tree type specific. Figure 7.8b and c shows such examples for the throughfall dataset. Both lowest height of the crown and crown radius are very specific for beech and spruce trees, which may also result from inter- and intra-specific competition. It is therefore hard to interpret an effect of crown radius or lowest height of the crown, when the model already accounts for the grouping into beech and spruce trees that have very different lowest height of the crown and crown radius characteristics, at least in this data.

7.8 Transformations as Potential Actions

Often assumptions of a statistical procedure can only be justified after applying transformations to the variables or parameters. For example, linear regression assumes additive effects of the regressors and of the error terms and, in addition, the normal distribution of the error terms, needed for statistical inference in the form of tests or confidence and prediction intervals, is itself often justified by assuming the error terms arise additively from many independent effects with small range (central limit theorem; cf. Thm. 5.12 in Kallenberg 2002). Thus, in a situation where effects are known to be multiplicative, a linear regression model can only be applied after transforming the response variables by taking logarithms.

Of course, when applying such transformations, any conclusions also need to be interpreted in terms of the transformed variables. In particular, expected values computed after a transformation need not be equal to the transformed expected values. They will typically be smaller in the case of concave transformations such as logarithms, cf. Jensen's inequality.

Less strictly, transformation of variables can be considered for various reasons: (i) to get rid of outliers, (ii) achieve normality and homogeneity (e.g., for the residuals of a regression analysis) and/or (iii) linearize relationships (Zuur and Ieno 2015). The methods that can be applied depend on the details of the specific pattern that is wished to be removed.

Such transformations, when not based on theoretical considerations or prior knowledge, can be problematic. It is generally recommended to avoid unnecessary transformations. Especially for the response variable, it is worth considering alternative statistical models that are able to deal with greater variation (e.g., using a gamma distribution for a continuous response variable that has a large mean value) (Zuur et al. 2010).

Nonetheless, when the sample size is small and observations cannot be dropped, transformations can help to close gaps in inadequately sampled data ranges along an environmental gradient (Zuur and Ieno 2015). Variables containing outliers on the right side of the majority of observations can be transformed by a square root (see Fig. 7.7f) or by a much stronger logarithmic transformation (Zuur and Ieno 2015). If outliers are present on the left side of the majority of observations, the data has to be reflected prior to the transformation. This is achieved by subtracting the variables from a constant, which represents, for instance, the maximum observed value or the result of 1 added to the largest value (which also allows log transformations when zeros are present) (Zuur and Ieno 2015). Similarly, dimensionless quantities are obtained subtracting the variable mean μ from the variable x and dividing it by its standard deviation σ ($z = (x\text{-}\mu)/\sigma$). This standardizing yields variables with mean 0 and standard deviations of 1, which are especially important for techniques such as principal component analysis (PCA) that are sensitive to the dimension of the data. Furthermore, standardizing can be used to identify outliers, since z-scores of more than 2.5–3 refer to potential outliers, although robust z-scores derived from the median and pair-wise differences between observations are more reliable (Daszykowski et al. 2007).

Parameter values obtained by a statistical analysis using transformed variables might be transformed back to the original scale in order to simplify the interpretation of the results. A variable $\boldsymbol{x}$ that was log transformed (log (x)) can be transformed back using the anti-log ($x = e^{\log(x)}$) as well as a square root transformation (x) can be transformed back using the square (x^2; more examples are demonstrated in our *R* code in the electronic supplement).

In any case, care has to be taken when interpreting the results of tests using transformed data. For example, interaction terms in linear regression models can be significant using the original data, but not for the transformed data (Zuur and Ieno 2015). It is therefore crucial to think about any transformation and how this affects the analysis.

7.9 Future Opportunities/Directions

We provided an outline and introduction to basic tasks and tools on data exploration and transformation. This is a primary step in analysing data. Many different statistical methods can be applied to address a certain research question. If you are not sure about the right method, Dytham (2006) provides a key that helps to choose a test. We believe that research benefits from diversity, since the combination of various approaches can provide additional insight into the reliability of results. This can include the consideration of simple non-parametric tests such as the chi-square test, others of Kruskal-Wallis type (e.g., Chenouri and Small 2012) or more advanced parametric analysis such as mixed-effects modelling (e.g., Pinheiro and Bates 2010) or the information-theoretic approach (e.g., Burnham and Anderson 2002). Non-parametric tests have less demanding assumptions and can represent a useful alternative, e.g., when the data are of unknown distribution type. Although they are less powerful, the effects are likely strong, if they are indicated by a non-parametric test.

Regarding to the aspect of collinearity, Dormann et al. (2013) list a series of future research directions: (i) Developing regression techniques that are able to balance model complexity and fit in a way that is less sensitive to collinearity; (ii) improving the implementation of correlated explanatory variables that are non-linearly related to each other ('concurvity') in an exploratory data analysis; (iii) assessing the role of the positive and negative of correlations on collinearity diagnostics more comprehensively.

In the end, we believe that the rise of new interactive graphical methods has the ability to foster both data exploration and transparency in science. We recommend that datasets as well as the source code that were used for the analysis are always provided as supplement or as an independent publication in data-focused journals such as 'Data in Brief'. This allows research to be fully replicable, which has never been more important than in the modern era of data science.

References

Bivand RS, Pebesma EJ, Gomez-Rubio V (2008) Applied spatial data analysis with R. Use R series, Springer, New York. https://doi.org/10.1007/978-1-4614-7618-4

Burnham KP, Anderson DR (2002) Model selection and multimodel inference: a practical information-theoretic approach. Springer, New York. https://doi.org/10.1007/b97636

Carlyle-Moses DE, Lishman CE, McKee AJ (2014) A preliminary evaluation of throughfall sampling techniques in a mature coniferous forest. J For Res 25:407–413. https://doi.org/10.1007/s11676-014-0468-8

Chenouri S, Small CG (2012) A nonparametric multivariate multisample test based on data depth. Electron J Stat 6:760–782. https://doi.org/10.1214/12-EJS692

Daszykowski M, Kaczmarek K, Vander Heyden Y, Walczak B (2007) Robust statistics in data analysis – a review. basic concepts Chemometrics Intell Lab Syst 85:203–219. https://doi.org/10.1016/j.chemolab.2006.06.016

Dytham C (2006) Choosing and using statistics: a biologist's guide. 2nd edn (Repr.), Blackwell Publishing., Malden, p 248

Dormann CF, Elith J, Bacher S, Buchmann C, Carl G, Carré G et al (2013) Collinearity: a review of methods to deal with it and a simulation study evaluating their performance. Ecography 36:027–046. https://doi.org/10.1111/j.1600-0587.2012.07348.x

Fox J, Weisberg S (2011) An R companion to applied regression, 2nd edn. Sage Publications, Thousand Oaks. http://tinyurl.com/carbook

Freckleton RP (2011) Dealing with collinearity in behavioural and ecological data: model averaging and the problems of measurement error. Behav Ecol Sociobiol 65:91–101. https://doi.org/10.1007/s00265-010-1045-6

Frischbier N (2012) Study on the single-tree related small-scale variability and quantity-dependent dynamics of net forest precipitation using the example of two mixed beech-spruce stands. TUDpress, Dresden. (Dissertation). http://nbn-resolving.de/urn:nbn:de:bsz:14-qucosa-94870

Frischbier N, Wagner S (2015) Detection, quantification and modelling of small-scale lateral translocation of throughfall in tree crowns of European beech (Fagus sylvatica L.) and Norway spruce (Picea abies (L.) karst.). J Hydrol 522:228–238. https://doi.org/10.1016/j.jhydrol.2014.12.034

Hurlbert SH (1984) Pseudoreplication and the design of ecological field experiments. Ecol Monogr 54:187–211. https://doi.org/10.2307/1942661

Joliffe IT, Cadima J (2016) Principal component analysis: a review and recent developments. Phil Trans R Soc A 374:20150202. https://doi.org/10.1098/rsta.2015.0202

Kallenberg O (2002) Foundations of modern probability, 2nd edn. Springer, New York, p 638

Keim RF, Skaugset AE, Weiler M (2005) Temporal persistence of spatial patterns in throughfall. J Hydrol 314:263–274. https://doi.org/10.1016/j.jhydrol.2005.03.021

Pinheiro J, Bates D (2010) Mixed-effects models in S and S-PLUS. Springer, Dordrecht. ISBN: 9781441903181. https://doi.org/10.1007/b98882

Quinn GP, Keough MJ (2002) Experimental design and data analysis for biologists. Repr. With corr. 2003. Cambridge University Press, Cambridge, p 537

Schielzeth H, Forstmeier W (2009) Conclusions beyond support: overconfident estimates in mixed models. Behav Ecol 20:416–420. https://doi.org/10.1093/beheco/arn145

Schielzeth H (2010) Simple means to improve the interpretability of regression coefficients. Methods Ecol Evol 1:103–113. https://doi.org/10.1111/j.2041-210X.2010.00012.x

Schielzeth H, Nakagawa S (2013) Nested by design: model fitting and interpretation in a mixed model era. Methods Ecol Evol 4:14–24. https://doi.org/10.1111/j.2041-210x.2012.00251.x

Sievert C (2018) Plotly for R. https://plotly-book.cpsievert.me

Sun F, Roderick ML, Farquhar GD (2018) Rainfall statistics, stationarity, and climate change. P Natl Acad Sci USA 115:2305–2310. https://doi.org/10.1073/pnas.1705349115

Tischer A, Zwanzig M, Frischbier N (2019) Spatiotemporal statistics: analysis of spatially and temporally-correlated throughfall data: exploring and considering dependency and heterogeneity. In: Levia DF, Carlyle-Moses DE, Iida S, Michalzik B, Nanko K, Tischer A (eds) Forest-water interactions. Ecological studies series, No. 240. Springer, Heidelberg. https://doi.org/10.1007/978-3-030-26086-6_8

Townend J (2008) Practical statistics for environmental and biological scientists. Wiley, Chichester, p 276. ISBN: 978-0-471-49665-6

Unwin A (2018). OutliersO3: draws overview of outliers (O3) Plots. R package version 0.5.4. https://CRAN.R-project.org/package=OutliersO3

Wickham H (2016) ggplot2: elegant graphics for data analysis. Springer-Verlag, New York. https://doi.org/10.1007/978-0-387-98141-3

Wilks DS (2006) Statistical methods in the atmospheric sciences. Second edition. Elsevier, Amsterdam, p 676

Zuur AF, Ieno EN, Elphick CS (2010) A protocol for data exploration to avoid common statistical problems. Methods Ecol Evol 1:3–14. https://doi.org/10.1111/j.2041-210X.2009.00001.x

Zuur AF, Ieno EN (2015) A beginner's guide to data exploration and visualisation with R. Highland Statistics Ltd.

Chapter 8
Spatiotemporal Statistics: Analysis of Spatially and Temporally Correlated Throughfall Data: Exploring and Considering Dependency and Heterogeneity

Alexander Tischer, Martin Zwanzig, and Nico Frischbier

8.1 Introduction

Ecosystems consist of all the organisms in an area and the physical environment with which they interact (Chapin et al. 2011). From this simple definition alone, it is clear that ecosystems have a spatial dimension. Inherent to the nature of organisms, ecosystems also have a temporal dimension. In particular, for long-lived forest ecosystems, space and time are of relevance for interactions between organisms and, for example, the redistribution of resources such as solar and photosynthetic active radiation, water, and nutrients (Reifsnyder et al. 1971; Zimmermann et al. 2008; Zimmermann and Elsenbeer 2008; Wälder et al. 2008; Frischbier and Wagner 2015; Keim et al. 2005). It has long been recognized that single trees change the distribution of resources due to the abundance and the structure of their canopies (Ford and Deans 1978; Wu et al. 1985; Canham et al. 1994). This includes both the vertical and horizontal dimensions of plant parts, such as the canopy, the stem, and the root system, as well as the changes in spatial properties during the vegetation period and tree life (Marin et al. 2000; Carlyle-Moses et al. 2004; Müller 2009;

Electronic Supplementary Material The online version of this chapter (https://doi.org/10.1007/978-3-030-26086-6_8) contains supplementary material, which is available to authorized users.

A. Tischer (✉)
Institute of Geography, Friedrich Schiller University Jena, Jena, Germany
e-mail: alexander.tischer@uni-jena.de

M. Zwanzig
Institute of Forest Growth and Forest Computer Sciences, Technische Universität Dresden, Tharandt, Germany

N. Frischbier
Forestry Research and Competence Centre, ThüringenForst, Gotha, Germany

D. F. Levia et al. (eds.), *Forest-Water Interactions*, Ecological Studies 240,
https://doi.org/10.1007/978-3-030-26086-6_8

Metzger et al. 2017; Carlyle-Moses et al. 2018). Such properties relate, for example, to canopy storage capacity (Frischbier and Wagner 2015) with effects on water redistribution processes (Allen et al. 2014; Allen et al. 2017). Spatial and/or temporal variations in resource distribution affect patterns of plant communities (Saetre et al. 1997; Okland et al. 1999), tree seedling establishment (Pukkala and Kolström 1992; Kellner and Swihart 2017), soil water availability (Schume et al. 2003; Metzger et al. 2017), as well as organic matter and nutrient turnover (Wälder et al. 2009; Saetre and Bååth 2000; Weber and Bardgett 2011). Thus, hypotheses on ecosystem processes and ecosystem engineering in forests are strongly affected by the spatial structures and the temporal dynamics of the respective ecosystem under investigation (Dale and Fortin 2015). One relevant scale that structures the temporal and spatial aspects of forest ecosystems are single trees (Wu et al. 1985; Mou et al. 1993; Peters et al. 2016; Engel et al. 2018).

Studies on ecosystem processes are therefore directly or indirectly related to the consideration of spatial and or temporal dependencies of observations (Dale 2007; Zuur et al. 2009; Dale and Fortin 2015; Zuur et al. 2017). The problem with dependent observations is one about pseudo-replication (Hurlbert 1984; Webster 2001; Zuur et al. 2017). Zuur et al. (2017) defined pseudoreplication as: "the use of inferential statistics to test for treatment effects with data from experiments where either treatments are not replicated (. . .) or replicates are not statistically independent." One relevant point is treating data as independent when they do not violate the assumption on independence of the covariances (Zuur et al. 2017). Using covariance structures that account for dependencies is important for the calculation of standard errors of model estimates and therefore affect confidence intervals and p-values for treatment effects and thus conclusions on experimental results (Zuur et al. 2009; Zuur et al. 2017). Several applications of incorporation of temporal and/or spatial correlated observations are available in various fields of ecological research (see, e.g., Carl and Kühn (2007), Kissling and Carl (2008), Legendre and Gauthier (2014), and Hefley et al. (2017)).

In this chapter, the data exploration of Zwanzig et al. (2019; see Chap. 7) is continued in order to investigate how strong the dependency among observations in space and time is and if temporal and spatial patterns in throughfall data exist. Frischbier (2012) tested the hypothesis that interception and lateral water translocation (see Frischbier and Wagner (2015)) within the canopy are affected by the tree species canopy structure. In order to investigate this hypothesis, Frischbier (2012) continuously monitored throughfall at defined measurement points below the canopy by taking single-tree characteristics such as species and distance from a particular stem into account. Consequently, the sampling design was optimized regarding the observation of throughfall at varying time-points of ideal rainfall events and at multiple spatial measurement points underneath the canopy of a mixed beech-spruce stand and within forest gaps.

Some of the main objectives of the study of Frischbier (2012) were related to the following questions:

- Does forest structure (i.e., crown position and tree species, mixture, gap—encoded as "canopy" within the dataset) affect throughfall redistribution?

- Does an effect of forest structure vary with space (e.g., distance from tree stem—encoded as "absolute distance" and "relative distance")? This is a question on the existence of spatial patterns.
- Does any spatial pattern vary with time (i.e., encoded as "obs.date" and "[foliage] period") throughout the period of observation from May 2006 to April 2007? In other words, are spatial patterns temporally persistent?

From these questions several consequences for the structure of models of inferential statistics such as linear mixed effects models and nonlinear mixed effects models arise:

- Variance of model residuals among observational groups such as canopy type and period may vary, and this heterogeneity must be considered.
- Temporally and spatially near observations might be strongly correlated, and these correlations must be considered in order to obtain appropriate estimates for the effects of factors with temporal (event gross precipitation) and spatial (relative distance, distance to the nearest tree) dimensions.

If these consequences are not considered, model outputs, interpretations, and conclusions drawn from the final models might be misleading due to underestimation of standard errors of estimated model parameters (Haining 2009; Zuur et al. 2009).

The throughfall dataset contains temporally repeated observations of throughfall and synchronized gross precipitation collected by 175 funnels on 33 rainfall events (i.e., ideal rainfall events with wind speeds <6 [m s^{-1}]; see Frischbier and Wagner (2015)). The funnels are either placed in forest gaps of the studied forest stand or under the canopy of Norway spruce (*Picea abies* (L.) Karst.), European beech (*Fagus sylvatica* L.), and conditions of intensive mixture of both tree species. All funnels were placed in the same 0.75 ha stand along tree-to-tree transects.

The aim of this chapter is to provide guidance on the analysis of spatial and temporal dependencies and to demonstrate how spatially/temporally correlated covariance structures can be assessed and included in inferential statistics such as linear and nonlinear mixed effects models in order to derive more appropriate standard errors, confidence intervals, and *p*-values from model output. For more details on underlying mathematics and further examples on environmental datasets, we recommend the textbooks by Zuur et al. (2009), Pinheiro and Bates (2000), Dale and Fortin (2015), and Zuur et al. (2017). This chapter does not aim to demonstrate how throughfall patterns can be extrapolated in space and at larger scales (for that interested readers are directed, e.g., to Dale and Fortin (2015)).

The chapter begins with a section that aims to continue the data exploration of Chap. 7 in order to learn more about the dataset. The other sections are then dedicated to show which possibilities exist to consider heterogeneity of model residuals and correlations among repeated and spatially related observations for inferential statistics. We will apply linear as well as nonlinear mixed effects models in order to test for effects of the main explanatory variables, to include heterogeneity among observational groups and correlations of observations. We will demonstrate how the several model adjustments affect model output and we will show how model improvement can be assessed. We will briefly discuss the model output and the respective indicators. For interested readers, the real-world dataset and the R code are included online as electronic supplementary material to this chapter (see Extras Springer online). The R code is stimulated by examples included in the textbooks of Pinheiro and Bates (2000) and Zuur et al. (2009) as well as by the broad diversity of tutorials available in the Internet.

8.2 Data Exploration

Along with a comprehensive exploratory data analysis, the main features of the dataset on throughfall, including scales of variables, have been introduced in Chap. 7. Here, the spatial and temporal patterns of throughfall and relevant covariates, such as event gross precipitation (EGP), will be demonstrated.

8.2.1 Properties of Event Gross Precipitation and Throughfall

It can be assumed that throughfall amount at a single event is not random and at least influenced by the characteristics and cumulated amount of the particular rainfall event, i.e., the duration and intensity of gross precipitation. Figure 8.1 (panels **a–c**) shows the magnitude, temporal pattern, and frequency of event gross precipitation of the 33 events. The range of gross precipitation spans from close to 0 mm up to around 50 mm (equivalent to $l\ m^{-2}$). From the density plots in Fig. 8.1 (panels **c** and **d**), it is clear that most observations of EGP and of throughfall are in the lower range (<10 mm). However, the density plot of EGP demonstrates a more differentiated curve with several peaks until ~25 mm. In contrast, throughfall shows in general lower amounts compared to EGP due to interception within the canopy of trees. Gross precipitation will be included as an explanatory variable in inferential statistics later in this chapter.

As already demonstrated in Chap. 7 (Fig. 7.7e, f), throughfall amount increases with increasing EGP. However, this increase is accompanied by an increase in variance of the residuals of the linear model at higher levels of EGP (i.e., heteroscedasticity or heterogeneity of model residuals). Variable transformation, in this case the square root transformation, partially solved the problem of heteroscedasticity. However, as indicated by lower distances between quantiles/whiskers unequal variances still exist between canopy types with higher variances under the canopy of European beech and Norway spruce as compared to mixtures of both species and canopy gaps (Chap. 7, Fig. 7.6b). This observation might be the norm rather than the exception in ecological research and points to the need of

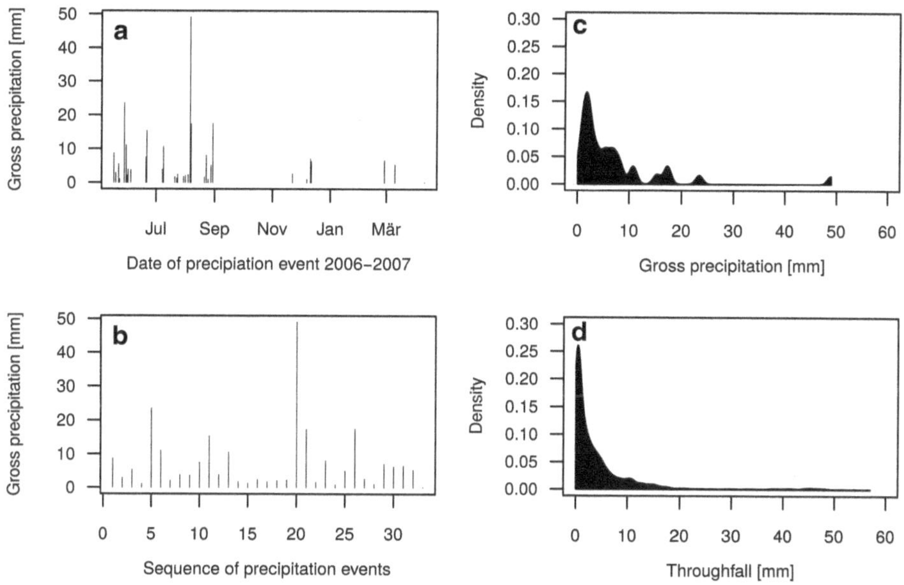

Fig. 8.1 Properties of gross precipitation and throughfall among 33 rainfall events of a mixed European beech-Norway spruce forest in Germany. Panels **a** and **b** show absolute gross precipitation amounts based on date (**a**) and in a temporal order (**b**, rainfall events during periods of frost and with wind speeds >6 m s^{-1} are not included). Panels **c** and **d** summarize event gross precipitation (**c**) and throughfall (**d**) as density plots indicating frequency distributions. The total sum of the area below the curve is 1 (Panels **c** and **d**)

including unequal variances between experimental groups in inferential statistics, which is possible in mixed effects models.

8.2.2 Dependencies Between Throughfall Observations: Spatial, Temporal, and Spatial-Temporal Correlations

The concept of autocorrelation refers to correlations among values of a single variable (Dale and Fortin 2015). It is used to quantitatively test whether subsequent observations are more similar than temporally distant observations (Zuur et al. 2009). Chapter 7 already presented a test for temporal autocorrelation of individual throughfall samples applying the "acf"-function in R. Figure 7.4 shows that throughfall events are not very strongly correlated from the perspective of temporal patterns (<0.4), while more distant events are close to the threshold value for relevant autocorrelation (at lag of 4 and of lag 10). One explanation might be that the included rainfall events are selected in accordance with the definition of an ideal rainfall event, which led to an inconsistent time series. Another explanation is that the observed wavelike pattern may indicate that the properties of the single rainfall event (e.g., intensity, duration) are more relevant than the temporal order of the rainfall events, thereby confirming the observation that the temporal correlation between rainfall events is often low. This also points to the relevance of EGP as explanatory variable for inferential statistics. It is therefore possible to include temporal correlation by adding the amount of event gross precipitation as covariate within explanatory models and thus to account for temporal correlation within the throughfall model. Nonetheless, including many fixed effect covariates and nested factors poses high demands on the dataset and can lead to models that cannot be fitted. For example, including both EGP as fixed covariate and observation date as nested factor for the random term within the model led to error messages. Another question is if continuous covariates should be transformed or not. Zuur et al. (2009) as well as Schielzeth and Nakagawa (2013) recommend standardization of continuous covariates within the inferential model. One reason is to avoid convergence problems during model parameter estimation (Zuur et al. 2009), another reason is to allow cross-study comparisons of model estimates (Schielzeth and Nakagawa 2013). One has to keep in mind that the consequence of standardization is that model estimates for the continuous covariate represent the deviations from the average conditions of the respective study. The information, for example, on the absolute amount of EGP that is needed to saturate canopy storage capacity is then masked.

Figure 8.2 shows how throughfall amounts and relative throughfall (expressed as a proportion of event gross precipitation: TF (%)) are distributed along the *X*- and the *Y*-range of the sample grid. The observations are grouped according to the canopy type. European beech is located more to the lower values of the *Y*-range while Norway spruce and gap are located to higher values of the *Y*-range, which indicates spatial clustering. The canopy types are more evenly distributed along the *X*-range. Panels **c** and **d** indicate regular patterns of TF (%) along both the *X*- and the *Y*-range. Since the selection of the *X*- and *Y*-range is somehow artificial, a calculation of the distance to the nearest tree for each measurement point permits a more reliable consideration of the spatial influence of single tree individuals on throughfall patterns.

The equivalent of a temporal auto-correlogram for spatial pattern analysis is an experimental variogram (Zuur et al. 2017). The underlying question is whether values of a particular variable observed on neighboring sample units, such as throughfall measurement points, are more similar than those from sample units far apart. Variograms can also be used for analysis of temporal patterns in cases of irregularly spaced sampling dates (Zuur et al. 2017). As already indicated by Fig. 8.2 (panels **c** and **d**), a kind of wavelet pattern is partly mirrored by the semivariogram in Fig. 8.3 (panels **a** and **b**). The semivariograms of panel **a** (throughfall amount) and panel **b** (TF, as proportion of EGP) were calculated using the "Variogram"-function of the nlme package for a gls model (i.e., generalized least squares fit according to Crawley (2013)). The semivariogram of the TF (as proportion of EGP) data indicates that spatial dependencies do exist (increasing semivariance up to *x*-axis values (range) of 10), although the *y*-axis intercept is at a high level, which refers to a large nugget effect indicating noise, i.e., variance that is not explained.

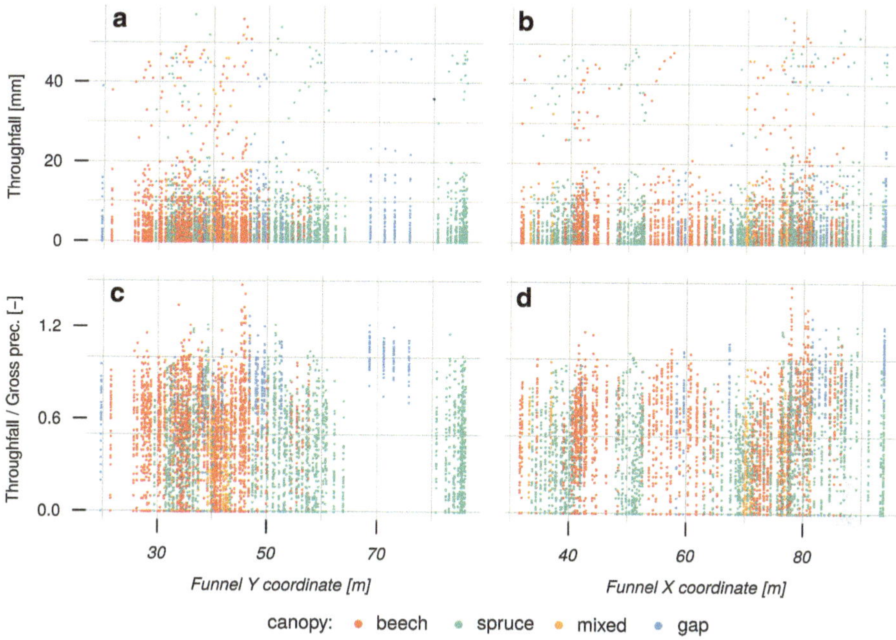

Fig. 8.2 The spatial dimension of throughfall. Panels **a** and **b** show throughfall amounts along X and Y coordinates of the sample grid, respectively. Panels **c** and **d** show the proportion of throughfall relative to event gross precipitation. Different colors represent the canopy types. The sample grid is referenced to the investigated forest stand and is therefore an artificial grid. However, it is possible from this grid to calculate funnel distances and distances to, for example, the nearest tree by using the functionality of respective R packages such as raster (an example for the dataset is included as electronic supplementary material)

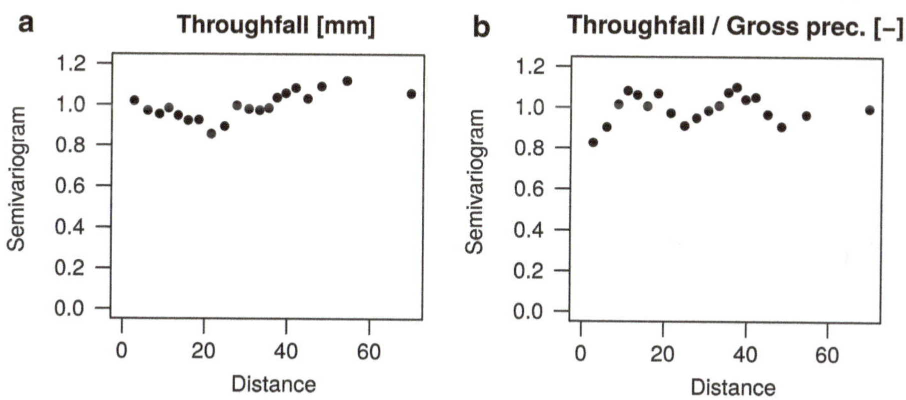

Fig. 8.3 Spatial analysis of throughfall data. Panels **a** and **b** show semivariograms of the absolute throughfall (**a**) and relative throughfall (as a proportion of event gross precipitation) in order to test for spatial and temporal dependencies between rainfall events

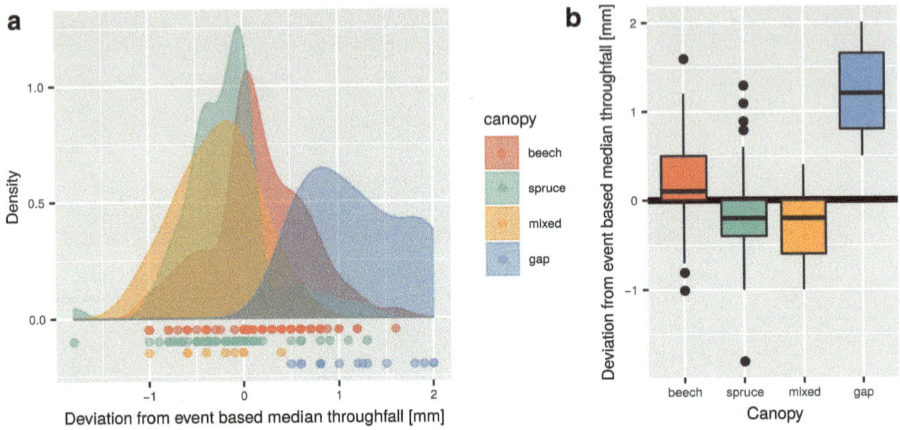

Fig. 8.4 Throughfall deviations from event-based median. (**a**) Density curve and points per canopy type and (**b**) boxplots grouped according to canopy type. Lines indicate median deviation; boxes indicate first and third quantiles; whiskers indicate the ranges ≤1.5∗SD; dots indicate deviations >3∗SD

One abundant question in throughfall research is whether observed spatial patterns are persistent among several rainfall events (i.e., temporal persistence). Several authors investigated temporal patterns of throughfall (Keim et al. 2005; Zimmermann et al. 2008; Zimmermann and Elsenbeer 2008; Keim and Link 2018). Keim et al. (2005), among others, calculated temporal persistence by subtracting the mean or median of throughfall amounts of a particular rainfall event from the amount observed at a particular throughfall measurement point. Positive values indicate higher throughfall than average and vice versa at a particular measurement point. These deviations from average throughfall are then summarized and plotted at a spatial scale by box-whisker plots, for example, indicating whether a single measurement point is systematically above or below average. Figure 8.4 shows the density plots for median deviations (i.e., single value at a collector minus median of all measurement points at an event) of all 33 events (panel **a**) that are summarized in the form of box-whisker plots (panel **b**). Observations were grouped according to canopy type. As indicated by both figures, the density curves and the boxplots, the canopy type "gap" is characterized by persistently higher than average throughfall amounts. In addition, throughfall below European beech appears to be persistently close to or higher than average event-based throughfall. Collectors below Norway spruce and in mixtures are below average. This indicates an effect of canopy type on throughfall amount among the 33 rainfall events. However, all box-whisker plots show substantial variability within canopy types pointing to the fact that other sources of variability are of additional relevance. In addition, one has to keep in mind that such indicators of temporal persistence depend on sample size and represent a first tool for data exploration of throughfall or other temporal repeated datasets.

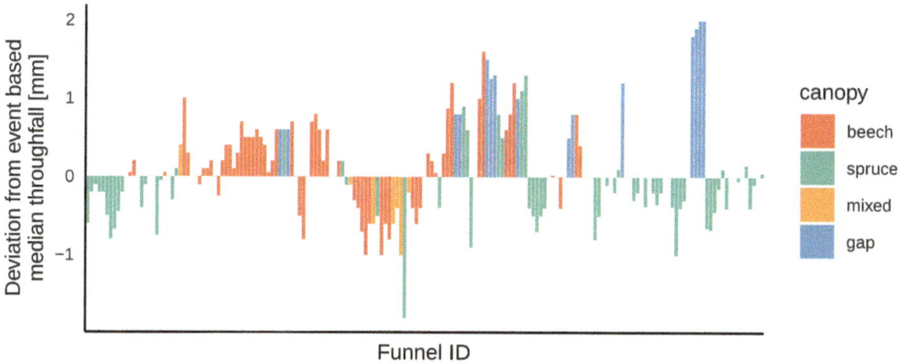

Fig. 8.5 Spatial pattern of temporal stability of throughfall amounts. *X*-axis is order by funnel ID. The *y*-axis shows the median of the deviation of throughfall collector from the median of the rainfall event summarized for 33 events. Funnel ID is related to the spatial position of a measurement point but does not represent an absolutely strict spatial sequence along one single transect

In Fig. 8.5, median deviations are given along the sequence of funnel IDs, which indicates a spatial explicit pattern of temporal persistence, since the designation of the funnel ID and, thus, the order of funnels in the dataset of Frischbier (2012) partly include spatial dependency. Although in particular European beech has a wavelike pattern, one has to keep in mind the possibility that funnel ID is not a strict spatial gradient, since it includes measurement points from different not directly connected transects. Additionally, funnel IDs are linked to single tree individuals (with the exception of canopy type "gap") and correlations between neighbored funnels might be due to the effect of the particular tree individual and its tree and crown characteristics.

8.3 Inferential Statistics

8.3.1 General Aspects

In the section above, a set of spatial and temporal analyses of the throughfall dataset was applied. One of the main objectives of the study of Frischbier (2012) was to test for differences between canopy structures of observed tree species. The challenge in this chapter is to consider:

- Heterogeneity of model residuals among groups (i.e., levels of a factor).
- Spatial and temporal dependencies/correlations of observations for inferential statistics.

One approach is the use of mixed effects models (Pinheiro and Bates 2000; Zuur et al. 2009; Zuur et al. 2017). A mixed effects model contains fixed effects as well as random effects. The fixed part of the model consists of the explanatory variables

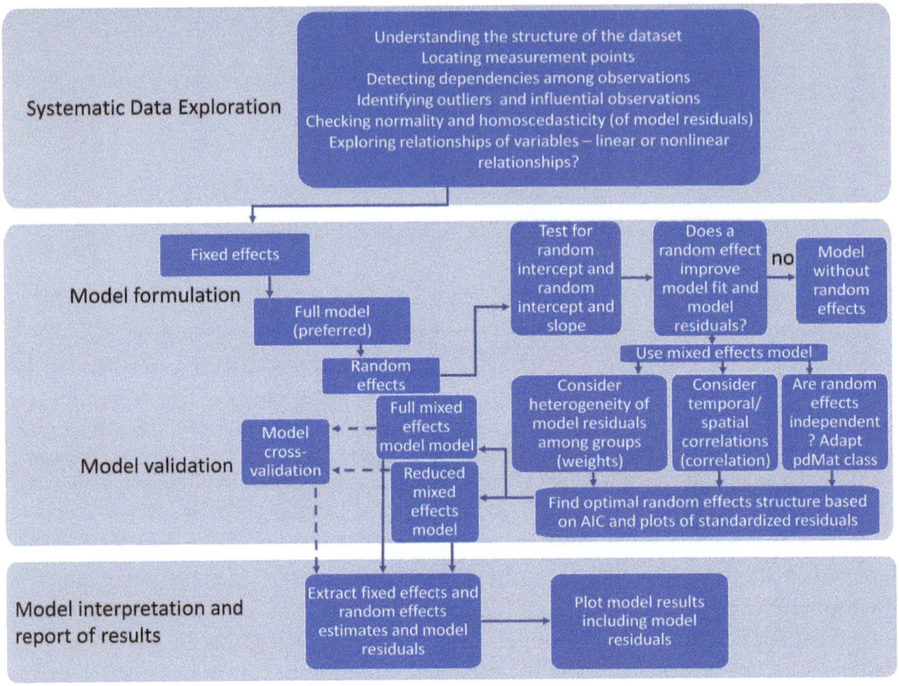

Fig. 8.6 Flowchart on the main steps for model formulation and validation based on Pinheiro and Bates (2000) and Zuur et al. (2009). The nlme and lme4 packages in R can be used for the main steps of model building and testing. See Chap. 7 of this book for more information on systematic data exploration

such as tree species or canopy structure and period in our case. The random part of the model consists of a real random term and components that include heterogeneity, nesting of data, and temporal as well as spatial correlations of observations (Zuur et al. 2009). With random effects it is possible to consider heteroscedasticity of within-group errors and dependencies, i.e., spatial and temporal correlations, as well as deviations from the normal distribution (Pinheiro and Bates 2000). This offers clear advantages for model parameter estimation and robustness in comparison to classical analysis of variance procedures (Pinheiro and Bates 2000; Schielzeth and Nakagawa 2013).

The definition of an appropriate model is sometimes challenging. Zuur et al. (2009) proposed a very helpful ten-step protocol that structures a systematic procedure of model development and validation of assumptions. In Fig. 8.6, we take up some of these steps and combine them with further recommendations by Pinheiro and Bates (2000). A first question is on which factors and covariates are available and should be included. When all factors and covariates are defined (ideally before an experiment is conducted), the question is whether full models (including all predictors and relevant interactions among them) or reduced models have to be selected. Based on model simulations, Schielzeth and Nakagawa (2013) recommended the selection of full models since they best represent the dataset and are less sensitive to biased standard errors, confidence intervals, and p-values. In contrast, Zuur et al. (2009) proposed the use of reduced, simple, but sufficient models with the final selection of factors and covariates based on information criteria, such as Akaike's Information Criterion (AIC) (Akaike 1998) or the Bayesian Information Criterion (BIC). Both criteria measure the model fit and consider the complexity of the tested model with the lower the values for AIC and BIC the more information is contained in the model (Zuur et al. 2009). The latter strategy of defining reduced models may result in easier interpretations of model output but is prone to biased model estimates (Schielzeth and Nakagawa 2013; Zuur et al. 2017). However, it appears to be a good choice to start with full models or at least with a model that includes all relevant explanatory variables and plausible interactions (Zuur et al. 2009). Sometimes full models do not run due to numerical and convergence problems (Zuur et al. 2009). Full or close to full models are used for finding the optimal random effects structure and are then reduced or not reduced based on AIC or BIC and residual patterns.

Another question is whether a linear or a nonlinear model approach is more suitable regarding the data structure. The decision on the shape of the relationships can be done by insights from systematic data exploration or from expected theoretical relationships between response and explanatory variables. In this chapter we will apply both linear mixed effects model (LMM) and nonlinear mixed effects model (NLMM) to the throughfall dataset. NLMMs extend LMMs by allowing the regression function to depend nonlinearly on fixed and random effects and are generally better interpretable (Pinheiro and Bates 2000).

8.3.2 Model Formulation and Model Validation

The main predictors of the throughfall study of Frischbier (2012) are canopy type (with four levels), period (two levels with leafed and non-leafed period), some tree metrics such as tree diameter and height, event gross precipitation, and variables related to the spatial position of a single throughfall collector (i.e., coordinates of funnels within the X-Y-sample grid, distances of funnel from the tree stem). Repeated measures are included in some of the variables of the available dataset, i.e., period (leafed and non-leafed), date or series ID ($n = 33$), as well as event gross precipitation amount ($n = 33$). Not all of them can be included in the full model, since the canopy type "gap" is not assigned to a specific tree with respective tree metrices. However, since the study is a spatial explicit investigation, distances of gap funnels to the nearest tree can be calculated from the X-Y sample grid, and thus, the potential effect of a spatial covariate can be included.

For demonstration issues, it is useful to start with a simple linear model (lm) that includes throughfall, square root transformed EGP according to the data exploration in Chap. 7 (as well as z-transformed using the scales function with center = TRUE and scales = TRUE), canopy type, and period as factors. It reveals strong and significant effects of EGP, canopy type, period, as well as the period-specific differences between canopy types as indicated by a significant interaction term. The R^2 of the model is high (0.9118). However, we have to keep in mind one key massage from Pinheiro and Bates (2000)—that is to examine that the assumptions have been followed once we have derived a model.

The diverse statistical software packages within the R environment (R core team 2013) provides graphical methods to test for violations of the model assumptions. Classical tools are the:

- plot(*model*)-function drawing the standardized model residuals as a function of the fitted values (standardized residuals are best for assessing heteroscedasticity).
- qqnorm(*model*)-function drawing a quantile–quantile plot of the normalized residuals.
- qqnorm(*model*, ~ranef(.)) command drawing the respective plots to test for normality of residuals of random effects in mixed effects models.
- plot(Variogram(*model*)) command to illustrate autocorrelation within the model (semivariogram of the within-group residuals).
- plot(*model,* resid(., type = "pearson") ~ *variable X*)-function allows for testing if patterns in model residuals are affected by individual explanatory variables.

For the linear model estimated above, the plot of model residuals against event gross precipitation (Fig. 8.7a) reveals significant patterns, such as the increase of the

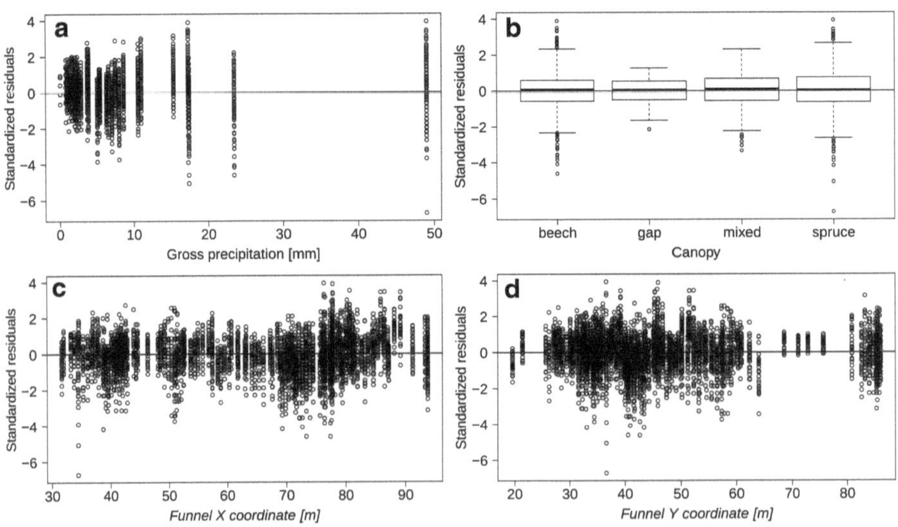

Fig. 8.7 Scatterplots of standardized residuals of a linear model. The model ignores the spatial–temporal dependencies of the data. In panel **a** the relationship between the standardized residuals and the main covariate EGP is investigated. Panel **b** shows the standardized residuals as a function of canopy type (a grouping factor) in order to test for heteroscedasticity. Panels **c** and **d** investigate the spatial scale of variability of standardized residuals of the respective linear model in order to indicate whether the position within the sample grid affects the model residuals

variance of residuals as EGP increases (ideally an evenly spread of residuals among the *x*-axis is preferred). In addition, variances of model residuals are not homogeneous among canopy types (Fig. 8.7b). Both observations indicate violations of important model assumptions. Thus, this linear model, which does not account for heterogeneity of model residuals among groups as well as spatial and temporal patterns, should not be used. An additional check can be done by plotting standardized model residuals along the spatial dimensions of the study, that is, the *X*- and *Y*-coordinates of the throughfall funnels (Fig. 8.7c, d). This indicates unevenly distributed residuals in space such as a tendency for patterned deviations along the *X*-coordinates.

These observations lead to the conclusion that we have to consider heteroscedasticity and dependencies of within-group errors in the model building process. Two sources of heterogeneity of model residuals in the dataset are:

- Canopy type.
- Period.

The nlme package (Pinheiro et al. 2013) offers a great variety of possibilities for extending linear and nonlinear models for random effects in order to account for and to explore the above mentioned violations of model assumptions, respectively inherent important ecological properties of the dataset. The tool for including heterogeneity of model residuals in mixed models is:

- To define the variance structure by the *weights* argument (to deal with heterogeneity; see Table 8.1 and cited references).

Inherent to our dataset is temporal and spatial correlation of observations. Two sources for temporal and spatial correlation in the dataset are:

- Temporally repeated observations at the same funnel.
- Observations along transects (and below single trees and among neighbored trees).

The nlme package (Pinheiro et al. 2013) offers several possibilities for including temporal and spatial correlated observations in the model. The tool for including correlation among observations in mixed models is:

- To define the correlation structure by the *correlation* argument (to account for correlations among observations; see Table 8.2 and cited references).

Another task is to find a suitable variable that can be attributed as a random effect. This is done before variance–covariance structure and the correlation between observations of the model is optimized in terms of AIC or BIC (Zuur et al. 2009). The random effects structure usually results from the design of the experiment. For example, taking independent samples at different locations to investigate a particular treatment effect results in a random effect of location that is not of primary interest to the study. Data derived from experiments in which the single observations are grouped according to predefined classifications can be stored in R a "groupedData" object available from the nlme package (Pinheiro and Bates 2000). This is

Table 8.1 Overview on the various variance structures that can be specified in order to account for heteroscedasticity

Variance function in R (nlme) specified for weights argument	Variance assumption	Description
varFixed (~variable)	Variance is proportional to the covariate	Fixed variance.
varExp (form = ~variable)	Variance is proportional to the exponential of a covariate	Exponential of the variance covariate; instead of using covariates the fitted values of the model can be used.
varPower (form = ~variable)	Variance is proportional to the power of a covariate	Power of the variance covariate; instead of using covariates the fitted values of the model can be used.
varConstPower (form = ~variable)	Variance is proportional to the power of a covariate plus a constant	
varIdent (form = ~1| A∗B)	A (and B) is a factor, the variance is allowed to vary for each level of the factor	Different variances per group or stratum or factor combinations
varComb()	Combinations of the other variance functions	

The table is adapted from Pinheiro and Bates (2000), Zuur et al. (2009) and http://www.flutterbys.com.au/stats/tut/tut8.2a.html. (Assessed on February 27, 2019).

particularly useful as it greatly simplifies the plotting and analysis of data from such experiments. Details on "groupedData" objects can be found in Chap. 3 of Pinheiro and Bates (2000).

One candidate for the grouping variable and the random effect in the models investigated here is the funnel ID ($n = 175$). Multiple observations of throughfall are nested within the funnel and each funnel represents a distinct spatial entity as defined by its coordinates and its relationship to all other funnels and tree individuals in the forest stand. We used funnel ID as grouping variable that divides the observations into the distinct groups of observations. Throughfall is the response variable and EGP was used as primary explanatory variable, while canopy type and period are defined as further experimental factors. The plot() function allows for rapid assessment whether funnel ID has to be considered as random effect within mixed effects models. Figure 8.8 indicates a large variability in regression slopes among individual throughfall funnels and the variability in regression slopes exceeds the magnitude of variability of canopy type and period. We can conclude that funnel ID is suitable as random effect variable to account for nested sampling. Another observation with relevance for model setup and interpretation is the difference in magnitudes of EGP between leafed and non-leafed period, given by a much narrower distribution of EGP in the lower range of absolute values in the non-leafed season of the year.

The plot of funnel-specific regression parameters in Fig. 8.9 supports the abovementioned observation. The slopes and the respective confidence intervals are higher than zero indicating an effect of throughfall.

Table 8.2 Overview on the various correlation structures that can be specified in order to account for autocorrelation among observations

Predefined classes for correlation structures in R (nlme)	Name	Description/Application
corAR1	Autoregressive of order 1	Commonly used for time-series analyses; parameter phi indicates correlations between neighbored observations. Degree of correlation decreases, that is, exponential decay of correlation with temporal distance.
corCAR1	Continuous-time AR(1)	Commonly used for time-series analyses; parameter phi indicates correlations between neighbored observations. Degree of correlation decreases exponentially with temporal distance (time variable does not have to be integers).
corARMA	Autoregressive-moving average	Commonly used for time-series analyses; parameters phi and theta indicate correlations between neighbored observations (first lag or first two lags). Higher order than AR1 and/or moving average.
corCompSymm	Compound symmetry	All observations of the same group are correlated. Correlation does not exponentially decay with greater temporal distance.
corSymm	General correlation structure	Each value in the correlation matrix can have separate values.
corLin	Linear correlation structure	Semivariance increases linear until it reaches maximum (=distance of uncorrelated observations). Do not accept observations that have identical *X* and *Y* coordinates (zero distance) in the dataset and cannot be applied for datasets with temporally repeated measures at the same point.
corExp	Exponential correlation structure	Semivariance increases exponential until it reaches maximum (=distance of uncorrelated observations). Do not accept observations that have identical *X* and *Y* coordinates (zero distance) in the dataset and cannot be applied for datasets with temporally repeated measures at the same point.
corGaus	Gaussian correlation structure	Semivariance increases following a Gaussian model until it reaches maximum (=distance of uncorrelated observations). Do not accept observations that have identical *X* and *Y* coordinates (zero distance) in the dataset and cannot be applied for datasets with temporally repeated measures at the same point.
corSpher	Spherical correlation structure	Semivariance increases following a Gaussian model until it reaches maximum (=distance of uncorrelated observations). Do not accept observations that have identical *X* and

(continued)

Table 8.2 (continued)

Predefined classes for correlation structures in R (nlme)	Name	Description/Application
		Y coordinates (zero distance) in the dataset and cannot be applied for datasets with temporally repeated measures at the same point.
corRatio	Rational quadratic correlation structure	Semivariance increases following a quadratic model until it reaches maximum (=distance of uncorrelated observations). Do not accept observations that have identical *X* and *Y* coordinates (zero distance) in the dataset and cannot be applied for datasets with temporally repeated measures at the same point.

The table is adapted from Pinheiro and Bates (2000) and *http://plantecology.syr.edu/fridley/bio793/mixed2.html.* (Assessed on February 27, 2019)

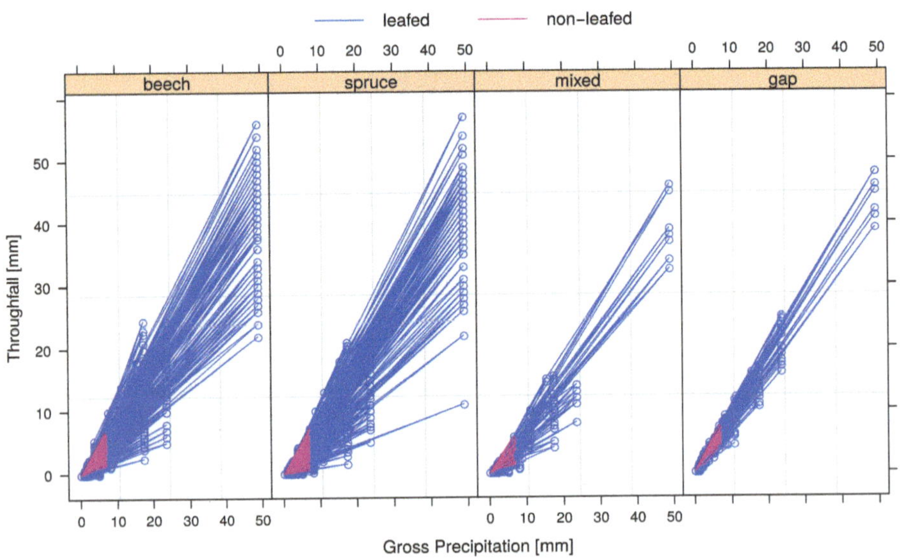

Fig. 8.8 Throughfall as a function of event gross precipitation, canopy type, and period grouped within funnel ID (model similar to fm1a.lme)

In Table 8.3, the main characteristics of a set of LMMs are listed. We started with a model that implies that throughfall is modeled as an intercept plus a random term (fm1.lme). This model serves as a baseline model (e.g., for comparing AIC and residual patterns). Model fm1a.lme includes all fixed factors (full model) and allows for random intercept of each funnel. Model fm1b.lme extends the previous model and allows that the slope of EGP can vary among funnels. This is a clear improvement (see AIC) and we will use this random model for the next steps that consider heterogeneity among groups. Model fm2.lme and model fm2a.lme consider

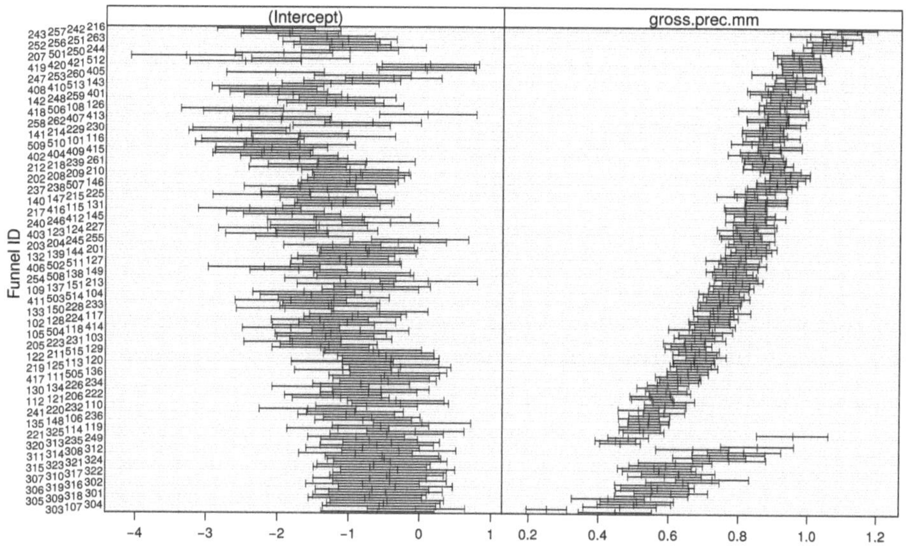

Fig. 8.9 Ninety-five percent confidence intervals on intercept and slope for each funnel of the throughfall dataset. Event gross precipitation represent the slope of the linear mixed effects model (model similar to fm1a.lme)

heterogeneity of model residuals among canopy types or period, respectively. Both models improve model fit and changed patterns of standardized residuals. However, the best adaption is to consider both, heterogeneity of residual variances among canopy types and period (fm2b.lme). In models fm2c.lme and fm2d.lme we tested whether the inclusion of the fitted values or the residuals of the model as variance covariates improved model fit. Indeed, the AIC decreased strongly and the patterns of residuals changed substantially. However, the plot of throughfall as a function of the fitted values of both models indicate larger variability as for model fm2b.lme. Therefore, we proceeded with this model to adapt the correlation structure. Another option for model improvement is to adapt the variance–covariance structure of the random effects, that is a question on how the random effects are correlated (pdMat classes, see Pinheiro and Bates (2000)). In the case of the tested LMMs, the default setting “pdSymm” was best.

In the next step, we aimed to include correlations among observations. We used two different options of autoregressive correlations (corAR1 and corARMA). The spatial correlation cannot be included in the model since in the dataset are repeated observations at the same location. Both variants improved the AIC and estimates for phi and theta indicated correlation between observations. For model comparisons, the “anova” command can be used to test for differences between models estimated by Maximum likelihood (ML). The model parameter estimation of the final model should be done by Restricted Maximum likelihood estimation (REML, see Zuur et al. (2009) for a discussion).

Table 8.3 Overview of main characteristics of tested linear mixed effects models

Model number (abbreviation in text)	Type of model	Adjustments/ structure	AIC	Residual patterns/ notes
1 (fm1.lme)	LMM	Fixed effects: No Random effects: 1|funnel.id Estimation by maximum likelihood (ML)	37,891.1	Strong patterns in standardized residuals and positive relation to EGP.
1a (fm1a.lme)	LMM (random intercept)	Fixed effects: EGP∗canopy type∗period∗nearest tree distance Random effects: 1|funnel.id Estimation by ML	23,801.1	Spread of standardized residuals is reduced but new patterns exist. Variance of residuals increases with increasing fitted values. Higher variance of standardized residuals with lower values for nearest tree distance. Higher variance of standardized residuals with higher values for EGP.
1b (fm1b.lme)	LMM (random intercept and slope)	Fixed effects: EGP∗canopy type∗period∗nearest tree distance Random effects: EGP|funnel.id Estimation by ML	21,317.6	Large spread of residuals for single trees with some trees showing larger variances than others (tree.id is not included as factor since funnels in gaps are not assigned to a particular tree). Mainly positive residuals at high fitted values. As suggested by the reduction of AIC, model 1b represents the "best" model applying the anova test for model comparison ($p < 0.0001$).

(continued)

Table 8.3 (continued)

Model number (abbreviation in text)	Type of model	Adjustments/ structure	AIC	Residual patterns/ notes
2 (fm2.lme)	LMM (including heteroscedasticity)	Fixed effects: EGP∗canopy type∗period∗nearest tree distance Random effects: EGP|funnel.id Weights = varIdent (1|canopy) Estimation by ML	20,950.7	Close to the pattern of model 1b with a slight reduction in magnitude of standardized residuals. Model allows for different standard deviations per canopy type with spruce as reference (1), beech (79% of spruce), and gap (48%) showed lower standard deviations (mixed with 1.01 resample that of spruce).
2a (fm2a.lme)	LMM	Fixed effects: EGP∗canopy type∗period∗nearest tree distance Random effects: EGP|funnel.id Weights = varIdent (1|period) Estimation by ML	20,362.5	Close to the pattern of model 1b with a slight reduction in magnitude of standardized residuals. Model allows for different standard deviations per period with nonleafed period characterized by 39.5% of the standard deviation of the leafed period.
2b (fm2b.lme)	LMM	Fixed effects: EGP∗canopy type∗period∗nearest tree distance Random effects: EGP|funnel.id Weights = varIdent (form = ~1| canopy∗period) Estimation by ML	20,038.3	Only minor changes in residual patterns. Variogram shows autocorrelation of within-group residuals. Reduction of standardized residuals of period and canopy. Same residual patterns for nearest tree distance and EGP as above.

(continued)

Table 8.3 (continued)

Model number (abbreviation in text)	Type of model	Adjustments/ structure	AIC	Residual patterns/ notes
2c (fm2c.lme)	LMM	Fixed effects: EGP∗canopy type∗period∗nearest tree distance Random effects: EGP|funnel.id Weights = varExp (form = ~resid(.)) Estimation by ML	16,270.3	Completely different pattern of model residuals (L-shaped). The plot of observed throughfall over fitted values indicate poorer fit as compared to fm2b. lme, despite AIC indicate improvement of model fit.
2d (fm2d.lme)	LMM (combined variance structure considering variability per canopy type and period as well as increasing variance with fitted values)	Fixed effects: EGP∗canopy type∗period∗nearest tree distance Random effects: EGP|funnel.id Weights = varComb (VarIdent(form = ~ 1 | period∗canopy), varExp(form = ~fitted(.))) Estimation by ML	15,022.8	Comparable pattern as observed forfm2c.lme. Pattern within plot of standardized residuals and EGP changed and tend to disappear. The plot of observed throughfall over fitted values indicate poorer fit as compared to fm2b. lme, despite AIC indicates improvement of model fit. Residual plots for canopy and period indicating lower suitability of model fm2c.lme and fm2d.lme. Note: Final model should be estimated by REML.
3 (fm3.lme)	LMM (including correlation among observations)	Same as model 2b plus correlation = corAR1 (form = ~obs.date| funnel.id)	19,875.8	Compared to fm2b. lme model residuals only slightly changed. Parameter phi = 0.19, indicating (weak) correlations between

(continued)

Table 8.3 (continued)

Model number (abbreviation in text)	Type of model	Adjustments/ structure	AIC	Residual patterns/ notes
				consecutive observations. Note: Final model should be estimated by REML
3a (fm3a.lme)	LMM	corARMA does not work for the full model. Adapted fixed effects: EGP∗canopy type∗period + nearest tree distance Correlation = corARMA (form = ~1|funnel. id, $p = 1$, $q = 2$) Estimation by ML	19,048.6	Compared to fm2b. lme model residuals only slightly changed. Parameter phi = 0.79, indicating strong correlations between consecutive observations. Theta2 (−0.57) indicates the wave-like pattern

Note: not all possible alternative models in terms of weights argument are listed, but were tested. Only the models with lowest AICs are reported

However, the intercepts of the linear regressions are almost negative suggesting negative throughfall amounts when event gross precipitation is zero. This points to another important issue that is masked by the high R^2 of the linear model applied earlier in this chapter. The linear model probably does not reflect the mechanism in the relationship between EGP and throughfall: If a rain event starts, it is more likely that rain drops are partly retained within the canopy on the surface of leaves and twigs and throughfall just starts to significantly set in linear according to EGP if this storage is saturated (Allen et al. 2014; Frischbier and Wagner 2015). This canopy storage capacity can be included in a nonlinear model (formula 8.1) as described in Frischbier and Wagner (2015).

$$TF = \text{beta} * \text{EGP} * \left(1 - \exp^{-(1/\text{alpha}*\text{EGP})}\right) \tag{8.1}$$

where TF is the throughfall (mm) as an exponential function of event gross precipitation (mm) and two model parameters (alpha and beta). Of both estimated parameters, alpha is of particular interest since it is an estimate on the amount of event gross precipitation that is required to saturate the canopy storage capacity (Frischbier and Wagner 2015). The nlme package provides possibilities for the formulation of nonlinear models and subsequent estimation of model parameters, including self-starting functions, that better mirror mechanistic relationships in environmental datasets (Pinheiro and Bates 2000; Pinheiro et al. 2013). Table 8.4 provides an overview of the main characteristics on NLMMs. An additional benefit of nonlinear

Table 8.4 Overview of the main characteristics of tested nonlinear mixed effects models (NLMMs). (Note: not all possible alternative models in terms of weights argument are listed, but were tested. Only the models with lowest AICs are reported)

Model number (abbreviation in text)	Type of model	Adjustments/structure	AIC	Residual patterns/notes
4 (fm4.nlme)	NLMM	Fixed effects: EGP, beta and alpha according to formula (8.1) Random effects: Alpha ~ 1 | funnel.id Estimation by maximum Likelihood (ML)	22,838.6	Increasing standardized residuals with increasing fitted values and EGP. Tendency for heterogeneity of residual variances among canopy types and between periods. Higher variances of standardized residuals at lower tree distances and underestimation at higher nearest tree distances.
5 (fm5.nlme)	NLMM	Fixed effects: EGP, beta and alpha according to formula (8.1) Random effects: Alpha ~ 1 | funnel.id; pdDiag(beta + alpha ~ 1) Estimation by ML	20,837.1	Substantial change in residual patterns, now close to fm2b.lme.
6 (fm6.nlme)	NLMM	Fixed effects: EGP, beta and alpha according to formula (8.1) plus alpha∗canopy and beta∗canopy Random effects: Alpha ~ 1 | funnel.id; pdDiag(beta + alpha ~ 1) Estimation by ML	20,774.1	Close to fm5.nlme.
7 (fm7.nlme)	NLMM	Fixed effects: EGP, beta and alpha according to formula (8.1) plus alpha∗canopy and beta∗canopy∗period Random effects: Alpha ~ 1 | funnel.id; pdDiag(beta + alpha ~ 1) Estimation by ML	20,590.3	Close to fm6.nlme. Residual patterns for canopy type and period were improved. Intercept of alpha tend to deviate from normal distribution.
8 (fm8.nlme)	NLMM	Fixed effects: Same as fm7.nlme Random effects: Alpha ~ 1 | funnel.Id; pdDiag(beta + alpha ~ 1); Weights = varIdent (1|canopy) Estimation by ML	20,415.4	Close to fm7.nlme. Note: Final model should be estimated by REML.

(continued)

Table 8.4 (continued)

Model number (abbreviation in text)	Type of model	Adjustments/structure	AIC	Residual patterns/notes
8a (fm8a. nlme)	NLMM	Fixed effects: Same as fm7.nlme Random effects: Alpha ~ 1 | funnel.id; pdDiag(beta + alpha ~ 1); Weights = varIdent (1|period*canopy) Estimation by ML	19,524.8	Close to fm7.nlme. Magnitude of residual variances decreased. However, variogram indicates increased correlation between observations. In addition, residual patterns for groups (canopy type and period) shifted in an unwanted manner. Note: Final model should be estimated by REML.
9 (fm9.nlme)	NLMM	Fixed effects: Same as fm7.nlme Random effects: Alpha ~ 1 | funnel.id; pdDiag(beta + alpha ~ 1); Weights = varIdent (1|period*canopy) Correlation = corARMA (c(0.1, 0.1, 0.1), form = ~1| funnel.id, p = 1, q = 2) Estimation by ML	18,714.6	Close to fm7.nlme. Magnitude of residual variances decreased. However, variogram indicates increased correlation between observations. In addition, residual patterns for groups (canopy type and period) shifted in an unwanted manner.

(mixed) models is that the original scales of the data are easier to implement and additional data transformations (and respective back transformations of model output) are not required (see also Chap. 7 for a discussion of data transformation). Frischbier and Wagner (2015) extended the model expressed in formula (8.1), however, for the sake of simplicity we keep on working with this formula in the course of this chapter.

Applying the "nls" function to formula (8.1) estimates alpha and beta for each funnel ID. Without considering canopy type or period, the average value for alpha is 6.70 ± 0.17 SE ($p < 0.001$) indicating that a mean event gross precipitation of 6.70 mm is needed until throughfall begins (for this study). The diagnostic plots on the model residuals indicate, heteroscedasticity along both the EGP and the canopy types (Fig. 8.10a and b).

It seems to be reasonable to assume that at least one of the estimated parameters may vary among measurement points. Therefore, the next step in the model building process is to extend the nonlinear model to a first nonlinear mixed effects model that allows the parameter alpha, as an indirect estimate of canopy storage capacity, to vary with funnel ID. This model (fm4.nlme) now considers repeated measures at a single measurement point for several observation dates.

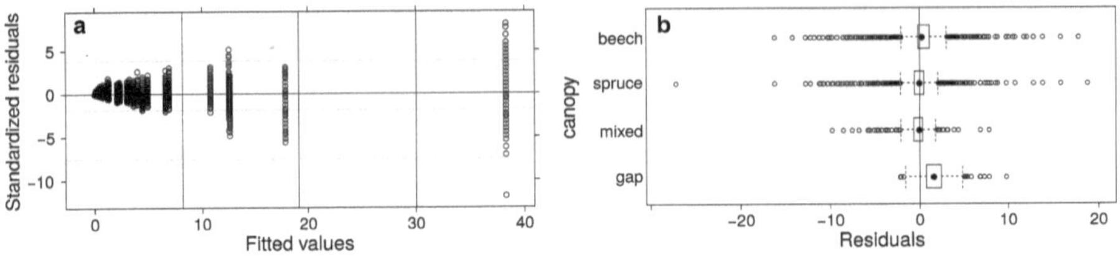

Fig. 8.10 Scatterplot (**a**) and box-whisker plot (**b**) of residuals of a nonlinear model. The model ignores the spatial–temporal dependencies of the data

Model estimates significantly changed as compared to the nonlinear model without a random effect. Average alpha increased to 13.61 ± 0.87 SE ($p < 0.001$) suggesting that a higher average amount of event gross precipitation is needed to generate throughfall linearity in the forest stand than previously estimated. Next, tasks in stepwise model improvement are to check whether estimated model parameters (alpha and beta) are correlated (= 0.14) and whether heteroscedasticity for canopy types can be taken into account. The later model indicates that the within-group variance of model residuals of beech and gap are both 90.8 and 58.3% of spruce, respectively, while variability of model residuals of mixed canopy strata is close to that of spruce (99.2% of spruce). A further step is to test a model that allows for period and canopy type specific beta estimates and canopy type specific variation of alpha (fm6.nlme). Compared to the first nonlinear mixed effects model, this further improved the throughfall model as suggested by the lower AIC and the results of the model comparison by the log-likelihood ratio test (L. ratio = 2451.2, $p < 0.0001$), as called by the "anova" command. Model estimates indicate significant interactions between beta and period, beta and canopy type as well as between beta, period, and canopy type. Alpha values significantly differ between beech (5.79 ± 0.44), gap (1.15 ± 1.06), spruce (10.78 ± 0.66), and mixed (10.66 ± 1.45), confirming model estimates by Frischbier and Wagner (2015). The model has an R^2 of 0.9654, meaning that the model accounts for nearly all the given variation.

Diagnostic plots for this model (Fig. 8.11a–c) indicate an underestimation of throughfall in the range of high throughfall values (related to high gross precipitation events) and an overestimation in the intermediate throughfall range. Model residuals among canopy types now center around zero indicating no violation of the respective model assumptions. Quantile–quantile plots of both model estimates indicate a normal distribution of model residuals (with a tendency of violation for alpha). Panel **d** in Fig. 8.11, a variogram, indicates autocorrelation between the grouping factors for observations, that is, funnel ID. These diagnostic plots lead to the conclusion that other factors have to be included, which account for spatial effects of canopy structure and position of the funnel. Further model improvement can be achieved by allowing for heterogeneity of model residuals among canopy types and periods (fm8.nlme and fm8a.nlme). Model (fm9.nlme) included corARMA and further led to a decrease of AIC, however, it showed negative impacts on residual

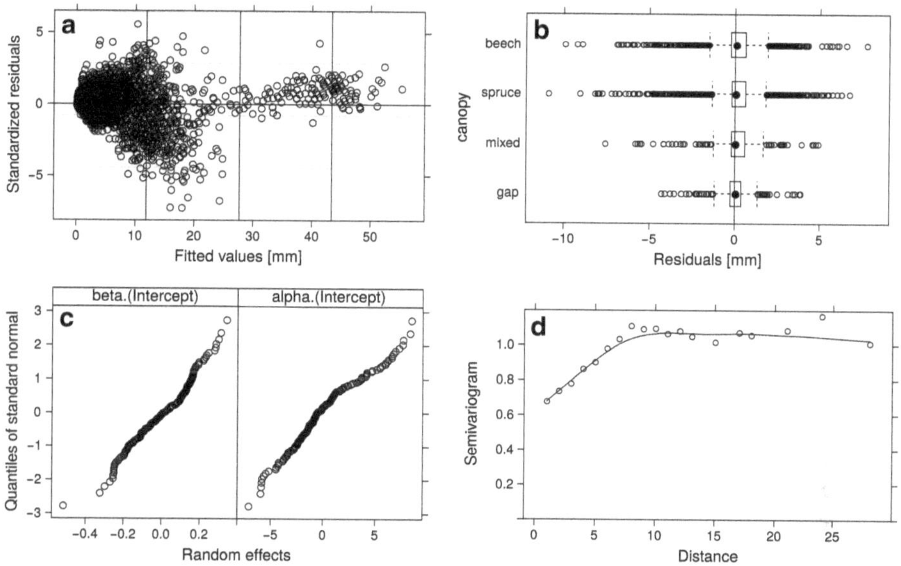

Fig. 8.11 Scatterplot (**a**) and box-whisker plot (**b**) of standardized residuals of a nonlinear mixed effects model. The model allows for heteroscedasticity between canopy types (**c**), canopy type and period specific parameters alpha and beta but it does not explicitly include dependencies of the observations. Panel **c** shows quantile–quantile plots for the random effects in order to check for normality of residuals of random effects. Panel **d** is the semivariogram of the respective model and shows the correlations among funnels

patterns. Therefore, we would prefer the NLMM without an additional correlation structure.

In the previous section, we tested several steps of model adaptation to the structure of the dataset. We applied linear, linear mixed effects model as well as the corresponding nonlinear models. Strong improvements of model in terms of model fit (indicated by AIC) and in terms of patterns of standardized residuals were achieved by including random effects. In particular, the consideration of heterogeneity among canopy types and periods were suitable modifications. The additional consideration of the correlation of observations was found to further decrease AIC and hence, improved a measure of model fit. However, inconsistent effects on patterns of model residuals and on the modeled throughfall were the consequence. The nonlinear mixed effects models offered the possibility to obtain model estimates that can be better explained in terms of ecological and ecohydrological theory (compare, for example, Fig. 8.12c and d with focus on the shape of the relationships at lower range of values). The estimated values of the parameter alpha, an indirect indication for canopy storage capacity, were sensible to model improvement. With the current model approach, it was not directly possible to include spatial correlation models for our dataset with repeated measures at the same location. We included variables such as distance to the nearest tree to explicitly consider spatial information, which improved the information contained within the model. However, the

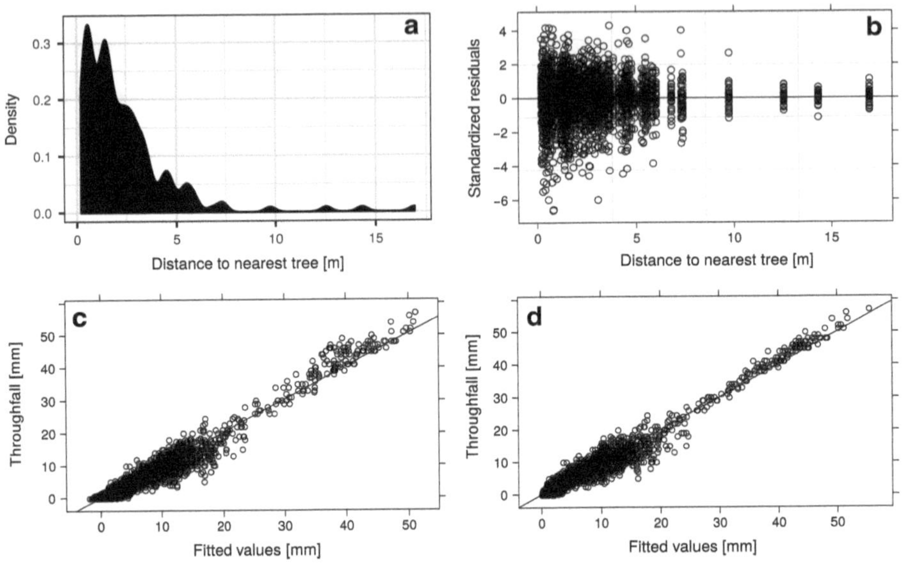

Fig. 8.12 Density plot (**a**) of nearest tree distance calculated for all 175 throughfall collectors. Scatterplots of standardized residuals of a nonlinear mixed effects model (fm8.nlme) and distance to the nearest tree (**b**). The model allows for heteroscedasticity between canopy types and period, canopy type and period specific parameters alpha and beta but it does not explicitly include dependencies of the observations. Panel **c** compares the fitted values of a linear mixed effects model (fm3.lme) to the observed throughfall values. Panel **d** compares the fitted values of a nonlinear mixed effects model (fm8.nlme) to the observed throughfall values

final models still showed residual patterns correlated with nearest tree distance (Fig. 8.12b). This might be related to some properties of the dataset such as a high degree of clumping of funnels with low distances to the nearest tree (Fig. 8.12a) and the abundance of low-intensity rainfall events during non-leafed period (Fig. 8.8). Applying R^2, as a common indicator of model fit, generally revealed values >0.90 for the models, and the fitted values generally well reflect the observed throughfall data (Fig. 8.12c and d). It is likely that the models might be overfitted, which can be resolved in detail using cross-validation strategies. For this, datasets are split in a test and a training dataset. Models that have been fitted with the training data are then evaluated in their performance on predicting the test data. The simplest approach is to extract random observations from the full dataset to generate the test dataset. However, for structured data such as the throughfall dataset, a nonrandom, blocked cross-validation approach has been demonstrated to be universally more appropriate, e.g., providing reliable error estimates and selecting better predictive models (Roberts et al. 2017). This blocked cross-validation evaluates a model's performance based on a test dataset that is similarly structured as the training dataset and thus also reflects a grouping of the data according to spatial, temporal, or hierarchical factors. Roberts et al. (2017) provide a guide on how to block. However, if data are scarce but highly structured, such as in our example, it might not be possible to achieve the required independence between training and test datasets.

8.4 Outlook and Recommendations for Future Work

A key message of Pinheiro and Bates (2000) is that the model building processes in the course of empirical data analysis have to be accompanied by tests whether models violate important model assumptions and properly consider correlations among observations. Dale and Fortin (2015) pointed out that almost any system inherently has spatial and temporal correlations; thus, the appropriate test and, if possible, consideration for such types of autocorrelation are necessary.

- In this spirit, we call for the report of diagnostic plots of model residuals, at least in supplementary materials as this is possible for almost all journals.

It is not a shame to find that final models still show patterns in residuals. It bears great potential for further experiments since this points to the fact that other explanatory variables have to be included or further shifts in model structure that were not previously intended have to be uncovered (Zuur et al. 2017). These steps can stimulate further research in order to find unexplored factors, inter- and extrapolate ecological phenomena (Dale and Fortin 2015) and more important, to define new, previously not intended and emergent research questions with potentially better answers (Popper 2002).

- In our opinion, it is better to keep the dataset as complete and connected as it is compiled.

Splitting datasets can be sometimes useful to apply particular types of statistics or adjustments of model structure (as in our case, we cannot apply spatial correlation structures due to temporally repeated observations). However, as with the approach to start with full models regarding the fixed effects, the complete dataset represents the view of the respective researcher on the research question and we tend to favor this over applicability of a specific type of statistical analysis.

- Other approaches can be used for analysis of spatiotemporal datasets.

Of course, linear and nonlinear mixed effects models are only one possibility for analyzing spatiotemporal datasets. For example, Zuur et al. (2017) explained and applied the R-INLA (Integrated Nested Laplace Approximation) approach, a Bayesian statistics, for spatiotemporal datasets. It seems that this is a very powerful way to overcome problems of common methods and more importantly to adequately describe and consider autocorrelations in datasets of ecological studies.

- Model evaluation by randomized procedures or by including generated random variables.

In order to test the robustness of the final model, it is possible to take random samples from the entire dataset, analyze these sets with the final model, and to systematically compare the "new" model estimates with the complete dataset (Roberts et al. 2017). In addition, adding random generated variables to the model can give information whether the obtained model estimates are close to what can be expected from randomness or not.

References

Akaike H (1998) Information theory and an extension of the maximum likelihood principle. In: Akaike H, Parzen E, Tanabe K, Kitagawa G (eds) Selected papers of Hirotugu Akaike. Springer, New York, pp 199–213

Allen ST, Brooks JR, Keim RF, Bond BJ, McDonnell JJ (2014) The role of pre-event canopy storage in throughfall and stemflow by using isotopic tracers. Ecohydrology 7:858–868. https://doi.org/10.1002/eco.1408

Allen ST, Keim RF, Barnard HR, McDonnell JJ, Brooks JR (2017) The role of stable isotopes in understanding rainfall interception processes: a review. WIREs Water 4:1–17. https://doi.org/10.1002/wat2.1187

Canham CD, Finzi AC, Pacala SW, Burbank DH (1994) Causes and consequences of resource heterogeneity in forests: interspecific variation in light transmission by canopy trees. Can J For Res 24:337–349. https://doi.org/10.1139/x94-046

Carl G, Kühn I (2007) Analyzing spatial autocorrelation in species distributions using Gaussian and logit models. Ecol Model 207:159–170. https://doi.org/10.1016/j.ecolmodel.2007.04.024

Carlyle-Moses DE, Iida S, Germer S, Llorens P, Michalzik B, Nanko K et al (2018) Expressing stemflow commensurate with its ecohydrological importance. Adv Water Resour 121:472–479. https://doi.org/10.1016/j.advwatres.2018.08.015

Carlyle-Moses DE, Laureano JSF, Price AG (2004) Throughfall and throughfall spatial variability in Madrean oak forest communities of northeastern Mexico. J Hydrol 297:124–135. https://doi.org/10.1016/j.jhydrol.2004.04.007

Chapin FS, Matson PA, Vitousek PM (2011) Principles of terrestrial ecosystem ecology, 2nd edn. Springer, New York

Crawley MJ (2013) The R book, 2nd edn. Wiley, Chichester

Dale MRT (2007) Spatial pattern analysis in plant ecology. Cambridge University Press, Cambridge

Dale MRT, Fortin M-J (2015) Spatial analysis: a guide for ecologists, 2nd edn. Cambridge University Press, Cambridge

Engel M, Körner M, Berger U (2018) Plastic tree crowns contribute to small-scale heterogeneity in virgin beech forests—an individual-based modeling approach. Ecol Model 376:28–39. https://doi.org/10.1016/j.ecolmodel.2018.03.001

Ford ED, Deans JD (1978) The effects of canopy structure on stemflow, throughfall and interception loss in a young Sitka spruce plantation. J Appl Ecol 15:905–917. https://doi.org/10.2307/2402786

Frischbier N (2012) Study on the single-tree related small-scale variability and quantity-dependent dynamics of net forest precipitation using the example of two mixed beech-spruce stands. Dissertation. TUD Press, Dresden, Germany

Frischbier N, Wagner S (2015) Detection, quantification and modelling of small-scale lateral translocation of throughfall in tree crowns of European beech (*Fagus sylvatica* L.) and Norway spruce (*Picea abies* (L.) Karst.). J Hydrol 522:228–238. https://doi.org/10.1016/j.jhydrol.2014.12.034

Haining RP (2009) Spatial data analysis: theory and practice, 6th edn. Cambridge University Press, Cambridge

Hefley TJ, Broms KM, Brost BM, Buderman FE, Kay SL, Scharf HR et al (2017) The basis function approach for modeling autocorrelation in ecological data. Ecology 98:632–646. https://doi.org/10.1002/ecy.1674

Hurlbert SH (1984) Pseudoreplication and the design of ecological field experiments. Ecol Monogr 54:187–211. https://doi.org/10.2307/1942661

Keim RF, Link TE (2018) Linked spatial variability of throughfall amount and intensity during rainfall in a coniferous forest. Agric For Meteorol 248:15–21. https://doi.org/10.1016/j.agrformet.2017.09.006

Keim RF, Skaugset AE, Weiler M (2005) Temporal persistence of spatial patterns in throughfall. J Hydrol 314:263–274. https://doi.org/10.1016/j.jhydrol.2005.03.021

Kellner KF, Swihart RK (2017) Simulation of oak early life history and interactions with disturbance via an individual-based model, SOEL. PLoS One 12:e0179643. https://doi.org/10.1371/journal.pone.0179643

Kissling WD, Carl G (2008) Spatial autocorrelation and the selection of simultaneous autoregressive models. Glob Ecol Biogeogr 17:59–71. https://doi.org/10.1111/j.1466-8238.2007.00334.x

Legendre P, Gauthier O (2014) Statistical methods for temporal and space-time analysis of community composition data. Proc R Soc B Biol Sci 281:20132728. https://doi.org/10.1098/rspb.2013.2728

Marin CT, Bouten W, Sevink J (2000) Gross rainfall and its partitioning into throughfall, stemflow and evaporation of intercepted water in four forest ecosystems in western Amazonia. J Hydrol 237:40–57. https://doi.org/10.1016/S0022-1694(00)00301-2

Metzger JC, Wutzler T, Dalla Valle N, Filipzik J, Grauer C, Lehmann R et al (2017) Vegetation impacts soil water content patterns by shaping canopy water fluxes and soil properties. Hydrol Process 31:3783–3795. https://doi.org/10.1002/hyp.11274

Mou P, Mitchell RJ, Jones RH (1993) Ecological field theory model: a mechanistic approach to simulate plant–plant interactions in southeastern forest ecosystems. Can J For Res 23:2180–2193. https://doi.org/10.1139/x93-271

Müller J (2009) Forestry and water budget of the lowlands in northeast Germany — consequences for the choice of tree species and for forest management. J Water Land Dev 13a:133–148. https://doi.org/10.2478/v10025-010-0024-7

Okland RH, Rydgren K, Okland T (1999) Single-tree influence on understorey vegetation in a Norwegian boreal spruce forest. Oikos 87:488–498. https://doi.org/10.2307/3546813

Peters R, Lin Y, Berger U (2016) Machine learning meets individual-based modelling. Self-organising feature maps for the analysis of below-ground competition among plants. Ecol Model 326:142–151. https://doi.org/10.1016/j.ecolmodel.2015.10.014

Pinheiro J, Bates D, DebRoy S, Sarkar D (2013) nlme: linear and nonlinear mixed effects models. R core team

Pinheiro JC, Bates DM (2000) Mixed-effects models in sand S-PLUS. Springer, New York

Popper KR (2002) Conjectures and refutations: the growth of scientific knowledge. Routledge, London

Pukkala T, Kolström T (1992) A stochastic spatial regeneration model for *Pinus sylvestris*. Scand J For Res 7:377–385. https://doi.org/10.1080/02827589209382730

R core team (2013) R: a language and environment for statistical computing. R Foundation for Statistical Computing, Vienna

Reifsnyder WE, Furnival GM, Horowitz JL (1971) Spatial and temporal distribution of solar radiation beneath forest canopies. Agric For Meteorol 9:21–37. https://doi.org/10.1016/0002-1571(71)90004-5

Roberts DR, Bahn V, Ciuti S, Boyce MS, Elith J, Guillera-Arroita G et al (2017) Cross-validation strategies for data with temporal, spatial, hierarchical, or phylogenetic structure. Ecography 40:913–929. https://doi.org/10.1111/ecog.02881

Saetre P, Bååth E (2000) Spatial variation and patterns of soil microbial community structure in a mixed spruce-birch stand. Soil Biol Biochem 32:909–917. https://doi.org/10.1016/S0038-0717(99)00215-1

Saetre P, Saetre LS, Brandtberg P-O, Lundkvist H, Bengtsson J (1997) Ground vegetation composition and heterogeneity in pure Norway spruce and mixed Norway spruce – birch stands. Can J For Res 27:2034–2042. https://doi.org/10.1139/x97-177

Schielzeth H, Nakagawa S (2013) Nested by design: model fitting and interpretation in a mixed model era. Methods Ecol Evol 4:14–24. https://doi.org/10.1111/j.2041-210x.2012.00251.x

Schume H, Jost G, Katzensteiner K (2003) Spatio-temporal analysis of the soil water content in a mixed Norway spruce (*Picea abies* (L.) Karst.)–European beech (*Fagus sylvatica* L.) stand. Geoderma 112:273–287. https://doi.org/10.1016/S0016-7061(02)00311-7

Wälder K, Frischbier N, Bredemeier M, Näther W, Wagner S (2008) Analysis of O_f-layer humus mass variation in a mixed stand of European beech and Norway spruce: an application of structural equation modelling. Ecol Model 213:319–330. https://doi.org/10.1016/j.ecolmodel.2007.12.014

Wälder K, Näther W, Wagner S (2009) Improving inverse model fitting in trees—anisotropy, multiplicative effects, and Bayes estimation. Ecol Model 220:1044–1053. https://doi.org/10.1016/j.ecolmodel.2009.01.034

Weber P, Bardgett RD (2011) Influence of single trees on spatial and temporal patterns of belowground properties in native pine forest. Soil Biol Biochem 43:1372–1378. https://doi.org/10.1016/j.soilbio.2011.03.015

Webster R (2001) Statistics to support soil research and their presentation. Eur J Soil Sci 52:331–340. https://doi.org/10.1046/j.1365-2389.2001.00383.x

Wu H-I, Sharpe PJH, Walker J, Penridge LK (1985) Ecological field theory: a spatial analysis of resource interference among plants. Ecol Model 29:215–243. https://doi.org/10.1016/0304-3800(85)90054-7

Zimmermann A, Germer S, Neill C, Krusche AV, Elsenbeer H (2008) Spatio-temporal patterns of throughfall and solute deposition in an open tropical rain forest. J Hydrol 360:87–102. https://doi.org/10.1016/j.jhydrol.2008.07.028

Zimmermann B, Elsenbeer H (2008) Spatial and temporal variability of soil saturated hydraulic conductivity in gradients of disturbance. J Hydrol 361:78–95. https://doi.org/10.1016/j.jhydrol.2008.07.027

Zuur AF, Ieno EN, Saveliev AA (2017) Beginner's guide to spatial, temporal, and spatial-temporal ecological data analysis with R-INLA, vol 7. Highland Statistics Ltd, Newburgh

Zuur AF, Ieno EN, Walker N, Saveliev AA, Smith GM (2009) Mixed effects models and extensions in ecology with R. Springer, New York

Zwanzig M, Schlicht R, Frischbier N, Berger U (2019) Primary steps in analyzing data: tasks and tools for a systematic data exploration. In: Levia DF, Carlyle-Moses DE, Iida S, Michalzik B, Nanko K and Tischer A (Eds.), Forest-Water Interactions. Ecological Studies Series, No. 240, Springer-Verlag, Heidelberg, Germany. https://doi.org/10.1007/978-3-030-26086-6_7

Chapter 9
Analysis of Vegetation-Water Interactions: Application and Comparison of Maximum-Likelihood Estimation and Bayesian Inference

Istem Fer

9.1 Introduction

In ecology and environmental science, we often want to understand and predict complex systems with many processes and interactions. Doing so generally requires models, and regardless of whether those are statistical or mechanistic in nature, this also requires measurement data to inform model parameters and their uncertainties. Occasionally individual model parameters can be constrained directly with observations, but increasing data from field experiments, observatory networks and remote sensing missions can be mapped to the outputs that we are trying to predict with our models. In that case, such data can be used to indirectly inform model parameters through likelihood-based estimation methods.

Likelihood functions constitute the basis for both Bayesian inference and maximum-likelihood estimation. These functions quantify the probability of the observed data, D, given a model with parameter vector Φ, $P(D|\Phi)$ (Hartig et al. 2011; for a more detailed statistical background, please see below). In other words, likelihoods allow us to compare different model parameter values in terms of their ability to reproduce the observed data, D (Hartig et al. 2012). Putting this in a framework where we repeatedly compare model predictions with data as we change parameters, we can (either analytically or numerically) find parameter values with higher likelihoods and reduce the uncertainty around model parameters (Williams et al. 2009). Accordingly, both maximum-likelihood estimation (Richardson and Hollinger 2005; Brookshire et al. 2011; Kim et al. 2012; Smith et al. 2016) and

Electronic Supplementary Material The online version of this chapter (https://doi.org/10.1007/978-3-030-26086-6_9) contains supplementary material, which is available to authorized users.

I. Fer (✉)
Department of Earth and Environment, Boston University, Boston, MA, USA
e-mail: istfer@bu.edu

D. F. Levia et al. (eds.), *Forest-Water Interactions*, Ecological Studies 240,
https://doi.org/10.1007/978-3-030-26086-6_9

Table 9.1 Maximum-likelihood estimation (MLE) optimization vs Bayes

MLE optimization	Bayes
Given my parameters, what is the probability of observing this data? $P(D\|\theta)$	Given the data, what is the probability of my parameters? $P(\theta\|D)$
95% confidence interval: If we were to calculate the 95% intervals many times using the data sets generated by the same process, true parameter value would be in the 95% CI, 95% of those times	95% credible interval: Value falls into the interval with 95% probability
Point estimate of parameter values	Distribution of parameter values
Could be easier to implement for simple problems	Easier than MLE for modelling complex systems
Treat uncertainties as known quantities	Identify, quantify, partition and propagate uncertainties

Bayesian inference (van Oijen et al. 2005; Keenan et al. 2011; MacBean et al. 2018; Fer et al. 2018) have been widely used in the analysis of environmental data.

Although both approaches use likelihoods as a statistical measure of fit between model and data, and can often give very similar results, maximum-likelihood estimation and Bayesian inference fundamentally differ in theory and application (Ellison 2004). While I will discuss these differences in more detail in the following sections, Table 9.1 highlights the most striking and relevant points. These differences arise from two essentially different views of probability and, in a way, boils down to treatment of the likelihood. Likelihood, by definition, expresses information about the data and carries no absolute information about a particular parameter (Hartig et al. 2011). In the maximum-likelihood approach, likelihood is used to search for the parameter whose probability of producing the observed data is the highest without saying anything about the probability of the parameter, whereas in Bayesian approach, likelihood is evaluated together with our prior information about the parameter values, $P(\Phi)$, e.g., from previous measurements and analyses, to solve for $P(\Phi|D)$, the probability of the model's parameters given the data, which is called the posterior. Instead of a single value, the posterior is interpreted as a probability distribution which encapsulates information about the uncertainty of the parameter.

In the following sections, I first give a more detailed statistical background on likelihood-based approaches. However, this quick refresher still assumes a certain level of familiarity with statistical and computational methods. For additional background and more in-depth discussion of these concepts you may want to consult other textbooks (Clark 2007; Bolker 2007; McElreath 2015; Kruschke 2015; Kurt 2019). Then, I provide scientific background on vegetation-water interactions, which I use as an example for contrasting maximum-likelihood estimate and Bayesian inference. Finally, I finish by discussing future opportunities and directions. For interested readers, the example data and the full R code used in the analyses are provided as online electronic supplementary material to this chapter (see Extras Springer online).

9.2 Statistical Background on Likelihood-Based Approaches

This section provides an overview of the notation used throughout this chapter and the basic concepts, with an emphasis on the differences between maximum-likelihood and Bayesian approaches. As already mentioned above, an overarching premise of this chapter is on likelihood-based methods where likelihood is essentially a conditional probability. So, let us talk about probability first.

The contrast between the two inference methods begins with their definition of probability. Traditionally, maximum-likelihood inference sees probability as long-term frequencies of random variables whereas Bayesian inference sees it as plausible reasoning about random variables (Jaynes 2003). In a broad sense, random variables (are functions that) represent events (e.g., making a specific observation) and can take on multiple values as there are multiple outcomes of the event each with a certain probability of occurring. Notationally, random variables are typically expressed using capital letters, e.g., *A*, which can be mapped to specific (lower case) events, a_i, with some probability p_i, e.g., $P(A = a_i) = p_i$ (Kurt 2019). While it is possible to specify these probabilities individually, it is more common to describe them using functions, known as probability distributions, which formulate how the values of a random variable is distributed.

When we start describing events with relation to another, probabilities such as joint, marginal and conditional allow us to reason about the interactions between those events (Dietze 2017). The joint probability of random variables *A* and *B*, $P(A, B)$, is the probability of their associated events occurring simultaneously. Marginal probabilities, $P(A)$ and $P(B)$, describe the probability of an event happening irrespective of other events. From another point of view, the marginal probability of an event *x* becomes the sum of all the ways *x* can happen, expressed as its joint probabilities with all events in the event space (using the law of total probability). Conditional probability, $P(A|B)$, is the probability of random variable *A*, conditioned on random variable *B* taking on a specific, fixed value (i.e., given the event associated with *B* took place).

9.2.1 Likelihood

Once we start applying the basic concepts of probability to the elements of inference, likelihood becomes the probability of the data (*D*) under the assumption that the model (*M*) is true, $P(D|M)$. In this framework, we can evaluate different models as well as different parameter values for a given model, $M(\Phi)$. Then the notation becomes $P(D|\Phi)$, read as the probability that *D* would be observed given $M(\Phi)$ (Hartig et al. 2011). It then follows that a parameter value would be a better estimate than another if it is the one for which the conditional probability of the data, $P(D|\Phi)$, is higher (Clark 2007). Using this likelihood principle, we can search for a parameter value for which the data are more probable. The function that is obtained by viewing $P(D|\Phi)$ as a function of the parameter Φ is called the likelihood function:

$$L(\Phi) \propto P(D|\Phi) \tag{9.1}$$

To work our way through writing down the likelihood, let us imagine that we have decided simple linear regression as our model, $f(x) = \beta_0 + \beta_1 * x$. We will call this model the "process model" because it is where we model the real-world process (hence the name) to the best of our knowledge. Every time we feed data set x and a parameter vector $\{\beta_0, \beta_1\}$ into this model, we will obtain a prediction. Because we are not able to take measurements or make observations perfectly, the data will scatter around our predictions. Therefore, the probability of data that we want to calculate as the likelihood will also depend on how we expect this scatter to be. In other words, we cannot ignore this variability because it is part of the data. We want to model this mismatch separately, because we only want the process to influence the estimation of the parameters of our process model, not the observation noise. In this simple example, we can assume that the observed data will be normally distributed around the model predictions with observation error. This part where we model the residuals is typically called the "data model" (Fig. 9.1) in the literature (Clark 2007). Note how we assumed that all the error is only in the observations. Likewise, we could easily imagine that there is non-negligible error in the measurement of x as well (typically called "errors in variables"). Data model is also where we can model errors in variables and more. For the sake of this simple example, we will only consider observation error. Then, the likelihood we will calculate becomes $y \sim N(\beta_0 + \beta_1 * x, \sigma^2)$. Every time we plug-in the same data sets $\{x, y\}$ and a different parameter vector $\{\beta_0, \beta_1, \sigma\}$ into this likelihood, we will get a different value for the probability of the data. Then, the values of $\{\beta_0, \beta_1, \sigma\}$ that maximize this value, the likelihood function, would become our best estimates for these parameters.

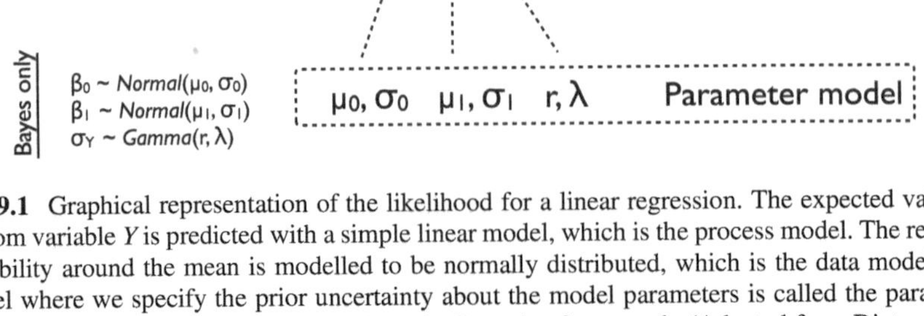

Fig. 9.1 Graphical representation of the likelihood for a linear regression. The expected value of random variable Y is predicted with a simple linear model, which is the process model. The residual variability around the mean is modelled to be normally distributed, which is the data model. The model where we specify the prior uncertainty about the model parameters is called the parameter model (a.k.a. the priors) and applies only to the Bayesian framework. (Adapted from Dietze 2017)

9.2.2 Maximum-Likelihood Estimation

The approach used for finding the parameter value that maximizes $L(\Phi)$ is called the maximum-likelihood estimation (MLE). We can maximize the likelihood function, like we would maximize any function: by taking its derivative with respect to each parameter, setting the equations to zero and solve for the values that maximize it. However, as you might have guessed, taking the derivative of the likelihood function may not always be easy or even possible (e.g., when the $M(\Phi)$ is a complex simulation model, with no open form). In that case, we can turn to numerical methods, in which we can plug in different values for Φ, evaluate the $M(\Phi)$ and calculate $P(D|\Phi)$ for each of these values and search for the parameter value that maximizes it. Numerical optimization algorithms (Arora 2015) with the following steps are the standard approaches:

1. Start from some initial value.
2. Evaluate $M(\Phi)$ and $L(\Phi)$ for the initial value.
3. Propose a new parameter value.
4. Evaluate $M(\Phi)$ and $L(\Phi)$ for the new value.
5. Decide whether or not to accept the new value.
6. Repeat 3–5 until you cannot find a better value.

Numerical optimization algorithms mainly differ in the third and fifth steps (in the way they propose a new value and then decide whether to accept it). Traditionally, there are many algorithms that are designed for minimizing functions; therefore, it is typical to work with negative likelihood functions and minimize them to find the best estimates for parameters. It is also worth noting that, for numerical stability reasons, we almost always work with (negative) log-likelihoods.

9.2.3 Bayesian Statistics

The superlative adjective "best estimate" in the previous sentences reminds us that until now, we have only been talking about a parameter value being a better estimate relative to another. This is because the MLE framework only allows us to evaluate the probability of getting the data given a fixed value for a parameter. However, broadly speaking, we are trying to come up with an understanding about the parameters or the models. In other words, what we are scientifically interested in is the probability of the parameter or the model, not the probability of the data. Mathematically, reversing the $P(D|\Phi)$ into $P(\Phi|D)$ is only possible through the Bayes' theorem:

$$P(\Phi|D) = \frac{P(D|\Phi)P(\Phi)}{P(D)} \tag{9.2}$$

where $P(D|\Phi)$ is the likelihood, $P(\Phi)$ is the prior and $P(\Phi|D)$ is the posterior. In the representation of the Bayes' theorem above, the term in the denominator, $P(D)$, becomes the probability of the data which does not make intuitive sense. However, going back to the probability rules, we can write marginal probabilities using the law of total probability, which becomes the probability of D having integrated over all possible values Φ can take; $P(D) = \int P(D|\Phi)P(\Phi)\mathrm{d}\Phi$. This is just the integral of everything in the numerator, thus acting as a normalizing constant that ensures $P(\Phi|D)$ is a valid probability distribution (integrates to 1). Overall, this leads to a more intuitive interpretation of Bayes' theorem: combine your prior knowledge about a parameter $P(\Phi)$, with the support given to it by the data, $P(D|\Phi)$, and scale this by all the ways data could occur, $\int P(D|\Phi)P(\Phi)\mathrm{d}\Phi$, to update your (posterior) understanding about the parameter $P(\Phi|D)$:

$$P(\Phi|D) = \frac{P(D|\Phi)P(\Phi)}{\int P(D|\Phi)P(\Phi)d\Phi} \tag{9.3}$$

The $P(D|\Phi)$ term, which we know can be hard or intractable to take its derivative of, now appears in an integral in the denominator. But, once calculated, the term in the denominator will be a constant and will be the same for all the Φ values we are testing. Therefore, it is often named as the "normalizing constant" in the literature and left out of calculations, making the posterior proportional to likelihood times the prior:

$$P(\Phi|D) \propto P(D|\Phi)P(\Phi) \tag{9.4}$$

9.2.4 *Numerical Methods for Bayes*

Similar to numerical optimization approach, where we were drawing parameter values, evaluating their likelihoods and comparing them, we can now perform the search for "better" parameter values by additionally multiplying the likelihood by the prior. As such, Bayesian numerical calibration algorithms generally follow these analogous steps to numerical optimization algorithms:

1. Start from some initial value, Φ.
2. Evaluate the (unnormalized) posterior, $P(D|\Phi)P(\Phi)$.
3. Propose a new parameter value, $\Phi*$.
4. Evaluate the new (unnormalized) posterior, $P(D|\Phi*)P(\Phi*)$.
5. Decide whether or not to accept the new value according to a rule.
6. Repeat 3–5 until you converge to a posterior distribution.

Algorithms that aim at solving complex probability problems through repeated stochastic simulations are called Monte Carlo (MC) algorithms. While different flavours of MC algorithms have their own pros and cons, one particular algorithm is worth mentioning here: the Markov-chain Monte Carlo (MCMC) algorithm. The main characteristic of MCMC is the fifth step:

5. Decide whether or not to accept the new value according to a rule
   [accept proportional to $P(D|\Phi*)P(\Phi*)$ / $P(D|\Phi)P(\Phi)$]

This transition rule (acceptance criteria) is essential to MCMC functionality, which makes the algorithm explore the parameter space by avoiding the less likely regions while ensuring convergence to the posterior (i.e., if you use another transition rule, your algorithm might converge to something that is not the posterior). For more information about different algorithms, see Hartig et al. (2011) and references therein.

Although there are notable similarities, MCMC is not just optimization. First, we evaluate the (unnormalized) posterior in MCMC rather than just the likelihood. Second, MCMC has a stricter criterion for accepting a proposed value. Third, MCMC converges to a probability distribution rather than a single point. Finally, because MCMC converges to a distribution, the stopping condition depends on the design of its application, e.g., pre-determined number of iterations, proposal distribution, number of chains and convergence diagnostics. To learn about these design issues and diagnostics, see Andrieu et al. (2003) and Gelman et al. (2003).

9.3 Bayes vs MLE

Obtaining a posterior distribution instead of a point estimate is the key difference between MLE and Bayesian analysis. Theoretically, this quantity [posterior] now directly tells us about the probability of our parameter having certain values, which really is what we want in most of such analyses (as opposed to quantities about the probability of the observed data). Practically, it also immediately provides a readily interpretable uncertainty statement about these values. For example, 95% credible interval (analogue of confidence interval in Bayesian statistics) simply becomes the central 95% of the posterior density, and it is a quantity about the probability of a parameter being in the credible interval.

It is possible to get uncertainty estimates from the MLE optimization as well, but doing so always requires making additional assumptions and performing additional calculations (e.g., via bootstrapping; see the example below for a bootstrapped confidence interval for MLE and also the accompanying code for the steps of a parametric bootstrapping algorithm). However, it is not as straightforward to interpret MLE confidence intervals. For example, 95% confidence interval would mean if you run your models repeatedly on new data, the true value of your parameter will be

Table 9.2 Some of the Bayesian softwares/packages

Language	Package	Url	Citation
JAGS		mcmc-jags.sourceforge.net	Plummer (2003)
BUGS		www.mrc-bsu.cam.ac.uk/software/bugs	Lunn et al. (2012)
STAN		mc-stan.org	Carpenter et al. 2017
R	NIMBLE	r-nimble.org	De Valpine et al. (2017)
R	BayesianTools	CRAN.R-project.org/ package=BayesianTools	Hartig et al. (2018)
Python	PyMC3	docs.pymc.io	Martin (2018)
R	INLA	r-inla.org	Rue et al. (2009)
R	Greta	greta-stats.org	Golding (2018)
Julia	Mamba	mambajl.readthedocs.io	Smith et al. (2014)
R	PEcAn	pecanproject.github.io	Fer et al. (2018)

Note that BUGS (Bayesian Inference Using Gibbs Sampling), JAGS (Just Another Gibbs Sampler) and STAN (Named after Stanislaw Ulam, co-inventor of Monte Carlo methods) are standalone languages that can be interfaced from other languages such as R, Python, MATLAB, Julia and Stata

in the 95% CI in 95% of your trials. Furthermore, there is no way to tell if your current analysis is one of those 95%, but a certain error rate is guaranteed in the long term. Overall, frequentist measures such as p-values and confidence intervals are not quantities about probabilities of a hypothesis being true or that a parameter being in a certain range. Such statements are only possible through Bayesian approach (Ellison 2004).

That being said, the statistical inference drawn from Bayesian and MLE approach are practically the same for relatively simple (and well-behaving) problems (Clark 2005). Moreover, for simple problems, application of the Bayesian approach might be harder, as some of the mentioned techniques require a certain level of expertise. However, the pace of software development to overcome such technical difficulties of Bayesian approach is at its fastest (see Table 9.2 for some of these softwares). Also, Bayesian approach might still be desirable over MLE as it provides a formal way of incorporating prior knowledge in the analysis. For example, we can reduce parameter trade-off and unidentifiability issues by putting an informative prior on parameters for which we have more prior information. Many optimization algorithms also allow setting boundaries for parameters, which produces a similar effect, although some argue that this is same as using uniform priors without admitting it. Furthermore, when such trade-offs exist, *numerical methods for Bayes* (e.g., *MCMC) are far more robust than the numerical methods to find the MLE* as we will see in an example later in the chapter.

Another reason to choose Bayes over MLE is the iterative nature of the Bayesian approach. The posterior of one analysis naturally becomes the prior of your next analysis (the sample size used in the previous analyses etc. are all encapsulated in the posterior) whereas you would need to repeat your MLE analysis as new data become available. Repeating MLE analyses might not be too inconvenient if your problem is simple enough, but can become computationally prohibitive when working with large data volumes or complex models. Indeed, when it becomes too computationally expensive to use the entirety of the data all at once, Bayesian analysis can be done in a sequential mode (Dietze 2017).

The power of Bayesian approach becomes even more apparent as we move on to modelling more complex systems. As we begin to break down the complex problem into multiple levels, we can begin to account for random effects (effects that would be different if the study were to be replicated), hidden variables (variables that we cannot directly observe but estimate with uncertainty), errors in variables and integrating multiple data sets at different scales with different uncertainties (see Clark (2005) and Figure 1 therein for examples of incorporating complexity into Bayesian models). This type of modelling structure, where there are multiple stages to reflect and treat complexity of the system components, is called a Hierarchical Bayesian framework. Hierarchical Bayes allows us to account for the sources of variability and uncertainty throughout the layers of complexity between what we can measure/observe and what we actually want to model/predict. Ignoring this intervening complexity, and bringing data to models rather than models to data, could result in inaccurate inference (Clark 2007). In the following section, we will explore examples to contrast (Hierarchical) Bayes and maximum-likelihood approaches.

9.4 Vegetation-Water Relationships at Flux Sites

In this example, we will use eddy flux data to infer two important ecosystem properties, canopy and aerodynamic conductances, that ultimately govern the relationship between vegetation and the water cycle. I will first provide an overview of this relationship; however, I invite the interested reader to check the following studies for a more detailed scientific background: Medlyn et al. 2011, 2017; De Kauwe et al. 2013, 2017; and Knauer et al. 2018a.

9.4.1 Scientific Background

Earth's carbon and water cycles are intrinsically coupled through plant stomata where CO_2 uptake (photosynthesis) is accompanied by water loss (transpiration) during gas exchange. The rate of CO_2 entering and water vapor exiting the stomata is regulated by a leaf property called stomatal conductance. Accordingly, the behaviour of stomatal conductance is modelled to have effects on transpiration at global scales (McNaughton and Jarvis 1991; De Kauwe et al. 2013). However, the overall effect depends on how stomatal conductance is upscaled to the ecosystem level, which is measured as canopy conductance, G_c, as well as how effectively the ecosystem can

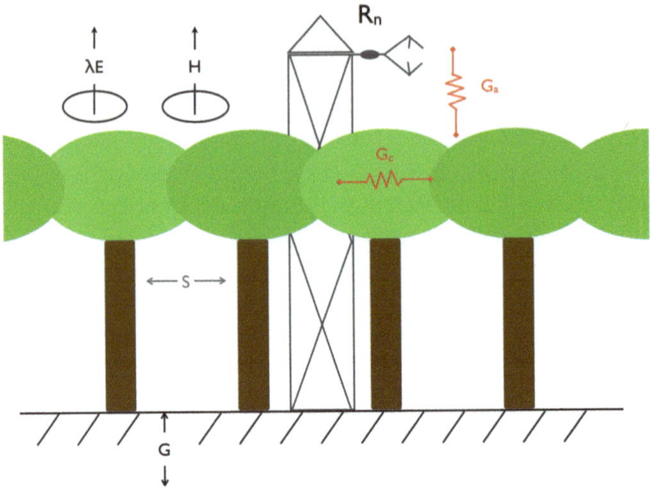

Fig. 9.2 A conceptual diagram on canopy (G_c) and aerodynamic conductance (G_a). R_n net radiation, S storage heat flux (faded because it is usually not available in flux data sets), G ground heat flux, λE latent heat, H sensible heat (Adapted from Knauer et al. 2018a)

transfer mass and energy fluxes to the atmosphere, which is a function of aerodynamic conductance, G_a (Fig. 9.2). Below we will explore how flux data provide a way to study these two conductances (G_c and G_a) and obtain a more comprehensive understanding of vegetation-water interactions (Knauer et al. 2018a).

The latent heat flux measured by flux towers can be modelled by the Penman-Monteith equation:

$$\lambda E = \frac{\Delta(R_n - G - S) + \rho_a c_p (VPD) G_a}{\Delta + \gamma\left(1 + \frac{G_a}{G_s}\right)} \tag{9.5}$$

where λE is the latent heat flux (W m^{-2}), Δ is the slope of the saturation vapor pressure curve (kPa K^{-1}), ρ_a is air density (kg m^{-3}), c_p is the specific heat of air for constant pressure (J K^{-1} kg^{-1}), VPD is the vapor pressure deficit (kPa), γ is the psychrometric constant (kPa K^{-1}), G_a is aerodynamic conductance to heat/water vapor (m s^{-1}) and G_s is the surface conductance to water vapor. The term $(R_n - G - S)$ refers to the available energy, where R_n is the net radiation (W m^{-2}), G is the ground heat flux (W m^{-2}) and S is the sum of all storage fluxes (W m^{-2}).

The latent heat flux has two main components: plant transpiration and evaporation from surfaces. We can treat the canopy transpiration as the sole evaporative source when (i) vegetation cover is dense (closed canopy) and (ii) canopy and soil surfaces are not wet. In other words, assuming a dry canopy carpeting the surface below like a single big-leaf, the only water exchange from the land to the atmosphere would happen through canopy transpiration (Fig. 9.2). In that case, we can approximate the plant transpiration using the Penman-Monteith equation above, in which canopy conductance (G_c) can be interpreted as a proxy for the surface conductance to water vapor (G_s) term:

$$\lambda E = \frac{\Delta(R_n - G - S) + \rho_a c_p (VPD) G_a}{\Delta + \gamma\left(1 + \frac{G_a}{G_c}\right)} \tag{9.6}$$

The available energy[1] ($R_n - G - S$), the latent heat flux (λE) and the environmental variables (VPD, air pressure and temperature—which are used in derivation of other terms) in Eq. (9.6) are being measured by flux networks such as Ameriflux (http://ameriflux.lbl.gov/) and FluxNet (http://fluxnet.fluxdata.org/). Therefore, with the assumptions mentioned above, finding the canopy conductance (G_c) and the aerodynamic conductance (G_a) becomes an inverse modelling problem (Hartig et al. 2012). As flux data are available from many sites, spanning many years, we can perform spatial and temporal analyses about the physiological and aerodynamic ecosystem quantities of interest, G_c and G_a, and in return characterize the coupling between carbon and water cycles under environmental change.

9.5 Parameter Estimation Using Flux Data

For the sake of clarity and reproducibility, I use the "bigleaf" R-package (v0.6.1) as described in detail by Knauer et al. (2018b), because this package comes with filtering functions and sample flux data that I use in the first example (see below). In addition, I use "dfoptim" R-package for MLE optimization (v2-1, Varadhan et al. 2018) and "rjags" R-package (v4-6, Plummer 2016) for MCMC to be able to focus on the modelling rather than algorithm implementations. *Rjags* package provides an interface from R to JAGS (Just Another Gibbs Sampler) language. Therefore, the accompanying code involves JAGS syntax, which has a lot of similarities to more familiar languages like R. A brief introduction on JAGS syntax is still provided in Box 9.1. I start with the simplest case possible and add increasing complexity.

Box 9.1: A Short Introduction on JAGS

We will work with JAGS through R for Bayesian analysis because it helps us focus on our modelling instead of getting lost in the details of algorithm implementations. The most crucial part of JAGS language is the usage of the distributions. Except a few distributions, JAGS has the same naming convention as R. However, the parameterization of these distributions is not always the same as R (see LeBauer et al. 2013 for translation tables and functions).

(continued)

[1] *G* and *S* are usually absent from flux data sets, though.

Box 9.1 (continued)
For example, in R the Gaussian distribution is parameterized with mean and standard deviation (σ), while in JAGS it is parameterized with mean and precision ($1/\sigma^2$).

R: dnorm(mean, sd).

JAGS: dnorm(mean, tau).

All distribution names and their parameterizations can be found in JAGS user manual (last accessed 2019/02/18). Another (possibly the most) important feature to note about JAGS is its polymorphic behaviour. While running JAGS models, if you provide values for a variable, it is treated as known and other random variables will be evaluated conditioned on these fixed values. If you do not provide the values, the variable is treated as unknown and values are drawn from its distribution. As a reflection of this behaviour, all distributions in JAGS are in "d" form even when you use them to generate random numbers (i.e., there is no "rnorm" in JAGS).

To mention a few technical details in JAGS: Results from deterministic calculations are assigned by using the arrow (<−) operator, whereas tilde (~) operator is used for assignment of random variables. The equal sign (=) cannot be used for either.

Y ~ dnorm(mu, tau) # data model, probabilistic.

mu < − b0 + b1∗x # process model, deterministic.

Finally, the order of instructions is not important for JAGS (i.e., you do not need to define a variable before using it in a calculation) which could be confusing to many programmers who are used to the sequential order of commands. JAGS reads the instructions as a whole, figures out the relations between them based on the conditional probabilities described, parses the problem into a graph representation and then analyses how to best sample from it.

For a more hands-on JAGS primer, please also see existing online primers: https://github.com/EcoForecast/EF_Activities/blob/master/Exercise_05_JAGS.Rmd.

9.5.1 Data Description

The data in the first example are half-hourly eddy covariance from a spruce forest, Tharandt (DE-Tha), Germany, for June 2014. See package documentation and Knauer et al. (2018b) for more details about this data. The second data set is hourly eddy covariance from a deciduous forest site, Harvard Forest (US-Ha1), USA, from 1991 to 2012 (although *LE* values from years 1991 and 2010 are completely missing from this data set, we will be excluding those years, which leaves us with 20 years in total). Variable names and units are used as provided by the *bigleaf* package and are listed in Table 9.2.

Table 9.3 Constants, (derived) variables and units used in this chapter

Abbreviation	Description	Units
G_a	Aerodynamic conductance	m s^{-1}
G_c	Canopy conductance	m s^{-1}
LE (λE)	Latent heat flux	W m^{-2}
Tair	Air temperature	°C
VPD	Vapor pressure deficit	kPa
PPFD	Photosynthetic photon flux density	μmol m^{-2} s^{-1}
G	Ground heat flux	W m^{-2}
Pressure	Atmospheric pressure	kPa
Ca	CO_2 concentration	Ppm
Ustar	Friction velocity	m s^{-1}
R_n	Net radiation	W m^{-2}
Δ	The slope of the saturation vapor pressure curve	kPa K^{-1}
γ	Psychrometric constant	kPa K^{-1}
ρ_a	Air density	Kg m^{-3}
c_p	Specific heat of air for constant pressure	J K^{-1} kg^{-1}

For more details see *bigleaf* documentation

As mentioned above, to approximate the surface conductance to water vapor (G_s) term in the Penman-Monteith equation with the canopy conductance (G_c), certain assumptions need to be made. Therefore, to meet these assumptions, the *LE* data are initially filtered[2] as follows:

(i). Data points with photosynthetic photon flux density (PPFD) values bigger than 200 (μmol m^{-2} s^{-1}) were chosen to account for periods when the vegetation is photosynthesizing.
(ii). Data points where the friction velocity is lower than 0.4 (m s^{-1}) are eliminated to avoid conditions of low turbulence.
(iii). Data points with precipitation and the subsequent 24 h were excluded to minimize influence of evaporation from wet surfaces.

VPD, R_n, *G* and *LE* (see Eq. (9.6) and Table 9.3) were extracted from the flux data. Specific heat of air for constant pressure, c_p, is provided by the bigleaf package as 1004.834 (J K^{-1} kg^{-1}). The psychrometric constant, γ, is calculated from c_p, atmospheric pressure, latent heat of vaporization and ratio of the molecular weight of water vapor to dry air by the bigleaf::psychrometric.constant[3] function. The slope of the saturation vapor pressure curve, Δ, is calculated from *Tair* by the bigleaf::Esat.slope function. Air density, ρ_a, (kg m^{-3}), is calculated from *Tair*, air pressure and gas constant of dry air by the bigleaf::air.density function. For more information, see the accompanying code and *bigleaf* package documentation.

[2]Additional filtering can be done by using the quality flags when it is provided in the data set.

[3]package.name::function.name

9.5.2 Statistical Inference

Throughout the examples, *LE* is treated as the observed variable with a Gaussian likelihood. Errors of *LE* data have shown to have asymmetric heteroskedastic Laplacian distribution (Richardson et al. 2006), however, for simplicity, here I use a Gaussian likelihood. Therefore, the inverse modelling is initially set up for estimating three parameters, G_a, G_c and σ_{LE}, with the last parameter being the standard deviation of the Gaussian likelihood (a.k.a. the residual standard error). To avoid confusion, it is worth noting that it would be more accurate to treat G_a and G_c as fluxes. For simplicity, in these examples, they are treated as parameters of the Penman-Monteith equation and estimated accordingly. However, the approach described here can be extended to estimate time-series values of G_a and G_c. All the model and code used to produce this section can be found in the accompanying scripts.

$$LE_{obs} \sim Normal\left(LE_{Penman-Monteith}, \sigma_{LE}\right) \tag{9.7}$$

After filtering according to the assumptions listed above, we are left with 381 data points (26% of the total). Note that filtering is only applied to the *LE* data, and not the meteorological data, since the specific conditions listed above influences the quality of the eddy covariance fluxes but not the micrometeorology data.

I use the same uniform priors, *unif(0,10)*, for the G_c and G_a parameters, and *Gamma(1, 100)* for the precision (as noted in Box 9.1, JAGS uses precision for normal distribution instead of standard deviation) in the Bayesian approach. These priors inform the framework that both parameters are zero-bound, but otherwise they are wide. For the MLE, the null hypothesis can be thought of as having no conductance ($G_c = G_a = 0$) as we deliberately chose conditions we expect nonzero conductances. However, the same way I constrain the priors in the Bayesian approach, I use constrained optimization algorithm for the MLE solution and provide a lower bound of zeros for both parameters. Figure 9.3 shows the estimated values through MLE and Bayes with associated CIs *(CI: bootstrapped confidence interval for MLE and credible interval for Bayes for the rest of the chapter)*. As

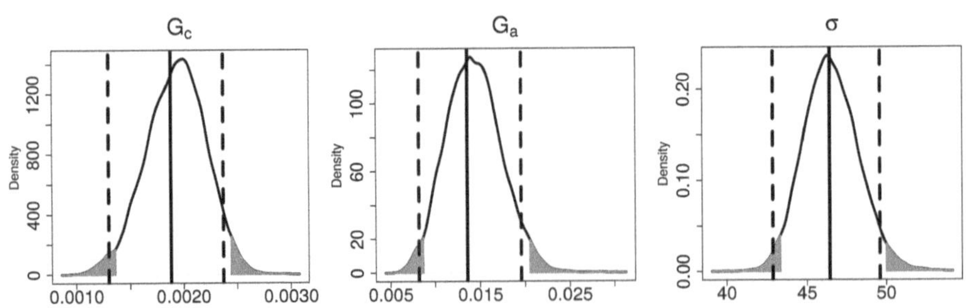

Fig. 9.3 Bayes vs MLE analysis gives similar results. Posterior densities are shown for the three estimated parameters. Unshaded central densities are the 95% Bayesian credible intervals. Vertical lines are the maximum-likelihood approach results (solid MLE, dashed 95% confidence intervals)

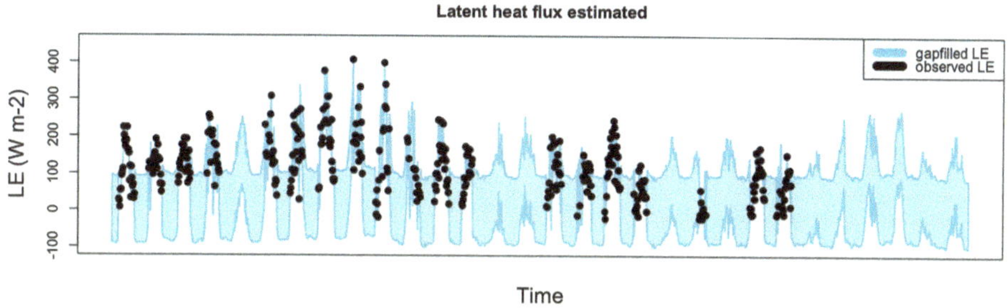

Fig. 9.4 *LE* data (predicted) is gap filled with uncertainties as a by-product in the Bayesian approach

expected, for this simple case the estimated values for G_a, G_c and σ_{LE}, are similar with both approaches. However, the CIs of the MLE bootstrapping approach are slightly narrower than the Bayesian CIs, which is a known tendency that occurs because bootstrapping is limited to "in-sample" inference (i.e., it can only resample the variability in the observed data, which is typically lower than the possible variability of a system).

The Bayesian approach also predicts the missing data points in the response variable with an estimation of uncertainty. As a by-product of the Bayesian approach, the *LE* data "becomes" gap filled (Fig. 9.4). The estimated missing data points unrealistically extend to large negative night-time values which is a result of our choice of the likelihood as a Gaussian distribution where variance is constant. If we had used an asymmetric heteroskedastic Laplacian likelihood (which allows variance to vary with flux magnitude, hence the heteroscedasticity), narrower uncertainty bands would have been estimated as flux values get smaller.

9.5.3 Unidentifiability

There is one more point I would like to highlight about our results. In Eq. (9.6), although G_a appears in the numerator alone, both parameters that we are trying to estimate also appear as G_a/G_c in the denominator which leads to an equifinality problem. That is to say, combinations of G_a and G_c are sharing a similar likelihood which makes it difficult to estimate their values individually. This manifests itself as an almost perfect correlation ($r \approx 1$) in the MCMC samples (Fig. 9.5).

High correlation is not necessarily an issue for prediction. We basically need to take the correlation into account while making predictions (i.e., we need to use the joint posterior probability, instead of marginal probability density distributions for each parameter). However, if the aim is to make inferences about each parameter then equifinality would become an issue. In such cases, the ability to construct informative priors, where appropriate could help reduce the equifinality problem in the Bayesian approach. For example, if I had prior information that suggested

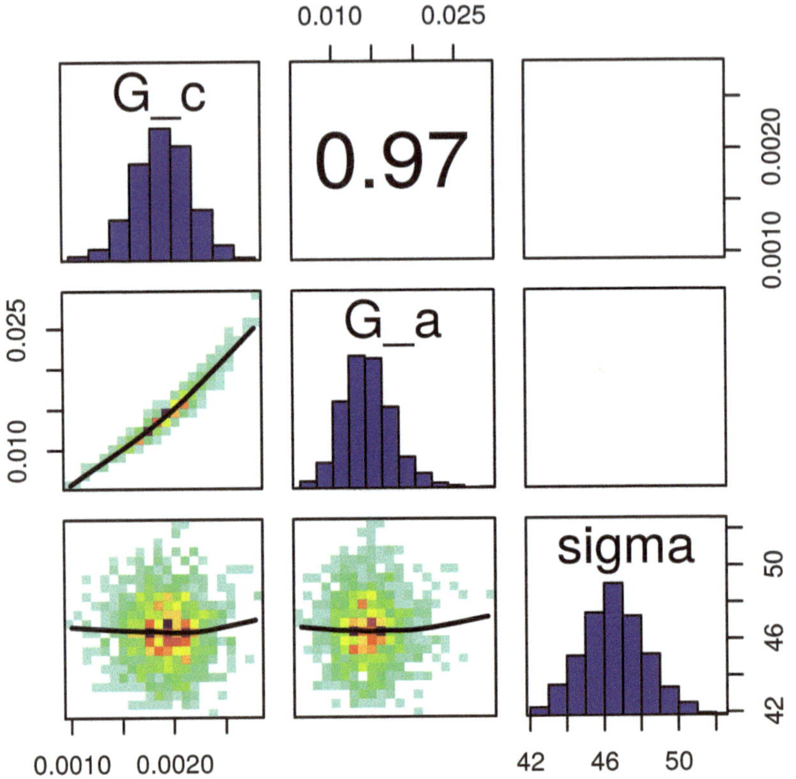

Fig. 9.5 Equifinality in parameters. High correlation between G_a and G_c in the MCMC samples

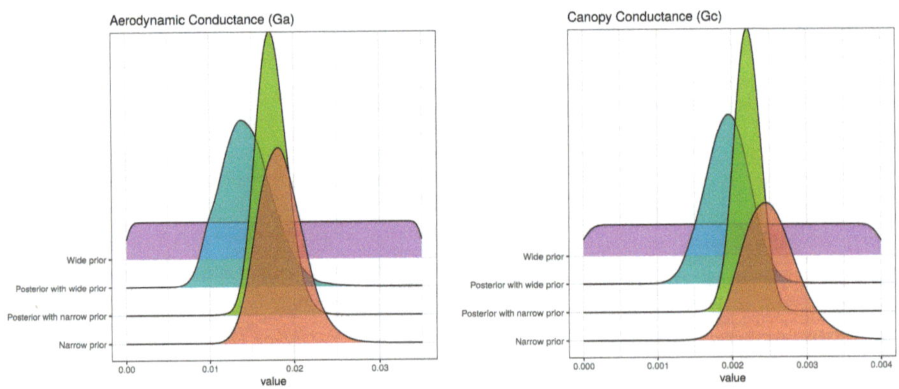

Fig. 9.6 Differences in posteriors when a narrower prior is used

narrower log-normal priors for G_a and G_c, *Lognorma(−4, 0.14)* and *Lognormal(−6, 0.14)*, respectively, and repeat the parameter estimation, this leads to narrower posteriors (Fig. 9.6). These priors were chosen solely to demonstrate the effects of using informative priors. In a real analysis, priors must be formed based on prior information (e.g., literature research, posteriors from a previous analysis, expert elicitation[4]), and *cannot be constructed based on the data being fit* or as a "solution" to posteriors being wider than you would have liked. Repeated analysis results show that MCMC samples of G_a and G_c now have reduced, although still high, correlation ($r \approx 0.9$), and the posterior distributions are influenced by the narrow priors (Fig. 9.6). The more priors reduce the possible combinations, the more identifiable the parameters will be. This is not possible in MLE.

Let us look at another example of unidentifiability from vegetation models. Dynamic process-based computer simulators often have many parameters. It is also very common that there are trade-offs between these parameters which results in unidentifiability issues as different combinations of parameter values would share similar likelihoods. In these cases, the optimizers could fail to search the parameter space exhaustively, then the associated CI techniques could fail as well.

The Very Simple Ecosystem Model (VSEM) implementation in BayesianTools package presents an opportunity to demonstrate this. VSEM, as the name suggests, is a simple "toy" model consisting of a few formulations for photosynthesis and carbon pools (for details see BayesianTools package documentation). One such example of the trade-off situation presents itself between two parameters of this model: the light use efficiency (*LUE*) parameter and a parameter called *GAMMA* which determines the fraction of Gross Primary Productivity (*GPP*) that is autotrophic respiration. Following the package examples, I first simulated synthetic data from the model and added normal noise on it to obtain pseudo-observations. Then, I tried to find back the parameter values of *LUE* and *GAMMA* that created this data using both MCMC and optimization approaches. The advantages of doing this test with synthetic data is that, we now know that we have the perfect model, we know the true parameter values (which are 0.002 and 0.4 for *LUE* and *GAMMA*, respectively, in this example) and we know which likelihood we exactly need to write. We can then inspect the likelihood profile over the range of values for our parameters of interest. The trade-off between parameters would appear as a "ridge" in the likelihood surface (Fig. 9.7).

Such structures in the parameter space makes it hard for MCMC to converge in a small region. However, MCMC would explore this ridge fully and allow better interpretation of the results. Although the posteriors would not resolve finely around the true values, they would not exclude the true values either (Fig. 9.8).

While numerical methods for Bayes (e.g., MCMC) are often more robust to such structures in the parameter space, it is not uncommon that different optimization algorithms give inconsistent results. Moreover, depending on the algorithm used, MLE optimization and the following CI method could fail to capture the true values as optimization algorithms return a point estimate instead of a distribution of parameter values. Figure 9.9 shows how the results returned by a commonly used numerical optimization algorithm, Nelder-Mead, are sensitive to the starting point in the parameter space for this example. Admittedly, there are more robust global

[4]See O'Hagan (2019) and references therein.

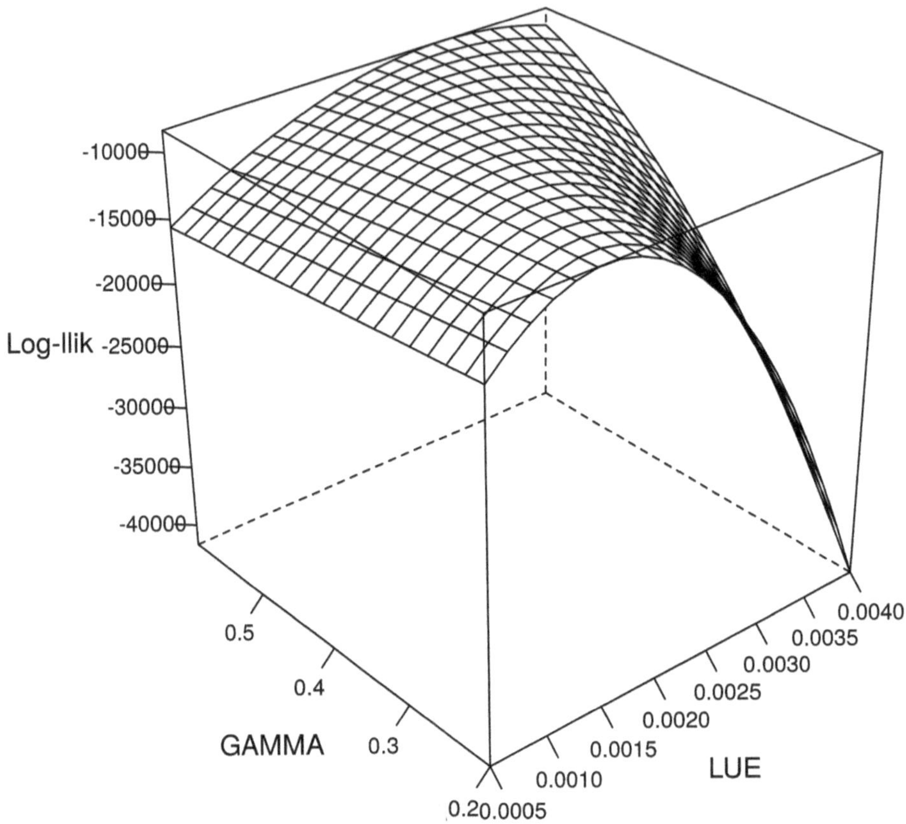

Fig. 9.7 Ridge in the parameter space. Different combinations of parameter values have similar likelihood values

optimization algorithms such as the Differential Evolution algorithm.[5] However, if we were to use numerical methods to find the MLE without checking the likelihood surface first and without comparing different algorithms, we might have been falsely over-confident in the estimates of our parameters. Furthermore, this was a simple low-dimensional problem (and we were using synthetic data). It would be particularly hard to diagnose problems relating to unidentifiability in higher dimensions (and when we do not know the true values).

[5]Even then, joint distributions could be overconfident. Please see the accompanying code for DEoptim application to the same problem and comparison of its bootstrapped samples to the MCMC samples.

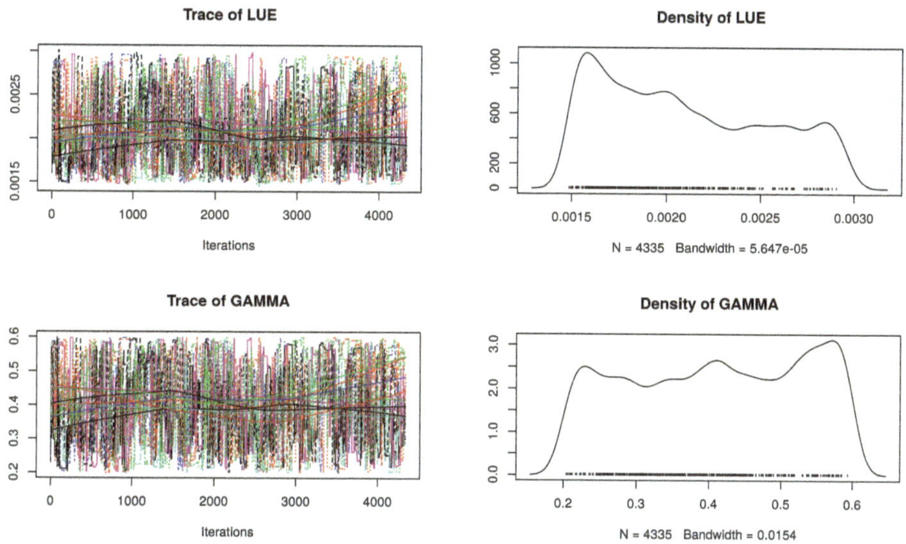

Fig. 9.8 MCMC chains exploring the parameter space (left) and marginal posterior probability distributions (right) are wide but they include the true values (0.002 and 0.4 for *LUE* and *GAMMA*, respectively)

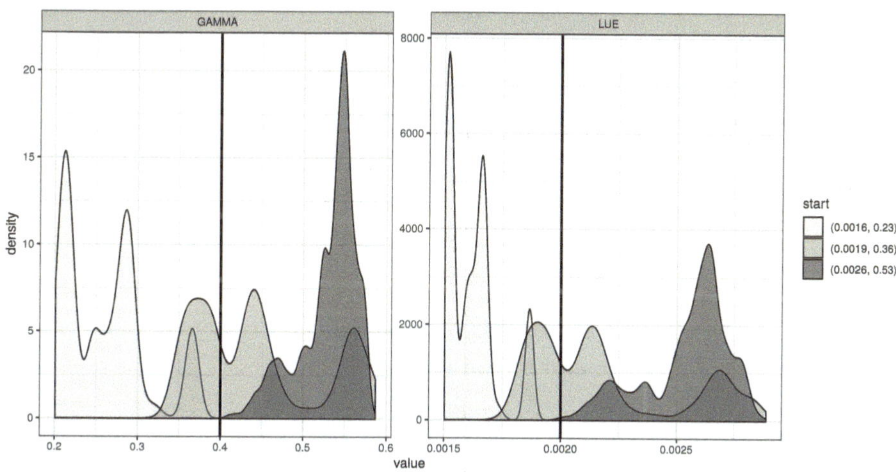

Fig. 9.9 (Parametric) Bootstrapped parameter samples using the MLE values that are found by the Nelder-Mead optimization algorithm. Parameter samples are distributed around values depending on where the MLE is, and MLE values are sensitive to the starting points in the parameter space for Nelder-Mead algorithm when there is a strong trade-off between the parameters. As a result, bootstrapped samples could fail to capture the true values (vertical lines) unless the MLE optimization (with Nelder-Mead) algorithm starts exploring the parameter space from a close point

9.5.4 Hierarchical Bayes

As a final activity, we will begin exploring a hierarchical Bayesian (HB) modelling framework. HB allows accounting for uncertainty and variability at different levels in our models. In the previous examples, we had a nonhierarchical fixed effect (Penman-Monteith equation) in our model. Now we will add a hierarchical random effect to it. More specifically, we will include a year-effect in the model to account for year-to-year variability in the model.

As the sample data set that comes with the *bigleaf* package is very limited (only a month of data), for the HB example, I use flux data from Harvard Forest, US-Ha1, which is available for download from the FluxNet website (https://fluxnet.fluxdata.org/). I applied trivial amount of reformatting to US-Ha1 data to have the same structure and the units as the DE-Tha sample used in the previous examples (this data is also provided together with the code of this book). The same functions are used to calculate the air density, the slope of the saturation vapor pressure curve, the psychrometric constant. In terms of data filtering, I first excluded the months November, December, January, February and March during which the deciduous trees are dormant. This both helped reducing the dimensionality of the data set and made the data set easier to manage because otherwise further processing would be necessary to handle different sizes of leap and non-leap years. The remaining portions of the data are again screened to meet the aforementioned assumptions using the bigleaf::filter.data function.

Initial exploration of the data (e.g., if you plot different years on the same plot) shows that there is year-to-year variability in the latent heat fluxes from this site. Assuming we do not know what causes this variability, we can add a random effect for year in our model. This requires adding a year effect (α_t) to the process model[6] and then specifying a prior on the random effects variance.

$$
\begin{aligned}
&\sigma_\alpha \sim Gamma\,(k, \theta)\\
&\alpha_t \sim Normal\,(0, \sigma_\alpha)\\
&LE_{E,\,t} \leftarrow LE_{Penman-Monteith,\,t} + \alpha_t\\
&LE_{obs,\,t} \sim Normal\big(LE_{E,\,t}, \sigma_{LE}\big)
\end{aligned}
\tag{9.8}
$$

Note how we decided to add a hierarchical random effect. If your data shows other types of variability in the exploratory data analysis (please see Chap. 7), such as site to site, species to species or experimental setting to experimental setting variability, you might want to add a different type of random effect.

Another difference in this exercise from the previous one is that we now have gaps in the independent variable R_n. As we gapfilled the *LE* data in the previous example by treating it as a random variable in the Bayesian framework, we can treat

[6]In this example, I modeled this as an additive effect (such that some years would be consistently higher or lower all year) rather than putting the random effect on our parameters of interest (G_c, G_a).

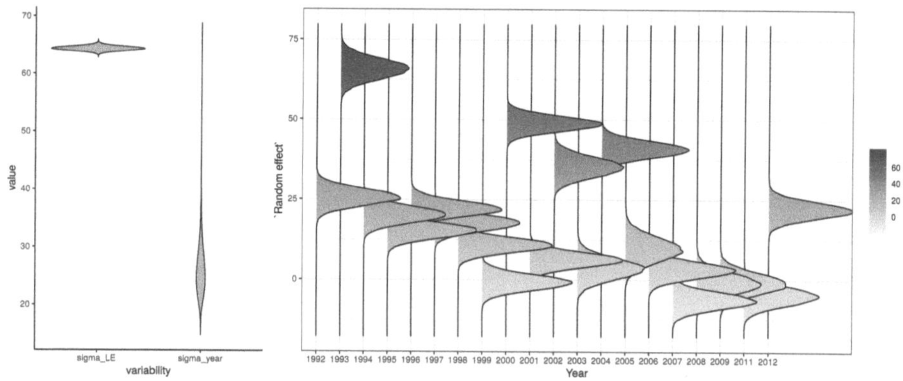

Fig. 9.10 Variance partitioning after adding year effects (left) and the year effect posteriors through time (right). Some years (e.g., 1993) seem particularly different

this variable the same way. This requires specifying prior for this variable by considering its individual characteristics. However, this prior is also chosen for demonstration only and is Gaussian (for actual values used, please see the accompanying code).

$$R_{n,\,obs} \sim Normal\,(R_n, \sigma_r) \tag{9.9}$$

After fitting the model, we can analyse the σ_α and α_t terms in Eq. (9.8) to find out approximately how much of the variance in the latent heat flux is explained by the year effects and what the year-to-year variability looks like, respectively. Notice that now we have two terms that the overall variability is partitioned into: the σ_{LE} in Eq. (9.7) and the σ_α in Eq. (9.8). σ_{LE} is the residual, unexplained variance, whereas σ_α is the variance due to year to year variability. Looking at their posterior distributions (Fig. 9.10), approximately 30% of the overall variability is explained by year effects, even though we may not know what particularly causes this year-to-year variability.

But, the good thing about partitioning the overall variability is that now we can begin to explore the reasons behind additional effects (Dietze 2017). We can now ask questions such as "Were there particularly different years? What could explain this year to year variability?" Next, similar to how we approach all of our scientific questions, we can start generating hypotheses and conduct exploratory analyses. For example, in our case, we can start exploring the factors that affect our ecosystem properties, G_a and G_c. G_a is influenced by wind speed and friction velocity, whereas G_c is influenced by *GPP*, *Ca*, *VPD* and G_1 (stomatal slope parameter) (Medlyn et al. 2011; Knauer et al. 2018a). G_1 is further influenced by CO_2 compensation point ($\Gamma*$) and marginal water cost of carbon (λ) (Medlyn et al. 2011). λ and $\Gamma*$ are further influenced by water stress and growth temperatures, respectively. Therefore, to find out what could be driving this year to year variability, we can explore the correlations between the posterior mean of our year effect and annual mean values of the candidates: friction velocity, wind speed, *GPP*, *Ca*, *VPD* and air temperature. Among these, friction velocity and *VPD* show the only significant correlations ($p < 0.05$) with our year effect, their correlation coefficients being 0.46 and −0.55, respectively. It is also worth noting that *VPD* and friction velocity were not collinear (please see the accompanying code). These results are not unexpected as energy balance closure at flux sites has been shown to increase with friction velocity (Stoy et al. 2013) and *VPD* has been shown to be an important driver of surface conductance in forest ecosystems (Novick et al. 2016) (also see discussion below). Overall, our preliminary analysis suggests that the relationships between *LE* and *VPD* and friction velocity are worth exploring further. Our future data collection and modelling efforts will most likely benefit from focusing on relationships that encompass these variables. Finally, in our HB example, we used one flux site (US-Ha1), whereas data are available from FluxNet network for many other sites. One can easily think of a multisite analysis where site-to-site variability is included in the hierarchical framework and variability is partitioned further.

9.6 Discussion

In the first example, we contrasted and compared MLE and Bayesian approaches on a simple inverse parameter estimation problem. However, even this simple exercise surfaced many important and scientifically relevant dissimilarities between the two approaches. The most important is that Bayesian approach allows us to treat all terms in models as probability distributions. Both the model parameters, the state we are estimating, and the covariates are all treatable with uncertainty. In fact, in the examples above, we have left many sources of uncertainties unexplored. For example, flux towers often do not close the energy balance (i.e., $R_n - G - S \neq H + LE$, left-hand side represents available energy, right-hand side represents turbulent fluxes (Twine et al. 2000), and estimation of G_c ($\approx G_s$) is sensitive to violations of closed energy budget assumption (Knauer et al. 2018a)). Accordingly, not only the energy budget terms, but all of the terms in Eq. (9.6) (including air temperature and pressure terms which are used in derivation of psychrometric constant and air density) are treatable with uncertainty, similar to the approach we took in the HB example. As demonstrated, building and fitting such complex models is generally easier in the Bayesian framework than the maximum-likelihood approach, and this trend is likely to grow, with ever improving statistical and computational tools such as development of more efficient sampler algorithms, open-source user-friendly software packages.

Likewise, we filtered out the data for meeting certain assumptions, such as excluding the data points during and following rainfall to minimize the effects of evaporation from wet surfaces. If the screening was not sufficient and evaporation is still a component of the *LE* flux, it would introduce further errors to our estimate of G_c (as an approximation of G_s) (Knauer et al. 2018a). Instead, evaporation could be modelled as a function of environmental variables, such as *VPD* and wind speed, and estimated explicitly within the overall framework. Indeed, our preliminary analysis of the year effects in the hierarchical Bayesian example showed that we can make better predictions by accounting for further processes involving *VPD* in our model. Similarly, other assumptions (e.g., closed energy budget) can also be relaxed one by one, each becoming their own level in the hierarchical model and associated with their own uncertainties.

Although the flexibilities listed above could be thought as the most convincing arguments for Bayes over MLE, the Bayesian approach also has other scientific advantages that are worth emphasizing. The more philosophical advantage is that Bayes allows us to talk about the probability of our parameters/model/hypotheses being true, which is often what we, as scientists, are interested in and want to communicate in the first place. Frequentist measures (p-values, confidence intervals) can often be misinterpreted as factual statements about parameters/model/hypotheses and this miscommunication could be slowing down the progress of environmental science, if not misleading it altogether (see comment in *Nature* by Amrhein et al. 2019, and the Forum discussions, P values and model selection in *Ecology* 2014). This situation of course could be remedied with better statistical education. The more practical argument is that, Bayesian approach naturally allows us to update our analyses as new data becomes available. This simply involves using posteriors from previous analyses in the new analyses, and repeat this cycle iteratively. Unlike Bayes, we would need to re-do the analyses in the MLE approach every time from the beginning when new data is obtained (or shared between researchers). The ability to update our understanding as new evidence becomes available is a critical aspect of science and Bayesian approach provides a formal framework to do so, in a faster way.

9.7 The Future of Statistical Inference in Environmental Science

One of the main takeaways of this chapter is about the way we treat uncertainties in our models. The notion of treating all model terms as probabilistic is independent of whether we approach the world as a Frequentist or as a Bayesian. Even if we are living in a deterministic world (i.e., there are fixed true values as Frequentism assumes), we only have an imperfect perception of that reality. Therefore, our models should reflect this ignorance to the best of our abilities. As we start incorporating multiple uncertainty sources in our modelling effort, we might realize that we lack data to inform each and every one of these terms, which might result in inconclusive inference (e.g., uncertainty intervals so wide that our inference is basically no better than by chance). However, traditional methods that ignore these uncertainties do not mean these uncertainties do not exist, but that our previous conclusions have been overconfident and prone to false-positives. Indeed, hierarchical Bayesian frameworks that allows uncertainty and variability into our models from many sources do have higher data (and computational) needs. However, this should only encourage our community to move towards identifying, quantifying and partitioning the sources of uncertainty in our study systems, not to shy away from them. Once we identify our data needs to reduce those uncertainties (e.g., do we need more frequent data, data from more sites or more treatments, do we need better equipments or new technology), we can invest in future research efforts accordingly (Dietze 2017).

One of the biggest strengths of Bayesian statistical inference that we have not touched upon explicitly in this chapter is its proficiency in model averaging. When we report our model results as probability distributions, we can seamlessly integrate these results with probabilistic results from other models. In environmental science, where we have many models (hypotheses), this is a powerful method to reconcile and synthesize different predictions. Bayesian approach provides a formal way of combining models where their probability distributions (i.e., their degree of plausibility) dictate how much weight they would contribute to the overall statistical inference. Although Bayesian model averaging requires its own expertise, it is much more advanced than Frequentist model averaging as a field (Ellison 2004; Link and Barker 2006; Hooten and Hobbs 2015; Dormann et al. 2018).

Bayesian inference is becoming fundamental to environmental science all the more. Environmental scientists are increasingly feeling the pressure to provide actionable information to decision makers (Clark et al. 2001; Clark 2005; Luo et al. 2011). This is achievable by putting our models into the iterative machinery of the Bayesian approach (Dietze et al. 2018). As we make probabilistic forecasts, we can confront and combine our predictions with data as they become available and subsequently improve both our models and predictions which makes environmental science more reliable and relevant (Dietze et al. 2018). This is not only a way to meet societal needs but also one of the fastest ways of learning which makes the future of statistical inference in environmental sciences Bayesian (Dietze and Lynch 2019).

Acknowledgements I thank Michael Dietze, Florian Hartig, Martin De Kauwe, Jürgen Knauer and all members of the Boston University Ecological Forecasting Lab for their helpful comments and discussions.

References

Amrhein A, Greenland S, McShane B et al (2019) Scientists rise up against statistical significance. Nature 567:305–307. https://doi.org/10.1038/d41586-019-00857-9

Andrieu C, de Freitas N, Doucet A, Jordan MI (2003) An introduction to MCMC for machine learning. Mach Learn 50:5–43. https://doi.org/10.1023/A:1020281327116

Arora RK (2015) Optimization: algorithms and applications. Chapman and Hall, London

Bolker BM (2007) Ecological models and data in R. Princeton University Press, Princeton

Brookshire ENJ, Gerber S, Webster JR, Vose JM, Swank WT (2011) Direct effects of temperature on forest nitrogen cycling revealed through analysis of long-term watershed records. Glob Chang Biol 17:297–308. https://doi.org/10.1111/j.1365-2486.2010.02245.x

Carpenter B, Gelman A, Hoffman MD, Lee D, Goodrich B, Betancourt M et al (2017) Stan: a probabilistic programming language. J Stat Softw 76:1–32. https://doi.org/10.18637/jss.v076.i01

Clark JS (2005) Why environmental scientists are becoming Bayesians. Ecol Lett 8:2–14. https://doi.org/10.1111/j.1461-0248.2004.00702.x

Clark JS (2007) Models for ecological data. Princeton University Press, Princeton

Clark JS, Carpenter SR, Barber M, Collins S, Dobson A, Foley JA et al (2001) Ecological forecasts: an emerging imperative. Science 293:657–660. https://doi.org/10.1126/science.293.5530.657

De Kauwe MG, Medlyn BE, Zaehle S, Walker AP, Dietze MC, Hickler T et al (2013) Forest water use and water use efficiency at elevated CO2: a model-data intercomparison at two contrasting temperate forest FACE sites. Glob Chang Biol 19:1759–1779. https://doi.org/10.1111/gcb.12164

De Kauwe MG, Medlyn BE, Knauer J, Williams CA (2017) Ideas and perspectives: how coupled is the vegetation to the boundary layer? Biogeosciences 14:4435–4453. https://doi.org/10.5194/bg-14-4435-2017

de Valpine P, Turek D, Paciorek CJ, Anderson-Bergman C, Temple Lang D, Bodik R (2017) Programming with models: writing statistical algorithms for general model structures with NIMBLE. J Comput Graph Stat 26:403–413. https://doi.org/10.1080/10618600.2016.1172487

Dietze MC (2017) Ecological forecasting. Princeton University Press, Princeton

Dietze MC, Fox A, Beck-Johnson LM, Betancourt JL, Hooten MB, Jarnevich CS et al (2018) Iterative near-term ecological forecasting. Proc Natl Acad Sci U S A 115:1424–1432. https://doi.org/10.1073/pnas.1710231115

Dietze M, Lynch H (2019) Forecasting a bright future for ecology. Front Ecol Environ 17:3. https://doi.org/10.1002/fee.1994

Dormann CF, Calabrese JM, Guillera-Arroita G, Matechou E, Bahn V, Bartoń K, Beale CM, Ciuti S, Elith J, Gerstner K, Guelat J, Keil P, Lahoz-Monfort JJ, Pollock LJ, Reineking B, Roberts DR, Schröder B, Thuiller W, Warton DI, Wintle BA, Wood SN, Wüest RO, Hartig F (2018) Model averaging in ecology: a review of Bayesian, information-theoretic, and tactical approaches for predictive inference. Ecol Monogr 88:485–504. https://doi.org/10.1002/ecm.1309

Ellison AM (2004) Bayesian inference in ecology. Ecol Lett 7:509–520. https://doi.org/10.1111/j.1461-0248.2004.00603.x

Fer I, Kelly R, Moorcroft PR, Richardson AD, Cowdery EM, Dietze MC (2018) Linking big models to big data: efficient ecosystem model calibration through Bayesian model emulation. Biogeosciences 15:5801–5830. https://doi.org/10.5194/bg-15-5801-2018

Forum – P values and model selection (2014) Ecology 95:609–653. https://esajournals.onlinelibrary.wiley.com/toc/19399170/2014/95/3

Gelman A, Carlin JB, Stern HS, Rubin DB (2003) Bayesian data analysis. Chapman and Hall, London

Golding N (2018) Greta: simple and scalable statistical modelling in R. R package version 0.3.0. https://CRAN.R-project.org/package=greta

Hartig F, Calabrese JM, Reineking B, Wiegand T, Huth A (2011) Statistical inference for stochastic simulation models – theory and application. Ecol Lett 14:816–827. https://doi.org/10.1111/j.1461-0248.2011.01640.x

Hartig F, Dyke J, Hickler T, Higgins SI, O'Hara RB, Scheiter S et al (2012) Connecting dynamic vegetation models to data – an inverse perspective. J Biogeogr 39:2240–2252. https://doi.org/10.1111/j.1365-2699.2012.02745.x

Hartig F, Minunno F, Paul S (2018) BayesianTools: general-purpose MCMC and SMC samplers and yools for Bayesian statistics. R package version 0.1.5. https://CRAN.R-project.org/package=BayesianTools

Hooten MB, Hobbs NT (2015) A guide to Bayesian model selection for ecologists. Ecol Monogr 85:3–28. https://doi.org/10.1890/14-0661.1

Jaynes ET (2003) Probability theory: the logic of science. Cambridge University Press, Cambridge

Keenan TF, Carbone MS, Reichstein M, Richardson AD (2011) The model-data fusion pitfall: assuming certainty in an uncertain world. Oecologia 167:587–597. https://doi.org/10.1007/s00442-011-2106-x

Kim Y, Knox RG, Longo M, Medvigy D, Hutyra LR, Pyle EH et al (2012) Seasonal carbon dynamics and water fluxes in an Amazon rainforest. Glob Chang Biol 18:1322–1334. https://doi.org/10.1111/j.1365-2486.2011.02629.x

Knauer J, Zaehle S, Medlyn BE et al (2018a) Towards physiologically meaningful water-use efficiency estimates from eddy covariance data. Glob Change Biol 24:694–710. https://doi.org/10.1111/gcb.13893

Knauer J, El-Madany TS, Zaehle S, Migliavacca M (2018b) Bigleaf – an R package for the calculation of physical and physiological ecosystem properties from eddy covariance data. PLoS One 13:e0201114. https://doi.org/10.1371/journal.pone.0201114

Kruschke JK (2015) Doing Bayesian data analysis: a tutorial with R, JAGS, and Stan, 2nd edn. Academic Press, Cambridge, MA

Kurt W (2019) Bayesian statistics the fun way. No Starch Press, San Francisco

LeBauer DS, Dietze MC, Bolker BM (2013) Translating probability density functions: from R to BUGS and back again. R Journal 5:207–209. https://doi.org/10.32614/RJ-2013-020

Link WA, Barker RJ (2006) Model weights and the foundations of multimodel inference. Ecology 87:2626–2635. https://doi.org/10.1890/0012-9658(2006)87[2626:MWATFO]2.0.CO;2

Luo Y, Ogle K, Tucker C, Fei S, Gao C, LaDeau S, Clark JS, Schimel DS (2011) Ecological forecasting and data assimilation in a data-rich era. Ecol Appl 21:1429–1442. https://doi.org/10.1890/09-1275.1

Lunn D, Jackson C, Best N, Thomas A, Spiegelhalter DJ (2012) The BUGS book – a practical introduction to Bayesian analysis. Chapman and Hall, London

MacBean N, Maignan F, Bacour C, Lewis P, Peylin P, Guanter L et al (2018) Strong constraint on modelled global carbon uptake using solar-induced chlorophyll fluorescence data. Sci Rep 8:2045–2322. https://doi.org/10.1038/s41598-018-20024-w

Martin O (2018) Bayesian analysis with Python, 2nd edn. Packt Books, Birmingham

McElreath R (2015) Statistical rethinking: a Bayesian course with examples in R and Stan. Chapman and Hall, London

McNaughton K, Jarvis P (1991) Effects of spatial scale on stomatal control of transpiration. Agric For Meteorol 54:279–302. https://doi.org/10.1016/0168-1923(91)90010-N

Medlyn BE, Duursma RA, Eamus D, Ellsworth DS, Prentice IC, Bartons CVM et al (2011) Reconciling the optimal and empirical approaches to modelling stomatal conductance. Glob Chang Biol 17:2134–2144. https://doi.org/10.1111/j.1365-2486.2010.02375.x

Medlyn BE, De Kauwe MG, Lin Y, Knauer J, Duursma RA, Williams CA et al (2017) How do leaf and ecosystem measures of water-use efficiency compare? New Phytol 216:758–770. https://doi.org/10.1111/nph.14626

Novick KA, Ficklin DL, Stoy PC, Williams CA, Bohrer G, Oishi AC et al (2016) The increasing importance of atmospheric demand for ecosystem water and carbon fluxes. Nat Clim Chang 6:1023–1027. https://doi.org/10.1038/nclimate3114

O'Hagan A (2019) Expert knowledge elicitation: subjective but scientific. Am Stat 73:69–81. https://doi.org/10.1080/00031305.2018.1518265

Plummer M (2003) JAGS: a program for analysis of Bayesian graphical models using Gibbs sampling. In: Hornik K, Leisch F, Zeileis A (eds) Proceedings of the 3rd international workshop on distributed statistical computing (DSC 2003). Technische Universität Wien, Vienna. https://www.r-project.org/conferences/DSC-2003/Proceedings/Plummer.pdf

Plummer M (2016) rjags: Bayesian graphical models using MCMC. R package version 4-6. https://CRAN.R-project.org/package=rjags

Richardson AD, Hollinger DY (2005) Statistical modeling of ecosystem respiration using eddy covariance data: maximum likelihood parameter estimation, and Monte Carlo simulation of model and parameter uncertainty, applied to three simple models. Agric For Meteorol 131:191–208. https://doi.org/10.1016/j.agrformet.2005.05.008

Richardson AD, Braswell BH, Hollinger DY, Burman P, Davidson EA et al (2006) Comparing simple respiration models for eddy flux and dynamic chamber data. Agric For Meteorol 141:219–234. https://doi.org/10.1016/j.agrformet.2006.10.010

Rue H, Martino S, Chopin N (2009) Approximate Bayesian inference for latent Gaussian models using integrated nested Laplace approximations (with discussion). J Roy Stat Soc B 71:319–392. https://doi.org/10.1111/j.1467-9868.2008.00700.x

Smith BJ, Deonovic B et al (2014) Mamba: Markov chain Monte Carlo for Bayesian analysis in julia. 2014. julia software package. https://github.com/brian-j-smith/Mamba.jl

Smith WK, Reed SC, Cleveland CC, Ballantyne AP, Anderegg WRL, William WR et al (2016) Large divergence of satellite and Earth system model estimates of global terrestrial CO_2 fertilization. Nat Clim Chang 6:306–310. https://doi.org/10.1038/nclimate2879

Stoy PC, Mauder M, Foken T, Marcolla B, Boegh E, Ibrom A et al (2013) A data-driven analysis of energy balance closure across FLUXNET research sites: the role of landscape scale heterogeneity. Agric For Meteorol 171-172:137–152. https://doi.org/10.1016/j.agrformet.2012.11.004

Twine TE, Kustas WP, Norman JM, Cook DR, Houser PR, Meyers TP, Prueger JH, Starks PJ, Wesely ML (2000) Correcting eddy-covariance flux underestimates over a grassland. Agric For Meteorol 103(3):279–300. https://doi.org/10.1016/S0168-1923(00)00123-4

van Oijen M, Rougier J, Smith R (2005) Bayesian calibration of process-based forest models: bridging the gap between models and data. Tree Physiol 25:915–927. https://doi.org/10.1093/treephys/25.7.915

Varadhan R, Borchers HW, ABB Corporate Research (2018) dfoptim: derivative-free optimization. R package version 2018.2-1. https://CRAN.R-project.org/package=dfoptim

Williams M, Richardson AD, Reichstein M, Stoy PC, Peylin P, Verbeeck H et al (2009) Improving land surface models with FLUXNET data. Biogeosciences 6:1341–1359. https://doi.org/10.5194/bg-6-1341-2009

Chapter 10
Machine Learning Applications in Hydrology

H. Lange and S. Sippel

10.1 Introduction

Hydrological processes operate on vastly different spatiotemporal scales (Fatichi et al. 2016). From water transport in the pore space to continental runoff, from rapid infiltration and diffusion at the scale of seconds to multiyear phenomena, the hydrosphere exhibits a bewildering diversity. It is without question that very different processes are the most relevant at the different scales. Therefore, process-oriented modeling approaches which assume the same set of processes across all scales clearly imply inherent limitations and are often inadequate (Sivapalan 2003, 2006; Clark et al. 2015). However, this assumption is often made in hydrological modeling. Arguably the best known example is the application of the Richards equation (Richards 1931), which is basically a nonlinear transport equation derived from averaging microscopic processes within the pore volume, and thus operates in pore space at typical spatial scales of some centimeters or decimeters, to a hillslope, whole catchments or even continental scales in the context of Earth system models. Problems induced by this issue are basically known (Blöschl 2001; Sivapalan et al. 2004; Viney and Sivapalan 2004); see, however, Ma et al. (2017) for an attempt to bridge the scale gaps by linking mapping, monitoring, modeling, and management and by "comprehensive integration."

Electronic Supplementary Material The online version of this chapter (https://doi.org/10.1007/978-3-030-26086-6_10) contains supplementary material, which is available to authorized users.

H. Lange (✉)
Norwegian Institute of Bioeconomy Research, Ås, Norway
e-mail: Holger.Lange@nibio.no

S. Sippel
Norwegian Institute of Bioeconomy Research, Ås, Norway

Department of Environmental System Science, ETH Zürich, Zürich, Switzerland

D. F. Levia et al. (eds.), *Forest-Water Interactions*, Ecological Studies 240,
https://doi.org/10.1007/978-3-030-26086-6_10

It is a common feature of process-oriented hydrological models that they require values for parameters which are difficult to obtain through measurements or are almost unconstrained by observations. They are often fitted or at least constrained through *inverse* modeling: the hydrological system is supposed to "have" the parameter values which lead to the best fit of the model to the data available. However, since the fitting routines seeking the parameter estimates often necessarily operate in high-dimensional spaces, no unique minimum of the objective function is obtained, and the problem of "equifinality" occurs (Beven and Freer 2001).

In general, *process-oriented* modeling tries to *explain* the observed behavior and can be characterized as a bottom-up approach where phenomena at a higher level (e.g., runoff from a catchment) are explained through processes at a lower level (water transport through soil, groundwater movement, etc.). Equifinality and over-parametrization are two severe caveats impeding progress along that route. It is an open and rather difficult question whether this is due to our lack of system understanding, scale errors, too little detailed or simply not enough measurements, or whether focus is on variables unsuitable to foster proper understanding, to name a few of the basic obstacles one could think of.

Alternatively, a *data-oriented* approach seeks to detect and describe patterns in univariate or multivariate hydrological datasets, tries to generalize them, and then makes *predictions* on the system state. Hydrology is particularly well-suited to this line of research since in many cases, it deals with input–output systems and transport processes (i.e., water flow) along gravitational gradients. The attribution of a variable as input or output is in many cases obvious. We are mentioning this since in other contexts, e.g., the biochemistry of ecosystems, the direction of causality is not intuitive and might even change over time (Sugihara et al. 2012).

"Process-oriented" and "data-oriented" modeling approaches are not complementary per se, but in many cases, the combination of both entails a large potential for hydrological modeling. For example, data-oriented models occasionally provide a detailed description of the input–output mapping and thereby support process-oriented model building and process understanding. Furthermore, *hybrid modeling*, i.e., the combination of both approaches within one modular hydrological model, can yield improved forecasts (Corzo Perez 2009). Hybrid modeling tries to combine the advantages of both process-oriented and data-oriented modeling approaches. For example, a key process-oriented modeling principle is that the energy and water balances across the entire model remain preserved, while determining model parameters might be achieved in a parsimonious way with data-driven approaches. Earlier studies have shown that such pathways are indeed feasible, for instance, for the parametrization of clouds within Earth system models (Rasp et al. 2018), and in hydrology automated machine learning (ML)-based upscaling of streamflow observations to the grid cell level for Earth system modeling has been conducted (Gudmundsson and Seneviratne 2015). ML is also used to improve the understanding and analysis of process-oriented models (Peters et al. 2016).

Table 10.1 Examples for methods used in data analysis in four different categories

	Unsupervised	Supervised
Manual/ non-iterative	Time series analysis, frequency decomposition, principal component analysis	ESLLE[a] GLMs, GAMs, GAMMs[b]
Automatic/ iterative	Clustering, autoencoders, data mining, dimensionality reduction	Neural networks, decision trees, random forests, support vector machines, and many more; proper machine learning

[a]Enhanced Supervised Locally Linear Embedding (Zhang 2009)
[b]Generalized Linear Models, Generalized Additive Models, Generalized Additive Mixed Models

Data-oriented detection and description of patterns can be done in a variety of ways. Two basic distinctions are "manual" (or one-time, non-learning) versus "automatic" (iterative, learning) algorithms on one hand and supervised versus nonsupervised ones on the other (Table 10.1).

As can be seen from Table 10.1, machine learning methods (as we use the term) belong to the class of supervised iterative methods, where the iteration is the process of learning. Thus, in most applications, the algorithm gets training and validation datasets presented, learns on them, and is afterward tested on the remaining (or independent) datasets. For hydrological applications, a well-designed cross-validation scheme is of utmost importance since, as the underlying data are highly structured and spatiotemporally correlated, the estimation of the generalization error can be severely distorted (see, e.g., Roberts et al. (2017); a simple, tutorial-style introduction into this topic is given in Sect. 10.2 from a bias-variance trade-off perspective where the R code is provided as online electronic supplementary material [see Extras Springer online]).

Moreover, modelers are often afraid that the training part could not contain all situations which occurred throughout the history of the system; proper extrapolation beyond the range observed (or "hydrological regime") in the training sample is a difficult issue as even the most sophisticated machine learning techniques might fail outside the training range. In many cases less flexible, more parsimonious methods might actually extrapolate better, and causal learning techniques that would allow to incorporate the correct causal structures are still in their infancy (Peters et al. 2017).

The present review of (supervised) ML in the context of hydrological applications is structured as follows: in the next section, we provide a short overview of the origin and rapid expansion of machine learning in popular as well as scientific publications and outline basic statistical principles of machine learning from a practitioner's perspective. In Sect. 10.3, some of the key machine learning algorithms that are already in frequent use in hydrology are described in a nontechnical manner, broadly along a gradient of more parsimonious to more flexible methods. Sect. 10.4 then highlights recent notable cases of ML application in hydrology; a brief outlook is given in Sect. 10.5.

10.2 Machine Learning: Historical Overview and Statistical Basis

10.2.1 The Origin and Rapid Expansion of ML

The term "Machine Learning" was coined already 60 years ago (Samuel 1959) in the context of game-learning computers (in that case, checkers). The fundamental and then-new insight was that software is able to learn how to play strategic games, with a performance superior to that of the human programmer.

From the start, machine learning was closely connected to Artificial Intelligence (AI). However, since probabilistic and iterative methods are notoriously data-thirsty and data availability was always an issue prior to the ability to collect, store and distribute large datasets automatically, machine learning faded away in favor of expert systems and knowledge-based AI software.

However, with the advent of the Internet, the corresponding availability of digitized information and with increasing computing power, machine learning experienced a revival, now as its own field separated from AI. It has strong roots in statistics and probability theory, and many applications of machine learning, including those fashionable in hydrology, are indeed tools based on and designed for regression and/or classification.

A closely related concept is that of *data mining*. In this review, we consider data mining and machine learning as separate approaches under the umbrella of data-oriented modeling: whereas data mining explores unknown datasets for knowledge discovery, the typical machine learning algorithms has prediction as its target; to achieve it, it is first fed with training data to learn patterns and dynamics and then validated on hitherto unseen datasets. Still, the two concepts are interwoven: some unsupervised learning methods are taken from data mining as a preprocessing step of machine learning; some data mining approaches utilize machine learning methods but with a different goal. Data mining is often concerned with dimension reduction, providing an efficient (low-dimensional) representation of the main patterns present in high-dimensional data, unraveling redundancies, and so on. Whereas some machine learning methods perform dimension reduction as preparatory step, it is not part of machine learning per se.

It is surely no overstatement that we are living in the era of machine learning since a few years. Their applications span a wide range, including not only scientific domains but also industry and commercial businesses. Examples are automatic text and speech recognition, translation from one natural language to another or from specialist to laymen language, health care and medical data analysis, stock market predictions, spotting fraud and plagiarism in science and elsewhere, optimizing search engines like Google, autonomous driving vehicles, or autonomous self-exploration and social interactions of robots. This list is far from being exhaustive, and there are many job announcements and project applications mentioning machine learning in their title. Overall, the scientific and commercial potential is considered as huge.

A word of caution might be necessary at this stage. The rise of machine learning is deeply connected to data availability, which includes the ability to automatically collect, store, and distribute data—we also live in the era of "big data." The titanic flow of data presents a challenge in its own, not only to computer storage and processing speed but also to human comprehension. However, large classes of methods used in machine learning are not genuinely new; others are mere refinements of existing approaches. It is possible that there is a mismatch between data volume and sophistication of our tools; the latter is always in danger to lag behind.[1] And as today, skeptics could consider machine learning as just nonlinear fitting in big datasets, vulnerable to wild errors when extrapolating from the domain of training sets provided to the "machine" and (as a consequence) difficult to apply in areas where huge amounts of reliable data are difficult or impossible to get or where the test datasets differ in some underlying aspects from the training datasets.

A recent poll among chemists[2] indicated that 45% of the participants agreed or even strongly agreed to the statement that machine learning is overhyped. An extreme example of hyping is that of the director of Artificial Intelligence at Tesla calling machine learning (and neural nets in particular) "Software2.0."[3] While overrating or overselling are common side effects of new technologies and emerging markets, one could as well relax and just add machine learning approaches to the toolbox of the modeler, recognizing that machine learning is good at things where machines are good at learning and where the appropriate training data are available, no more no less.

The increasing popularity of machine learning in hydrology originates also in the availability of data sources other than traditional ones like precipitation, runoff, groundwater height, and so on. Remote sensing from satellites or airplanes, embedded sensor networks, drones and even internet-based social networks (*sensu* citizen science) all contribute to ever-increasing data streams, creating a rich playground for machine learning approaches.

10.2.2 The Bias-Variance Trade-Off and Cross-Validation of Spatiotemporally Correlated Data

A key principle of machine learning methods is that unknown functional relationships are approximated via the iterative tuning of hyperparameters. This leads directly to the bias variance trade-off (see, e.g., the longer introduction in Hastie et al. (2008)) with important implications for hydrological ML applications where

[1] https://insidebigdata.com/2018/10/19/report-depth-look-big-data-trends-challenges/ (accessed January 30th, 2019).

[2] Found on https://cen.acs.org/physical-chemistry/computational-chemistry/machine-learning-overhyped/96/i34 (accessed January 30th, 2019).

[3] https://medium.com/@karpathy/software-2-0-a64152b37c35 (accessed January 30th, 2019).

substantial spatial or temporal dependencies are crucial data properties. Here, we illustrate the bias-variance trade-off with the approximation of an "unknown" (simple third-order polynomial) function f via non-parametric k-nearest neighbor (kNN) regression. kNN has one tuning parameter, which is the number of nearest neighbors k (in feature space, i.e., X) that are averaged to obtain the prediction $\widehat{Y}$ (see Sect. 10.3.1 for a more detailed introduction); i.e.,

$$\widehat{Y}(x) = \frac{1}{k} \sum_{x_i \in N_k(x)} y_i$$

where $N_k(x)$ is the set of the k points x_i in the training data which are closest to the point x. Mean squared error (MSE) curves for approximating f are shown as a function of k in Fig. 10.1a for the training and test datasets. For large k, kNN regression "predictions" are close to the training sample mean; hence, a bias arises toward the extreme cases of X resulting in underfitting (e.g., $k = 50$ in Fig. 10.1b where the prediction tends toward a constant at both edges). In contrast, for small k, kNN "predictions" follow (arbitrarily) closely the *training* dataset (Fig. 10.1b). Hence, by successively decreasing k, the *training set* MSE reduces monotonically as flexibility is added to the model. This continuously reduces bias and leads eventually to a perfect fit of the training dataset (Fig. 10.1b). However, a "perfect" fit on the training dataset (such as shown for $k = 1$ in Fig. 10.1b) is clearly undesirable, as the model will generalize poorly to a different ("test set") realization of f (Fig. 10.1c), i.e., leading to high variance of the fit. Hence, generalization ability to unseen data is a crucial property of any machine learning model, and consequently training set MSE is not a reliable measure for model performance. MSE on an unseen test (or validation) dataset is a sum of bias (resulting from too little model flexibility, i.e., underfitting, such as for $k = 50$ in Fig. 10.1b/c), variance (resulting from too much model flexibility, i.e., overfitting, e.g., for $k = 1$ in Fig. 10.1b/c), and irreducible error (see for more details, e.g., Hastie, Tibshirani, and Friedman (2008)); and hence from Fig. 10.1a, the modeler would choose a k-value that minimizes test MSE.

In most practical cases, hyperparameters are tuned through *cross-validation*, i.e., through a systematic partitioning of the training dataset into internal training/ validation sets.

An important property of hydrological datasets is that they are highly structured and typically involve substantial spatial and temporal correlations and nonstationarities (Roberts et al. 2017) that might be induced either through missing predictors or through underlying correlated noise. This property implies that a random partitioning of data into folds for cross-validation, which is the default in many standard software cross-validation routines, is prone to fail: The reason is of course that data used internally for training and validation in cross-validation are correlated independent, and hence the resulting MSE curve would resemble the "training set MSE" in Fig. 10.1a; resulting in an underestimation of generalization error of the model and leading to a model fit that is tuned too closely to the

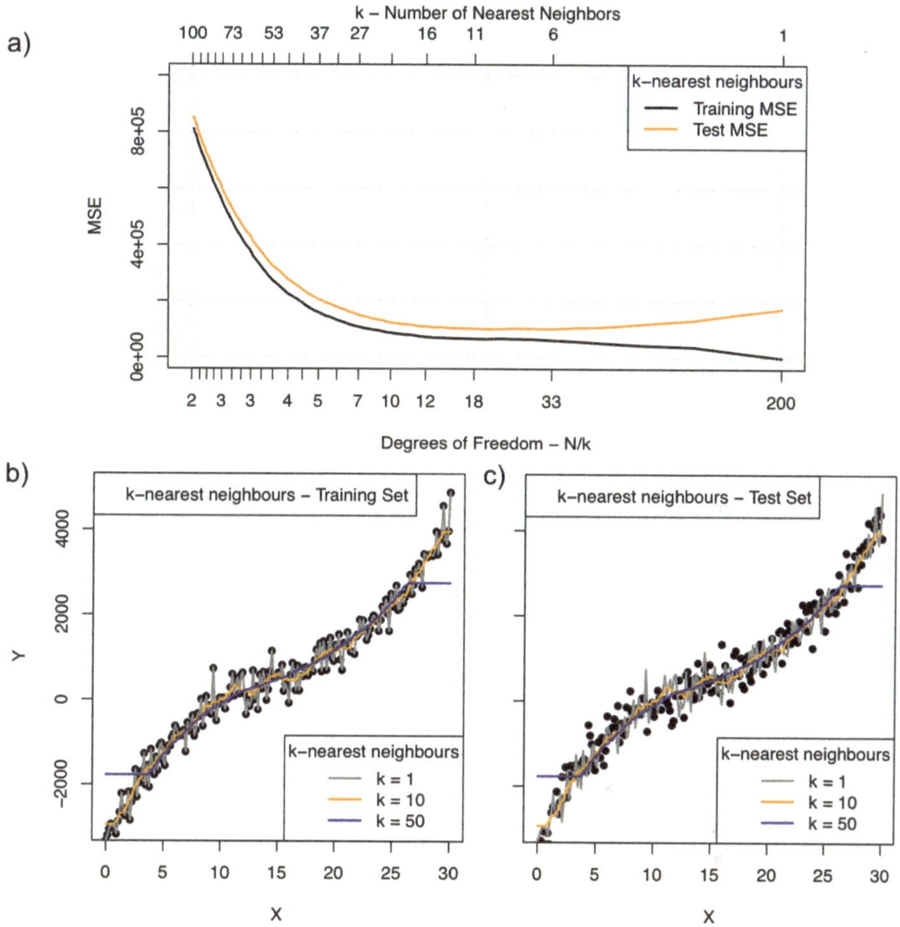

Fig. 10.1 Illustration of the bias-variance trade-off as a key concept in machine learning that is crucially relevant for hydrological applications. (**a**) Training and test set mean squared error of k-nearest neighbor (kNN) regression with various tuning parameter values k used to approximate an unknown function (third-degree polynomial). (**b**, **c**) kNN regression fits for the (**b**) training and (**c**) test datasets. k-nearest neighbor regression fits are based on the "kknn" R-package (Schliep and Hechenbichler 2016)

available data. Hence, a smart partitioning of the data into training and test datasets is crucial particularly in real-world hydrological applications (Roberts et al. 2017). For practical applications, the issue of correlations might be alleviated by choosing block-wise folds (in space and/or time, i.e., eliminating information leakage between folds) for cross-validation (Roberts et al. 2017). Although the problem of correlated errors in prediction problems is known since quite some time (see Schoups and Vrugt (2010) for hydrology, and Bergmeir et al. (2018) for an overview), it has been argued that some predictions that failed gracefully such as the outcome of the 2017 US presidential elections or the onset of the 2008 financial crisis had been at least partly caused by ignoring correlated errors among the instances used for prediction (Silver 2012).

10.3 Machine Learning Algorithms Used in Hydrological Research

In this section, we provide a short overview on some important machine learning techniques used in hydrology. Given the pace of development in this field, the overview is necessarily incomplete and might also be imbalanced; however, each one of the methods to follow has had an application in the hydrological literature. Moreover, following up on the bias-variance trade-off considerations in the previous section and for practical considerations, we present the methods broadly in a sequence from more parsimonious methods to more flexible methods.

We also add references to some notable algorithms and software implementations. However, because machine learning algorithms and implementations have become so widespread and diverse, it is beyond the scope of this chapter to recommend specific software packages. Therefore, we refer the interested reader to dedicated overview sites targeted at machine learning applications and algorithm implementations, for instance, *mloss.org*[4] ("machine learning open source software") or the CRAN task view "Machine Learning & Statistical Learning" (Hothorn 2019) that provides a comprehensive overview of packages and implementations that are available within the R programming environment.

10.3.1 k-Nearest Neighbors (KNN)

This is a classic algorithm for both classification and regression (Altman 1992) – and has been used for illustration in Sect. 10.2. In *feature space*, i.e., the space spanned by pairs of input and output vectors, one must find the k-nearest neighbors of any sample point, which obviously requires a distance measure. Among the common choices, the Minkowski distance (with distance parameter q) as a generalization of the Euclidean distance is the most flexible one. In KNN regression, the output (prediction) for the sample point is the weighted average of the values for the k-nearest neighbors, where the weights are inversely proportional to the distances. It might be necessary to reduce the dimensions of the problem first.

[4]https://mloss.org/software/

10.3.2 Regularized Linear Models: Lasso, Ridge, and Elastic Net Regression

Regularized regression methods have been developed to account for cases in which the number of predictors is relatively large compared to the number of samples, i.e., where multiple linear regression might overfit (Hastie et al. 2008).[5] The idea is to reduce the regression model's flexibility by shrinking the regression coefficients, such as to avoid overfitting and to allow interpretability of the model. Shrinkage is done by adding a constraint that encapsulates the size of the regression coefficients to the least squares optimization function. Lasso and ridge regression are two related regression methods that differ in the nature of shrinkage (Hastie et al. 2008), where ridge regression shrinks coefficients toward zero based on the L2-norm, whereas Lasso regression performs some kind of subset selection by preferably shrinking coefficients *exactly* to zero via the L1-norm penalty. The degree of shrinkage is regulated by a hyperparameter (λ in the equation below) that is typically determined by cross-validation.

Elastic net models (Zou and Hastie 2005) represent a blending of Lasso and ridge regression, and uses an additional hyperparameter allowing to switch continuously between the two regression methods. The vector of regression coefficients β is obtained in elastic nets as

$$\widehat{\beta} = \underset{\beta}{\operatorname{argmin}}\left(\|y - X\beta\|^2 + \lambda\left[\alpha\|\beta\|^2 + (1-\alpha)\|\beta\|_1\right]\right)$$

where X is the matrix of predictors, λ is the penalty strength, and α is the new hyperparameter ($\alpha = 0$ is Lasso, $\alpha = 1$ is ridge regression).

Elastic nets are implemented in the glmnet package in R (Friedman et al. 2010).

10.3.3 Artificial Neural Networks (ANNs)

These are the most well-known learning algorithms, probably also among the oldest and with applications to hydrology since several decades (Maier and Dandy 1995; Kuligowski and Barros 1998; Zealand et al. 1999; Lischeid 2001; Parasuraman et al. 2006; Wang et al. 2006; Daliakopoulos and Tsanis 2016). There are also many excellent textbooks on the subject, so we provide only some key elements here. The interested reader might consult Chapter 11 of (Hastie et al. 2008) or Chapter 5 in Bishop (2006) for a thorough overview on ANNs.

---

[5]This book is freely available as pdf document from https://web.stanford.edu/~hastie/ElemStatLearn/

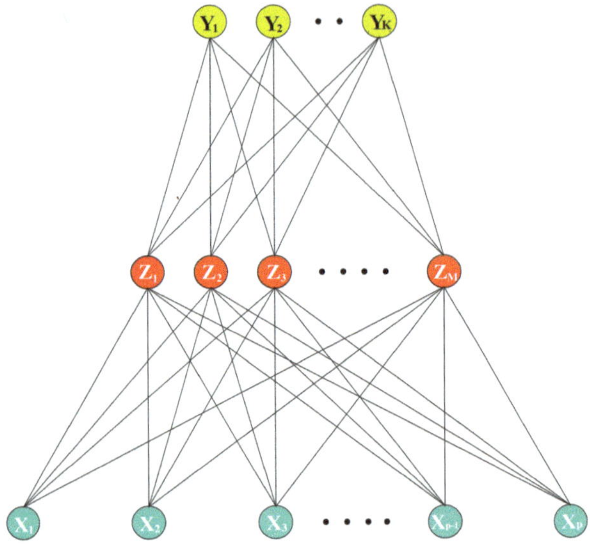

Fig. 10.2 A typical three-layer feed-forward neural network. The X_i are input and the Y_i output nodes; between them, there is a "hidden" layer with nodes Z_i. The links between the nodes are weighted (not shown). (Source: Hastie et al. 2008; reproduced with permission of Springer)

ANNs are motivated by an analogy to the physiology of the human brain, where neurons are represented as nodes in a network, whereas the connections between neurons (synapses) are links between them. Each neuron is equipped with an activation function, determining according to its input value whether it "fires" (produces output) or not (or rather, the "firing" is a continuous process between none and maximal firing). The prototypical architecture of an ANN is shown in Fig. 10.2. It should be obvious from the figure that the analogy with the human brain should not be taken too far. In the latter, there are no immediately identifiable "input" and "output" neurons, let alone hidden layers between them.

In many applications, there will be only one output node Y; it is also important to note that foreseeing more than one hidden layer does not necessarily improve model performance, consistent with the universal approximation theorem (Cybenko 1989) which states that a feed-forward network with a single hidden layer containing a finite number of neurons can approximate continuous functions, under mild assumptions on the activation function; therefore, one hidden layer is the most common choice.

ANNs have a long history, where the invention of the perceptron (Rosenblatt 1958) was a particular important milestone. Plagued by the bottleneck of insufficient computing power and some theoretical problems, the proper breakthrough came with the invention of the backpropagation algorithm (Werbos 1975), later made popular in particular through the seminal work of (Rumelhart et al. 1986), which is until today by far the most common method to determine the weights on the links through training.

There are some serious issues with ANNs in practical circumstances. The learning rate can be quite slow, preventing an efficient online updating, where the networks learns continuously while new data are streaming in, rather than working fully through an offline training set. There is also a severe danger of overparameterization, unless the training sample is huge. Letting the backpropagation run for a long time on the same sample bears the danger of overfitting (or overlearning): the network gets to "understand" tiny insignificant details of that data sample, loses generalization power and thus is performing worse in the subsequent validation phase compared to learning shorter steps. As Hastie et al. (2008) phrase it, "there is quite an art in training neural networks."

There is one additional problem which is, strictly speaking, outside proper quality assessments for machine learning, but it has been subject of intense debate since a long time: trained neural networks which perform well are usually opaque and not open to interpretation. The matching between input and output, presented as weights between (many) input neurons, the nodes in the hidden layer(s) and from there to the output neuron(s), is eluding human comprehension; developed ANN architectures are nontransparent to the data analyst. One might wonder whether the incomprehensibility of the network architecture is a relevant issue for systems doing analysis and prediction largely automatic and without human interference, but it has limited the spread of ANNs; the target of scientific applications (rather than, say, industrial or engineering applications) is often to understand *why* things work.

10.3.3.1 What Is Deep Learning?

The adjective "deep" in front of a learning algorithm is a fashionable attribute. A machine learning algorithm is called deep when it comprises a multitude of layers of nonlinear processing units, where each successive layer uses the output of the previous layer. The canonical example for deep learners are ANNs (Schmidhuber 2015), but not every ANN is necessarily deep; a special variant of them, Recurrent ANNs, have potentially unlimited depth (layers are added during the learning process) and are thus (very) deep. The term deep learning was brought to the machine learning community by Dechter (1986). The most general framework is Deep Reinforcement Learning (DRL). DRL agents are forced to learn how to interact with an initially unknown and only partially observable environment, unassisted, in order to maximize their cumulative reward signals (Schultz 2007). Reward signals are produced by reward neurons and are used to influence brain activity that controls actions, decisions and choices. In the human brain, an important example of reward neurons are dopamine neurons. DRL became popular recently as Silver et al. (2016) managed to construct DRL systems beating world elite players in the board game Go. Nowadays, deep learning is a very active research area with a loose recognition of the term; it refers to any algorithm capable of extracting knowledge from big data in an efficient manner (Najafabadi et al. 2015). Implementations of deep learning algorithms are available for all major programming languages.[6]

[6]see e.g., https://en.wikipedia.org/wiki/Comparison_of_deep_learning_software

10.3.4 Convolutional Neural Networks (CNNs)

For the ANNs discussed up to here, a spatial neighborhood of input or output neurons (or more generally, a geodesic distance between them) is ignored, i.e., does not enter the network architecture. For some tasks, with automatic image analysis as foremost example, this is unfortunate since local information (the surrounding pixels) is more important than remote one. This flaw is overcome by CNNs (LeCun et al. 1989), designed to emulate the behavior of the visual cortex. The neurons in a CNN are organized in three dimensions (height, width and depth) (Fig. 10.3), and this spatial locality is utilized through a local connectivity pattern between neurons of adjacent layers. The neurons in a given layer (to the right in Fig. 10.3) are connected to only a few neurons or a small region within the layer before it, called the receptive field (the left and middle box in Fig. 10.3 show two iterations of this principle).

The core building element of a CNN is the convolutional layer, whose parameters are changeable kernels (filters) with a small viewing field (focusing on close neighbors). Each filter is "convolved" through the width and height of the layer, which means that the scalar products between the inputs and the entries of the filter are taken. It is therefore more correct to speak of a "cross-correlation neural network," but this is not the common term for this type of networks.

10.3.5 Support Vector Machines (SVMs)

The starting point for support vector machines (SVMs) (Boser et al. 1992) is the observation that there is always noise in any set of observations and that a

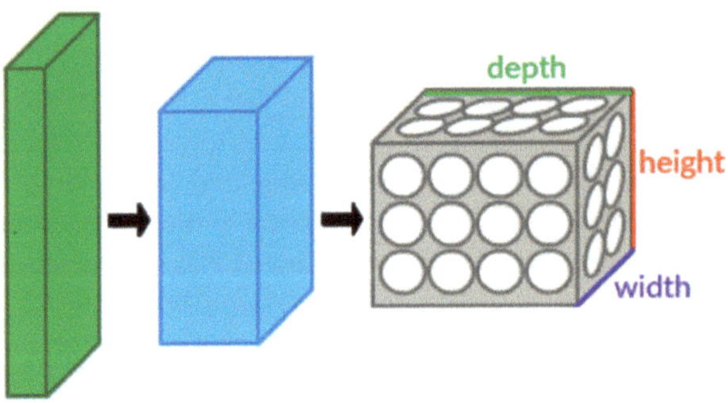

Fig. 10.3 The typical architecture of a convolutional neural network. The convolutional layer to the right is a three-dimensional object which is connected to a small region of the layer before it (box in the middle), the receptive field, which, in turn, has few connections with the layer before it (left). (Source: Wikipedia (https://en.wikipedia.org/wiki/Convolutional_neural_network); Reproduced under Creative Commons license type Attribution-Share Alike 4.0 International)

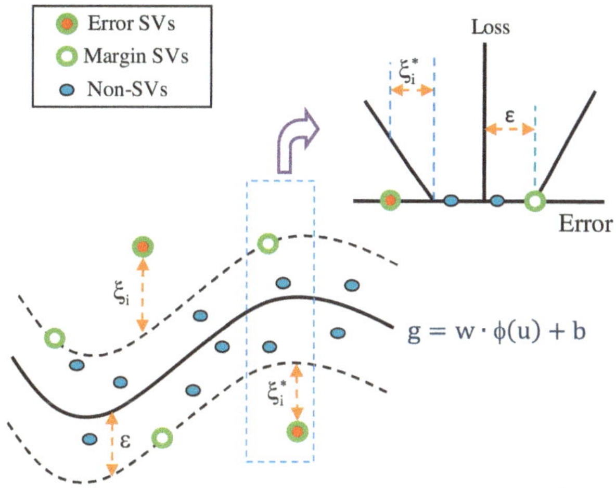

Fig. 10.4 Support vector regression. Points inside the ε boundaries are not used as support vectors (non-SVs). (Source: Raghavendra and Deka (2014); Reprinted with permission from Elsevier)

(generalized) regression should not run through data points which differ less than an assumed threshold ε; SVM regression is sometimes also called ε-insensitive regression. Only the data points outside of the threshold are used for the regression; they constitute the support vectors (Fig. 10.4). Therefore, SVMs are especially suitable when analyzing noisy data. Arbitrary nonlinear regression is possible when using kernel functions (the "kernel trick" (Hofmann et al. 2008)). Typical kernels are Gaussian (in this context often called "Radial Basis Functions" (RBF)) or polynomial ones, but also sigmoidal kernels have been tried. This flexibility comes also with a price, however: the selection of the kernel function is largely heuristic and based on trial and error.

The best-known application of SVMs is the automatic identification of handwritten digits, but there are also other important areas like bioinformatics, biochemistry, and not the least environmental sciences, including hydrology.

10.3.6 Decision Tree Learning

One of the key ingredients of many machine learning algorithms are *decision trees*. These are (graphical) presentations of conditional statements about the outcome of decisions; depending on weights (probabilities) of the input features, one branch or the other in a decision tree is taken. Observations are contained in the branches, whereas target values (predictions) are contained in the leaves. Thus, decision trees are examples of prediction models. If the target values are continuous, decision trees are often called regression trees. The prototype and arguably the oldest regression tree procedure is the Classification and Regression Tree (CART) introduced by

Breiman et al. (1984). An implementation of CART in R is provided by the rpart package.[7]

A decision tree learning algorithm usually starts with a tree with a simple structure (in terms of branches and leaves), and recursively splits the observations set into subsets, in many cases using just thresholds for the value of the observations (take one branch if the observation is smaller than a prescribed threshold value, and the other one otherwise). The process of re-branching is called recursive partitioning. The process stops when further splitting does no longer improve the predictions. In many algorithms, a secondary step is performed where insignificant branches are removed again to end up with a less complex model providing virtually identical prediction performance, an operation called pruning.

10.3.7 Random Forests

It is a common property of decision trees to be sensible to overtraining, in particular if the training dataset is not providing all features of the datasets which can occur. It is generally believed that increasing the depth of decision trees (or the complexity of many other learning algorithms for that matter) comes at the price of overfitting (see, e.g., Fig. 10.1); deep trees tend to learn highly irregular patterns. This disadvantage can be overcome to a certain extent by working with multiple decision trees at once; the individual trees are generated using a random selection (without replacement) of the training set. This ensemble of trees is called a random forest (Breiman 2001). Each tree is making a prediction of the target variable, and the average of all predictions is then the resulting prediction of the random forest.

A crucial performance improvement is obtained when, repeatedly, a random sample of the training set (now with replacement) is presented to the generated trees, and the trees are fitted to these subsamples, a process called "bagging" (Breiman 2001) which is short for "bootstrap aggregating." If this is done B times (say), one has B different regression trees trained. For the prediction of hitherto unseen samples, one takes the arithmetic average of the predictions of all trained trees. This has a noise cancelling effect, and as a consequence, the model performance is improved in the sense of decreased variance while at the same time the bias is not increased.

The number of trees B is a parameter of the approach, and may be adjusted by considering the *out-of-bag error*, i.e., the prediction performance for samples not contained in the training set of the individual trees. This error usually flattens out once B exceeds a couple of hundreds to a few thousands. The limit of B growing arbitrary large (*infinite* random forests) is an active area of research in the theory of random forests.

[7]https://cran.r-project.org/web/packages/rpart/index.html

The learning algorithm of random forests involves the invention of candidate splits (branching points) in the trees. At each candidate split, the learner is presented with a fraction (the recommended value being 1/3) of the features, randomly chosen. This procedure minimizes correlations among the trees which would otherwise occur when there are strong predictors in the input dataset.

Random forests show a relatively strong resistance to overfitting (Kleinberg 1996), arguably the most decisive reason for its popularity in recent years. Further details on the theory of random forests can be found in Hastie et al. (2008), and in the R programming environment, the randomforest package (Liaw and Wiener 2002) may be used.

The number of articles in hydrology using random forests has rapidly increased recently. A first review of applications of random forests in the water sciences (Tyralis et al. 2019) counts more than 200 papers on that method in water journals; more than 60% of them have been published since 2016.

A further step, closely connected to random forests, consists in presenting the whole learning sample (rather than just a subsample) to each tree, and assign splits in the trees at random many times; the split which yields the highest score is finally chosen to split the node. This variation is called extremely randomized trees (Geurts et al. 2006) or ExtraTrees for short.

10.3.8 Gradient Boosting Machine

The generalization of gradient-descent algorithms to (stochastic) gradient boosting, introduced by (Friedman 2002), consists in providing an iterative algorithm with a set of "base learners"—simple predictor functions which may be non-adapted to the problem or are in other senses "weak." The most extreme example of a weak base learner would be the mean of the whole dataset. Other, more sophisticated examples include regression trees. The model error is then used in the next iteration to build a set of new learners, whose predictions are added to the prediction of the base learners (this is the boosting step), resulting in a new set of prediction functions. This process is repeated until a user-defined number of trees is reached. The gradient boosting can be fully automatized using software like the gbm3 package in R.[8]

10.3.9 M5 and M5-cubist

These two algorithms also belong to the family of decision/regression tree approaches; they are sometimes also called model tree approach. Contrary to usual regression trees, they contain regression functions at their leaves, not single

[8]https://github.com/gbm-developers/gbm3

(threshold) values. The M5 model tree treats the standard deviation of the class values that reach a node as the node error and calculates the expected reduction in this error by splitting the node and testing each attribute at that node. Among all splits, M5 choses the one which maximizes the expected error reduction. In most cases, a subsequent pruning step is required. M5 was introduced by Quinlan (1993).

Cubist or M5-cubist (Loh 2011) is a further extension of M5 avoiding sudden "jumps" or discontinuities potentially arising in M5 when changing the decision node, if the corresponding linear models have very different coefficients. M5-cubist is available through the Cubist R package.[9]

10.3.10 Stack Generalization

As the diversity of methods discussed in this chapter indicate, it is not always easy to decide for the "optimal" approach when confronted with a specific modeling problem or dataset. Meta approaches (or ensemble models) combine the predictions of a suite of (machine learning) algorithms (level 0 methods). These predictions are input (predictor variables) to another layer of machine learning (level 1). This procedure resembles boosting in the GBM or bagging in random forests and is referred to as stacked regression (Wolpert 1992). It can also be iterated (level 2, i.e., meta-meta models and so on), although this has been rarely seen until now.

10.4 Examples of Machine Learning Applications from Hydrology

10.4.1 A Note on Quality Metrics

As machine learning applications are still relatively new in the hydrological literature, the performance is usually compared to more conventional approaches, or against each other. This raises the question how to compare observations and predictions (or simulations). There are many ways to do this; it is not obvious which metric serves the purpose best, and it also depends on the focus of the data analyst: is reproduction of the mean and the variance most important? Should the autocorrelation function be reproduced (or the power spectrum)? Should the prediction do well only on short time scales, or also on longer ones (statistically)? Is the reproduction of the seasonal cycle (phase, asymmetry) an issue? There are many more aspects of the time series to be considered. However, the vast majority of papers restrict themselves to very basic metrics, usually delivering just one number for the whole record indicating the data-model mismatch: the correlation coefficient, Root Mean Squared

[9]https://cran.r-project.org/web/packages/Cubist/index.html

Error (RMSE), Mean Absolute Error (MAE) or the Nash-Sutcliffe Efficiency (Nash and Sutcliffe 1970). There is a risk that according to these simple metrics, the methods do not differ much from each other, and their ranking is much more a random function of the case study at hand rather than a typical one for the research area. In some cases, it might also not do justice to some methods which have a very good description of the general dynamics, but simply fail to reproduce the correct scale (or the mean value) accurately enough.

There are lots of more sophisticated model-data comparisons available; two examples are (Lange et al. 2013) and (Tongal and Berndtsson 2017), focusing on the complexity of streamflow. In general, we advocate to optimize for more than just one metric (or several, but completely independent from each other), i.e., a multi-objective optimization (Miettinen 1999), to proper evaluate the different machine learning techniques and to create structure in the zoo of algorithms and their performance in hydrological applications.

10.4.2 ANNs in Hydrology

As discussed in the Introduction, the focus of machine learning in hydrology (and mostly elsewhere) is on prediction. The most obvious variable to predict in hydrology is streamflow. Here, applications of ANNs have a long history, extending well before the period where machine learning was a buzzword. In Karunanithi et al. (1994), the network architecture was determined by a so-called cascade-correlation algorithm, and it was observed that the network was capable of adapting to changes in flow history, contrary to analytical models. Also, (Zealand et al. 1999), working on eight years of streamflow from a larger catchment in Canada, observe the ability of ANNs to model complex nonlinear relationships and to predict flow a couple of time steps ahead. On very short time scales (few hours), it was also attempted to predict precipitation using ANNs (Kuligowski and Barros 1998), based on wind direction and antecedent precipitation data from a weather gauge.

The literature on ANNs and streamflow exploded with the start of the new millennium: an ISI Web of Science search lists 4 papers with publication date prior to 2000, and 602 after 2000. We can only sketch some representative examples.

The lack of transparency of resulting ANN architecture and the weight matrix was the concern of Kingston et al. (2005). These authors tried to implement constraints in the objective function, no longer seeking the global optimum but physically plausible input conditions. The obtained predictive performance was still comparable to that of the unconstrained fit. Many examples have shown that ANNs (and many other machine learning techniques for that matter) have difficulties to reproduce isolated short-term peaks in runoff records; the summation over different neurons from the hidden layer(s) to produce the output (often a single neuron) has a notorious smoothing effect. To mitigate this smoothing, (Parasuraman et al. 2006) added a "spiking" layer to the hidden layer which is active in particular for high flow. A similar layout could also successfully be used for Eddy Covariance gas fluxes (in the same paper). A strictly univariate approach (called "top-down black box"), i.e., using only streamflow to predict streamflow, was used in Wang et al. (2006). It turned out that careful preprocessing, in particular deseasonalization without normalization, was the clue to obtain good performances; consistent with that, periodic ANNs performed best in their case study.

The excellent performance of ANNs for real-time forecasting purposes was documented in Toth and Brath (2007); however, these authors point to the importance of extensive hydro-meteorological datasets which the ANNs need for training.

Daliakopoulos and Tsanis (2016) compare an ANN with a conceptual model for rainfall-runoff modeling. They conclude that, on average, their ANN variant ("input delay neural network") is superior to the conceptual model, but it is outmatched for low flow conditions. It could be that extrapolation to yet unseen conditions (not contained in the training set) is a problem for ANNs. A comparison of ANNs to a newer version of it, the extreme learning machine (ELM) (Huang et al. 2006), was performed in Yaseen et al. (2018). ELMs are single-hidden layer feedforward neural networks; however, the output weights are calculated analytically rather than iteratively using gradient-based methods: they are non-tuned and tremendously faster than ordinary ANNs. In the paper, it was also shown that they exhibit slightly better performance on different time scales. Another paper on ELM expands the method to incorporate online changes in the network structure (number of hidden nodes), which the authors call variable complexity online sequential ELM (Lima et al. 2017). For daily streamflow prediction, this extension is clearly outperforming the standard ELM.

In an applied approach (Bozorg-Haddad et al. 2018), ANNs and SVMs are used to give advice for the optimal operation of water reservoirs. In their case, SVMs outperform ANNs in the forecasting scenarios. In a similar vein, Siqueira et al. (2018) investigate forecasting for Brazilian hydroelectric power plants using "unorganized" machines—ELM and so-called echo state networks—in comparison with standard ANNs. Concerning quality metrics, they present an exception from the rule as they consider the partial autocorrelation function, mutual information and maximum relevance/minimum common redundancy as evaluation criteria. Still another exception is the already mentioned paper (Tongal and Berndtsson 2017) which conclude for daily streamflow data, based on entropy and complexity analysis, that ANNs are well suited for 1-day forecasting, but should be used with care beyond 2-day forecasting (their favorite algorithm for longer forecasting horizons, the Self-Exciting Threshold AutoRegressive Model (SETAR), is not in the class of machine learning techniques).

10.4.3 SVMs in Hydrology

The application of support vector regression (SVR) to hydrological data seems to have started around 15 years ago. One of the earliest examples is (Lin et al. 2006), pointing to the important fact that over-fitting is rarely an issue for SVMs, contrary to

ANNs. For rainfall forecasting, a rather advanced setup of learning techniques is exploited in Hong (2008), combining SVMs with recurrent ANNs, and using a particular methodology from the field of genetic algorithms, the chaotic particle swarm optimization (Kennedy and Eberhart 1995), to choose the parameters for the SVR. The resulting model framework (called RSVRCPSO in the paper) is shown to yield well forecasting performance.

A comparison of SVR and Bayesian ANNs for prediction of daily streamflow can be found in Rasouli et al. (2012). Interestingly, they show that climate indices like the North Atlantic Oscillation (NAO) or the Pacific-North American teleconnection (PNA) contribute to forecast scores in the case of longer lead times (5-7 days), for a small watershed in British Columbia, Canada.

In the context of detecting Land Use and Land Use Changes (LULC), (Nourani et al. 2018) combine remote sensing data and a conceptual rainfall-runoff model generating outflow, using a Storage Coefficient (SC) as parameter representing the LULC. The relation between SC and the streamflow time series is then simulated by SVM and ANNs combined.

The performance of SVR seems to depend on the preprocessing of the hydrological data, an observation made also for other analysis methods. In Yu et al. (2018), SVR is combined with Fourier transformation to generate independent forecasting models according to the frequency of the components. This FT-SVR appears to yield remarkably high performance for inflow prediction of the Three Gorges Dam in China.

For further examples, consider an existing review of SVM applications in hydrological literature as of 2014 (Raghavendra and Deka 2014).

10.4.4 Papers Combining and Comparing Machine Learning Methods in Hydrology

It should be obvious at this stage that there is no single best machine learning technique for hydrological applications. The request for analysis and modeling is spread over different temporal and spatial scales, the amount of (training) data is very different, and even if we focus here on prediction tasks, comparisons with conceptual or process-based models is occasionally desired. This can't be covered by a single framework. Thus, an increasing number of papers is comparing several methods and elucidate their respective strengths and weaknesses for the application at hand.

Groundwater potential mapping is the focus of Naghibi and Pourghasemi (2015). Here, boosted regression trees, CART (which is not necessary in the class of machine learning algorithms), and random forests were trained on spring locations, using 14 predictors. These three were all performing very well, but almost indistinguishable, and were outperforming conventional methods easily. The heterogeneity of the predictors makes the task a typical application domain for machine learning.

Using wavelets as base functions for ANNs, Shafaei and Kisi (2017) compare ANNs to SVM for daily streamflow prediction. For short prediction times (up to 3 days), the wavelet-based ANN was outperforming both ordinary ANN as well as SVM.

In the context of daily streamflow from semiarid mountainous regions, Yin et al. (2018) compare SVR, a multivariate adaptive spline algorithm (MARS) and the M5 tree discussed above, where M5 turned out to be the winner for short-term prediction up to 3 days.

A new technique coined selected model fusion is developed in (Modaresi et al. 2018a, b). This is an example where the individual methods are not performed individually and then compared against each other; rather, the output of all of them (ANN, SVR, K-NN, and ordinary multiple linear regression) are fused together with an ordered selection algorithm. The fusion of the outputs is superior to even the best of the individual methods for the case of monthly streamflow prediction, demonstrating that it is feasible to combine the respective strengths of single algorithms.

Finally, Worland et al. (2018) set up most of the methods presented here (i.e., elastic nets, gradient boosting machine, KNN, M5-cubist, and SVM) to predict a peculiar extreme statistics, the 7-day mean streamflow which corresponds to the 10% quantile ("7Q10"), i.e., an indicator for low flow. They tackle the problem of generalizing these 7Q10 values obtained from gauged sites also to ungauged ones. They also exploit stack generalization, using a M5-cubist as metamodel, resembling a Leave-One-Out Cross-Validation, but now on the level of machine learning techniques (plus three baseline models not related to machine learning). The meta-model ("meta cubist") outcompetes each individual method in terms of the standard quality metrics RMSE and Nash-Sutcliffe.

10.5 Outlook

Machine learning and deep learning are on their way to becoming the key empirical analysis techniques in hydrology; and increasingly ML applications in hydrological studies are made reproducible through code sharing (e.g., Peters et al. 2016). According to Shen (2018), however, we are still in the early "value discovery" stage of deep learning. One can expect synergies between deep learning/machine learning methods and process-based models; it is possible that patterns detected by the automatic methods initiate new questions on the relevance and nature of processes at different scales, leading to new routes in mechanistic modeling. The bet is open on whether this is going to happen. So far, process-based modeling and ML approaches are more in competition to each other, and their proponents often belong to different scientific communities.

However, the interplay between process understanding and ML application is a more complex one. Broadly speaking, ML results are based solely on data presented to the algorithm, and do not come with any interpretation of what is going on in the system. It is rather easy to produce colorful results devoid of any meaning unless combined with expert knowledge. In return, this knowledge can be sharpened, confirmed, or revised with the aid of pattern detection based on ML. It is this feedback loop which should be pursued further, be it in the field of hydrology proper or as an interdisciplinary effort.

Arguably the central benefit of deep learning is the learning of huge amounts of unsupervised (or unlabeled, uncategorized) data. The pattern extraction, information retrieval, classification, and prediction abilities of deep learning algorithms indicate their suitability for "Big Data Hydrology" (Irving et al. 2018). Still, the low maturity of deep learning warrants extensive further research (Najafabadi et al. 2015). A prerequisite of big data analysis methods to be successful is the presence of big data in the first place. The monitoring of hydrological systems has to be continued and extended. This is a challenging and long-lasting task, since some patterns are only apparent in time series extending over decades.

We expect the field of ML and deep learning to expand rapidly, with a proliferation of publications also in the field of hydrology. Currently, SVMs, CNNs, and random forests appear to be the most actively investigated algorithms, but new ones are forthcoming. Thus, it is likely that any future chapter on ML in hydrology in the coming years would probably focus on new additional (or different) methods as well as those that are now being widely utilized.

At the time of writing, a special collection of Water Resources Research on "Big Data and Machine Learning in Water Sciences: Recent Progress and Their Use in Advancing Science"[10] is being compiled, where seven articles already have been published. We are waiting with excitement to the rest of the contributions and their discussion within the community of hydrology researchers.

References

Altman NS (1992) An introduction to kernel and nearest-neighbor nonparametric regression. Am Stat 46:175–185. https://doi.org/10.2307/2685209

Bergmeir C, Hyndman RJ, Koo B (2018) A note on the validity of cross-validation for evaluating autoregressive time series prediction. Comput Stat Data Anal 120:70–83. https://doi.org/10.1016/j.csda.2017.11.003

Beven K, Freer J (2001) Equifinality, data assimilation, and uncertainty estimation in mechanistic modelling of complex environmental systems. J Hydrol 249:11–29. https://doi.org/10.1016/S0022-1694(01)00421-8

Bishop C (2006) Pattern recognition and machine learning. Springer, New York. 738 p

Blöschl G (2001) Scaling in hydrology. Hydrol Process 15:709–711. https://doi.org/10.1002/hyp.432

Boser BE, Guyon IM, Vapnik VN (1992) A training algorithm for optimal margin classifiers. In: Proceedings of the fifth annual workshop on computational learning theory. ACM, Pittsburgh, pp 144–152. https://doi.org/10.1145/130385.130401

[10]https://agupubs.onlinelibrary.wiley.com/doi/toc/10.1002/(ISSN)1944-7973.MACHINELEARN

Bozorg-Haddad O, Aboutalebi M, Ashofteh PS, Loaiciga HA (2018) Real-time reservoir operation using data mining techniques. Environ Monit Assess 190:594. https://doi.org/10.1007/s10661-018-6970-2
Breiman L (2001) Random forests. Mach Learn 45:5–32. https://doi.org/10.1023/A:1010933404324
Breiman L, Friedman JH, Stone CJ, Olshen RA (1984) Classification and regression trees. Chapman & Hall, Boca Raton. 368 p
Clark MP, Nijssen B, Lundquist JD, Kavetski D, Rupp DE, Woods RA et al (2015) A unified approach for process-based hydrological modeling: 1. Modeling concept. Water Resour Res 51:2498–2514. https://doi.org/10.1002/2015WR017198
Corzo Perez GA (2009) Hybrid models for hydrological forecasting: Integration of data-driven and conceptual modelling techniques. Doctoral thesis, TU Delft. 215 p
Cybenko G (1989) Approximation by superpositions of a sigmoidal function. Math Control Signal 2:303–314. https://doi.org/10.1007/BF02551274
Daliakopoulos IN, Tsanis IK (2016) Comparison of an artificial neural network and a conceptual rainfall-runoff model in the simulation of ephemeral streamflow. Hydrol Sci J 61:2763–2774. https://doi.org/10.1080/02626667.2016.1154151
Dechter R (1986) Learning while searching in constraint-satisfaction problems. In: AAAI '86 Proceedings of the Fifth AAAI national conference on artificial intelligence. Pennsylvania, Philadelphia, pp 178–183
Fatichi S, Pappas C, Valeriy IY (2016) Modeling plant–water interactions: an ecohydrological overview from the cell to the global scale. WIRES Water 3:327–368. https://doi.org/10.1002/wat2.1125
Friedman JH (2002) Stochastic gradient boosting. Comput Stat Data An 38:367–378. https://doi.org/10.1016/S0167-9473(01)00065-2
Friedman JH, Hastie T, Tibshirani R (2010) Regularization paths for generalized linear models via coordinate descent. J Stat Softw 33:1–22. https://doi.org/10.18637/jss.v033.i01
Geurts P, Ernst D, Wehenkel L (2006) Extremely randomized trees. Mach Learn 63:3–42. https://doi.org/10.1007/s10994-006-6226-1
Gudmundsson L, Seneviratne SI (2015) Towards observation-based gridded runoff estimates for Europe. Hydrol Earth Syst Sci 19:2859–2879. https://doi.org/10.5194/hess-19-2859-2015
Hastie T, Tibshirani R, Friedman JH (2008) The elements of statistical learning. Springer, New York. 745 p
Hofmann T, Schölkopf B, Smola AJ (2008) Kernel methods in machine learning. Ann Stat 36:1171–1220. https://doi.org/10.1214/009053607000000677
Hong W-C (2008) Rainfall forecasting by technological machine learning models. Appl Math Comput 200:41–57. https://doi.org/10.1016/j.amc.2007.10.046
Hothorn T (2019) CRAN task view: machine learning and statistical learning. R-project.org. Accessed 27 Feb 2019. https://cran.r-project.org/web/views/MachineLearning.html
Huang G-B, Zhu Q-Y, Siew C-K (2006) Extreme learning machine: theory and applications. Neurocomputing 70:489–501. https://doi.org/10.1016/j.neucom.2005.12.126
Irving K, Kuemmerlen M, Kiesel J, Kakouei K, Domisch S, Jähnig SC (2018) A high-resolution streamflow and hydrological metrics dataset for ecological modeling using a regression model. Sci Data 5:180224. https://doi.org/10.1038/sdata.2018.224
Karunanithi N, Grenney WJ, Whitley D, Bovee K (1994) Neural networks for river flow prediction. J Comput Civil Eng 8:201–220. https://doi.org/10.1061/(ASCE)0887-3801(1994)8:2(201)
Kennedy J, Eberhart R (1995) Particle swarm optimization. In: Proceedings of the ICNN'95 international conference on neural networks, vol 4, pp 1942–1948. https://doi.org/10.1109/ICNN.1995.488968
Kingston GB, Maier HR, Lambert MF (2005) Calibration and validation of neural networks to ensure physically plausible hydrological modeling. J Hydrol 314:158–176. https://doi.org/10.1016/j.jhydrol.2005.03.013

Kleinberg EM (1996) An overtraining-resistant stochastic modeling method for pattern recognition. Ann Stat 24:2319–2349

Kuligowski RJ, Barros AP (1998) Experiments in short-term precipitation forecasting using artificial neural networks. Mon Weather Rev 126:470–482. https://doi.org/10.1175/1520-0493(1998)126<0470:EISTPF>2.0.CO;2

Lange H, Rosso OA, Hauhs M (2013) Ordinal pattern and statistical complexity analysis of daily stream flow time series. Eur Phys- J Spec Top 222:535–552. https://doi.org/10.1140/epjst/e2013-01858-3

LeCun Y, Boser B, Denker JS, Henderson D, Howard RE, Hubbard W et al (1989) Backpropagation applied to handwritten zip code recognition. Neural Comput Appl 1:541–551. https://doi.org/10.1162/neco.1989.1.4.541

Liaw A, Wiener M (2002) Classification and regression by randomForest. R News 2:18–22

Lima AR, Hsieh WW, Cannon AJ (2017) Variable complexity online sequential extreme learning machine, with applications to streamflow prediction. J Hydrol 555:983–994. https://doi.org/10.1016/j.jhydrol.2017.10.037

Lin JY, Cheng CT, Chau KW (2006) Using support vector machines for long-term discharge prediction. Hydrol Sci J 51:599–612. https://doi.org/10.1623/hysj.51.4.599

Lischeid G (2001) Investigating short-term dynamics and long-term trends of SO_4 in the runoff of a forested catchment using artificial neural networks. J Hydrol 243:31–42. https://doi.org/10.1016/S0022-1694(00)00399-1

Loh W-Y (2011) Classification and regression trees. WIRES Data Min Knowl 1:14–23. https://doi.org/10.1002/widm.8

Ma Y, Li XY, Guo L, Lin H (2017) Hydropedology: interactions between pedologic and hydrologic processes across spatiotemporal scales. Earth-Sci Rev 171:181–195. https://doi.org/10.1016/j.earscirev.2017.05.014

Maier HR, Dandy GC (1995) Comparison of the Box-Jenkins procedure with artificial neural network methods for univariate time series modelling. Research Report No R 127, June 1995. Department of Civil and Environmental Engineering, University of Adelaide, Adelaide, Australia

Miettinen K (1999) Nonlinear multiobjective optimization. Springer, New York., 298 p. https://doi.org/10.1007/978-1-4615-5563-6

Modaresi F, Araghinejad S, Ebrahimi K (2018a) A comparative assessment of artificial neural network, generalized regression neural network, least-square support vector regression, and K-nearest neighbor regression for monthly streamflow forecasting in linear and nonlinear conditions. Water Resour Manag 32:243–258. https://doi.org/10.1007/s11269-017-1807-2

Modaresi F, Araghinejad S, Ebrahimi K (2018b) Selected model fusion: an approach for improving the accuracy of monthly streamflow forecasting. J Hydroinform 20:917–933. https://doi.org/10.2166/hydro.2018.098

Naghibi SA, Pourghasemi HR (2015) A comparative assessment between three machine learning models and their performance comparison by bivariate and multivariate statistical methods in groundwater potential mapping. Water Resour Manag 29:5217–5236. https://doi.org/10.1007/s11269-015-1114-8

Najafabadi MM, Villanustre F, Khoshgoftaar TM, Seliya N, Wald R, Muharemagic E (2015) Deep learning applications and challenges in big data analytics. J Big Data 2(1). https://doi.org/10.1186/s40537-014-0007-7

Nash JE, Sutcliffe V (1970) River flow forecasting through conceptual models, I. A discussion of principles. J Hydrol 10:282–290. https://doi.org/10.1016/0022-1694(70)90255-6

Nourani V, Roushangar K, Andalib G (2018) An inverse method for watershed change detection using hybrid conceptual and artificial intelligence approaches. J Hydrol 562:371–384. https://doi.org/10.1016/j.jhydrol.2018.05.018

Parasuraman K, Elshorbagy A, Carey SK (2006) Spiking modular neural networks: a neural network modeling approach for hydrological processes. Water Resour Res 42:W05412. https://doi.org/10.1029/2005WR004317

Peters J, Janzing D, Schölkopf B (2017) Elements of causal inference, Foundations and learning algorithms. MIT Press, Cambridge, MA. 288 p

Peters R, Lin Y, Berger U (2016) Machine learning meets individual-based modelling: self-organising feature maps for the analysis of below-ground competition among plants. Ecol Model 326:142–151. https://doi.org/10.1016/j.ecolmodel.2015.10.014

Quinlan JR (1993) Combining instance-based and model-based learning. In: Proceedings of the tenth international conference on machine learning. Morgan Kaufmann, Amherst, MA, pp 236–243

Raghavendra SN, Deka PC (2014) Support vector machine applications in the field of hydrology: a review. Appl Soft Comput 19:372–386. https://doi.org/10.1016/j.asoc.2014.02.002

Rasouli K, Hsieh WW, Cannon AJ (2012) Daily streamflow forecasting by machine learning methods with weather and climate inputs. J Hydrol 414–415:284–293. https://doi.org/10.1016/j.jhydrol.2011.10.039

Rasp S, Pritchard MS, Gentine P (2018) Deep learning to represent subgrid processes in climate models. Proc Natl Acad Sci USA 115:9684–9689. https://doi.org/10.1073/pnas.1810286115

Richards LA (1931) Capillary conduction of liquids in porous mediums. Physics 1:318–333. https://doi.org/10.1063/1.1745010

Roberts DR, Bahn V, Ciuti S, Boyce MS, Elith J, Guillera-Arroita G et al (2017) Cross-validation strategies for data with temporal, spatial, hierarchical, or phylogenetic structure. Ecography 40:913–929. https://doi.org/10.1111/ecog.02881

Rosenblatt F (1958) The perceptron: a probabilistic model for information storage and organization in the brain. Psychol Rev 65:386–408. https://doi.org/10.1037/h0042519

Rumelhart DE, Hinton GE, Williams RJ (1986) Learning representations by back-propagating errors. Nature 323:533–536. https://doi.org/10.1038/323533a0

Samuel AL (1959) Some studies in machine learning using the game of checkers. IBM J Res Dev 3:210–229. https://doi.org/10.1147/rd.33.0210

Schliep K, Hechenbichler K (2016) kknn: Weighted k-Nearest Neighbors. R package version 1.3.1. https://CRAN.R-project.org/package=kknn

Schmidhuber J (2015) Deep learning in neural networks: an overview. Neural Netw 61:85–117. https://doi.org/10.1016/j.neunet.2014.09.003

Schoups G, Vrugt JA (2010) A formal likelihood function for parameter and predictive inference of hydrologic models with correlated, heteroscedastic, and non-Gaussian errors. Water Resour Res 46:W10531. https://doi.org/10.1029/2009WR008933

Schultz W (2007) Reward signals. Scholarpedia 2:2184. https://doi.org/10.4249/scholarpedia.2184

Shafaei M, Kisi O (2017) Predicting river daily flow using wavelet-artificial neural networks based on regression analyses in comparison with artificial neural networks and support vector machine models. Neural Comput Appl 28:S15–S28. https://doi.org/10.1007/s00521-016-2293-9

Shen C (2018) Deep learning: a next-generation big-data approach for hydrology. EOS Trans 99. https://doi.org/10.1029/2018EO095649

Silver D, Huang A, Maddison CJ, Guez A, Sifre L, van den Driessche G et al (2016) Mastering the game of Go with deep neural networks and tree search. Nature 529:484–489. https://doi.org/10.1038/nature16961

Silver N (2012) The signal and the noise: why so many predictions fail--but some don't. Penguin Books, New York. 560 p

Siqueira H, Boccato L, Luna I, Attux R, Lyra C (2018) Performance analysis of unorganized machines in streamflow forecasting of Brazilian plants. Appl Soft Comput 68:494–506. https://doi.org/10.1016/j.asoc.2018.04.007

Sivapalan M (2003) Process compexity at hillslope scale, process simplicity at the watershed scale: is there a connection? Hydrol Process 17:1037–1041. https://doi.org/10.1002/hyp.5109

Sivapalan M (2006) Pattern, process and function: elements of a unified theory of hydrology at the catchment scale. Encycl Hydrol Sci. https://doi.org/10.1002/0470848944.hsa012

Sivapalan M, Grayson R, Woods R (2004) Scale and scaling in hydrology. Hydrol Process 18:1369–1371. https://doi.org/10.1002/hyp.1417

Sugihara G, May R, Ye H, Hsieh C-H, Deyle E, Fogarty M et al (2012) Detecting causality in complex ecosystems. Science 338:496–500. https://doi.org/10.1126/science.1227079

Tongal H, Berndtsson R (2017) Impact of complexity on daily and multi-step forecasting of streamflow with chaotic, stochastic, and black-box models. Stoch Environ Res Risk Assess 31:661–682. https://doi.org/10.1007/s00477-016-1236-4

Toth E, Brath A (2007) Multistep ahead streamflow forecasting: role of calibration data in conceptual and neural network modeling. Water Resour Res 43:W11405. https://doi.org/10.1029/2006WR005383

Tyralis H, Papacharalampous G, Langousis A (2019) A brief review of Random Forests for water scientists and practitioners and their recent history in water resources. Water 11:910. https://doi.org/10.3390/w11050910

Viney NR, Sivapalan M (2004) A framework for scaling of hydrologic conceptualizations based on a disaggregation-aggregation approach. Hydrol Process 18:1395–1408. https://doi.org/10.1002/hyp.1419

Wang W, Van Gelder P, Vrijling JK, Ma J (2006) Forecasting daily streamflow using hybrid ANN models. J Hydrol 324:383–399. https://doi.org/10.1016/j.jhydrol.2005.09.032

Werbos PJ (1975) Beyond regression: new tools for prediction and analysis in the behavioral sciences. Harvard University Press, Cambridge, MA. 906 p

Wolpert DH (1992) Stacked generalization. Neural Netw 5:241–259. https://doi.org/10.1016/S0893-6080(05)80023-1

Worland SC, Farmer WH, Kiang JE (2018) Improving predictions of hydrological low-flow indices in ungaged basins using machine learning. Environ Model Softw 101:169–182. https://doi.org/10.1016/j.envsoft.2017.12.021

Yaseen ZM, Allawi MF, Yousif AA, Jaafar O, Hamzah FM, El-Shafie A (2018) Non-tuned machine learning approach for hydrological time series forecasting. Neural Comput Appl 30:1479–1491. https://doi.org/10.1007/s00521-016-2763-0

Yin ZL, Feng Q, Wen XH, Deo RC, Yang LS, Si JH et al (2018) Design and evaluation of SVR, MARS and M5Tree models for 1, 2 and 3-day lead time forecasting of river flow data in a semiarid mountainous catchment. Stoch Environ Res Risk Assess 32:2457–2476. https://doi.org/10.1007/s00477-018-1585-2

Yu X, Zhang XQ, Qin H (2018) A data-driven model based on Fourier transform and support vector regression for monthly reservoir inflow forecasting. J Hydro-Environ Res 18:12–24. https://doi.org/10.1016/j.jher.2017.10.005

Zealand CM, Burn DH, Simonovic SP (1999) Short term streamflow forecasting using artificial neural networks. J Hydrol 214:32–48. https://doi.org/10.1016/S0022-1694(98)00242-X

Zhang S-Q (2009) Enhanced supervised locally linear embedding. Pattern Recogn Lett 30:1208–1218. https://doi.org/10.1016/j.patrec.2009.05.011

Zou H, Hastie T (2005) Regularization and variable selection via the elastic net. J R Stat Soc B 67:301–320. https://doi.org/10.1111/j.1467-9868.2005.00503.x

Part III
Synthesizing Process-Based Understanding of Forest-Water Interactions: Some Recent Advances

Chapter 11
Advances and Future Research Directions in the Study of Leaf Water Repellency

Curtis D. Holder

11.1 Introduction

For more than a century, ecohydrologists reported on characteristics of vegetation that result in differences in interception during rainfall events. From these numerous studies, leaf area index emerged as one of the prominent variables influencing rainfall interception and evapotranspiration differences between vegetation types (Kittredge 1948; Sopper and Lull 1967; Hewlett 1982). Decreases in leaf area index, because of canopy fires or removal of forest cover because of land use changes, emerged as the proximate driver of increases in runoff and stream discharge in affected watersheds. The quantity of leaf surface area is only part of the explanation for differences in rainfall interception. Now ecohydrologists are exploring the nuances of vegetation canopy differences in a search to explain how vegetation from species with a similar leaf area index can intercept different quantities of rainfall. Vegetation surface characteristics also produce differences in rainfall interception between species (Garcia-Estringana et al. 2010; Holder 2013). For example, in recent times, several articles examined the role of leaf surfaces in repelling water drops from the canopy (Holder 2007).

The study of the repellency of water drops from surfaces goes back more than a couple of centuries, but in recent times, materials science engineers have emerged to develop products and applications from the study of water repellency. For example, engineers continue to develop water-repellent fabrics, computer components, and other consumer products. Chemical engineers are interested in the adhesion of industrial pesticides on leaves of agricultural crops to maximize the persistency of pesticides on foliage. Biologists began to explore leaf water repellency in the 1980s (Smith and McClean 1989; Brewer et al. 1991; Barthlott and Neinhuis 1997).

C. D. Holder (✉)
Leaf Biomechanics and Ecohydrology Research Group, Department of Geography and Environmental Studies, University of Colorado Colorado Springs, Colorado Springs, CO, USA
e-mail: cholder@uccs.edu

D. F. Levia et al. (eds.), *Forest-Water Interactions*, Ecological Studies 240,
https://doi.org/10.1007/978-3-030-26086-6_11

The initial ecological studies examined ecophysiological explanations for differences in leaf water repellency between species. Many of these studies explored the epicuticular structure of the leaf surface as a mechanism for shedding water (Koch et al. 2009). These early studies focused on the significance of leaf water repellency to promote leaf gas exchange and carbon assimilation and to minimize the growth of pathogens on leaf surfaces (Smith and McClean 1989).

The increase in leaf water repellency literature over the past decade is a testament to the growing importance of this research. During the past decade, several leaf water repellency studies emerged with new information and perspectives. The purpose of this chapter is to review the latest knowledge of the ecohydrological significance of leaf water repellency with the objective of framing underexplored questions for future research. Ten broad research questions are proposed to frame future leaf water repellency research to explore the significance of leaf water repellency in ecohydrological processes. The following ten sections of this chapter address each question systematically and provide the context for each question and a rationale for why each question should be explored in future studies.

11.2 What Does Leaf Water Repellency Mean?

Recent studies have used the following terms to describe the adhesion of water on a leaf surface: leaf contact angle, water drop behavior, leaf wettability, leaf water repellency, leaf hydrophobicity, leaf wetness traits, wetting properties, leaf wetness, water droplet retention, physicochemical properties, water storage retention, and water repellency (Holder 2007; Matos and Rosado 2016; Aparecido et al. 2017; Fernández et al. 2017; Goldsmith et al. 2017; Klamerus-Iwan and Witek 2018; Xiong et al. 2018). In this chapter, I refer to leaf water repellency as the general phenomenon in which water repels from a leaf surface. Leaf water repellency is measured either as leaf hydrophobicity (static repellency) or as water droplet retention (dynamic repellency). Leaf hydrophobicity is the degree to which water adheres to a leaf surface and is commonly measured as the contact angle of a water droplet on a leaf surface. Water droplet retention, which is another measure of leaf water repellency, is measured as the leaf angle at which a water droplet begins to move along the leaf surface.

Although the various terms describe a similar process, there still may be confusion with the use of these terms. The repellency of water from a leaf surface is often explained methodologically in a couple of different ways. Some studies emphasize the contact angle between a water droplet and the leaf surface as the primary definition of leaf water repellency. The higher the contact angle between a water droplet and the leaf surface, the more hydrophobic or repellent the leaf surface. Other studies emphasize the quantity of water that can be retained on a leaf surface. The less water that is retained on a leaf surface on a leaf area basis after a rainfall event or by immersion of the leaf in water in laboratory experiments, the more hydrophobic the leaf surface. Other studies emphasize the leaf angle at which water drops on leaf surface begin to move off the leaf. Leaves that are not greatly inclined from the horizontal and shed water from the leaf surface are categorized as more hydrophobic than leaves inclined closer to 90° from horizontal and retain water droplets on the leaf surface.

A term for leaf water repellency is often chosen depending on the emphasis of the study. In many cases, it can be confusing if the author is using leaf wettability interchangeably for leaf water repellency. Pérez-Harguindeguy et al. (2013) published an article to provide guidance to standardize the measurement of leaf functional traits, including leaf wettability and droplet retention ability. Pérez-Harguindeguy et al. (2013) stated the definition and procedures for measuring leaf wettability as "how easily a leaf can get wet is determined by measuring contact angles between water droplets and the leaf surface" (p. 185). This definition of leaf wettability appears to be the similar to other articles that use the terms leaf water repellency or leaf hydrophobicity. The problem with the label "leaf wettability" is that water films are largely ignored in the procedures to calculate leaf wettability (see Sect. 11.7). Additionally, water droplets of different sizes form on leaf surfaces during rainfall events (see Sect. 11.8), and the standard water droplet size to measure contact angle proposed by Pérez-Harguindeguy et al. (2013) is between 2 and 5 μL (1.56–2.12 mm in diameter). Della Volpe and Siboni (2015) recommend describing with explicit details any methods to obtain contact angle measurement. As a start to overcome confusion over different terms, I would suggest that any term used to express the repellency of water on a leaf surface be explicitly stated in research papers and include the explicit methods used to gather data on the chosen term. The use of different terminology can then be standardized to specific methodological approaches to measuring the repellency of water on a leaf surface.

11.3 How Is Leaf Water Repellency Measured?

There are numerous techniques in calculating leaf hydrophobicity and water droplet retention in the literature (Rosado and Holder 2013). The two most common methods to measure leaf hydrophobicity are with a goniometer and a digital camera. In both techniques, a contact angle is measured between the leaf surface and a sessile water droplet placed on a flattened, horizontal leaf surface (Fig. 11.1). Positioning the leaf surface horizontally standardizes the contact angle measurements for comparison between species. Goniometers are mostly used within laboratories because the instrumentation is not easily portable to the field. Leaf specimens are harvested in the field and then transported to the laboratory. Portability into the field is an advantage of using digital cameras. Specimens can be collected in the field and a photograph can be taken immediately to avoid the potential for leaf desiccation and wilting. Images taken of the water droplet on the leaf, either by the goniometer or by the camera in the field, are then processed to calculate the contact angle. Generally, goniometer manufacturers provide proprietary software to directly measure contact angles from digital images. Image processing software is easily accessible to

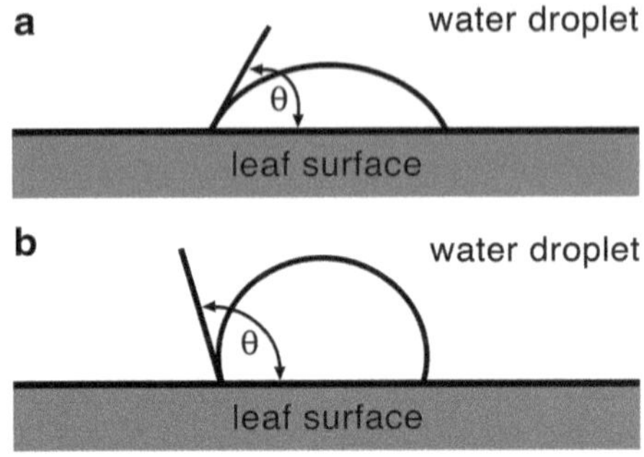

Fig. 11.1 Diagram of the contact angle between a sessile water droplet and a leaf surface for (**a**) a wettable leaf and (**b**) a non-wettable leaf

Table 11.1 Classification of leaf hydrophobicity based on Aryal and Neuner (2010)

Contact angle (°)	Classification
<40	Superhydrophilic
40–90	Highly wettable
90–110	Wettable
110–130	Non-wettable
130–150	Highly non-wettable
>150	Superhydrophobic

measure contact angles from images taken with a digital camera in the field. For example, contact angle measurements can be determined with Image J, a free software available for download from the National Institute of Health (https://imagej.nih.gov/ij/index.html). Aryal and Neuner (2010) provided a classification of leaf hydrophobicity from superhydrophilic to superhydrophobic based on the measured contact angles (Table 11.1). If the contact angle between the water droplet and the leaf surface is greater than 110°, then the leaf surface is classified as non-wettable (Fig. 11.1).

Water droplet retention is commonly measured by two techniques. The first technique is to use an angle protractor on a tilting platform, such as a fluid-head tripod with a small platform or lab jack. A second technique is to use a tilting platform on a goniometer. In both techniques, a horizontal platform is constructed to stabilize the leaf as it is being gradually tilted. The angle at which a water droplet begins to move freely as the platform is being tilted is the water droplet retention. The difficulty of water droplet retention measurement is determining when the water droplet begins to move freely along the leaf surface. In leaves with complex surface microstructures, such as trichomes, the water droplet may begin to advance and move for a short distance along a leaf surface but may cease movement because of blockage by a microstructure along the path of movement. Should the water droplet retention angle be recorded as the angle in which the water droplet moved the short distance, or should the water droplet retention angle be recorded as the angle in which the water droplet would freely flow off the entire leaf surface? Similar to the instruments for measuring leaf hydrophobicity, the goniometer is not easily transported from the laboratory to the field and the fluid-head tripod is more mobile for field-based form. Aryal and Neuner (2010) developed three classifications for water droplet retention angles. Leaves with low water retainability had water droplet retention angles less than 20°, and leaves with high water retainability had water droplet retention angles greater than 60°.

There may be a presumption that measurements with the goniometer are more accurate because variables, such as the horizontality of the goniometer platform, are easier to control in the lab. Also, the ease of using the goniometer instrumentation and software can remove some of the potential human errors in the contact angle and water droplet retention calculations in the field. However, to my knowledge, a study directly comparing these techniques has not been published. A direct comparison of different techniques to calculate leaf hydrophobicity and water droplet retention would be useful to determine if comparisons of results using different techniques are consistent. The variability of contact angle measurements can be large for the same species using the same technique. Does the goniometer produce a similar variability in contact angle measurements as the contact angles determined from processing images taken with a digital camera in the field?

Although two approaches to measure leaf water repellency (contact angle measurements and water droplet retention angle measurements) are common in the literature, the relationship between the two approaches has not been largely explored. Although Brewer and Nuñez (2007) found a relationship that highly non-wettable leaves (high leaf hydrophobicity) have water droplets that roll off leaves easily (low water droplet retention angle) in 37 species of Argentina, Matos and Rosado (2016) did not find a significant relationship between leaf hydrophobicity and water droplet retention for 14 species in Brazil. The differences in the complexity of leaf surface microstructures between species may influence this relationship (Sect. 11.4).

Della Volpe and Siboni (2015) provided a review of common misuses of contact angle measurements. They make multiple recommendations for contact angle research, including describing with details the methods used to calculate contact angle and providing the exact protocols applied in the measurement to enable the replication of the experiment. Although much of the critique of Della Volpe and Siboni (2015) involved differences in terminology between studies (see Sect. 11.2), the methods to calculate leaf water repellency are informed by the definitions researchers use.

Besides the specific instrumentation used to calculate leaf water repellency, the protocols to measure leaf water repellency in the lab or the field may produce different data. Papierowska et al. (2018) measured contact angles of a sessile water droplet at 1 second intervals for a total of 2 minutes. After an examination of 19 species, a general decrease in contact angle over time occurred from this method. In some species the difference between the initial contact angle and the final contact angle after 2 minutes exceeded 10°. Papierowska et al. (2018) speculate that the changes in contact angles over the 2-minute interval could be related to the microstructure of the leaf surfaces (e.g., trichomes and epicuticular waxes) examined. This study suggests that standard protocols should be determined in the calculation of contact angles on leaf surfaces after a specific, uniform time of initial placement of the water droplet. Differences in protocols for the calculation of leaf water repellency may make comparisons of results between various studies difficult.

11.4 In What Ways Do Leaf Surface Microstructures Influence Leaf Water Repellency?

Leaf surface structure and chemistry are the most significant traits that influence leaf water repellency. The leaf surface is not flat and homogeneous, but rather is composed of three-dimensional structures that may include trichomes, hairs, and epicuticular waxes. Della Volpe and Siboni (2015) recommend that any experiments using contact angles should describe with details the chemistry and geometrical structures of the surface. For composites, such as leaves, the heterogeneity of the leaf surface is problematic in linking directly to the position of the water droplet on the leaf surface.

The surface structure directly affects how water is retained on or repelled from the leaf surface and the drainage patterns in the canopy during precipitation events (Garcia-Estringana et al. 2010; Biddick et al. 2018). Leaf surface structures enable plants to acquire water from fog and dew in some plants in arid environments (Roth-Nebelsick et al. 2012; Tao and Zhang 2012; Konrad et al. 2015). During the past decade, numerous studies have investigated the relationship between leaf surface structure and the repellency of water using scanning electron microscopy (Tranquada and Erb 2014; Koukos et al. 2015; Revilla et al. 2016; Fernández et al. 2017; Kang et al. 2018). Voigt et al. (2012) used cryo-scanning electron microscopy to produce images of small water droplets (1–10 μm diameter) and water films on leaf surfaces. These microscopy studies revealed how differences in leaf surface microstructure and the presence of nanoscale wax features produce various degrees of leaf water repellency.

What are we able to generalize about heterogeneous leaf microsurfaces? Does the presence of specific microstructures, such as trichomes, produce greater probabilities that the leaf retains water on the surface? How does the development of epiphylls in humid environments or the accumulation of dust on the surface in rain-free arid locations add to the complexity of the leaf surface and the leaf's ability to retain water? Advances in microscopy may point to generalizable features of leaf microstructure that influence leaf water repellency.

11.5 Is Leaf Water Repellency a Beneficial or Costly Plant Trait?

Leaf water repellency is presented as a mechanism that helps the plant remove water to promote gas exchange and minimize the growth of pathogens (Aparecido et al. 2017; Goldsmith et al. 2017). In this context, leaf water repellency as a functional leaf trait is seen as a benefit. However, leaf wetness is not always negative, especially during times of drought stress in which the plant does not need to promote gas exchange to exacerbate the effects of drought stress. Evaporation of water on leaves can reduce leaf temperature and reduce heat stress to the plant (Katata et al. 2010).

Additionally, several studies found that plants can absorb water through their leaves to improve the water balance within the leaf (Goldsmith et al. 2013; Berry and Smith 2014). In this context, shedding water from the leaf surface is seen negatively.

Matos and Rosado (2016) identified positive and negative outcomes for leaves to retain or repel water in the introductory section of their research article. Dawson and Goldsmith (2018) presented the known costs and benefits of leaf wetness and argue that these costs and benefits are spatially and temporally scale-dependent. Examples of costs for wet leaves discussed in Dawson and Goldsmith (2018) include the development of pathogens on the leaf surface and a reduction in gas exchange for photosynthesis. Each world biome has a distinct position along the continua of annual precipitation and potential evapotranspiration. Tropical montane cloud forests are distinctly different from fog deserts even though in both environments there is a high persistency of fog. Plants adapt to the environments and occupy distinct niches within these environments. Although tropical montane cloud forests have a large annual precipitation (and fog precipitation) total compared with fog deserts, epiphytes in tropical montane cloud forests still need to adapt to drought-stress niches in the canopy compared with plants that extend their root systems into the moist, organic-rich soils of the cloud forests. Is leaf wettability a plant-specific benefit or an ecosystem-specific benefit? Do plants with highly hydrophobic leaves coexist within the same ecosystem with highly hydrophilic leaves because each plant adapts to diverse habitats along a water resource continuum within the ecosystem? What generalizations about leaf water repellency can be made between or within diverse ecosystems? Further exploration of leaf water repellency as a leaf functional trait would help address these ecohydrological questions.

11.6 How Does Seasonal Changes and Ambient Pollution Influence Leaf Water Repellency?

Precipitation has an erosive effect on leaf surface waxes (Reicosky and Hanover 1976; Baker and Hunt 1986). As a result, older leaves in the canopy can have more waxes eroded from the leaf surface than younger leaves. As leaves age and senesce, exposure to natural acid precipitation can erode many of the leaf waxes and change the three-dimensional properties and chemistry resulting in changes in the leaf water repellency. Tranquada and Erb (2014) examined the nanoscale wax morphologies of leaves from three species in Ontario, Canada, on a weekly basis during the growing season from May to October. They found that hot and humid weather conditions had the greatest impact on the degradation of the leaf waxes and this degradation resulted in a decrease in leaf hydrophobicity. A gradual increase in hydrophobicity occurs during early leaf development in aspen when the leaf surfaces develop wax from April to May (Tranquada et al. 2015). Degradation of leaf waxes during the growing season can make the leaves more wettable and increase canopy rainfall interception as the growing season progresses (Klamerus-Iwan and Blonska 2017; Klamerus-

Iwan and Kraj 2017). Kang et al. (2018) found that brown, senesced leaves of *Cercidiphyllum japonicum* in autumn (October to November) support up to 17 times greater water volume than summer leaves (June to September) and explain this finding based on the loss of leaf waxes resulting in a decrease in leaf hydrophobicity. Further investigations should explore temporal changes in leaf water repellency and direct measurements of interception and throughfall from different types of forest environments to compare deciduous and evergreen forests, arid and humid vegetation, and high-latitude and low-latitude forests.

Leaf life span also determines the amount of time leaves are exposed to ambient pollutants. Various particulate and aqueous pollutants can degrade epicuticular waxes on leaf surfaces (Burkhardt and Pariya 2014; Klamerus-Iwan et al. 2018a; Lu et al. 2018). Leaf hydrophobicity can decrease because of exposure to ambient pollutants and acid deposition that degrade the leaf surface waxes (Cape 1983; Kardel et al. 2012). In some cases, exposure to pollutants can weaken the plant to become more susceptible to disease and result in increases in leaf wettability (Klamerus-Iwan and Witek 2018). Klamerus-Iwan et al. (2018b) found that the content of polycyclic aromatic hydrocarbons (PAHs) in leaves increased the canopy water storage capacity of three coniferous tree species in Poland. Urban vegetation is exposed to a various mixture of airborne pollutants. Detailed studies on the effects of common airborne pollutants on leaf water repellency should be continued. Additional studies from different geographical settings would provide insight into the ecohydrological impacts of pollutants in urban watersheds.

11.7 Do Water Droplets or Water Films Develop on Wettable and Non-wettable Leaves During Rainfall Events?

At the beginning of a rainfall event, various sizes of raindrops splash onto leaf surfaces and the raindrops separate into various sizes. Some of these raindrops can coalesce on the leaf surfaces to form larger water droplets. Figure 11.2a shows an image of a leaf with clearly visible water droplets. The water droplets were produced from a raindrop generator splashing onto the leaf surface. These water droplets are comparably sized droplets measured in leaf water repellency studies. Water droplets of various sizes continue to form on leaf surfaces as rainfall events progress, but with long, persistent rainfall events, water films begin to form on leaf surfaces, especially on leaf surface that are not classified as wettable (i.e., $<110°$ based on Aryal and Neuner 2010).

Figure 11.2b shows an image of a leaf that has developed a water film on the surface after repeated raindrop impacts. Water films are very thin layers of water on the leaf surface. As the leaf surface becomes more wetted through raindrop splash, the water film develops. The quantity of water that composed a thin water film on an *Acer saccharinum* L. leaf represents approximately seven 10 μL water droplets on a leaf surface of 20 cm^2 or a 0.035 mm water film covering the leaf surface evenly.

Fig. 11.2 Photograph of two *Acer x freeman* "Jeffersred" leaves with (**a**) individual water droplets on the leaf surface and (**b**) a water film

Water films cover larger areas of the leaf surface than the area covered by the same quantity of water formed into water droplets. Water films do not bead up as a water droplet, and the measurement of contact angles is more problematic to calculate. Because water droplets are measured in leaf water repellency research, water films are largely ignored or assumed to have similar hydrophobic qualities as water droplets.

Future studies should explore differences in the repellency of water films and water droplets. Additionally, the ecohydrological processes that create water films and water droplets should be further examined. Are water films produced with greater frequency in persistently humid environments such as cloud forests? During rainfall events are water droplets or water films more present on leaf surfaces of native species in arid environments than on species from humid environments? Understanding the characteristics of leaf wetness between species in different environments can provide insight on ecophysiological responses of plants. If leaf water repellency is a functional leaf trait to maximize gas exchange and photosynthesis, leaves that enhance the development of water droplets may be advantageous over leaves that enhance the development of water films. The development of water films may promote water absorption by leaves to minimize water stress in some species over leaves that develop water droplets. Water films cover a larger leaf surface area providing more opportunities for water absorption. The evaporation rates between water droplets and water films also present an underexplored research topic. Do water films evaporate faster than water droplets because water films expose the water to a greater area of incident sunlight?

The dewetting process creates another research direction. Edwards et al. (2016) developed a model of a dewetting liquid film. Although Edwards et al. (2016) were not investigating water droplets or water films on leaves, their model of dewetting piques questions related to the evaporation process on leaves, such as do water films transition into water droplets during the evaporation process? Does the contact angle between a water film and a leaf surface change as water droplets transition from a water film? These questions would provide insight to better understand the evaporation process after a rainfall event.

11.8 Does the Water Droplet Size Matter in the Measurement of Leaf Water Repellency?

The droplet size used to measure contact angles on leaf surfaces ranged from 1 to 10 μL in research reviewed by Rosado and Holder (2013). Pérez-Harguindeguy et al. (2013) recommended droplets ranging from 2 to 5 μL. Schreiber (1996) determined that the contact angle would not be significantly different if the droplet size ranged from 1 to 10 μl. Also, Rosado and Holder (2013) found an even larger range of water droplet sizes (10–50 μL) used from studies that calculated water droplet retention. Since this review, Matos and Rosado (2016) conducted an experiment to determine if water droplet sizes influence leaf water repellency measurement. They concluded that the water droplet sizes used in leaf hydrophobicity and water droplet retention studies significantly affect the results. Matos and Rosado (2016) recommend a standard of 5 μL for leaf hydrophobicity measurements and 50 μL for water droplet retention measurements. Even since this study emerged (in less than 2 years!), numerous subsequent studies have been published that used a wide range of water droplet sizes in the experiments. If water droplet size matters, how can leaf water repellency studies using different sizes of water droplets be compared? In some cases, in particular coniferous species, needles are so small that a 10 μL water droplet could not be placed on the needle to measure the contact angle. Is there a method to compare leaf water repellency measurements between studies that use different sizes of water droplets? Afterall, during a rainfall event, different sizes of water droplets form and adhere to leaf surfaces. Also, during a rainfall event, water droplets change in size as water coalesces with other raindrops to form larger water droplets or evaporate to form smaller droplets. Should leaf water repellency studies have a standard droplet size? Konrad et al. (2012) and Nanko et al. (2013) examined the physical processes of how water droplets form and move off leaf and branch surfaces. Further development of these process-based theoretical models to explore the interconnections between water droplet size and leaf water repellency would provide more insight into rainfall interception studies.

11.9 Does Leaf Water Repellency Influence Canopy Storage Capacity?

Canopy storage capacity is the amount of water held on the canopy at the end of a rainfall event after drip has ceased and in the absence of wind. Ecohydrologists have addressed how the branch and canopy architecture influence the variability of canopy storage capacity between species (Garcia-Estringana et al. 2010). Because a large percentage of rainfall interception is explained by variables such as leaf area index, species with leaves that are more hydrophobic may intercept less rainfall than species with more hydrophilic leaves. A few studies have investigated the influence of leaf water repellency on canopy storage capacity and found a general trend of decreasing canopy storage capacity with increasing leaf hydrophobicity (Holder 2013; Holder and Gibbes 2017). Klamerus-Iwan and Blonska (2018) reported that the contact angle on leaf surfaces are dependent on the temperature of the water. Specifically, they found that a 1 °C increase in temperature decreases the contact angle of the water droplet on the leaf surface by greater than 3.4° (Klamerus-Iwan and Blonska 2018). Furthermore, temperature of the water explained a significant variation in canopy storage capacity. Nanko et al. (2013) found that the drop size distribution of throughfall changes due to variations in leaf structure and leaf hydrophobicity.

The importance of these and future studies investigating the relationship between leaf water repellency and canopy storage capacity is to understand how different species affect the rainfall interception process. Two species with the same leaf area index does not necessarily intercept the same quantity of rainfall; however, leaf area index is often reported as the most important vegetation variable in controlling interception in ecohydrological models. Species with highly repellent leaf surfaces may increase the quantities of throughfall at a site and may result in greater hydrological inputs beneath the canopy during a rainfall event. If the difference between leaf water repellency between species and between species in contrasting ecosystems is large, then hydrological inputs and hydrological processes operating in an ecosystem may be significantly different. Also, if species in an ecosystem are replaced by different species by either human activities, the site water balance may change not only because of leaf area index changes but because of changes in the hydrophobicity of the canopy. The significance of leaf water repellency as a mechanism that influences hydrological processes is an underexplored ecohydrological topic.

11.10 How Can Leaf Biomechanics Inform Leaf Water Repellency Studies?

Biomechanics is the study of the mechanical functions of living organisms. Studies within the discipline of biomechanics examine how internal or external forces act upon the biological organism. The significance of leaf water repellency and leaf angle as variables that influence the mechanics and process of rainfall interception is an underexplored research topic in ecohydrology. Information on the interaction between leaf water repellency and leaf angle and the mechanics of a raindrop impacting a leaf may enhance hydrological models of rainfall interception. Because leaf water repellency and leaf angle are important canopy parameters that explain differences in hydrologic fluxes under different canopies with constant leaf area index, future studies could provide linkages between biomechanics and ecohydrology.

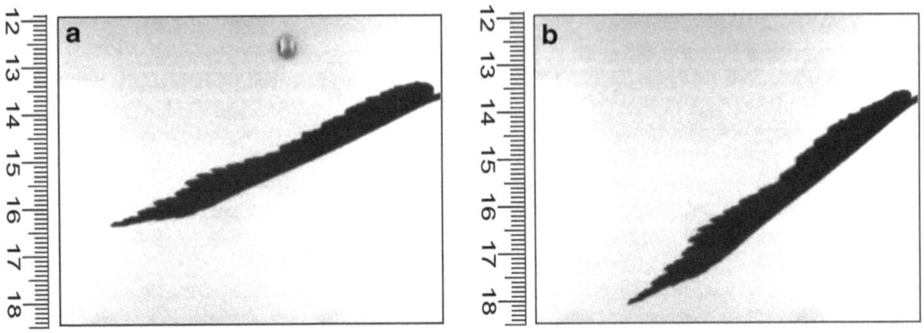

Fig. 11.3 Photograph of a 30 μL water droplet (**a**) before impact with an *Ulmus pumila* leaf and (**b**) immediately after impact. The time interval between each image is approximately 50 milliseconds. Notice that the leaf has deflected to a greater angle from the horizontal after impact with the water droplet. The scale of the ruler on the left of the image is in centimeters

I will provide an example to illustrate how biomechanics can be approached in the field of ecohydrology. Figure 11.3a shows a frame from a high-speed video of a 30 μL water droplet just before impacting a single *Ulmus pumila* leaf. The video was taken to provide a profile view of the water droplet upon impaction. In this image the leaf is inclined at an angle similar to the angle that the leaf would appear in the canopy. Also, the leaf's petiole is attached to the branch. Figure 11.3b shows the same leaf less than a second after the raindrop impacted on the leaf surface. In Fig. 11.3b the leaf angle from horizontal is visibly larger than the leaf angle in Fig. 11.3a. If the kinetic energy of the water droplet forces the leaf angle to surpass the water droplet retention angle produced by the leaf water repellency, then the water droplet will drain off the leaf upon impaction decreasing the water storage on the leaf during a rainfall event. Other questions that arise from this type of research are the following: do the average initial leaf angles on the canopy of a species surpass the average water droplet retention angle produced by the leaf surface and does the kinetic energy of a fall raindrop trigger the leaf to incline at a steeper angle allowing the water droplet to easily drain off the leaf surface? Wind may also influence rainfall interception processes by changing leaf inclination angles (Tadrist et al. 2014). The influence of the biomechanics of the leaves, branches, and the entire canopy on interception totals is an unexplored research topic.

The complexity of the dynamic process of rainfall interception has made it difficult to develop analytical models of leaf water repellency. Current approaches rely heavily on experimental data and provide predictions with a moderate degree of uncertainty, due to the variation in the experimental data. Furthermore, while this experimental approach provides insight into the impact of the individual physical and chemical processes, it is difficult to extend these results to describe the interactions of these factors or to extend the results to another species. Exploring and integrating connections between biomechanics and canopy variable that influence rainfall interception, such as leaf water repellency, can provide more nuanced understandings of the dynamic process of rainfall interception.

11.11 From Leaf to Landscape: Does Leaf Water Repellency Influence Water Resources?

Providing adequate supplies of water resources is an issue that affects the majority of people regardless of whether a municipal water system depends on groundwater or surface water supplies in rural or urban areas. In practically every urban area, municipal water departments are struggling to find adequate, healthy, and sustainable sources of water to support growing urban populations. The acquisition of new water rights for municipal water delivery systems is also increasingly contested by downstream water users often across state boundaries. Within this context, ecohydrologists need to scale leaf surface studies to the landscape or broader watershed. These leaf-to-landscape studies can provide the basis for informing water resources planners and policy-makers to secure sustainable supplies of water resources through effective management of municipal watersheds and aquifers.

According to the World Bank, the global population living in urban areas increased from 46.7% in 2000 to 54.8% in 2017, and the trend of an urbanizing population is expected to increase in the future. Urban watersheds represent a complex mosaic of land uses, including buildings (residential, commercial, and industrial), infrastructure (roads, sidewalks, drainage outlets, etc.), landscaped vegetation, and undeveloped open spaces (forests, grasslands, parks, etc.). Urban land use changes through time affect the hydrology of the watershed. Information on how changes in vegetation influence hydrological processes such as rainfall interception differences and runoff is important for urban planners as the population of urban citizens continues to increase and urban land cover continues to change. Citizens in urban areas, either directly or through representative municipal officials, decide which species to plant in urban areas. This individual choice can have profound impacts on the hydrological processes within urban areas. Leaf water repellency between species in these watersheds is significantly different as previous studies have indicated. Scaling up the results of these studies is an important step in identifying potential landscape impacts of our individual choices. Future research should examine linkages between individual plant-level studies to the watershed. Geographic information science may provide a useful tool to model ecohydrological data, including leaf water repellency, at multiple levels (leaf, canopy, vegetative stand, and watershed) to examine the hydrological impacts of land use changes in urbanized areas.

Additionally, investigations of the geography of leaf water repellency at the landscape level could determine whether leaf water repellency is a species-specific variable or a biome-specific variable (Aryal and Neuner 2010). For example, do plants with more hydrophobic leaves occur in larger proportions in one type of ecosystem versus another type of ecosystem? Because of the rise of recent publications with leaf water repellency data from various species and different biomes, we are approaching a time when meta-analyses could be conducted to determine whether leaf water repellency varies between physical environments. Any meta-analysis would need to examine studies that explored the dominant species within different environments and, also, control for such factors as leaf phenology. Holder (2007), Aryal and Neuner (2010), Aparecido et al. (2017), Goldsmith et al. (2017), and Xiong et al. (2018) investigated comparisons of leaf water repellency or water storage retention of species from different environments. These studies are first steps in understanding the geographic variability of leaf water repellency and how species with different leaf water repellency respond to their hydrological environment. These studies have generally found that arid environments tend to have a larger proportion of species with hydrophobic leaves. However, more research in other locations needs to be conducted to systematically corroborate this finding.

11.12 Synthesis and Future Research

This chapter explored recent advances in leaf water repellency research with the intention of focusing future research to several broad questions. Ten questions were provided to direct future research in leaf water repellency research. The questions posed in the book chapter were not all-encompassing. The purpose of proposing these questions was to promote a dialogue within the ecohydrology discipline for future research in leaf water repellency.

The study of the role of leaf water repellency in rainfall interception processes occurs within a nexus area of the disciplines of hydrology, meteorology, ecology, and botany. Because of this, scientists from various disciplines have examined leaf water repellency through different lenses to focus on different aspects of leaf water repellency. The interdisciplinarity of the study of leaf water repellency provides excitement to this field of research, especially if scientists from the various disciplines communicate research findings across disciplinary boundaries. However, the disciplinary boundaries can create separate vocabularies and procedures for the study and measurement of leaf water repellency. In many examples, the definition and measurement of leaf water repellency can vary between plant physiologists, forest hydrologists, plant pathologists, botanists, geographers, materials engineers, and physicists. I hope this chapter will call attention to the various definitions (Sect. 11.2) and measurements (Sects. 11.3 and 11.8) of leaf water repellency to bring together scientists from disparate disciplines to discuss the use of standard definitions and procedures in leaf water repellency studies. Perhaps an interdisciplinary symposium is needed to further refine definitions and measurements.

Within specific disciplines, there are clear avenues of future research directions. For example, botanists can explore the role of variations in leaf surface structures and epiphylls on leaf water repellency using advanced techniques in microscopy (Sect. 11.4). Plant physiologists can assess leaf water repellency as a beneficial or costly functional trait for plants to determine if generalizable patterns exist between biomes (Sects. 11.5 and 11.11). Ecologists can explore the influence of leaf phenology and atmospheric pollution on leaf water repellency (Sect. 11.6). Ecohydrologists can explore differences between water drops and water films on wettable and non-wettable leaf surfaces (Sect. 11.7) and the influence of leaf water repellency on canopy storage capacity (Sect. 11.9) at various geographic scales from leaf to landscape (Sect. 11.11). Engineers and botanists interested in leaf biomechanics can explore the influence of leaf water repellency, leaf angle, and leaf biomechanics as raindrops splash on leaf surfaces during rainfall events (Sect. 11.10). Although this chapter emphasized leaf water repellency, the vegetation canopy consists of other surfaces besides leaves, including bark, branches, and stems. The repellency of these woody surfaces is an additional research question that is largely unexplored in ecohydrology.

Several opportunities exist to formulate experiments and provide further insights to understand the role of canopy surface water repellency in ecohydrological processes. Throughout this chapter, I provided several questions for clarification and further exploration. I would encourage colleagues to use these questions and formulate additional questions, to advance our knowledge of the ecohydrological significance of leaf water repellency.

References

Aparecido LMT, Miller GR, Cahill AT, Moore GW (2017) Leaf surface traits and water storage retention affect photosynthetic responses to leaf surface wetness among wet tropical forest and semiarid savanna plants. Tree Physiol 37:1285–1300. https://doi.org/10.1093/treephys/tpx092

Aryal B, Neuner G (2010) Leaf wettability decreases along an extreme altitudinal gradient. Oecologia 162:1–9. https://doi.org/10.1007/s00442-009-1437-3

Baker EA, Hunt GM (1986) Erosion of waxes from leaf surfaces by simulated rain. New Phytol 102:161–173. https://doi.org/10.1111/j.1469-8137.1986.tb00807.x

Barthlott W, Neinhuis C (1997) Purity of the sacred lotus, or escape from contamination in biological surfaces. Planta 202:1–8. https://doi.org/10.1007/s004250050096

Berry ZC, Smith WK (2014) Experimental cloud immersion and foliar water uptake in saplings of *Abies fraseri* and *Picea rubens*. Trees Struct Funct 28:115–123. https://doi.org/10.1007/s00468-013-0934-5

Biddick M, Hutton I, Burns KC (2018) An alternative water transport system in land plants. Proc R Soc B Biol Sci 285:20180995. https://doi.org/10.1098/rspb.2018.0995

Brewer CA, Nuñez CI (2007) Patterns of leaf wettability along an extreme moisture gradient in western Patagonia, Argentina. Int J Plant Sci 168:555–562. https://doi.org/10.1086/513468

Brewer CA, Smith WK, Vogelmann TC (1991) Functional interaction between leaf trichomes, leaf wettability and the optical properties of water droplets. Plant Cell Environ 14:955–962. https://doi.org/10.1111/j.1365-3040.1991.tb00965.x

Burkhardt J, Pariya S (2014) Particulate pollutants are capable to 'degrade' epicuticular waxes and to decrease the drought tolerance of Scots pine (*Pinus sylvestris* L.). Environ Pollut 184:659–667. https://doi.org/10.1016/j.envpol.2013.04.041

Cape JN (1983) Contact angles of water droplets on needles of Scots pine (*Pinus sylvestris*) growing in polluted atmospheres. New Phytol 83:293–299. https://doi.org/10.1111/j.1469-8137.1983.tb03432.x

Dawson TE, Goldsmith GR (2018) The value of wet leaves. New Phytol 219:1156–1169. https://doi.org/10.1111/nph.15307

Della Volpe C, Siboni S (2015) Use, abuse, misuse and proper use of contact angles: a critical review. Rev Adhesion Adhesives 3:365–385. https://doi.org/10.7569/RAA.2015.097310

Edwards AMJ, Ledesma-Aguilar R, Newton MI, Brown CV, McHale G (2016) Not spreading in reverse: the dewetting of a liquid film into a single drop. Sci Adv 2:e1600183. https://doi.org/10.1126/sciadv.1600183

Fernández V, Bahamonde HA, Peguero-Pina JJ, Gil-Pelegrín E, Sancho-Knapik D, Gil L et al (2017) Physico-chemical properties of plant cuticles and their functional and ecological significance. J Exp Bot 68:5293–5306. https://doi.org/10.1093/jxb/erx302

Garcia-Estringana P, Alonso-Blázquez N, Alegre J (2010) Water storage capacity, stemflow and water funneling in Mediterranean shrubs. J Hydrol 389:363–372. https://doi.org/10.1016/j.jhydrol.2010.06.017

Goldsmith GR, Matzke NJ, Dawson TE (2013) The incidence and implications of clouds for cloud forest plant water relations. Ecol Lett 16:307–314. https://doi.org/10.1111/ele.12039

Goldsmith GR, Bentley LP, Shenkin A, Salinas N, Blonder B, Martin RE et al (2017) Variation in leaf wettability traits along a tropical montane elevation gradient. New Phytol 214:989–1001. https://doi.org/10.1111/nph.14121

Hewlett JD (1982) Principles of forest hydrology. University of Georgia Press, Athens

Holder CD (2007) Leaf water repellency of species in Guatemala and Colorado (USA) and its significance to forest hydrology studies. J Hydrol 336:147–154. https://doi.org/10.1016/j.jhydrol.2006.12.018

Holder CD (2013) Effects of leaf hydrophobicity and water droplet retention on canopy storage capacity. Ecohydrology 6:483–490. https://doi.org/10.1002/eco.1278

Holder CD, Gibbes C (2017) Influence of leaf and canopy characteristics on rainfall interception and urban hydrology. Hydrolog Sci J 62:182–190. https://doi.org/10.1080/02626667.2016.1217414

Kang H, Graybill PM, Fleetwood S, Boreyko JB, Jung S (2018) Seasonal changes in morphology govern wettability of Katsura leaves. PLoS One 13:e020290. https://doi.org/10.1371/journal.pone.0202900

Kardel F, Wuyts K, Babanezhad M, Wuytack T, Adriaenssens S, Samson R (2012) Tree leaf wettability as passive bio-indicator of urban habitat quality. Environ Exp Bot 75:277–285. https://doi.org/10.1016/j.envexpbot.2011.07.011

Katata G, Nagai H, Kajino M, Ueda H, Hozumi Y (2010) Numerical study of fog deposition on vegetation for atmosphere-land interactions in semi-arid and arid regions. Agr Forest Meteorol 150:340–353. https://doi.org/10.1016/j.agrformet.2009.11.016

Kittredge J (1948) Forest influences. McGraw-Hill Book Company Inc., New York

Klamerus-Iwan A, Blonska E (2017) Seasonal variability of interception and water wettability of common oak leaves. Ann For Res 60:73–73. https://doi.org/10.15287/afr.2016.706

Klamerus-Iwan A, Kraj W (2017) Wettability and interception in relationship with the seasonal changes on the *Fagus sylvatica* leaf surface. For Res Pap 78:210–217. https://doi.org/10.1515/frp-2017-0023

Klamerus-Iwan A, Blonska E (2018) Canopy storage capacity and wettability of leaves and needles: the effect of water temperature changes. J Hydrol 559:534–540. https://doi.org/10.1016/j.jhydrol.2018.02.032

Klamerus-Iwan A, Blonska E, Lasota J, Waligórski P, Kalandyk A (2018a) Seasonal variability of leaf water capacity and wettability under the influence of pollution in different city zones. Atmos Pollut Res 9:455–463. https://doi.org/10.1016/j.apr.2017.11.006

Klamerus-Iwan A, Emanuel G, Sadowska-Rociek A, Blonska E, Lasota J, Lagan S (2018b) Linking the contents of hydrophobic PAHs with the canopy water storage capacity of coniferous trees. Environ Pollut 242:1176–1184. https://doi.org/10.1016/j.envpol.2018.08.015

Klamerus-Iwan A, Witek W (2018) Variability of the wettability and water storage capacity of common oak leaves (*Quercus robur* L.). Water 10:695. https://doi.org/10.3390/w10060695

Koch K, Bohn HF, Barthlott W (2009) Hierarchically sculptured plant surfaces and superhydrophobicity. Langmuir 25:14116–14120. https://doi.org/10.1021/la9017322

Konrad W, Ebner M, Traiser C, Roth-Nebelsick A (2012) Leaf surface wettability and implications for drop shedding and evaporation from forest canopies. Pure Appl Geophys 169:835–845. https://doi.org/10.1007/s00024-011-0330-2

Konrad W, Burkhardt J, Ebner M, Roth-Nebelsick A (2015) Leaf pubescence as a possibility to increase water use efficiency by promoting condensation. Ecohydrology 8:480–492. https://doi.org/10.1002/eco.1518

Koukos D, Meletiou-Christou M-S, Rhizopoulou S (2015) Leaf surface wettability and fatty acid composition of *Arbutus unedo* and *Arbutus andrachne* grown under ambient conditions in a natural macchia. Acta Bot Gallea Bot Lett 162:225–232. https://doi.org/10.1080/12538078.2015.1039579

Lu S, Yang X, Li S, Chen B, Jiang Y, Wang D et al (2018) Effects of plant leaf surface and different pollution levels on $PM_{2.5}$ adsorption capacity. Urban For Urban Green 34:64–70. https://doi.org/10.1016/j.ufug.2018.05.006

Matos IS, Rosado BHP (2016) Retain or repel? Droplet volume does matter when measuring leaf wetness traits. Ann Bot 117:1045–1052. https://doi.org/10.1093/aob/mcw025

Nanko K, Watanabe A, Hotta N, Suzuki M (2013) Physical interpretation of the difference in drop size distributions of leaf drips among tree species. Agric For Meteorol 169:74–84. https://doi.org/10.1016/j.agrformet.2012.09.018

Papierowska E, Szporak-Wasilewska S, Szewinska J, Szatylowicz J, Debaene G, Utratna M (2018) Contact angle measurements and water drop behavior on leaf surface for several deciduous shrub and tree species from a temperate zone. Trees Struct Funct 31:1253–1266. https://doi.org/10.1007/s00468-018-1707-y

Pérez-Harguindeguy N, Díaz S, Garnier E, Lavorel S, Poorter H, Jaureguiberry P et al (2013) New handbook for standardized measurement of plant functional traits worldwide. Aust J Bot 61:167–234. https://doi.org/10.1071/BT12225

Reicosky DA, Hanover JW (1976) Seasonal change in leaf surface waxes of *Picea pungens*. Am J Bot 63:449–456. https://doi.org/10.2307/2441912

Revilla P, Fernández V, Álvarez-Iglesias L, Medina ET, Cavero J (2016) Leaf physico-chemical and physiological properties of maize (*Zea mays* L.) populations from different origins. Plant Physiol Bioch 107:319–325. https://doi.org/10.1016/j.plaphy.2016.06.017

Rosado BHP, Holder CD (2013) The significance of leaf water repellency in ecohydrological research: a review. Ecohydrology 6:150–161. https://doi.org/10.1002/eco.1340

Roth-Nebelsick A, Ebner M, Miranda T, Gottschalk V, Voigt D, Gorb S et al (2012) Leaf surface structures enable the endemic Namib desert grass *Stipagrostis sabulicola* to irrigate itself with fog water. J R Soc Interface 9:1965–1974. https://doi.org/10.1098/rsif.2011.0847

Schreiber L (1996) Wetting of the upper needle surface of *Abies grandis*: influence of pH, wax chemistry and epiphyllic microflora on contact angles. Plant Cell Environ 19:455–463. https://doi.org/10.1111/j.1365-3040.1996.tb00337.x

Smith WK, McClean TM (1989) Adaptive relationship between leaf water repellency, stomatal distribution, and gas exchange. Am J Bot 76:465–469. https://doi.org/10.1002/j.1537-2197.1989.tb11335.x

Sopper WE, Lull HW (1967) International symposium on forest hydrology. Pergamon Press, New York

Tadrist L, Sandreua M, de Langre E (2014) Wind and gravity mechanical effects on leaf inclination angles. J Theor Biol 341:9–16. https://doi.org/10.1016/j.jtbi.2013.09.025

Tao Y, Zhang YM (2012) Effects of leaf hair points of a desert moss on water retention and dew formation: implications for desiccation tolerance. J Plant Res 125:351–360. https://doi.org/10.1007/s10265-011-0449-3

Tranquada GC, Erb U (2014) Morphological development and environmental degradation of superhydrophobic aspen and black locust leaf surfaces. Ecohydrology 7:1421–1436. https://doi.org/10.1002/eco.1468

Tranquada GC, Victor JJ, Erb U (2015) Early season development of micro/nano-morphology and superhydrophobic wetting properties on aspen leaf surfaces. Am J Plant Sci 6:2197–2208. https://doi.org/10.4236/ajps.2015.613222

Voigt D, Roth-Nebelsick A, Miranda T, Ebner M, Gorb S (2012) Visualization of small water droplets on surfaces with different degree of wettability by using cryo-scanning electron microscopy. J Adv Microsc Res 7:64–67. https://doi.org/10.1166/jamr.2012.1091

Xiong P, Chen Z, Jia Z, Wang Z, Palta JA, Xu B (2018) Variability in leaf wettability and surface water retention of main species in semiarid Loess Plateau of China. Ecohydrology 11(e):2021. https://doi.org/10.1002/eco.2021

Chapter 12
Throughfall Erosivity in Relation to Drop Size and Crown Position: A Case Study from a Teak Plantation in Thailand

K. Nanko, N. Tanaka, M. Leuchner, and D. F. Levia

12.1 Introduction

Soil is an essential resource for plant growth, water conservation, stocks of organic carbon and nutrients, and sustainable forest management (Miura et al. 2015; Goebes et al. 2015b). Multidisciplinary interest in the role of the soil in topics such as biodiversity, ecosystem services, land degradation, and water security has been growing (Brevik et al. 2015). Since soil loss is irrecoverable on the scale of human decision making, the prediction of soil erosion is important. Water-driven soil erosion is governed by two principal forces: rain splash and water flow (Meyer and Wischmeier 1969; Morgan 2005). Raindrop impact is the initial process in water erosion and splash detachment (Ellison 1947; Kinnell 1973; Lal 1976; Fernández-Raga et al. 2017). The ground cover protects the soil surface from raindrops impact (Chapman 1948; Calder et al. 1993; Liu et al. 2016). There is generally an inverse relationship between the extent of protective ground cover and soil erosion rates in forests (Li et al. 2015; Miura et al. 2015). In fact, forests are generally considered an effective land use to protect against soil erosion (Cerdà 1999; Sidle et al. 2006).

K. Nanko (✉)
Department of Disaster Prevention, Meteorology and Hydrology, Forestry and Forest Products Research Institute, Tsukuba, Ibaraki, Japan
e-mail: knanko@ffpri.affrc.go.jp

N. Tanaka
Ecohydrology Research Institute, The University of Tokyo Forests, Graduate School of Agricultural and Life Sciences, The University of Tokyo, Seto, Aichi, Japan

M. Leuchner
Physical Geography and Climatology, RWTH Aachen University, Aachen, Germany

D. F. Levia
Departments of Geography and Plant & Soil Sciences, University of Delaware, Newark, DE, USA

D. F. Levia et al. (eds.), *Forest-Water Interactions*, Ecological Studies 240,
https://doi.org/10.1007/978-3-030-26086-6_12

However, some monospecific forests or plantations have suffered from soil erosion because of little protective ground cover, including unmanaged Japanese cypress (*Chamaecyparis obtusa* Sieb. & Zucc.) plantations (Miura et al. 2002; Nanko et al. 2006; Onda et al. 2010), eucalyptus (*Eucalyptus exserta* and *Eucalyptus globulus*) plantations (Calder et al. 1993; Terry 1996; Zhou et al. 2002), teak (*Tectona grandis* Linn. f.) plantations (Calder 2001), coffee plantations (Hanson et al. 2004), rubber (*Hevea brasiliensis*) plantations (Ziegler et al. 2009; Liu et al. 2016), and invasive species stands (Nanko et al. 2015). Throughfall has different characteristics from open rainfall (Levia et al. 2017). Rainfall erosivity is usually represented by kinetic energy and momentum (Nanko et al. 2008a; Goebes et al. 2014) which are determined by rainfall intensity (and amount), drop size distributions (DSDs), and drop velocity (Shinohara et al. 2018). Therefore, estimation of throughfall erosivity is necessary for the prediction of soil erosion in forests.

The DSD of open rainfall has been studied in meteorological and soil erosion contexts to improve the accuracy of radar rain gauges (Atlas et al. 1984; Brandes et al. 1999) and more accurately estimate rainfall erosivity (Kinnell 1973; Wischmeier and Smith 1978; Nanko et al. 2016b), respectively. Open rainfall DSD is correlated with rainfall intensity, type of rainfall, and wind speed (Marshall and Palmer 1948; Joss and Waldvogel 1967; Sempere-Torres et al. 1994; Erpul et al. 1998). Nowadays, an exponential relation (Marshall and Palmer 1948) and a general gamma relation (Ulbrich 1983) between raindrop size and drop number density are widely used. In contrast, throughfall has a different DSD from open rainfall due to rainfall interception by canopies. Once intercepted by an aboveground vegetative surface (e.g., leaf, branch, cone), raindrops begin to wet the canopy and are routed to the forest floor as different types of throughfall. The canopy sometimes generates larger throughfall drops by the coalescence of raindrops on the plant surface (i.e., canopy drip) (Chapman 1948; Nanko et al. 2013), and sometimes spawns smaller droplets as raindrops impact the plant surface with high kinetic energy creating splash droplets (i.e., splash throughfall) (Yang and Madden 1993; Saint-Jean et al. 2004). Other raindrops pass through the canopy directly without contacting any aboveground plant surface (i.e., free throughfall or direct throughfall). As a result of interactions with plant canopies, throughfall has a distinctly different DSD than that of open rainfall which can change as a function of various biotic and abiotic factors (Levia et al. 2017). Briefly, plant surface characteristics (e.g., plant surface wettability) (Nanko et al. 2013), canopy structure (e.g., canopy thickness and canopy phenophase) (Nanko et al. 2008b, 2016a), and meteorological conditions (e.g., rainfall intensity and wind speed) (Nanko et al. 2006) all have detectable and substantial effects on throughfall drop size. The myriad of factors determining throughfall DSD are summarized in Levia et al. (2017). Readers interested in the theory and methods of throughfall DSD studies and current knowledge gaps are referred to Levia et al. (2017).

The raindrops of open rainfall usually have a certain fall velocity, called terminal velocity. The fall velocity is determined between raindrop weight and friction of air. Bigger raindrops have higher fall velocity than smaller raindrops. Empirical and theoretical equations for terminal velocity related to raindrop diameter have been widely used (Laws 1941; Gunn and Kinzer 1949; Atlas et al. 1973). Throughfall, however, has a different fall velocity, in particular canopy drip. Drops need sufficient fall distance to gain terminal velocity, for example, for raindrops with diameters > 3 mm, it should be at least 12 m (Wang and Pruppacher 1977). In juvenile forests and mature forest with low canopy layers, throughfall drop velocity is lower than terminal velocity (Nanko et al. 2008b, 2011).

As demonstrated by a litany of studies, throughfall is characterized by high spatial variability compared to open rainfall (Robson et al. 1994; Keim et al. 2005; Staelens et al. 2006; Zimmermann et al. 2009; Zimmermann and Zimmermann 2014). Nevertheless, it is important to clarify the spatial variation of throughfall DSD and velocity in relation to throughfall erosivity. To date, only one study is known to examine the spatial variability of throughfall DSD in relation to throughfall erosivity. Nanko et al. (2011) investigated the spatial variation of throughfall DSD and kinetic energy with repetition of measuring throughfall drops using a single Japanese cypress in a large-scale rainfall simulator. More studies are necessary. Oppositely, the number of recent studies examining the spatial variability of throughfall erosivity has increased. Scholten et al. (2011) developed sand-filled cups to estimate the erosivity of rainfall or throughfall, called the Tübingen splash cup. The Tübingen splash cup enables the measurement of throughfall kinetic energy at many points in the forest at low cost. Spatial variation of throughfall kinetic energy has been investigated by Tübingen splash cups in subtropical forest plantations in southeast China (Geißler et al. 2013; Goebes et al. 2015a, b; Song et al. 2018), rubber plantations and rubber-based agroforestry in southwest China (Liu et al. 2018a, b), and Japanese cypress plantations (Shinohara et al. 2018). Goebes et al. (2016) developed a model to estimate throughfall kinetic energy from biotic and abiotic factors. The model shows that leaf area, tree height, leaf area index, and crown area are the most prominent biotic factors to estimate the kinetic energy of throughfall.

In this chapter, we introduce throughfall drop measurements in a teak plantation in Thailand. We examine the interrelationships among throughfall DSD and throughfall kinetic energy to provide a better understanding of canopy–soil interactions and provide other researchers with another window to conceptualize soil losses under trees. Teak is commercially important hardwood among tropical countries and is often cultivated in plantations (Krishnapillay 2000). The most prominent morphological feature of teak may be its large-sized mature leaves (Jarvis and McNaughton 1986; Pandy and Brown 2000), which could produce large throughfall drops. It was assumed that such large-sized throughfall drops could be a reason for the often-criticized high rate of soil loss in teak plantations (Bell 1973; Tangtham 1992). Because the magnitude of soil erosion losses in teak plantations is still under debate (Calder 2001; Fernández-Moya et al. 2014), detailed measurements of throughfall drops of teak can provide clearer insights into soil erosion losses in teak plantations. As such, the primary objective of this research is to report the effects of throughfall drop type (canopy drip, splash throughfall, free throughfall) and crown position (near-stem, one crown, two overlapping crowns, canopy gap) on the kinetic energy of throughfall in relation to that of open rainfall. Knowledge gaps of throughfall and its effects on soil erosion are also identified in an effort to stimulate further research to reduce soil erosion in teak plantations.

12.2 Methodology

12.2.1 Site Description

Field work was conducted in a teak plantation of uniform age in Mae Mo district, Lampang province, northern Thailand (18°25′N, 99°43′E; 380 m asl), where teak saplings were planted in 1968 by the Thai Forest Industry Organization. Mean (± 1 SD) stem diameters at breast height and tree height in the plantation were 22.4 ± 7.3 cm and 19.9 ± 3.5 m ($n = 308$), respectively, with a stand density of 440 trees ha^{-1} as of November 2005.

The Asian Monsoon is the dominant feature of the regional climate with a wet season, occurring from approximately May to October, and a dry season, roughly covering November to April. Mean annual rainfall was 1230.9 mm (1986–2010). Mean annual air temperature was 25.4 °C (2001–2010), with a minimum monthly mean of 21–22 °C in December–January and a maximum monthly mean of 29.5 °C in April. Detailed information is shown in Tanaka et al. (2015).

Throughfall drops measurements were made in a plot (Fig. 12.1) in July 2004 when the teak trees were fully leaved. Teak trees in the plot had a mean (± 1 SD) canopy projection area of 13.7 ± 7.1 m^2 ($n = 22$), with a plot-scale canopy gap of 31.8%. Simultaneous measurements of drop size and velocity of throughfall were conducted at four crown positions beneath the canopy with laser disdrometers: canopy gap, beneath one crown, beneath two crowns, and near-stem, hereafter the positions are called TF_{CG}, TF_{1C}, TF_{2C}, and TF_{NS}, respectively (Fig. 12.1). In addition to the throughfall drop measurements, raindrops were also measured at an open site about 500 m away from the throughfall measurement plot, hereafter this position is referred to as OP. All of the disdrometers were mounted 0.8 m above the ground. A splash screen was placed beneath the disdrometer, halfway to the surface, to prevent back splatter from drops striking the ground. The canopy openness at each crown position was determined using CanopOn 2 software (http://takenaka-akio.org/etc/canopon2/) from hemispherical photographs (Fig. 12.2A).

Data acquired from three small rain events were utilized in the analysis, which included data from the five disdrometers for all rain events. The rainfall and duration for each event were as follows: 2.8 mm in 1.4 h, 13.5 mm in 3.2 h, and 3.6 mm in 6.6 h. In 2004, there were 120 rain events from April to November in this site. Event precipitation ranged from 0.5 to 90.1 mm, with a median of 3.3 mm and a 75th percentile of 15.2 mm.

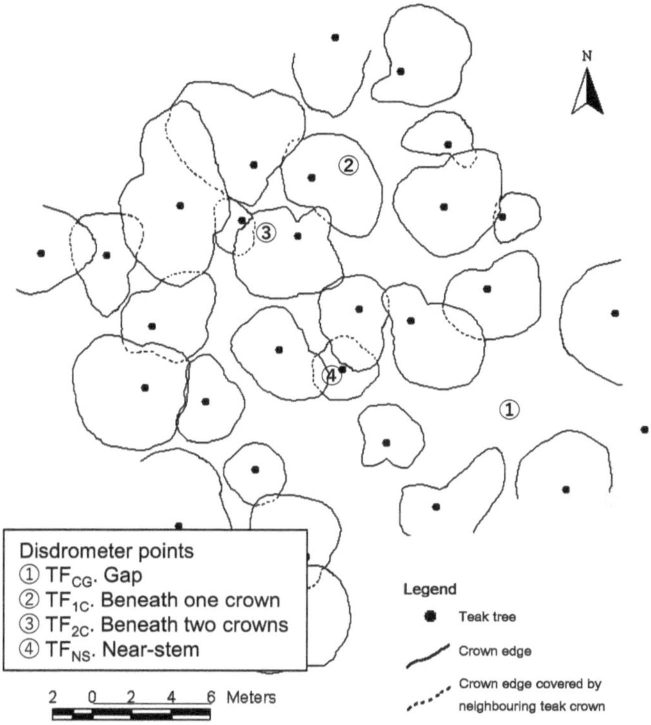

Fig. 12.1 (**a**) Site map and (**b**) throughfall drop measurement points with tree positions and canopy projection areas

12.2.2 Measuring Raindrops with Disdrometers

There are various methods to manually and automatically measure raindrops and throughfall drops. One useful automatic instrument is an optical or laser disdrometer, which measures the drop size distribution and velocity of falling hydrometeors. Optical or laser disdrometers have a pair of optical or laser transmitters and receivers. When a raindrop passes through the optical sheet, analog output voltage from the receiver is reduced in proportion to the intercepted area. Larger raindrops attenuate the voltage signal more so than smaller raindrops. Drop size is calculated (or estimated) by the maximum intercepted area. With high sampling speeds (>10 kHz), the intercepted time imposed by a passing raindrop is detected. The fall velocity of the raindrop is calculated using raindrop size, laser sheet width, and the intercepted time (Nanko et al. 2008a). Most recent meteorological studies use laser disdrometers either the OTT PARSIVEL^2 or the Thies CLIMA Laser Precipitation Monitor (Löffler-Mang and Joss 2000; Angulo-Martínez et al. 2018). Laser disdrometers have the advantage to measure drop size and velocity directly with an assumption concerning raindrop shape (i.e., an oblate spheroid).

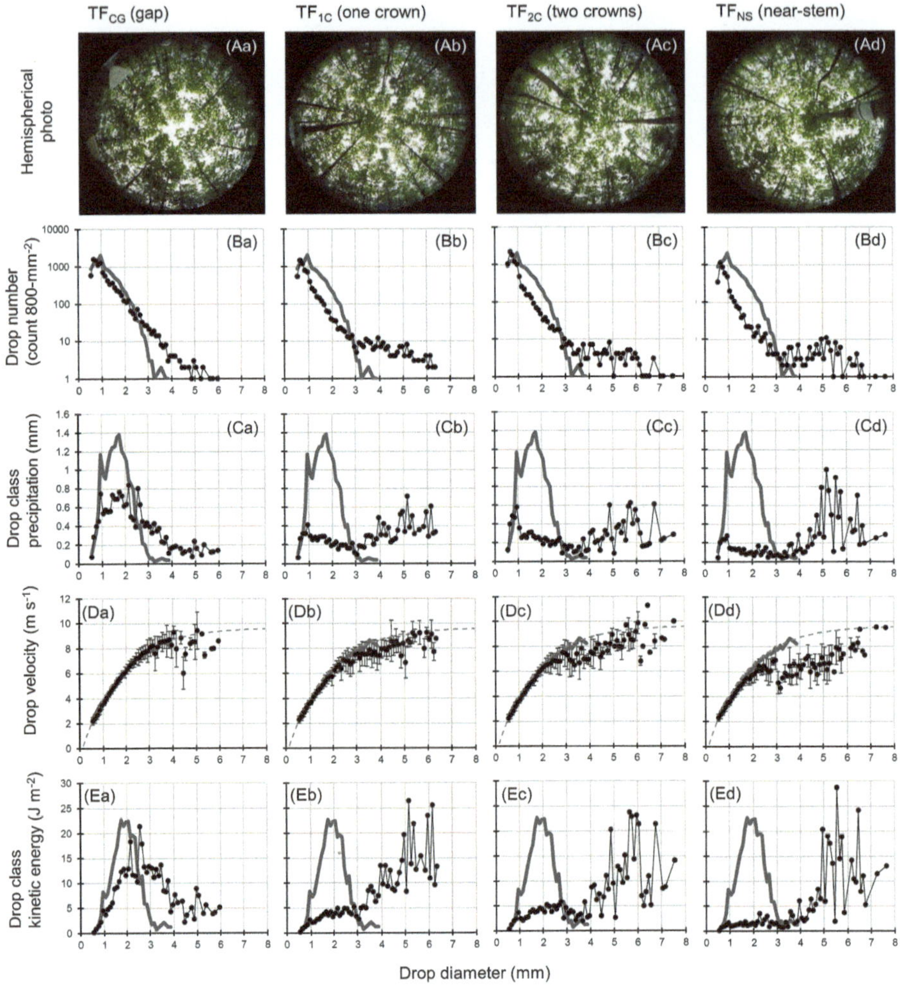

Fig. 12.2 (**A**) Hemispherical photos, (**B**) drop size distribution based on the number of drops (number-based DSD), (**C**) drop size distribution based on precipitation (volume-based DSD), (**D**) drop velocity, and (**E**) kinetic energy at TF$_{CG}$, TF$_{1C}$, TF$_{2C}$, and TF$_{NS}$ (gap, beneath one crown, beneath two crowns, and near-stem, respectively). All drop data used in the study was summed up. The *gray* line denotes open rainfall and the *black line with circles* denotes throughfall. *Error bars* in drop velocity indicate ± 1 standard deviation. The *dashed gray line* in (**D**) denotes the Gunn and Kinzer (1949) terminal fall velocity curves, Eq. (12.5)

The raindrop and throughfall drop data introduced in the chapter were collected by laser disdrometers developed by Nanko et al. (2006, 2008a). The output voltage variation from the disdrometer was digitized and stored on a laptop PC. Sampling speed was 12.5 kHz. The sampling area of the laser disdrometers was 800 mm^2.

The shape of falling drops is not spherical because of the friction of air (Beard 1976). Each falling raindrop was assumed to be an oblate spheroid whose axis ratio (f, dimensionless) was determined using the equation given by Andsager et al. (1999):

$$f = \frac{b}{a} \tag{12.1}$$

$$f = \min\left\{1.0, 1.0048 + 0.0057(D_{\mathrm{cm}}) - 2.628(D_{\mathrm{cm}})^2 + 3.682(D_{\mathrm{cm}})^3 - 1.677(D_{\mathrm{cm}})^4\right\} \tag{12.2}$$

where a is the major axis of an oblate spheroid (mm), b is the minor axis of an oblate spheroid (mm), and D_{cm} is an equivalent spherical diameter of a given drop (cm). For example, f is 0.98 for drops with 1 mm diameter, 0.78 for 4 mm diameter, and 0.52 for 8 mm diameter. The fall velocity of a raindrop (v, m s^{-1}) was calculated from the laser disdrometer data as

$$v = \frac{b + L - d_{\mathrm{initial}} - d_{\mathrm{last}}}{t} \tag{12.3}$$

where L is the width of laser sheet (mm, 1 mm for the laser disdrometer here), t is the intercepted time (msec) measured by the laser disdrometer, and d_{initial} (and d_{last}) is the intercepted distance by a passed raindrop calculated by initial (and last) output voltage reductions (mm). The disdrometers used here were capable of measuring drops with diameters > 0.5 mm. Calculated drop diameters > 8.0 mm were filtered in this study. All of the drops were classified into 0.1 mm diameter classes.

The kinetic energy (e in Joule) of a raindrop was calculated using the following equation:

$$e = \frac{1}{2}mv^2 = \frac{1}{2}\rho\left(\frac{\pi}{6}D^3\right)v^2 \tag{12.4}$$

where m is raindrop mass (g), D is drop diameter (mm), and ρ is raindrop density (1×10^{-6} gm^{-3}).

12.2.3 *Throughfall Partitioning*

Rainfall is partitioned by forest canopy into throughfall, stemflow, and interception loss (Carlyle-Moses and Gash 2011). The throughfall component may be divided into free and release throughfall (Dunkerley 2000), with release throughfall subsequently further categorized into splash throughfall and canopy drip (Moss and Green 1987; Nanko et al. 2006; Levia et al. 2017).

Throughfall partitioning was conducted using simultaneous measurements of both open rainfall and throughfall (and their corresponding DSDs) in accord with Levia and Nanko et al. (2019). The DSD of free throughfall was estimated first, followed by splash throughfall, and canopy drip. Throughfall partitioning was based on the following assumptions:

1. Throughfall in each drop size class was the sum of free throughfall, splash throughfall, and canopy drip.
2. The DSD of free throughfall had the same shape as that of open rainfall.
3. The fraction of free throughfall to open rainfall was assigned as the maximum value under the condition that free throughfall did not exceed throughfall in all drop size classes.
4. Splash throughfall DSD was based on the Weibull distribution, with fluctuations of the Weibull coefficients and D_{MAX} for splash throughfall with diameters < 2 mm.

These assumptions were based on the theories and experimental results of Yang and Madden (1993), Nakaya et al. (2011), and Frasson and Krajewski (2013). The detailed procedure is shown in Levia and Nanko et al. (2019). From the throughfall partitioning procedure, the number of drops, precipitation, and kinetic energy of each throughfall type in each drop size class were quantified.

12.3 Results

12.3.1 Throughfall Drops: Diameter and Kinetic Energy in Relation to Crown Position

Throughfall drops had different characteristics compared to raindrops. Usually, throughfall is composed of fewer but larger sized drops than open rainfall (Nanko et al. 2004; Frasson and Krajewski 2011). We showed two different types of DSDs in Fig. 12.2. Figure 12.2B and 12.2C illustrates DSD based on the drop number (number-based DSD) and the drop volume (volume-based DSD), respectively. In Fig. 12.2C, precipitation in each drop diameter class is shown as the cumulative volume of drops in each drop diameter class (mm^3) divided by the sampling area (mm^2).

At first, throughfall precipitation was lower than open rainfall. The amount of rainfall was 19.9 mm at OP and 18.9, 17.7, 17.0, and 15.1 mm at TF_{CG}, TF_{1C}, TF_{2C}, and TF_{NS}, respectively. The total number of measured drops was 13,869 for OP and 10,348, 6,660, 9,257, and 4,411 for TF_{CG}, TF_{1C}, TF_{2C}, and TF_{NS}, respectively. The maximum drop diameter (D_{MAX}) was 3.81 mm for OP and 5.98, 6.37, 7.56, and 7.61 mm for TF_{CG}, TF_{1C}, TF_{2C}, and TF_{NS}, respectively. The median volume drop diameter, such as the diameter of respective cumulative volume 50th percentiles (D_{50}) was 1.62 mm for OP and 2.15, 4.06, 4.25, and 5.12 mm for TF_{CG}, TF_{1C}, TF_{2C}, and TF_{NS}, respectively. Some raindrops coalesced into larger drops in the canopy and were released as canopy drip. TF_{NS} had the least number of drops but were of the largest size, whereas TF_{CG} had the highest number of drops and the smallest size. TF_{2C} had higher drop numbers and larger sizes than TF_{1C}.

Volume-based DSD (Fig. 12.2C) shows what size of drops contributed to total precipitation. OP had a peak between 1.7 and 1.8 mm (Fig. 12.2C). TF_{CG} had some peaks < 2.6 mm in diameter, whereas the other three crown positions had some peaks > 3.8 mm in diameter. These larger drops were the main contributors of total throughfall precipitation. Throughfall precipitation at TF_{NS} was composed by larger drops than at the other crown positions.

Throughfall drops had different drop diameter–velocity relationships compared to OP (Fig. 12.2D). The present data showed that OP had similar drop diameter–velocity relationship as suggested by Atlas et al. (1973) based on Gunn and Kinzer (1949) (Fig. 12.2D) with

$$v_t = 9.65 - 10.3\exp(-0.6D) \tag{12.5}$$

where v_t is terminal velocity (m s^{-1}). Throughfall drops > 2 mm in diameter had velocities lower than that of terminal velocity (Fig. 12.2D). The difference between the terminal velocities and the measured velocities was smaller at TF_{CG} and increased in order of TF_{1C}, TF_{2C}, and TF_{NS}. Most or all drops with > 3 mm in diameter were produced by canopy drip because the number of drops > 3 mm in diameter in OP was lower (Fig. 12.2B). The fall distance from the canopy and disdrometer was insufficient to reach terminal velocity, which for raindrops with diameters > 3 mm should be at least 12 m (Wang and Pruppacher 1977). Although mean tree height (19.9 m) was sufficient to reach terminal velocity, crown base height was insufficient. From TF_{CG} to TF_{NS}, the crown base height decreases.

The kinetic energy distribution based on drop size was different between OP and throughfall (Fig. 12.2E). Larger drops with > 3 mm in diameter had higher velocity than smaller drops of < 3 mm in diameter (Fig. 12.2D), which lacked a sufficient fall distance to gain terminal velocities. Therefore, larger drops with > 3 mm in diameter had higher kinetic energy than smaller drops. The effect of larger drops on the distribution shape was more drastic for kinetic energy (Fig. 12.2E) than precipitation (Fig. 12.2C). Throughfall had higher total kinetic energy than open rainfall, 313 J m^{-2} for OP and 410, 458, 470, and 351 J m^{-2} for throughfall at TF_{CG}, TF_{1C}, TF_{2C}, and TF_{NS}, respectively. Normalized kinetic energy by unit rainfall was 15.7 J m^{-2} mm^{-1} for OP and 21.6, 26.9, 26.6, and 23.3 J m^{-2} mm^{-1} for throughfall at TF_{CG}, TF_{1C}, TF_{2C}, and TF_{NS}, respectively. The measured normalized kinetic energy of throughfall was close to the maximum value of open rainfall, which is 28.3 J m^{-2} mm^{-1} (van Dijk et al. 2002). These normalized kinetic energies of throughfall were realized by rainfall intensity of open rainfall with 19, 52, 56, and 26 mm h^{-1}, respectively (van Dijk et al. 2002). Mid-canopy positions (TF_{1C} and TF_{2C}) had higher kinetic energy than TF_{CG} and TF_{NS}. TF_{CG} had larger throughfall precipitation but smaller drop size (Fig. 12.2Ca). TF_{NS} had larger drops (Fig. 12.2Cd); however, both throughfall precipitation and drop velocity were smaller (Fig. 12.2Dd).

12.3.2 Throughfall Partitioning by DSD in Relation to Crown Position

Figure 12.3 is an example of the throughfall partitioning at TF_{1C} shown in Fig. 12.2Cb. The throughfall partitioning was conducted based on a 5-min time step and summed for the whole dataset. Figure 12.3 indicates that the throughfall drops of < 1 mm in diameter were mainly composed of splash throughfall and the drops with > 3 mm in diameter were mostly composed of canopy drip.

Throughfall partitioning was conducted for all four crown positions (Fig. 12.4). As for drop number (Fig. 12.4b), the number of splash throughfall droplets was higher than that of any other throughfall type at all crown positions, contributing 54%, 67%, 77%, and 75% at TF_{CG}, TF_{1C}, TF_{2C}, and TF_{NS}, respectively. The number of canopy drip drops was the lowest, contributing 18%, 14%, 9%, and 11% at TF_{CG}, TF_{1C}, TF_{2C}, and TF_{NS}, respectively.

Regarding throughfall amount, canopy drip was the major contributor for throughfall precipitation (Fig. 12.4c). Canopy drip contributed 66%, 81%, 77%, and 88% of throughfall precipitation for TF_{CG}, TF_{1C}, TF_{2C}, and TF_{NS}, respectively. Free throughfall precipitation was larger at TF_{CG} (21%) than the other crown positions (10%, 10%, and 5%, respectively); however, canopy drip was more than free throughfall even at TF_{CG}. TF_{CG} was located completely beyond the canopy projection area (Fig. 12.1). The mean tree height was 19.9 m. The nearest distance from TF_{NS} to the crown projection area was 2.2 m. Canopy openness was low (Figs. 12.2Aa and 12.4a). If all of raindrops and throughfall drops fell completely

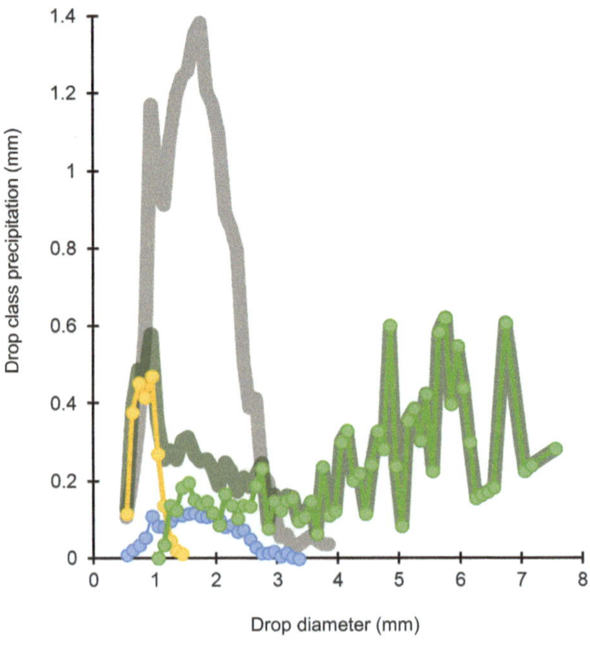

Fig. 12.3 An example of throughfall partitioning at the one crown point in Fig. 12.2Cb. The *gray* line denotes open rainfall, the *thick green* line denotes throughfall, the *blue line with circles* denotes free throughfall, the *yellow line with circles* denotes splash throughfall, and the *green line with circles* denotes canopy drip

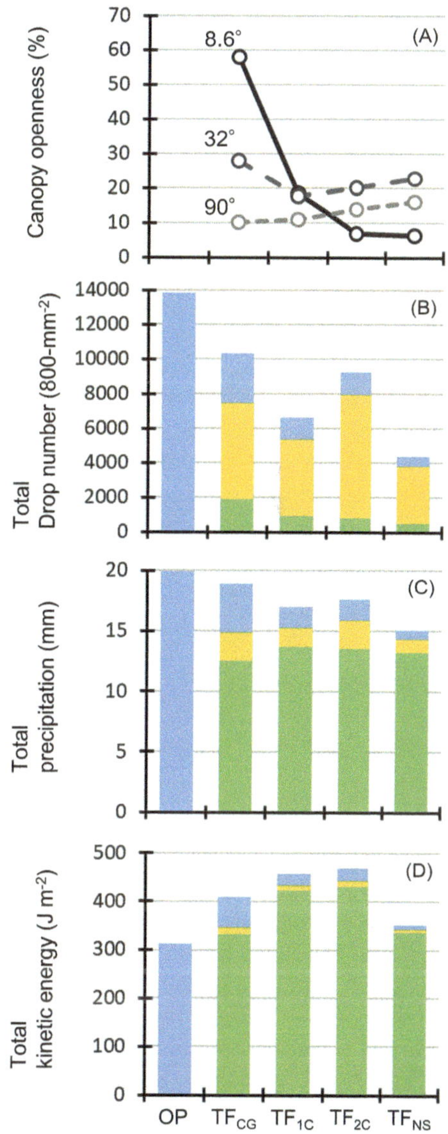

Fig. 12.4 (**a**) Canopy openness with three zenith angles; (**b**) drop number; (**c**) precipitation; and (**d**) kinetic energy of throughfall types at OP (open space) and TF_{CG}, TF_{1C}, TF_{2C}, and TF_{NS} (gap, beneath one crown, beneath two crowns, and near-stem, respectively). The *blue* bar denotes open rainfall or free throughfall, with the *yellow* and *green* bars denoting splash throughfall and canopy drip, respectively

vertically, TF_{CG} would receive only free throughfall. However, the throughfall partitioning based on DSD measurements revealed that the gap area received a combination of splash throughfall, canopy drip, and free throughfall. This is likely attributable to the fact that splash throughfall and canopy drip are introduced to the canopy gap due to the angle of emergence of release throughfall by rainwater being routed toward the canopy gap (Nanko et al. 2008b) or the redistribution by wind under conditions of wind-driven rainfall. In fact, wind-driven rainfall has been found to change the effective area of crown interception (Herwitz and Slye 1995).

The drop size of canopy drip increased from TF_{CG} towards TF_{NS}. Using canopy drip DSD, the D_{50} of canopy drip was calculated as 2.67, 4.58, 4.98, and 5.21 mm for TF_{CG}, TF_{1C}, TF_{2C}, and TF_{NS}, respectively. The canopy drip at TF_{CG} was mainly generated by leaf tips at the edge of canopy projection area. More leaves and branches were above the other crown positions. Nanko et al. (2013) conducted an indoor experiment to create leaf drips by applying spray mist to different canopy species. The D_{50} of leaf drips ranged from 3.7 to 5.3 mm. The experiment showed potential maximum canopy drip size because the foliage was stable inside the rainfall simulator without wind (Nanko et al. 2013). Canopy vibration reduced the amount of water coalescence on the canopy and forced water to drip from the canopies, and thus the drop size of canopy drip decreased (Nanko et al. 2006). Teak generated larger canopy drips on average, possibly due to its large leaves. Moreover, the amount of bark surface over the crown positions may influence the canopy drip size. The bark surface produced larger canopy drips than foliar surface for yellow poplar (*Liriodendron tulipifera* L.) when potential drip points were unmasked following leaf fall (Nanko et al. 2016a).

The kinetic energy of throughfall was highly influenced by canopy drip (Fig. 12.4d). Canopy drip contributed 81%, 92%, 92%, and 96% of the total kinetic energy at TF_{CG}, TF_{1C}, TF_{2C}, and TF_{NS}, respectively. Together, the amount of canopy drip (Fig. 12.3c), the DSD of canopy drip (Figs 12.2C and 12.4), and the fall distance of canopy drip (Fig. 12.2D) determine the total throughfall energy and its potential erosivity.

12.3.3 Temporal Variation of Throughfall-Type Generation

Figure 12.5 shows the temporal variation of throughfall as a function of throughfall type. The temporal variations of total throughfall were different at each crown position. TF_{CG} had similar variation with OP; however, TF_{NS} exhibited a different variation compared to OP. Throughfall generation is later at TF_{NS} than the other crown positions. The timing of throughfall inputs depends on the canopy openness (Figs 12.2A and 12.4a) and the amount of foliage over the crown positions.

At TF_{CG} (Fig. 12.5a), throughfall generated immediately after the onset of open rainfall. However, most of the initial throughfall was composed of canopy drip, surprisingly. Canopy drip with diameter > 2 mm was easily produced by teak canopy. But, the maximum size of canopy drip rarely exceeded 4 mm in diameter. The temporal variation of open rainfall was well correlated with free throughfall (Fig. 12.5a).

In contrast, at TF_{NS} (Fig. 12.5d), throughfall generation was initiated later in the rain events. Most of throughfall was made by canopy drip. There was more foliage and thick primary branches from the stem over TF_{NS} than the other crown positions, making it difficult for free throughfall to reach the disdrometer at TF_{NS}. More

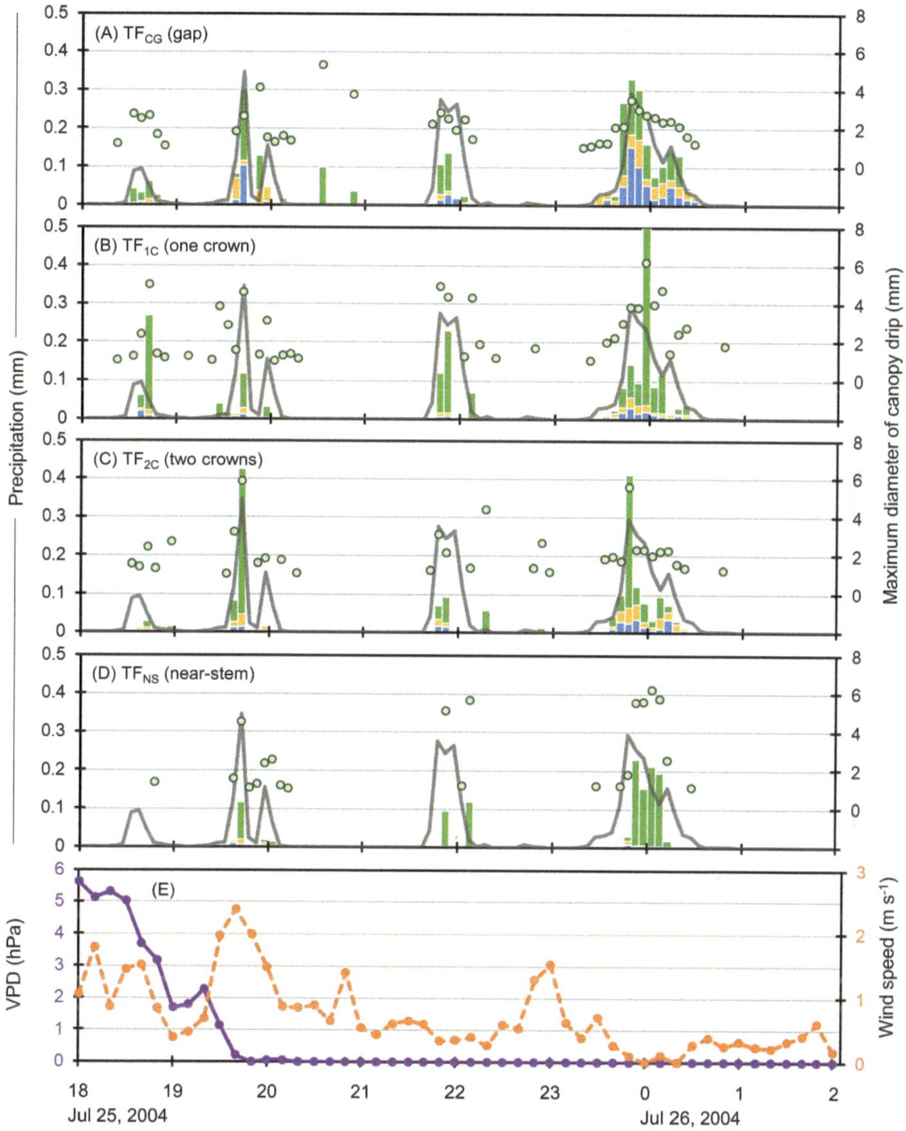

Fig. 12.5 Temporal variation of precipitation, and VPD, vapor pressure deficit, and wind speed at OP every 5 min. Throughfall was measured at (**a**) TF_{CG}, gap; (**b**) TF_{1C}, beneath one crown; (**c**) TF_{2C}, beneath two crowns; and (**d**) TF_{NS} near-stem, respectively. The *blue* bar denotes free throughfall, the *yellow* bar denotes splash throughfall, and the *green* bar denotes canopy drip. The *green* circle denotes the maximum diameter of canopy drip. The *purple* and *orange* lines denote VPD and wind speed, respectively

rainwater was necessary to saturate the canopy compared to other crown positions. After saturation, throughfall was mainly composed of canopy drip and the size of canopy drip was larger, exceeding 4 mm in diameter.

At TF_{1C} and TF_{2C}, the correlation between throughfall and open rainfall varied temporally. Sometimes, these crown positions receive concentrated throughfall with more canopy drip. Dynamic changes in the incident angle of open rainfall, canopy sway, and flowpaths of water through the canopy and changes in structurally created dripping points with differences in wind-driven rainfall can all result in changes in the magnitude and timing of canopy drip. Once there is sufficient data building upon the smaller magnitude rain events analyzed here, a more comprehensive DSD dataset at various crown positions for different tree species can be assembled to better document water flow paths in canopies.

12.4 Directions for Future Work on Throughfall Drop Studies and Canopy–Soil Interactions

It should now be clear that throughfall type, DSD, and the fall velocity of throughfall drops affect the kinetic energy of throughfall impacting the ground in teak plantations. We found a spatial heterogeneous pattern of throughfall kinetic energy at different crown positions underneath the teak trees examined that are at least partly attributable to differences in throughfall DSD and drop fall velocity.

Prior work examining the spatial variations of throughfall kinetic energy using Tübingen splash cups in mixed-species forest (Geißler et al. 2013; Goebes et al. 2015b), plantations (Goebes et al. 2015a; Liu et al. 2018a; Song et al. 2018; Shinohara et al. 2018), and agroforestry systems (Liu et al. 2018b) have identified the following:

1. At first, the variation of crown base height increases spatial variation of throughfall kinetic energy, with a lower crown base height decreasing throughfall kinetic energy due to the lower fall velocity of canopy drip (Goebes et al. 2015b; Shinohara et al. 2018). Mature monospecific forest with crown base height >10 m has a lower spatial variability of throughfall kinetic energy, where throughfall amount is the dominant factor and positively correlated with throughfall kinetic energy (Liu et al. 2018a, b; Shinohara et al. 2018).
2. Local LAI changes throughfall kinetic energy due to differences in the proportions of throughfall type. Higher LAI decreases free throughfall fraction and throughfall amount but increases the canopy drip fraction (Geißler et al. 2012; Goebes et al. 2015a). Vertical and horizontal distribution of LAI affect throughfall kinetic energy due to the fall distance of canopy drips (Song et al. 2018).

3. Greater tree species diversity increases spatial variability of throughfall kinetic energy (Geißler et al. 2013). Consequently, differences in tree traits and forest structures among species induce spatial variability of throughfall kinetic energy, which is spawned by different throughfall amount, DSD, and drop fall velocity.

Building upon the insights gained from earlier work (as summarized above), future studies must simultaneously measure throughfall drops using both splash cups and laser disdrometers to characterize the spatial variations of throughfall kinetic energy in detail. One solution is the development of low-cost disdrometers. The deployment of hundreds of low-cost disdrometers would permit us to gain a comprehensive understanding of the spatial variability and uncertainty of throughfall kinetic energies at different crown positions beneath various forests. Ideally, one could couple LiDAR measurements of the forest canopy with estimates of throughfall DSD, fall velocity, and kinetic energy at hundreds of points beneath the canopy per square meter to produce high-resolution maps of soil erosion risk, such as Song et al. (2018). With a greater number of point measurements, the kriged values between point measurements would be greatly improved. It is our contention that such precision maps of throughfall inputs (and kinetic energy inputs thereof) would assist forest managers in making better decisions to improve timber yields by possibly altering tree spacing patterns and tree species selections to minimize soil losses.

The influence of woody surface drip points on throughfall erosivity is largely unknown. While we know that woody surface drip points generate larger canopy drip than conifers or leafed broadleaved and leafless broadleaved trees (Levia and Nanko et al. 2019), we lack sufficient understanding as to how structurally created drip points created by irregularities in branch architecture affect throughfall erosivity. Changes in the location or magnitude of potential drip points with fluctuations in meteorological conditions or across seasons are also unknown. If the woody surface is high enough above the ground surface, the dripping point could create a much higher throughfall erosivity than from other plant surfaces. Further, more detailed understanding of the interplay between throughfall DSD and kinetic energy is necessary to fully envisage soil erosion in forests as well as litter decomposition by drop impact (Tashiro-Uchimura and Mizunaga 2017).

Future throughfall drop studies seeking to better understand the spatiality of throughfall kinetic energy might, in relation to canopy phenophase, also benefit from better prediction of seasonal soil loss. Based on results of Tanaka et al. (2015), it is known that throughfall inputs vary with canopy phenophase. As such, further work should better synchronize the interplay among canopy phenophase, throughfall DSD, and throughfall kinetic energy. What canopy phenophase (leafed, leafless, leafing, senescing) experiences the highest throughfall kinetic energy inputs? Does the spatial positioning of throughfall kinetic inputs change with canopy phenophase? Answers to such questions will provide a deeper understanding of canopy–soil interactions in teak plantations.

12.5 Summary

We have introduced the usefulness of throughfall drop measurement in the study of forest–water interactions with examples from a teak plantation in Thailand. Throughfall was partitioned into free throughfall, splash throughfall, and canopy drip using DSDs of both open rainfall and throughfall. Throughfall exhibited the following characteristics compared with open rainfall:

- Throughfall drops were lower in number but larger in size due to the coalescence of raindrops on canopies.
- Throughfall drops, especially canopy drip, had lower velocity due to insufficient fall distance from the canopy to the forest floor to reach terminal velocity, which partly depends on crown base height and the vertical distribution of foliage.
- Throughfall usually had higher kinetic energy due to larger drop size, which depends on the amount and size of canopy drip and the crown base height.

In addition, comparison of throughfall drop characteristics among four crown positions revealed throughfall kinetic energy was higher in mid-crown positions than in the gap or near-stem positions. Compared to mid-crown positions, the gap position had smaller drops and less canopy drip, while the near-stem position had lower drop fall velocity. The erosivity of throughfall with respect to crown position should now be examined in other types of plantations and forests to gain a more holistic understanding on canopy–water–soil interactions and develop high-resolution maps of throughfall kinetic energy that could help maintain forest productivity.

Acknowledgments The work was funded by JSPS KAKENHI Grant numbers JP17780119, JP15H05626, and JP17KK0159 and the CREST Program of JST (Japan Science and Technology Agency).

References

Andsager K, Beard KV, Laird NF (1999) Laboratory measurements of axis ratios for large raindrops. J Atmos Sci 56:2673–2683. https://doi.org/10.1175/1520-0469(1999)056<2673:LMOARF>2.0.CO;2

Angulo-Martínez M, Beguería S, Latorre B, Fernández-Raga M (2018) Comparison of precipitation measurements by OTT Parsivel 2 and Thies LPM optical disdrometers. Hydrol Earth Syst Sci 22:2811–2837. https://doi.org/10.5194/hess-22-2811-2018

Atlas D, Srivastava RC, Sekhon RS (1973) Doppler radar characteristics of precipitation at vertical incidence. Rev Geophys Space Phys 11:1–35. https://doi.org/10.1029/RG011i001p00001

Atlas D, Ulbrich CW, Meneghini R (1984) The multi parameter remote measurement of rainfall. Radio Sci 19:3–21. https://doi.org/10.1029/RS019i001p00003

Beard KV (1976) Terminal velocity and shape of cloud and precipitation drops aloft. J Atmos Sci 33:851–864. https://doi.org/10.1175/1520-0469(1976)033<0851:TVASOC>2.0.CO;2

Bell TIW (1973) Erosion in the Trinidad teak plantations. Commonw For Rev 52:223–233

Brandes EA, Vivekanandan J, Wilson JW (1999) A comparison of radar reflectivity estimates of rainfall from collocated radars. J Atmos Ocean Technol 16:1264–1272. https://doi.org/10.1175/1520-0426(1999)016<1264:ACORRE>2.0.CO;2

Brevik EC, Cerdà A, Mataix-Solera J, Pereg L, Quinton JN, Six J et al (2015) The interdisciplinary nature of SOIL. SOIL 1:117–129. https://doi.org/10.5194/soil-1-117-2015

Calder IR (2001) Canopy processes: implications for transpiration, interception and splash induced erosion, ultimately for forest management and water resources. Plant Ecol 153:203–214. https://doi.org/10.1023/A:1017580311070

Calder IR, Hall RL, Prasanna KT (1993) Hydrological impact of Eucalyptus plantation in India. J Hydrol 150:635–648. https://doi.org/10.1016/0022-1694(93)90129-W

Cerdà A (1999) Parent material and vegetation affect soil erosion in eastern Spain. Soil Sci Soc Am J 63:362–368. https://doi.org/10.2136/sssaj1999.03615995006300020014x

Carlyle-Moses DE, Gash JHC (2011) Rainfall interception loss by forest canopies. In: Levia DF, Carlyle-Moses DE, Tanaka T (eds) Forest hydrology and biogeochemistry, Ecological studies (Analysis and synthesis), vol 216. Springer, Dordrecht, pp 407–423. https://doi.org/10.1007/978-94-007-1363-5_20

Chapman G (1948) Size of raindrops and their striking force at the soil surface in a red pine plantation. Trans Am Geophys Union 29:664–670. https://doi.org/10.1029/TR029i005p00664

Dunkerley D (2000) Measuring interception loss and canopy storage in dryland vegetation: a brief review and evaluation of available research strategies. Hydrol Process 14:669–678. https://doi.org/10.1002/(SICI)1099-1085(200003)14:4<669::AID-HYP965>3.0.CO;2-I

Ellison WD (1947) Soil erosion studies – Part I. Agric Eng 28:145–146

Erpul G, Gabriels D, Janssens D (1998) Assessing the drop size distribution of simulated rainfall in a wind tunnel. Soil Tillage Res 45:455–463. https://doi.org/10.1016/S0933-3630(97)00030-5

Fernández-Moya J, Alvarado A, Forsythe W, Ramírez L, Algeet-Abarquero N, Marchamalo-Sacristán M (2014) Soil erosion under teak (*Tectona grandis* L.f.) plantations: general patterns, assumptions and controversies. Catena 123:236–242. https://doi.org/10.1016/j.catena.2014.08.010

Fernández-Raga M, Palencia C, Keesstra S, Jordán A, Fraile R, Angulo-Martínez M et al (2017) Splash erosion: a review with unanswered questions. Earth Sci Rev 171:463–477. https://doi.org/10.1016/j.earscirev.2017.06.009

Frasson RPM, Krajewski WF (2011) Characterization of the drop-size distribution and velocity–diameter relation of the throughfall under the maize canopy. Agric For Meteorol 151:1244–1251. https://doi.org/10.1016/j.agrformet.2011.05.001

Frasson RPM, Krajewski WF (2013) Rainfall interception by maize canopy: development and application of a process-based model. J Hydrol 489:246–255. https://doi.org/10.1016/j.jhydrol.2013.03.019

Geißler C, Nadrowski K, Kühn P, Baruffol M, Bruelheide H, Schmid B et al (2013) Kinetic energy of throughfall in subtropical forests of SE China – effects of tree canopy structure, functional traits, and biodiversity. PLoS One 8:1–8. https://doi.org/10.1371/journal.pone.0049618

Geißler C, Lang AC, von Oheimb G, Härdtle W, Baruffol M, Scholten T (2012) Impact of tree saplings on the kinetic energy of rainfall-The importance of stand density, species identity and tree architecture in subtropical forests in China. Agric For Meteorol 156:31–40. https://doi.org/10.1016/j.agrformet.2011.12.005

Goebes P, Bruelheide H, Härdtle W, Kröber W, Kühn P, Li Y, Seitz S et al (2015a) Species-specific effects on throughfall kinetic energy in subtropical forest plantations are related to leaf traits and tree architecture. PLoS ONE 10:e0128084. https://doi.org/10.1371/journal.pone.0128084

Goebes P, Seitz S, Geißler C, Lassu T, Peters P, Seeger M et al (2014) Momentum or kinetic energy – how do substrate properties influence the calculation of rainfall erosivity? J Hydrol 517:310–316. https://doi.org/10.1016/j.jhydrol.2014.05.031

Goebes P, Seitz S, Kühn P, Li Y, Niklaus PA, von Oheimb G et al (2015b) Throughfall kinetic energy in young subtropical forests: investigation on tree species richness effects and spatial variability. Agric For Meteorol 213:148–159. https://doi.org/10.1016/j.agrformet.2015.06.019

Goebes P, Schmidt K, Härdtle W, Seitz S, Stumpf F, von Oheimb G et al (2016) Rule-based analysis of throughfall kinetic energy to evaluate biotic and abiotic factor thresholds to mitigate erosive power. Prog Phys Geogr 40:431–449. https://doi.org/10.1177/0309133315624642
Gunn R, Kinzer GD (1949) The terminal velocity of fall for water droplets in stagnant air. J Meteorol 6:243–248. https://doi.org/10.1175/1520-0469(1949)006<0243:TTVOFF>2.0.CO;2
Hanson DL, Steenhuis TS, Walter MF, Boll J (2004) Effects of soil degradation and management practices on the surface water dynamics in the Talgua River Watershed in Honduras. Land Degrad Dev 15:367–381. https://doi.org/10.1002/ldr.603
Herwitz SR, Slye RE (1995) Three-dimensional modeling of canopy tree interception of wind-driven rainfall. J Hydrol 168:205–226. https://doi.org/10.1016/0022-1694(94)02643-P
Jarvis PG, McNaughton KG (1986) Stomatal control of transpiration: scaling up from leaf to region. Adv Ecol Res 15:1–49. https://doi.org/10.1016/S0065-2504(08)60119-1
Joss J, Waldvogel A (1967) Ein Spektrograph für Niederschlagstropfen mit automatischer Auswertung. Pure Appl Geophys 68:240–246. https://doi.org/10.1007/BF00874898
Keim RF, Skaugset AE, Weiler M (2005) Temporal persistence of spatial patterns in throughfall. J Hydrol 314:263–274. https://doi.org/10.1016/j.jhydrol.2005.03.021
Kinnell PIA (1973) The Problem of Assessing the Erosive Power of Rainfall from Meteorological Observations 1. Soil Sci Soc Am J 37:617. https://doi.org/10.1016/j.neurenf.2013.02.002
Krishnapillay B (2000) Silviculture and management of teak plantations. Unasylva 201:14–21
Lal R (1976) Soil erosion on Alfisols in Western Nigeria III. Effects of rainfall characteristics. Geoderma 16:389–401. https://doi.org/10.1016/0016-7061(76)90003-3
Laws JO (1941) Measurements of the fall-velocity of water-drops and raindrops. Trans Am Geophys Union 22:709–721. https://doi.org/10.1029/TR022i003p00709
Levia DF, Hudson SA, Llorens P, Nanko K (2017) Throughfall drop size distributions: a review and prospectus for future research. WIRES Water 4(e):1225. https://doi.org/10.1002/wat2.1225
Levia DF*, Nanko K*, Amasaki H, Giambelluca TW, Hotta N, Iida S, Mudd RG, Nullet MA, Sakai N, Shinohara Y, Sun X, Suzuki M, Tanaka N, Tantasirin C, Yamada K (2019) Throughfall partitioning by trees. Hydrol Process 33:1698–1708. https://doi.org/10.1002/hyp.13432. [*denotes equal contributors]
Li Y, Yu HQ, Zhou N, Tian G, Poesen J, Zhang ZD (2015) Linking fine root and understory vegetation to channel erosion in forested hillslopes of southwestern China. Plant Soil 389:323–334. https://doi.org/10.1007/s11104-014-2362-8
Liu W, Zhu C, Wu J, Chen C (2016) Are rubber-based agroforestry systems effective in controlling rain splash erosion? Catena 147:16–24. https://doi.org/10.1016/j.catena.2016.06.034
Liu J, Liu W, Li W, Jiang X, Wu J (2018a) Effects of rainfall on the spatial distribution of the throughfall kinetic energy on a small scale in a rubber plantation. Hydrol Sci J 63:1078–1090. https://doi.org/10.1080/02626667.2018.1473580
Liu J, Liu W, Zhu K (2018b) Throughfall kinetic energy and its spatial characteristics under rubber-based agroforestry systems. Catena 161:113–121. https://doi.org/10.1016/j.catena.2017.10.014
Löffler-Mang M, Joss J (2000) An optical disdrometer for measuring size and velocity of hydrometeors. J Atmos Ocean Technol 17:130–139. https://doi.org/10.1175/1520-0426(2000)017<0130:AODFMS>2.0.CO;2
Marshall JS, Palmer WMK (1948) The distribution of raindrops with size. J Meteorol 5:165–166. https://doi.org/10.1175/1520-0469(1948)005<0165:TDORWS>2.0.CO;2
Meyer LD, Wischmeier WH (1969) Mathematical simulation of the process of soil erosion by water. Trans ASAE 12:754–758. https://doi.org/10.13031/2013.38945
Miura S, Hirai K, Yamada T (2002) Transport rates of surface materials on steep forested slopes induced by raindrop splash erosion. J For Res 7:201–211. https://doi.org/10.1007/BF02763133
Miura S, Ugawa S, Yoshinaga S, Yamada T, Hirai K (2015) Floor cover percentage determines splash erosion in *Chamaecyparis obtusa* forests. Soil Sci Soc Am J 79:1782–1791. https://doi.org/10.2136/sssaj2015.05.0171
Morgan RPC (2005) Soil erosion and conservation, 3rd edn. Blackwell, Oxford

Moss AJ, Green TW (1987) Erosive effects of the large water drops (gravity drops) that fall from plants. Aust J Soil Res 25:9–20. https://doi.org/10.1071/SR9870009

Nakaya K, Wakamatsu T, Ikeda H, Abe S, Toyoda Y (2011) Development of raindrop kinetic energy model under canopy for the estimation of soil erosion in forest (in Japanese with English summary). CRIEPI Res Rep 2011: V11001

Nanko K, Giambelluca TW, Sutherland RA, Mudd RG, Nullet MA, Ziegler AD (2015) Erosion potential under *Miconia calvescens* stands on the Island of Hawai'i. Land Degrad Dev 26:218–226. https://doi.org/10.1002/ldr.2200

Nanko K, Hotta N, Suzuki M (2004) Assessing raindrop impact energy at the forest floor in a mature Japanese cypress plantation using continuous raindrop-sizing instruments. J For Res 9:157–164. https://doi.org/10.1007/s10310-003-0067-6

Nanko K, Hotta N, Suzuki M (2006) Evaluating the influence of canopy species and meteorological factors on throughfall drop size distribution. J Hydrol 329:422–431. https://doi.org/10.1016/j.jhydrol.2006.02.036

Nanko K, Hudson SA, Levia DF (2016a) Differences in throughfall drop size distributions in the presence and absence of foliage. Hydrol Sci J 61:1–8. https://doi.org/10.1080/02626667.2015.1052454

Nanko K, Mizugaki S, Onda Y (2008a) Estimation of soil splash detachment rates on the forest floor of an unmanaged Japanese cypress plantation based on field measurements of throughfall drop sizes. Catena 72:348–361. https://doi.org/10.1016/j.catena.2007.07.002

Nanko K, Moskalski SM, Torres R (2016b) Rainfall erosivity-intensity relationships for normal rainfall events and a tropical cyclone on the US southeast coast. J Hydrol 534:440–450. https://doi.org/10.1016/j.jhydrol.2016.01.022

Nanko K, Onda Y, Ito A, Moriwaki H (2008b) Effect of canopy thickness and canopy saturation on the amount and kinetic energy of throughfall: an experimental approach. Geophys Res Lett 35: L05401. https://doi.org/10.1029/2007GL033010

Nanko K, Onda Y, Ito A, Moriwaki H (2011) Spatial variability of throughfall under a single tree: experimental study of rainfall amount, raindrops, and kinetic energy. Agric For Meteorol 151:1173–1182. https://doi.org/10.1016/j.agrformet.2011.04.006

Nanko K, Watanabe A, Hotta N, Suzuki M (2013) Physical interpretation of the difference in drop size distributions of leaf drips among tree species. Agric For Meteorol 169:74–84. https://doi.org/10.1016/j.agrformet.2012.09.018

Onda Y, Gomi T, Mizugaki S, Nonoda T, Sidle RC (2010) An overview of the field and modelling studies on the effects of forest devastation on flooding and environmental issues. Hydrol Process 24:527–534. https://doi.org/10.1002/hyp.7548

Pandy D, Brown C (2000) Teak: a global overview. Unasylva 201:3–13

Robson JJ, Neal C, Ryland GPP, Harrow M (1994) Spatial variations in throughfall chemistry at the small plot scale. J Hydrol 158:107–122. https://doi.org/10.1016/0022-1694(94)90048-5

Saint-Jean S, Chelle M, Huber L (2004) Modelling water transfer by rain-splash in a 3D canopy using Monte Carlo integration. Agric For Meteorol 121:183–196. https://doi.org/10.1016/j.agrformet.2003.08.034

Scholten T, Geißler C, Goc J, Kühn P, Wiegand C (2011) A new splash cup to measure the kinetic energy of rainfall. J Plant Nutr Soil Sci 174:596–601. https://doi.org/10.1002/jpln.201000349

Sempere-Torres D, Porra JM, Creutin JD (1994) A general formulation for raindrop size distribution. J Appl Meteorol 33:1494–1502. https://doi.org/10.1175/1520-0450(1994)033<1494:AGFFRS>2.0.CO;2

Shinohara Y, Ichinose K, Morimoto M, Kubota T, Nanko K (2018) Factors influencing the erosivity indices of raindrops in Japanese cypress plantations. Catena 171:54–61. https://doi.org/10.1016/j.catena.2018.06.030

Sidle RC, Ziegler AD, Negishi JN, Nik AR, Siew R, Turkelboom F (2006) Erosion processes in steep terrain – truths, myths, and uncertainties related to forest management in Southeast Asia. For Ecol Manage 224:199–225. https://doi.org/10.1016/j.foreco.2005.12.019

Song Z, Seitz S, Zhu P, Goebes P, Shi X, Xu S et al (2018) Spatial distribution of LAI and its relationship with throughfall kinetic energy of common tree species in a Chinese subtropical forest plantation. For Ecol Manage 425:189–195. https://doi.org/10.1016/j.foreco.2018.05.046

Staelens J, De Schrijver A, Verheyen K, Verhoest NEC (2006) Spatial variability and temporal stability of throughfall water under a dominant beech (*Fagus sylvatica* L.) tree in relationship to canopy cover. J Hydrol 330:651–662. https://doi.org/10.1016/j.jhydrol.2006.04.032

Tanaka N, Levia D, Igarashi Y, Nanko K, Yoshifuji N, Tanaka K et al (2015) Throughfall under a teak plantation in Thailand: a multifactorial analysis on the effects of canopy phenology and meteorological conditions. Int J Biometeorol 59:1145–1156. https://doi.org/10.1007/s00484-014-0926-1

Tangtham N (1992) Soil erosion problem in teak plantation. Proceedings of the seminar on 50th anniversary of Huay-Tak teak plantation: 60th birthday celebration of her Majesty the Queen of Thailand. Royal Forestry Department, Bangkok, pp 247–259

Tashiro-Uchimura Y, Mizunaga H (2017) Dynamics of remaining amount and vertical distribution of a *Cryptomeria japonica* needle litter created by non-commercial thinning (in Japanese with English summary). Jpn J For Environ 59:13–25. https://doi.org/10.18922/jjfe.59.1_13

Terry JP (1996) Erosion pavement formation and slope process interactions in commercial forest plantations, northern Portugal. Zeitschrift für Geomorphol Suppl Issues 40:97–115

Ulbrich CW (1983) Natural variations in the analytical form of the raindrop size distribution. J Clim Appl Meteorol 22:1764–1775. https://doi.org/10.1175/1520-0450(1983)022<1764:NVITAF>2.0.CO;2

van Dijk AIJM, Bruijnzeel LA, Rosewell CJ (2002) Rainfall intensity–kinetic energy relationships: a critical literature appraisal. J Hydrol 261:1–23. https://doi.org/10.1016/S0022-1694(02)00020-3

Wang PK, Pruppacher HR (1977) Acceleration to terminal velocity of cloud and raindrops. J Appl Meteorol 16:275–280. https://doi.org/10.1175/1520-0450(1977)016<0275:ATTVOC>2.0.CO;2

Wischmeier WH, Smith DD (1978) Predicting rainfall erosion. (Agricultural handbook no. 537) United States Department of Agriculture, Washington, DC

Yang X, Madden LV (1993) Effect of ground cover, rain intensity and strawberry plants on splash of simulated raindrops. Agric For Meteorol 65:1–20. https://doi.org/10.1016/0168-1923(93)90035-G

Zhou G, Wei X, Yan J (2002) Impacts of eucalyptus (*Eucalyptus exserta*) plantation on sediment yield in Guangdong Province, Southern China—a kinetic energy approach. Catena 49:231–251. https://doi.org/10.1016/S0341-8162(02)00030-9

Ziegler AD, Fox JM, Xu J (2009) The rubber juggernaut. Science 324(80):1024–1025. https://doi.org/10.1126/science.1173833

Zimmermann A, Zimmermann B (2014) Requirements for throughfall monitoring: the roles of temporal scale and canopy complexity. Agric For Meteorol 189–190:125–139. https://doi.org/10.1016/j.agrformet.2014.01.014

Zimmermann A, Zimmermann B, Elsenbeer H (2009) Rainfall redistribution in a tropical forest: spatial and temporal patterns. Water Resour Res 45:W11413. https://doi.org/10.1029/2008WR007470

Chapter 13
Assessing the Ecological Significance of Throughfall in Forest Ecosystems

William H. McDowell, Katherine X. Pérez-Rivera, and Meaghan E. Shaw

13.1 Introduction

13.1.1 Throughfall: The Hydrologic Context

Throughfall is defined as precipitation that reaches the ground surface following passage through a vegetative canopy, or "the portion of incoming gross precipitation that penetrates or drips through the canopy" (Hewlett 1982; Fig. 13.1). Throughfall is an important hydrologic flow path in many ecosystems. Although shrublands and grasslands produce throughfall, relatively few papers assess this topic (e.g., Vega et al. 2005; Levia and Germer 2015). In this chapter, we will thus focus on the better-studied forest environment, while also recognizing that throughfall may play an equally important role in other ecosystems. The quantity of throughfall (TF) in a forested ecosystem is determined by the balance between precipitation inputs, the production of stemflow (SF), and return of precipitation to the atmosphere due to evaporation following canopy interception (I) (Fig. 13.1; Crockford and Richardson 2000). Stemflow is produced when precipitation is delivered to the forest floor by flow down boles or stems. It is a considerably smaller hydrologic flux than throughfall. For a given precipitation regime, interactions among the quantity and timing of atmospheric inputs, micrometeorology of the site, and architecture of the forest canopy all play a role in determining the quantity of SF, I, and TF that occur during an individual storm or an entire year at a forested site (Crockford and Richardson 2000; Levia et al. 2011).

W. H. McDowell (✉) · K. X. Pérez-Rivera · M. E. Shaw
Department of Natural Resources and the Environment, University of New Hampshire, Durham, NH, USA
e-mail: Bill.McDowell@unh.edu

D. F. Levia et al. (eds.), *Forest-Water Interactions*, Ecological Studies 240,
https://doi.org/10.1007/978-3-030-26086-6_13

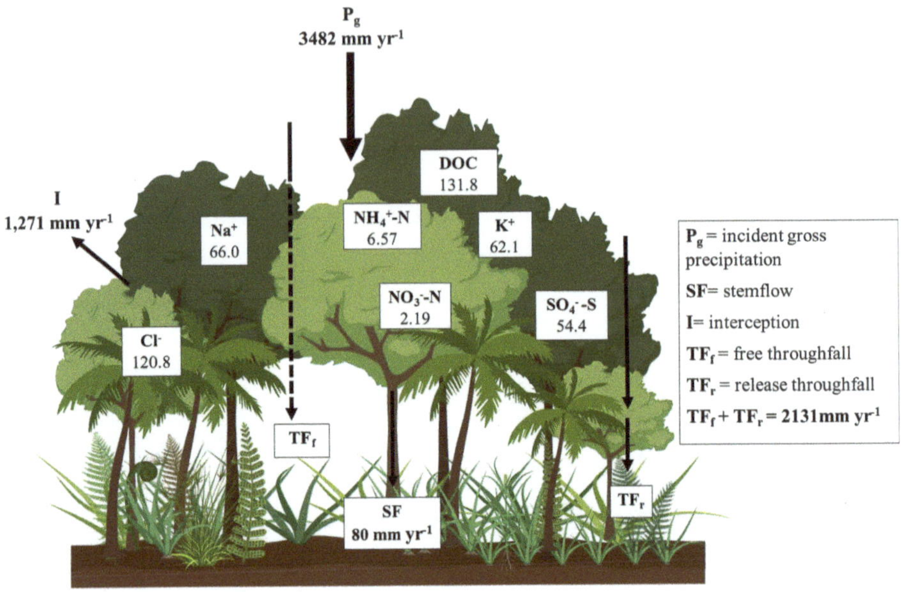

Fig. 13.1 Mean throughfall fluxes for solutes and chemical species (kg ha^{-1} y^{-1}) based on weekly samples in a tropical rainforest during a 15-year period (1988–2002). Data taken from Heartsill-Scalley et al. (2007). Values for incident gross precipitation (P_g), throughfall (TF_f + TF_r), and interception (I) were taken from Heartsill-Scalley et al. (2007) after estimating stemflow (SF) from precipitation (Scatena 1990). This figure was created with BioRender software (BioRender.com)

13.1.2 Mechanisms Generating Throughfall

The processes that result in throughfall generation have been studied in considerable detail. Throughfall can be classified into two types: free and release. Free throughfall refers to precipitation that passes through the canopy without interacting with any vegetation; release throughfall refers to incoming precipitation that is first intercepted and then drips from the plant (Levia and Frost 2006; Fig. 13.1).

In forested ecosystems, most of the incoming precipitation that is transferred from the canopy to the soil is deposited as throughfall (Lawson 1967; Henderson et al. 1977; Cape et al. 1991; Crockford and Khanna 1997; Huber and Iroumé 2001; Levia and Frost 2003; Carlyle-Moses et al. 2004; Lilienfein and Wilcke 2004; Levia and Frost 2006). The amount of precipitation that reaches the forest floor is determined by the canopy coverage, total leaf area, the number of layers of vegetation, tree species, and the intensity as well as the duration of rainfall events (Brooks et al. 2012). The total amount of throughfall produced in an individual storm varies considerably as a function of precipitation received (Fig. 13.2), because once the canopy is wetted and throughfall begins to be produced little additional canopy storage and subsequent evaporation can be expected (Carlyle-Moses and Gash 2011). This principle has been invoked to explain the high interception loss at

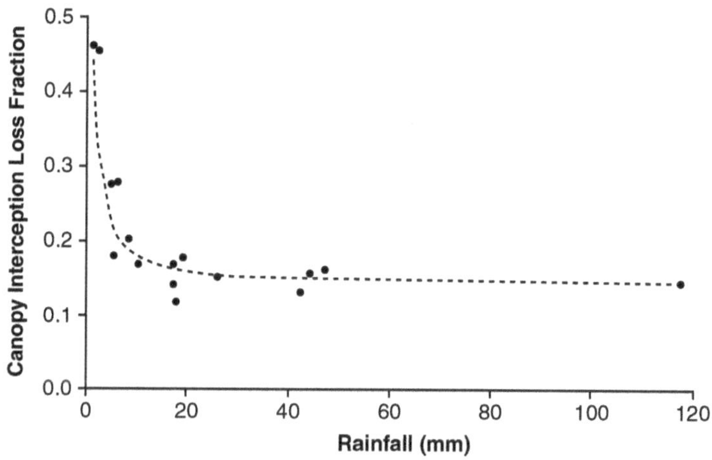

Fig. 13.2 Canopy interception (proportion) as a function of rainfall during a storm (mm). (From Carlyle-Moses and Gash (2011). Reproduced with permission of Springer)

tropical forest sites in Puerto Rico, where steady winds, consistently warm temperatures, and small average size of an individual rainstorm lead to ideal conditions for canopy interception, resulting in a watershed-scale estimate of throughfall that is only 59% of precipitation (Scatena 1990). Because the size and shape of the canopy also directly affect the amount, intensity and distribution of throughfall, it shows high temporal and spatial variability.

13.1.3 Field Collection Methods

Many localized factors within the forest canopy influence the amount of throughfall that is generated within a given forested stand, making quantification of throughfall at the watershed scale challenging. A wide variety of methods have been developed to quantify throughfall and collect it for additional analyses. Usually, throughfall is collected in forests using individual funnels or troughs that are installed suspended above the forest floor in order to limit any overestimation due to splash from the forest floor (Levia and Frost 2006). Both of these collection techniques consist of the same principle, with throughfall collected from a known surface area in order to estimate overall inputs for an entire plot (Levia and Frost 2006). However, the sampling strategy applied and the number of collectors installed are important elements that affect the accuracy of throughfall measurements in forested ecosystems (Helvey and Patric 1965; Lloyd and Marques 1988; Thimonier 1998; Price and Carlyle-Moses 2003; Levia and Frost 2006).

Lloyd and Marques (1988) showed that switching the position of throughfall collectors randomly can increase the accuracy of throughfall estimates compared to those made with collectors at fixed position. Holwerda et al. (2006) compared these two approaches for sampling throughfall (fixed and roving collectors) in a tropical rainforest. They found that both fixed and roving deployments can give similar values, but at least 100 fixed gauges are required to obtain values within the 95% confidence interval of those obtained from 30 roving gauges (Holwerda et al. 2006). Therefore, fewer throughfall collectors are needed when using the roving method. Rotating collectors are not always desirable, however, as they preclude comparisons over time under a given set of canopy conditions, and might limit ability to understand throughfall dynamics and drivers of biogeochemical fluxes when canopy conditions change at a given location (Levia and Frost 2006).

Another aspect of the hydrologic regime that influences the quantity of collectors needed is the magnitude of rainfall during individual precipitation events. During a given storm, as precipitation input increases, variability in the volume of throughfall among individual collectors decreases, meaning that throughfall is more spatially consistent (and a larger proportion of incoming precipitation; Fig. 13.2) when greater inputs of precipitation are entering the canopy (Price and Carlyle-Moses 2003). In fact, Price and Carlyle-Moses (2003) determined that more gauges are needed to adequately sample a precipitation event of less than 2 mm than for one that has an input of >4 mm. Zimmermann et al. (2010) developed a model to evaluate the performance of various sampling schemes. They found that a greater sample size is required to be able to quantify small events, due mainly to increases in the coefficient of variation for throughfall during small events. In general, a longer period of record and as many collectors as is practical are recommended to characterize throughfall quantity adequately (Zimmermann et al. 2010).

13.1.4 Global Variation in Quantity of Forest Throughfall

The structure of the canopy, specifically the arrangement, size and angle of the leaves plays an important role in determining the amount of throughfall that is generated because a greater individual leaf size and canopy surface area results in more interception and less throughfall reaching the forest floor (Brooks et al. 2012). Due to differences in canopy architecture, leaf area index (LAI) as well as physical positioning of leaf area are likely to be important in determining throughfall quantity. The relationship between throughfall and precipitation has been established for a variety of forest types in different parts of the world. These relationships are based on the dependence throughfall has on the canopy surface, area, and cover (Brooks et al. 2012; Table 13.1) and may be best addressed in future studies through the use of LIDAR to quantify canopy structure (e.g., Roth et al. 2007). In regressions that describe the relationship between throughfall and precipitation, the slope typically is near 90%, and depending on the size of canopy storage, the intercept (related to canopy storage) is several mm (Table 13.1). Alteration of the forest canopy by hurricane defoliation (Heartsill-Scalley et al. 2007) or spatial variability in forest canopy structure from other, more subtle causes (Konishi et al. 2006) can lead to increased throughfall and decreased interception. In the case of Puerto Rico,

Table 13.1 Relationship between throughfall and precipitation for different vegetative canopies[a]

Vegetative canopy	Throughfall relationship	Notes
Eastern Hardwood Forest, USA (Helvey and Patric 1965)	Growing Season	T_h is throughfall (in); *P* is total precipitation (in); and n is number of storms
	$T_h = 0.901\ (P) - 0.031\ (n)$	
	Dormant Season	
	$T_h = 0.914\ (P) - 0.015\ (n)$	
Southern Pine Forests, USA (Roth and Chang 1981)	Longleaf Pine	T_h is throughfall (mm) and *P* is total rainfall (mm)
	$T_h = 1.002(P) - 0.0008\ (P)^2 - 1.397$	
	Loblolly Pine	
	$T_h = 0.930\ (P) - 0.0011(P)^2 - 0.610$	
New Zealand Vegetation Communities (Blake 1975)	Kauri Forest	T_h is throughfall (mm) and *P* is total rainfall (mm)
	$T_h = 0.60\ (P) - 3.71$	
	Manuka Shrub	
	$T_h = 0.44\ (P) - 0.10$	
	Mountain Beech Forest	
	$T_h = 0.69\ (P) - 1.90$	

[a]Modified from Brooks et al. (2012)

revegetation returned relationships between P and I to near pre-hurricane levels within a year following Hurricane Hugo in 1989 (Heartsill-Scalley et al. 2007).

13.2 Throughfall as a Biogeochemical Flux

Throughfall chemistry integrates the effects of multiple biogeochemical drivers. These include the chemistry of incoming precipitation, the transfer of materials deposited on leaf surfaces by dry deposition, insect activity, the leaching of intracellular materials from canopy leaves or needles, and the net uptake or release of solutes by epiphytic organisms such as lichens, mosses, bromeliads, and algae. Because throughfall represents the largest internal transfer of water within many forest ecosystems, the solutes that it contains can represent significant biogeochemical fluxes from forest canopy to forest floor (e.g., McDowell 1998).

Fluxes of throughfall are often remarkably enriched in K^+, at levels 10–40 times higher than precipitation (e.g., Table 13.2), and often enriched in dissolved organic carbon (DOC), dissolved organic nitrogen (DON) and NH_4^+ but much more variable in terms of NO_3^- concentrations and fluxes (e.g., Filoso et al. 1999; Hervé-Fernández et al. 2016). The literature contains many examples of specific studies of individual elements or groups of ions in throughfall, such as extensive work on DOC in throughfall (summarized in Table 13.3), but it is relatively sparse in

Table 13.2 Solute fluxes in throughfall (kg ha^{-1} yr^{-1}) at several sites with relatively complete major ion, nutrient, and organic matter fluxes

	Site			
Solute	Beddgelert (Stevens 1987)	Luquillo (Heartsill-Scalley et al. 2007)	Brazil São Paulo (Forti et al. 2005)	Brazil Negro River (Filoso et al. 1999)
NO_3^--N	6.51	2.19	9.09	0.62
NH_4^+-N	4.47	6.57	6.56	0.51
DOC	–	131.8	–	158.5
Na^+	75.8	66.1	12.4	3.83
K^+	15.1	62.1	110	27.7
Ca^{2+}	9.14	20.8	10.6	5.54
Mg^{2+}	9.72	13.1	5.86	2.70
H^+	1.02	0.015	0.01	0.05
SO_4^{2-}-S	26.5	54.4	30.1	1.71
Cl^-	136.1	120.8	61.3	5.38

complete descriptions of major anions, cations, nutrients, and organic matter for individual sites that, taken together, would provide a global understanding of throughfall across biomes. This is a major shortcoming in understanding the biogeochemistry of throughfall at the global scale and leads us to recommend that more comprehensive studies on the full range of throughfall biogeochemistry be undertaken in the future.

13.2.1 Net Uptake and Release of Solutes

Determination of net uptake and release of solutes from the forest canopy as throughfall is generated requires measurement of solute fluxes, which are the product of volume and solute concentrations. Because the quantity of liquid water returned to the atmosphere through evaporation can be substantial (Fig. 13.1), evaporative concentration of solutes can result in an increase in solute concentrations even when canopy interactions result in no net uptake or generation of solutes (Hansen et al. 1994).

Many sources within the forest canopy can contribute to the increases in solute concentrations and fluxes that are frequently observed when comparing precipitation and throughfall. Comparison of deposition (incoming precipitation) and throughfall yields a measurement sometimes referred to as "net throughfall" (Lovett and Lindberg 1984). The concept of net throughfall is particularly useful in addressing the sources of solutes entering a forest ecosystem. In one of the simpler applications of the concept, net throughfall can be used to estimate dry deposition of sea salt aerosols onto the forest canopy by comparing Na^+ and Cl^- fluxes in throughfall and

Table 13.3 Ranges of dissolved organic carbon (DOC) concentrations in throughfall (TF) for different forest types[1]

Forest Type	Range of [DOC] (mg C L^{-1}) in TF	Reference
Tropical rain	6–17	McDowell (1998)
		Tobón et al. (2004)
		Goller et al. (2006)
		Wanek et al. (2007)
		Heartsill-Scalley et al. (2007)
Tropical moist	5–9	Möller et al. (2005)
		Fujii et al. (2009)
		Hofhansl et al.(2012)
Tropical dry and seasonally dry	7–12	Laclau et al.(2003)
		Schrumpf et al.(2006)
		Germer et al. (2007)
Subtropical humid	5–11	Liu and Sheu (2003)
		Wang et al. (2004)
		Jian-fen et al. (2005)
Subtropical steppe	5–11	Ciglasch et al.(2004)
Temperate oceanic	14–17	Moore (1987)
		Solinger et al.(2001)
		Levia et al. (2012)
Temperate continental	9–29	Moore (1987)
		McDowell and Likens (1988)
		Qualls et al. (1991)
		Currie et al. (1996)
		Chang and Matzner (2000)
		Hagedorn et al. (2000)
		Guggenberger and Zech (1994)
Temperate steppe	18–20	Kolka et al. (1999)
Boreal	9–20	Dalva and Moore (1991)
		Moore (2003)
		Turgeon and Courchesne (2008)
Boreal tundra	49–57	Koprivnjak and Moore (1992)

[1]Modified from Van Stan and Stubbins (2018)

precipitation. Using this principle, for example, McDowell (1998) showed that dry deposition of marine aerosols provides a measurable but relatively small increase (15%) in total atmospheric deposition to a tropical forest with considerable rainfall (3.5 m). Because Na^+ and Cl^- are relatively inert under most circumstances (little biotic uptake or release), their quantification in net throughfall provides an effective measure of inputs of marine aerosols and other sources of dry deposition. In a more complex application of the same principle, use of net precipitation was pioneered by Lovett and Lindberg (1984) to estimate dry deposition of atmospheric pollutants with somewhat limited potential for biotic uptake. By measuring SO_4^{2-} fluxes in net throughfall, they inferred that dry deposition of SO_4^{2-} aerosols provided a significant input to forest ecosystems in much of the eastern USA and provided very important early estimates of the magnitude of dry deposition as a component of acidic atmospheric deposition (Ollinger et al. 1993). This use of net throughfall to estimate dry deposition is problematic, however, for more biologically active elements such as nitrogen. Lovett and Lindberg (1984) concluded that canopy interactions precluded use of net throughfall N concentrations to estimate dry deposition of nitric acid. Net throughfall below surrogate surfaces (artificial trees; Davidson et al. 1985) provides another way to estimate dry deposition, although little research on artificial collector surfaces has been conducted recently.

13.2.2 *Spatial and Temporal Variation in Throughfall Chemistry*

The chemical composition of throughfall varies considerably among biomes (Lovett and Lindberg 1984; Stevens 1987; Ollinger et al. 1993; Lovett et al. 1996; Li et al. 1997; Filoso et al. 1999; Liu et al. 2002; Tobón et al. 2004). A variety of factors influence throughfall chemistry, including dry deposition of various solutes to the forest canopy, intracellular foliar uptake or release, uptake or release by epiphytic canopy communities, and insect infestations. Dry deposition affects throughfall chemistry primarily when aerosol particles settle on canopy leaves and are subsequently solubilized in throughfall (Lovett and Lindberg 1984; McDowell 1998). Canopy exchange processes include both active exchange across the cell walls of the leaves or needles that make up the forest canopy and the uptake or release of solutes by associated epiphytic communities. Foliar uptake of nutrients (Wittwer and Teubner 1959) and nutrient loss from leaf tissue (Tukey Jr 1970) were both recognized over 50 years ago as important processes that can affect throughfall chemistry, but the importance of epiphytic processes has only been widely documented more recently (e.g., Clark et al. 1998; Filoso et al. 1999). Because of the large effects of epiphytes on soluble organic matter, Van Stan and Pypker (2015) hypothesized that canopy epiphytes may play a particularly important role in the organic matter cycles of forest ecosystems. Insect infestations have also been recognized as contributing to variation in throughfall chemistry, with particularly large effects on DOC and DON fluxes. Insect infestations can damage leaves or result in the deposition of waste materials on leaf surfaces that can contribute to throughfall solute fluxes. Le Mellec et al. (2011), in a study of throughfall in oak stands, found that fluxes of DOC increased by a factor of 2.5 with insect infestation, resulting in a flux of 166 kg/ha during the growing season. This flux is similar to or higher than those observed on an annual basis in wet tropical forests (Table 13.2).

Spatial variation in the two primary regulatory processes (dry deposition and canopy exchange) varies by geographic location, extent of pollution sources, and biome. Indeed, even variations within a forest can have an influence on the chemical composition of throughfall. In forests with a wide range of canopy species such as tropical rainforests, the chemical composition of throughfall can differ between neighboring transects in an otherwise similar environment. In the Rio Negro basin of Brazil, for example, phosphate flux was more than double in the stand with the highest flux, compared to the stand with the lowest, apparently due to differences in canopy composition among the stands (Filoso et al. 1999). For base cations the range among stands was lower, with differences of 10–25% depending on the cation. Other small-scale spatial variation in throughfall has been attributed to differential exposure to pollution sources. In Japan, Chiwa et al. (2003) found that pine forest sites with exposure to urban atmospheric inputs had higher solute fluxes, while nearby sites facing away from the urban area had lower NO_3^-, NH_4^+, and SO_4^{2-} net throughfall deposition.

Throughfall solute concentrations can vary during an individual storm as well as by season, providing potential insights into the drivers of temporal variation in throughfall chemistry and elemental fluxes. One of the most thorough examinations of variation in throughfall concentrations during storms found that the concentrations of most ions peaked early in a storm, as sources (dry deposition, internal pools of readily leachable solutes) tended to become depleted following the initial flushing associated with the start of a rainstorm (Hansen et al. 1994). Seasonal variation in throughfall chemistry has been described in the Rio Negro basin, Brazil by Filoso et al. (1999). They found that dissolved NO_3^- in throughfall was lower than incoming precipitation during the wet season but not during the dry season, suggesting enhanced foliar or epiphytic uptake of nitrate during the wet season. In contrast, the concentration of NH_4^+ in throughfall was higher than the concentration in precipitation during the dry season, suggesting the possibility of higher rates of dry deposition of ammonium aerosols during the wet season (Filoso et al. 1999). In Puerto Rico, seasonality in solute concentrations has been reported for rain (McDowell et al. 1990; Gioda et al. 2011, 2013), with highest NO_3 concentrations in April and May, but throughfall at the site shows few seasonal patterns (Heartsill-Scalley et al. 2007). More long-term studies are needed to assess the seasonality of throughfall chemistry in multi-year studies across the globe.

Sampling interval can affect throughfall chemistry, as solute concentrations can change while samples are held in collection vessels prior to removal and processing for chemical analysis (Thimonier 1998). Biologically active solutes such as nutrients or organic matter, and the solutes that can be leached from particulate matter (dust, pollen) are susceptible to potential changes due to collection frequency. The influence of collection frequency on the chemical composition of throughfall samples is variable, with reports of both increases and decreases in the concentrations of various solutes (reviewed by Thimonier 1998). Early assessments of the stability of bulk and wet-only deposition chemistry reported little influence of collection interval on chemical concentrations (Galloway and Likens 1976; Madsen 1982; Thimonier 1998) but the influence of collection interval on throughfall chemistry is much less well documented. Van der Mass and Valent (1989) found an increase in the concentration of ammonium when throughfall samples were collected at longer intervals, which they attributed to mineralization of organic matter in the collector prior to sample retrieval. Decreases in the concentration of throughfall ammonium as a function of collection interval have also been observed, and pH has been observed to both increase and decrease with collection interval (Liechty and Mroz 1991; Ferm 1993; Thimonier 1998). Routine sampling programs for throughfall are often weekly (Heartsill-Scalley et al. 2007), although biweekly (e.g., US National Ecological Observatory Network) and monthly (e.g., Oulehle et al. 2017) collection intervals have been employed. Interpretation of throughfall chemistry should include assessment of the possible impacts of collection interval on solute concentrations, particularly in regions where acid deposition is not driving throughfall chemistry to low pH values.

13.2.3 Dissolved Organic Matter in Throughfall

Dissolved organic matter (DOM) is a particularly important and enigmatic solute flux in forest throughfall. On a mass basis, it represents one of the dominant solutes in throughfall. Even when measured by its carbon content alone (as DOC), which explicitly underestimates the overall solute load found in DOM by excluding the nitrogen, hydrogen, and oxygen content of organic compounds, DOC is typically dominant (e.g., Table 13.2). Little of the DOC in throughfall, which can range up to an average of 57 mg L^{-1} at individual sites (Table 13.3), is from precipitation, which globally averages only about 1–2 mg L^{-1} (Willey et al. 2000). The recent compilation of throughfall DOC concentrations by Van Stan and Stubbins (2018) shows a strong latitudinal trend in DOC concentrations, suggesting that there are fundamental differences in the production of throughfall from the tropics to the boreal zone. This broad global generalization includes many different forest types, as well as latitudinal patterns in the length of the growing season, precipitation, and temperature. All of these co-varying factors should be investigated in order to understand what is driving spatial variability in this important carbon flux among forest ecosystems. Biome-scale patterns may also be obscured by regional factors. In a study of similar Canadian northern hardwoods forest communities, DOC concentration was 30% higher at a northern than a southern site (in keeping with the global pattern) but was attributed to higher rainfall in the southern site, rather than any difference in DOC flux as a function of latitude (Liechty et al. 1995).

Compared to the other dominant solutes in throughfall (Na^+, Cl^-, or SO_4^{2-}), DOC is the only major solute contributed almost entirely by biotic processes, as high SO_4^{2-} is usually from dry deposition of anthropogenic pollutants (e.g., Lovett and Lindberg 1984) and Na^+ and Cl^- are largely from marine aerosols when found in high concentrations in precipitation (e.g., McDowell et al. 1990). Whether from leaf leaching, aphid excretion, insect feces, or epiphytic sources rich in DOC such as tank bromeliads (Richardson et al. 2000), DOC in throughfall is derived almost exclusively from sources that originate in the forest canopy, rather than materials deposited on the canopy by atmospheric processes. Potassium is the only other solute in throughfall that is almost exclusively from vegetative sources (rather than deposition) and is thought to be derived largely from leaching of intracellular sources within the canopy (Tukey Jr 1970; Hansen et al. 1994). Examination of possible mechanistic links between production of K^+ and organic solutes in throughfall has not yet been undertaken.

13.3 Ecological Impacts of Throughfall

13.3.1 Effects on Forest Floor Respiration, Greenhouse Gas Production, and Microbial Activity

It is difficult to disentangle the effects of throughfall (water and solutes) from the effects of water alone on the biogeochemistry of the forest floor. Observational studies can only be used to make inferences about the possible roles of water, nutrients, or organic matter that are all delivered to the forest floor by throughfall. Multiple experiments have been conducted to examine the effects of both solute additions (e.g., NITREX and other nitrogen saturation experiments; Lamersdorf et al. 1998; Aber et al. 1998) and water additions or removals (e.g., Lamersdorf et al. 1998; Nepstad et al. 2007) on plant growth and forest floor processes. The design of most experimental additions or removals of throughfall rarely considers the role of water alone, and thus it is difficult to untangle the effects of added water and solutes on any forest floor response. Typical responses to manipulations of incoming throughfall or water input in tropical forests include modest changes in soil greenhouse gas fluxes in a wet tropical forest subjected to additional water input (Hall et al. 2013), reduced tree growth in Amazonian forest subjected to decreased water inputs (Nepstad et al. 2007), and increased greenhouse gas production when additional water input was provided in a seasonally dry Amazonian forest (Vasconcelos et al. 2004). These manipulative experiments suggest that the spatial and temporal heterogeneity of throughfall might drive hot spots and hot moments of biogeochemical fluxes in the forest floor (McClain et al. 2003; Siegert et al. 2017; Van Stan and Stubbins 2018).

In temperate forests, Kalbitz et al. (2007) found that doubling the amount of throughfall that reaches the forest floor of a Bavarian forest affects the microbial transformation of mineral nitrogen and production of DON. They found that when the amount of throughfall increased long-term, there was a significant decline in the concentrations of DOC and DON in the forest floor (Kalbitz et al. 2007). This experiment did not separate the importance of water fluxes and other biogeochemical inputs on the production of dissolved organic matter (DOM). Another long-term throughfall manipulation shows that significant changes in deep rooting can occur during a 12-year manipulation (Johnson et al. 2008). Johnson et al. (2002)

studied the effects of manipulating throughfall fluxes on soil leaching by decreasing (−33%) or increasing (+33%) total throughfall. By contrasting these treatments to the unmanipulated control, they found that nitrogen fluxes in soil solution increased in the wet treatment and decreased in the dry treatment (Johnson et al. 2002). Ion concentrations increased in the dry treatment but were unchanged in the wet treatment. These differences were attributed primarily to the differences in soil water flux with increasing depth (Johnson et al. 2002). After 12 years of throughfall manipulation, however, relatively few long-term biogeochemical changes resulted from either of the experimental treatments (Johnson et al. 2008).

13.4 Evolutionary Implications of Throughfall Biogeochemistry

The loss of nutrients and organic matter from canopy foliage and other components of the canopy ecosystem such as epiphytic bromeliads has uncertain implications for the evolution of forest vegetation and forest canopies. The measurement of throughfall biogeochemistry at the stand level has been driven largely by consideration of the impacts of atmospheric pollution and the extent to which vegetation takes up, responds to, or transforms atmospheric pollutants such as acidic deposition (e.g., Lovett and Lindberg 1984). This focus emphasizes the important role of throughfall as a responsive component of the forest ecosystem that can buffer the underlying forest floor from various inputs.

Early work on throughfall in the flux of DOC through forest ecosystems (e.g., McDowell and Likens 1988) did emphasize the importance of throughfall as a distinctive flux of organic matter with characteristic functional groups (more carbohydrates, less phenolics) than those fluxes deeper in the solum, but the implications of this distinctive chemical signature have not been fully explored. Throughfall can be metabolized as an energy source by microbes, as was established decades ago by seminal papers such as Qualls and Haines (1992). Nevertheless, there has been a surprising lack of investigation into the specific physiological role that throughfall may play in fueling or inhibiting microbial communities and seedling success on the forest floor (Levia et al. 2017), although some research groups have addressed this topic recently (Bischoff et al. 2015).

When compared to the role that other organic compounds are known to play in the ecology of forest communities and ecosystems, the lack of information on the function and evolutionary significance of organic matter in throughfall is striking. Decades of research have focused on the role of volatile compounds that are released by forest canopies under attack by herbivores (Baldwin and Schultz 1983; Dicke and Baldwin 2010) and that appear to serve as a form of communication among plants. To what extent does dissolved organic matter production in throughfall represent simply an unfortunate waste of energy, or purposeful release of organic compounds that drive a specific response in the forest floor? The example of juglone provides some evidence that throughfall release of DOC may be more than an accident. The throughfall and foliage of black walnut trees, *Juglans nigra* L., are known to release the compound juglone in throughfall, which is phytotoxic to all but black walnut seedlings (Gabriel 1975; Von Kiparski et al. 2007). These early observations of what is termed "allelopathy" have led to dozens of studies of plant interactions, which typically provide experimental evidence of the phytotoxic or insecticidal properties of various foliage. Although originally published as a contribution to the field of chemical ecology, interest in the topic has now contributed to explosive growth in the field of alternative agriculture and organic farming using natural plant defense mechanisms (Macías et al. 2019). In a similar vein, allelopathy is now being used in attempts to control harmful algal blooms through the use of allelochemicals (Tan et al. 2019). Despite this successful translation from forest ecology to organic agriculture and aquatic ecology, understanding of potential allelopathy and the ecological role of organic matter in throughfall has lagged behind. The strong latitudinal variation in throughfall DOC concentrations (Table 13.3) suggests that there may be different drivers controlling DOC leaching from plant canopies among biomes, with uncertain implications for ecosystem function and evolution of forest plants.

13.5 Future Directions

Future research on the role of throughfall in forested ecosystems should focus in three primary areas: temporal dynamics, elemental interactions, and functional significance. These focal areas are not meant to be mutually exclusive, and in fact a robust research portfolio on throughfall should include all three. Our review shows that there is remarkably little long-term data on throughfall chemistry. Throughfall biogeochemistry is likely to change as a function of changing climate, changing rates and nature of atmospheric deposition, and changing stand age or composition. Despite these well-recognized changes in the forest environment, throughfall response to long-term drivers has typically been limited to sites documenting dramatic changes in atmospheric deposition (e.g., Oulehle et al. 2017) and forest dynamics in response to major canopy disturbance by hurricanes (e.g., Heartsill-Scalley et al. 2007). Other more subtle or gradual changes that might be occurring as a result of drivers such as increased atmospheric CO_2 concentrations or warming remain unexplored. Three of the best-studied forest ecosystems in the USA, for example, the Hubbard Brook, Coweeta, and HJ Andrews Experimental Forests, have ongoing hydrological and biogeochemical measurements of forest function, but no long-term sampling of throughfall chemistry.

Understanding the elemental interactions that drive biogeochemical fluxes from the forest canopy is another area that has received relatively little study in the past few decades. After a flurry of intensive work on the effects of acidic atmospheric deposition including some studies that spanned several decades (e.g., Matzner and Meiwes 1994), interest in the biogeochemistry of throughfall has waned at most sites in Europe and North America. In regions where acidic deposition has declined, significant opportunities now exist for assessing the fundamental controls on production and alteration of the most important solutes generated by canopy processes (DOC, DON, and K^+). With declines in acidic deposition, before-after studies might provide useful insights into the interactions among these solutes as well as their response to changes in acidity. Similarly, in regions now undergoing extensive acidification of deposition due to increased industrial output, an emphasis on understanding how fundamental changes in canopy biogeochemistry occur during acidification is warranted. To what extent are production of DOC, DON, and K^+ responding to similar or different drivers across a range of acidic deposition? Is there any evidence that production of these solutes is linked physiologically? Another impetus for focusing on the organic matter in throughfall is the widespread increase in DOC in some surface waters in northeastern USA and western Europe (e.g., Gavin et al. 2018, Monteith et al. 2007). Understanding the importance of throughfall as a potential source of DOC to surface waters should take on new importance with the ongoing increases in DOC now observed at a wide range of sites.

Generating new insights into the ecological and evolutionary significance of throughfall, especially the role of specific organic compounds in throughfall, is now possible with recent advances in analytical instrumentation and growing interest in DOC over the last thirty years. Recent work in Germany shows that additional analytical chemistry might help to provide insights into the role of DOC in throughfall. Bischoff et al. (2015) studied beech and spruce throughfall with solid-state ^{13}C nuclear magnetic resonance spectroscopy with cross-polarization and magic-angle spinning (CPMAS-NMR). They found large differences by species, with much greater proportions of phenolic and presumably recalcitrant molecules in beech throughfall and suggested that this production of aromatic materials might have implications for allelopathic interactions with other tree species. They also make the case that fine particulate organic matter (0.45 to 500 μm) should be studied, as very little is known about its origin or ecological significance in throughfall. Given the extensive body of literature related to allelopathy in crops, ecosystem ecologists should be examining insights from this literature to better understand controls on forest community structure, soil respiration, microbial greenhouse gas production from the forest floor, and carbon accumulation in forest soils.

References

Aber J, McDowell W, Nadelhoffer K, Magill A, Berntson G, Kamakea M et al (1998) Nitrogen saturation in temperate forest ecosystems-hypotheses revisited. Bioscience 48:921–934. https://doi.org/10.2307/1313296

Baldwin I, Schultz J (1983) Rapid changes in tree leaf chemistry induced by damage: evidence for communication between plants. Science 221:227–280. https://doi.org/10.1126/science.221.4607.277

Bischoff S, Schwarz MT, Siemens J, Thieme L, Wilcke W, Michalzik B (2015) Properties of dissolved and total organic matter in throughfall, stemflow and forest floor leachate of central European forests. Biogeosciences 12:2695–2706. https://doi.org/10.5194/bg-12-2695-2015

Blake GJ (1975) The interception process. In: Chapman TG, Dunin FX (eds) Prediction in catchment hydrology, national symposium on hydrology. Melbourne, Australian Academy of Science, pp 59–81

Brooks KN, Ffolliott PF, Magner JA (2012) Hydrology and the management of watersheds, 4th edn. Wiley-Blackwell, Ames

Cape JN, Brown AHF, Robertson SMC, Howson G, Paterson IS (1991) Interspecies comparisons of throughfall and stemflow at three sites in northern Britain. For Ecol Manage 46:165–177. https://doi.org/10.1016/0378-1127(91)90229-O

Carlyle-Moses DE, Gash JHC (2011) Rainfall interception loss by forest canopies. In: Levia DF, Carlyle-Moses DE, Tanaka T (eds) Forest hydrology and biogeochemistry, Ecological studies (Analysis and Synthesis), vol 216. Springer, Dordrecht, pp 407–423. https://doi.org/10.1007/978-94-007-1363-5_20

Carlyle-Moses DE, Laureano JSF, Price AG (2004) Throughfall and throughfall spatial variability in Madrean oak forest communities of northeastern Mexico. J Hydrol 297:124–135. https://doi.org/10.1016/j.jhydrol.2004.04.007

Chang SC, Matzner E (2000) The effect of beech stemflow on spatial patterns of soil solution chemistry and seepage fluxes in a mixed beech/oak stand. Hydrol Process 14:135–144. https://doi.org/10.1002/(SICI)1099-1085(200001)14:1<135::AID-HYP915>3.0.CO;2-R

Chiwa M, Kim D, Sakugawa H (2003) Rainfall, stemflow, and throughfall chemistry at urban-and mountain-facing sites at Mt. Gokurakuji Hiroshima, Western Japan. Water Air Soil Pollut 146:93–109. https://doi.org/10.1023/A:1023946603217

Çiglasch H, Lilienfein J, Kaiser K, Wilcke W (2004) Dissolved organic matter under native Cerrado and *Pinus caribaea* plantations in the Brazilian savanna. Biogeochemistry 67:157–182. https://doi.org/10.1023/B:BIOG.0000015281.74705.f8

Clark KL, Nadkarni NM, Schaefer D, Gholz HL (1998) Atmospheric deposition and net retention of ions by the canopy in a tropical montane forest. J Trop Ecol 14:27–45. https://doi.org/10.1017/S0266467498000030

Crockford RH, Khanna PK (1997) Chemistry of throughfall, stemfall and litterfall in fertilized and irrigated *Pinus radiata*. Hydrol Process 11:1493–1507. https://doi.org/10.1002/(sici)1099-1085(199709)11:11<1493::aid-hyp475>3.3.co;2-g

Crockford RH, Richardson DP (2000) Partitioning of rainfall into throughfall, stemflow and interception: effect of forest type, ground cover and climate. Hydrol Process 2903–2920. https://doi.org/10.1002/1099-1085(200011/12)14:16/17<2903::AID-HYP126>3.0.CO;2–6

Currie WS, Aber JD, McDowell WH, Boone RD, Magill AH (1996) Vertical transport of dissolved organic C and N under long-term N amendments in pine and hardwood forests. Biogeochemistry 35:471–505. https://doi.org/10.1007/BF02183037

Dalva M, Moore TR (1991) Sources and sinks of dissolved organic carbon in a forested swamp catchment. Biogeochemistry 15:1–19. https://doi.org/10.1007/BF00002806

Davidson CI, Lindberg SE, Schmidt JA, Cartwright LG, Landis LR (1985) Dry deposition of sulfate and nitrate onto surrogate surfaces. J Geophys 90:2123–2130. https://doi.org/10.1029/JD090iD01p02123

Dicke M, Baldwin I (2010) The evolutionary context for herbivore-induced plant volatiles: beyond the "cry for help.". Trends Plant Sci 15:167–175. https://doi.org/10.1016/j.tplants.2009.12.002

Ferm M (1993) Throughfall measurements of nitrogen and sulphur compounds. Int J Environ Anal Chem 50:29–43. https://doi.org/10.1080/03067319308027581

Filoso S, Williams MR, Melack JM (1999) Composition and deposition of throughfall in a flooded forest archipelago (Negro River, Brazil). Biogeochemistry 45:169–195. https://doi.org/10.1023/A:1006108618196

Forti MC, Bicudo DC, Bourotte C, de Cicco V, Arcova FCS (2005) Rainfall and throughfall chemistry in the Atlantic Forest: a comparison between urban and natural sites (São Paulo State, Brazil). Hydrol Earth Syst Sci 9:570–585. https://doi.org/10.5194/hess-9-570-2005

Fujii JL, Uemura M, Hayakawa C, Funakawa S, Sukartiningsih KT et al (2009) Fluxes of dissolved organic carbon in two tropical forest ecosystems of East Kalimantan, Indonesia. Geoderma 152:127–136. https://doi.org/10.1016/j.geoderma.2009.05.028

Gabriel WJ (1975) Allelopathic effects of black walnut on white birches. J For 73:234–237. https://doi.org/10.1093/jof/73.4.234

Galloway JN, Likens GE (1976) Calibration of collection procedures for the determination of precipitation chemistry. Water Air Soil Pollut 6:241–258. https://doi.org/10.1007/BF00182868

Gavin AL, Nelson SJ, Klemmer AJ, Fernandez IJ, Strock KE, McDowell WH (2018) Acidification and climate linkages to increased dissolved organic carbon in high-elevation lakes. Water Resour Res 54:5376–5393. https://doi.org/10.1029/2017WR020963

Germer S, Neill C, Krusche AV, Gouveia Neto SC, Elsenbeer H (2007) Seasonal and within-event dynamics of rainfall and throughfall chemistry in an open tropical rainforest in Rondônia, Brazil. Biogeochemistry 86:155–174. https://doi.org/10.1007/s10533-007-9152-9

Gioda A, Reyes-Rodríguez GJ, Santos-Figueroa G, Collett JL Jr, Decesari S, Ramos MCKV et al (2011) Speciation of water-soluble inorganic, organic, and total nitrogen in a background marine environment: cloud water, rainwater, and aerosol particles. J Geophys Res Atmos 116. https://doi.org/10.1029/2010JD015010

Gioda A, Mayol-Bracero OL, Scatena FN, Weathers KC, Mateus VL, McDowell WH (2013) Chemical constituents in clouds and rainwater in the Puerto Rican rainforest : potential sources and seasonal drivers. Atmos Environ 68:208–220. https://doi.org/10.1016/j.atmosenv.2012.11.017

Goller R, Wilcke W, Fleischbein K, Valarezo C, Zech W (2006) Dissolved nitrogen, phosphorous, and sulfur forms in the ecosystem fluxes of a montane forest in Ecuador. Biogeochemistry 77:57–89. https://doi.org/10.1007/s10533-005-1061-1

Guggenberger G, Zech W (1994) Composition and dynamics of dissolved carbohydrates and lignin-degradation products in two coniferous forests, N.E. Bavaria, Germany. Soil Biol Biochem 26:19–27. https://doi.org/10.1016/0038-0717(94)90191-0

Hagedorn F, Schleppi P, Waldner P, Flühler H (2000) Export of dissolved organic carbon and nitrogen from Gleysol dominated catchments – the significance of water flow paths. Biogeochemistry 50:137–161. https://doi.org/10.1023/A:1006398105953

Hall SJ, McDowell WH, Silver WL (2013) When wet gets wetter: decoupling of moisture, redox biogeochemistry, and greenhouse gas fluxes in a humid tropical forest soil. Ecosystems 16:576–589. https://doi.org/10.1007/s10021-012-9631-2

Hansen K, Draaijers G, Ivens W, Gundersen P, van Leeuwen N (1994) Concentration variations in rain and canopy throughfall collected sequentially during individual rain events. Atmos Environ 28:3195–3205. https://doi.org/10.1016/1352-2310(94)00176-L

Heartsill-Scalley T, Scatena FN, Estrada C, McDowell WH, Lugo AE (2007) Disturbance and long-term patterns of rainfall and throughfall nutrient fluxes in a subtropical wet forest in Puerto Rico. J Hydrol 333:472–485. https://doi.org/10.1016/j.jhydrol.2006.09.019

Helvey JD, Patric JH (1965) Canopy and litter interception of rainfall by hardwoods of eastern United States. Water Resour Res 1:193–206. https://doi.org/10.1029/WR001i002p00193

Henderson GS, Harris WF, Todd DE, Grizzard T (1977) Quantity and chemistry of throughfall as influenced by forest-type and season. J Ecol 65:365–374. https://doi.org/10.2307/2259488

Hervé-Fernández P, Oyarzún C, Woelfl S (2016) Throughfall enrichment and stream nutrient chemistry in small headwater catchments with different land cover in southern Chile. Hydrol Process 30:4944–4955. https://doi.org/10.1002/hyp.11001

Hewlett JD (1982) Principles of forest hydrology. University of Georgia Press, Athens

Hofhansl F, Wanek W, Drage S, Huber W, Weissenhofer A, Richter A (2012) Controls of hydrochemical fluxes via stemflow in tropical lowland rainforests: effects of meteorology and vegetation characteristics. J Hydrol 452–453:247–258. https://doi.org/10.1016/j.jhydrol.2012.05.057

Holwerda F, Scatena FN, Bruijnzeel LA (2006) Throughfall in a Puerto Rican lower montane rain forest: A comparison of sampling strategies. J Hydrol 327:592–602. https://doi.org/10.1016/j.jhydrol.2005.12.014

Huber A, Iroumé A (2001) Variability of annual rainfall partitioning for different sites and forest covers in Chile. J Hydrol 248:78–92. https://doi.org/10.1016/S0022-1694(01)00394-8

Jian-fen G, Yu-Sheng Y, Guang-shi C, Peng L (2005) Dissolved organic carbon and nitrogen in precipitation, throughfall and stemflow from *Schima superba* and *Cunninghamia lanceolata* plantations in subtropical China. J For Res 16:19–22. https://doi.org/10.1007/BF02856847

Johnson DW, Hanson PJ, Todd DE (2002) The effects of throughfall manipulation on soil leaching in a deciduous forest. J Environ Qual 31:204–216. https://doi.org/10.2134/jeq2002.0204

Johnson DW, Todd Jr. DE, Hanson PJ (2008) Effects of throughfall manipulation on soil nutrient status: results of 12 years of sustained wet and dry treatments. Global Change Biology 14:1661–1675. https://doi.org/10.1111/j.1365-2486.2008.01601.x

Kalbitz K, Meyer A, Yang R, Gerstberger P (2007) Response of dissolved organic matter in the forest floor to long-term manipulation of litter and throughfall inputs. Biogeochemistry 86:301–318. https://doi.org/10.1007/s10533-007-9161-8

Kolka RK, Nater EA, Grigal DF, Verry ES (1999) Atmospheric inputs of mercury and organic carbon into a forested upland/bog watershed. Water Air Soil Pollut 113:273–294. https://doi.org/10.1023/A:1005020326683

Konishi S, Tani M, Kosugi Y, Takanashi S, Sahat MM, Nik AR et al (2006) Characteristics of spatial distribution of throughfall in a lowland tropical rainforest, Peninsular Malaysia. For Ecol Manage 224:19–25. https://doi.org/10.1016/j.foreco.2005.12.005

Koprivnjak J-F, Moore TR (1992) Sources, sinks, and fluxes of dissolved organic carbon in subarctic fen catchments. Arct Alpine Res 24:204–210. https://doi.org/10.2307/1551658

Laclau J-P, Ranger J, Bouillet J-P, de Dieu NJ, Deleporte P (2003) Nutrient cycling in a clonal stand of *Eucalyptus* and an adjacent savanna ecosystem in Congo 1. Chemical Composition of rainfall, throughfall and stemflow solutions. For Ecol Manage 176:105–119. https://doi.org/10.1016/S0378-1127(02)00280-3

Lamersdorf N, Beier C, Blanck K, Bredemeier M, Cummins T, Farrell EP et al (1998) Effect of drought experiments using roof installations on acidification/nitrification of soils. For Ecol Manage 101:95–109. https://doi.org/10.1016/S0378-1127(97)00128-X

Lawson ER (1967) Throughall and stemflow in a pine-hardwood stand in the Ouachita Mountains of Arkansas. Water Resour Res 3:731–735. https://doi.org/10.1029/WR003i003p00731

le Mellec A, Gerold G, Michalzik B (2011) Insect herbivory, organic matter decomposition and effects on belowground organic matter fluxes in a central European oak forest. Plant Soil 342:393–403. https://doi.org/10.1007/s11104-010-0704-8

Levia DF, Frost EE (2003) A review and eveluation of stemflow literature in the hydrologic and biogeochemical cycles of forested and agricultural ecosystems. J Hydrol 274:1–29. https://doi.org/10.1016/S0022-1694(02)00399-2

Levia DF, Frost EE (2006) Variability of throughfall volume and solute inputs in wooded ecosystems. Prog Phys Geogr 30:605–632. https://doi.org/10.1177/0309133306071145

Levia DF, Germer S (2015) A review of stemflow generation-environment interactions in forests and shrublands. Rev Geophys 53:673–714. https://doi.org/10.1002/2015RG000479

Levia DF, Keim RF, Carlyle-Moses DE, Frost EE (2011) Throughfall and stemflow in wooded ecosystems. In: Levia DF, Carlyle-Moses DE, Tanaka T (eds) Forest hydrology and biogeochemistry, Ecological studies (Analysis and Synthesis), vol 216. Springer, Dordrecht, pp 425–443. https://doi.org/10.1007/978-94-007-1363-5_21

Levia DF, Van Stan IIJT, Inamdar SP, Jarvis MT, Mitchell MJ, Mage SM et al (2012) Stemflow and dissolved organic carbon cycling: temporal variability in concentration, flux, and UV-Vis spectral metrics in a temperate broadleaved deciduous forest in the eastern United States. Can J For Res 42:207–216. https://doi.org/10.1139/x11-173

Levia DF, Hudson SA, Llorens P, Nanko K (2017) Throughfall drop size distributions: a review and prospectus for future research. WIRES Water 4:e1225. https://doi.org/10.1002/wat2.1225

Li YC, Alva K, Calvert DV, Zhang M (1997) Chemical composition of throughfall and stemflow from citrus canopies. J Plant Nutr 20:1351–1360. https://doi.org/10.1080/01904169709365339

Liechty HO, Mroz GD (1991) Effects of collection interval on quality of throughfall samples in two northern hardwood stands. J Environ Qual 20:588–591. https://doi.org/10.2134/jeq1991.00472425002000030014x

Liechty HO, Kuuseoks E, Mroz GD (1995) Dissolved organic carbon in northern hardwood stands with differing acidic inputs and temperature regimes. J Environ Qual 24:927–933. https://doi.org/10.2134/jeq1995.00472425002400050021x

Lilienfein J, Wilcke W (2004) Water and element product input into native, agri- and silvicultural ecosystems of Brazilian savanna. Biogeochemistry 67:183–212. https://doi.org/10.1023/b:biog.0000015279.48813.9d

Liu CP, Sheu BH (2003) Dissolved organic carbon in precipitation, throughfall, stemflow, soil solution, and stream water at the Guandaushi subtropical forest in Taiwan. For Ecol Manage 172:315–325. https://doi.org/10.1016/S0378-1127(01)00793-9

Liu W, Fox JED, Xu Z (2002) Nutrient fluxes in bulk precipitation, throughfall and stemflow in montane subtropical moist forest on Ailao Mountains in Yunnan, south-west China. J Trop Ecol 18:527–548. https://doi.org/10.1017/S0266467402002353

Lloyd CR, Marques DO (1988) Spatial variability of throughfall and stemflow measurements in Amazonian rainforest. Agr Forest Meteorol 42:63–73. https://doi.org/10.1016/0168-1923(88)90067-6

Lovett GM, Lindberg SE (1984) Dry deposition and canopy exchange in a mixed oak forest as determined by analysis of throughfall. J Appl Ecol 21:1013–1027. https://doi.org/10.2307/2405064

Lovett GM, Nolan SS, Driscoll CT, Fahey TJ (1996) Factors regulating throughfall flux in a New Hampshire forested landscape. Can J For Res 26:2134–2144. https://doi.org/10.1139/x26-242

Macías F, Mejías F, Molinillo J (2019) Recent advances in allelopathy for weed control: from knowledge to applications. Pest Manag Sci: Early view. https://doi.org/10.1002/ps.5355

Madsen BC (1982) An evaluation of sampling interval length on the chemical composition of wet-only deposition. Atmos Environ 16:2515–2519. https://doi.org/10.1016/0004-6981(82)90143-3

Matzner E, Meiwes KJ (1994) Long-term development of element fluxes with bulk precipitation and throughfall in two German forests. J Env Qual 23:162–166. https://doi.org/10.2134/jeq1994.00472425002300010025x

McClain ME, Boyer EW, Dent LC, Gergel SE, Grimm NB, Groffman PM et al (2003) Biogeochemical hot spots and hot moments at the interface of terrestrial and aquatic ecosystems. Ecosystems 6:301–312. https://doi.org/10.1007/s10021-003-0161-9

McDowell WH (1998) Internal nutrient fluxes in a Puerto Rican rain forest. J Trop Ecol 14:521–536. https://doi.org/10.1017/S0266467498000376

McDowell WH, Likens GE (1988) Origin, composition, and flux of dissolved organic carbon in the Hubbard Brook valley. Ecol Monogr 58:177–195. https://doi.org/10.2307/2937024

McDowell WH, Sanchez CG, Asbury CE, Perez CRR (1990) Influence of sea salt aerosols and long-range transport on precipitation chemistry at El Verde, Puerto Rico. Atmos Environ 24:2813–2821. https://doi.org/10.1016/0960-1686(90)90168-M

Möller A, Kaiser K, Guggenberger G (2005) Dissolved organic carbon and nitrogen in precipitation, throughfall, soil solution, and stream water of the tropical highlands in northern Thailand. J Plant Nutr Soil Sci 168:649–659. https://doi.org/10.1002/jpln.200521804

Monteith DT, Stoddard JL, Evans CD, de Wit HA, Forsius M, Høgåsen T et al (2007) Dissolved organic carbon trends resulting from changes in atmospheric deposition chemistry. Nature 450:537–540. https://doi.org/10.1038/nature06316

Moore TR (1987) An assessment of a simple spectrophotometric method for the determination of dissolved organic carbon in freshwaters. New Zeal J Mar Freshw Res 21:585–589. https://doi.org/10.1080/00288330.1987.9516262

Moore TR (2003) Dissolved organic carbon in a northern boreal landscape. Global Biogeochem Cycles 17:1109. https://doi.org/10.1029/2003GB002050

Nepstad DC, Tohver IM, Ray D, Moutinho P, Cardinot G (2007) Mortality of large trees and lianas following experimental drought in an Amazon forest. Ecology 88:2259–2269. https://doi.org/10.1890/06-1046.1

Ollinger SV, Aber JD, Lovett GM, Millham SE, Lathrop LG, Ellis JM (1993) A spatial model of atmospheric deposition for the Northeastern U.S. Ecol Appl 3:459–472. https://doi.org/10.2307/1941915

Oulehle F, Chuman T, Hruška J, Krám P, McDowell WH, Myška O et al (2017) Recovery from acidification alters concentrations and fluxes of solutes from Czech catchments. Biogeochemistry 132:251–272. https://doi.org/10.1007/s10533-017-0298-9

Price AG, Carlyle-Moses DE (2003) Measurement and modelling of growing-season canopy water fluxes in a mature mixed deciduous forest stand, southern Ontario, Canada. Agr Forest Meteorol 119:69–85. https://doi.org/10.1016/S0168-1923(03)00117-5

Qualls R, Haines B (1992) Biodegradability of dissolved organic matter in forest throughfall, soil solution, and stream water. Soil Sci Society Am J 56:578–586. https://doi.org/10.2136/sssaj1992.03615995005600020038x

Qualls RG, Haines BL, Swank WT (1991) Fluxes of dissolved organic nutrients and humic substances in a deciduous forest. Ecology 72:254–266. https://doi.org/10.2307/1938919

Richardson BA, Richardson MJ, Scatena FN, McDowell WH (2000) Effects of nutrient availability and other elevational changes on bromeliad populations and their invertebrate communities in a humid tropical forest in Puerto Rico. J Trop Ecol 16:167–188. https://doi.org/10.1017/S0266467400001346

Roth FA II, Chang M (1981) Throughfall in planted stands of four southern pine species in east Texas. J Am Water Resour Assoc 17:880–885. https://doi.org/10.1111/j.1752-1688.1981.tb01312.x

Roth BE, Slatton KC, Cohen MJ (2007) On forest the potential rainfall ecosystems for high-resolution interception estimates lidar in to improve. Front Ecol Environ 5:421–428. https://doi.org/10.1890/060119.1

Scatena FN (1990) Watershed scale rainfall interception on two forested watersheds in the Luquillo Mountains of Puerto Rico. J Hydrol 113:89–102. https://doi.org/10.1016/0022-1694(90)90168-W

Schrumpf M, Zech W, Lehmann J, Lyaruu HVC (2006) TOC, TON, TOS and TOP in Rainfall, Throughfall, Litter Percolate and Soil Solution of a Montane Rainforest Succession at Mt. Kilimanjaro, Tanzania. Biogeochemistry 78:361–387. https://doi.org/10.1007/s10533-005-4428-4

Siegert CM, Levia DF, Leathers DJ, Van Stan JT, Mitchell MJ (2017) Do storm synoptic patterns affect biogeochemical fluxes from temperate deciduous forest canopies? Biogeochemistry 132:273–292. https://doi.org/10.1007/s10533-017-0300-6

Solinger S, Kalbitz K, Matzner E (2001) Controls on the dynamics of dissolved organic carbon and nitrogen in a Central European deciduous forest. Biogeochemistry 55:327–349. https://doi.org/10.1023/A:1011848326013

Stevens PA (1987) Throughfall chemistry beneath Sitka spruce of four ages in Beddgelert Forest, North Wales, UK. Plant Soil 101:291–294. https://doi.org/10.1007/BF02370658

Tan K, Huan Z, Ji R et al (2019) A review of allelopathy on microalgae. Microbiology. https://doi.org/10.1099/mic.0.000776

Thimonier A (1998) Measurement of atmospheric deposition under forest canopies: Some recommendations for equipment and sampling design. Environ Monit Assess 52:353–387. https://doi.org/10.1023/A:1005853429853

Tobón C, Sevink J, Verstraten JM (2004) Solute fluxes in throughfall and stemflow in four forest ecosystems in northwest Amazonia. Biogeochemistry 70:1–25. https://doi.org/10.1023/B:BIOG.0000049334.10381.f8

Tukey H Jr (1970) The leaching of substances from plants. Annu Rev Plant Physiol 21:305–324. https://doi.org/10.1146/annurev.pp.21.060170.001513

Turgeon JML, Courchesne F (2008) Hydrochemical behaviour of dissolved nitrogen and carbon in a headwater stream of the Canadian Shield: relevance of antecedent soil moisture conditions. Hydrol Process 22:327–339. https://doi.org/10.1002/hyp.6613

van der Mass MP, Valent A (1989) In situ conservation of throughfall samples. Air Pollut Rep Ser 21:137

Van Stan J, Pypker T (2015) A review and evaluation of forest canopy epiphyte roles in the partitioning and chemical alteration of precipitation. Sci Total Environ 536:813–824. https://doi.org/10.1016/j.scitotenv.2015.07.134

Van Stan JT, Stubbins A (2018) Tree-DOM: dissolved organic matter in throughfall and stemflow. Limnol Oceanogr Lett 3:199–214. https://doi.org/10.1002/lol2.10059

Vasconcelos SS, Zarin DJ, Capanu M, Littell R, Davidson EA, Ishida FY et al (2004) Moisture and substrate availability constrain soil trace gas fluxes in an eastern Amazonian regrowth forest. Global Biogeochemical Cycles 18(2):GB2009. https://doi.org/10.1029/2003GB002210

Vega JA, Fernández C, Fonturbel T (2005) Throughfall, runoff and soil erosion after prescribed burning in gorse shrubland in Galicia (NW Spain). Land Degrad Dev 16:37–51. https://doi.org/10.1002/ldr.643

Von Kiparski GR, Lee LS, Gillespie AR (2007) Occurrence and fate of phytotoxin juglone in alley soils under black walnut trees. J Environ Qual 36:709–717. https://doi.org/10.2134/jeq2006.0231

Wanek W, Hofmann J, Feller IC (2007) Canopy interactions of rainfall in an offshore mangrove ecosystem dominated by Rhizophora mangle. J Hydrol 345:70–79. https://doi.org/10.1016/j.jhydrol.2007.07.012

Wang MC, Liu CP, Sheu BH (2004) Characterization of organic matter in rainfall, throughfall, stemflow, and streamwater from three subtropical forest ecosystems. J Hydrol 289:275–285. https://doi.org/10.1016/j.jhydrol.2003.11.026

Willey J, Kieber R, Eyman M, Avery JG (2000) Rainwater dissolved organic carbon: Concentrations and global flux. Glob Biogeochem Cycl 14:139–148. https://doi.org/10.1029/1999GB900036

Wittwer SH, Teubner FG (1959) Foliar absorption of mineral nutrients. Annu Rev Plant Physiol 10:13–30. https://doi.org/10.1146/annurev.pp.10.060159.000305

Zimmermann B, Zimmermann A, Lark RM, Elsenbeer H (2010) Sampling procedures for throughfall monitoring: a simulation study. Water Resour Res 46:W01503. https://doi.org/10.1029/2009WR007776

Chapter 14
Root-Water Relations and Interactions in Mixed Forest Settings

Anke Hildebrandt

14.1 Introduction

Besides light and nutrients, water is one of the most important resources that determines ecosystem productivity (Reichstein et al. 2013). Most plants obtain their water from belowground and root-water relations are essential for maintaining a positive carbon balance in most ecosystems. Forests are long-living ecosystems with potentially substantial structural diversity, a combination of different species, sizes and ages, that harbor a variety of habitats at comparatively small scales within a heterogeneous environment (Puettmann et al. 2009). Longer-lived organisms have a special capacity for creating rooting systems with sufficient lateral as well as vertical extent (Schenk and Jackson 2002) to bridge that heterogeneity which allows complex belowground interactions.

The changing climate and expected shifts in precipitation regimes pose challenges to forest management, and requires to aim toward adaptive, more drought-resilient, and potentially more diverse forest structures (Forrester and Bauhus, 2016). Intelligent decision-making on forest management as well as understanding the effect of climate change on forest ecosystems requires basic understanding on belowground water-related interactions in mixed forests, such as the effect of specific assemblies or diversity per se on drought vulnerability (Broedel et al. 2017; Manoli et al. 2017), shifts in competitive advantages (Goisser et al. 2016; Vitali et al. 2018), and expected productivity gains or losses (Gutsch et al. 2015).

A. Hildebrandt (✉)
Helmholtz Centre for Environmental Research – UFZ, Leipzig, Germany

Institute of Geoscience, Friedrich Schiller University Jena, Jena, Germany

German Centre for Integrative Biodiversity Research (iDiv) Halle-Jena-Leipzig, Leipzig, Germany
e-mail: anke.hildebrandt@ufz.de

D. F. Levia et al. (eds.), *Forest-Water Interactions*, Ecological Studies 240,
https://doi.org/10.1007/978-3-030-26086-6_14

Table 14.1 Overview of keyword search (October 2018) results using the indicated search terms in the Web of Science. Hits refer to the actual number of indexed references returned. Validated hits are based on additional inspection, removing all references mainly dealing with modeling and having their actual focus in a different field (like erosion, urban hydrology, field crops)

Search item	Hits	Validated hits
Forest AND "root water relation"	0	
Forest AND "root water uptake"	91	
Forest AND "root water uptake" AND (experiment OR observ∗ or meas∗)	75	40
Forest AND drought	11,003	
Forest AND drought AND soil	4,264	
Forest AND drought AND "soil water"	2,897	
Forest AND drought AND root∗	1,548	
Forest AND drought AND root∗ (transpiration or water use)	403	
Forest AND "soil water"	6,178	
Forest AND "soil water" AND root∗	1,404	
Forest AND "soil water" AND root∗ AND (transpiration OR "water use")	427	
Forest AND hydrol∗	12,308	
Forest AND hydrol∗ AND soil	6,398	
Forest AND hydrol∗ AND soil AND root∗	743	
Forest AND hydrol∗ AND soil AND root∗ AND (transpiration OR uptake OR "water use" OR "evapo∗")	314	
Forest AND hydrol∗ AND soil AND root∗ AND (transpiration OR uptake OR "water use" OR "evapo∗") AND (experiment OR observ∗ OR meas∗)	219	95

Knowledge of belowground processes and specifically on the capacity of ecosystems to access water is also crucial for models assessing land surface-atmosphere exchange and vegetation dynamics (Smithwick et al. 2014), like transpiration and carbon assimilation, in hydrological and climate prediction (Kuhl et al. 2018), drainage patterns for hydrological prediction (Guswa 2012), or forest management decision-making (Pretzsch et al. 2015). The quality of such assessments hinges upon faithful representation of flow paths through the soil-plant-atmosphere continuum. In forests, this continuum is characterized by substantial internal heterogeneity above and below the surface, which also may propagate to affect water fluxes at larger scales (i.e., Guswa 2012; Hentschel et al. 2013).

Compared to aboveground processes, belowground interactions are generally much more challenging to observe. This is arguably one of the primary reasons why our knowledge on belowground interactions is still comparatively small despite their acknowledged central importance. A Web of Science search (summarized in Table 14.1) for *forest AND "root water uptake"* yielded surprisingly few results (91), while a search for *forest AND "drought"* yielded an amazing 11,000 hits, which shrunk to roughly 4300 and 2300 by additionally including the search terms *soil* or *"soil water"*, respectively. Including *root∗ AND "water use" OR transpiration* decreased the initially substantial number to only 400, although one would think that those terms are strongly related to drought processes. This might either suggest that a great deal of published drought research on forests is unrelated to hydrology or has been conducted aboveground. A second search on *forest AND hydrol***, filtered for "soil" yields very similar results, suggesting the accuracy of the second explanation. This illustrates not only the greater focus on the more accessible aboveground parts of forests, but may also suggest it provides a great deal of information on soil hydrological processes. When narrowing down the search further to observational studies and hand-selecting to exclude modeling studies and those mainly related to other processes, like nutrient uptake, the number of available references is comparatively small (only 40 in the case of root water uptake).

This review chapter aims at collecting the advances made on understanding root-water relations and interactions in mixed forests across disciplines. It starts with a short recap of the process of root water uptake at the individual plant level and connects it to aboveground and other plant functions. Next, it reviews methods for assessing root-water relations. The last part deals with root water uptake at the ecosystem scale, as well as a discussion of neighborhood interactions and their relation with basic ecological functions, such as productivity and resilience, as well as their effects on hydrologic patterns. It closes with a short outlook.

14.2 Background: Below- and Aboveground Aspects of Root Water Uptake and Abiotic Limits

14.2.1 The Process of Root Water Uptake

Root systems supply the transpiration flux, which results from the loss of leaf water through stomata during carbon uptake. It is therefore almost impossible to consider root water uptake independently of the remainder of the plant hydraulic system and the boundary conditions imposed by the soil and the atmosphere (Ivanov et al. 2012; Brum et al. 2019). In order for the leaf to remain hydrated and functional, the lost water needs to be replaced. The water flows along the water potential gradient from the soil (Fig. 14.1, No. 1, least negative potential) though the rhizosphere (Fig. 14.1, No. 2), which is the plant-influenced space around the roots, radially into the roots and is transported along the xylem conduit system from the root to the leaf (Fig. 14.1, No. 4), where the water potential is the most negative. At the plant scale, the water flux between soil and the leaf is often summarized in the following conceptual equation (i.e., Wullschleger et al. 1998; Brodribb 2009; David et al. 2016):

$$E_T = K_p(\psi_L - \psi_S) \tag{14.1}$$

where E_T refers to the transpirational water loss in $m^3\ s^{-1}$, K_p to the total plant hydraulic conductivity (in $m^2\ s^{-1}$) between the soil and the leaf, and ψ_L and ψ_s to the leaf and soil water potentials (given here in meters of water column). The water potential includes several components, most importantly potential energy, capillary suction, and osmotic potential. All have negative signs indicating that the potential is

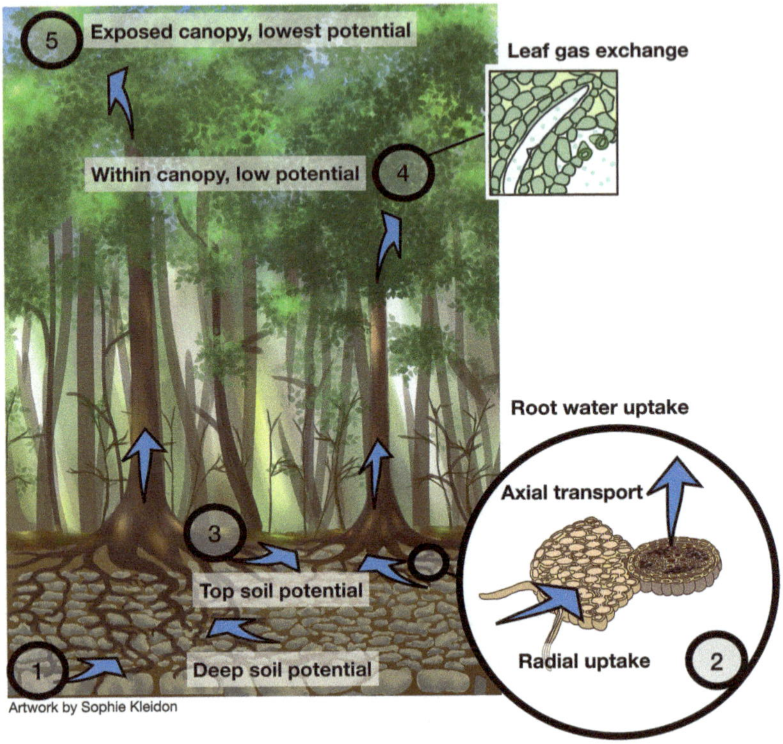

Fig. 14.1 Flow path from the bulk soil to the atmosphere for over and understorey trees with coordination of above- and belowground hydraulic traits in a forest community. (1) Water uptake from deep storage, potentially including rock; (2) cross-section of a root, illustrating the porous pipe, with radial and axial flow paths; (3) hydraulically redistributed water by deep rooted vegetation to the upper soil may benefit shallow rooting species; (4) leaf cross-section with water vapor release during carbon uptake; (5) radiation-exposed tall trees shading the lower canopy. Inlets image source: Modified from Taiz and Zeiger (2002), Sinauer Associates

lower than that of free water. The values drop to be more negative from the unsaturated soil (some meters of suction) to the atmosphere (several 1000 of meters of suction). The potentials within the xylem itself are in between these values, which means typically far below the water vapor pressure (Tyree and Zimmermann 2002). The water uptake path and physiology of xylem vessels are adapted in multi-faceted ways to both withstand collapse and avoid cavitation and embolism (see below, Tyree and Sperry 1989).

The conceptual Eq. 14.1 shows the greatest spread in water potentials along the soil-plant pathway, which is the driver for the overall flow along a connected water pipeline from the root surface to the leaf stomata (Tyree and Sperry 1989). Stomata are considered the greatest resistors along that flow path (Brodribb 2009), operating to avoid xylem cavitation (Sperry et al. 1998). Along the entire pathway, however, a multitude of mechanisms exist that affect the overall plant hydraulic conductivity.

Given that they act in series, each segment may become limiting (Hildebrandt et al. 2016).

Within the plant hydraulic system, roots act like porous pipes (Zwieniecki et al. 2003), which both transport water and take up more along the way. Within an individual root segment, the hydraulic conductance is separated into the axial and radial parts (Landsberg and Fowkes 1978) (Fig. 14.1, No. 2). The axial pathway is responsible for transporting the water within and along the root system to the collar. Its hydraulic conductance depends on the number and size of the xylem conduits and is smallest in the youngest roots (Vetterlein and Doussan 2016) where the xylem vessels only start to be developed (Steudle and Peterson 1998). The radial pathway leads from the root surface toward the xylem vessels. Root radial resistance increases during root maturation as the cells surrounding the root xylem develop (Steudle and Peterson 1998). This leads to root tips having the highest radial, but lowest axial conductance, and makes the apical parts of the root system prominent places for water uptake (Rewald et al. 2011). On the other hand, mature roots with lower radial and higher axial conductance function rather as transport roots. Along the entire root system, the radial conductance is much smaller than the axial one (Frensch and Steudle 1989). Radial conductance can, however, be increased by activating aquaporins, which act as water channels shortcutting the most resistant root tissue (Steudle and Peterson 1998). Aquaporins can be activated relatively quickly, which theoretically allows the plant to regulate the radial uptake area within the root system (Johnson et al. 2014).

Using the porous pipe concept allows one to evaluate the location of water uptake within the root system (Bechmann et al. 2014). The process is nicely explained by Zwieniecki et al. (2003), making reference to a perforated drinking straw in a martini. The ratio of the radial and axial conductance determines uptake location along the porous pipe. When the root radial conductance is high compared to the axial one (many holes in the straw), most of the uptake occurs near the root collar (the top of the straw), while it extends longer along the straw when the radial conductance is lower (fewer holes in the straw) (Zwieniecki et al. 2003), which is also comparable to dry soil. This leads to water uptake occurring preferentially near the root collar (or top soil) in well-watered conditions, but moves down and away to more apical parts of the root system as the soil dries, thus creating "moving uptake fronts". Observations show uptake shifting between shallow and deep soil, depending on water availability (Schwendenmann et al. 2015; Broedel et al. 2017; Brinkmann et al. 2019). The swiftness of the process suggests that this relies mainly on basic root infrastructure and less so on root growth. Also, at any given moment, large parts of the root system are not involved in root water uptake (Hodge et al. 2009), as they are activated sequentially.

Furthermore, the porous pipe concept explains the phenomenon of hydraulic redistribution, by which water is redistributed via the root system from wet to dry soil locations (Neumann and Cardon 2012). In doing so, hydraulic redistribution is directed to equilibrate the water potential differences within the soil with the root system acting as a shortcut (see below).

14.2.2 Abiotic Limits

The most prevalent factor limiting root water uptake is lack of or limited access to soil water. Although waterlogging and soil water salinity are also prominent causes for decreases in transpiration and root water uptake, these are special situations and are not considered here.

At the ecosystem scale, the water balance equations reads:

$$W_{max}\frac{dS}{dt} = n_{PAW}\cdot z_r\frac{dS}{dt} = I - D - E_{ET} \quad (14.2)$$

where W_{max} is the maximal available soil water storage, which is the product of the plant accessible soil porosity (n_{PAW}, unitless or m^3 water per m^3 soil) and the rooting depth (z_R in m). The plant accessible soil porosity encompasses the intermediate pore size classes, which are still accessible to the roots, but small enough to retain water against gravitation. S is the unitless saturation, referenced to range from 0 to 1 between residual water content and field capacity. The right-hand side describes the fluxes (all in m m s^{-1}) in and out of the rooting zone: I is the infiltration and D the drainage out of the root zone. E_{ET} is the total evapotranspiration consisting of soil evaporation and transpiration, and is dominated by transpiration in forest environments (Miralles et al. 2011). Transpiration depends foremost on the atmospheric demand and leaf area, but can be further be adjusted, most importantly by stomatal regulation.

According to Equation (14.2), the soil water storage acts as a buffer supplying water to satisfy atmospheric demand for evaporation between precipitation events. The total plant available water storage depends both on hydraulic soil properties (reflected by n_{PAW}), as well as the rooting depth (z_R). Both are coordinated with the abiotic environment and affected plant traits.

Rooting depth coordinates with climate and soil conditions. Roots should extend deep enough to withstand a regularly reoccurring water deficit (Collins and Bras 2007; Guswa 2010), but not be deeper (Milly and Dunne 1994) in order to avoid inefficient investment. Research on the optimal rooting depth stems mainly from modeling based on optimal carbon balances (e.g., Kleidon and Heimann 1998; Collins and Bras 2007; Schymanski et al. 2008; Guswa 2010) but agrees well with observations on global patterns on root extent (Schenk and Jackson 2002). Typically, at the optimum, productive water loss (transpiration) would be maximized and the unproductive one (evaporation, drainage) minimized (Hildebrandt and Eltahir 2007; Collins and Bras 2007). Rooting depth relative to canopy size is expected to increase in climates with a strong seasonal cycle in water availability, which agrees well with observations (Canadell et al. 1996; Schenk and Jackson 2002). For example, very deep roots have not only been reported from schlerophyllous forests, but also from the Amazon (Canadell et al. 1996). More shallow rooting is reported from temperate environments, where evaporation is energy limited and deeper roots would not increase transpiration.

Rooting depth can, however, be confined to less than the climatic optimum due to weak penetrability of the subsurface. Root growth entails a small displacement of the solid material (Aravena et al. 2014), which is strongly hampered in compacted soils or bedrock. Penetrating solid material requires substantial root pressure, which is arguably not sufficient in most plant species to penetrate rocks or extremely fine pores (Brantley et al. 2017). Smaller pores can then only be accessed by limited adaption in root morphology (Zwieniecki and Newton 1995). Regardless, some bedrock formations have substantial storage capacity, which have repeatedly been shown to be tapped by plants (Schwinning 2010). Mycorrhizal fungi are suggested as key facilitators, allowing rock penetration and water uptake (Allen 2007; Brantley et al. 2017). Their finer hyphae enter small rock fractures inaccessible to roots (Allen 2007), and at the same time decrease tortuosity. Additionally, mycorrhizae are weathering pioneers and are expected to enhance both the storage and accessibility of the bedrock with time (Brantley et al. 2017).

The prevalence of water uptake from bedrock still needs to be tested and the magnitude of uptake determined. For example, Heilman et al. (2009) found that a shallow karst bedrock was no source for root water uptake in a drought setting, despite its storage potential. Also, Hassler et al. (2018) found that geology was one of the most important factors determining tree transpiration along a geological gradient. Finally, patchy access to groundwater (depending on the type of rock or the depth to groundwater) has been suggested as a reason for patchy drought mortality (Barbeta et al. 2015). In a cross-cutting study, O'Brien et al. (2017) determined soil depth and texture as important factors affecting drought mortality patterns (next to plant traits) and potentially a cause of the substantial within-species variation. This suggests that penetrability of the subsurface is an important abiotic limit of root water uptake.

Next to rooting depth, the plant available soil porosity (n_{paw}) determines the overall plant water storage (Eq. 14.2). Low soil water holding capacity (that is low n_{paw}) may be due to a high drainage capacity associated with larger pores, e.g., in coarse textures. Alternatively, a high proportion of small pores (due to fine texture or compaction) retains water so strongly that it cannot be removed by roots, also limiting n_{PAW}. Lower water-holding and higher drainage capacity of coarse soils would need to be compensated by a greater rooting depth in order to allow access to the same soil water storage (Collins and Bras 2007; Hildebrandt and Eltahir 2007). Overall, both vertical and horizontal gradients in soil hydraulic properties have been shown to affect to root water uptake and productivity (Hacke et al. 2000a; Markewitz et al. 2010; Hertel et al. 2013; Kirchen et al. 2017).

Additionally, n_{PAW} can also be affected by the plant, most prominently by adjustments in root length density. Besides soil water retention, also soil hydraulic conductivity changes when the soil dries between precipitation events. With only smaller and smaller pores still filled with water, the cross-sectional area for water flow decreases and friction increases (i.e., Hillel 1998). This has an effect on the small-scale water flow toward individual roots. Soil hydraulic conductivity drops below the radial conductance of the root cortex in drier soils (Draye et al. 2010), and this small-scale process may represent a second impediment of root water access (Sperry et al. 1998). Technically, this problem can be compensated for by a plant increasing its root length density (more root length per soil volume) and therefore decreasing the distance for soil water flow toward the root. This compensation was, for example, observed along a precipitation gradient, where mature beech (*Fagus sylvatica*) root biomass increased from wet to dry sites. Additionally, roots were stretched longer per biomass (higher specific root area) in sandy soils, where the hydraulic conductivity drop per loss of water content is the steepest (Hertel et al. 2013).

Overall, plant available storage is affected by abiotic factors that coordinate with plant traits. That is, on the one hand, rooting depth, a plant trait, adjusts with abiotic factors, namely climate conditions and root penetrable soil depth. On the other hand, plant accessible porosity, n_{PAW}, depends foremost on the mostly abiotic soil water retention function but can be modified by plant traits, i.e., rooting density as well as drought resistance (capacity of plants to take up water from dry soil; see below).

14.2.3 Resulting Interaction of Above- and Belowground Functions at the Individual Scale

Leaf carbon uptake in plants relies delicately on sustained root water uptake. Therefore, above- and belowground traits need to be coordinated and adapted to specific environmental variations in order to maintain a connected flow path between root and leaf during prolonged dry periods, and to avoid xylem cavitation (Choat et al. 2012). Coordinated hydraulic adaptations at the individual plant scale are targeted at maintaining a positive carbon balance and avoiding desiccation and risk of death.

One way to avoid cavitation is by enhancing cavitation resistance (Tyree and Zimmermann 2002; Zanne and Falster 2010). It is noteworthy that this comes with lower vessel hydraulic conductance, requiring more vessels (Höllta et al. 2011) and eventually higher wood density (Poorter et al. 2010). This higher carbon investment is for the long term as, once built, the properties of the non-living xylem vessels cannot easily be changed (Brodribb 2009). Besides building xylem with higher cavitation resistance, more dynamic regulation is achieved either by controlling the transpiration loss aboveground by stomata (Sperry et al. 1998) or by adjusting shoot and/or belowground biomass (Bréda et al. 2006; David et al. 2016).

Stomatal regulation strategy changes between species, with some species operating stomata opening to maintain constant leaf water potentials even in varying environmental conditions (isohydric behavior), while others allow leaf water potentials to vary substantially as soil water status changes (anisohydric behavior) (Tardieu and Simonneau 1998). Interestingly, most trees operate stomata with very limited safety margins to cavitation (Choat et al. 2012). Thus, while cavitation resistance per se changes with climate (decreasing with higher mean annual precipitation; observations by Choat et al. 2012, agreeing with theory by Sperry et al. 1998), the remainder of the flow path organization seems to be highly optimized and suggests that control mechanisms or additional investments are only applied when necessary.

Considering individual life forms, specifically a more water-conserving or water-spending transpiration strategy, also affects the individual optimal rooting depth.

Modeling studies (Guswa 2010; Ivanov et al. 2012) suggest that deeper roots might be associated with a more water-conserving transpiration strategy (more anisohydric), whereas shallow roots may be more efficient for a drought-resistant strategy. This agrees with observations (e.g., Hacke et al. 2000b; Brum et al. 2019), which indicate that rooting depth is related to cavitation resistance and numerous other publications indicating that deeper rooting is associated with drought avoidance (i.e., Maeght et al. 2013; David et al. 2016; O'Brien et al. 2017). On the other hand, validating those models with hydrological data has been challenging (Speich et al. 2017), and a meta-analysis suggests that in the extreme case drought mortality is strongly related to hydraulic traits, especially the safety margin to cavitation (Anderegg et al. 2016) and less strongly to rooting depth (O'Brien et al. 2017).

Root available water can also be adapted via tree spacing, or increasing root spread compared to canopy size in environments requiring greater water storage, such as more arid climates or in coarser texture soils (Casper et al. 2003; Hacke et al. 2000a; Hertel et al. 2013). For example, lower plant available porosity can be compensated by shifting root to shoot ratios, that is, increasing root length (Sperry et al. 1998) or rooting depth (Hacke et al. 2000a) at the expense of leaf area. At the same time, this decreases water demand and increases the apparent hydraulic conductivity (see above). Those plastic adaptations illustrate the strong coordination of the above- and belowground plant organs with soil conditions.

Taken together, coordination between above- and belowground function is omnipresent, potentially revealing a multitude of different strategies to adapt to the specific environment. Unfortunately, many studies investigating, e.g., belowground water foraging do not consider aboveground data relating to transpiration demand such as leaf area index, stand density, or canopy position. Considering the above- and belowground traits together is a promising way forward to gaining a mechanistic understanding on belowground interactions, such as those recently demonstrated by Brum et al. (2019).

14.3 Assessing Root Water Uptake

Most of the available process understanding described in the preceding section relies either on research at the individual plant or the entire ecosystem scale. However, especially for the most pressing questions regarding climate change mitigation or assessing the consequences of biodiversity loss, adaptation and interaction effects are essential (Forrester and Bauhus 2016). For example, understanding which species or functional composition promotes ecosystem functions such as carbon sequestration or resilience relies on information on belowground water foraging at the individual plant scale. However, assessing root water uptake patterns is challenging, not only because the subsurface is difficult to access, but also because of the limitations of techniques applicable in field conditions (Rothfuss and Javaux 2017). Thus, there is no single preferable method available. This section gives a short introduction into common methods used for assessing water foraging at a sufficiently small scale to evaluate individual uptake strategies of trees and resulting interactions.

14.3.1 Indirect Assessment of Water Sourcing Based on Root Abundance and Root Morphology

A great deal of studies investigating forest species interactions for resources, including water, rely on root morphology parameters, which relate to the potential of root water uptake. Those include rooting depth, root length, root length density, root surface area, as well as lateral extent, branching and root growth (Atkinson 2000). Root biomass is often invoked as a suitable parameter, but needs to be interpreted with care, since root extent rather than biomass affects access to resources (see above), and in stress situations, root length growth may be achieved at lower root biomass (reviewed in Brunner et al. 2015).

Most commonly, reported studies obtained a representative number of rooted soil cores. From those, the roots are washed out, dried and weighed, thus obtaining root biomass. Most studies apply additional sorting, most commonly separating live from dead roots as well as differentiating root size classes. Some studies even identify species or functional groups based on root morphological traits (e.g., Brassard et al. 2011) in order to reveal the overlap of zones of influence. Recently, DNA barcoding techniques have been introduced for the same purpose. Their use is likely to increase with the development of faster workflows for genomic sequencing and establishing a global DNA database (Kress 2017). Root length can be obtained from images of the washed and spread out roots using analysis software (Richner et al. 2000).

The relatively small soil samples can only give reasonable information for smaller-sized roots, while very coarse roots are not well captured (Oliveira et al. 2000). Also, root length density varies considerably in forests, and representative sampling requires a substantial number of replicates (Bengough et al. 2000). Thus, non-destructive methods for assessing root system structure of entire trees have also been promoted (Nadezhdina and Čermák 2003; Jayawickreme et al. 2014). The use of geophysical techniques, such as ground-penetrating radar (Guo et al. 2013) and electrical resistivity tomography (Amato et al. 2012), has revealed coarse root structure and biomass of the root system. These techniques are capable of revealing the root system extent in well-drained soils, but do not resolve the size class of fine roots (<2 mm, Amato et al. 2012; Guo et al. 2013), which are most active in root water uptake. With further development, they are thought to become increasingly available for ecological studies of the subsurface in the near future (Jayawickreme et al. 2014).

For assessing root competition and reaction to stress, information about root growth, mortality, and the resulting turnover have been collected. Root growth can be tracked visually *in situ* using rhizotrons (Vamerali et al. 2012) by repeatedly recording roots on transparent tubes or windows embedded in the soil, and by calculating growth and mortality based on the difference between observations. This method runs the risk of bias since root growth is affected by the observation windows (Hajek et al. 2014). Alternatively, in-growth cores are used to assess root growth. Those are root-free soil cores that are inserted in the soil for a given time (Oliveira et al. 2000). The contained roots are all newly formed during the time the soil core resides in the field, which allows the comparison of root growth, e.g., between locations or treatments.

Root morphological studies have been extremely useful in fostering understanding of belowground zones of influence and niche filling. Results need to be considered with some caution because: (1) root properties vary a great deal within the same plant and size class and their activity can differ substantially between morphologically similar roots (Atkinson 2000; Leuschner et al. 2004); and (2) root uptake location changes dynamically due to the porous pipe nature of the root system, with only a part of the root system being active at any given moment (Hodge et al. 2009). Therefore, root morphology provides primarily information on root extent (Atkinson 2000) and potential for water access.

14.3.2 Assessment of Water Sourcing Based on Tracer Studies

Environmental tracers are prominent tools to trace water through the hydrological system and have been used to investigate the sources (pools) of transpiration water. Natural gradients in isotopic composition of the stable water isotopes have been a way for discerning water pools since the 1960s (Zimmermann et al. 1967). Stable water isotopes have been a prominent means of investigating root water sources for ecological applications (e.g., studies listed by Silvertown et al. 2015), especially in trees (Rothfuss and Javaux 2017). Isotopic composition depends on the sequence of fractionation during phase changes and equilibration processes related to evaporation from the sea surface and subsequent transformations by the hydrological cycle. This may lead to discernible differences in isotopic signatures of alternative root water uptake locations. For example, groundwater signatures can often be discerned from soil water, since evaporation fractionates in favor of the lighter isotopes, leading to enriched water in the soil. The same process also leads to enrichment profiles near the soil surface with the stronger influence of direct soil evaporation (Zimmermann et al. 1967; Allison 1982). This permits a derivation of vertical uptake profiles in combination with inverse modeling (Rothfuss and Javaux 2017), based on samples of soil and sap flow water.

Since their first applications to root water uptake studies, isotope tracers have allowed valuable mechanistic insights into plant hydraulics (Meinzer et al. 2001; Sprenger et al. 2016). For example, they revealed the distinction between root abundance and root activity (Ehleringer and Dawson 1992). Interpretation of the isotopic signal is not straightforward and associated with numerous challenges, including the choice of sampling techniques and simplifications, some of which interfere strongly with the interpretation (Sprenger et al. 2016). For example, applied soil water extraction methods, soil water content and soil type (Orlowski et al. 2016) as well as the method applied for mixing model inversion or the actual isotopic gradients (Rothfuss and Javaux 2017) all affect the estimated water fluxes. Also, stable isotope studies rely often on destructive sampling of the aquifer or soil and plant tissues. This prevents repeated sampling at the exact same spot, which necessarily increases uncertainty. Most results are therefore snapshots of water sourcing of specified sampled individual plants at a given point in time.

Because of their great popularity in various disciplines, the methods are quickly advancing, i.e., by using more sophisticated statistics (Parnell et al. 2010). Coupling those with process models (Ogle et al. 2004; Rothfuss et al. 2010) further decreases the chances of misinterpretation when root water uptake is dynamic, i.e., in situations with active hydraulic redistribution (Rothfuss and Javaux 2017), which are particularly interesting from the perspective of root water uptake in forests.

14.3.3 Direct Assessment Based on Root and Stem Sap Flow as well as Stem Diameter

The most direct method to assess water fluxes within the plant and root is by measuring sap flux velocity, either near the root collar (within the stem) or on single roots (Coners and Leuschner 2002, Nadezhdina et al. 2012b; Vandegehuchte and Steppe 2013). Most methods rely on measuring the heat transport within the sap wood, which increases proportionally with sap velocity. For this, a heat signal is applied, either by inserting needles into the sap wood, or by heating from the outside, and the temperature around the heat source is assessed. A large number of measurement setups and methods have been developed translating the temperature signal into flow velocities or total sap flux, and the methods have been increasingly refined (Vandegehuchte and Steppe 2013). For example, some setups allow analyzing flow direction, others use the radial sap flow profiles in the stem, partly combined with partial irrigation, to discern locations of root water uptake around single trees (Nadezhdina et al. 2012a). A considerable degree of uncertainty is attached to the methods (Steppe et al. 2010; Forster 2017), potentially leading to substantial underestimation of water fluxes (Forster 2017). Species-specific calibration is required to yield comparable results between species, which is not always economical. Chapter 1 of this volume covers the calibration of sap flow sensors.

Sap flow studies on roots have been instrumental in elucidating the processes of hydraulic redistribution (see below, Burgess et al. 1998). Several recent studies have combined sap flow measurements with other techniques, such as isotopes (Jia et al. 2017) or changes in soil water content (Schwärzel et al. 2009; Domec et al. 2010), for ecological interpretations.

14.3.4 Interpreting Soil Moisture Signals

Soil moisture patterns are, among other factors, affected by root water uptake, and soil moisture patterns have been repeatedly interpreted to reveal uptake distribution (Schume et al. 2004; Schwärzel et al. 2009). However, soil water patterns are the result of several water fluxes (Eq. 14.2), that is, infiltration, root water uptake and soil water flow, all of which show comparatively time-stable patterns. Therefore, singling out the role of root water uptake on soil water patterns is challenging (Vanderlinden et al. 2012). Also, deriving root water uptake from inverse modeling of the soil water balance is not trivial (Hupet et al. 2002) since both root water uptake and soil water redistribution are of the same order of magnitude. Deducing soil water flow and root water uptake profiles inversely based on observed soil water content is therefore not universally possible (Musters et al. 2000), but only in specific soil conditions and ideally requires additional information, like transpiration or drainage, which are typically not available (Hupet et al. 2003; Guderle and Hildebrandt 2015). One way forward is to separate the time series into day and night cycles, and assuming dominance of soil water flow during the night (Guderle and Hildebrandt 2015). Those approaches have been validated against lysimeters in grasslands (Guderle et al. 2018) or against eddy covariance data (Breña Naranjo et al. 2011) and yielded plausible results in forest settings (Breña Naranjo et al. 2011; Renner et al. 2016).

Relatively dense and spatially highly-resolution soil water content monitoring (Schume et al. 2004; Schwärzel et al. 2009) allows tracking of the most prevalent features of soil moisture dynamics and to draw first conclusions about root water uptake. However, due to the substantial small-scale heterogeneity, caused by soil properties (Metzger et al. 2017) and heterogeneous input patterns (Levia et al. 2011), sensor spacing can pose a limit to successfully observing all necessary hotspots (Schwärzel et al. 2009). Remote sensing techniques, in particular electrical resistance measurements, are promising ways to fill the spatial gap (Fan et al. 2015; Kuhl et al. 2018), although their temporal resolution is often much smaller than classical soil moisture measurements (i.e., based on frequency or time domain reflectometry).

14.4 Water Sourcing and Interaction of Root Systems within Forest Communities

Soil water availability changes in space and time. Interannual changes in meteorological conditions can expose an ecosystem to dramatically different conditions between years. Resilience and productivity therefore depend on the capacity to withstand those changing conditions. At the same time, interaction of species with different resource-use strategies potentially allows for reducing stress and maintaining higher productivity of diverse forest communities. This section takes a closer look at adaptation (plasticity) of water uptake location as well as species interaction processes related to root water uptake.

14.4.1 Plasticity of Root Water Uptake Location

Ideally, root uptake would be centered in the most prolific soil regions, especially during drought. With soil moisture distribution changing continuously with time, the capacity to shift root activity to prolific soil zones is an essential plant trait to avoid stress (Brinkmann et al. 2019). Indeed, so-called moving uptake fronts have been observed repeatedly, both at the individual and ecosystem scales in a wide range of climates during periods of water deficit (e.g., Ehleringer and Dawson 1992; Schwendenmann et al. 2015; Brinkmann et al. 2019).

Most reported shifts in root activity relate to shifts between shallow and deep soils, with root uptake preferably situated in the top soil when water is available, but moving down during time of temporary or prolonged drought. Most of the studies reporting shifts in root water uptake are based on isotopic tracer observations, partly on soil water balance approaches and combined with sap flow (Ehleringer and Dawson 1992; Schwendenmann et al. 2015; Brinkmann et al. 2019). This research shows that uptake depth and its plasticity depends strongly on the species or functional group (Schwendenmann et al. 2015; Brinkmann et al. 2019; Guderle et al. 2018), and aligns with traits regarding canopy position and drought tolerance (Brum et al. 2019). For example, some species, like *Fagus sylvatica*, have repeatedly been reported to shifting water uptake to greater depth during drought, while others, like *Picea abies,* confine water uptake to shallow depth, even in severe stress conditions (Goisser et al. 2016; Brinkmann et al. 2019). This suggests that climatically optimal root uptake profiles only emerge at the ecosystem scale.

Dynamic shifts of root water uptake to greater depth with more reliable soil water storage during drought have been encountered in a great variety of ecosystems around the world, including temperate regions (Goisser et al. 2016) and the Amazon (Broedel et al. 2017). This is intuitive, since environmental variation is omnipresent, and many plants act with little safety margin to cavitation (Choat et al. 2012). In other words, while "drought" may relate to different magnitudes of water deficit in each specific environment, drought stress is prone to occur in almost any environment.

As described by the porous pipe concept, root water uptake can shift along the root system to moister layers, provided the root system includes appropriate infrastructure such as transport roots, like tap roots, in so-called dimorphic root systems (David et al. 2016; Brum et al. 2017). Suitable root morphologies can be species-specific, starting with seedlings developing sinker roots and having a much greater belowground compared to aboveground extent (David et al. 2016; Matheny et al. 2017). However, drought can also trigger plastic changes in root system architecture allowing for deeper uptake. For example, mesquite trees developed larger and deeper roots when in rain exclusion treatments (Ansley et al. 2007).

In addition to adjusting coarse roots, adaptation in the location of fine roots also has been observed. Since young apical roots have higher uptake capacity, shedding old and growing young roots (increasing turnover) in a slight drought may help maintain a positive carbon balance, despite the additional investment in young roots (Gaul et al. 2008). This also adjusts the location of uptake according to water availability, resulting in shifts to greater depth (Ansley et al. 2007). In severe drought, however, root shedding seems to prevail, and fine root biomass decreases (Brunner et al. 2015). Overall root turnover seems to depend little on precipitation (Gill and Jackson 2000; Joslin et al. 2000). It is noteworthy that changes in fine root biomass may be disjointed from root length, as specific root weight can be adjusted to grow more roots of the same length at lower carbon cost, and probably higher vulnerability (Hertel et al. 2013; Brunner et al. 2015).

Although shifting root activity has mainly been reported in the vertical direction, there is no reason to assume that similar effects are not encountered laterally, for example, in response to variation in soil properties or water input leading to stable soil moisture patterns (Vanderlinden et al. 2012; Metzger et al. 2017). Dynamic lateral shifts in root water uptake have been demonstrated using partial irrigation experiments on trees (Green and Clothier 1999), and spatial throughfall variation has been hypothesized to have similar effects (Schwärzel et al. 2009), with roots taking up water preferably at locations with higher infiltration. Therefore, lateral adaptation of root water uptake patterns in forests is plausible, and would agree with theory of root water uptake equalizing infiltration heterogeneity in forests (Guswa 2012). There is, however, little systematic experimental research on matching root water uptake with the substantial soil water heterogeneity in forests.

14.4.2 Belowground Interactions – Background

Forests, no matter how complex they may be, are composed of trees living together in comparatively close proximity, creating a specific, sometimes very heterogeneous neighborhood for each individual. Although evapotranspiration scales with leaf area index at the ecosystem scale, water fluxes vary substantially between neighborhoods and individuals (Aranda et al. 2012). At the same time, belowground zones of influence of different individuals and species strongly overlap (e.g., Brunner et al. 2004; Göttlicher et al. 2008), and individual root distributions respond to the presence of a neighbor (Rewald and Leuschner 2009; Valverde-Barrantes et al. 2013; Laclau et al. 2013; Goisser et al. 2016). Questions about the strength and outcome of those belowground interactions, either positive or negative, are therefore close at hand. Two ecological strains of research have contributed a great deal to the current understanding of interactions regarding root water uptake—research dealing with understanding coexistence mechanisms and research investigating the relationship between ecosystem diversity and function (like ecosystem productivity and stress resilience).

Interactions may be positive or negative, meaning they either increase or decrease the function of one of the partners. For example, resource competition is a typically two-sided negative interaction. On the other hand, given the multitude of water use strategies (e.g., levels of isohydricity, shallow to deep rooting, canopy position or phenology, etc.), functional diversity could also reduce competition compared to monocultures (Forrester 2015) or even promote ecosystems for mutual and one-sided facilitation (Forrester and Bauhus 2016). The latter refers to interactions that improve the functions of at least one partner without detrimental effects for neither. For example, moderate shading can decrease water stress in the shaded species.

Forest communities interact in several ways, with water being only one aspect (Forrester et al. 2014). Thus, strong interaction effects between species with regard to water use and uptake are only expected in regions and/or at times where water is the most limiting resource (Forrester and Bauhus 2016). Intuitively, when development of leaf area is, for example, nutrient limited, optimizing ecosystem water use strategies will affect neither coexistence nor productivity. Accordingly, the stress-gradient hypothesis (Bertness and Callaway 1994) formulates that facilitation is to be expected most in locations where resource stress is extreme. Forrester and Bauhus (2016) generalized this by stating that interaction effects targeted at any given resource should increase when the resource is most limiting. In order to increase productivity for a given neighborhood compared to another, interactions would have to increase the total availability of the resource (Forrester and Bauhus 2016). At the same time, the water balance (Eq. 14.2) still holds. Increasing water availability would either require increasing soil water storage by: (1) extending uptake to greater depth (presuming regular replenishment of that added soil storage, Schwinning 2010) or increasing the plant available porosity (n_{PAW} in Eq. 14.2), for example by including species capable of drying the soil more profoundly (higher drought tolerance); or (2) increasing water use efficiency (being more productive with the same amount of water).

14.4.3 Competition and Competitive Reduction

Vertical niche partitioning for root water uptake is one of the most investigated mechanisms (Silvertown et al. 2015) falling in the category of complementarity (Forrester and Bauhus 2016). It implies that coexisting species tap water at different soil depth, thus reducing competition for water. Silvertown et al. (2015) specifically reviewed “hydrologic niche segregation” strategies, one of which is strictly related to belowground space partitioning. They found that most of the identified studies on forests, including semiarid to temperate sites, confirmed vertical spatial hydrologic niche partitioning of co-occurring tree species to some degree. More recent results are in line with this finding and suggest that niche overlap decreases with increasing stress (Guo et al. 2018), thus supporting the stress gradient hypothesis. Belowground niche separation has been derived from indirect inference, based on soil moisture measurements (e.g., Schume et al. 2004; Fan et al. 2015) or root distributions (Valverde-Barrantes et al. 2013).

More recently, Brum et al. (2019) combined isotope-derived root water uptake with above- and belowground plant trait measurements and suggest that the hydraulic niche partitioning extends to the aboveground. They show that overstorey drought-tolerant species take up water from greater depth, compared to smaller understorey species, which are shallow rooted and less drought tolerant (Fig. 14.1). Thus, shading of smaller individuals is an aboveground processes with belowground consequences. This is congruent with ecohydrological modeling resolving canopy layers and rooting depths (Ivanov et al. 2012). Invoking the same mechanism, the model also predicts higher drought mortality of tall exposed trees in exceptionally dry periods, which agrees with observations (Bennett et al. 2015).

Complementarity is not only possible in space, but also in time. Coexistence of trees with different phenology or water use strategies (e.g., evergreen and deciduous) allows for sequential access to the water storage (Goisser et al. 2016; Schwendenmann et al. 2015) which increases total annual transpiration compared to, e.g., the deciduous species growing in monoculture. Taken together, the available studies demonstrate that belowground resource partitioning for root water uptake in mixed forests is a prevalent feature, and coordinated with aboveground traits.

Most of those previously mentioned studies did not address in how far the observed niche separation contributed to ecosystem processes, such as maintenance of coexistence or improved ecosystem functions (like productivity), or whether it was coordinated with neighbors. This is because these studies were conducted in mixed species environments, without means for comparing how species act without neighbors. Some experimental manipulation approaches have attempted to close this gap. Root distributions of single species have been shown to change, depending on whether the species is grown in mixture or monoculture in a range of climates (Brassard et al. 2011; Goisser et al. 2016; Laclau et al. 2013). However, for a temperate forest, Laclau et al. (2013) found root water uptake depth (determined from isotopes) differed little between monoculture and mixture, despite significant root adaptations in mixtures. Similar results were obtained by Goisser et al. (2016) in a temperate environment. Also in the Amazon, Schwendenmann et al. (2015) found that uptake depth depended mainly on species identity, which nevertheless seemed to support complementarity in severe drought, given the right pairings. Comparative studies between monocultures and mixed species based on the soil water balance approach suggested deeper uptake in mixtures compared to monocultures in a temperate forest (Schume et al. 2004).

The differences between neighborhood effects on root abundance (derived from root sampling) versus root activity (derived from isotope measurements) may partly be due to roots being deployed for foraging nutrients instead of water, and the uncertainty attached to the isotope methods masking subtle differences. Thus, more advanced isotopic methods may still reveal adaptations in the future. Overall, current research indicates that despite the substantial plasticity in water uptake, species assembly plays a substantial role for complementarity effects. In other words, increasing complementarity in drier sites goes hand in hand with adaptations in ecosystem structure (Guo et al. 2018, based on isotopes), while structural adaptation to new conditions is a comparatively slow process.

Given the great horizontal heterogeneity of soil water content in forest ecosystems (Schume et al. 2004; Metzger et al. 2017), lateral niche partitioning also would be an interesting subject of study in mixed forests. However, only few such studies are available (Pretzsch et al. 2015), and most studies are based on morphological data. There is a general tendency toward suggesting more complete horizontal belowground space filling in diverse mixtures (Pretzsch et al. 2015) and very adaptable root structures. One study specifically concluded that more complete belowground space filling was responsible for over-yielding in mixed compared to single species stands (Brassard et al. 2011). However, this study was based on root biomass and not root length or root activity, and other factors but water may have affected the result.

It is worth noting that interactions can proceed at the expense of one or more less competitive neighbor species (Vitali et al. 2018). For example, both root extent and colonization with mycorrhizae was decreased in northern pine in the presence of a European beech (Goisser et al. 2016), with potential detrimental consequences for drought resilience. Also, in regions with shallow rooting depth where space for vertical niche partitioning lacks, the presence of a water spending species can exhaust the limited water storage at the expense of all neighbors (Gebauer et al. 2012), and increase water stress in dry years. Thus, whether mixtures successfully enhance ecosystem function depends on the specific species assemblages (Forrester et al. 2016; Vitali et al. 2018) and site conditions.

14.4.4 Facilitation – Hydraulic Redistribution

Facilitation describes interactions, which positively affect at least one of the interacting partners while at the same time causing no negative effect on any of them. Arguably the most prominent mechanism for belowground facilitation is hydraulic redistribution, which has been observed in a wide range of ecosystems (Neumann and Cardon 2012). Hydraulic redistribution is the passive transport of water from moist to dry soil regions through the root system. It occurs when the suction at the root collar is small, mainly during the night, and when the potential differences within the soil parts connected by roots become considerable. Most observations report on distribution from deeper moist to the shallow dry soil layers, but top-down and lateral directions are possible, depending mainly on the distribution of soil water potential. The water flow through the root system can be much quicker than bulk soil water flow (Domec et al. 2010), and facilitate equilibration of soil water potentials (Katul and Siqueira 2010). The release of water helps avoiding severe water stress in shallow rooted neighbors (Katul and Siqueira 2010), but also maintaining nutrient supply and nurturing microbial activity in dry soil conditions (Prieto et al. 2012).

Hydraulic redistribution is species-specific and associated with a combination of functional shallow and deep roots in the redistributing species (Bleby et al. 2010). Depending on species and condition, redistribution can substantially contribute to overall plant and ecosystem transpiration (on average 15%, maximum 80% reported, Neumann and Cardon 2012), but even small amounts of redistributed water may contribute to improving the water status in the receiving plant (e.g., Domec et al. 2004; Warren et al. 2008). Hydraulic redistribution has entered the common knowledge of the root water uptake modeling community, although the number of studies measuring it directly, either with isotope tracers (Domec et al. 2004; Warren et al. 2008) or sap flux sensors on roots (Burgess et al. 1998; Bleby et al. 2010), is comparatively small (Neumann and Cardon 2012). This is partly due to the difficulty with accessing roots, but also models allow pinpointing the combined benefits of hydraulic redistribution with a systemic approach. They allow, for example, coupling soil-root interactions with the plant carbon balance, which reveals the benefit of hydraulic lift for ecosystem drought resilience in overstorey and understorey mixtures (Manoli et al. 2017).

In addition to root activity, mycorrhizae are increasingly acknowledged in playing a role in hydraulic redistribution, especially with regard to supporting tree seedlings. Common mycorrhizal networks were shown to contribute to transpiration of seedlings, both in the same and other species, in a mesocosm experiment with labeled water (Egerton-Warburton et al. 2007). Field studies confirm the potentially very important role of common mycorrhizal networks for hydrating seedlings (Warren et al. 2008; Booth and Hoeksema 2009). For example, based on a systematic root and mycorrhizal exclusion study, Booth and Hoeksema (2009) showed that shared mycorrhizal networks between overstorey trees and seedlings significantly increased their survival in the otherwise harsh, low-light, and drought-prone understorey.

Overall, hydraulic redistribution is a reoccurring phenomenon across climates, with positive effects on ecosystem drought resilience, recruitment and sometimes considerable contribution to ecosystem transpiration.

14.5 Effect of Root-Water Relations on Hydrological Patterns

Interaction within tree neighborhoods is also interesting from the perspective of the expected spatial variation of soil water content, resulting microhabitats and affecting soil percolation and runoff. Water input at the forest floor is characterized by substantial spatial heterogeneity (Levia et al. 2011), induced by throughfall and stemflow hotspots. Those input patterns are at the same time temporally quite stable. The porous pipe nature of the extensive root system and its capacity for adaptations should act to smoothen that heterogeneous input. Based on high-resolution soil moisture monitoring, it has been suggested that root water uptake is increased at locations with enhanced input (Schwärzel et al. 2009). However, regardless of the larger potential of tree root systems for equalizing soil water, forests showed higher variation in soil water content compared to cropland in a study across several datasets (Korres et al. 2015). Moreover, Zehe et al. (2010) observed that variation was enhanced in forests compared to grassland. This conflicting result may be attributed to variation in net precipitation heterogeneity but may also be due to soil properties which are systematically modified by trees (Paluch and Gruba 2012; Archer et al. 2013), even at a small scale (Metzger et al. 2017).

Moreover, within-stand heterogeneity of sap flow, that is, the variation of root water uptake by different tree individuals, varies strongly, depending on species (Gebauer et al. 2012), tree age (Aranda et al. 2012), or location (Renner et al. 2016). Models suggest that small-scale transpiration varies substantially due to self-shading (Hentschel et al. 2013, He et al. 2014). Those differences would act to increase soil water heterogeneity at the meter scale (He et al. 2014), but do not explain why they are not equalized by overlapping root systems.

To illustrate this, Fig. 14.2 shows gradients in soil water potentials at 27 cm soil depth, which are based on nine transect measurements in a beech dominated forest plot (see Metzger et al. 2017, for site description) during a dry period following a rain event (Huss 2017). Transects point away from the tree with one sensor near the trunk (10 cm distance) and one further away (0.9 m to 2.9 m distance, median distance 0.9 m). We expected potentials to equalize when progressing into the dry period, but the opposite was observed.

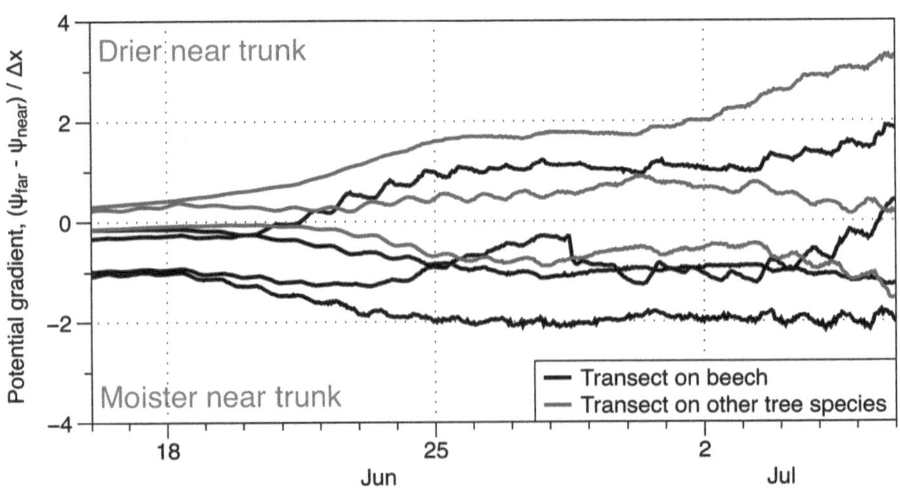

Fig. 14.2 Observed gradients in soil matric potential over transects starting at 0.1 m distance from the tree trunk pointed outward, maximum distance (Δx) is 2.9 m. (Data source: Huss 2017)

Gradients were comparatively equalized after the rain, but increased during dry down in both directions, i.e., leading to both drier and moister areas near the stem. At the end of the period, gradients are steeper in the dry compared to the wettest state, which is not the expected result considering the potentially equalizing role of root water uptake. This example is to show that understanding the dynamics of spatial soil water patterns still requires further investigation, and promises further insights into the formation of hydrological niches in forests.

Roots have also been strongly associated with the change of soil hydraulic properties of forest soils of different ages (Archer et al. 2013), increasing infiltrating capacity, as well as with preferential flow (e.g., Anderson et al. 2009). Based on soil water observations combined with a sprinkling experiment, Jost et al. (2012) proposed that deeper coarse roots of beech allow for preferential flow to deeper layers, whereas water does not infiltrate as deep in the spruce forest with a shallower rooting depth. This is in line with earlier work, demonstrating substantial preferential flow along suberized beech roots (Schwärzel et al. 2012).

Currently, our understanding of the consequences of tree interaction on soil water patterns and fluxes in forests is still incomplete. More work is required to evaluate the consequences of forest structural changes on preferential flow and precipitation water partitioning.

14.6 Outlook

Root systems support the canopy needs to supply water and nutrients. In order to provide the water needs, the root systems' extent has to be adapted to the demand of the canopy, with abiotic limits imposed by the length and severity of droughts as well as the accessibility of the deep subsurface. Root water relations are therefore a result of tightly coupled above- and belowground processes.

Because of the limited access to the belowground, the subsurface complexity of mixed forests is far less investigated than the aboveground. Also, unfortunately, too few studies take into account both above- and belowground parts of the plant when discussing differences in belowground water relations or root morphology. Instead it would be required to discuss Eqs. 14.1 and 14.2 together. Several modeling studies have moved ahead along those lines and shown that aboveground interactions affecting shading, combination of water use strategies, and species interactions for water and light can improve predictions of transpiration and explain coexistence.

The reviewed literature suggests that roots are flexible. Mechanisms of adaptation are multifold and allow dealing with typical water deficits that occur temporally in almost any ecosystem. Numerous adaptation mechanisms allow trees to adjust the uptake locations to the local conditions, and coordinate them with overall hydraulic strategies. Associations with mycorrhizae allow plant roots to push the border of the rooting zone into the bedrock.

Despite this flexibility, plant water foraging strategies are nevertheless quite species-specific. Recent research indicates that species traits align along a hydrological niche axis that encompasses both notions of light and water use strategies. A great deal of research suggests that mixing hydraulic strategies allows for coexistence and increased wood productivity in mixed forests compared to monocultures, but also that the positive effects depend on the specific neighborhood assembly.

Mycorrhizae play a key role in belowground resource trading in forests. The exact quantitative contribution of mycorrhizae for forest hydrology is not well investigated, but their role for drought mitigation and seedling survival is more and more established. Mycorrhizae may be affected by climate change (Walter 2018) and decrease in biodiversity. There is a need to consider the potential hydrological consequences independently of the potential role of biodiversity for increased productivity.

From a hydrological perspective, forest ecosystems at the plot scale are comparatively predictable, with transpiration and ecosystem root water uptake comparatively well correlated with leaf area index. However, very little is known about the smaller scale processes at the scale of several meters. Despite substantial research invested in learning about root system properties within the last years, the insight into consequences of individual interactions on soil hydrology has moved very slowly ahead. Upcoming isotope sensing techniques for online monitoring of tracers in soil and sap are a promising way forward. This would greatly help understanding overlap of tree species zones of influence.

An important role of roots may also lie in their effect on soil properties, preferential flow and runoff routing at the hillslope scale. Very few studies address the hydrological consequences of both, heterogeneous infiltration and soil properties being shaped by the same organisms for substantial time periods. For example, we still know little about the origin of the increased heterogeneity in soil water content in forests compared to, e.g., grasslands. This observation is in contradiction with the commonly held assumption that large and overlapping root systems should equilibrate soil water contents. Resolving this inconsistency would reveal new process insights. Current monitoring techniques do not yet allow attributing soil water content patterns to differences in soil properties or root water uptake. Combining processes models of plant water uptake with soil and tracer observations is a promising way forward. Soil moisture sensing has become cheaper and more accessible in recent years, which contributes to an ever-increasing database for future research. Yet more insight could be gained by focusing besides on soil water content, also on variables and parameters that are directly related to the process of root water uptake, such as the soil water potential or soil hydraulic properties.

Overall, root water relations in forests have great potential to affect not only ecosystem functions, but also fast and slow water fluxes. Understanding root interactions provides a promising way forward to learning how forest management and related changes in forest structure affect hydrological dynamics. While some work has been dedicated to revealing ecological consequences of tree interactions, still more is needed to reveal the effects on hydrology. This requires taking a literally deeper look into the subsurface and a focus on process-based, quantitative approaches.

Acknowledgment Data in Fig. 14.2 were collected in the Critical Zone Exploratory AquaDiva funded by the German Research Foundation (DFG grant CRC 1076) with additional funding from the DFG funded Centre for Integrative Biodiversity Research (iDiv) Halle-Jena-Leipzig.

References

Allen MF (2007) Mycorrhizal fungi: highways for water and nutrients in arid soils. Vadose Zo J 6:291. https://doi.org/10.2136/vzj2006.0068

Allison GB (1982) The relationship between 18O and deuterium in water in sand columns undergoing evaporation. J Hydrol 55:163–169. https://doi.org/10.1016/0022-1694(82)90127-5

Amato M, Lapenna V, Rossi R, Bitella G (2012) Multi-electrode resistivity imaging. In: Measuring roots: an updated approach. Springer-Verlag, Berlin Heidelberg, pp 189–211

Anderegg WRL, Klein T, Bartlett M et al (2016) Meta-analysis reveals that hydraulic traits explain cross-species patterns of drought-induced tree mortality across the globe. Proc Natl Acad Sci 113:5024–5029. https://doi.org/10.1073/pnas.1525678113

Anderson AE, Weiler M, Alila Y, Hudson RO (2009) Dye staining and excavation of a lateral preferential flow network. Hydrol Earth Syst Sci 13:935–944. https://doi.org/10.5194/hess-13-935-2009

Ansley RJ, Boutton TW, Jacoby PW (2007) Mesquite root distribution and water use efficiency in response to long-term soil moisture manipulations. In: Proc Shrubl Dyn Water. U.S. Department of Agriculture, Forest Service, Rocky Mountain Research Station, Fort Collins, pp 96–103

Aranda I, Forner A, Cuesta B, Valladares F (2012) Species-specific water use by forest tree species: from the tree to the stand. Agric Water Manag 114:67–77. https://doi.org/10.1016/j.agwat.2012.06.024

Aravena JE, Berli M, Ruiz S et al (2014) Quantifying coupled deformation and water flow in the rhizosphere using X-ray microtomography and numerical simulations. Plant Soil 376:95–110. https://doi.org/10.1007/s11104-013-1946-z

Archer NAL, Bonell M, Coles N et al (2013) Soil characteristics and landcover relationships on soil hydraulic conductivity at a hillslope scale: a view towards local flood management. J Hydrol 497:208–222. https://doi.org/10.1016/j.jhydrol.2013.05.043

Atkinson D (2000) Root characteristics: why and what to measure. In: Smit AL, Bengough AG, Engels C et al (eds) Root methods. Springer, Berlin Heidelberg, pp 1–32. https://doi.org/10.1007/978-3-662-04188-8_1

Barbeta A, Mejía-Chang M, Ogaya R et al (2015) The combined effects of a long-term experimental drought and an extreme drought on the use of plant-water sources in a Mediterranean forest. Glob Chang Biol 21:1213–1225. https://doi.org/10.1111/gcb.12785

Bechmann M, Schneider C, Carminati A et al (2014) Effect of parameter choice in root water uptake models – the arrangement of root hydraulic properties within the root architecture affects dynamics and efficiency of root water uptake. Hydrol Earth Syst Sci 18:4189–4206. https://doi.org/10.5194/hess-18-4189-2014

Bengough AG, Catrignano A, Pages L, van Noordwijk M (2000) Root methods. Springer, Berlin Heidelberg, Berlin, Heidelberg

Bennett AC, Mcdowell NG, Allen CD, Anderson-Teixeira KJ (2015) Larger trees suffer most during drought in forests worldwide. Nat Plants 1:1–5. https://doi.org/10.1038/nplants.2015.139

Bertness MD, Callaway R (1994) Positive interactions in communities. Trends Ecol Evol 9:187–191. https://doi.org/10.1016/0169-5347(94)90087-6

Bleby TM, Mcelrone AJ, Jackson RB (2010) Water uptake and hydraulic redistribution across large woody root systems to 20 m depth. Plant Cell Environ 33:2132–2148. https://doi.org/10.1111/j.1365-3040.2010.02212.x

Booth MG, Hoeksema JD (2009) Mycorrhizal networks counteract effects of canopy trees on seedlings. Ecology 91:2294–2302. https://doi.org/10.1890/09-1139.1

Brantley SL, Eissenstat DM, Marshall JA et al (2017) Reviews and syntheses: on the roles trees play in building and plumbing the critical zone. Biogeosciences 14:5115–5142. https://doi.org/10.5194/bg-14-5115-2017

Brassard BW, Chen HYH, Bergeron Y, Pare D (2011) Differences in fine root productivity between mixed- and single-species stands. Funct Ecol 25:238–246. https://doi.org/10.1111/j.1365-2435.2010.01769.x

Bréda N, Huc R, Granier A, Dreyer E (2006) Temperate forest trees and stands under severe drought: a review of ecophysiological responses, adaptation processes and long-term consequences. Ann For Sci 63:625–644. https://doi.org/10.1051/forest:2006042

Breña Naranjo JA, Weiler M, Stahl K (2011) Sensitivity of a data-driven soil water balance model to estimate summer evapotranspiration along a forest chronosequence. Hydrol Earth Syst Sci 15:3461–3473. https://doi.org/10.5194/hess-15-3461-2011

Brinkmann N, Eugster W, Buchmann N, Kahmen A (2019) Species-specific differences in water uptake depth of mature temperate trees vary with water availability in the soil. Plant Biol 21:71–81. https://doi.org/10.1111/plb.12907

Brodribb TJ (2009) Xylem hydraulic physiology: the functional backbone of terrestrial plant productivity. Plant Sci 177:245–251. https://doi.org/10.1016/j.plantsci.2009.06.001

Broedel E, Tomasella J, Cândido LA, von Randow C (2017) Deep soil water dynamics in an undisturbed primary forest in central Amazonia: differences between normal years and the 2005 drought. Hydrol Process 31:1749–1759. https://doi.org/10.1002/hyp.11143

Brum M, Teodoro GS, Abrahão A, Oliveira RS (2017) Coordination of rooting depth and leaf hydraulic traits defines drought-related strategies in the campos rupestres, a tropical montane biodiversity hotspot. Plant Soil 420:467–480. https://doi.org/10.1007/s11104-017-3330-x

Brum M, Vadeboncoeur MA, Ivanov V et al (2019) Hydrological niche segregation defines forest structure and drought tolerance strategies in a seasonal Amazon forest. J Ecol 107:318–333. https://doi.org/10.1111/1365-2745.13022

Brunner I, Ruf M, Lüscher P, Sperisen C (2004) Molecular markers reveal extensive intraspecific below-ground overlap of silver fir fine roots. Mol Ecol 13:3595–3600. https://doi.org/10.1111/j.1365-294X.2004.02328.x

Burgess SSO, Adams MA, Turner NC, Ong CK (1998) The redistribution of soil water by tree root systems. Oecologia 115:306–311. https://doi.org/10.1007/s004420050521

Canadell JG, Jackson RB, Ehleringer JR et al (1996) Maximum rooting depth of vegetation types at the global scale. Oecologia 108:583–595. https://doi.org/10.1007/BF00329030

Casper BB, Schenk HJ, Jackson RB (2003) Defining a plant's belowground zone of influence. Ecology 84:2313–2321. https://doi.org/10.1890/02-0287

Choat B, Jansen S, Brodribb TJ et al (2012) Global convergence in the vulnerability of forests to drought. Nature 491:752–755. https://doi.org/10.1038/nature11688

Collins DBG, Bras RL (2007) Plant rooting strategies in water-limited ecosystems. Water Resour Res 43:W06407. https://doi.org/10.1029/2006WR005541

Coners H, Leuschner C (2002) In situ water absorption by tree fine roots measured in real time using miniature sap-flow gauges. Funct Ecol 16:696–703. https://doi.org/10.1046/j.1365-2435.2002.00665.x

David TS, Pinto CA, Nadezhdina N, David JS (2016) Water and forests in the Mediterranean hot climate zone: a review based on a hydraulic interpretation of tree functioning. For Syst 25:eR02. https://doi.org/10.5424/fs/2016252-08899

Domec JC, Warren JM, Meinzer FC et al (2004) Native root xylem embolism and stomatal closure in stands of Douglas-fir and ponderosa pine: mitigation by hydraulic redistribution. Oecologia 141:7–16. https://doi.org/10.1007/s00442-004-1621-4

Domec JC, King JS, Noormets A et al (2010) Hydraulic redistribution of soil water by roots affects whole-stand evapotranspiration and net ecosystem carbon exchange. New Phytol 187:171–183. https://doi.org/10.1111/j.1469-8137.2010.03245.x

Draye X, Kim Y, Lobet G, Javaux M (2010) Model-assisted integration of physiological and environmental constraints affecting the dynamic and spatial patterns of root water uptake from soils. J Exp Bot 61:2145–2155. https://doi.org/10.1093/jxb/erq077

Egerton-Warburton LM, Querejeta JI, Allen MF (2007) Common mycorrhizal networks provide a potential pathway for the transfer of hydraulically lifted water between plants. J Exp Bot 58:1473–1483. https://doi.org/10.1093/jxb/erm009

Ehleringer JR, Dawson TE (1992) Water uptake by plants: perspectives from stable isotope composition. Plant Cell Environ 15:1073–1082. https://doi.org/10.1111/j.1365-3040.1992.tb01657.x

Fan J, Scheuermann A, Guyot A et al (2015) Quantifying spatiotemporal dynamics of root-zone soil water in a mixed forest on subtropical coastal sand dune using surface ERT and spatial TDR. J Hydrol 523:475–488. https://doi.org/10.1016/j.jhydrol.2015.01.064

Forrester DI (2014) The spatial and temporal dynamics of species interactions in mixed-species forests: from pattern to process. For Ecol Manag 312:282–292. https://doi.org/10.1016/j.foreco.2013.10.003

Forrester DI (2015) Transpiration and water-use efficiency in mixed-species forests versus monocultures: effects of tree size, stand density and season. Tree Physiol 35:289–304. https://doi.org/10.1093/treephys/tpv011

Forrester DI, Bauhus J (2016) A review of processes behind diversity—productivity relationships in forests. Curr For Rep 2:45–61. https://doi.org/10.1007/s40725-016-0031-2

Forrester DI, Bonal D, Dawud S et al (2016) Drought responses by individual tree species are not often correlated with tree species diversity in European forests. J Appl Ecol 53:1725–1734. https://doi.org/10.1111/1365-2664.12745

Forster MA (2017) How reliable are heat pulse velocity methods for estimating tree transpiration? Forests 8:1–11. https://doi.org/10.3390/f8090352

Frensch J, Steudle E (1989) Axial and radial hydraulic resistance to roots of maize (*Zea mays* L.). Plant Physiol 91:719–726. https://doi.org/10.1104/pp.91.2.719

Gaul D, Hertel D, Borken W et al (2008) Effects of experimental drought on the fine root system of mature Norway spruce. For Ecol Manag 256:1151–1159. https://doi.org/10.1016/j.foreco.2008.06.016

Gebauer T, Horna V, Leuschner C (2012) Canopy transpiration of pure and mixed forest stands with variable abundance of European beech. J Hydrol 442–443:2–14. https://doi.org/10.1016/j.jhydrol.2012.03.009

Gill RA, Jackson RB (2000) Global patterns of root turnover for terrestrial ecosystems. New Phytol 147:13–31. https://doi.org/10.1046/j.1469-8137.2000.00681.x

Goisser M, Geppert U, Rötzer T et al (2016) Does belowground interaction with *Fagus sylvatica* increase drought susceptibility of photosynthesis and stem growth in *Picea abies*? For Ecol Manag 375:268–278. https://doi.org/10.1016/j.foreco.2016.05.032

Göttlicher SG, Taylor AFS, Grip H et al (2008) The lateral spread of tree root systems in boreal forests: estimates based on 15N uptake and distribution of sporocarps of ectomycorrhizal fungi. For Ecol Manag 255:75–81. https://doi.org/10.1016/j.foreco.2007.08.032

Green S, Clothier B (1999) The root zone dynamics of water uptake by a mature apple tree. Plant Soil 206:61–77. https://doi.org/10.1023/A:1004368906698

Guderle M, Hildebrandt A (2015) Using measured soil water contents to estimate evapotranspiration and root water uptake profiles – a comparative study. Hydrol Earth Syst Sci 19:409–425. https://doi.org/10.5194/hess-19-409-2015

Guderle M, Bachmann D, Milcu A et al (2018) Dynamic niche partitioning in root water uptake facilitates efficient water use in more diverse grassland plant communities. Funct Ecol 32:214–227. https://doi.org/10.1111/1365-2435.12948

Guo L, Chen J, Cui X et al (2013) Application of ground penetrating radar for coarse root detection and quantification: a review. Plant Soil 362:1–23. https://doi.org/10.1007/s11104-012-1455-5

Guo JS, Hungate BA, Kolb TE, Koch GW (2018) Water source niche overlap increases with site moisture availability in woody perennials. Plant Ecol 219:719–735. https://doi.org/10.1007/s11258-018-0829-z

Guswa AJ (2010) Effect of plant uptake strategy on the water – optimal root depth. Water Resour Res 46:1–5. https://doi.org/10.1029/2010WR009122

Guswa AJ (2012) Canopy vs. roots: production and destruction of variability in soil moisture and hydrologic fluxes. Vadose Zo J 11. https://doi.org/10.2136/vzj2011.0159

Gutsch M, Lasch-Born P, Suckow F, Reyer CPO (2015) Modeling of two different water uptake approaches for mono-and mixed-species forest stands. Forests 6:2125–2147. https://doi.org/10.3390/f6062125

Hacke UG, Sperry JS, Ewers BE et al (2000a) Influence of soil porosity on water use in Pinus taeda. Oecologia 124:495–505. https://doi.org/10.1007/PL00008875

Hacke UG, Sperry JS, Pittermann J (2000b) Drought experience and cavitation resistance in six shrubs from the Great Basin, Utah. Basic Appl Ecol 1:31–41. https://doi.org/10.1078/1439-1791-00006

Hajek P, Hertel D, Leuschner C (2014) Root order- and root age-dependent response of two poplar species to belowground competition. Plant Soil 377:337–355. https://doi.org/10.1007/s11104-013-2007-3

Hassler SK, Weiler M, Blume T (2018) Tree-, stand- and site-specific controls on landscape-scale patterns of transpiration. Hydrol Earth Syst Sci 22:13–30. https://doi.org/10.5194/hess-22-13-2018

He L, Ivanov VY, Bohrer G et al (2014) Effects of fine-scale soil moisture and canopy heterogeneity on energy and water fluxes in a northern temperate mixed forest. Agric For Meteorol 184:243–256. https://doi.org/10.1016/j.agrformet.2013.10.006

Heilman JL, McInnes KJ, Kjelgaard JF et al (2009) Energy balance and water use in a subtropical karst woodland on the Edwards Plateau, Texas. J Hydrol 373:426–435. https://doi.org/10.1016/j.jhydrol.2009.05.007

Hentschel R, Bittner S, Janott M et al (2013) Simulation of stand transpiration based on a xylem water flow model for individual trees. Agric For Meteorol 182–183:31–42. https://doi.org/10.1016/j.agrformet.2013.08.002

Hertel D, Strecker T, Müller-Haubold H, Leuschner C (2013) Fine root biomass and dynamics in beech forests across a precipitation gradient – is optimal resource partitioning theory applicable to water-limited mature trees? J Ecol 101:1183–1200. https://doi.org/10.1111/1365-2745.12124

Hildebrandt A, Eltahir EAB (2007) Ecohydrology of a seasonal cloud forest in Dhofar: 2. Role of clouds, soil type, and rooting depth in tree-grass competition. Water Resour Res 43:1–13. https://doi.org/10.1029/2006WR005262

Hildebrandt A, Kleidon A, Bechmann M (2016) A thermodynamic formulation of root water uptake. Hydrol Earth Syst Sci 20:3441–3454. https://doi.org/10.5194/hess-20-3441-2016

Hillel D (1998) Environmental soil physics. Academic Press, Boston

Hodge A, Berta G, Doussan C et al (2009) Plant root growth, architecture and function. Plant Soil 321:153–187. https://doi.org/10.1007/s11104-009-9929-9

Höllta T, MENCUCCINI M, NIKINMAA E (2011) A carbon cost-gain model explains the observed patterns of xylem safety and efficiency. Plant Cell Environ 34:1819–1834. https://doi.org/10.1111/j.1365-3040.2011.02377.x

Hupet F, Lambot S, Javaux M, Vanclooster M (2002) On the identification of macroscopic root water uptake parameters from soil water content observations. Water Resour Res 38:1300. https://doi.org/10.1029/2002WR001556

Hupet F, Lambot S, Feddes RA et al (2003) Estimation of root water uptake parameters by inverse modeling with soil water content data. Water Resour Res 39:1312. https://doi.org/10.1029/2003WR002046

Huss J (2017) Untersuchung der Abhängigkeit des Bodenwasserpotentials von der Entfernung zu einem Baum in einem Buchenmischwald. Unpbublished BSc theisis. Friedrich Schiller University, Jena

Ivanov VY, Hutyra LR, Wofsy SC et al (2012) Root niche separation can explain avoidance of seasonal drought stress and vulnerability of overstory trees to extended drought in a mature Amazonian forest. Water Resour Res 48:W12507. https://doi.org/10.1029/2012WR011972

Jayawickreme DH, Jobbágy E, Jackson RB (2014) Geophysical subsurface imaging for ecological applications. New Phytol 201:1170–1175. https://doi.org/10.1111/nph.12619

Jia G, Liu Z, Chen L, Yu X (2017) Distinguish water utilization strategies of trees growing on earth-rocky mountainous area with transpiration and water isotopes. Ecol Evol 7:10640–10651. https://doi.org/10.1002/ece3.3584
Johnson DM, Sherrard ME, Domec J-C, Jackson RB (2014) Role of aquaporin activity in regulating deep and shallow root hydraulic conductance during extreme drought. Trees 28:1323–1331. https://doi.org/10.1007/s00468-014-1036-8
Joslin JD, Wolfe MH, Hanson PJ (2000) Effects of altered water regimes on forest root systems. New Phytol 147:117–129. https://doi.org/10.1046/j.1469-8137.2000.00692.x
Jost G, Schume H, Hager H et al (2012) A hillslope scale comparison of tree species influence on soil moisture dynamics and runoff processes during intense rainfall. J Hydrol 420–421:112–124. https://doi.org/10.1016/j.jhydrol.2011.11.057
Katul GG, Siqueira MB (2010) Biotic and abiotic factors act in coordination to amplify hydraulic redistribution and lift. New Phytol 187:3–6. https://doi.org/10.1111/j.1469-8137.2010.03306.x
Kirchen G, Calvaruso C, Granier A et al (2017) Local soil type variability controls the water budget and stand productivity in a beech forest. For Ecol Manag 390:89–103. https://doi.org/10.1016/j.foreco.2016.12.024
Kleidon A, Heimann M (1998) A method of determining rooting depth from a terrestrial biosphere model and its impacts on the global water and carbon cycle. Glob Chang Biol 4:275–286. https://doi.org/10.1046/j.1365-2486.1998.00152.x
Korres W, Reichenau TG, Fiener P et al (2015) Spatio-temporal soil moisture patterns – a meta-analysis using plot to catchment scale data. J Hydrol 520:326–341. https://doi.org/10.1016/j.jhydrol.2014.11.042
Kress WJ (2017) Plant DNA barcodes: applications today and in the future. J Syst Evol 55:291–307. https://doi.org/10.1111/jse.12254
Kuhl AS, Kendall AD, Van Dam RL, Hyndman DW (2018) Quantifying soil water and root dynamics using a coupled hydrogeophysical inversion. Vadose Zo J 17:170154. https://doi.org/10.2136/vzj2017.08.0154
Laclau JP, Nouvellon Y, Reine C et al (2013) Mixing Eucalyptus and Acacia trees leads to fine root over-yielding and vertical segregation between species. Oecologia 172:903–913. https://doi.org/10.1007/s00442-012-2526-2
Landsberg J, Fowkes N (1978) Water movement through plant roots. Ann Bot 42:493–508
Leuschner C, Coners H, Icke R (2004) In situ measurement of water absorption by fine roots of three temperate trees : species differences and differential activity of superficial and deep roots †. Tree Physiol 24:1359–1367. https://doi.org/10.1093/treephys/24.12.1359
Levia DF, Keim RF, Carlyle-Moses DE, Frost EE (2011) Througfall and stemflow in wooded ecosystems. In: Levia DF, Carlyle-Moses DE, Tanaka T (eds) Forest hydrology and biogeochemistry, ecological studies, vol 216. Springer Netherlands, Dordrecht, pp 425–433. https://doi.org/10.1007/978-94-007-1363-5_21
Maeght J, Rewald B, Pierret A (2013) How to study deep roots—and why it matters. Front Plant Sci 4:1–14. https://doi.org/10.3389/fpls.2013.00299
Manoli G, Huang CW, Bonetti S et al (2017) Competition for light and water in a coupled soil-plant system. Adv Water Resour 108:216–230. https://doi.org/10.1016/j.advwatres.2017.08.004
Markewitz D, Devine S, Davidson EA et al (2010) Soil moisture depletion under simulated drought in the Amazon: impacts on deep root uptake. New Phytol 187:592–607. https://doi.org/10.1111/j.1469-8137.2010.03391.x
Matheny AM, Fiorella RP, Bohrer G et al (2017) Contrasting strategies of hydraulic control in two codominant temperate tree species. Ecohydrology 10:1–16. https://doi.org/10.1002/eco.1815
Meinzer FC, Clearwater MJ, Goldstein G (2001) Water transport in trees: current perspectives, new insights and some controversies. Environ Exp Bot 45:239–262. https://doi.org/10.1016/S0098-8472(01)00074-0
Metzger JC, Wutzler T, Valle ND, Filipzik J, Grauer C, Lehmann R et al (2017) Vegetation impacts soil water content patterns by shaping canopy water fluxes and soil properties. Hydrol Process 31:3783–3795. https://doi.org/10.1002/hyp.11274

Milly PCD, Dunne KA (1994) Sensitivity of the global water cycle to the water-holding capacitiy of land. J Clim 7:506–526. https://doi.org/10.1175/1520-0442(1994)007<0506

Miralles DG, De Jeu RAM, Gash JH et al (2011) Magnitude and variability of land evaporation and its components at the global scale. Hydrol Earth Syst Sci 15:967–981. https://doi.org/10.5194/hess-15-967-2011

Musters PAD, Bouten W, Verstraten JM (2000) Potentials and limitations of modelling vertical distributions of root water uptake of an Austrian pine forest on a sandy soil. Hydrol Process 14:103–115. https://doi.org/10.1002/(SICI)1099-1085(200001)14:1<103::AID-HYP913>3.0.CO;2-5

Nadezhdina N, Čermák J (2003) Instrumental methods for studies of structure and function of root systems of large trees. J Exp Bot 54:1511–1521. https://doi.org/10.1093/jxb/erg154

Nadezhdina N, David TS, David JS et al (2012a) Root structure: in situ studies through sap flow research. In: Mancuso S (ed) Measuring roots: an updated approach. Springer-Verlag, Berlin, Heidelberg, pp 247–266

Nadezhdina N, David TS, David JS et al (2012b) Root function: in situ studies through sap flow research. In: Mancuso S (ed) Measuring roots: an updated approach. Springer-Verlag, Berlin, Heidelberg, pp 267–290

Neumann RB, Cardon ZG (2012) The magnitude of hydraulic redistribution by plant roots: a review and synthesis of empirical and modeling studies. New Phytol 194:337–352. https://doi.org/10.1111/j.1469-8137.2012.04088.x

O'Brien MJ, Engelbrecht BMJ, Joswig J et al (2017) A synthesis of tree functional traits related to drought-induced mortality in forests across climatic zones. J Appl Ecol 54:1669–1686. https://doi.org/10.1111/1365-2664.12874

Ogle S, Wolpert RL, Reynolds JF (2004) Reconstructing plant root area and water uptake profiles. Ecology 85:1967–1978. https://doi.org/10.1002/jhrc.1240160411

Oliveira MDRG, van Noordwijk M, Gaze SR et al (2000) Auger sampling, ingrowth cores and pinboard methods. In: Smit AL, Bengough AG, Engels C et al (eds) Root methods. Springer, Berlin Heidelberg, pp 175–210. https://doi.org/10.1007/978-3-662-04188-8_6

Orlowski N, Pratt DL, McDonnell JJ (2016) Intercomparison of soil pore water extraction methods for stable isotope analysis. Hydrol Process 30:3434–3449. https://doi.org/10.1002/hyp.10870

Paluch JG, Gruba P (2012) Effect of local species composition on topsoil properties in mixed stands with silver fir (*Abies alba* Mill.). Forestry 85:413–426. https://doi.org/10.1093/forestry/cps040

Parnell AC, Inger R, Bearhop S, Jackson AL (2010) Source partitioning using stable isotopes: coping with too much variation. PLoS One 5:1–5. https://doi.org/10.1371/journal.pone.0009672

Poorter L, McDonald I, Alarcón A et al (2010) The importance of wood traits and hydraulic conductance for the performance and life history strategies of 42 rainforest tree species. New Phytol 185:481–492. https://doi.org/10.1111/j.1469-8137.2009.03092.x

Pretzsch H, Forrester DI, Rötzer T (2015) Representation of species mixing in forest growth models: a review and perspective. Ecol Model 313:276–292. https://doi.org/10.1016/j.ecolmodel.2015.06.044

Prieto I, Armas C, Pugnaire FI (2012) Water release through plant roots : new insights into its consequences at the plant and ecosystem level. New Phytol 193:830–841. https://doi.org/10.1111/j.1469-8137.2011.04039.x

Puettmann KJ, Coates DK, Messier C (2009) A critique of silviculture : managing for complexity. Island Press, Washington, DC

Reichstein M, Bahn M, Ciais P et al (2013) Climate extremes and the carbon cycle. Nature 500:287–295. https://doi.org/10.1038/nature12350

Renner M, Hassler SK, Blume T et al (2016) Dominant controls of transpiration along a hillslope transect inferred from ecohydrological measurements and thermodynamic limits. Hydrol Earth Syst Sci Discuss 1–30. https://doi.org/10.5194/hess-2015-535

Rewald B, Leuschner C (2009) Does root competition asymmetry increase with water availability? Plant Ecol Divers 2:255–264. https://doi.org/10.1080/17550870903022865

Rewald B, Ephrath JE, Rachmilevitch S (2011) A root is a root is a root? Water uptake rates of *Citrus* root orders. Plant Cell Environ 34:33–42. https://doi.org/10.1111/j.1365-3040.2010.02223.x

Richner W, Liedgens H, Bürgi H et al (2000) Root image analysis and interpretation. In: Smit AL, Bengough AG, Engels C et al (eds) Root methods. Springer, Berlin Heidelberg, pp 305–341. https://doi.org/10.1007/978-3-662-04188-8_10

Rothfuss Y, Javaux M (2017) Isotopic approaches to quantifying root water uptake and redistribution: a review and comparison of methods. Biogeosciences 14:2199–2224. https://doi.org/10.5194/bg-14-2199-2017

Rothfuss Y, Biron P, Braud I et al (2010) Partitioning evapotranspiration fluxes into soil evaporation and plant transpiration using water stable isotopes under controlled conditions. Hydrol Process 24:3177–3194. https://doi.org/10.1002/hyp.7743

Schenk HJ, Jackson RB (2002) Rooting depths, lateral root spreads and belowground aboveground allometries of plants in water limited ecosystems. J Ecol 480–494. https://doi.org/10.1046/j.1365-2745.2002.00682.x

Schume H, Jost G, Hager H (2004) Soil water depletion and recharge patterns in mixed and pure forest stands of European beech and Norway spruce. J Hydrol 289:258–274. https://doi.org/10.1016/j.jhydrol.2003.11.036

Schwärzel K, Menzer A, Clausnitzer F et al (2009) Soil water content measurements deliver reliable estimates of water fluxes: a comparative study in a beech and a spruce stand in the Tharandt forest (Saxony, Germany). Agric For Meteorol 149:1994–2006. https://doi.org/10.1016/j.agrformet.2009.07.006

Schwärzel K, Ebermann S, Schalling N (2012) Evidence of double-funneling effect of beech trees by visualization of flow pathways using dye tracer. J Hydrol 470–471:184–192. https://doi.org/10.1016/j.jhydrol.2012.08.048

Schwendenmann L, Pendall E, Sanchez-Bragado R et al (2015) Tree water uptake in a tropical plantation varying in tree diversity: interspecific differences, seasonal shifts and complementarity. Ecohydrology 8:1–12. https://doi.org/10.1002/eco.1479

Schwinning S (2010) The ecohydrology of roots in rocks. Ecohydrology 130:238–245. https://doi.org/10.1002/eco.134

Schymanski SJ, Sivapalan M, Roderick ML et al (2008) An optimality-based model of the coupled soil moisture and root dynamics. Hydrol Earth Syst Sci 12:913–932. https://doi.org/10.5194/hess-12-913-2008

Silvertown J, Araya Y, Gowing D (2015) Hydrological niches in terrestrial plant communities: a review. J Ecol 103:93–108. https://doi.org/10.1111/1365-2745.12332

Smithwick EAH, Lucash MS, McCormack ML, Sivandran G (2014) Improving the representation of roots in terrestrial models. Ecol Model 291:193–204. https://doi.org/10.1016/j.ecolmodel.2014.07.023

Speich MJR, Lischke H, Zappa M (2017) Testing an optimality-based model of rooting zone water storage capacity in temperate forests. Hydrol Earth Syst Sci Discuss 1–33. https://doi.org/10.5194/hess-2017-723

Sperry JS, Adler FR, Campbell GS, Comstock JP (1998) Limitation of plant water use by rizosphere and xylem conductance: results of a model. Plant Cell Environ 21:347–359. https://doi.org/10.1046/j.1365-3040.1998.00287.x

Sprenger M, Leistert H, Gimbel K, Weiler M (2016) Illuminating hydrological processes at the soil-vegetation-atmosphere interface with water stable isotopes. Rev Geophys 54:674–704. https://doi.org/10.1002/2015RG000515

Steppe K, De Pauw DJW, Doody TM, Teskey RO (2010) A comparison of sap flux density using thermal dissipation, heat pulse velocity and heat field deformation methods. Agric For Meteorol 150:1046–1056. https://doi.org/10.1016/j.agrformet.2010.04.004

Steudle E, Peterson CA (1998) How does water get through roots? J Exp Bot 49:775–788. https://doi.org/10.1093/jxb/49.322.775

Taiz L, Zeiger E (2002) Plant physiology, 3rd edn. Sinauer Associates, Sunderland

Tardieu F, Simonneau T (1998) Variability among species of stomatal control under fluctuating soil water status and evaporative demand: modelling isohydric and anisohydric behaviours. J Exp Bot 49:419–432. https://doi.org/10.1093/jxb/49.Special_Issue.419
Tyree MT, Sperry JS (1989) Vulnerability of xylem to cavitation and embolism. Annu Rev Plant Physiol Plant Mol Biol 40:19–36. https://doi.org/10.1146/annurev.pp.40.060189.000315
Tyree MT, Zimmermann MH (2002) Xylem structure and the ascent of sap. Springer, Berlin, Heidelberg
Valverde-Barrantes OJ, Smemo KA, Feinstein LM et al (2013) The distribution of below-ground traits is explained by intrinsic species differences and intraspecific plasticity in response to root neighbours. J Ecol 101:933–942. https://doi.org/10.1111/1365-2745.12087
Vamerali T, Bandiera M, Mosca G (2012) Minirhizotrons in modern root studies. In: Measuring roots: an updated approach, pp 341–361
Vandegehuchte MW, Steppe K (2013) Sap- flux density measurement methods : working principles and applicability. Funct Plant Biol 40:213–223. https://doi.org/10.1071/FP12233
Vanderlinden K, Vereecken H, Hardelauf H et al (2012) Temporal stability of soil water contents: a review of data and analyses. Vadose Zo J 11. https://doi.org/10.2136/vzj2011.0178
Vetterlein D, Doussan C (2016) Root age distribution: how does it matter in plant processes? A focus on water uptake. Plant Soil 407:145–160. https://doi.org/10.1007/s11104-016-2849-6
Vitali V, Forrester DI, Bauhus J (2018) Know your neighbours: drought response of Norway spruce, silver fir and Douglas fir in mixed forests depends on species identity and diversity of tree neighbourhoods. Ecosystems 21:1215–1229. https://doi.org/10.1007/s10021-017-0214-0
Walter J (2018) Effects of changes in soil moisture and precipitation patterns on plant-mediated biotic interactions in terrestrial ecosystems. Plant Ecol 219:1449–1462. https://doi.org/10.1007/s11258-018-0893-4
Warren JM, Brooks JR, Meinzer FC, Eberhart JL (2008) Hydraulic redistribution of water from Pinus ponderosa trees to seedlings: evidence for an ectomycorrhizal pathway. New Phytol 178:382–394. https://doi.org/10.1111/j.1469-8137.2008.02377.x
Wullschleger SD, Meinzer FC, Vertessy RA (1998) A review of whole-plant water use studies in trees. Tree Physiol 18:499–512. https://doi.org/10.1093/treephys/18.8-9.499
Zanne AE, Falster DS (2010) Plant functional traits – linkages among stem anatomy, plant performance and life history. New Phytol 185:343–348. https://doi.org/10.1111/j.1469-8137.2009.03138.x
Zehe E, Graeff T, Morgner M et al (2010) Plot and field scale soil moisture dynamics and subsurface wetness control on runoff generation in a headwater in the Ore Mountains. Hydrol Earth Syst Sci 14:873–889. https://doi.org/10.5194/hess-14-873-2010
Zimmermann U, Ehhalt D, Muennich KO (1967) Soil-water movement and evapotranspiration: Changes in the isotopic composition of the water. In Symposium on isotopes in hydrology; Vienna (Austria); 14–18 Nov 1966. International Atomic Energy Agency (IAEA), Vienna (Austria), pp 567–584
Zwieniecki MA, Newton M (1995) Roots growing in rock fissures: their morphological adaptation. Plant Soil 172:181–187. https://doi.org/10.1007/BF00011320
Zwieniecki MA, Thompson MV, Holbrook NM (2003) Understanding the hydraulics of porous pipes: tradeoffs between water uptake and root length utilization. J Plant Growth Regul 21:315–323. https://doi.org/10.1007/s00344-003-0008-9

Chapter 15
Effects of Stemflow on Soil Water Dynamics in Forest Stands

Wei-Li Liang

15.1 Introduction

Soil moisture in a forested stand varies in time and space and is controlled by meteorological (e.g., rainfall amount and intensity), geological (e.g., landform, soil depth, and soil texture), and vegetation (e.g., species, density, and morphologic appearance) factors. The factors controlling soil moisture behavior can also be classified into dynamic and static factors. They are synonymous with the concept of local or nonlocal controls proposed by Grayson et al. (1997), indicating that soil water responses are dominated by local environmental features or contributed by surface/subsurface lateral water movement elsewhere, respectively. Thus, soil moisture behavior is not necessarily controlled by a limited set of factors, but rather a combination of various controlling factors is important (Western et al. 2002; Robinson et al. 2008; Vereecken et al. 2008; Lin 2010). For example, interactions between meteorological and vegetation factors influence rainfall redistribution processes aboveground, and interactions between geological and vegetation factors control water storage and movement via infiltration and water redistribution processes underground. Although the main controlling factors are site-dependent, it would be of great interest for ecohydrologists to clarify the generality of the relationships between soil water dynamics and their controlling factors (Western et al. 1999; Tromp-van Meerveld and McDonnell 2006; Gwak and Kim 2017; Liang et al. 2017).

With regard to rainfall redistribution processes above forested ground, precipitation in forests is first intercepted by canopies and then partitioned into throughfall and stemflow components as diffuse and point inputs. Thus, precipitation that reaches the forest floor is considerably uneven. Rainfall redistribution processes have been broadly reviewed in the previous literature (e.g., Crockford and

W.-L. Liang (✉)
School of Forestry and Resource Conservation, National Taiwan University, Taipei, Taiwan
e-mail: liangwl@ntu.edu.tw

D. F. Levia et al. (eds.), *Forest-Water Interactions*, Ecological Studies 240,
https://doi.org/10.1007/978-3-030-26086-6_15

Richardson 2000; Levia and Frost 2003, 2006; Levia and Germer 2015). In addition to rainfall redistribution processes aboveground, trees can also increase the heterogeneity of the rainwater infiltration and redistribution processes underground due to root-induced preferential flow or root uptake. Root-induced channels could be preferential flow pathways, and stemflow tends to follow these channels within the soil (Voigt 1960; Martinez-Meza and Whitford 1996; Schwärzel et al. 2012). Thus, stemflow contributes to an uneven water input to the forest floor and preferential percolation and flow in the subsurface that increases the heterogeneity of soil water dynamics in forested stands. Johnson and Lehmann (2006) referred to these aboveground and underground effects as the double-funneling effect of a tree. The double-funneling effect may have implications for hydrological responses including soil water dynamics (Metzger et al. 2017), soil erosion (Tanaka et al. 1991), subsurface saturation (Germer 2013), slope stability (Liang et al. 2010), groundwater recharge (Taniguchi et al. 1996), and runoff generation (Neave and Abrahams 2002) in space and time. A comprehensive and integrated understanding of the double-funneling effect, involving two heterogeneous processes of the stemflow supply aboveground and root-induced preferential flow belowground, is needed. This chapter reviews the current understanding of how stemflow affects soil water dynamics in both wetting and drying processes. After discussing field evidence of the effects of stemflow on soil water responses, their modeling and issues requiring future study are highlighted.

15.2 Stemflow Infiltration in the Wetting Process

15.2.1 Effects of Throughfall and Stemflow on Soil Water Dynamics in Forests

Many studies have examined the effects of rainfall redistribution processes on soil water dynamics based on field observations or numerical experiments (e.g., Durocher 1990; Bouten et al. 1992a; Keim et al. 2006; Coenders-Gerrits et al. 2013). However, most of these studies minimized stemflow, dealt with interception loss as the difference between gross and net rainfall, and emphasized that the spatial distribution of throughfall mainly affects soil water dynamics. Throughfall that reaches the forest floor would either fall from foliage as foliage drip or pass directly through gaps within the canopy as direct throughfall, with both pathways mainly determined by the leaf traits or canopy architecture (Ford and Deans 1978; Carleton and Kavanagh 1990; Loustau et al. 1992; Beier et al. 1993; Nanko et al. 2013). Higher throughfall has been observed at the canopy perimeter in a Norway spruce (*Picea abies* [L.] Karst.) forest (Beier et al. 1993) and a juvenile lodgepole pine (*Pinus contorta*) forest (Carlyle-Moses and Lishman 2015), in the middle position within the canopy of a black spruce (*Picea mariana*) forest (Carleton and Kavanagh 1990), and near tree stems in a Sitka spruce (*Picea sitchensis*) forest (Ford and Deans 1978). Loustau et al. (1992) measured throughfall in a maritime pine (*Pinus pinaster*

Ait.) forest using 52 rain gauges and concluded that there was no spatial distribution pattern with distance from the tree stem. These inconsistencies indicate large inter-species variation in throughfall, even within the same genus of *Picea* or *Pinus*. The inconsistent distribution patterns of throughfall were also attributed to local drip points of throughfall that might deliver greater amounts than gross rainfall (Lloyd and Marques 1988; Kellman and Roulet 1990; Robson et al. 1994; Keim et al. 2005).

Although spatial variation in throughfall would contribute to spatial variation in soil water, the diffuse pattern of throughfall patterns may not necessarily be reflected in the soil moisture patterns (Pressland 1976; Raat et al. 2002; Metzger et al. 2017). Excluding the potential influences of geological factors and soil properties, one explanation for this pattern would be the influence of local rainwater inputs from stemflow concentrated around the base of a tree (Voigt 1960; Durocher 1990; Liang et al. 2007). Previous studies of soil water dynamics in forested stands have given little attention to the effects of stemflow due to the low ratio of stemflow to observed precipitation. When considering stemflow in rainfall redistribution processes, the total volume of stemflow is usually divided by the canopy projection area or observed stand area. For example, Huber and Iroume (2001) conducted long-term observations of precipitation, throughfall, and stemflow at 29 research plots covering a wide range of rainfall zones and forest types, species, ages, and densities. The ratio of throughfall to total precipitation ranged from 55% to 86%, with a larger ratio in coniferous stands than in broadleaved forests. The ratio of stemflow to total precipitation ranged from 1% to 13% in coniferous stands and from 1% to 8% in broadleaved forests. However, it is important to evaluate the point input characteristics of stemflow when considering soil water dynamics, even for tree species with a low ratio of stemflow to precipitation. Aboal et al. (1999) quantified the stemflow of 30 sample trees from six different species in a laurel forest and found that precipitation could be concentrated by up to 12.8 times in infiltration areas of trees due to stemflow, although the annual stemflow only represented 6.9% of the gross precipitation. Similarly, Liang et al. (2007) reported that the stemflow per infiltration area on the downslope side of a tall stewartia (*Stewartia monadelpha*) stem was 18.9 times the open area rainfall, while the stemflow per canopy projection area was only 38.9% of the open area rainfall for a heavy storm event. Thus, the funneling ratio (F) has been proposed to evaluate the concentrated nature of stemflow (Herwitz 1986a; Levia and Frost 2003):

$$F = V/(BG) \tag{15.1}$$

where V is the stemflow volume, B is the basal area of the trunk, and G is the depth of the gross precipitation.

Levia and Germer (2015) further reviewed several indices related to the quantification of stemflow, in which the key parameters of basal area, stemflow infiltration area, or canopy projection area were combined into several patterns to represent the funneling characteristics of chemical or physical fluxes in stemflow. Building on the work of Levia and Germer (2015), Carlyle-Moses et al. (2018) provided a comprehensive examination of funneling ratios per unit trunk basal area and per unit stemflow infiltration area at the tree and stand scales to enable accurate scaling of stemflow commensurate with its ecohydrological importance.

The point inputs of stemflow in forested stands may affect not only the local soil water dynamics but also the physical and chemical properties of the soil (Gersper and Holowaychuk 1970) and root development (Herwitz and Levia 1997). Ford and Deans (1977) found a greater concentration of fine roots close to tree stems and noted that this might reflect the spatial pattern of the water flux in forest soil. In addition, Herwitz (1986b) and Tanaka et al. (1991) observed soil erosion scars around the bases of some trees and attributed them to the infiltration excess produced by concentrated stemflow inputs to the forest floor, presumably during extreme events. The influences of stemflow have not only been found on a small scale at the tree base, but also at the catchment scale influencing hydrological processes such as runoff generation (Neave and Abrahams 2002), groundwater recharge (Taniguchi et al. 1996), and the response of chemicals in solution (Chang and Matzner 2000).

15.2.2 General Features of Stemflow for Trees Growing on a Hillslope

Stemflow characteristics vary with species, particularly for deciduous trees in which the canopy architecture changes between leafed and leafless conditions. Previous studies of deciduous beech and oak trees have reported that stemflow in the leafless period tends to be greater than in the leafed period (Helvey and Patric 1965; Giacomin and Trucchi 1992; Neal et al. 1993; Staelens et al. 2008) or shows no seasonal variation (Deguchi et al. 2006; Sraj et al. 2008). Although Liang et al. (2009a) observed greater stemflow in the leafed period than in the leafless period for tall stewartia trees, negative effects of leaves on stemflow yield are commonly recognized (Levia and Frost 2003; Levia et al. 2015). Greater stemflow in the leafless period is commonly explained by the largely exposed branches that could capture more rainfall (Helvey and Patric 1965). Thus, leaves may prevent the branches from becoming wet and conducting water down the trunk (Giacomin and Trucchi 1992). The branch architecture is also a main factor affecting intraspecific variability in stemflow production rates (Levia and Frost 2003; Levia et al. 2015). Herwitz (1987) clarified the strong positive relationship between branch flow and branch inclination in a laboratory experiment. Additionally, smooth bark has lower water storage capacity and contributes to the generation of a continuous flow path, which would increase stemflow yield (Levia and Herwitz 2005). Helvey and Patric (1965) attributed the greater stemflow in young stands than in old stands to a smoother bark with branches tending to grow upward rather than outward in young stands.

In addition to the general controlling factors discussed above, the morphological features of trees growing on hillslopes may differ from those on areas of gentle relief, even for the same species. Unlike trees growing on an area of relatively flat terrain, those on a steep hillslope generally incline toward the downslope and are more or less S-shaped (Schweingruber 1996), particularly for broad-leafed trees. This may cause uneven canopy architecture and flow pathways between the upslope and downslope sides and result in different stemflow yields on the upslope and

Fig. 15.1 (**a**) Separate stemflow collection systems along the upslope and downslope sides of a tree trunk (from Liang et al. 2009a) and (**b**) the asymmetrical canopy area and stemflow pathways for a tree growing on a hillslope

downslope sides of a tree trunk. Liang et al. (2009a) used two tubes cut longitudinally and wrapped spirally around the upslope and downslope sides of the trunk to separately collect stemflow from the upslope (SF-up) and downslope (SF-down) sides (Fig. 15.1a). Their observations showed that the amount of SF-down was 12–132 times greater than SF-up for six tall stewartia on a hillslope with a mean gradient of 28°. In addition to the asymmetrical canopy projection area, the greater stemflow volume along the downside of the trunk was attributable to asymmetrical stemflow pathways between the upslope and downslope sides of the trunk due to the tilt of the tree trunk toward the downslope direction (Fig. 15.1b). Although some previous studies have measured stemflow for trees on forested hillslopes (e.g., Park and Cameron 2008), the asymmetrical generation of stemflow upslope and downslope of tree stems has rarely been reported. Such asymmetrical generation of stemflow may be a typical feature of trees growing on a hillslope, which enlarges the point input of water into soil layers and therefore has significant implications for hillslope hydrological processes.

15.2.3 Root-Induced Preferential Flow

In addition to the asymmetrical distribution of water supply, preferential infiltration pathways around trees also increase the heterogeneity of soil water dynamics. Rock fractures and macropores around tree roots are primary pathways for preferential flow (Gaiser 1952; Dasgupta et al. 2006). Beven and Germann (1982) indicated that macropores in soil may be associated with either living or decayed tree roots and that the structure of macropore systems derived from roots may be very effective at channeling water through soil layers. Beven and Germann (1982) also indicated that the decayed tree roots may facilitate the formation of hose-type macropores sealed by the remaining bark; due to the fact that the bark of tree roots sometimes resists decay longer than the xylem. Based on sprinkling experiments, root observations, and numerical simulations, Lange et al. (2009) provided evidence that root density is closely related to the occurrence of preferential infiltration. Weiler (2017) indicated that preferential flow is a normal phenomenon rather than an exception in the field. The preferential pathway formed by tree roots is wider and deeper than those of crops or pasture plants (Yunusa et al. 2002). Thus, the preferential flow pathways created by roots can have important implications for soil water dynamics in forested hydrological processes. Such complex networks of interconnected preferential flow pathways could affect subsurface flow, hillslope stability, and stormflow responses even at the catchment scale, which has been revealed by both field observations and conceptual models (Noguchi et al. 1999; Sidle et al. 2001; Lin 2006; Ghestem et al. 2011; Jost et al. 2012; Sidle and Bogaard 2016).

Preferential pathways have been identified using dye tracers and image analyses (Noguchi et al. 1999; Kulli et al. 2003; Sander and Gerke 2007), data analyses of soil water responses (Liang et al. 2007; Lin and Zhou 2008; Lozano-Parra et al. 2016), or combinations of both tracers and field datasets (Liang et al. 2011). Noguchi et al. (1999) conducted a dye experiment and showed that flow impeded by living roots can be diverted around the perimeter of roots, generating subsurface flow and saturated zones in the soil matrix. The application of dye tracers enables infiltration pathways to be visualized and the presence of preferential and matrix flows to be distinguished in excavated soil profiles. However, Beven and Germann (2013) indicated that dyes require destructive sampling and may not reveal all of the pathways of moving water. Using datasets acquired by nondestructive techniques to evaluate soil water behavior or to identify preferential flow is preferable for ongoing monitoring in the field. The occurrence of preferential flow is recognized when soil water responds to a rainwater input earlier in deeper soil layers than in shallower layers (Liang et al. 2007; Lin and Zhou 2008). Moreover, Lozano-Parra et al. (2016) assumed that the slow and rapid responses of soil water content are caused by matrix and preferential flows, respectively, and used the maximum increase in soil water content during rainfall events to determine the hydrological response due to matrix or preferential flows in several plots within a catchment. Van Stiphout et al. (1987) demonstrated that bypass flow reached depths of 60–135 cm, whereas the wetting front in the soil matrix expanded only to depths of 7–12 cm. Thus, unlike the slow expansion of the wetting front in matrix flow, preferential flow could result in rapid soil water responses.

Stemflow tends to enter soil layers along the network of root channels and serves as a principal source of the rainwater input around a tree. Martinez-Meza and Whitford (1996) and Li et al. (2009) sprinkled a dye tracer on the soil surface around

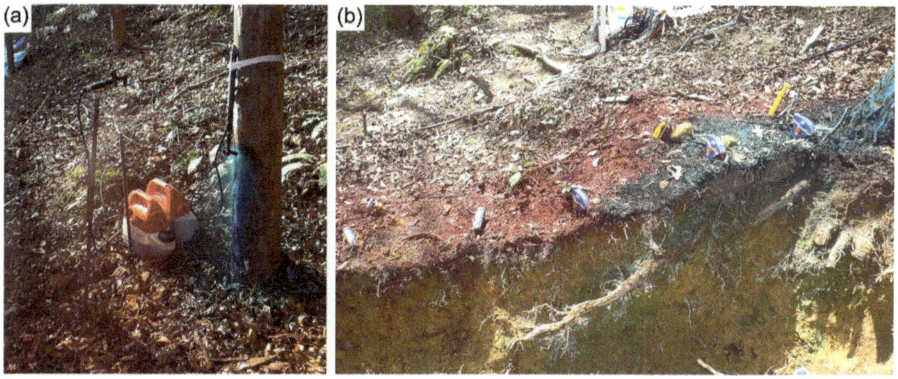

Fig. 15.2 (**a**) A sprinkling experiment using two dye solutions applied toward the ground (red) and tree stem (greenish blue) to represent throughfall and stemflow surface infiltration areas, respectively. (**b**) Stained areas in the downslope soil profile, depicting prefential channelization of stemflow along roots. (Modified from Liang et al. 2011)

the base of a trunk and demonstrated that the root channels of desert shrubs were preferential pathways for the movement of stemflow water into soil. Schwärzel et al. (2012) conducted a dye experiment and performed detailed analyses that demonstrated the double-funneling effect of beech trees. They also revealed differences in the double-funneling effect between spruce and beech trees and suggested that stemflow and root-induced preferential flow in a beech forest could trigger fast lateral subsurface flow bypassing large parts of the soil matrix, at least during rainstorm events. However, the application of only one dye tracer would not necessarily reveal the entire infiltration pattern in a forested stand because both throughfall and stemflow contribute rainwater inputs to the soil around a tree. Liang et al. (2011) separately identified infiltration areas derived from throughfall and stemflow using two dye solutions that were sprinkled toward the ground and a tree stem, respectively (Fig. 15.2a). Infiltration and percolation from throughfall input was limited to shallow soil layers, but stemflow inputs, infiltrating near the trunk, bypassed shallow soil layers and flowed horizontally along thick roots (Fig. 15.2b). Using event-based analyses, the authors also conducted an experiment that monitored soil water dynamics around a tree on a hillslope based on 2 years of field observations before (SF period) and after (non-SF period) intercepting the stemflow of the tree. The maximum increase in soil water storage was greater than rainfall supply and there was irregular variation in soil water content downslope from the tree, both of which were attributed to the concentration of stemflow on the downslope side of the tree during rainfall events in the SF period. When focusing on the downslope region from the SF to non-SF periods, they found that the frequency of occurrence of this phenomenon decreased from 43% to 10%, but there was little change in the frequency of bypass flow. The results revealed that stemflow could cause asymmetrical soil water dynamics between regions downslope and upslope from a tree growing on a hillslope. In addition, stemflow contributed to soil water storage but did not initiate bypass flow. In other words, stemflow enlarged the magnitude or extent of preferential flow but did not generate preferential flow. Preferential flow is mainly induced by the root architecture.

15.2.4 Stemflow-Induced Perched Water Tables

At the vegetation community scale, Glover et al. (1962) observed deeper wetting fronts underneath taller grassland plants in Maasailand, Kenya, and indicated that the depth of rainwater percolation was approximately equal to the depth of the normal rainfall percolation into the bare soil plus the height of the plants in dry grassland communities. Their finding was likely due to the funneling effect of the grass leaves, which concentrate and accumulate rainwater to become stemflow that infiltrates around the plants. Similarly, Ludwig et al. (2005) reported that considerable soil water was stored in small groves of mulga trees, causing a deeper wetting front under mulga groves than in inter-grove areas without trees. Taniguchi et al. (1996) calculated groundwater recharge rates using a mass balance method with chloride in subsurface waters in a pine forest. They found that the ratio of stemflow to net precipitation was small, ranging from 0.5% to 1.2%, but the ratio of the recharge by stemflow to the total groundwater recharge was relatively large, ranging from 10.9% to 19.1%. They concluded that the effects of stemflow on groundwater recharge cannot be disregarded, even for tree species with a low ratio of stemflow to precipitation.

At the single tree scale, Durocher (1990) observed very rapid soil water movement and the generation of saturated zones above the bedrock surface underneath sweet chestnut trees recorded by a monitoring network of tensiometers during a 7 mm rainfall event, while there were no significant changes in soil water potential at a location 3.7 m from the trees. The inconsistent responses of soil water potential were attributed to small-scale water inputs from stemflow that rapidly infiltrated into and deeply percolated within the soil matrix. Liang et al. (2007) observed an asymmetric saturated zone at the soil–bedrock interface around a tall stewartia tree. Due to the generation of large amounts of stemflow on the downslope side of the tree, a realtively small area of stemflow infiltration into the subsurface, and rapid stemflow channelization along roots, perched water tables were generated more rapidly at the soil–bedrock interface and to a greater degree in the region downslope from the tree stem than in the upslope region (Fig. 15.3).

In addition to subsurface saturation or perched water table that tends to be generated at the soil–bedrock interface due to the sharp changes and clear differences in hydraulic properties between soil and bedrock layers (Weiler et al. 2006; Katsura et al. 2009; Masaoka et al. 2016; Liang and Chan 2017), Germer (2013) provided evidence that stemflow-induced saturation could also be generated in a shallow soil layer above an impeding layer. Based on a monitoring network of piezometers at soil depths of 20 and 50 cm, Germer (2013) demonstrated that stemflow can flow through the root system, induce initial horizontal water flows through the soil, and contribute to perched water table development above a shallow impeding layer in a

Fig. 15.3 A schematic showing the evolution of a stemflow-induced perched water table at the soil–bedrock interface

babassu palm forest. The perched water tables were observed to be more frequently located near palm stems than at locations further away. Although the above studies clearly indicate that the generation of perched water tables is related to concentrated rainwater inputs, it is still not clear if perched water tables can be attributed to the point input of stemflow or to root-induced preferential flow, which is a consequence of the double-funneling effect of a tree. Based on a comparison before and after the interception of stemflow from a tree, Liang et al. (2011) found that the frequency of saturated zones at the soil–bedrock interface decreased from the condition with stemflow supply (58%) to that without stemflow supply (16%), and there was a lesser extent of saturation without stemflow supply. Liang et al. (2011) also provided evidence that both stemflow and root-induced preferential flow contribute to subsurface saturation and that the point input of stemflow can enlarge the frequency and amount of subsurface saturation underneath a tree.

Concentrated rainwater infiltration from stemflow has significant implications for other hydrological processes. Germer (2013) indicated that high-intensity water input to the soil surface due to stemflow might lead to the following three responses: Horton overland flow, subsurface saturation above an impeding soil layer, or fast, deep percolation as preferential flow along thick roots and subsequent saturated zone generation at the soil–bedrock interface. In addition, the formation of perched water tables in the soil mantle generally controls rainwater discharge and slope instability on steep landscapes (Anderson and Sitar 1995; Wang and Sassa 2003). The generation of a saturated zone at the soil–bedrock interface contributes to runoff or deep percolation (Montgomery and Dietrich 2002; Kosugi et al. 2006; Jost et al. 2012). Information regarding the spatial and temporal extent of saturated zones at the soil–bedrock interface can be used to help predict the location and timing of shallow landslides (Montgomery et al. 2002; Dhakal and Sidle 2004). Thus, in addition to the hydrologically active areas referred to by Bachmair and Weiler (2011), trees are also key triggers of the generation of subsurface saturated zones.

15.3 Effects of Stemflow on the Soil Water Redistribution Process

15.3.1 General Focus on the Soil Water Redistribution Process

Soil water behavior could differ between the wetting and drying processes in forested stands. In addition to meteorological and geological factors, many previous studies have considered the effects of vegetation on water redistribution processes by different species, evapotranspiration, or the combination of these variables (Tromp-van Meerveld and McDonnell 2006). The focus has generally been on the species or density of vegetation (Garcia-Estringana et al. 2012; Emanuel et al. 2014), root system (Archer et al. 2012; Robinson et al. 2012), and evapotranspiration (Schume et al. 2004; Krämer and Hölscher 2010). Krämer and Hölscher (2010) found that soil water was extracted at a higher rate in species-rich plots at the beginning of the drying process, but only beech-dominated stands extracted large amounts of soil water at later stages of the drying process when the atmospheric evaporative demand was higher. Robinson et al. (2012) investigated the spatiotemporal variability of hydraulic redistribution within an oak–pine forest and found that soil moisture demand and usage was greater in areas with trees than in areas without them, suggesting that root zones likely retain and regulate moisture through hydraulic redistribution or hydraulic lift. Schume et al. (2004) reported higher water content in the deep soil layers of a beech stand than in a spruce stand, due to root systems producing different evapotranspiration rates in the stands. Schwärzel et al. (2009) demonstrated that spatial and temporal variability in soil water depletion and transpiration in beech and spruce stands is mainly due to differences in soil properties and root density. According to Schwärzel et al. (2012), the different root architecture and depth of root development between beech and spruce stands not only influenced transpiration but also affected soil water redistribution. Based on these studies, it is clear that soil water use by vegetation is strongly related to the root system, which then affects evapotranspiration and soil water redistribution in forested stands.

15.3.2 Effects of Stemflow and Roots on Soil Water Depletion

The soil water redistribution process is mainly controlled by upward evapotranspiration and downward rainwater drainage. Generally, soil water depletion is expected to be greater in locations near a tree than in other areas due to root uptake and upward transpiration in the drying process. According to Jackson et al. (2000), soil water inputs and depletion were more active at a distance of 0.3 m from a tree than at 2.5 m, which was attributed to stemflow inputs and root uptake. According to Schwärzel et al. (2009), the differences in evapotranspiration estimated using the eddy covariance and soil water balance methods were often large around a beech tree, due to a lack of information about stemflow-induced wetting and subsequent drying around the trunk.

Although many previous studies have indicated the effects of root uptake or evapotranspiration on soil water redistribution, soil water depletion from vertical drainage has rarely been discussed, particularly with regard to the decrease in soil water following an extra increase in soil water due to stemflow. Germer (2013) found that root distributions were related to soil water redistribution and that the development of perched water tables due to the infiltration and percolation of stemflow would still occur even after small rainfall events. This suggests that soil conditions should remain wet close to a tree after a rainwater supply ceases. However, Durocher (1990) found that the rapid and deep increases in soil water potential generated by stemflow following a small rainfall event (total amount of 7 mm) were quickly and spatially redistributed after the event, resulting in a near-uniform water potential distribution. This raises the questions of what controls soil water depletion around a tree and how long the stemflow-induced soil water recharge influences soil water redistribution processes. Liang et al. (2015) attempted to separate rainwater drainage and evapotranspiration based on the rate of instantaneous soil water depletion in the drying durations before and after the interception of stemflow from a tree on a hillslope. They found that the contribution of soil water drainage was much greater than evapotranspiration in the 10 days without a rainfall supply. According to Liang et al. (2015), the effects of stemflow on soil water depletion were only significant for a short period of time (0.5 h) after the end of rainfall events, although stemflow increased the initial soil water content at the start of the drying process. In addition, the effects of root-induced pathways enhanced drainage during the drying process and resulted in irregular vertical distributions of changes in soil water depletion. This suggests that the double-funneling effect of a tree influences not only the wetting process but also the drying process. Thus, greater soil water depletion should occur at locations near a tree under natural conditions with stemflow (Metzger et al. 2017). The spatial pattern of soil water would be more even during the drying process if stemflow was removed.

15.4 Modeling Stemflow Infiltration and Redistribution

Numerical models based on Darcy's law or the Richards' equation have extensively been used to simulate soil water dynamics in one dimension (Van Dam and Feddes 2000; Belk et al. 2007; Rassam et al. 2018), two dimensions (Gardenas et al. 2006; Keim et al. 2006), and three dimensions (Hopp and McDonnell 2011; Liang and Uchida 2014). For physically-based predictive models, field information regarding surface/subsurface topography and soil/bedrock hydraulic properties is the main determinant of model performance. Numerical simulations allow for more extensive analyses to explore the first-order controls on hydrologic dynamics. For example, Weiler and McDonnell (2004) used detailed data of surface and bedrock topography in a simple physically-based model to explore the first-order controls of runoff generation on a wet, steep hillslope with high- and low-drainable porosity. They called their approach a "virtual experiment." Similar approaches have been used to study other issues, such as the effects of the canopy on subsurface flow (Keim et al. 2006; Hopp and McDonnell 2011), the capillary fringe groundwater ridging hypothesis (Cloke et al. 2006), the effects of bedrock fractures (Ebel et al. 2008), and the interactions between storm size, bedrock permeability, slope angle, soil depth, and hydrological connectivity (Hopp and McDonnell 2009).

In addition to the topographic or hydraulic parameters, the methods used to determine water inputs have implications for modeling soil/bedrock flows in forested stands. Although many numerical models consider the division of precipitation into interception and throughfall components, rainwater concentrated by stemflow has generally been disregarded. Bouten et al. (1992b) simulated soil water dynamics in a forested lowland catchment using the Soil Water in Forested Ecosystems (SWIF) model, in which throughfall and stemflow were treated as net precipitation and added evenly to the forest floor. In a one-dimensional model developed by Belk et al. (2007) to evaluate the variation in soil water content in a tropical forest, the rainfall entering the forest system was only partitioned into throughfall and canopy interception. Similarly, Keim et al. (2006) simulated subsurface flow generation on a hillslope during storm conditions with a two-dimensional model (HYDRUS-2D) in which only throughfall and canopy interception were considered in the rainfall redistribution processes. Evaporative loss from canopy interception was responsible for delaying the onset of subsurface stormflow and lowering stormflow peaks, total flow, and the runoff ratio. Hopp and McDonnell (2011) examined the role of throughfall patterns on subsurface stormflow generation with a three-dimensional model (HYDRUS-3D) and found that the effects of spatial input variability of throughfall on lateral subsurface stormflow generation were small on their study hillslope. Keim and Skaugset (2003) investigated the potential effects of rainfall intensity smoothed by forest canopies on slope stability based on the modeling of soil water responses to measured rainfall and throughfall during high-intensity rainfall. They concluded that precipitation intensities smoothed by forest canopies may cause greater slope stability compared to a hillslope without forest canopies. Although the contribution of stemflow was ignored in their model, they noted that preferential-flow pathways in soil layers may develop at the location of stemflow inputs and therefore affect the prediction of slope stability. Although these studies examined the effects of canopy interception or throughfall on hydrological responses, models disregarding the contribution of stemflow may generate potential errors in their simulation results. How to express the point-input feature and root-induced bypass flow process of stemflow are the key challenges in the modeling of stemflow infiltration and subsequent redistribution.

Among stemflow input methods, Tanaka et al. (1996) developed a primary model based on a cylindrical infiltration area of stemflow-induced water to determine stemflow inputs to the soil surface. They assumed that the relationship between the radius of the infiltration area of stemflow inputs and the diameter of the tree base is as follows:

$$R_{\mathrm{inf}} = a \times \ln \mathrm{D}_{\mathrm{tree}} - b \tag{15.2}$$

where D_{tree} is the diameter of the tree base, R_{inf} is the radius of the infiltration area of stemflow inputs from the center of the trunk, and *a* and *b* are constants. Assuming that the infiltration area is a ring, the area of stemflow inputs (A_s) can be calculated as follows:

$$A_s = \pi\left[R_{inf}^2 - (D_{tree}/2)^2\right] \tag{15.3}$$

The same concept of infiltration area for stemflow input was used by Sansoulet et al. (2008), in which water fluxes around a banana plant were simulated with the HYDRUS-3D software and the measured volume of stemflow was applied on a small area around the plant stem. Although Tanaka et al. (1996) and Sansoulet et al. (2008) successfully simulated local water infiltration by stemflow, they did not present the detailed stemflow transport process in soil layers. In comparisons between field observations and numerical simulations, Liang et al. (2009b) showed evidence that inputting stemflow to the surface infiltration area only represented the point-input feature, not the rapid transport of stemflow that bypasses soil layers along roots. Considering the bypass characteristic of stemflow channelization, Liang et al. (2009b) proposed a new subsurface stemflow transport model based on the three-dimensional Richards' equation in pressure head (ψ) form:

$$C(\psi)\frac{\partial\psi}{\partial t} = \frac{\partial}{\partial x}\left[K(\psi)\frac{\partial\psi}{\partial x}\right] + \frac{\partial}{\partial y}\left[K(\psi)\frac{\partial\psi}{\partial y}\right] + \frac{\partial}{\partial z}\left[K(\psi)\left(\frac{\partial\psi}{\partial z} + 1\right)\right] - S \tag{15.4}$$

where *t* is time; *x*, *y*, and *z* are the longitudinal, lateral, and vertical directions, respectively; and $C(\psi)$ and $K(\psi)$ are the soil water capacity function and hydraulic conductivity function, respectively. The *S* is a source–sink term that is positive as a sink and negative as a source. The *S* term represents water supply (e.g., precipitation) or water loss (e.g., evaporation or root uptake), and the other terms comprehensively describe water conservation and movement of matrix flow in soil. In the proposed model, stemflow is parameterized as a spring source flux in soil source regions close to a tree and separately inputted to each 10 cm depth by different ratios. The model results indicated an irregular distribution of vertical soil water content changes and a rapid response in the deeper soil layer in the area close to the tree, in agreement with field observations. Thus, considering stemflow as a source flux in the soil layers and applying the variable source term in the Richards' equation is an effective way to represent both the characteristics of the point input of stemflow infiltration and preferential flow of stemflow in the subsurface. Liang et al. (2010) further combined slope stability analyses to test the proposed model (i.e., stemflow assigned as a source flux), with a conventional model in which stemflow was incorporated into net precipitation and added evenly to the forest floor. The proposed model resulted in the timing of a minimum safety factor < 1.0 at an earlier stage than in the conventional model. Furthermore, the proposed model indicated a risk of slope failure throughout all parts of the slope, but the conventional model only showed a risk of slope failure in the lower part of the slope. Thus, the treatment of stemflow input has significant implications for the representation of soil water dynamics and the evaluation of slope stability.

15.5 Future Research Directions

An integrated understanding of hydrological dynamics from canopies to roots is needed at the single-tree scale (Fig. 15.4). Jost et al. (2012) and Schwärzel et al. (2012) demonstrated how the differences in the crown and root architecture between spruce and beech may result in different soil water responses to rainstorms. This kind of comprehensive investigation can help elucidate the hydrological linkage from canopies to roots for different species and predict the hydrological responses in different types of forested stands. There is also a need to better understand the functions of stemflow and roots in the drying process compared to their functions in the wetting process. Brooks et al. (2010) and Evaristo et al. (2015) demonstrated that the soil water used by vegetation is largely decoupled from the soil water entering streams and proposed "two water worlds" of tightly bound water represented by the plant water and mobile water expressed in the stream. It implies that stemflow would be of little use for root uptake. Stemflow supplies a large amount of water, but root uptake depletes the soil water around a tree. Future studies should separately evaluate their positive and negative impacts on soil water resources.

The double-funneling effect of a tree causing a point input of stemflow infiltration and transport along preferential pathways has been qualitatively demonstrated for various species with links to soil erosion, subsurface flow, slope stability,

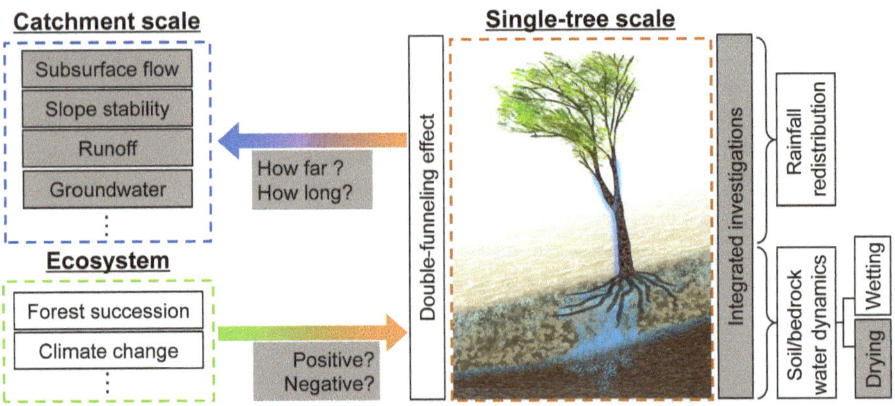

Fig. 15.4 A diagram of the future directions of studies on stemflow and soil water dynamics. White denotes areas where research has been conducted. Gray shading denotes areas where future research is needed

groundwater recharge, and runoff generation at the local, plot, and hillslope scales. Rather than a qualitative understanding, more studies that provide quantitative evidence on the effects of stemflow on the spatial and temporal variation in hydrological responses are needed. It is still not clear whether the double-funneling effect on hydrological responses is diluted or smoothed at a larger scale, such as the catchment or watershed scale (Fig. 15.4). Thus, there is a need to quantify the extent of the double-funneling effect. One fruitful way forward would be to better couple stemflow infiltration areas, expressed as a stand-scale infiltration area funneling ratio (Carlyle-Moses et al. 2018), with percolation and preferential flow of stemflow in the subsurface. This would permit a more definitive linkage between infiltration areas and the magnitude of root-induced preferential flow in the subsurface and the scaling of the double-funneling effect at the stand (and ultimately catchment) scale. Similarly, quantifying the duration of the effect is also important, particularly for the drying process. Both Durocher (1990) and Liang et al. (2015) reported that stemflow only contributed to soil water responses within a short period after rainfall events. To address these issues, both numerical virtual experiments and field observations with isotope/chemical tracing techniques in a large catchment would be useful.

The impacts of global climate change on forest succession, natural disasters, and ecosystem functions are critical issues that require study. For example, Iida et al. (2005) observed a substantial increase in stemflow, essentially no change in throughfall, and a substantial decrease in interception during a period of forest stand succession from Japanese red pine to oak trees. Tanaka et al. (2008) predicted that changes in forest stands from coniferous trees to broadleaved evergreen trees due to both natural thinning and global warming would cause significant changes in water movement and groundwater recharge processes. Although positive effects of forested catchments on water resources and disaster mitigation have been acknowledged or are expected, trees may also contribute to landslide hazards due to the concentrated input of stemflow in a local region (Liang et al. 2010; Liang et al. 2011). Therefore, it is also important to clarify how forest succession and climate change alter the double-funneling effect to evaluate the positive and negative impacts on water resource management and disaster mitigation in changing ecosystems (Fig. 15.4). This information could be used to identify strategies for adapting to climate change.

Acknowledgments This chapter was supported by the grant from the Ministry of Science and Technology, Taiwan (107-2313-B-002-016). The author thanks Dr. Takahisa Mizuyama and Dr. Ken'ichirou Kosugi at Kyoto University, Japan, for providing valuable suggestions. Special thanks are given to Dr. Delphis Levia, an Editor of the book, for his encouragement on writing this chapter.

References

Aboal JR, Morales D, Hernandez M, Jimenez MS (1999) The measurement and modelling of the variation of stemflow in a laurel forest in Tenerife, Canary Islands. J Hydrol 221:161–175. https://doi.org/10.1016/S0022-1694(99)00086-4

Anderson SA, Sitar N (1995) Analysis of rainfall induced debris flows. J Geotech Eng-Asce 121:544–552. https://doi.org/10.1061/(ASCE)0733-9410(1995)121:7(544)
Archer NAL, Quinton JN, Hess TM (2012) Patch vegetation and water redistribution above and below ground in south-east Spain. Ecohydrology 5:108–120. https://doi.org/10.1002/eco.210
Bachmair S, Weiler M (2011) New dimensions of hillslope hydrology. In: Levia DF, Carlyle-Moses D, Tanaka T (eds) Forest hydrology and biogeochemistry: synthesis of past research and future directions. Springer, Dordrecht, pp 455–481. https://doi.org/10.1007/978-94-007-1363-5_23
Beier C, Hansen K, Gundersen P (1993) Spatial variability of throughfall fluxes in a spruce forest. Environ Pollut 81:257–267. https://doi.org/10.1016/0269-7491(93)90208-6
Belk EL, Markewitz D, Rasmussen TC, Maklouf Carvalho EJ, Nepstad DC et al (2007) Modeling the effects of throughfall reduction on soil water content in a Brazilian Oxisol under a moist tropical forest. Water Resour Res 43:W08432. https://doi.org/10.1029/2006WR005493
Beven K, Germann P (1982) Macropores and water flow in soils. Water Resour Res 18:1311–1325. https://doi.org/10.1029/WR018i005p01311
Beven K, Germann P (2013) Macropores and water flow in soils revisited. Water Resour Res 49:3071–3092. https://doi.org/10.1002/wrcr.20156
Bouten W, Heimovaara TJ, Tiktak A (1992a) Spatial patterns of throughfall and soil water dynamics in a Douglas fir stand. Water Resour Res 28:3227–3233. https://doi.org/10.1029/92WR01764
Bouten W, Schaap MG, Bakker DJ, Verstraten JM (1992b) Modeling soil water dynamics in a forested ecosystem.1. A site specific evaluation. Hydrol Process 6:435–444. https://doi.org/10.1002/hyp.3360060405
Brooks JR, Barnard HR, Coulombe R, McDonnell JJ (2010) Ecohydrologic separation of water between trees and streams in a Mediterranean climate. Nat Geosci 3:100–104. https://doi.org/10.1038/ngeo722
Carleton TJ, Kavanagh T (1990) Influence of stand age and spatial location on throughfall chemistry beneath black spruce. Can J For Res 20:1917–1925. https://doi.org/10.1139/x90-257
Carlyle-Moses DE, Lishman CE (2015) Temporal persistence of throughfall heterogeneity below and between the canopies of juvenile lodgepole pine (*Pinus contorta*). Hydrol Process 29:4051–4067. https://doi.org/10.1002/hyp.10494
Carlyle-Moses DE, Iida S, Germer S, Llorens P, Michalzik B, Nanko K, Tischer A, Levia DF (2018) Expressing stemflow commensurate with its ecohydrological importance. Adv Water Resour 121:472–479. https://doi.org/10.1016/j.advwatres.2018.08.015
Chang SC, Matzner E (2000) The effect of beech stemflow on spatial patterns of soil solution chemistry and seepage fluxes in a mixed beech/oak stand. Hydrol Process 14:135–144. https://doi.org/10.1002/(SICI)1099-1085(200001)14:1<135::AID-HYP915>3.3.CO;2-I
Cloke HL, Anderson MG, McDonnell JJ, Renaud JP (2006) Using numerical modelling to evaluate the capillary fringe groundwater ridging hypothesis of streamflow generation. J Hydrol 316:141–162. https://doi.org/10.1016/j.jhydrol.2005.04.017
Coenders-Gerrits AMJ, Hopp L, Savenije HHG, Pfister L (2013) The effect of spatial throughfall patterns on soil moisture patterns at the hillslope scale. Hydrol Earth Syst Sci 17:1749–1763. https://doi.org/10.5194/hess-17-1749-2013
Crockford RH, Richardson DP (2000) Partitioning of rainfall into throughfall, stemflow and interception: effect of forest type, ground cover and climate. Hydrol Process 14:2903–2920. https://doi.org/10.1002/1099-1085(200011/12)14:16/17<2903::AID-HYP126>3.0.CO;2-6
Dasgupta S, Mohanty BP, Kohne JM (2006) Impacts of juniper vegetation and karst geology on subsurface flow processes in the Edwards Plateau, Texas. Vadose Zone J 5:1076–1085. https://doi.org/10.2136/vzj2005.0073
Deguchi A, Hattori S, Park H-T (2006) The influence of seasonal changes in canopy structure on interception loss: application of the revised Gash model. J Hydrol 318:80–102. https://doi.org/10.1016/j.jhydrol.2005.06.005
Dhakal AS, Sidle RC (2004) Pore water pressure assessment in a forest watershed: simulations and distributed field measurements related to forest practices. Water Resour Res 40:W02405. https://doi.org/10.1029/2003WR002017

Durocher MG (1990) Monitoring spatial variability of forest interception. Hydrol Process 4:215–229. https://doi.org/10.1002/hyp.3360040303

Ebel BA, Loague K, Montgomery DR, Dietrich WE (2008) Physics-based continuous simulation of long-term near-surface hydrologic response for the Coos Bay experimental catchment. Water Resour Res 44:W07417. https://doi.org/10.1029/2007wr006442

Emanuel RE, Hazen AG, McGlynn BL, Jencso KG (2014) Vegetation and topographic influences on the connectivity of shallow groundwater between hillslopes and streams. Ecohydrology 7:887–895. https://doi.org/10.1002/eco.1409

Evaristo J, Jasechko S, McDonnell JJ (2015) Global separation of plant transpiration from groundwater and streamflow. Nature 525:91–94. https://doi.org/10.1038/nature14983

Ford ED, Deans JD (1977) Growth of a Sitka spruce plantation: spatial distribution and seasonal fluctuations of lengths, weights and carbohydrate concentrations of fine roots. Plant Soil 47:463–485. https://doi.org/10.1007/BF00011504

Ford ED, Deans JD (1978) Effects of canopy structure on stemflow, throughfall and interception loss in a young Sitka spruce plantation. J Appl Ecol 15:905–917. https://doi.org/10.2307/2402786

Gaiser RN (1952) Root channels and roots in forest soils. Soil Sci Soc Am Proc 16:62–65. https://doi.org/10.2136/sssaj1952.03615995001600010019x

Garcia-Estringana P, Latron J, Llorens P, Gallart F (2012) Spatial and temporal dynamics of soil moisture in a Mediterranean mountain area (Vallcebre, NE Spain). Ecohydrology 6:741–753. https://doi.org/10.1002/eco.1295

Gardenas AI, Simunek J, Jarvis N, van Genuchten MT (2006) Two-dimensional modelling of preferential water flow and pesticide transport from a tile-drained field. J Hydrol 329:647–660. https://doi.org/10.1016/j.jhydrol.2006.03.021

Germer S (2013) Development of near-surface perched water tables during natural and artificial stemflow generation by babassu palms. J Hydrol 507:262–272. https://doi.org/10.1016/j.jhydrol.2013.10.026

Gersper PL, Holowaychuk N (1970) Effects of stemflow water on a Miami soil under a beech tree. 2. Chemical properties. Soil Sci Soc Am Proc 34:786–794. https://doi.org/10.2136/sssaj1970.03615995003400050033x

Ghestem M, Sidle RC, Stokes A (2011) The influence of plant root systems on subsurface flow: implications for slope stability. Bio Science 61:869–879. https://doi.org/10.1525/bio.2011.61.11.6

Giacomin A, Trucchi P (1992) Rainfall interception in a beech coppice (Acquerino, Italy). J Hydrol 137:141–147. https://doi.org/10.1016/0022-1694(92)90052-w

Glover PE, Glover J, Gwynne MD (1962) Light rainfall and plant survival in E. Africa. 2. Dry grassland vegetation. J Ecol 50:199–206. https://doi.org/10.2307/2257204

Grayson RB, Western AW, Chiew FHS, Blöschl G (1997) Preferred states in spatial soil moisture patterns: local and nonlocal controls. Water Resour Res 33:2897–2908. https://doi.org/10.1029/97wr02174

Gwak Y, Kim S (2017) Factors affecting soil moisture spatial variability for a humid forest hillslope. Hydrol Process 31:431–445. https://doi.org/10.1002/hyp.11039

Helvey JD, Patric JH (1965) Canopy and litter interception of rainfall by hardwoods of eastern United States. Water Resour Res 1:193–206. https://doi.org/10.1029/WR001i002p00193

Herwitz SR (1986a) Episodic stemflow inputs of magnesium and potassium to a tropical forest floor during heavy rainfall events. Oecologia 70:423–425. https://doi.org/10.1007/BF00379506

Herwitz SR (1986b) Infiltration-excess caused by stemflow in a cyclone-prone tropical rainforest. Earth Surf Process Landf 11:401–412. https://doi.org/10.1002/esp.3290110406

Herwitz SR (1987) Raindrop impact and water flow on the vegetative surfaces of trees and the effects on stemflow and throughfall generation. Earth Surf Process Landf 12:425–432. https://doi.org/10.1002/esp.3290120408

Herwitz SR, Levia DF (1997) Mid-winter stemflow drainage from bigtooth aspen (*Populus grandidentata* Michx) in Central Massachusetts. Hydrol Process 11:169–175. https://doi.org/10.1002/(SICI)1099-1085(199702)11:2<169::AID-HYP428>3.0.CO;2-J

Hopp L, McDonnell JJ (2009) Connectivity at the hillslope scale: identifying interactions between storm size, bedrock permeability, slope angle and soil depth. J Hydrol 376:378–391. https://doi.org/10.1016/j.jhydrol.2009.07.047

Hopp L, McDonnell JJ (2011) Examining the role of throughfall patterns on subsurface stormflow generation. J Hydrol 409:460–471. https://doi.org/10.1016/j.jhydrol.2011.08.044

Huber A, Iroume A (2001) Variability of annual rainfall partitioning for different sites and forest covers in Chile. J Hydrol 248:78–92. https://doi.org/10.1016/S0022-1694(01)00394-8

Iida S, Tanaka T, Sugita M (2005) Change of interception process due to the succession from Japanese red pine to evergreen oak. J Hydrol 315:154–166. https://doi.org/10.1016/j.jhydrol.2005.03.024

Jackson NA, Wallace JS, Ong CK (2000) Tree pruning as a means of controlling water use in an agroforestry system in Kenya. Forest Ecol Manag 126:133–148. https://doi.org/10.1016/S0378-1127(99)00096-1

Johnson MS, Lehmann J (2006) Double-funneling of trees: stemflow and root-induced preferential flow. Ecoscience 13:324–333. https://doi.org/10.2980/i1195-6860-13-3-324.1

Jost G, Schume H, Hager H, Markart G, Kohl B (2012) A hillslope scale comparison of tree species influence on soil moisture dynamics and runoff processes during intense rainfall. J Hydrol 420–421:112–124. https://doi.org/10.1016/j.jhydrol.2011.11.057

Katsura S, Kosugi K, Mizutani T, Mizuyama T (2009) Hydraulic properties of variously weathered granitic bedrock in headwater catchments. Vadose Zone J 8:557–573. https://doi.org/10.2136/vzj2008.0142

Keim RF, Skaugset AE (2003) Modelling effects of forest canopies on slope stability. Hydrol Process 17:1457–1467. https://doi.org/10.1002/hyp.5121

Keim RF, Skaugset AE, Weiler M (2005) Temporal persistence of spatial patterns in throughfall. J Hydrol 314:263–274. https://doi.org/10.1016/j.jhydrol.2005.03.021

Keim RF, Tromp-van Meerveld HJ, McDonnell JJ (2006) A virtual experiment on the effects of evaporation and intensity smoothing by canopy interception on subsurface stormflow generation. J Hydrol 327:352–364. https://doi.org/10.1016/j.jhydrol.2005.11.024

Kellman M, Roulet N (1990) Stemflow and throughfall in a tropical dry forest. Earth Surf Process Landf 15:55–61. https://doi.org/10.1002/esp.3290150106

Kosugi K, Katsura S, Katsuyama M, Mizuyama T (2006) Water flow processes in weathered granitic bedrock and their effects on runoff generation in a small headwater catchment. Water Resour Res 42:W02414. https://doi.org/10.1029/2005WR004275

Krämer I, Hölscher D (2010) Soil water dynamics along a tree diversity gradient in a deciduous forest in Central Germany. Ecohydrology 3:262–271. https://doi.org/10.1002/eco.103

Kulli B, Stamm C, Papritz A, Fluhler H (2003) Discrimination of flow regions on the basis of stained infiltration patterns in soil profiles. Vadose Zone J 2:338–348. https://doi.org/10.2113/2.3.338

Lange B, Luescher P, Germann PF (2009) Significance of tree roots for preferential infiltration in stagnic soils. Hydrol Earth Syst Sci 13:1809–1821. https://doi.org/10.5194/hess-13-1809-2009

Levia DF, Frost EE (2003) A review and evaluation of stemflow literature in the hydrologic and biogeochemical cycles of forested and agricultural ecosystems. J Hydrol 274:1–29. https://doi.org/10.1016/S0022-1694(02)00399-2

Levia DF, Frost EE (2006) Variability of throughfall volume and solute inputs in wooded ecosystems. Prog Phys Geogr 30:605–632. https://doi.org/10.1177/0309133306071145

Levia DF, Germer S (2015) A review of stemflow generation dynamics and stemflow-environment interactions in forests and shrublands. Rev Geophys 53:673–714. https://doi.org/10.1002/2015RG000479

Levia DF, Herwitz SR (2005) Interspecific variation of bark water storage capacity of three deciduous tree species in relation to stemflow yield and solute flux to forest soils. Catena 64:117–137. https://doi.org/10.1016/j.catena.2005.08.001

Levia DF, Michalzik B, Näthe K, Bischoff S, Richter S, Legates DR (2015) Differential stemflow yield from European beech saplings: the role of individual canopy structure metrics. Hydrol Process 29:43–51. https://doi.org/10.1002/hyp.10124

Li X-Y, Yang Z-P, Li Y-T, Lin H (2009) Connecting ecohydrology and hydropedology in desert shrubs: stemflow as a source of preferential flow in soils. Hydrol Earth Syst Sci 13:1133–1144. https://doi.org/10.5194/hess-13-1133-2009
Liang W-L, Chan M-C (2017) Spatial and temporal variations in the effects of soil depth and topographic wetness index of bedrock topography on subsurface saturation generation in a steep natural forested headwater catchment. J Hydrol 546:405–418. https://doi.org/10.1016/j.jhydrol.2017.01.033
Liang W-L, Uchida T (2014) Effects of topography and soil depth on saturated-zone dynamics in steep hillslopes explored using the three-dimensional Richards' equation. J Hydrol 510:124–136. https://doi.org/10.1016/j.jhydrol.2013.12.029
Liang W-L, Kosugi K, Mizuyama T (2007) Heterogeneous soil water dynamics around a tree growing on a steep hillslope. Vadose Zone J 6:879–889. https://doi.org/10.2136/vzj2007.0029
Liang W-L, Kosugi K, Mizuyama T (2009a) Characteristics of stemflow for tall stewartia (*Stewartia monadelpha*) growing on a hillslope. J Hydrol 378:168–178. https://doi.org/10.1016/j.jhydrol.2009.09.027
Liang W-L, Kosugi K, Mizuyama T (2009b) A three-dimensional model of the effect of stemflow on soil water dynamics around a tree on a hillslope. J Hydrol 366:62–75. https://doi.org/10.1016/j.jhydrol.2008.12.009
Liang W-L, Kosugi K, Mizuyama T, Musashi Y (2010) Field observations and numerical experiments to assess the effect of trees on slope stability. In: Chen S-C (ed) INTERPRAEVENT 2010, The International Research Society INTERPRAEVENT, pp 476–485
Liang W-L, Kosugi K, Mizuyama T (2011) Soil water dynamics around a tree on a hillslope with or without rainwater supplied by stemflow. Water Resour Res 47:W02541. https://doi.org/10.1029/2010wr009856
Liang W-L, Kosugi K, Mizuyama T (2015) Soil water redistribution processes around a tree on a hillslope: the effect of stemflow on the drying process. Ecohydrology 8:1381–1395. https://doi.org/10.1002/eco.1589
Liang W-L, Li S-L, Hung F-X (2017) Analysis of the contributions of topographic, soil, and vegetation features on the spatial distributions of surface soil moisture in a steep natural forested headwater catchment. Hydrol Process 31:3796–3809. https://doi.org/10.1002/hyp.11290
Lin H (2006) Temporal stability of soil moisture spatial pattern and subsurface preferential flow pathways in the shale hills catchment. Vadose Zone J 5:317–340. https://doi.org/10.2136/vzj2005.0058
Lin H (2010) Earth's Critical Zone and hydropedology: concepts, characteristics, and advances. Hydrol Earth Syst Sci 14:25–45. https://doi.org/10.5194/hess-14-25-2010
Lin H, Zhou X (2008) Evidence of subsurface preferential flow using soil hydrologic monitoring in the Shale Hills catchment. Eur J Soil Sci 59:34–49. https://doi.org/10.1111/j.1365-2389.2007.00988.x
Lloyd CR, Marques AD (1988) Spatial variability of throughfall and stemflow measurements in Amazonian rainforest. Agric For Meteorol 42:63–73. https://doi.org/10.1016/0168-1923(88)90067-6
Loustau D, Berbigier P, Granier A, Moussa FEH (1992) Interception loss, throughfall and stemflow in a maritime pine stand I Variability of throughfall and stemflow beneath the pine canopy. J Hydrol 138:449–467. https://doi.org/10.1016/0022-1694(92)90130-n
Lozano-Parra J, van Schaik NLMB, Schnabel S, Gómez-Gutiérrez Á (2016) Soil moisture dynamics at high temporal resolution in a semiarid Mediterranean watershed with scattered tree cover. Hydrol Process 30:1155–1170. https://doi.org/10.1002/hyp.10694
Ludwig JA, Wilcox BP, Breshears DD, Tongway DJ, Imeson AC (2005) Vegetation patches and runoff-erosion as interacting ecohydrological processes in semiarid landscapes. Ecology 86:288–297. https://doi.org/10.1890/03-0569
Martinez-Meza E, Whitford WG (1996) Stemflow, throughfall and channelization of stemflow by roots in three Chihuahuan desert shrubs. J Arid Environ 32:271–287. https://doi.org/10.1006/jare.1996.0023

Masaoka N, Kosugi K, Yamakawa Y, Tsutsumi D (2016) Processes of bedrock groundwater seepage and their effects on soil water fluxes in a foot slope area. J Hydrol 535:160–172. https://doi.org/10.1016/j.jhydrol.2016.01.081

Metzger JC, Wutzler T, Dalla Valle N, Filipzik J, Grauer C et al (2017) Vegetation impacts soil water content patterns by shaping canopy water fluxes and soil properties. Hydrol Process 31:3783–3795. https://doi.org/10.1002/hyp.11274

Montgomery DR, Dietrich WE (2002) Runoff generation in a steep, soil-mantled landscape. Water Resour Res 38:1168. https://doi.org/10.1029/2001WR000822

Montgomery DR, Dietrich WE, Heffner JT (2002) Piezometric response in shallow bedrock at CB1: implications for runoff generation and landsliding. Water Resour Res 38:1274. https://doi.org/10.1029/2002WR001429

Nanko K, Watanabe A, Hotta N, Suzuki M (2013) Physical interpretation of the difference in drop size distributions of leaf drips among tree species. Agric For Meteorol 169:74–84. https://doi.org/10.1016/j.agrformet.2012.09.018

Neal C, Robson AJ, Bhardwaj CL, Conway T, Jeffery HA, Neal M et al (1993) Relationships between precipitation, stemflow and throughfall for a lowland beech plantation, Black Wood, Hampshire, southern England: findings on interception at a forest edge and the effects of storm damage. J Hydrol 146:221–233. https://doi.org/10.1016/0022-1694(93)90277-g

Neave M, Abrahams AD (2002) Vegetation influences on water yields from grassland and shrubland ecosystems in the Chihuahuan Desert. Earth Surf Process Landf 27:1011–1020. https://doi.org/10.1002/esp.389

Noguchi S, Tsuboyama Y, Sidle RC, Hosoda I (1999) Morphological characteristics of macropores and the distribution of preferential flow pathways in a forested slope segment. Soil Sci Soc Am J 63:1413–1423. https://doi.org/10.2136/sssaj1999.6351413x

Park A, Cameron JL (2008) The influence of canopy traits on throughfall and stemflow in five tropical trees growing in a Panamanian plantation. Forest Ecol Manag 255:1915–1925. https://doi.org/10.1016/j.foreco.2007.12.025

Pressland AJ (1976) Soil moisture redistribution as affected by throughfall and stemflow in an arid zone shrub community. Aust J Bot 24:641–649. https://doi.org/10.1071/bt9760641

Raat KJ, Draaijers GPJ, Schaap MG, Tietema A, Verstraten JM (2002) Spatial variability of throughfall water and chemistry and forest floor water content in a Douglas fir forest stand. Hydrol Earth Syst Sci 6:363–374. https://doi.org/10.5194/hess-6-363-2002

Rassam D, Šimůnek J, Mallants D, van Genuchten MT (2018) The HYDRUS-1D software package for simulating the one-dimensional movement of water, heat, and multiple solutes in variably-saturated media: tutorial. CSIRO Land and Water, Adelaide

Robinson DA, Campbell CS, Hopmans JW, Hornbuckle BK, Jones SB, Knight R et al (2008) Soil moisture measurement for ecological and hydrological watershed-scale observatories: a review. Vadose Zone J 7:358–389. https://doi.org/10.2136/vzj2007.0143

Robinson JL, Slater LD, Schäfer KVR (2012) Evidence for spatial variability in hydraulic redistribution within an oak–pine forest from resistivity imaging. J Hydrol 430–431:69–79. https://doi.org/10.1016/j.jhydrol.2012.02.002

Robson AJ, Neal C, Ryland GP, Harrow M (1994) Spatial variations in throughfall chemistry at the small plot scale. J Hydrol 158:107–122. https://doi.org/10.1016/0022-1694(94)90048-5

Sander T, Gerke HH (2007) Preferential flow patterns in paddy fields using a dye tracer. Vadose Zone J 6:105–115. https://doi.org/10.2136/vzj2006.0035

Sansoulet J, Cabidoche Y-M, Cattan P, Ruy S, Simunek J (2008) Spatially distributed water fluxes in an andisol under banana plants: experiments and three-dimensional modeling. Vadose Zone J 7:819–829. https://doi.org/10.2136/vzj2007.0073

Schume H, Jost G, Hager H (2004) Soil water depletion and recharge patterns in mixed and pure forest stands of European beech and Norway spruce. J Hydrol 289:258–274. https://doi.org/10.1016/j.jhydrol.2003.11.036

Schwärzel K, Menzer A, Clausnitzer F, Spank U, Häntzschel J, Grünwald T, Köstner B, Bernhofer C, Feger K-H (2009) Soil water content measurements deliver reliable estimates of water fluxes: a comparative study in a beech and a spruce stand in the Tharandt forest (Saxony, Germany). Agric For Meteorol 149:1994–2006. https://doi.org/10.1016/j.agrformet.2009.07.006

Schwärzel K, Ebermann S, Schalling N (2012) Evidence of double-funneling effect of beech trees by visualization of flow pathways using dye tracer. J Hydrol 470–471:184–192. https://doi.org/10.1016/j.jhydrol.2012.08.048

Schweingruber FH (1996) Influence of mass movement. In: Schweingruber FH (ed) Tree rings and environment dendroecology. Haupt Publ, Bern, pp 271–287

Sidle RC, Bogaard TA (2016) Dynamic earth system and ecological controls of rainfall-initiated landslides. Earth-Sci Rev 159:275–291. https://doi.org/10.1016/j.earscirev.2016.05.013

Sidle RC, Noguchi S, Tsuboyama Y, Laursen K (2001) A conceptual model of preferential flow systems in forested hillslopes: evidence of self-organization. Hydrol Process 15:1675–1692. https://doi.org/10.1002/hyp.233

Sraj M, Brilly M, Mikos M (2008) Rainfall interception by two deciduous Mediterranean forests of contrasting stature in Slovenia. Agric For Meteorol 148:121–134. https://doi.org/10.1016/j.agrformet.2007.09.007

Staelens J, De Schrijver A, Verheyen K, Verhoest NEC (2008) Rainfall partitioning into throughfall, stemflow, and interception within a single beech (Fagus sylvatica L.) canopy: influence of foliation, rain event characteristics, and meteorology. Hydrol Process 22:33–45. https://doi.org/10.1002/hyp.6610

Tanaka T, Tsujimura M, Taniguchi M (1991) Infiltration area of stemflow-induced water. Annu Rep Inst Geosci, Univ Tsukuba 17:30–32

Tanaka T, Taniguchi M, Tsujimura M (1996) Significance of stemflow in groundwater recharge.2. A cylindrical infiltration model for evaluating the stemflow contribution to groundwater recharge. Hydrol Process 10:81–88. https://doi.org/10.1002/(sici)1099-1085(199601)10:1<81::aid-hyp302>3.0.co;2-m

Tanaka T, Iida S, Kakubari J, Hamada Y (2008) Effect of forest stand succession from conifer trees to broad-leaved evergreen trees on infiltration and groundwater recharge processes. In: Abesser C, Wagener T, Nuetzmann G (eds) Groundwater-surface water interaction: process understanding, conceptualization and modelling, vol 321. IAHS, Wallingford, pp 54–60

Taniguchi M, Tsujimura M, Tanaka T (1996) Significance of stemflow in groundwater recharge.1. Evaluation of the stemflow contribution to recharge using a mass balance approach. Hydrol Process 10:71–80. https://doi.org/10.1002/(sici)1099-1085(199601)10:1<71::aid-hyp301>3.3.co;2-h

Tromp-van Meerveld HJ, McDonnell JJ (2006) On the interrelations between topography, soil depth, soil moisture, transpiration rates and species distribution at the hillslope scale. Adv Water Resour 29:293–310. https://doi.org/10.1016/j.advwatres.2005.02.016f

Van Dam JC, Feddes RA (2000) Numerical simulation of infiltration, evaporation and shallow groundwater levels with the Richards equation. J Hydrol 233:72–85. https://doi.org/10.1016/S0022-1694(00)00227-4

Van Stiphout TPJ, Van Slanen HAJ, Boersma OH, Bouma J (1987) The effect of bypass flow and internal catchment of rain on the water regime in a clay loam grassland soil. J Hydrol 95:1–11. https://doi.org/10.1016/0022-1694(87)90111-9

Vereecken H, Huisman JA, Bogena H, Vanderborght J, Vrugt JA, Hopmans JW (2008) On the value of soil moisture measurements in vadose zone hydrology: a review. Water Resour Res 44: W00D06. https://doi.org/10.1029/2008wr006829

Voigt GK (1960) Distribution of rainfall under forest stands. For Sci 6:2–10. https://doi.org/10.1093/forestscience/6.1.2

Wang GH, Sassa K (2003) Pore-pressure generation and movement of rainfall-induced landslides: effects of grain size and fine-particle content. Eng Geol 69:109–125. https://doi.org/10.1016/S0013-7952(02)00268-5

Weiler M (2017) Macropores and preferential flow-a love-hate relationship. Hydrol Process 31:15–19. https://doi.org/10.1002/hyp.11074

Weiler M, McDonnell J (2004) Virtual experiments: a new approach for improving process conceptualization in hillslope hydrology. J Hydrol 285:3–18. https://doi.org/10.1016/s0022-1694(03)00271-3

Weiler M, McDonnell JJ, Tromp-van Meerveld I, Uchida T (2006) Subsurface stormflow. In: Anderson MG, JJ MD (eds) Encyclopedia of hydrological sciences. Wiley, Hoboken, 14 p. https://doi.org/10.1002/0470848944.hsa119

Western AW, Grayson RB, Blöschl G, Willgoose GR, McMahon TA (1999) Observed spatial organization of soil moisture and its relation to terrain indices. Water Resour Res 35:797–810. https://doi.org/10.1029/1998wr900065

Western AW, Grayson RB, Blöschl G (2002) Scaling of soil moisture: a hydrologic perspective. Annu Rev Earth Planet Sci 30:149–180. https://doi.org/10.1146/annurev.earth.30.091201.140434

Yunusa I, Mele P, Rab M, Schefe C, Beverly C (2002) Priming of soil structural and hydrological properties by native woody species, annual crops, and a permanent pasture. Aust J Soil Res 40:207–220. https://doi.org/10.1071/sr01038

Chapter 16
Radiocesium Cycling in the Context of Forest-Water Interactions

Hiroaki Kato

16.1 Radiocesium Cycling in Wooded Ecosystems: An Overview

Cesium (Cs) is an alkali metal, which has physical and chemical properties similar to those of rubidium and potassium. Natural cesium occurs only in one stable form as ^{133}Cs in rocks, soil, and dust at low concentrations in the environment. ^{137}Cs and ^{134}Cs are artificial radionuclides with half-lives of 30.2 years and 2.1 years, respectively. Both are products of past atmospheric testing of thermonuclear weapons and nuclear accidents during the late 1950s and 1960s (Cambray et al. 1989). The radiocesium released by nuclear weapons testing was dispersed into the stratosphere, distributed globally, and deposited as fallout, usually in association with precipitation (Tamura 1964). The amount of fallout is dependent on latitude and amount of precipitation, with deposition being greater in the Northern Hemisphere than the Southern Hemisphere, which experienced less weapons testing and lower amounts of fallout (Walling 2002). Accidents in nuclear facilities, such as nuclear power plants, release radiocesium into the atmosphere in particulate or gaseous forms, which result in significant deposition of radiocesium to local areas with decreasing deposition with distance from the accident source.

Most cesium compounds are readily soluble in water, and the chemical properties of ionic Cs (Cs^+) generally dictate its high degree of mobility and bioavailability (Ashraf et al. 2014). However, in soils, it strongly binds to common clay minerals and is therefore rather uncommon in groundwater. The relatively long half-life of ^{137}Cs provides an opportunity to conduct a detailed study of its behavior in terrestrial, aquatic, and marine ecosystems following historical nuclear events.

The forest canopy serves as an efficient filter of atmospheric contaminants due to its large surface area and aerodynamic roughness, thereby acting as an important

H. Kato (✉)
University of Tsukuba, Tsukuba, Ibaraki, Japan
e-mail: kato.hiroaki.ka@u.tsukuba.ac.jp

D. F. Levia et al. (eds.), *Forest-Water Interactions*, Ecological Studies 240,
https://doi.org/10.1007/978-3-030-26086-6_16

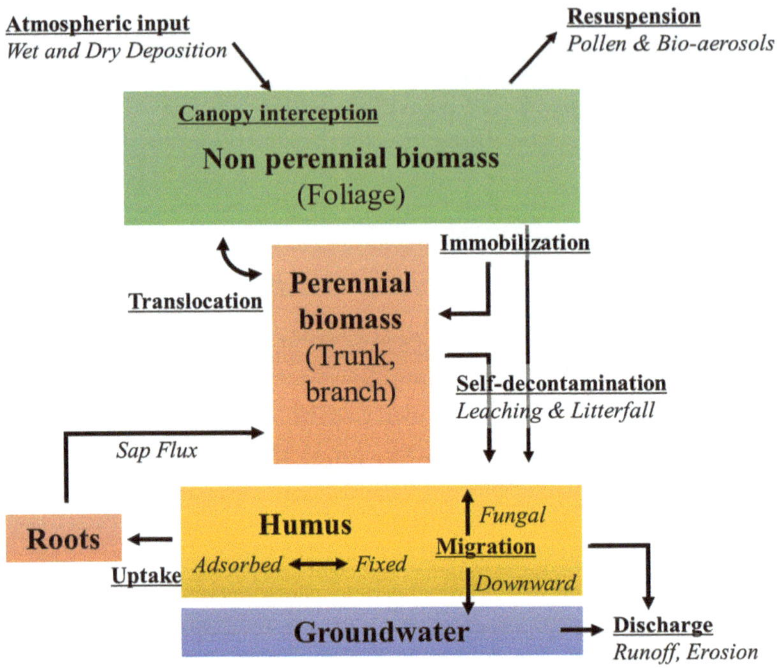

Fig. 16.1 Biogeochemical cycling of radiocesium in forest ecosystems. (Adapted from IAEA 2002)

interface connecting the atmosphere and soil surface. Following the atmospheric deposition of radiocesium onto a forested area, the primary source of tree contamination is direct interception of aerosol-associated radiocesium by the canopy (Pröhl 2009; IAEA 2002, 2010).

Biogeochemical cycling of radiocesium in forested environments has been investigated following past nuclear accidents, such as Chernobyl in 1986 and Fukushima in 2011. The transport and storage of radiocesium through various compartments of forests (e.g., tree trunk, roots) has been described well in the existing literature (Fig. 16.1) (e.g., IAEA 2002). The capture and interception of particulate or dissolved radiocesium in rainwater by the canopy is followed by foliar uptake, with further translocation from foliar surfaces to structural components of the tree in due course (Myttenaere et al. 1993; Nimis 1996; Thiry et al. 2016). Further changes in tree contamination after the initial fallout are due to two major processes. The first of these is a dominant and relatively rapid self-decontamination process of the tree canopy, in association with precipitation washoff (e.g., via throughfall and stemflow) and leaf fall due to leaf senescence and physical removal by heavy storms (e.g., greenfall) (Bunzl et al. 1989; Bonnett and Anderson 1993; Pröhl 2009; Kato et al. 2012, 2017, 2018a). Once the canopy radiocesium reaches the forest floor, it migrates into the soil profile over the longer term. This is followed by root uptake because radiocesium is a nutrient analogue to potassium (Ronneau et al. 1991).

Therefore, the rate of radiocesium cycling within forests is considered to be relatively rapid with a quasi-equilibrium of its distribution being reached in a few years to a decade after atmospheric fallout. Output from the watershed system via stream water and groundwater is generally limited due to the fixation of radiocesium on micaceous clay minerals. In fact, the transport of radiocesium from the watershed in association with sediment discharge is less than 1% of total inventory in the drainage area (Iwagami et al. 2017a).

Trees serve an important role in the recycling of radiocesium in the forest. Partial and transient storage of radiocesium occurs particularly in perennial woody components with large biomass, such as tree trunks and branches. However, the major portion of radiocesium accumulated by vegetation from foliar surfaces and the soil is recycled annually through litterfall and leaching from the canopy, resulting in the long-lasting biological availability of radiocesium in surface soil (Myttenaere et al. 1993; Schell et al. 1996; IAEA 2002). Internal translocation of radiocesium within vegetation from foliar to growing organs also occurs immediately after the initial contamination, possibly masking the contribution of root uptake until the radiocesium recycling in the soil-plant system is in equilibrium or quasi-equilibrium states (e.g., Thiry et al. 2016).

Like radiocesium, some scientists have examined the cycling of radioiodine (^{131}I), a largely beta-emitting radioisotope (^{131}I) with half-life of 8.02 days. ^{131}I is a by-product of the oxidation of irradiated fissile material in chemical processing plants, in reactor accidents, and in nuclear explosions. Iodine is naturally present in the atmosphere and an essential dietary nutrient for mammals, although it is one of the less abundant elements in the environment. The deposition of ^{131}I at the Earth's surface following nuclear accidents have been rarely clarified due to the technical difficulties involved in measuring the short-lived radioactivity. Thus, we include data on the cycling of ^{131}I in this chapter where data are available to compare and contrast to radiocesium. Elemental iodine does exist near the Earth's surface in multiple oxidation states, which form inorganic (e.g., iodide, I-, and iodate IO_3^-) and organic (e.g., CH_3I) species in gaseous (e.g., iodine, I_2), particle-reactive, or soluble phases (Landis et al. 2012). When iodine is present in particulate form, the diffusion to surfaces is controlled by the mechanisms which determine the transport of particles across boundary layers.

Even though the transfer process of radiocesium in and among forest compartments has been well documented following the global fallout and past nuclear accidents, there is insufficient monitoring and data available for modeling the long-term dynamics of radiocesium cycling in forest ecosystems. This is particularly true with respect to the initial partitioning of atmospherically deposited radiocesium by forest canopies and its influences on the spatiotemporal distribution and long-term recycling of radiocesium in the soil-plant system. Accordingly, the purpose of this chapter is to (1) provide an overview of radiocesium dynamics in forests affected by past nuclear accidents; (2) introduce the latest monitoring results of radiocesium cycling during the early phase of the Fukushima accident; and (3) discuss the potential suitability of radiocesium as a tracer of water and element cycling in forest ecosystems.

16.2 Atmospheric Deposition and Initial Canopy Interception of Radiocesium

Canopy interception of atmospherically deposited radiocesium was observed in a Japanese cedar (*Cryptomeria japonica* D. Don) plantation following the Fukushima Dai-ichi Nuclear Power Plant (FDNPP) accident. The rainwater volume and its radiocesium inventory were compared between open rainwater (RF) and throughfall (TF) and stemflow (SF) for the first rain event (in the 11 to 28 March 2011 sampling period) following the accident (Table 16.1, Kato et al. 2012). Even though the canopy interception of rainwater was 31% of total precipitation, 93% of total ^{137}Cs activity was retained by the canopy biomass after the water passed through the forest canopy. In contrast, 50% of total ^{131}I activity in rainwater was captured by the canopy. These data indicated that the atmospherically deposited radiocesium was efficiently captured and retained by forest canopies, whereas the retention of ^{131}I by the canopy was less effective (e.g., Hoffman et al. 1995; Pröhl 2009; Kato et al. 2012; Onda et al. 2015).

The interception fraction, f (dimensionless), which is defined as the ratio of the radionuclide inventory (Bq m^{-2}) initially retained by the trees A_i (Bq m^{-2}), immediately subsequent to the deposition event, to the total deposition density (A_t, Bq m^{-2}) is defined as:

$$f = \frac{A_i}{A_t} \tag{16.1}$$

(e.g., IAEA 2002; Pröhl 2009)

The interception factor has been reported by several researchers for Japanese forests in the aftermath of the Fukushima reactor accident (Table 16.2). Although the methodology varied among studies (e.g., soil-litter sampling, mass balance, modeling, and experimental approaches), the existing studies indicated that evergreen conifers tended to show higher interception fractions for atmospherically deposited radiocesium as compared to broadleaved deciduous species with lower f values.

The f values for various forest canopies have been reported for some forests affected by Chernobyl Nuclear Power Plant accident in 1986 (e.g., Bunzl et al. 1989; Sombré et al. 1990; Schimmack et al. 1991; Melin et al. 1994). The observed interception factor varied significantly among different tree species (Table 16.2). Similar to trees in Japan, evergreen conifers (e.g., Norway spruce, red pine) tended to

Table 16.1 Mass balance of rainwater and atmospherically deposited radionuclides during the first rain event following the Fukushima accident in a Japanese cedar plantation

Hydrological component	Amount (mm)	^{137}Cs (Bq m^{-2})	^{131}I (Bq m^{-2})
Rainfall	21.2	5420	28,400
Throughfall	13.9	389	13,800
Stemflow	0.56	12.5	527

Modified from Kato et al. (2012)

Table 16.2 The observed interception fractions for the Chernobyl and Fukushima accident-derived radionuclide inputs

Forest species	Radionuclides	*f* value (%)	Methodology	Note	References
Japanese cedar	Cs-137, Cs-134, I-131	93 51	Mass balance ($n = 20$)	Fukushima accident	Kato et al. (2012)
Japanese hinoki cypress	Cs-137, Cs-134, I-131	92 25	Mass balance ($n = 20$)	Fukushima accident	Kato et al. (2012)
Japanese cedar	Cs-137, Cs-134, I-131	69 29	Soil and litter sampling ($n = 5$)	Fukushima accident	Onda et al. (2015)
Japanese cedar	Cs-137	70	Field loss model ($n = 7$)	Fukushima accident	Kato et al. (2017)
Japanese hinoki cypress	Cs-137, Cs-134	44–45	Mass balance ($n = 3$)	Fukushima accident	Itoh et al. (2015)
Japanese cedar	Cs-137, Cs-134	33–90	Mass balance ($n = 3$)	Fukushima accident	Itoh et al. (2015)
Japanese hinoki cypress	Cs-137	31	Mass balance ($n = 1$)	Fukushima accident	Nishikiori et al. (2015)
Konara oak and red pine	Cs-137	23	Field loss ($n = 6$)	Fukushima accident	Kato et al. (2017)
Deciduous broadleaved	Cs-137, Cs-134	34	Mass balance ($n = 3$)	Fukushima accident	Itoh et al. (2015)
Evergreen broadleaved	Cs-137, Cs-134	34–44	Mass balance ($n = 3$)	Fukushima accident	Itoh et al. (2015)
Coniferous	Not specified	70–90	–	Chernobyl accident	Tikhomirov and Shcheglov (1991)
Coniferous	Cs-137, Cs-134	76–79	–		Ronneau et al. (1987)
Coniferous	Cs-137, Cs-134	70	Soil and litter sampling		Bunzl et al. (1989)
Norway spruce	Cs-137, Cs-134	70	–		Schimmack et al. (1991)
Coniferous	Cs-137	80–100	–		Melin et al. (1994)
Coniferous	Cs-137	80	–		Sombre et al. (1990)
Norway spruce	Cs-134	79–86	Experiment		Thiry (1997)
Norway spruce	Cs-134	34	Mass balance (experiment)		Thiry et al. (2016)
Deciduous	Cs-137	10–40	–		Melin et al. (1994)
Beech	Cs-137, Cs-134	20	–		Schimmack et al. (1991)

have higher *f* values than broadleaved deciduous (e.g., beech, oak) forests. This is because the initial fallout from Chernobyl occurred in the early spring when many deciduous forests are in a leafless state (e.g., IAEA 2006). The situation was similar for the Fukushima accident where the initial atmospheric fallout occurred during the late winter to early spring (e.g., Kato et al. 2017, 2018b). Therefore, it should be noted that the observed *f* values for deciduous forest species shown in Table 16.2 probably represented a lower bounds as to the amount of radiocesium interception by the canopy that broadleaved deciduous trees could capture in full leaf conditions. In fact, the canopy interception of radiocesium by broadleaved deciduous forests remains insufficiently understood since initial fallout conditions have not occurred during the growing season when trees are full leafed.

16.3 Radiocesium Transfer from the Canopy to the Forest Floor: A Case Study

Initial canopy interception of atmospherically deposited radiocesium is followed by removal of contamination due to self-decontamination processes, such as rain wash and mechanical breakdown of plant bodies as litterfall. Transfer of canopy-intercepted radiocesium during the early phase of the atmospheric deposition is a key process governing its spatiotemporal distribution in forests and uptake by plants and animals; however, such data have been rarely collected from initial fallout conditions to the longer term. Nonetheless, the transfer of radiocesium from the canopy to forest floor has been monitored intensively in Japanese forest environments ever since the Fukushima accident (from initial fallout to present). The following data represents a case study from coniferous and mixed deciduous-coniferous forests of Japan affected by the Fukushima accident.

The study sites are located 40 km northwest of the FDNPP and are highly contaminated by deposited radionuclides (Fig. 16.2; Kato et al. 2017). The plume released by the FDNPP from approximately 12:00 to 15:00 JST (Japan Standard Time) on 15 March flowed northwestward, and wet deposition via precipitation occurred on the same night (Chino et al. 2011). In this area, the total atmospheric deposition of ^{137}Cs following the FDNPP accident was estimated to be 300-–600 kBq m^{-2} based on results of the Third Airborne Monitoring Survey of radioactive contamination (MEXT 2011). Three forest stands were selected as experimental sites, namely, a mature (31-year-old) Japanese cedar stand, a younger (15-year-old) Japanese cedar stand, and a mixed broadleaved stand of konara oak (*Quercus serrata* Murray) and Japanese red pine (*Pinus densiflora* Siebold & Zucc.), a typical secondary forest in the Fukushima area (Table 16.3; Kato et al. 2017, 2018a, b). Cesium-137 concentrations were monitored and quantified both in the vegetal components (e.g., foliage, outer bark, litterfall) and in open rainfall, throughfall, and stemflow (Fig. 16.3).

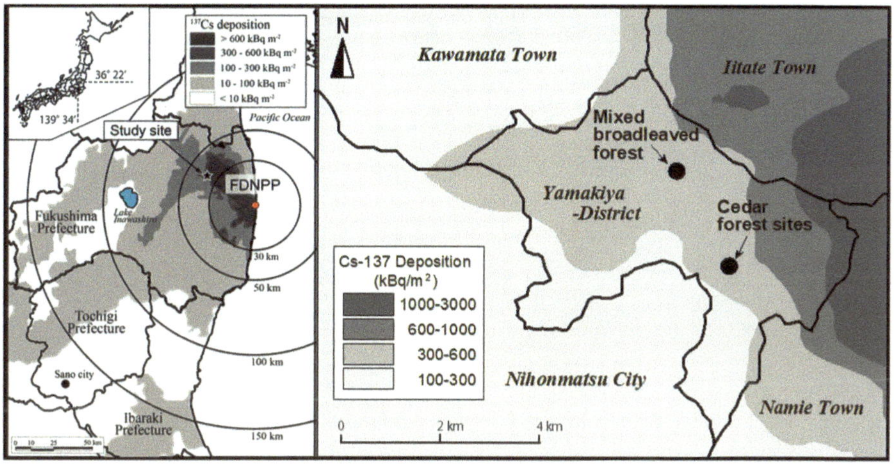

Fig. 16.2 Location of the study site. The ^{137}Cs deposition map is based on the Third Airborne Monitoring Survey of radioactivity by MEXT (2011). (The values on the ^{137}Cs inventory map were corrected to those of 2 July 2011). (From Kato et al. 2018a)

Table 16.3 Stand properties of the experimental forest sites in Fukushima Prefecture

Stand properties	Mature cedar (MC)	Young cedar (YC)	Mixed broadleaved (BL)
Stand age (year)	31	18	–
Stem density (stems ha^{-1})	1250	2600	2500 Konara oak 60%, Red pine 40%
Tree height (m)	25 <	< 20	25 <
Slope (°)	20	33	15
LAI (m^2 m^{-2})	4.2	10.3	–
^{137}Cs fallout (kBq m^{-2})	442		451

Markedly high ^{137}Cs concentrations were detected for the old Japanese cedar needles and the outer bark of both the Japanese cedar and konara oak, indicating contamination by direct deposition of atmospheric ^{137}Cs onto the plant surfaces (canopy interception) following the Fukushima accident (Table 16.4). The new needles/leaves of the Japanese cedar and konara oak were not affected by direct deposition from initial atmospheric fallout, but they showed high ^{137}Cs concentrations. The existence of ^{137}Cs in the new needles developed in 2011 indicated the direct incorporation of ^{137}Cs from the foliar and bark surfaces was followed by rapid translocation within the Japanese cedar trees (e.g., Yoshihara et al. 2014; Kanasashi et al. 2015; Nishikiori et al. 2015; Calmon et al. 2015; Thiry et al. 2016; Kato et al. 2017; Imamura et al. 2017a). On the other hand, foliar uptake had only a minor influence on ^{137}Cs concentrations in the new leaves of the konara oak because these leaves had not developed at the time of the initial atmospheric fallout. Therefore, the

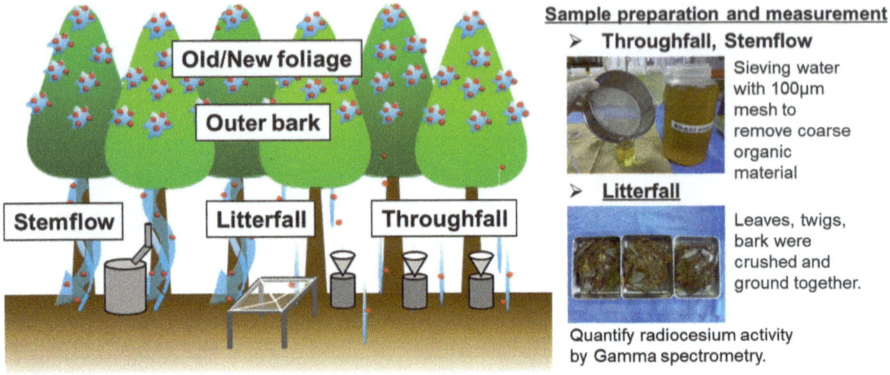

Fig. 16.3 Monitoring of radiocesium transfer via hydrological and biological pathways

^{137}Cs uptake into the newly developed konara oak leaves possibly occurred through the outer bark surfaces of the stems, branches, and twigs, rather than through foliar uptake (Kato et al. 2018b).

Cesium-137 concentrations in the Japanese cedar needles showed a sharp decreasing trend, indicating quick self-decontamination of ^{137}Cs over time (Table 16.4). In contrast to the cedar needles, ^{137}Cs concentrations in the konara oak leaves were more or less constant over time (Imamura et al. 2017a; Kato et al. 2018b). Cesium-137 concentrations in the outer bark samples showed concentrations staying more or less constant in the Japanese cedar and significantly decreasing for konara oak (Imamura et al. 2017a; Kato et al. 2018b).

The ^{137}Cs concentration in litterfall exponentially decreased over the 6-year observation period (Fig. 16.4a). The litterfall collected in the mixed broadleaved forest was a mixture of Japanese konara oak leaves, red pine needles, and twigs and bark from both species. Hisadome et al. (2013) reported the contribution of konara oak and red pine litterfall to the observed litterfall mass (kg m^{-2}) and ^{137}Cs deposition (kBq m^{-2}) in the same mixed broadleaved forest for an observation period from 31 July 2011 to 25 May 2012. The litterfall mass originating from the konara oak tree accounted for 74% of the total litterfall; however, contributions of these two tree species to the ^{137}Cs depositional flux via litterfall were nearly equal (Hisadome et al. 2013). A marked seasonal variability in ^{137}Cs concentrations in konara oak leaves has been reported (e.g., Hisadome et al. 2013; Okada et al. 2015). The ^{137}Cs concentration in konara oak leaves peaked during the growing season between May and September; however, the concentration significantly decreased during autumn (Okada et al. 2015). It is speculated that resorption of the ^{137}Cs from senescing leaves may have occurred before leaf fall during autumn, as has been reported for potassium, for example (e.g., Ninemets and Tamm 2005). Nevertheless, ^{137}Cs concentrations in the newly developed konara leaves have remained high for the last 6 years even though the canopy ^{137}Cs inventory has been reported to have decreased with time because of self-decontamination processes (e.g., Imamura et al. 2017a; Kato et al. 2017, 2018b). This discrepancy suggests that the contribution of

Table 16.4 Cesium-137 concentration in foliage and outer bark, kBq kg^{-1} (the ^{137}Cs concentration data were derived from Kato et al. (2018b))

Sampling year	Mature cedar			Young cedar			Mixed broadleaved (konara oak)		
	Old needle	New needle	Outer bark	Old needle	New needle	Outer bark	Old foliage	New foliage	Outer bark
< 200 d	119 ± 10	46 ± 5	–	45 ± 5	53 ± 4	–	–	15 ± 3	–
2011	75 ± 44	33 ± 13	–	33 ± 12	34 ± 19	–	–	12 ± 3	–
2012	39 ± 16	15 ± 5	12 ± 0	37 ± 6	10 ± 3	31 ± 0	–	15 ± 6	61 ± 0
2013	23 ± 6	8 ± 3	8 ± 3	30 ± 9	7 ± 4	25 ± 13	–	13 ± 3	26 ± 17
2014	18 ± 9	5 ± 0.5	8 ± 1	24 ± 12	4 ± 2	22 ± 4	–	11 ± 0	14 ± 2
2015	12 ± 0.3	5 ± 0.2	14 ± 0	11 ± 0	4 ± 0	28 ± 0	–	9 ± 0	–
2016	10 ± 4	2 ± 0.2	8 ± 1	12 ± 1	1 ± 1	16 ± 4	–	9 ± 0.2	6 ± 1

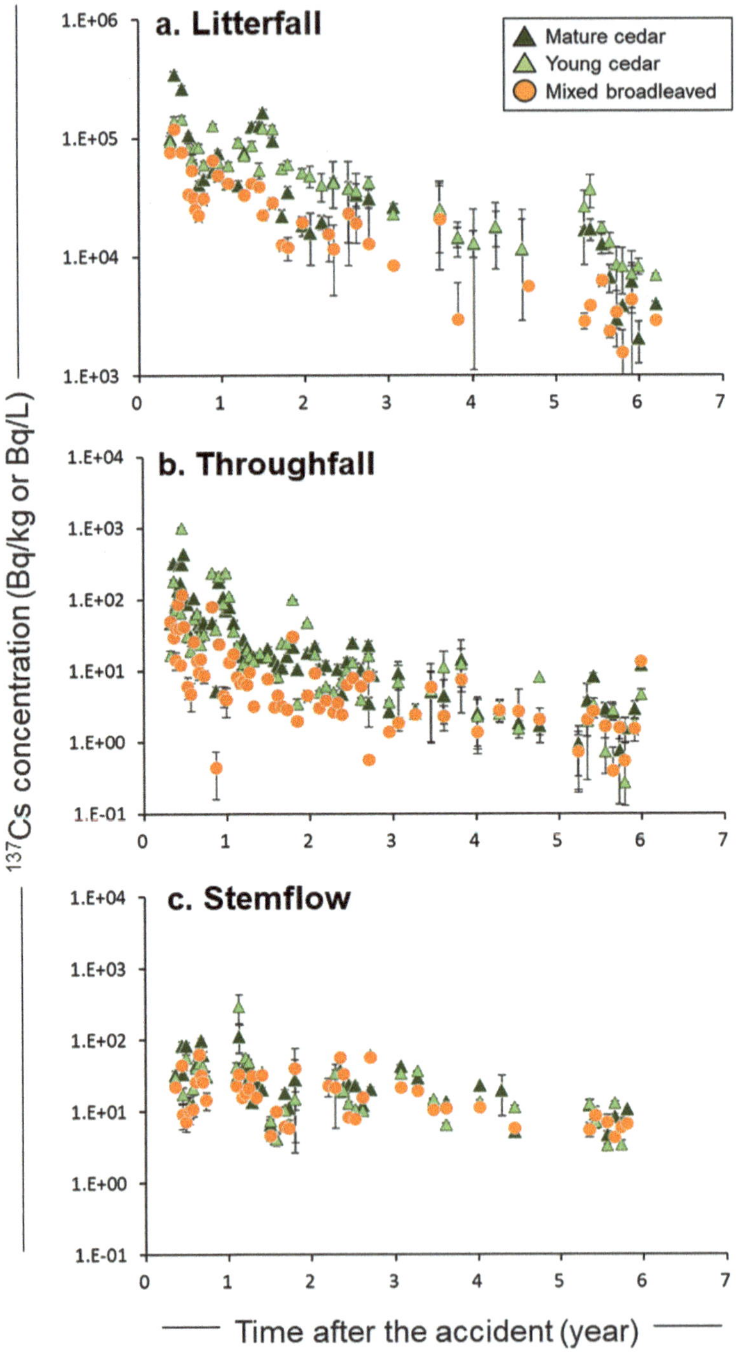

Fig. 16.4 Temporal changes of ^{137}Cs concentration in litterfall, throughfall, and stemflow. (The ^{137}Cs concentration data were derived from Hisadome et al. 2013 and Kato et al. 2017, 2018a)

root uptake of ^{137}Cs from the soil profile compensates the decrease in the ^{137}Cs stock in the trees and maintains the constant ^{137}Cs concentration in newly developed leaves.

Cesium-137 was detected in throughfall and stemflow samples collected from cedar and mixed broadleaved forest stands (Figs. 16.4b,c). Significant variability in ^{137}Cs concentrations was observed among the different sampling periods; however, relatively high concentrations exceeding 10 Bq L^{-1} were observed during the early phase of the contamination. The ^{137}Cs concentration in the throughfall exponentially decreased, and the rapid decrease during the early phase of the accident was followed by a slower decrease (Kato et al. 2012, 2017, 2018b; Loffredo et al. 2014). Such double-exponential declining trends have been used for simulating the loss of canopy radiocesium due to water movement through vegetation (whether trees or grass) (e.g., Kinnersley et al. 1996; Madoz-Escande et al. 2005; Kato et al. 2017). Furthermore, a mulch-exponential declining trend is observed for river water in the watershed affected by the Chernobyl accident (Smith et al. 2000, 2002, 2004) as well as stream and soil water in the headwater forest catchment in Fukushima Prefecture (Iwagami et al. 2017b, 2017c). These commonalities suggest that similar mechanisms in the fixation and leaching of radiocesium among plant, soil, and water determine the temporal evolution of ^{137}Cs concentration in the environment.

Cesium-137 concentrations in the foliage, bark, litterfall, and throughfall and stemflow were compared during each sampling year to clarify the key processes involved in ^{137}Cs transfer and leaching from the tree crowns. Mean ^{137}Cs concentrations during each sampling year were calculated for throughfall, stemflow, needles/leaves, and outer bark samples (Fig. 16.5).

Temporal changes in the ^{137}Cs concentration ratio of throughfall to old cedar needles/new konara oak leaves are shown in Fig. 16.5a (Kato et al. 2018b). It should be noted that the data for the early phase of the accident (<200 days), when the deposited radiocesium is readily mobile through hydrological pathways, is plotted

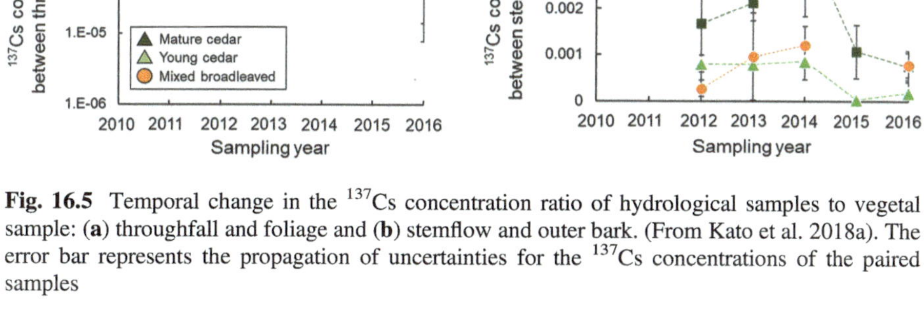

Fig. 16.5 Temporal change in the ^{137}Cs concentration ratio of hydrological samples to vegetal sample: (**a**) throughfall and foliage and (**b**) stemflow and outer bark. (From Kato et al. 2018a). The error bar represents the propagation of uncertainties for the ^{137}Cs concentrations of the paired samples

separately. In both the cedar and mixed broadleaved forests, the ^{137}Cs concentration ratio decreased exponentially with time (Fig. 16.5a), suggesting that the leachability of ^{137}Cs from the tree canopy to throughfall has been decreasing over the last 6 years. Furthermore, similar temporal trends in the concentration ratios were detected for both the cedar and mixed broadleaved forest stands. For the mixed broadleaved forest, the ^{137}Cs leaching from the pine canopies could have affected the observed decreasing trend of ^{137}Cs concentration in throughfall. However, the canopy coverage area of the konara oak accounted for more than 75% of the total plot area. Therefore, we considered that the majority of the ^{137}Cs in the throughfall originated from the canopy of the konara oak trees, particularly during the leafed season (May to late November). The decrease in the ^{137}Cs concentration ratio can be attributed to its migration from exterior vegetative surfaces to internal sinks over time (e.g., Thiry et al. 2016). This case study supports this claim as leaching has been found to decrease over time.

Temporal changes in the ^{137}Cs concentration ratio of stemflow to outer bark are shown in Fig. 16.5b. In the mature cedar stand, an increasing trend in the ^{137}Cs concentration ratio from 2012 to 2014 was followed by a marked decrease during 2015 and 2016. In the young cedar stand, the ^{137}Cs concentration ratio was nearly constant from 2012 to 2014, but it tended to decrease during 2015 and 2016. However, the ^{137}Cs concentration ratio for the konara oak trees showed an increasing trend during 2013–2016 with a lower ratio during 2012. In contrast to throughfall, there is limited knowledge concerning the mechanisms of radiocesium adsorption and leaching between the outer bark and rainwater via stemflow. Because we only collected outer bark samples from breast height (1.3 m), further investigation is required to clarify the mechanisms of ^{137}Cs entrainment from the outer bark surface to stemflow, especially in relation to bark thickness and texture which change with tree height.

Cumulative ^{137}Cs deposition fluxes to the forest floor were determined based on water inputs via throughfall and stemflow as well as litterfall inputs (Fig. 16.6). In the mature and young Japanese cedar stands, more than half of the total deposition flux occurred via throughfall, with the remainder mostly accounted for by litterfall. The mixed broadleaved stand had the largest deposition flux, with more than half of the total deposition occurring via litterfall and most of the remaining via throughfall. The ^{137}Cs deposition flux via stemflow was less than 5% of the total in all forest stands.

The contribution of stemflow on total depositional flux of radiocesium from the canopy to forest floor was rather small at forest stand scale; however, Imamura et al. (2017b) emphasized the importance of stemflow preferential flowpaths to the spatiality of ^{137}Cs concentrations and stocks in the soil of the proximal stem area in a konara oak forest. They found that trees with larger crown areas yielded higher ^{137}Cs concentrations and stocks in the near-trunk soils. With a similar focus on the spatiality of subcanopy radiocesium, Loffredo et al. (2015) investigated the relationship between the leachable stock of canopy-intercepted ^{137}Cs and canopy

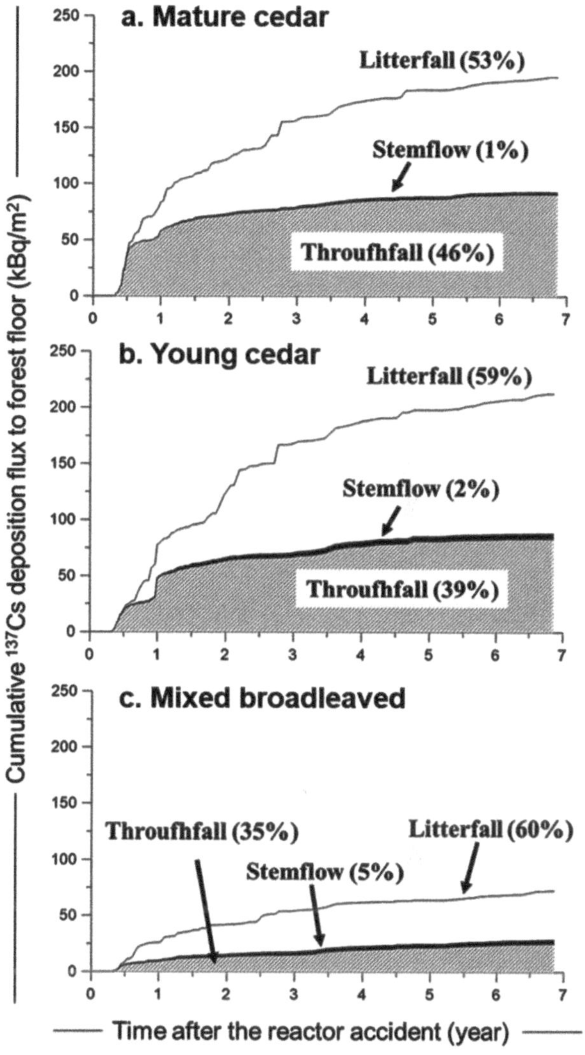

Fig. 16.6 Cumulative ^{137}Cs deposition onto forest floor via throughfall, stemflow, and litterfall during the 6-year observation period. (The ^{137}Cs depositional flux data was derived from Kato et al. 2017, 2018a)

closure based on intensive measurements of ^{137}Cs deposition flux to the forest floor in association with throughfall, stemflow, and litterfall in Japanese cedar stands after the FDNPP accident. They concluded that canopy closure (e.g., canopy density), representing the leachable ^{137}Cs stock in the canopy, could explain the horizontal variability of additional ^{137}Cs deposition fluxes to the forest floor in evergreen coniferous forests. On the other hand, the inter-canopy area is subjected to higher radiocesium deposition at the time of initial atmospheric deposition (Kato et al. 2018b; see next section). Further investigation is required to understand the complexities of radiocesium inputs to forests in relation to canopy structure both after the initial atmospheric fallout and over the longer term.

16.4 Effects of Canopy Interception on the Spatial Patterning of Radiocesium on the Forest Floor

Radiocesium stocks in forest soils are known to be highly heterogeneous (e.g., Guillitte et al. 1990; Takenaka et al. 1998; Shcheglov et al. 2001; Khomutinin et al. 2004; Korobova and Romanov 2011). Khomutinin et al. (2004) investigated the horizontal distribution of Chernobyl-derived ^{137}Cs in areas with various land uses and found that the horizontal variability in ^{137}Cs under forest canopies was greater than that of other land use types. Likewise, Shcheglov et al. (2001) found that the horizontal variability of the ^{137}Cs content in forest soil in the Russian Federation had a coefficient of variation ranging from 22% to 30%. They also reported an order of magnitude higher variation range in ^{137}Cs content in one near-site area within the 30-km exclusion zone of the Chernobyl Nuclear Power Plant, implying that the difference in the chemical form of fallout was affected by the spatial heterogeneity of atmospherically deposited radionuclides within the forest environment. Understanding the spatial heterogeneity of ^{137}Cs inputs to forest environments after nuclear events will not only help with understanding ^{137}Cs dynamics in forest environments but also other air pollutants. Moreover, the identification of underlying spatial patterns of radiocesium sinks and sources will be instrumental to understanding the long-term temporal dynamics and effects of radionuclide biogeochemical cycling on forest ecosystems (IAEA 2002; Goor and Thiry 2004).

Kato et al. (2018b) investigated the horizontal distribution of atmospherically deposited ^{137}Cs on the forest floor of a Japanese cedar plantation and a mixed broadleaved forest (konara oak and red pine, a typical secondary forest in Fukushima area) and its temporal evolution in the early phase of the FDNPP accident (Table 16.3).

Repeated in situ measurements of ^{137}Cs gamma count rates for 49 locations of the forest floor revealed the effects of initial canopy interception of the atmospherically deposited radiocesium and redistribution from the tree canopy. The Fukushima accident-derived ^{137}Cs was distributed heterogeneously over evergreen cedar and secondary mixed broadleaved forest floors (Fig. 16.7). In the cedar stands, the inter-canopy area had a relatively higher ^{137}Cs gamma count rate than the under-canopy area, whereas no clear relationship was found between the radiocesium pattern and canopy cover in the mixed broadleaved forest (Fig. 16.8). The lower ^{137}Cs count rates under the cedar canopy indicated the influence of more canopy interception by evergreen conifers, whereas such a trend was unclear in the mixed broadleaved forest, which was leafless season at the time of atmospheric fallout.

Rain was the dominant self-decontamination process for the canopy ^{137}Cs inventory, which led to additional ^{137}Cs inputs to the forest floor during the early phase of contamination (e.g., Hisadome et al. 2013; Loffredo et al. 2014; Kato et al. 2017, 2018b). The total depositional flux during the observation period of this study in the mature cedar, young cedar, and mixed broadleaved stands was 61 kBq m^{-2},

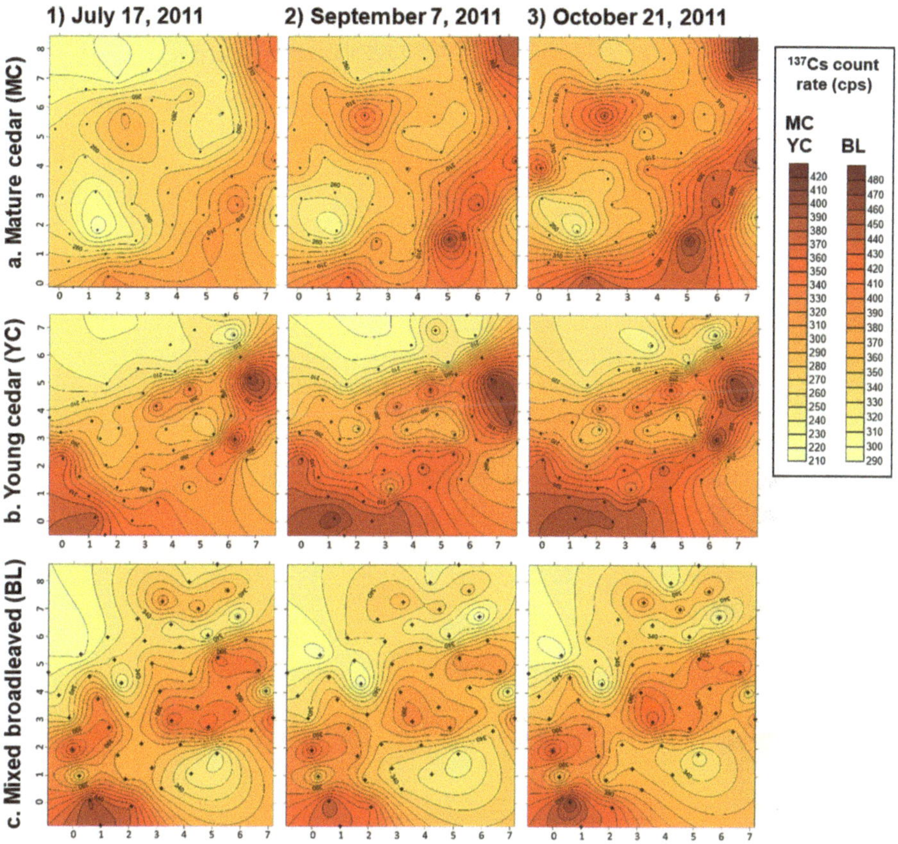

Fig. 16.7 Spatial pattern of the 137Cs gamma count rate in the experimental plot. (From Kato et al. 2018b)

33 kBq m^{-2}, and 22 kBq m^{-2}, respectively (Fig. 16.9). These depositional fluxes corresponded to 13.8% (mature cedar), 7.5% (young cedar), and 4.9% (mixed broadleaved) of the total atmospheric input following the Fukushima accident. Repeated ^{137}Cs measurements indicated that the spatial pattern of radiocesium activity on the forest floor did not substantially change following additional deposition inputs (Fig. 16.7) which were primarily associated with rain wash from the canopy. Furthermore, the magnitude of canopy covers partially explained the spatial variability of the ^{137}Cs on the forest floor in cedar stands. These results suggest that canopy structure affected the horizontal variability of atmospherically deposited radiocesium on the forest floor during the early phase of the Fukushima accident. Analysis of initial canopy interception and subsequent redistribution of accidentally released ^{137}Cs can yield new insights into the fate of atmospheric fallout in forest environments.

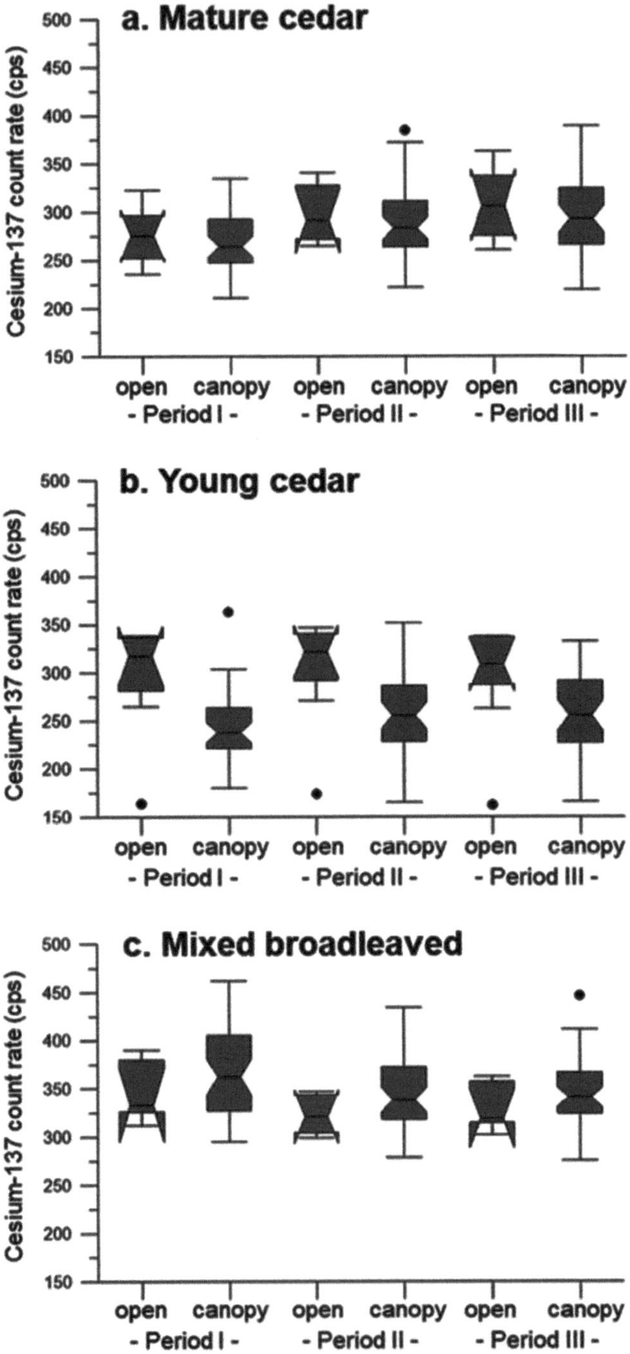

Fig. 16.8 Comparison of the measured ^{137}Cs count rate at locations under the forest canopy and open site (inter-canopy area) for different measurement periods. (The ^{137}Cs count rate data was derived from Kato et al. 2018b)

Fig. 16.9 Periodic and cumulative ^{137}Cs deposited onto forest floor. (The ^{137}Cs deposition flux data were derived from Kato et al. 2017, 2018a)

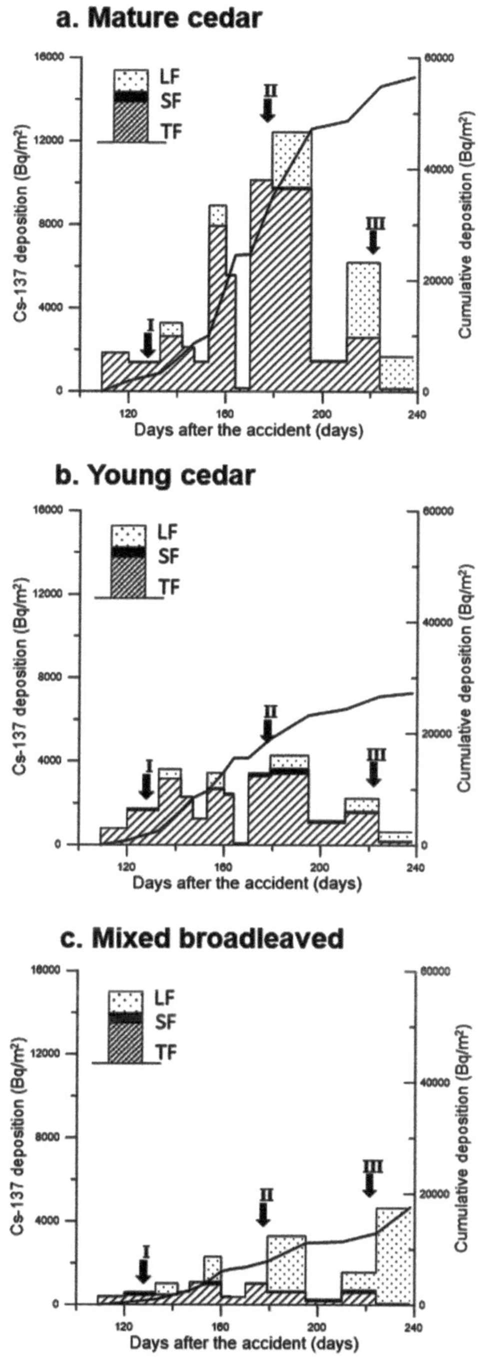

16.5 Future Direction of the Radiocesium and Hydrological Studies in Forest Ecosystems

Knowledge gaps and undefined or insufficiently understood processes of radiocesium cycling in forest ecosystems are highlighted in Fig. 16.10. Since the observed interception rate for deciduous forest species represented initial canopy interception during leafless season, canopy interception by broadleaved deciduous forests has not been sufficiently clarified due to limited data available during growing season. Furthermore, as demonstrated in this chapter, the initial canopy interception rates of radiocesium varied significantly, even for the same tree species. As such, further investigation is necessary to clarify the influences of stand properties, meteorological conditions, and deposition type (e.g., wet or dry dominated deposition) at the time of initial fallout on the efficiency of initial canopy interception.

Since the initial distribution of radiocesium in the forest soil is affected by canopy interception of the atmospheric fallout, it is necessary to more closely examine the mechanisms of adsorption and leaching of radiocesium among canopy components

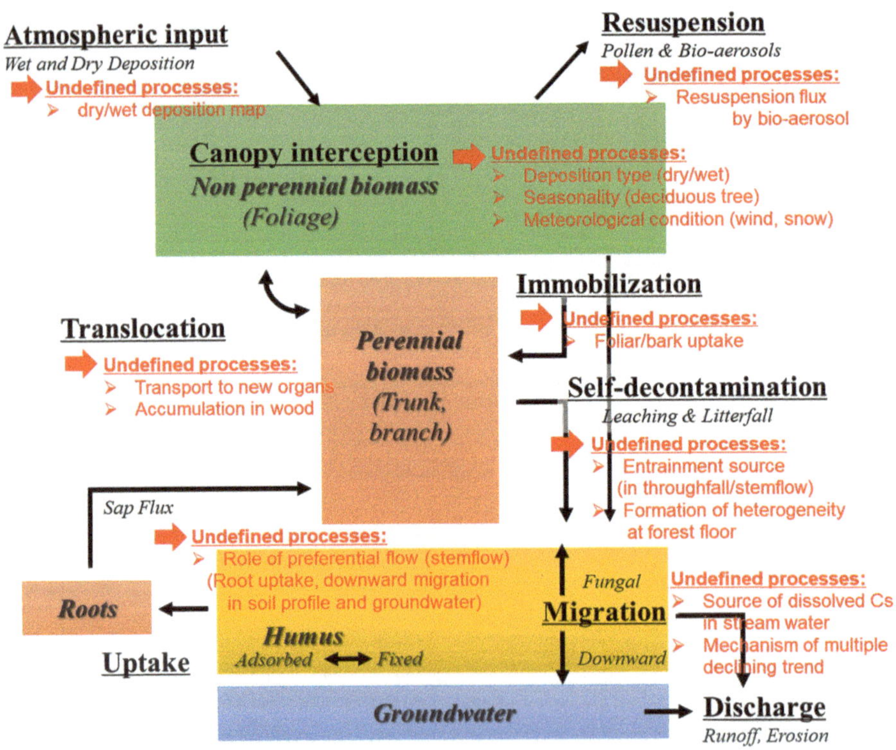

Fig. 16.10 Knowledge gaps and undefined processes for the cycling of radiocesium in forest ecosystems. (Builds on Fig. 16.1 which was adapted from IAEA 2002)

(e.g., foliage, twig, and branch) and between throughfall and stemflow, particularly the long-lasting source of radiocesium in stemflow. While the notable effects of stemflow on the spatial patterning of radiocesium inputs to the soil within the proximal area of konara oak tree trunks have been found (Imamura et al. 2017b), future work should examine the effects of these heterogeneous stemflow inputs (from preferential pathways on the tree trunk) on the ability of roots to uptake radiocesium. Linking the aboveground inputs of radiocesium in stemflow to the belowground functional ecology of roots would improve our understanding of the uneven spatial patterning of radiocesium stocks in forest soils. We recommend that researchers employ the stand-scale stemflow funneling ratios (Carlyle-Moses et al. 2018) and the infiltration area-based stemflow enrichment ratios (Levia and Germer 2015) to better quantify and understand the stemflow inputs of radiocesium. Coupling stemflow funneling and enrichment ratios with paired measurements of ground-penetrating radar and electrical resistivity tomography on subsurface features (such as roots) and soil moisture is one possible avenue to investigate stemflow-soil interactions to advance our knowledge of radionuclide cycling.

The use of multi-isotope tracer techniques using radiocesium, naturally occurring fallout radionuclides (e.g., unsupported ^{210}Pb, ^{7}Be), and stable isotopes is another avenue to improve our understanding of key hydrological processes at various temporal and spatial scales. It is recommended that a multi-isotope tracer approach be employed to better understand the interrelationships among the sources and sinks of radiocesium in forests and the processes and catalysts responsible for its transport. From Fig. 16.10, a number of new research questions could be formulated to advance the state of knowledge of radiocesium cycling in forests. Here are just a few examples: What is the precise role of bioaerosols in enhancing the deposition of radiocesium? Is there a seasonal dimension to the efficacy of bioaerosols in enhancing radiocesium deposition? To what extent are fungi and bacteria cycling radiocesium? Do they play a role in the transport and transfer of radiocesium via throughfall and stemflow? Further investigations also should examine the interaction of radionuclides with other pollutants on forest canopies and their transport and fate via throughfall and stemflow.

16.6 Summary

Canopy interception of atmospherically deposited radiocesium varied significantly among different tree species. The evergreen conifers tend to show high canopy interception rate, often exceeding 70% of atmospheric input. The high interception fraction of the deposited radiocesium by coniferous canopies indicated that the canopy will act as a secondary source of radioactive contamination of the forest floor. On the other hand, broadleaved deciduous tree species showed relatively low interception rates of less than 40% of atmospheric input. These observations appear to hold for forests affected by the Chernobyl and Fukushima accidents, although there were significant differences in physicochemical form of initial fallout and conditions of climate between those accidents.

Canopy interception of atmospherically deposited radiocesium is followed by hydrological and biological self-decontamination processes. Understanding the temporal trends of canopy radiocesium removal is important for determining the evolution of contamination to forest soils and the subsequent cycling of radiocesium within soils (and back to vegetation via uptake) of forest ecosystems. The monitoring of radiocesium concentrations in hydrological and biological components was an effective means to determine the transport of canopy radiocesium inventory to forest floor. Uncovering the spatial patterning of radiocesium stocks and its flux among different forest compartments can increase our knowledge of the hydrological and biogeochemical processes governing element cycling in forest ecosystems. Recent studies following the Fukushima accident have provided initial insights of the behavior of atmospheric radiocesium in the forest environment and have increased our process-based understanding of radiocesium cycling. Building upon the knowledge gleaned from earlier works, there are ample opportunities for new lines of research that further elucidate the role of forest-water interactions on the cycling of radiocesium.

References

Ashraf MA, Akib S, Maah MJ, Yusoff I, Balkhair KS (2014) Cesium-137: radio-chemistry, fate, and transport, remediation, and future concerns. Crit Rev Environ Sci Technol 44:1740–1793. https://doi.org/10.1080/10643389.2013.790753

Bonnett PJP, Anderson MA (1993) Radiocaesium dynamics in a coniferous forest canopy: a mid-Wales case study. Sci Total Environ 136:259–277. https://doi.org/10.1016/0048-9697(93)90314-V

Bunzl K, Schimmack W, Kreutzer K, Schierl R (1989) Interception and retention of Chernobyl-derived ^{134}Cs, ^{137}Cs and ^{106}Ru in a spruce stand. Sci Total Environ 78:77–87. https://doi.org/10.1016/0048-9697(89)90023-5

Calmon P, Gonze MA, Mourlon C (2015) Modeling the early-phase redistribution of radiocesium fallouts in an evergreen coniferous forest after Chernobyl and Fukushima accidents. Sci Total Environ 529:30–39. https://doi.org/10.1016/j.scitotenv.2015.04.084

Cambray RS, Playford K, Carpenter RC (1989) Radioactive fallout in air and rain: results to the end of 1988. In: UK atomic energy authority Rep., AERE-R 13575. HMSO, London

Carlyle-Moses DE, Iida S, Germer S, Llorens P, Michalzik B, Nanko K et al (2018) Expressing stemflow commensurate with its ecohydrological importance. Adv Water Resour 121:472–479. https://doi.org/10.1016/j.advwatres.2018.08.015

Chino M, Nakayama H, Nagai H, Terada H, Katata G, Yamazawa H (2011) Preliminary estimation of release amounts of ^{131}I and ^{137}Cs accidentally discharged from the Fukushima Daiichi nuclear power plant into the atmosphere. J Nucl Sci Technol 48:1129–1134. https://doi.org/10.1080/18811248.2011.9711799

Goor F, Thiry Y (2004) Processes, dynamics and modelling of radiocaesium cycling in a chronosequence of Chernobyl-contaminated scots pine (*Pinus sylvestris* L.) plantations. Sci Total Environ 325:163–180. https://doi.org/10.1016/j.scitotenv.2003.10.037

Guillitte O, Andolina J, Koziol M, Debauche A (1990) Plant-cover influence on the spatial distribution of radiocesium deposits in forest ecosystems. In: Desmet G, Nassimbeni P, Belli M (eds) Transfer of radionuclides in natural and semi-natural environments. Elsevier Applied Science, London, pp 441–449

Hisadome K, Onda Y, Kawamori A, Kato H (2013) Migration of radiocesium with litterfall in hardwood-Japanese red pine mixed forest and sugi plantation. J Jpn Forest Soc 95:267–274. (in Japanese with English summary). https://doi.org/10.4005/jjfs.95.267

Hoffman FO, Thiessen KM, Rael RM (1995) Comparison of interception and initial retention of wet-deposited contaminants on leaves of different vegetation types. Atmos Environ 29:1771–1775. https://doi.org/10.1016/1352-2310(95)00099-K

IAEA (2002) Modelling the migration and accumulation of radionuclides in forest ecosystems. International Atomic Energy Agency, Vienna

IAEA (2006) Environmental consequences of the Chernobyl accident and their remediation: twenty years of experience; report of the Chernobyl forum expert group 'environment. In: Radiological assessment reports series. International Atomic Energy Agency, Vienna, p 166

IAEA (2010) Handbook of parameter values for the prediction of radionuclide transfer in terrestrial and freshwater environments. IAEA Tech Rep Ser 472:191

Imamura N, Komatsu M, Ohashi S, Hashimoto S, Kajimoto T, Kaneko S et al (2017a) Temporal changes in the radiocesium distribution in forests over the five years after the Fukushima Daiichi nuclear power plant accident. Sci Rep- UK 7:8179. https://doi.org/10.1038/s41598-017-08261-x

Imamura N, Levia DF, Toriyama J, Kobayashi M, Nanko K (2017b) Stemflow-induced spatial heterogeneity of radiocesium concentrations and stocks in the soil of a broadleaved deciduous forest. Sci Total Environ 599-600:1013–1021. https://doi.org/10.1016/j.scitotenv.2017.05.017

Itoh Y, Imaya A, Kobayashi M (2015) Initial radiocesium deposition on forest ecosystems surrounding the Tokyo metropolitan area due to the Fukushima Daiichi nuclear power plant accident. Hydrol Res Lett 9:1–7. https://doi.org/10.3178/hrl.9.1

Iwagami S, Onda Y, Tsujimura M, Abe Y (2017a) Contribution of radioactive ^{137}Cs discharge by suspended sediment, coarse organic matter, and dissolved fraction from a headwater catchment in Fukushima after the Fukushima Dai-ichi Nuclear Power Plant accident. J Environ Radioact 166:466–474. https://doi.org/10.1016/j.jenvrad.2016.07.025

Iwagami S, Tsujimura M, Onda Y, Nishino M, Konuma R, Abe Y et al (2017b) Temporal changes in dissolved ^{137}Cs concentrations in groundwater and stream water in Fukushima after the Fukushima Dai-ichi Nuclear Power Plant accident. J Environ Radioact 166:458–465. https://doi.org/10.1016/j.jenvrad.2015.03.025

Iwagami S, Onda Y, Tsujimura M, Hada M, Pun I (2017c) Vertical distribution and temporal dynamics of dissolved ^{137}Cs concentrations in soil water after the Fukushima Dai-ichi Nuclear Power Plant accident. Environ Pollut 230:1090–1098. https://doi.org/10.1016/j.envpol.2017.07.056

Kanasashi T, Sugiura Y, Takenaka C, Hijii N, Uemura M (2015) Radiocesium distribution in sugi (*Cryptomeria japonica*) in eastern Japan: translocation from needles to pollen. J Environ Radioact 139:398–406. https://doi.org/10.1016/j.jenvrad.2014.06.018

Kato H, Onda Y, Gomi T (2012) Interception of the Fukushima reactor accident-derived ^{137}Cs, ^{134}Cs and ^{131}I by coniferous forest canopies. Geophys Res Lett 39:L20403. https://doi.org/10.1029/2012GL052928

Kato H, Onda Y, Loffredo N, Hisadome K, Kawamori A (2017) Temporal changes in radiocesium deposition in various forest stands following the Fukushima Dai-ichi Nuclear Power Plant accident. J Environ Radioact 116:449–457. https://doi.org/10.1016/j.jenvrad.2015.04.016

Kato H, Onda Y, Saidin ZH, Sakashita W, Hisadome K, Loffredo N (2018a) Six-year monitoring study of radiocesium transfer in forest environments following the Fukushima Nuclear Power Plant accident. J Environ Radioactiv In press:105817. https://doi.org/10.1016/j.jenvrad.2018.09.015

Kato H, Onda Y, Wakahara T, Kawamori A (2018b) Spatial pattern of atmospherically deposited radiocesium on the forest floor in the early phase of the Fukushima Daiichi Nuclear Power Plant accident. Sci Total Environ 615:187–196. https://doi.org/10.1016/j.scitotenv.2017.09.212

Khomutinin YV, Kashparov VA, Zhebrovska KI (2004) Sampling optimization when radioecological monitoring. Ukraine Institute for Agricultural Radiology, Kiev, p 137

Kinnersley RP, Shaw G, Bell JNB, Minski J, Goddard AJH (1996) Loss of particulate contaminants from plant canopies under wet and dry conditions. Environ Pollut 91:227–235. https://doi.org/10.1016/0269-7491(95)00047-X

Korobova E, Romanov S (2011) Experience of mapping spatial structure of Cs-137 in natural landscape and patterns of its distribution in soil toposequence. J Geochem Explor 109:139–145. https://doi.org/10.1016/j.gexplo.2011.02.006

Landis JD, Hamm NT, Renshaw CE, Dade WB, Magilligan FJ, Gartner JD (2012) Surficial redistribution of fallout 131iodine in a small temperate catchment. P Natl Acad Sci US 109:4064–4069. https://doi.org/10.1073/pnas.1118665109

Levia D, Germer S (2015) A review of stemflow generation dynamics and stemflow-environment interactions in forests and shrublands. Rev Geophys 53:673–714. https://doi.org/10.1002/2015RG000479

Loffredo N, Onda Y, Kawamori A, Kato H (2014) Modeling of leachable Cs-137 in throughfall and stemflow for Japanese forest canopies after Fukushima Daiichi Nuclear Power Plant accident. Sci Total Environ 493:701–707. https://doi.org/10.1016/j.scitotenv.2014.06.059

Loffredo N, Onda Y, Hurtevent P, Coppin F (2015) Equation to predict ^{137}Cs leaching dynamic from evergreen canopies after a radio-cesium deposit. J Environ Radioact 147:100–107. https://doi.org/10.1016/j.jenvrad.2015.05.018

Madoz-Escande C, Garcia-Sanchez L, Bonhomme T, Morello M (2005) Influence of rainfall characteristics on elimination of aerosols of cesium, strontium, barium and tellurium deposited on grassland. J Environ Radioact 84:1–20. https://doi.org/10.1016/j.jenvrad.2005.03.006

Melin J, Wallberg L, Suomela J (1994) Distribution and retention of cesium and strontium in Swedish boreal forest ecosystems. Sci Total Environ 157:93–105. https://doi.org/10.1016/0048-9697(94)90568-1

MEXT (2011) Results of the third Airborne monitoring survey by MEXT. radioactivity.nsr.go.jp/en/contents/5000/4182/24/1304797_0708e.pdf. Last Accessed on 1 July 2017

Myttenaere C, Schell WR, Thiry Y, Sombré L, Ronneau C, Van Der Stegen J (1993) Modelling of the Cs-137 cycling in forest: recent development and research needed. Sci Total Environ 136:77–91. https://doi.org/10.1016/0048-9697(93)90298-K

Nimis PL (1996) Radiocesium in plants of forest ecosystems. Stud Geobot 15:3–49

Nincmets Y, Tamm U (2005) Species differences in timing of leaf fall and foliage chemistry modify nutrient resorption efficiency in deciduous temperate forest stands. Tree Physiol 25:1001–1014. https://doi.org/10.1093/treephys/25.8.1001

Nishikiori T, Watanabe M, Koshikawa M, Takamatsu T, Ishii Y, Ito S et al (2015) Uptake and translocation of radiocesium in cedar leaves following the Fukushima nuclear accident. Sci Total Environ 502:611–616. https://doi.org/10.1016/j.scitotenv.2014.09.063

Okada N, Nakai W, Ohashi S, Tanaka A (2015) Radiocesium migration from the canopy to the forest floor in pine and deciduous forests. J Jpn Forest Soc 97:57–62. https://doi.org/10.4005/jjfs.97.57

Onda Y, Kato H, Hoshi M, Takahashi Y, Nguyen M (2015) Soil sampling and analytical strategies for mapping fallout in nuclear emergencies based on the Fukushima Dai-ichi Nuclear Power Plant accident. J Environ Radioact 139:300–307. https://doi.org/10.1016/j.jenvrad.2014.06.002

Pröhl G (2009) Interception of dry and wet deposited radionuclides by vegetation. J Environ Radioact 100:675–682. https://doi.org/10.1016/j.jenvrad.2008.10.006

Ronneau C, Cara J, Apers D (1987) The deposition of radionuclides from Chernobyl to a forest in Belgium. Atmos Environ 21:1467–1468. https://doi.org/10.1016/0004-6981(67)90094-7

Ronneau C, Sombré L, Myttenaere C, Andre P, Vanhouche M, Cara J (1991) Radiocaesium and potassium behavior in forest trees. J Environ Radioact 14:259–268. https://doi.org/10.1016/0265-931X(91)90032-B

Schell WR, Linkov I, Myttenaere C, Morel B (1996) A dynamic model for evaluating radionuclide distribution in forests from nuclear accidents. Health Phys 70:318–335. https://doi.org/10.1097/00004032-199603000-00002

Schimmack W, Bunzl K, Kreutzer K, Rondenkirchen E, Schierl R (1991) Einfluss von fichte (Picea abies L. Karst) und buche (*Fagus sylvatica* L.) auf die Wanderung von radiocasium im Boden. Fortwiss Forsch 39:242–251

Shcheglov AI, Tsvetnova OB, Klyashtorin AI (2001) Biogeochemical migration of technogenic radionuclides in forest ecosystems. NAUKA, Moscow, p 235

Smith JT, Comans RNJ, Beresford NA, Wright SM, Howard BJ, Camplin WC (2000) Chernobyl's legacy in food and water. Nature 405:141. https://doi.org/10.1038/35012139

Smith JT, Cross MA, Wright SM (2002) Predicting transfers of ^{137}Cs in terrestrial and aquatic environments: a whole-ecosystem approach. Radioprotection 37:37–42. https://doi.org/10.1051/radiopro/2002071

Smith JT, Wright SM, Cross MA, Monte L, Kudelsky AV, Saxén R et al (2004) Global analysis of the riverine transport of ^{90}Sr and ^{137}Cs. Environ Sci Technol 38:850–857. https://doi.org/10.1021/es0300463

Sombre L, Vanhouche M, Thiry Y, Ronneau C, Lambotte JM, Myttenaere C (1990) Transfer of radiocesium in forest ecosystems resulting from a nuclear accident. In: Desmet G et al (eds) Transfer of radionuclides in natural and semi-natural environments. Elsevier Applied Science, London, pp 74–83

Takenaka C, Onda Y, Hamajima Y (1998) Distribution of cesium-137 in Japanese forest soils: correlation with the contents of organic carbon. Sci Total Environ 222:193–199. https://doi.org/10.1016/S0048-9697(98)00305-2

Tamura T (1964) Selective sorption reaction of cesium with mineral soils. Nucl Saf 5:262–268

Thiry Y (1997) Etude du cycle du radiocesium en ecosysteme forestier: Distribution et facteurs de mobilité. Thesis, Université Catholique de Louvain, Louvain-la-Neuve, Belgium

Thiry Y, Garcia-Sanchez L, Hurtevent P (2016) Experimental quantification of radiocesium recycling in a coniferous tree after aerial contamination: field loss dynamics, translocation and final partitioning. J Environ Radioact 161:42–50. https://doi.org/10.1016/j.jenvrad.2015.12.017

Tikhomirov FA, Shcheglov AI (1991) The radiological consequences of the Kyshtym and Chernobyl radiation accidents for forest ecosystems. In: Proceedings of the seminar on comparative assessment of the environmental impact of radionuclides released during three major nuclear accidents, Kyshtym, Windscale, Chernobyl. CEC, Luxembourg, 1–5 October 1990, H, EUR 13574, pp 867–888

Walling DE (2002) Recent advances in the use of environmental radionuclides in soil erosion investigations. IAEA-SM-363/89, Report No IAEA-CSP--11/C, pp 279–301

Yoshihara T, Matsumura H, Tsuzaki M, Wakamatsu T, Kobayashi T, Hashida S, Nagaoka T, Goto F (2014) Changes in radiocesium contamination from Fukushima in foliar parts of 10 common tree species in Japan between 2011 and 2013. J Environ Radioact 138:220–226. https://doi.org/10.1016/j.jenvrad.2014.09.002

Part IV
Forest-Water Interactions in Urban Ecosystems

Chapter 17
Urban Trees as Green Infrastructure for Stormwater Mitigation and Use

Darryl E. Carlyle-Moses, Stephen Livesley, Mariana D. Baptista, Jasmine Thom, and Christopher Szota

17.1 Introduction

Globally, it is projected that five billion people will reside in cities by 2030, with urban land cover increasing by 200% over levels present in 2000 (Elmqvist et al. 2013). This increase in urban population and land use will require substantive ingenuity and planning to reduce the vulnerability of urban areas and their residents to hazards, especially in the face of extreme weather events associated with global climatic change (Voskamp and Van de Ven 2015; Green et al. 2016). Green infrastructure represents the management approach in urban and peri-urban landscapes of utilizing trees and other vegetation cover to provide community benefits and ecosystem services, including resiliency against various meteorological hazards. As a component of green infrastructure, trees in urbanised areas can diminish air pollution (Abhijith et al. 2017), including carbon dioxide concentrations (Nowak et al. 2013; Chen 2015), serve to mitigate heat stress caused by climate change and the urban heat island effect (Emmanuel and Loconsole 2015; Zölch et al. 2016), and provide habitat for wildlife (Stagoll et al. 2012). Additionally, trees can reduce urban stormwater runoff through canopy interception loss and water uptake (Chair 2000; Stovin et al. 2008; Livesley et al. 2016; Berland et al. 2017) with the potential

D. E. Carlyle-Moses (✉)
Department of Geography and Environmental Studies, Thompson Rivers University, Kamloops, BC, Canada
e-mail: dcarlyle@tru.ca

S. Livesley · J. Thom · C. Szota
School of Ecosystem and Forest Sciences, Faculty of Science, The University of Melbourne, Richmond, VIC, Australia

M. D. Baptista
Centre for Urban Research, School of Global, Urban and Social Studies, RMIT University, Melbourne, VIC, Australia

D. F. Levia et al. (eds.), *Forest-Water Interactions*, Ecological Studies 240,
https://doi.org/10.1007/978-3-030-26086-6_17

economic value associated with this stormwater reduction being considerable (Soares et al. 2011; McPherson et al. 2016; Nordman et al. 2018).

The densification and infilling of many urban centres has increased impervious surface cover such that runoff volumes and velocities can exceed the capacity of drainage infrastructure leading to local urban flooding and ecological damage to urban waterways (Shuster et al. 2005; Kaźmierczak and Cavan 2011; Ellis 2013). Given the increase in rainstorm frequency and intensity with climate change, this issue has the potential to intensify in many cities worldwide (Ashley et al. 2005; Neumann et al. 2015). There is great interest in urban stormwater control measures (SCMs) in addition to the traditional drainage infrastructure (grey infrastructure) to help lessen the magnitude of stormwater runoff. Many SCMs contain trees and other vegetation types, and tree canopies in their own right can provide runoff reduction benefits. Understanding the ways in which trees interact with rainfall, including the partitioning of rainfall by the canopy and tree water uptake (see Fig. 17.1) and the factors that control the magnitudes of these processes under urban conditions, can help guide municipal tree species selection and the design of planting initiatives to provide stormwater abatement (McPherson et al. 2005; Berland et al. 2017; Huang et al. 2017).

The goal of this chapter is to summarise the current state of knowledge of the impact of city trees on urban catchment hydrology and by extension the role of urban trees as components of SCMs. The chapter will include a review of rainfall partitioning by tree canopies into canopy interception loss, stemflow, and throughfall in urban areas and the potential these canopy water balance components have for stormwater mitigation. Tree water uptake and transpiration in the built environment are also evaluated with an emphasis on how these processes can contribute to SCMs. A discussion of future research opportunities and directions associated with trees as a component of green infrastructure to help address urban stormwater management issues is also provided.

17.2 Rainfall Redistribution by Urban Trees

17.2.1 Canopy Interception Loss

Where present in the urban landscape, trees and other vegetation alter the amount and spatial distribution of precipitation that reaches the impervious or permeable ground surfaces below. For example, a portion of the precipitation received by a tree canopy may be stored on and evaporated from that canopy and therefore never reach the ground. This *canopy interception loss* has been extensively studied in various natural and plantation forests that have near-continuous canopy cover (e.g., Helvey and Patric 1965; Toba and Ohta 2005; Šraj et al. 2008; Ghimire et al. 2017). In these forests, interception loss can account for 10–50% of rainfall, representing a sizable portion of the total evaporation from these environments (Carlyle-Moses and Gash 2011). However, the extrapolation of findings from natural and managed forests to

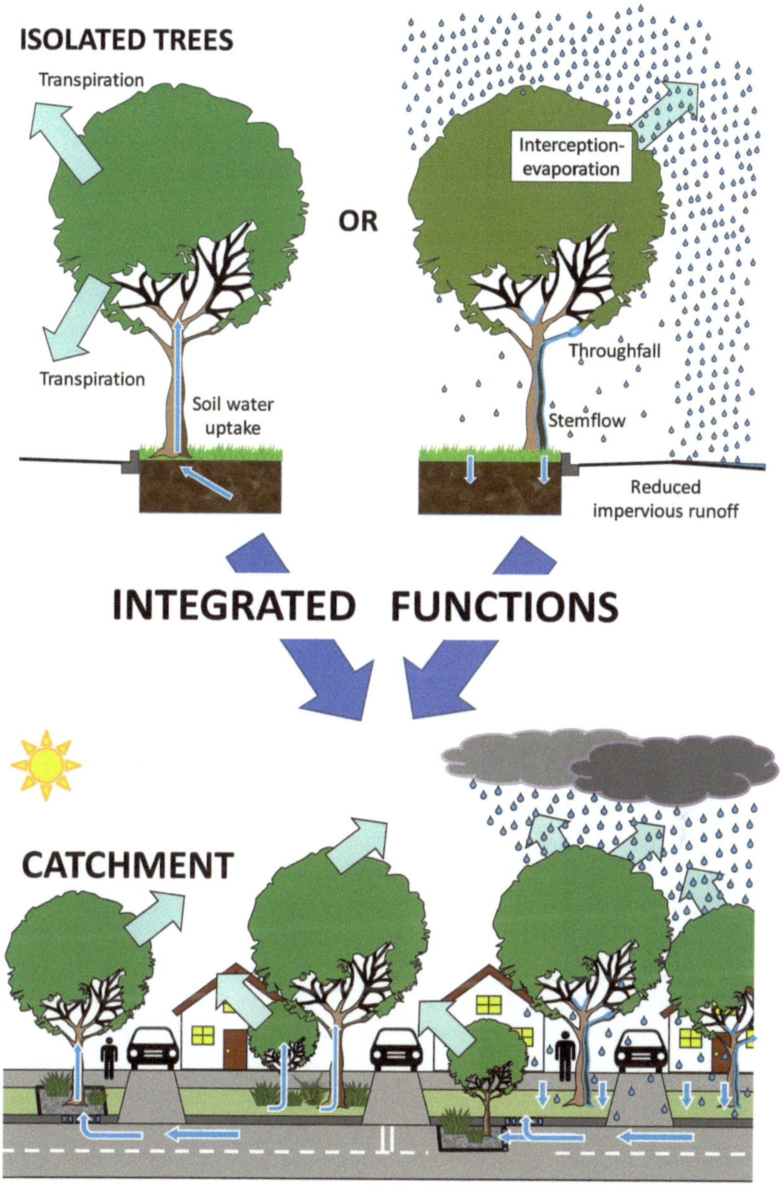

Fig. 17.1 Catchment-scale integration of tree interception-evaporation and transpiration water cycles

urban trees and the 'urban forest' may not be appropriate since single, isolated tree architecture differs from that of stand-grown trees and the urban microclimate is generally warmer and more strongly evaporative (Xiao et al. 1998; Guevara-Escobar et al. 2007; Kuehler et al. 2017). Although canopy interception studies in urban areas are limited, the available data suggest that interception loss is appreciable for individual trees in urban landscapes, as it is for urban woodlots and urban remnant forest patches (Table 17.1).

17.2.1.1 Canopy Water Storage Capacity

Of no small consequence to the volumetric importance of interception loss is the amount of water that the foliage and wood composing the canopy can detain on their surfaces. This canopy water storage capacity regulates the supply of water available for evaporation from the canopy while the storm event is in progress (Rutter et al. 1971; Klaassen et al. 1998) and once the rain event has ceased (Valente et al. 1997; Zeng et al. 2000). The canopy water storage capacity of a tree, or forest, is a function of both the canopy projection area and the canopy density (Holder 2013; Fathizadeh et al. 2017). Canopy density can be inferred from plant area index or more commonly leaf area index (LAI) (Livesley et al. 2014; Shahidan 2015).

Isolated, open-grown street and park trees that are healthy are likely to have more voluminous canopies with greater canopy densities per projection area and thus greater water storage and interception loss potential (Rhoades and Stipes 1999; Asadian and Weiler 2009; Pretzsch et al. 2015). However, many urban trees can be stressed and unhealthy because of restricted infiltration, contaminated soils, vandalism, excessive transpiration demand, air pollution, and/or insufficient nutrient availability, and, as a consequence, these trees may have limited canopy projection areas and canopy densities relative to their unstressed and healthy counterparts and therefore may not achieve their canopy water storage capacity potential (Nowak et al. 1990; Craul 1999; Zipper et al. 2017).

The ‘urban forest’ is very different from that of many natural and managed forests in that there can be hundreds of different tree species within a small geographical area. This highly diverse population of trees is subjected to large variations in light, water, and nutrient availability, as well as differing management interventions, including periodic pruning. Due to the vast array of species, site conditions, and management schemes, large differences in canopy water storage capacities amongst tree species planted throughout the urban landscape can be expected.

Canopy water storage capacity will be influenced by species-specific characteristics at the tree architecture level, as well as at the leaf and sub-leaf level. The fraction of the canopy that is comprised of wood (twigs, branches, and stem) and the surface texture and properties of the bark covering that wood are of importance for overall canopy water storage (Llorens and Gallart 2000; Livesley et al. 2014; Baptista et al. 2018a). The type of foliage (i.e., needle versus broadleaved), the size and shape, the hydrophobicity, the microstructure, and the angle and flexible movement of that foliage all play important roles in overall canopy water storage (Rosado and Holder 2013; Xiao and McPherson 2016).

Using branch samples under rainfall simulations, Xiao and McPherson (2016) determined the water storage capacity of 20 street trees common to Davis, California. The leaf and branch samples experienced simulated rainfall at different

Table 17.1 Mean study-period canopy interception loss (% of rainfall) from selected studies of isolated street and park trees, residential lawns, and urban landscape scale and for urban remnant forests

Species or forest type	Location	Mean interception loss (%)	References
Isolated street and park trees			
Broadleaved			
Albizia procera	San Juan, Puerto Rico	22.8	Nytch et al. (2018)
Calophyllum antillanum	Ibid.	16.4	Ibid.
Ficus benjamina	Queretaro City, Mexico	59.5	Guevara-Escobar et al. (2007)
Mangifera indica	Uruaçu, Brazil	47.9[a]	Alves et al. (2018)
Pachira aquatica	Ibid.	44.3[a]	Ibid.
Licania tomentosa	Ibid.	43.1[a]	Ibid.
Caesalpinia peltophoroides	Ibid.	28.7[a]	Ibid.
Caesalpinia pluviosa and *Tipuana tipu*	Sao Paulo, Brazil	60	Silva et al. (2010)
Eucalyptus saligna	Melbourne, Australia	29	Livesley et al. (2014)
Eucalyptus nicholii	Ibid.	44	Ibid.
Liquidambar styraciflua	Oakland, California	14.3	Xiao and McPherson (2011a)
Ginkgo biloba	Ibid.	25.2	Ibid.
Citrus limon	Ibid.	27.0	Ibid.
Pyrus calleryana 'Bradford'	Davis, California	15	Xiao et al. (2000b)
Quercus suber	Ibid.	27	Ibid.
Coniferous			
Pseudotsuga menziesii	North Shore (North Vancouver), Canada	49.1	Asadian and Weiler (2009)
Thuja plicata	Ibid.	60.9	Ibid.
Residential lawns			
	Raleigh, North Carolina	9.1–21.4	Inkiläinen et al. (2013)
City-wide			
	Santa Monica, California	27.3[b], 1.6[c]	Xiao and McPherson (2002)
	Modesto, California	11.9[b]	McPherson et al. (1999)
	Sacramento, California	11.1[b], 1.0[c]	Xiao et al. (1998)

(continued)

Table 17.1 (continued)

Species or forest type	Location	Mean interception loss (%)	References
Urban remnant forests			
Mixed forest (*Quercus petraea*, *Castanea sativa*, *Picea abies*)	Ljubljana, Slovenia	18.0	Kermavnar and Vilhar (2017)
Riparian forest (*Pinus sylvestris*, *Tilia cordata*, etc.)	Ibid.	3.9	Ibid.
Floodplain forest (*Acer pseudoplatanus*, *Alnus incana*, *Tilia cordata*, etc.)	Ibid.	7.1	Ibid.

[a]Derived from reported interception loss and rain depths
[b]Crown-level scale
[c]Landscape-level scale

intensities (3.6–139.7 mm h^{-1}), and mean water storage capacities ranged from 0.51 mm for Callery 'Bradford' pear (*Pyrus calleryana*) to 1.81 mm for blue spruce (*Picea pungens*). Mean water storage capacities were larger for coniferous canopies (1.25 ± 0.38 mm) than broadleaved deciduous (0.77 ± 0.19 mm) or broadleaved evergreen (0.78 ± 0.05 mm). Li et al. (2016), from rainfall simulation experiments of whole trees for four common species planted in Beijing, China, found that the two coniferous species had greater storage capacities compared to the two broadleaved deciduous trees. Xiao and McPherson (2016) attribute differences between the water storage capacities of coniferous and broadleaved species to dissimilarities in leaf and stem morphologies and surface roughness, while Li et al. (2016) suggest that the closer configuration of coniferous needles (fascicles) provides greater surface area and junction points where water can adhere under surface tension permitting greater water storage compared to broadleaved foliage.

The presence of coniferous cover has been attributed to greater interception loss in urban forest scenarios in Ljubljana, Slovenia (Kermavnar and Vilhar 2017), while Inkiläinen et al. (2013) found that the percentage of coniferous tree cover within urban lawns in Raleigh, North Carolina, was negatively correlated with throughfall amount and thus, by extension, positively correlated with canopy interception loss. Following this logic, broadleaved evergreen or deciduous tree genera that have large numbers of small, narrow, or elongated leaves (e.g., *Ailanthus*, *Callistemon*, *Gleditsia*, *Melaleuca*, *Salix*) could similarly provide greater canopy water storage potential.

From an urban stormwater management perspective, the benefits of coniferous and broadleaved evergreen trees may outweigh those of deciduous trees during the winter months. Although interception loss may still be appreciable from defoliated deciduous trees (e.g., Deguchi et al. 2006; Herbst et al. 2008) due, in part, to the large storage capacities of branches and the tree bole (Herwitz 1985), decreases in interception loss efficiency may be considerable during the leaf-off period. Xiao and McPherson (2002), for example, simulated interception loss from deciduous sweetgum (*Liquidambar styraciflua*, diameter at breast height (dbh) = 30.5–45.7 cm) in Santa Monica, California, at 70.5% for a 44-h duration, 20.3 mm rainfall event during the growing season, while only 5.5% was partitioned into interception loss by this same species for an event of similar duration and magnitude (41 h, 21.7 mm) but under leafless conditions. In contrast, the simulated interception loss from an evergreen camphor (*Cinnamomum camphora*) tree within the same dbh class as the sweetgum and for the same storm events decreased marginally from 24.4% under the growing season scenario to 19.4% in the dormant season. Xiao and McPherson (2002) attribute the lower interception loss during the winter storm from the *C. camphora* tree to lower evaporation rates during the rain event.

Livesley et al. (2014) evaluated the impact of canopy cover density on interception loss efficiency by comparing interception loss from two isolated street trees in Melbourne, Australia: a 30-m-tall Sydney sweet gum (*Eucalyptus saligna*) tree and a 35-m-tall narrow-leaved black peppermint (*Eucalyptus nicholii*) tree. Using regression functions based on 5 months of measurements, interception loss was estimated to account for approximately 44% of the 680.4 mm of annual rainfall for 2010 from the denser canopy of *E. nicholii* (plant area index = 3.88, canopy cover fraction = 86%). Interception loss over the same period from the *E. saligna* tree (plant area index = 3.03, canopy cover fraction = 80%) was considerably less than from *E. nicholii*, albeit still appreciable, at 28%. Similarly, Alves et al. (2018), comparing the four most commonly used tree species for urban afforestation projects in Uruaçu, Brazil, concluded that there was a tendency for interception loss to be greater from those trees with higher canopy cover fractions and larger canopy volumes and leaf area index values.

17.2.1.2 Influence of Rainfall Regime and Storm Meteorology on Interception Loss

Each time a rainfall occurs, the water held in storage at the end of the event evaporates. Therefore, if a fixed amount of rainfall is delivered over multiple events rather than just by one storm, a greater fraction of that rainfall will be partitioned into interception loss. As such, the rainfall regime of a city can influence the potential proportion of rainfall lost through canopy interception (Carlyle-Moses and Gash 2011). Xiao and McPherson (2011a, Table 2) demonstrated the importance of rainfall depth on interception loss for three rainfall depth classes for a lemon (*Citrus limon*) tree in Oakland, California. Mean interception losses for the evergreen *C. limon* were 67.0 ± 25.6%, 44.7 ± 32.6%, and 29.8 ± 17.0%, for <10.0, 10.0–19.9, and ≥ 20.0 mm rainfall classes, respectively.

The relationship between rainfall depth and the fraction of that rainfall partitioned into canopy interception loss is asymptotic, with interception loss as a percentage of rainfall decreasing rapidly with increasing rainfall depth until remaining quasi-constant for relatively large events (Gómez et al. 2001; Staelens et al. 2008; Nytch et al. 2018). The quasi-constant proportion of rainfall partitioned into interception loss for relatively large storms is an important consideration for SCMs in that trees having greater interception losses during these larger magnitude events would be more effective in mitigating against flood episodes. The quasi-constant percentage of relatively large rainfalls partitioned into interception loss may not be inconsequential. Staelens et al. (2008), for example, found that interception loss from a single European beech (*Fagus sylvatica*) canopy in full leaf averaged 18% for rain events ≥10.0 mm, albeit during the leafless periods, the *F. sylvatica* canopy only partitioned an average of 4% of large rainfalls into canopy interception loss.

In addition to rainfall depth, the duration, the intensity, the continuity, and the inclination angle of rainfall can modify the fraction of rainfall partitioned by a tree canopy as interception loss (Xiao et al. 2000a; Carlyle-Moses and Gash 2011; Nytch et al. 2018). Storm events associated with continuous, long-duration rainfall are subjected to greater evaporative losses from the canopy than short-duration events, while events populated by periods in which rain ceases or falls at a very low intensity (i.e., lower than the during-event evaporation rate) allow intermittent drying throughout the storm that increases interception loss through periodic replenishment of the canopy storage capacity (Zeng et al. 2000; Carlyle-Moses and Gash 2011). There is no consensus regarding the relationship between rainfall intensity and interception loss. Calder (1996) and Carlyle-Moses (2004), for example, suggest that under higher rainfall intensities, the storage capacity of a canopy is reduced due to the displacement of water stored on the canopy in response to relatively large raindrops striking the canopy. Murakami (2006), however, suggests that the higher kinetic energy associated with the large raindrop sizes results in greater splash and that this splash results in the formation of a multitude of smaller droplets. These smaller droplets are subjected to greater evaporation due to their smaller size since vapour pressures increase over convex surfaces with small radii of curvature and because a greater fraction of their total volume is exposed to the surrounding air (Murakami 2006; Dunkerley 2009). Thus, the role of rainfall intensity in the interception loss process is likely complex, with net increases or decreases in interception loss with increasing rain intensity being dependent on the net balance between a reduction in the canopy storage capacity and the increased during-event evaporation rate. As such, the influence of rainfall intensity on the magnitude of canopy interception loss is likely tree and storm specific.

Rainfall inclination angle, which is a function of wind speed and the intensity of rainfall, can also have an important influence on canopy interception loss, with the area of tree crowns, especially those of isolated street and park trees, exposed to direct rainfall being reduced or increased with increasing rainfall inclination angles. This *effective crown projection area* and its influence of rainfall interception loss and throughfall are discussed more fully in Sect. 17.2.2.1.

The evaporation of rainfall from tree canopies during storms follows a diffusive process, being a function of the vapour pressure deficit between the canopy surface and the overlying air and the aerodynamic conductance for water vapour integrated over the distance separating the canopy surface and the adjacent air (Monteith and Unsworth 2008; Pereira et al. 2009). As such, for a given canopy, the meteorological variables – apart from the aforementioned rainfall characteristics – influencing the evaporation of intercepted rain during an event are temperature, relative humidity, and wind speed (Toba and Ohta 2005; Carlyle-Moses and Gash 2011; Van Stan et al. 2017). Guevara-Escobar et al. (2007) suggest that the large fraction of rainfall partitioned into canopy interception loss by a lone ficus (*Ficus benjamina*) tree in Queretaro City, Mexico, was due to the relatively high temperatures and low humidity values during their study. The energy available to sustain evaporation from wetted canopies is attributed to advection energy with net radiation having little if any impact on during-event evaporation from wetted canopies (Stewart 1977; Wallace and McJanet 2006; Carlyle-Moses and Gash 2011). Cities themselves may represent a source of sensible heat for interception loss, especially due to the plethora of anthropogenic heat sources and the urban heat island effect (Asadian and Weiler 2009).

17.2.2 Canopy Throughfall

The canopy throughfall process ultimately determines the amount of rainfall that can potentially become runoff beneath or adjacent to an urban tree canopy, which is especially important in impervious urban landscapes, such as streetscapes and car parks (Livesley et al. 2014). As aforementioned, the intensity, frequency, and duration of rainfall are the primary drivers of canopy interception loss, and because throughfall is largely the inverse of canopy interception loss, those meteorological and tree factors that regulate interception loss also tend to influence the quantity of throughfall reaching the ground. Urban tree canopies will be more effective at intercepting rainfall and reducing throughfall and therefore reducing impervious asphalt runoff, in cities with a climate dominated by short, frequent, but low-intensity storms. In urban centres with a subtropical or tropical climate, the effectiveness of urban tree canopies to intercept rainfall and reduce throughfall and impervious runoff will be far less, as individual rainfall events will often be greater in intensity and duration (Kuehler et al. 2017; Nytch et al. 2018). Furthermore, throughfall amounts may significantly differ amongst tree species under low- or moderate-intensity storms but are likely to be more similar under high intensity (>50–70 mm) storms (Nytch et al. 2018).

17.2.2.1 Regulating Factors Influencing the Quantitative Importance of Throughfall

Unlike many natural forests and woodlands, single, isolated trees are very common in the urban landscape. As such, many urban studies of rainfall partitioning by trees have focused on measuring throughfall under isolated trees. Xiao et al. (2000a) captured all of the throughfall beneath a large *P. calleryana* and a large cork oak (*Quercus suber*) tree over the 1996/1997 winter and 1997/1998 winter periods, respectively. In this comprehensive study, throughfall represented 77% of gross rainfall for a single *P. calleryana* and 58% for a single *Q. suber*. Interestingly, Xiao et al. (2000a) concluded that rainfall frequency was more important than rainfall intensity or duration in determining interception loss, whereas throughfall increased with increasing event rainfall amount but decreased with increasing wind speed. Similarly, Zabret et al. (2018) measured throughfall under two urban trees in Ljubljana, Slovenia, using troughs and tipping gauges installed for more than 3 years. This study concluded that rainfall events of greater amount and intensity increased the proportion of throughfall as compared to interception-evaporation, whereas higher wind speeds decreased throughfall as a proportion. These observations substantiate the statement that tree canopies will reduce throughfall to a greater extent in climates where rainfall events may be frequent but low in intensity and duration.

Wind influences canopy interception and throughfall through complex interactions with rainfall (frequency, intensity, duration) and atmospheric properties (temperature and humidity). High wind speeds increase the potential for evaporative losses of intercepted rainfall, especially under warm and low humidity conditions. However, high wind speeds also change the incident angle at which rainfall intercepts a tree canopy potentially resulting in the generation of 'rain shadows' beyond the drip lines of single, isolated trees (Xiao et al. 2000b; Zabret et al. 2018). The importance of this rain-shadow effect for isolated trees was well illustrated in an urban study of throughfall under and around a large *F. benjamina* tree in Mexico (Guevara-Escobar et al. 2007). Throughfall was low at 38%, indicating the dominant role of interception loss for this *F. benjamina* tree. Kriging the spatial distribution of throughfall indicated that a 'rain-shadow' downwind of the tree represented almost 20% of the canopy interception loss of this lone tree, and thus, throughfall reduction beyond the drip line of this tree was substantive. The results of Guevara-Escobar et al. (2007) exemplify the need to consider the impact of wind direction and rain shadows beyond the tree canopy footprint when assessing the throughfall input associated with isolated trees in an urban landscape.

Simulating rainfall indoors provides a more controlled environment in which to investigate the relationship between tree- and leaf-level attributes and canopy throughfall (Aston 1979; Keim et al. 2006; Li et al. 2016; Xiao and McPherson 2016; Baptista et al. 2018a). Li et al. (2016) measured four species of contrasting needle- and broadleaved functional types along a simulated rainfall intensity gradient from 10 to 150 mm hr.$^{-1}$. Under these controlled conditions, throughfall in both needle- and broadleaved trees was related more to leaf and branch traits than rainfall intensity, which suggests that tree- and leaf-level characteristics may be more influential than rainfall characteristics in determining interception and throughfall. In a similar study using three tree species commonly used in urban streetscapes, Baptista et al. (2018a) manipulated tree canopy density to demonstrate the strong relationships between plant area index (i.e., LAI) and throughfall, again highlighting significant differences according to broadleaved canopy- and leaf-scale traits. This study suggests that mapping urban tree canopy cover alone will not accurately predict rainfall interception, reduced throughfall, and runoff. The health and quality of each tree canopy will significantly impact throughfall amounts.

17.2.2.2 Spatial Redistribution of Throughfall

At the scale of a single tree canopy, the spatial variability of throughfall can be very high. Tree-level and leaf-level parameters, such as canopy density or leaf and branch angles, can not only impact throughfall proportions but also the spatial redistribution of throughfall (King and Harrison 1998; Staelens et al. 2006; Nanko et al. 2011; Fathizadeh et al. 2014). Many tree canopies have naturally patchy or clumped foliage (e.g., many eucalypts), which can influence the spatial distribution of throughfall (Livesley et al. 2014). Variation in canopy throughfall can also develop because of natural gradients in foliage or branch density within a canopy but also because of urban tree management (e.g., pruning under or around power lines). Throughfall redistribution can be greatly influenced by the spatial distribution of major structural branches and as such may be more important in large limbed, mature trees (Fathizadeh et al. 2014). Staelens et al. (2006) studied the spatial redistribution of throughfall and noted that throughfall volume under a 30-m-high urban *F. sylvatica* tree was better correlated to branch area metrics alone, rather than the metrics of combined foliage and branch areas.

Under low-wind conditions, it is expected that throughfall volumes will be greater under less dense areas of the canopy. Baptista et al. (2018b) investigated the relationship between spatial variation in canopy density of individual trees and the spatial distribution of throughfall beneath three common urban tree species: London plane (*Platanus x acerifolia*), English elm (*Ulmus procera*), and spotted gum (*Corymbia maculata*). There was only a weak relationship between canopy density, as determined by terrestrial laser scanning point clouds, and throughfall amounts. Amongst the three species, coefficients of variation (CV) in throughfall redistribution were greatest (CV = 101%) for the large leafed *P. x acerifolia* and least for *U. procera* (CV = 50%) and *C. maculata* (CV = 51%) at 2.5 mm h^{-1} (Baptista et al. 2018b). This coefficient of variation in the spatial distribution of throughfall will generally decrease with increasing rainfall amounts. Beneath a solitary urban Persian oak (*Quercus brantii*) tree, the CV of throughfall decreased from >50% for rainfall events <5 mm to <25% when rainfall was >30 mm (Fathizadeh et al. 2014).

In the urban landscape, spatial redistribution of throughfall is not restricted to solitary tree canopies, because the matrix of overlapping and discontinuous canopies in fragmented private residential green spaces and public parks produces a large range of canopy conditions, from simple to structurally complex (Inkiläinen et al. 2013; Threlfall et al. 2016). The spatial redistribution of throughfall is important as it will impact how soil, or vegetation, surfaces below wet up, are eroded, or experience runoff (Ossola et al. 2015).

17.2.2.3 Temporal Delay of Throughfall

The time delay between the onset of rain and the initiation of throughfall can range from a few minutes to several hours – even days (Asadian and Weiler 2009; Nytch et al. 2018) – and is dependent on the storm conditions at the start of an event and on tree characteristics. Nytch et al. (2018) found that rainfall intensity during the early stages of an event had an important impact on throughfall delay times, with less intense rain resulting in greater time lags between rainfall and throughfall initiation. Although Nytch et al. (2018) did not find any difference in throughfall delay between two urban tree types – white siris (*Albizia procera*) and Antilles calophyllum (*Calophyllum antillanum*) – in San Juan, Puerto Rico, Asadian and Weiler (2009) found that western red cedar (*Thuja plicata*) had greater throughfall temporal delays than did Douglas fir (*Pseudotsuga menziesii*) trees in North Vancouver, Canada. Asadian and Weiler (2009) also observed that dominant and codominant trees that were in good health tended to have longer throughfall time delays. Although the temporal delay of throughfall has received little study, the postponement of canopy drainage reaching the ground may influence peak runoff characteristics, including delaying peak runoff itself (Xiao and McPherson 2002; Asadian and Weiler 2009; Kuehler et al. 2017).

17.2.2.4 Kinetic Energy and the Erosion Potential of Throughfall

Tree canopy and leaf-level traits can greatly modify throughfall drop size and velocity, which in turn changes the kinetic energy and erosive capacity of that throughfall (Goebes et al. 2015). Sand-filled splash cups provide a simple index of throughfall erosion potential and have been used to compare the erosive nature of throughfall under 11 different tree species monocultures in China (Geißler et al. 2012; Goebes et al. 2015). Leaf habit (deciduous/evergreen), leaf area, and tree canopy height were shown to be the most important factors influencing differences in throughfall kinetic energy amongst these 11 species. From the perspective of selecting urban tree species, the observation that evergreen, simple, and dentate leaved trees reduced throughfall kinetic energy more than deciduous, pinnate, or entire leaved species should be noted. The erosive potential of throughfall may be an important consideration when selecting urban tree species to be planted over bare soil or in green spaces on sloping land. The importance of erosion potential may not be as great when selecting trees to be planted over impervious surfaces (street trees) unless throughfall negatively impacts the movement of pedestrians on walkways under rainfall conditions.

17.2.3 Stemflow

The portion of rainfall that is not partitioned into canopy interception loss reaches the ground as either throughfall or stemflow, and, depending on site and rainfall characteristics, these components of understorey rainfall may contribute to urban stormwater runoff. Throughfall comprises rainfall that passes directly through gaps in the canopy (free throughfall) and release throughfall, which falls to the ground by either dripping or splashing from canopy surfaces. Stemflow represents the drainage from the canopy that reaches the ground by flowing down the tree bole. In urban scenarios, the fraction of rainfall partitioned into throughfall or stemflow may not be trivial, especially for trees whose canopies extend over impervious surfaces. Stemflow is a concentrated input of water, being delivered to the proximal area around the tree base, whereas throughfall is a diffuse input being delivered over a much larger area under the tree canopy and, in some cases, extending beyond the canopy area (see Sect. 17.2.2.2). Thus, if the surface area directly surrounding the tree base is sufficiently large and permeable to allow stemflow to infiltrate, selecting trees that favour diverting a greater amount of their drainage in the form of stemflow could be a useful stormwater management approach (Schooling and Carlyle-Moses 2015; Kuehler et al. 2017).

17.2.3.1 Quantitative Importance of Stemflow in Urban Areas

Stemflow from natural forests typically represents <5% of rainfall at the plot scale (Levia et al. 2011), albeit larger values have been reported (e.g., Wei et al. 2005; Johnson and Lehmann 2006). Citing the results from rural forests, some rainfall partitioning studies in urban environments have either ignored stemflow (Asadian and Weiler 2009; Alves et al. 2018; Nytch et al. 2018) or have held stemflow as a small fraction of rainfall (Xiao et al. 1998; Inkiläinen et al. 2013). However, because interception loss is estimated as the difference between rainfall and the sum of throughfall and stemflow (Helvey and Patric 1965; Carlyle-Moses and Gash 2011), large errors may accompany interception loss estimates, and the absolute stormwater retention services offered by urban trees may not be fully recognised if the assumptions of minimal stemflow are violated (Schooling and Carlyle-Moses 2015).

Studies of stemflow from city trees are scarce. However, the results presented to date suggest that although the proportion of rainfall partitioned into stemflow by the crowns of many isolated trees is similar to that by the canopies of rural forests, i.e., < 5% (e.g., Guevara-Escobar et al. 2007; Xiao and McPherson 2011a; Livesley et al. 2014), stemflow from certain urban trees may constitute a substantially greater fraction of rainfall. Xiao et al. (2000b), for example, measured understorey precipitation from two open-grown trees in Oakland, California: a *P. calleryana* (height = 8.5 m, dbh = 22 cm) and a *Q. suber* (height = 5.6 m, dbh = 12.5 cm). Stemflow, on a canopy area basis, from the pear tree represented 8% of the 441 mm of rainfall over the 1996–1997 winter season, while stemflow accounted for 15% of the >700 mm of the 1997–1998 winter rainfall that fell on the oak canopy.

Schooling and Carlyle-Moses (2015) measured stemflow over 18 months from 37 isolated deciduous broadleaved trees representing 21 cultivated species in an urban park in Kamloops, British Columbia, Canada. Twenty of the studied trees had a dominant main stem intersected by feeder branches, while the remaining 17 trees had multiple stem leaders converging at the base of the tree canopy, with each leader intersected by feeder branches. Stemflow over the study period averaged 0.9 ± 0.9,

2.8 ± 3.3, and 3.1 ± 2.8% of rainfalls 2 to <5, 5 to <10, and ≥ 10 mm, respectively, for the single-leader trees, while corresponding percentages for the multi-leader trees averaged 0.8 ± 1.2, 2.0 ± 2.5, and 2.7 ± 2.6%. However, as evidenced by the relatively large standard deviations, considerable variability in stemflow partitioning was found amongst the study trees. Stemflow was all but nonexistent for a large Russian olive (*Elaeagnus angustifolia*, dbh = 66.8 cm) tree with stemflow being generated only for rainfalls ≥10 mm and then only accounting for 0.1 ± 0.1% of rainfall depth, while a columnar English oak (*Quercus robur* 'Fastigiata', dbh = 23.5 cm) had mean stemflow percentages of 2.5 ± 5.9, 11.8 ± 9.1, and 12.3 ± 8.4 for 2 to <5, 5 to <10, and ≥ 10 mm rainfall depth classes, respectively. The largest single-event stemflow percentage from the columnar oak and the largest percentage observed from any tree during the study was 22.8% for a 25.6 mm rainfall.

17.2.3.2 Stemflow Funnelling Ratio and the Concentrated Nature of Stemflow

The ability of trees and other plants to concentrate canopy drainage in the form of stemflow has been quantified at the individual tree scale using the stemflow funnelling ratio, F_{P,B_t} (dimensionless) (Herwitz 1986):

$$F_{P,B_t} = \frac{S_{Y_t}}{P \cdot B_t} \tag{17.1}$$

where S_{Y_t} is the tree-scale stemflow yield (L) during a precipitation event of a given depth, P (mm) generated by a tree (t) having a given basal area of the stem, B_t (m^2).

When $F_{P,B_t} > 1$, the equivalent depth of stemflow delivered to the base of the tree exceeds the precipitation depth that would have been captured by a rain gauge with an opening area equivalent to the tree basal area. Under the different tree group and rainfall depth class combinations examined by Schooling and Carlyle-Moses (2015), F_{P,B_t} ranged from a mean of 8.5 ± 7.8 for 2 to <5 mm rainfall depth class for single-leader trees to 28.0 ± 20.7 for multi-leader trees under the relatively large >10 mm rainfalls. Certain trees had very large F_{P,B_t} values, including a mean of 50.7 ± 33.6 for a katsura tree (*Cercidiphyllum japonicum*, dbh = 10.2 cm) for 5 to <10 mm rainfalls and means of 58.6 ± 12.0 and 81.3 ± 64.9 for purple rain black locust (*Robinia pseudoacacia* 'Purple Rain', dbh = 14.6 cm) and River's purple European beech (*F. sylvatica* 'Riversii', dbh = 38.8 cm) trees, respectively. For a 25.6 mm rainfall event, the European beech cultivar had a F_{P,B_t} of 196.9; the highest F_{P,B_t} value of any tree for any storm during the study. Expressed as an equivalent depth of rain over the basal area of the beech tree, the F_{P,B_t} value equates to 5040 mm, or more than 18 times the annual precipitation depth of the study area.

Although the basal area does not necessarily equate with the area of stemflow water infiltration, if soils have associated surface hydraulic conductivities that are

Fig. 17.2 Stemflow water ponding at the base of a silver maple (*Acer saccharinum*) tree after a rainfall event on the campus of Thompson Rivers University in Kamloops, British Columbia, Canada. For this rain event, the available surface area of the moderately compacted soil was sufficiently large to contain all the stemflow; however, for events in which the stemflow rate may be higher, the available infiltration area may be insufficient to restrict stemflow from contributing to stormwater flow. Photo credit: DE Carlyle-Moses

relatively high, the area of infiltration is limited to a relatively small area around the base of the tree (Schwärzel et al. 2012; Carlyle-Moses et al. 2018). Thus, urban trees with high F_{P,B_t} and surface soil hydraulic conductivities favourable to infiltration could be important self-irrigators, delivering both relatively high volumes of water and nutrients (Abas et al. 1992; Schooling et al. 2017; see also Chap. 18 of this volume). If, however, soils are compacted or the area around the base is insufficiently large to accommodate the stemflow input, this understorey precipitation component has the potential to add to stormwater flow (Fig. 17.2).

17.2.3.3 Tree Characteristics and Meteorological Factors Influencing Stemflow

Schooling and Carlyle-Moses (2015) found that isolated deciduous trees with multiple leaders or many major branches converging at the base of the tree canopy favoured stemflow volume production, particularly for trees that exhibited high branch angles and smooth bark (e.g., *F. sylvatica*). Additionally, trees with smaller dbh values, which correlated with smaller canopy widths, tree heights, projected canopy areas, canopy volumes, wood cover fractions, and bark relief index values, tended to produce greater stemflow yields for relatively small rain events of <5 mm. However, for rain events ≥5 mm, and especially for events ≥10 mm that more completely saturated the tree canopy, it was found that more voluminous stemflow production was associated with broader, taller, and denser canopies due to their ability to capture more rainfall that could potentially be partitioned into stemflow.

Although trees with greater bark relief index values are normally associated with smaller stemflow volumes (Xiao et al. 2000a; Levia and Herwitz 2005; Van Stan and Levia 2010), Schooling and Carlyle-Moses (2015) found that if the bark was linearly furrowed – e.g., pin oak (*Quercus palustris*) and Norway maple (*Acer platanoides*) – stemflow yields were relatively high for relatively large rain depths ($\geq$ 10 mm), a possible consequence of enhanced flow conduction over a greater saturated surface area of the tree bole. Thus, as suggested by Schooling and Carlyle-Moses (2015), in cities with climates that have rainfall regimes dominated by relatively small rainfall events, stemflow will favour smaller trees with greater branch angles and smoother bark, while in climates where more modest rainfalls are more frequent, stemflow generation will favour trees with broader, more voluminous canopies with upright and numerous branches.

In addition to the importance of rainfall depth in generating stemflow yields, especially under saturated canopy conditions (Xiao et al. 2000a), Carlyle-Moses and Schooling (2015) found that, for the aforementioned 37 isolated trees in an urban park in British Columbia, Canada, stemflow yield was positively correlated with rainfall inclination and wind speed and negatively correlated with vapour pressure deficit and the duration of within-storm breaks in rainfall, in agreement with the findings of others (Xiao et al. 2000b; André et al. 2008; Van Stan et al. 2011). Rainfall intensity was not found to be consistently correlated to stemflow in the Carlyle-Moses and Schooling (2015) study. Although an inverse relationship between stemflow and rainfall intensity has been shown by some (e.g., Carlyle-Moses and Price 2006), others have found the opposite to be true (e.g., Van Stan et al. 2014). As Van Stan et al. (2014) state, tree size, bark roughness, and meteorological variables interact with rain intensity in complex ways, and thus the influence of rain intensity on stemflow production, as it is with canopy interception loss, can be expected to be tree and storm specific.

17.2.4 Modelling Tree Interception, Stemflow, and Throughfall

Modelling the combined processes of interception loss, stemflow, and throughfall is complex. Several models have been developed based on observed meteorological data and allometric relationships (Rutter et al. 1971; Gash et al. 1995). However, modelling rainfall partitioning for urban trees should take into account the different structure of isolated, open-grown trees and the environmental microclimate conditions (Xiao et al. 2000b; Asadian and Weiler 2009). A detailed urban tree hydrology model was developed by Xiao et al. (2000b) that includes many of the tree attributes that influence interception, stemflow, and throughfall, such as leaf angle, leaf area index, stem/branch area index, and leaf surface water storage capacity. The predictions from this detailed mechanistic model were most sensitive to input data on rainfall depth and leaf surface water storage capacity, followed by leaf area index and stem/branch area index. An interception model adapted for sparse canopy rainfall (Gash et al. 1995) was validated using empirical data from four urban tree species: white oak (*Quercus alba*), *A. platanoides*, green ash (*Fraxinus pennsylvanica*), and *Prunus* sp. (Huang et al. 2017). This model predicted throughfall and interception well, and sensitivity analyses indicated that evaporation and rainfall rates exhibited the greatest influence upon model outputs. As rainfall rates and evaporation rates are extremely variable within the complex structure of an urban landscape, direct measurement or careful consideration of these two factors is paramount (Huang et al. 2017).

17.3 Tree Water Uptake in Stormwater Control Measures

17.3.1 What Are Stormwater Control Measures?

Natural and agricultural landscapes are essentially pervious, allowing the majority of rainfall to infiltrate. These landscapes are dominated by vegetation such as trees that may intercept some rainfall but, perhaps more importantly, transpire large volumes of water back into the atmosphere. Conversely, urbanised landscapes are dominated by impervious surfaces which generate large volumes of stormwater runoff that is efficiently drained to prevent flooding (Chocat et al. 2001). Depending on the drainage infrastructure, runoff enters urban streams directly or during combined sewer overflows. Pollutant loads, as well as the volume and velocity of runoff, are highly damaging to urban stream ecosystems (Walsh et al. 2005), prompting cities around the world to adopt water-sensitive urban design (WSUD) philosophies (Fletcher et al. 2015) and reduce the impact of runoff with stormwater control measures (SCMs), examples of which are shown in Fig. 17.3. These SCMs are specifically designed to slow the velocity (detain), reduce the volume (retain), and reduce the pollutant load (treat) of stormwater runoff.

Stormwater control measures can be grouped according to their hydrological function (Fletcher et al. 2013): (i) infiltration-based systems (e.g., swales, infiltration trenches, basins, unlined biofiltration systems, and pervious pavements) can both infiltrate and retain runoff, and (ii) retention-based systems (e.g., green roofs, tanks and lined biofiltration systems, ponds and wetlands) reduce the overall runoff volume primarily through retention. Increasingly, SCMs are being designed to not simply reduce pollutant concentrations in runoff but also infiltrate and retain runoff to restore pre-development hydrological processes, particularly infiltration and evapotranspiration (Burns et al. 2012). As evapotranspiration can represent the greatest loss of water from natural landscapes (60–80%; Zhang et al. 2001), it has been suggested that integrating high water-using vegetation, such as trees, will improve the hydrological performance of SCMs and the urban landscape overall.

Fig. 17.3 Trees in stormwater control measures: a linear planting of trees in an SCM trench along a street in downtown Philadelphia, USA (**a**); a small SCM trench of gravel during construction (**b**) and after construction (**c**) in Melbourne, Australia; and a simple kerb-cut SCM where a tree is planted into the small biofilter sand rain garden in Melbourne, Australia (**d**). (Photo credits: DE Carlyle-Moses (**a**), C Szota (**b** and **c**), and J Thom (**d**))

17.3.2 Why Use Trees in Stormwater Control Measures?

Integrating trees into SCMs has the potential to significantly reduce the volumes of runoff generated in urban catchments. Aside from canopy interception and evaporation, urban trees have the potential to use significant volumes of water through transpiration (McCarthy et al. 2011; Pataki et al. 2011; Litvak et al. 2012), driven by high vapour pressure deficits (Coutts et al. 2013; Zipper et al. 2017). Trees can support a large transpiring leaf area from a small stem 'footprint' at ground level (Berland et al. 2017), requiring less space in highly urbanised areas (Grey et al. 2018a). Therefore, trees have significant potential to deplete captured runoff, creating storage before the next runoff event and therefore increasing the overall performance of SCMs (Mangangka et al. 2015; Wadzuk et al. 2015; Nocco et al. 2016; Hess et al. 2017; Yuan et al. 2017). As such, it has been suggested that the runoff retention performance of SCMs can be improved most by selecting trees with high transpiration rates (Szota et al. 2018). Trees are also likely to enhance the performance of SCMs if they possess traits that are linked with nutrient pollutant removal (Denman et al. 2016), including extensive root systems, high biomass, and rapid growth rates (Payne et al. 2018). Integrating trees with SCMs therefore has the potential to achieve retention of stormwater runoff, without jeopardising nutrient removal performance.

Integrating trees into SCMs has other potential benefits besides those associated with runoff. Trees are under pressure in urban streetscapes due to limited soil rooting volumes (Grabosky and Bassuk 1995) and high evaporative demands (Coutts et al. 2013; Zipper et al. 2017), both of which can lead to increased drought stress, reduced growth, and increased susceptibility to pathogens (Sjöman and Nielsen 2010; Mullaney et al. 2015; Savi et al. 2015). Supporting street trees, at a time when drought stress associated with climate change is likely, represents a serious challenge for city planners and urban forest managers (May et al. 2013; Nitschke et al. 2017; Bialecki et al. 2018). As cities become hotter and drier, trees are increasingly valued for their shading and cooling benefits, particularly increasing human thermal comfort (Livesley et al. 2016; Thom et al. 2016; Sanusi et al. 2017). Municipalities around the world have committed to not only retaining but significantly increasing urban tree canopy cover as a result. Providing appropriate root substrate volumes and supplying additional runoff water should not only decrease tree drought stress (Meineke and Frank 2018) but also increase tree growth (Grey et al. 2018a) to support an expanding urban forest.

17.3.3 How Can Trees Be Integrated with Stormwater Control Measures?

It is important to consider the design intent of integrating trees with SCMs. For example, 'passive irrigation', i.e., directing runoff to trees to provide supplementary irrigation to increase tree performance, is a different concept to integrating trees into the design of SCMs. This is because SCMs (with or without vegetation) are sized and designed to achieve specific targets, for example, to reduce total suspended solids and total nitrogen and total phosphorus concentrations (e.g., by 80, 45, and 45% (Melbourne Water 2005)) or to retain certain event sizes (e.g., 25 mm (NCDEQ 2017)). To achieve these targets, SCMs need to be sized to at least 2.5% of their impervious catchment area (Melbourne Water 2005; NCDEQ 2017). Such sizing is significantly more than that required to provide supplementary irrigation to support trees.

Trees have been successfully incorporated into SCMs such as biofilters without major modifications to their design (e.g., Li et al. 2009; Brown and Hunt 2011; Winston et al. 2016). However, increasing substrate depth has been recommended to ensure adequate soil volumes and soil moisture for tree growth and survival. For example, increasing the depth of the filter media from typical depths of 400–600 mm to >800 mm (e.g., Payne et al. 2015; NCDEQ 2017) can support deep-rooted tree species. This increased substrate depth returns to the original design guidelines of Clar and Green (1993), which recommended 1.2 m of substrate for adequate root development (Brown and Hunt 2011). Tree growth can increase by increasing the internal water storage of a SCM by producing a saturated subsoil volume via raised outflow or outlet control structures (Scharenbroch et al. 2016; Grey et al. 2018a).

Studies which report high runoff retention by SCMs planted with trees are often sized to represent a large proportion of their catchment area (e.g., Xiao and McPherson 2011a, b) and/or have a large storage volume (e.g., Li et al. 2009; DeBusk and Wynn 2011; Xiao and McPherson 2011b) or outlet control structures (Scharenbroch et al. 2016). Where streetscapes are being redesigned, trees could make use of runoff stored in reservoirs beneath the pavement surface: in structural soils (Bartens et al. 2009) or subsurface structures (Nichols and Lucke 2015; Page et al. 2015a, b; Ordóñez-Barona et al. 2018).

Regardless of SCM design, runoff needs to be captured as simply and efficiently as possible from the impervious catchment that supplies it so as to maximise system performance and minimise maintenance requirements (Szota et al. 2019). Bioswales that collect runoff from parking lots can avoid inlet restrictions by completely removing the kerb, such that runoff is delivered along the length of the system as a relatively uniform overland flow, rather than through an inlet (Xiao and McPherson 2011a, b; Scharenbroch et al. 2016). Kerb cuts have also been used to effectively maintain inflows for bioretention cells (Brown and Hunt 2011) and street tree pits (Grey et al. 2018a). Simple kerb cut inlets (Fig. 17.3d) can be cleared of sediment and debris relatively easily, easing the maintenance burden and cost (Houle et al. 2013). Although many of these studies have successfully incorporated trees into alternative SCM designs, very few have attempted to quantify how much water the trees in these SCMs transpire or what proportion of the water balance tree transpiration represents.

17.3.4 How Much Water Can Trees in the Urban Landscape Use?

Studies on urban tree transpiration point to the important role trees could play in the urban hydrological cycle, with reported values ranging from 3.2 kg day^{-1} to 260 kg day^{-1} (Pataki et al. 2011). This variability has been attributed to tree size (Riikonen et al. 2016; Litvak et al. 2017), wood anatomy (Bush et al. 2008; Litvak et al. 2012), and evaporative demand (Peters et al. 2010; Gotsch et al. 2018). A study in Salt Lake City demonstrated this variation for six tree species in well-watered urban landscapes (Bush et al. 2008). Transpiration estimated from sap flux varied from 14 kg day^{-1} for Gambel oak (*Quercus gambelii*, 13.5 cm diameter) to 200 kg day^{-1} for *P. x acerifolia* (25 cm diameter). In contrast, Ballinas and Barradas (2016) reported much lower transpiration, from similar-sized but water-limited trees of 3.7 kg day^{-1} for glossy privet (*Ligustrum lucidum*, 14.8 cm diameter) to 4.4 kg day^{-1} in paved streets during the dry season in Mexico City.

In addition to water availability, wood anatomy and transpiration sensitivity to evaporative demand influence the hydraulic transport of water through tree xylem. For example, Litvak et al. (2012) showed that diffuse-porous trees had significantly higher daily transpiration rates (260 kg day^{-1}, London sycamore (*Platanus hybrida*)) than semi ring-porous (175 kg day^{-1}, Chinese banyan (*Ficus microcarpa*)), ring-porous (150 kg day^{-1}, golden rain tree (*Koelreuteria paniculata*)), or coniferous species (100 L day^{-1}, Canary Island pine (*Pinus canariensis*)). These trees were grown in irrigated landscapes in Los Angeles and exposed to high evaporative demand, with average daily vapour pressure deficits >4 kPa (Litvak et al. 2012). In contrast, Peters et al. (2010) showed greater transpiration from ring-porous genera (~107 kg day^{-1}, *Ulmus*) than coniferous (~83 kg day^{-1}, *Picea*) or diffuse-porous (73 kg day^{-1}, *Juglans*) genera for residential urban trees experiencing lower evaporative demand (vapour pressure deficits <3 kPa). This is likely due to tighter stomatal control by ring-porous trees experiencing high evaporative demand (Bush et al. 2008; Litvak et al. 2012). Therefore, the potential volume of water that trees can utilise when they are grown in SCMs will depend not only on the local climatic conditions such as evaporative demand and water availability but also the characteristics of the tree itself, such as wood xylem anatomy and tree canopy size.

17.3.5 How Much Water Can Trees in Stormwater Control Measures Use?

A number of studies have directly measured transpiration in SCMs for grasses or shrubs (Denich and Bradford 2010; Wadzuk et al. 2015; Hess et al. 2017); however, only a few studies have quantified tree transpiration in SCMs. Of the three studies that have quantified transpiration by trees in SCMs, only two have measured transpiration rates (Bartens et al. 2009; Scharenbroch et al. 2016), while the other modelled transpiration from crop coefficients (Grey et al. 2018b). Both Bartens et al. (2009) and Grey et al. (2018b) reported reduced transpiration when water rapidly exfiltrated from SCMs into surrounding soil. Bartens et al. (2009) grew young trees (1–2 cm stem diameter) in structural soil in a nursery experiment which transpired 0.6–1.6 kg day^{-1} for swamp white oak (*Quercus bicolor*) to 1.2–2.1 kg day^{-1} for

F. pennsylvanica, depending on whether exfiltration rates were fast (20 mm h^{-1}) or slow (10 mm h^{-1}). Grey et al. (2018b) estimated transpiration from small deciduous trees (2–4 cm stem diameter) in individual tree pits that received runoff from road catchments via kerb cuts along a residential street, using a water balance model and crop coefficient of 1.3. Grey et al. (2018b) estimated similarly low transpiration from a rapid exfiltration design (~0.5 kg day^{-1} $tree^{-1}$; exfiltration 34 mm h^{-1}) that increased (~2.4 kg day^{-1} $tree^{-1}$) when exfiltration was slower, which increased water availability to the tree. While these values are relatively small, both studies assessed young trees planted in small individual tree pits which likely restricted access to infiltrated runoff.

Relatively high transpiration rates were recently reported for 63 trees (diameter at 1.3 m: 5.2–31.8 cm) planted in SCM bioswales capturing runoff from a car park (Scharenbroch et al. 2016). In this study, the bioswale outlet control structure meant a large proportion of captured runoff was accessible to trees and additional irrigation was also applied. Leaf-level measurements of stomatal conductance were scaled up from estimated canopy leaf area to calculate transpiration from the entire bioswale. Over a 4-month period over summer, total transpiration was estimated as 9194 Mg, or approximately 1196 kg day^{-1} $tree^{-1}$. These values are much higher than those reported for mature trees, even in irrigated landscapes with high evaporative demand (Pataki et al. 2011; Litvak et al. 2012). Although the high transpiration values may be due to design of the SCM, it is also possible these values were overestimated, given the challenges of scaling up leaf-level point measurements to continuous transpiration of a whole tree canopy. The large variation in reported transpiration values of trees in SCMs, and the methods used to quantify these values, highlights the need for further research in this area to better understand the influence of trees in SCMs on the urban hydrological cycle.

17.3.6 What Role Does Tree Transpiration Play in the Water Balance of SCMs?

Plants are generally considered to play a minor role in the water balance of SCMs, e.g., evapotranspiration typically represents <5% of runoff received for biofilters (Brown and Hunt 2011). The low evapotranspiration fraction is primarily due to the large volumes of runoff received by SCM systems sized to be 4–5% of their connected impervious catchment area (Brown and Hunt 2011). As we would expect, evapotranspiration can account for a significant proportion of runoff when systems are larger, e.g., 20% of impervious catchment area (Wadzuk et al. 2015; Hess et al. 2017). However, Li et al. (2009) estimated evapotranspiration by trees and shrubs to account for 19% of runoff for a system sized at only 4.5% of its impervious catchment. Trees therefore have the potential to increase evapotranspiration in the water balance of SCMs beyond that provided by sedges and rushes.

Per unit canopy area, trees may have very similar transpiration rates to commonly used biofiltration species, primarily sedges and rushes (Szota et al. 2018). However, tree canopy areas will likely extend beyond the area of a SCM as the tree is maturing, resulting in higher transpiration per unit treatment area. Therefore, if volumetric transpiration is expressed relative to the surface area of a SCM, it is likely that trees will use more water than an equivalent surface area planted with sedges or rushes (Szota et al. 2018). Several authors have now suggested that evapotranspiration can represent a substantial component of the water balance in SCMs (Wadzuk et al. 2015; Eger et al. 2017; Hess et al. 2017).

Unfortunately, few studies directly quantify evapotranspiration in SCMs (Denich and Bradford 2010; Wadzuk et al. 2015; Hess et al. 2017), as it is difficult to measure in situ (Hamel et al. 2014). Fewer studies have attempted to quantify transpiration by trees as a proportion of the water balance for SCMs. Scharenbroch et al. (2016) estimated that transpiration was 46–72% of runoff received, which is likely due to large storage volume and regulation of outflow via the outlet control structure in their design. In comparison, the tree pits evaluated by Grey et al. (2018b) were small (0.15–0.72% of their impervious catchment area), with little storage capacity. In this study, evapotranspiration was estimated as only 4% of runoff, likely due to the small storage volume and small size of the trees (Grey et al. 2018b). These studies demonstrate the significant potential for using trees to increase evapotranspiration in the water balance of SCMs.

17.4 Combined Impact on Urban Catchment Hydrology

Research into the role of vegetation in urban runoff mitigation plans and strategies has generally been limited to green roofs, biofilters, and engineered wetlands and has often focused on improved infiltration and water quality (Berland et al. 2017; Meerow and Newell 2017) rather than interception-evaporation and transpiration processes. Some cities do not consider green infrastructure, let alone trees, in their future spatial planning of runoff mitigation strategies (Dolowitz et al. 2018), while other cities are embracing the potential of green infrastructure and tree canopy cover to contribute to runoff reduction planning (Gotsch et al. 2018). Trees have a considerable impact upon urban water cycle processes and could play a far greater role in urban runoff mitigation strategies. However, to enable this role, there needs to be greater scientific evidence and predictive understanding of how, and to what degree, tree interception-evaporation and tree transpiration modify the urban water cycle. Furthermore, this understanding and model predictive capacity needs to expand from the individual tree canopy scale to a catchment and whole-of-city scale. From an urban planning perspective, we also need to better understand the potential of combining stormwater control measures (SCMs) with urban trees, so as to leverage the benefits of urban runoff retention, canopy interception, and urban tree transpiration (Berland et al. 2017; Kuehler et al. 2017; Grey et al. 2018a).

Research and modelling of the combined importance of tree interception-evaporation and tree transpiration has rarely been measured experimentally in the urban landscape. Similarly, it has rarely been simulated in detail by urban ecosystem models. Quantifying or accurately simulating urban evapotranspiration is essential to close the urban water balance and understanding interception-evaporation and transpiration implications for the design of stormwater runoff mitigation strategies and the overall management of urban catchments (Fletcher et al. 2013). There is a real need to increase research efforts into refining urban models to concurrently simulate the role of tree interception-evaporation and tree transpiration, and experimental data is required to validate and test those models. There are several types of hydrologic models and surface-atmosphere transfer (energy balance) models that can be applied to urban areas, and all of these models vary in their mechanistic complexity, process simulation capacity, development objective (Moors et al. 2014), and completeness in addressing the urban water cycle.

17.4.1 Urban Hydrologic Models

A simple hydrologic model can apply 'curve numbers' to different surface types within an urbanised catchment. As such, process understanding is limited, and the model can be context specific, but the curve number approach has been used to estimate a potential 7% reduction in urban runoff from canopy interception-evaporation in a city with 20% tree canopy cover (Sanders 1986). Advanced geospatial techniques mean that 'curve numbers' continue to be applied by researchers to understand and compare spatial contributions to urban runoff rates and amounts (Maragno et al. 2018). More mechanistic hydrologic models that vary in complexity and functional objective are WEP (water and energy transfer processes), SURF (semi-urbanised runoff flow), MIKE SHE, and UFORE-Hydro.

The UFORE-Hydro model (now i-Tree Hydro) was developed specifically to simulate the effects of urban trees on catchment hydrology and their potential role in runoff management (Wang et al. 2008; Liu et al. 2014). The UFORE-Hydro model requires input data for climate, elevation, land cover, and soil/vegetation parameters in order to simulate interception, evaporation, infiltration, and runoff. UFORE-Hydro has successfully shown that within an urbanised catchment, trees can significantly reduce runoff for low-intensity and short-duration events. The MIKE SHE model has been used to estimate runoff reduction at the catchment scale through canopy interception by various green infrastructure systems, including street trees (Zölch et al. 2017). As trees provide minimal water storage capacity in their canopies, MIKE SHE predicted greatest runoff reduction by green infrastructure infiltration systems (e.g., green roof, biofilter, rain garden), rather than trees. These hydrologic models can accurately estimate the role of tree canopy interception-evaporation but rarely provide the mechanistic functionality to simulate tree transpiration between rainfall events, which can be a major flux of the urban water cycle.

17.4.2 Urban Surface-Atmosphere Transfer Models

Urban surface-atmosphere transfer models differ in complexity but generally attempt to consider both the urban energy and the urban water balance, making use of high-temporal data input from climate or weather models. These models include the Town Energy Balance and Interaction Soil-Biosphere-Atmosphere (TEB-ISBA) and SUEWS. One of the earliest evaporation-interception models for urban areas was developed by Grimmond and Oke (1991) that has become known as the 'single-source urban evapotranspiration-interception scheme' (SUES). The SUES model uses both a water mass balance and an energy balance approach, enables a 5-min timestep, and can accommodate deciduous and evergreen vegetation. Mitchell et al. (2008) used SUES, and another model (Aquacycle), to explore the impact of different green infrastructure systems (none with trees) to reduce runoff and increase evapotranspiration in an urban catchment of Canberra, Australia. Green infrastructure could reduce urban stormwater runoff from 266 to 206 mm y^{-1} and concurrently increase evapotranspiration from 436 to 493 mm y^{-1}. It is this holistic level of vegetation impact upon urban water cycles that is lacking for urban trees and the 'urban forest' at a catchment or whole-of-city scale. Furthermore, few of these surface-atmosphere models consider canopy interception-evaporation, and those that do often do so using a static function of gross rainfall reduction. Given the limited research to date on the combined effect of tree interception and tree transpiration processes on urban water cycles, there remains great potential for improved simulation in mechanistic hydrology or surface-atmosphere models.

In recent years, urban field researchers have recognised the importance of measuring urban tree water uptake (Litvak et al. 2014, 2017), sometimes to assist in development and validation of urban energy balance or surface-atmosphere transfer models. Some recent urban tree transpiration studies have been motivated by understanding the role of trees in urban cooling (Ballinas and Barradas 2016; Konarska et al. 2016), whereas others have been motivated by understanding the role of tree transpiration in urban stormwater management (Gotsch et al. 2018). Despite this, the interaction between tree transpiration, climate, and soil conditions in a complex and heterogeneous urban landscape remains highly uncertain, largely due to the challenges of both measuring and scaling transpiration from the leaf to the tree and from the tree to the catchment (Litvak et al. 2017).

17.4.3 Comparative Impact of Urban Tree Interception-Evaporation and Transpiration

There have been a small number of urban hydrology simulations that have attempted to consider the proportional importance of canopy interception and transpiration. Using a hydrologic model (URBS-MO), Rodriguez et al. (2008) simulated 10 years of detailed urban sub-catchment rainfall, runoff, soil moisture, and stream discharge.

URBS-MO predicted that tree canopies reduced gross rainfall by 2.7% through interception-evaporation, whereas transpiration of the water that infiltrated into the soil represented a return of 43.5% of gross rainfall back to the atmosphere from a low-density urban sub-catchment. This hydrologic study estimates that urban tree transpiration is an order of magnitude greater than urban tree canopy interception-evaporation. Even if tree species were selectively planted to maximise canopy water storage capacity, it is likely that regardless of urban tree canopy cover, interception-evaporation will be an order of magnitude smaller than transpiration over an annual time scale and at a catchment scale. Regardless, from an urban future planning perspective, being able to accurately predict the role of all tree processes in reducing urban runoff and cooling the urban microclimate is of great importance. Furthermore, there is potential to increase urban tree transpiration by planting trees within SCM or WSUD systems that are engineered to receive and retain urban stormwater runoff (Berland et al. 2017; Kuehler et al. 2017). The ability of trees to transpire and remove the runoff water received can greatly improve overall SCM system performance by drying the infiltration substrate so that it is ready to receive as much runoff as possible from the next rainfall event (Grey et al. 2018a).

17.5 Conclusions and Future Opportunities

- Canopy interception loss in the urban environment is likely to be greater on a per unit canopy area basis than for rural forests. This is because healthy open-grown urban trees tend to be more voluminous than their counterparts in natural forests, and thus they possess a greater water storage capacity potential. Additionally, during-event evaporation rates may also be greater in the urban realm as a consequence of the potential for elevated sensible heat (artificial heat, urban heat island effect) and the increased air turbulence associated with the surface roughness of urban areas.
- The greatest interception loss values tend to be associated with larger coniferous trees compared to broadleaved species. Evergreen broadleaved trees, like coniferous trees, offer a distinct advantage with regard to interception loss compared with deciduous broadleaved trees during the dormant season.
- In addition to canopy interception loss, trees can also reduce the amount of rain researching impermeable surface by routing their drainage as stemflow. Although many trees may partition a relatively small fraction of rain striking their canopies into stemflow, for other trees, this fraction can be considerable. Even though the fraction of rain partitioned into stemflow on canopy-scale bases may be relatively small, the volume of water and the equivalent depth of that water delivered to the base of trees can be substantial. This can have important stormwater and tree water supply significance.

- For cities situated in regions where the rainfall regime is dominated by relatively small rain events, smaller trees will yield greater stemflow; however, for areas where larger rain events dominate, larger tree with more voluminous canopies will favour greater stemflow production. Trees with broad canopies and upright branching structure as well as smooth bark will favour stemflow generation. Furrowed bark may also conduct stemflow efficiently if the tree is thoroughly saturated.
- The redistribution of rainfall as throughfall is important both in time and space. The spatial variability of throughfall may have importance for how soil and understorey vegetation are wetted, how soils may be eroded, and the potential for runoff generation, while the temporal delay of throughfall reaching the ground compared to open rainfall may be of importance for stormwater flows, especially the timing of peak flows (Xiao and McPherson 2002; Asadian and Weiler 2009).
- Modelling rainfall partitioning by tree canopies in urban environments is made complex by the large variability of tree species and their condition. Models need to account for the differences in open-grown tree structure and microclimate compared to rural forests when assessing interception loss. Modelling procedures completed to date in cities have found that results are particularly sensitive to rainfall depth and rate as well as evaporation and canopy water storage capacity.
- Urban trees have the potential to transpire large volumes of water (e.g., captured runoff) due to the high evaporative demand in cities, and as such, stormwater control measures (SCMs) may be improved most by selecting those tree species with high transpiration rates.
- Evapotranspiration may represent a substantial component of SCM water balances, and trees may have similar transpiration rates (on a per unit canopy area basis) to sedges and rushes (commonly used biofiltration species). However, the extension of the canopy as a tree grows may mean that trees use more water than common biofiltration species planted over an equivalent surface area.
- The statements above regarding canopy interception loss, stemflow, throughfall, and transpiration in urban areas come from a very limited number of studies. Much more has yet to be learned. Research into the combined importance of interception-evaporation and tree transpiration in cities needs to be undertaken. For cities to embrace trees as part of their green infrastructure, greater scientific evidence and predictive understanding of the processes and the influence interception loss and transpiration have on the urban water balance and stormwater flows are required.
- Much work is needed to understand the complex interactions between transpiration, interception loss, climate, and soils in cities. As a greater understanding of tree-water interactions emerges with new research, the simulation results from urban hydrologic models and urban surface-atmosphere transfer models will be met with greater success.

References

Abas MR, Ahmad-Shah A, Awang MN (1992) Fluxes of ions in precipitation, throughfall and stemflow in an urban forest in Kuala Lumpur, Malaysia. Environ Pollut 75:209–213. https://doi.org/10.1016/0269-7491(92)90041-8

Abhijith KV, Kumar P, Gallagher J et al (2017) Air pollution abatement performances of green infrastructure in open road and built-up street canyon environments – a review. Atmos Environ 162:71–86. https://doi.org/10.1016/j.atmosenv.2017.05.014

Alves PL, Formiga KTM, Traldi MAB (2018) Rainfall interception capacity of tree species used in urban reforestation. Urban Ecosyst 21:697–706. https://doi.org/10.1007/s11252-018-0753-y

André F, Jonard M, Ponette Q (2008) Influence of species and rain event characteristics on stemflow volume in a temperate mixed oak-beech stand. Hydrol Process 22:4455–4466. https://doi.org/10.1002/hyp.7048

Asadian Y, Weiler M (2009) A new approach in measuring rainfall interception by urban trees in coastal British Columbia. Water Qual Res J Can 44:16–25. https://doi.org/10.2166/wqrj.2009.003

Ashley RM, Balmforth DJ, Saul AJ et al (2005) Flooding in the future-predicting climate change, risks and responses in urban areas. Water Sci Technol 52:265–273. https://doi.org/10.2166/wst.2005.0142

Aston AR (1979) Rainfall interception by eight small trees. J Hydrol 42:383–396. https://doi.org/10.1016/0022-1694(79)90057-X

Ballinas M, Barradas VL (2016) Transpiration and stomatal conductance as potential mechanisms to mitigate the heat load in Mexico City. Urban For Urban Green 20:152–159. https://doi.org/10.1016/j.ufug.2016.08.004

Baptista MD, Livesley SJ, Parmehr EG et al (2018a) Variation in leaf area density drives the rainfall storage capacity of individual urban tree species. Hydrol Process 32:3729–3740. https://doi.org/10.1002/hyp.13255

Baptista MD, Livesley SJ, Parmehr EG et al (2018b) Terrestrial laser scanning to predict canopy area metrics, water storage capacity and throughfall redistribution in urban trees. Remote Sens 10:1958. https://doi.org/10.3390/rs10121958

Bartens J, Day SD, Harris JR et al (2009) Transpiration and root development of urban trees in structural soil stormwater reservoirs. Environ Manag 44:646–657. https://doi.org/10.1007/s00267-009-9366-9

Berland A, Shiflett SA, Shuster WD et al (2017) The role of trees in urban stormwater management. Landsc Urban Plan 162:167–177. https://doi.org/10.1016/j.landurbplan.2017.02.017

Bialecki MB, Fahey RT, Scharenbroch B (2018) Variation in urban forest productivity and response to extreme drought across a large metropolitan region. Urban Ecosyst 21:157–169. https://doi.org/10.1007/s11252-017-0692-z

Brown RA, Hunt WF (2011) Impacts of media depth on effluent water quality and hydrologic performance of undersized bioretention cells. J Irrig Drain Eng 137:132–143. https://ascelibrary.org/doi/abs/10.1061/(ASCE)IR.1943-4774.0000167

Burns MJ, Fletcher TD, Walsh CJ et al (2012) Hydrologic shortcomings of conventional urban stormwater management and opportunities for reform. Landsc Urban Plan 105:230–240. https://doi.org/10.1016/j.landurbplan.2011.12.012

Bush SE, Pataki DE, Hultine KR et al (2008) Wood anatomy constrains stomatal responses to atmospheric vapor pressure deficit in irrigated, urban trees. Oecologia 156:13–20. https://doi.org/10.1007/s00442-008-0966-5

Calder IR (1996) Dependence of rainfall interception on drop size: 1. Development of the two-layer stochastic model. J Hydrol 185:363–376. https://doi.org/10.1016/0022-1694(95)02998-2

Carlyle-Moses DE (2004) A reply to R. Keim's comment on "Measurement and modelling of growing-season canopy water fluxes in a mature mixed deciduous forest stand, southern Ontario, Canada". Agric For Meteorol 124:281–284. https://doi.org/10.1016/j.agrformet.2004.02.004

Carlyle-Moses DE, Gash JHC (2011) Rainfall interception loss by forest canopies. In: Levia DF, Carlyle-Moses DE, Tanaka T (eds) Forest hydrology and biogeochemistry: synthesis of past research and future directions, Ecological Studies 216. Springer, Dordrecht, pp 407–423. https://doi.org/10.1007/978-94-007-1363-5_20

Carlyle-Moses DE, Price AG (2006) Growing-season stemflow production within a deciduous forest of southern Ontario. Hydrol Process 20:3651–3663. https://doi.org/10.1002/hyp.6380

Carlyle-Moses DE, Schooling JT (2015) Tree traits and meteorological factors influencing the initiation and rate of stemflow from isolated deciduous trees. Hydrol Process 29:4083–4099. https://doi.org/10.1002/hyp.10519

Carlyle-Moses DE, Iida S, Germer S et al (2018) Expressing stemflow commensurate with its ecohydrological importance. Adv Water Resour 121:472–479. https://doi.org/10.1016/j.advwatres.2018.08.015

Chair JT (2000) The hydrological effects of urban forests, with reference to the maritime Pacific Northwest. Landscape and Liveable Environments, Tech Bull, University of British Columbia No. 6 n/a-n/a

Chen WY (2015) The role of urban green infrastructure in offsetting carbon emissions in 35 major Chinese cities: a nationwide estimate. Cities 44:112–120. https://doi.org/10.1016/j.cities.2015.01.005

Chocat B, Krebs P, Marsalek J et al (2001) Urban drainage redefined: from stormwater removal to integrated management. Water Sci Technol 43:61–68. https://doi.org/10.2166/wst.2001.0251

Clar ML, Green R (1993) Design manual for use of bioretention in stormwater management. Prince George's County Govt., Watershed Protection Branch, Landover

Coutts AM, Tapper NJ, Beringer J et al (2013) Watering our cities: the capacity for Water Sensitive Urban Design to support urban cooling and improve human thermal comfort in the Australian context. Prog Phys Geogr 37:2–28. https://doi.org/10.1177/0309133312461032

Craul PJ (1999) Urban soils: applications and practices. Wiley, New York

da Silva LF, Lima AMLP, Filho DS et al (2010) Rainfall interception by two arboreal species in urban green area (In Portuguese). Cerne Lavras 16:547–555. https://doi.org/10.1590/S0104-77602010000400014

DeBusk KM, Wynn TM (2011) Storm-water bioretention for runoff quality and quantity mitigation. J Environ Eng 137:800–808. https://ascelibrary.org/doi/abs/10.1061/(ASCE)EE.1943-7870.0000388

Deguchi A, Hattori S, Park H-T (2006) The influence of seasonal changes in canopy structure on interception loss: application of the revised Gash model. J Hydrol 318:80–102. https://doi.org/10.1016/j.jhydrol.2005.06.005

Denich C, Bradford A (2010) Estimation of evapotranspiration from bioretention areas using weighing lysimeters. J Hydrol Eng 15:522–530. https://ascelibrary.org/doi/abs/10.1061/(ASCE)HE.1943-5584.0000134

Denman EC, May PB, Moore GM (2016) The potential role of urban forests in removing nutrients from stormwater. J Environ Qual 45:207–214. https://doi.org/10.2134/jeq2015.01.0047

Dolowitz DP, Bell S, Keeley M (2018) Retrofitting urban drainage infrastructure: green or grey? Urban Water J 15:83–91. https://doi.org/10.1080/1573062X.2017.1396352

Dunkerley DL (2009) Evaporation of impact water droplets in interception processes: historical precedence of the hypothesis and a brief literature overview. J Hydrol 376:599–604. https://doi.org/10.1016/j.jhydrol.2009.08.004

Eger CG, Chandler DG, Driscoll CT (2017) Hydrologic processes that govern stormwater infrastructure behaviour. Hydrol Process 31:4492–4506. https://doi.org/10.1002/hyp.11353

Ellis JB (2013) Sustainable surface water management and green infrastructure in UK urban catchment planning. J Environ Plan Manag 56:24–41. https://doi.org/10.1080/09640568.2011.648752

Elmqvist T, Fragkias M, Goodness J et al (2013) Stewardship of the biosphere in the urban era. In: Urbanization, biodiversity and ecosystem services: challenges and opportunities. Springer, Dordrecht, pp 719–746

Emmanuel R, Loconsole A (2015) Green infrastructure as an adaptation approach to tackling urban overheating in the Glasgow Clyde Valley Region, UK. Landsc Urban Plan 138:71–86. https://doi.org/10.1016/j.landurbplan.2015.02.012

Fathizadeh O, Attarod P, Keim RF et al (2014) Spatial heterogeneity and temporal stability of throughfall under individual Quercus brantii trees. Hydrol Process 28:1124–1136. https://doi.org/10.1002/hyp.9638

Fathizadeh O, Hosseini SM, Zimmerman A et al (2017) Estimating linkages between forest structural variables and rainfall interception parameters in semi-arid deciduous oak forest stands. Sci Total Environ 601–602:1824–1837. https://doi.org/10.1016/j.scitotenv.2017.05.233

Fletcher TD, Andrieu H, Hamel P (2013) Understanding, management and modelling of urban hydrology and its consequences for receiving waters: a state of the art. Adv Water Resour 51:261–279. https://doi.org/10.1016/j.advwatres.2012.09.001

Fletcher TD, Shuster W, Hunt WF et al (2015) SUDS, LID, BMPs, WSUD and more – the evolution and application of terminology surrounding urban drainage. Urban Water J 12:525–542. https://doi.org/10.1080/1573062X.2014.916314

Gash JHC, Lloyd CR, Lachaud G (1995) Estimating sparse forest rainfall interception with an analytical model. J Hydrol 170:79–86. https://doi.org/10.1016/0022-1694(95)02697-N

Geißler C, Lang AC, von Oheimb G et al (2012) Impact of tree saplings on the kinetic energy of rainfall-The importance of stand density, species identity and tree architecture in subtropical forests in China. Agric For Meteorol 156:31–40. https://doi.org/10.1016/j.agrformet.2011.12.005

Ghimire CP, Bruijnzeel LA, Lubczynski MW et al (2017) Measurement and modeling of rainfall interception by two differently aged secondary forests in upland eastern Madagascar. J Hydrol 545:212–225. https://doi.org/10.1016/j.jhydrol.2016.10.032

Goebes P, Bruelheide H, Härdtle W et al (2015) Species-specific effects on throughfall kinetic energy in subtropical forest plantations are related to leaf traits and tree architecture. PLoS One 10:1–13. https://doi.org/10.1371/journal.pone.0128084

Gómez JA, Giráldez JV, Fereres E (2001) Rainfall interception by olive trees in relation to leaf area. Agric Water Manag 49:65–76. https://doi.org/10.1016/S0378-3774(00)00116-5

Gotsch SG, Draguljić D, Williams CJ (2018) Evaluating the effectiveness of urban trees to mitigate storm water runoff via transpiration and stemflow. Urban Ecosyst 21:183–195. https://doi.org/10.1007/s11252-017-0693-y

Grabosky J, Bassuk N (1995) A new urban tree soil to safely increase rooting volumes under sidewalks. J Arboric 21:187–187

Green T, Kronenberg J, Andersson E et al (2016) Insurance value of green infrastructure in and around cites. Ecosystems 19:1051–1063. https://doi.org/10.1007/s10021-016-9986-x

Grey V, Livesley SJ, Fletcher TD et al (2018a) Establishing street trees in stormwater control measures can double tree growth when extended waterlogging is avoided. Landsc Urban Plan 178:122–129. https://doi.org/10.1016/j.landurbplan.2018.06.002

Grey V, Livesley SJ, Fletcher TD et al (2018b) Tree pits to help mitigate runoff in dense urban areas. J Hydrol 565:400–410. https://doi.org/10.1016/j.jhydrol.2018.08.038

Grimmond CSB, Oke TR (1991) An evapotranspiration-interception model for urban areas. Water Resour Res 27:1739–1755. https://doi.org/10.1029/91WR00557

Guevara-Escobar A, González-Sosa E, Véliz-Chávez C et al (2007) Rainfall interception and distribution patterns of gross precipitation around an isolated Ficus benjamina tree in an urban area. J Hydrol 333:532–541. https://doi.org/10.1016/j.jhydrol.2006.09.017

Hamel P, McHugh I, Coutts A et al (2014) An automated chamber system to measure field evapotranspiration rates. J Hydrol Eng 20:04014037. https://ascelibrary.org/doi/abs/10.1061/(asce)he.1943-5584.0001006

Helvey JD, Patric JH (1965) Canopy and litter interception of rainfall by hardwoods of eastern United States. Water Resour Res 1:193–206. https://doi.org/10.1029/WR001i002p00193

Herbst M, Rosier PTW, McNeil DD et al (2008) Seasonal variability of interception evaporation from the canopy of a mixed deciduous forest. Agric For Meteorol 148:1655–1667. https://doi.org/10.1016/j.agrformet.2008.05.011
Herwitz SR (1985) Interception storage capacities of tropical rainforest canopy trees. J Hydrol 77:237–252. https://doi.org/10.1016/0022-1694(85)90209-4
Herwitz SR (1986) Infiltration-excess caused by stemflow in a cyclone-prone tropical rainforest. Earth Surf Process Landf 11:401–412. https://doi.org/10.1002/esp.3290110406
Hess A, Wadzuk B, Welker A (2017) Evapotranspiration in rain gardens using weighing lysimeters. J Irrig Drain Eng 143:04017004. https://ascelibrary.org/doi/abs/10.1061/(ASCE)IR.1943-4774.0001157
Holder CD (2013) Effects of leaf hydrophobicity and water droplet retention on canopy storage capacity. Ecohydrology 6:483–490. https://doi.org/10.1002/eco.1278
Houle JJ, Roseen RM, Ballestero TP (2013) Comparison of maintenance cost, labor demands, and system performance for LID and conventional stormwater management. J Environ Eng 139:932–938. https://ascelibrary.org/doi/abs/10.1061/(ASCE)EE.1943-7870.0000698
Huang JY, Black TA, Jassal RS et al (2017) Modelling rainfall interception by urban trees. Can Water Res J 42:336–348. https://doi.org/10.1080/07011784.2017.1375865
Inkiläinen ENM, McHale MR, Blank GB et al (2013) The role of the residential urban forest in regulating throughfall: a case study in Raleigh, North Carolina, USA. Landsc Urban Plan 119:91–103. https://doi.org/10.1016/j.landurbplan.2013.07.002
Johnson MS, Lehmann J (2006) Double-funneling of trees: stemflow and root-induced preferential flow. Écoscience 13:324–333. https://doi.org/10.2980/i1195-6860-13-3-324.1
Kaźmierczak A, Cavan G (2011) Surface water flooding risk to urban communities: analysis of vulnerability, hazard and exposure. Landsc Urban Plan 103:185–197. https://doi.org/10.1016/j.landurbplan.2011.07.008
Keim RF, Skaugset AE, Weiler M (2006) Storage of water on vegetation under simulated rainfall of varying intensity. Adv Water Resour 29:974–986. https://doi.org/10.1016/j.advwatres.2005.07.017
Kermavnar J, Vilhar U (2017) Canopy precipitation interception in urban forests in relation to stand structure. Urban Ecosyst 20:1373–1387. https://doi.org/10.1007/s11252-017-0689-7
King BP, Harrison SJ (1998) Throughfall patterns under an isolated oak. Weather 53:111–121. https://doi.org/10.1002/j.1477-8696.1998.tb03973.x
Klaassen W, Bosweld F, de Water E (1998) Water storage and evaporation as constituents of rainfall interception. J Hydrol 212–213:36–50. https://doi.org/10.1016/S0022-1694(98)00200-5
Konarska J, Uddling J, Holmer B et al (2016) Transpiration of urban trees and its cooling effect in a high latitude city. Int J Biometeorol 60:159–172. https://doi.org/10.1007/s00484-015-1014-x
Kuehler E, Hathaway J, Tirpak A (2017) Quantifying the benefits of urban forest systems as a component of the green infrastructure stormwater treatment network. Ecohydrology 10:1–10. https://doi.org/10.1002/eco.1813
Levia DF, Herwitz SR (2005) Interspecific variation in bark water storage capacity of three deciduous tree species in relation to stemflow yield and solute flux to forest soils. Catena 64:117–137. https://doi.org/10.1016/j.catena.2005.08.001
Levia DF, Keim RF, Carlyle-Moses DE et al (2011) Throughfall and stemflow in wooded ecosystems. In: Levia DF, Carlyle-Moses DE, Tanaka T (eds) Forest hydrology and biogeochemistry: synthesis of past research and future directions, Ecological Studies 216. Springer, Dordrecht, pp 425–443. https://doi.org/10.1007/978-94-007-1363-5_21
Li H, Sharkey LJ, Hunt WF et al (2009) Mitigation of impervious surface hydrology using bioretention in North Carolina and Maryland. J Hydrol Eng 14:407–415. https://doi.org/10.1061/(ASCE)1084-0699(2009)14:4(407)
Li X, Xiao Q, Niu J et al (2016) Process-based rainfall interception by small trees in Northern China: the effect of rainfall traits and crown structure characteristics. Agric For Meteorol 218–219:65–73. https://doi.org/10.1016/j.agrformet.2015.11.017

Litvak E, McCarthy HR, Pataki DE (2012) Transpiration sensitivity of urban trees in a semi-arid climate is constrained by xylem vulnerability to cavitation. Tree Physiol 32:373–388. https://doi.org/10.1093/treephys/tps015

Litvak E, Bijoor NS, Pataki DE (2014) Adding trees to irrigation turfgrass lawns may be a water-saving measure in semi-arid environments. Ecohydrology 7:1314–1330. https://doi.org/10.1002/eco.1458

Litvak E, McCarthy HR, Pataki DE (2017) A method for estimating transpiration of irrigated urban trees in California. Landsc Urban Plan 158:48–61. https://doi.org/10.1016/j.landurbplan.2016.09.021

Liu W, Chen W, Peng C (2014) Assessing the effectiveness of green infrastructures on urban flooding reduction: a community scale study. Ecol Model 291:6–14. https://doi.org/10.1016/j.ecolmodel.2014.07.012

Livesley SJ, Baudinette B, Glover D (2014) Rainfall interception and stem flow by eucalypt street trees – the impacts of canopy density and bark type. Urban For Urban Green 13:192–197. https://doi.org/10.1016/j.ufug.2013.09.001

Livesley SJ, McPherson EG, Calfapietra C (2016) The urban forest and ecosystem services: impacts on urban water, heat, and pollution cycles at the tree, street, and city scale. J Environ Qual 45:119–124. https://doi.org/10.2134/jeq2015.11.0567

Llorens P, Gallart F (2000) A simplified method for forest water storage capacity measurement. J Hydrol 240:131–144. https://doi.org/10.1016/S0022-1694(00)00339-5

Mangangka IR, Liu A, Egodawatta P (2015) Performance characterisation of a stormwater treatment bioretention basin. J Environ Manag 150:173–178. https://doi.org/10.1016/j.jenvman.2014.11.007

Maragno D, Gaglio M, Robbi M et al (2018) Fine-scale analysis of urban flooding reduction from green infrastructure: an ecosystem services approach to the management of water flows. Ecol Model 386:1–10. https://doi.org/10.1016/j.ecolmodel.2018.08.002

May PB, Livesley SJ, Shears I (2013) Managing and monitoring tree health and soil water status during extreme drought in Melbourne, Victoria. Arboricult Urban For 39:136–145

McCarthy HR, Pataki DE, Jenerette GD (2011) Plant water-use efficiency as a metric of urban ecosystem services. Ecol Appl 21:3115–3127. https://doi.org/10.1890/11-0048.1

McPherson G, Simpson JR, Peper PJ et al (1999) Benefit-cost analysis of Modesto's municipal urban forest. J Arboric 25:235–248

McPherson G, Simpson JR, Peper PJ et al (2005) Municipal forest benefits and costs in five US cities. J For 103:411–416. https://doi.org/10.1093/jof/103.8.411

McPherson EG, van Doorn N, de Goede J (2016) Structure, function and value of street trees in California, USA. Urban For Urban Green 17:104–115. https://doi.org/10.1016/j.ufug.2016.03.013

Meerow S, Newell JP (2017) Spatial planning for multifunctional green infrastructure: growing resilience in Detroit. Landsc Urban Plan 159:62–75. https://doi.org/10.1016/j.landurbplan.2016.10.005

Meineke EK, Frank SD (2018) Water availability drives urban tree growth responses to herbivory and warming. J Appl Ecol 55:1701–1713. https://doi.org/10.1111/1365-2664.13130

Melbourne Water (2005) WSUD engineering procedures: stormwater. CSIRO Publishing, Melbourne

Mitchell VG, Cleugh HA, Grimmond CSB et al (2008) Linking urban water balance and energy balance models to analyse urban design options. Hydrol Process 22:2891–2900. https://doi.org/10.1002/hyp.6868

Monteith JL, Unsworth MH (2008) Principles of environmental physics, 3rd edn. Academic, London

Moors EJ, Grimmond CSB, Veldhuizen AA et al (2014) Urban water balance and hydrology models to support sustainable urban planning. In: Chrysoulakis N, de Castro EA, Moors EJ (eds) Understanding urban metabolism: a tool for urban planning. Routledge, New York, pp 106–117

Mullaney J, Lucke T, Trueman SJ (2015) A review of benefits and challenges in growing street trees in paved urban environments. Landsc Urban Plan 134:157–166. https://doi.org/10.1016/j.landurbplan.2014.10.013
Murakami S (2006) A proposal for a new forest canopy interception mechanism: splash droplet evaporation. J Hydrol 318:72–82. https://doi.org/10.1016/j.jhydrol.2005.07.002
Nanko K, Onda Y, Ito A et al (2011) Spatial variability of throughfall under a single tree: experimental study of rainfall amount, raindrops, and kinetic energy. Agric For Meteorol 151:1173–1182. https://doi.org/10.1016/j.agrformet.2011.04.006
NCDEQ (2017) C-0. Minimum design criteria for all SCMs. In: Stormwater design manual. North Carolina Department of Environmental Quality, Raleigh
Neumann JE, Price J, Chinowsky P et al (2015) Climate change risks to US infrastructure: impacts on roads, bridges, coastal development, and urban drainage. Clim Chang 131:97–109. https://doi.org/10.1007/s10584-013-1037-4
Nichols PWB, Lucke T (2015) Local level stormwater harvesting and reuse: a practical solution to the water security challenges faced by urban trees. Sustainability 7:8635–8648. https://doi.org/10.3390/su7078635
Nitschke CR, Nichols S, Allen K et al (2017) The influence of climate and drought on urban tree growth in southeast Australia and the implications for future growth under climate change. Landsc Urban Plan 167:275–287. https://doi.org/10.1016/j.landurbplan.2017.06.012
Nocco MA, Rouse SE, Balster NJ (2016) Vegetation type alters water and nitrogen budgets in a controlled, replicated experiment on residential-sized rain gardens planted with prairie, shrub, and turfgrass. Urban Ecosyst 19:1665–1691. https://doi.org/10.1007/s11252-016-0568-7
Nordman EE, Isely E, Isely P et al (2018) Benefit-cost analysis of stormwater green infrastructure practices for Grand Rapids, Michigan, USA. J Clean Prod 200:501–510. https://doi.org/10.1016/j.jclepro.2018.07.152
Nowak DJ, McBride JR, Beatty RA (1990) Newly planted street tree growth and mortality. J Arboric 16:124–130
Nowak DJ, Greenfield EJ, Hoehn RE et al (2013) Carbon storage and sequestration by trees in urban and community areas of the United States. Environ Pollut 178:229–236. https://doi.org/10.1016/j.envpol.2013.03.019
Nytch CJ, Melendez-Ackerman EJ, Perez M-E, Ortiz-Zayas JR (2018) Rainfall interception by six urban trees in San Juan, Puerto Rico. Urban Ecosyst 22:103–115. https://doi.org/10.1007/s11252-018-0768-4. Published online
Ordóñez-Barona C, Sabetski V, Millward AA et al (2018) De-icing salt contamination reduces urban tree performance in structural soil cells. Environ Pollut 234:562–571. https://doi.org/10.1016/j.envpol.2017.11.101
Ossola A, Hahs AK, Livesley SJ (2015) Habitat complexity influences fine scale hydrological processes and the incidence of stormwater runoff in managed urban ecosystems. J Environ Manag 159:1–10. https://doi.org/10.1016/j.jenvman.2015.05.002
Page JL, Winston RJ, Hunt WF (2015a) Soils beneath suspended pavements: an opportunity for stormwater control and treatment. Ecol Eng 82:40–48. https://doi.org/10.1016/j.ecoleng.2015.04.060
Page JL, Winston RJ, Mayes DB et al (2015b) Retrofitting with innovative stormwater control measures: hydrologic mitigation of impervious cover in the municipal right-of-way. J Hydrol 527:923–932. https://doi.org/10.1016/j.jhydrol.2015.04.046
Pataki DE, McCarthy HR, Litvak E et al (2011) Transpiration of urban forests in the Los Angeles metropolitan area. Ecol Appl 21:661–677. https://doi.org/10.1890/09-1717.1
Payne EGI, Hatt BE, Deletic A et al (2015) Adoption guidelines for stormwater biofiltration systems. Cooperative Research Centre for Water Sensitive Cities, Melbourne
Payne EGI, Pham T, Deletic A et al (2018) Which species? A decision-support tool to guide plant selection in stormwater biofilters. Adv Water Resour 113:86–99. https://doi.org/10.1016/j.advwatres.2017.12.022

Pereira FL, Gash JHC, David JS et al (2009) Evaporation of intercepted rainfall from isolated evergreen oak trees: do the crowns behave like wet bulbs? Agric For Meteorol 149:667–679. https://doi.org/10.1016/j.agrformet.2008.10.013

Peters EB, McFadden JP, Montgomery RA (2010) Biological and environmental controls on tree transpiration in a suburban landscape. J Geophys Res 15:G04006. https://doi.org/10.1029/2009JG001266

Pretzsch H, del Rio M, Ammer C et al (2015) Growth and yield of mixed versus pure stands of Scots pine (Pinus sylvestris L.) and European beech (*Fagus sylvatica* L.) analysed along a productivity gradient through Europe. Eur J For Res 134:927–947. https://doi.org/10.1007/s10342-015-0900-4

Rhoades RW, Stipes RJ (1999) Growth of trees on the Virginia Tech campus in response to various factors. J Arboric 25:211–217

Riikonen A, Järvi L, Nikinmaa E (2016) Environmental and crown related factors affecting street tree transpiration in Helsinki, Finland. Urban Ecosyst 19:1693–1715. https://doi.org/10.1007/s11252-016-0561-1

Rodriguez FR, Andrieu H, Morena F (2008) A distributed hydrological model for urbanized areas – model development and application to case studies. J Hydrol 351:268–287. https://doi.org/10.1016/j.jhydrol.2007.12.007

Rosado BHP, Holder CD (2013) The significance of leaf water repellency in ecohydrological research: a review. Ecohydrology 6:150–161. https://doi.org/10.1002/eco.1340

Rutter AJ, Kershaw KA, Robins PC, Morton AJ (1971) A predictive model of rainfall interception in forests, 1. Derivation of the model from observations in a plantation of Corsican pine. Agric Meteorol 9:367–384. https://doi.org/10.1016/0002-1571(71)90034-3

Sanders RA (1986) Urban vegetation impacts on the hydrology of Dayton, Ohio. Urban Ecol 9:361–376. https://doi.org/10.1016/0304-4009(86)90009-4

Sanusi R, Johnstone D, May P et al (2017) Microclimate benefits that different street tree species provide to sidewalk pedestrians relate to differences in plant area index. Landsc Urban Plan 157:502–511. https://doi.org/10.1016/j.landurbplan.2016.08.010

Savi T, Bertuzzi S, Branca S et al (2015) Drought-induced xylem cavitation and hydraulic deterioration: risk factors for urban trees under climate change? New Phytol 205:1106–1116. https://doi.org/10.1111/nph.13112

Scharenbroch BC, Morgenroth J, Maule B (2016) Tree species suitability to bioswales and impact on the urban water budget. J Environ Qual 45:199–206. https://doi.org/10.2134/jeq2015.01.0060

Schooling JT, Carlyle-Moses DE (2015) The influence of rainfall depth class and deciduous tree traits on stemflow production in an urban park. Urban Ecosyst 18:1261–1284. https://doi.org/10.1007/s11252-015-0441-0

Schooling JT, Levia DF, Carlyle-Moses DE et al (2017) Stemflow chemistry in relation to tree size: a preliminary investigation of eleven urban park trees in British Columbia, Canada. Urban For Urban Green 21:129–133. https://doi.org/10.1016/j.ufug.2016.11.013

Schwärzel K, Ebermann S, Schalling N (2012) Evidence of double-funneling effect of beech trees by visualization of flow pathways using dye tracer. J Hydrol 470–471:184–192. https://doi.org/10.1016/j.jhydrol.2012.08.048

Shahidan MF (2015) Potential of individual and cluster tree cooling effect performances through tree canopy density model evaluation in improving urban microclimate. Curr World Environ 10:398–413. https://doi.org/10.12944/CWE.10.2.04

Shuster WD, Bonta J, Thurston H et al (2005) Impacts of impervious surface on watershed hydrology: a review. Urban Water J 2:263–275. https://doi.org/10.1080/15730620500386529

Sjöman H, Nielsen AB (2010) Selecting trees for urban paved sites in Scandinavia – a review of information on stress tolerance and its relation to the requirements of tree planners. Urban For Urban Green 9:281–293. https://doi.org/10.1016/j.ufug.2010.04.001

Soares AL, Rego FC, McPherson EG et al (2011) Benefits and costs of street trees in Lisbon, Portugal. Urban For Urban Green 10:69–78. https://doi.org/10.1016/j.ufug.2010.12.001

Šraj M, Brilly M, Mikoš M (2008) Rainfall interception by two deciduous Mediterranean forests of contrasting stature in Slovenia. Agric For Meteorol 148:121–134. https://doi.org/10.1016/j.agrformet.2007.09.007

Staelens J, De Schrijver A, Verheyen K et al (2006) Spatial variability and temporal stability of throughfall water under a dominant beech (Fagus sylvatica L.) tree in relationship to canopy cover. J Hydrol 3–4:651–662. https://doi.org/10.1016/j.jhydrol.2006.04.032

Staelens J, De Schrijver A, Verheyen K et al (2008) Rainfall partitioning into throughfall, stemflow, and interception within a single beech (Fagus sylvatica L.) canopy: influence of foliation, rain event characteristics, and meteorology. Hydrol Process 22:22–45. https://doi.org/10.1002/hyp.6610

Stagoll K, Lindenmayer DB, Knight E et al (2012) Large trees are keystone structures in urban parks. Conserv Lett 5:115–122. https://doi.org/10.1111/j.1755-263X.2011.00216.x

Stewart JB (1977) Evaporation from the wet canopy of a pine forest. Water Resour Res 13:915–921. https://doi.org/10.1029/WR013i006p00915

Stovin VR, Jorgensen A, Clayden A (2008) Street trees and stormwater management. Arboricult J 30:297–310. https://doi.org/10.1080/03071375.2008.9747509

Szota C, McCarthy MJ, Sanders GJ et al (2018) Tree water-use strategies to improve stormwater retention performance of biofiltration systems. Water Res 144:285–295. https://doi.org/10.1016/j.landurbplan.2018.10.021

Szota C, Coutts AM, Thom JK et al (2019) Street tree stormwater control measures can reduce runoff but may not benefit established trees. Landsc Urban Plan 182:144–155

Thom JK, Coutts AM, Broadbent AM et al (2016) The influence of increasing tree cover on mean radiant temperature across a mixed development suburb in Adelaide, Australia. Urban For Urban Green 20:233–242. https://doi.org/10.1016/j.ufug.2016.08.016

Threlfall CG, Ossola A, Hahs AK et al (2016) Variation in vegetation structure and composition across urban green space types. Front Ecol Evol 4:66,. Article 6. https://doi.org/10.3389/fevo.2016.00066

Toba T, Ohta T (2005) An observational study of the factors that influence interception loss in boreal and temperate forests. J Hydrol 313:208–220. https://doi.org/10.1016/j.jhydrol.2005.03.003

Valente F, David JS, Gash JHC (1997) Modelling interception loss for two sparse eucalypt and pine forests in central Portugal using reformulated Rutter and Gash analytical models. J Hydrol 190:141–162. https://doi.org/10.1016/S0022-1694(96)03066-1

Van Stan JT, Levia DF (2010) Inter- and intraspecific variation of stemflow production from Fagus gradifolia Ehrh. (American beech) and Liriodendron tulipifera L. (yellow popular) in relation to bark microrelief in the eastern United States. Ecohydrology 3:11–19. https://doi.org/10.1002/eco.83

Van Stan JT, Siegert C, Levia DF et al (2011) Effects of wind-driven rainfall on stemflow generation between codominant tree species with differing crown characteristics. Agric For Meteorol 151:1277–1286. https://doi.org/10.1016/j.agrformet.2011.05.008

Van Stan JT, Van Stan JH, Levia DF (2014) Meteorological influences on stemflow generation across diameter size classes of two morphologically distinct deciduous species. Int J Biometeorol 58:2059–2069. https://doi.org/10.1007/s00484-014-0807-7

Van Stan JT, Norman Z, Meghoo A et al (2017) Edge-to-stem variability in wet-canopy evaporation from an urban tree row. Bound-Layer Meteorol 165:295–310. https://doi.org/10.1007/s10546-017-0277-7

Voskamp IM, Van de Ven FHM (2015) Planning support system for climate adaptation: composing effective sets of blue-green measures to reduce urban vulnerability to extreme weather events. Build Environ 83:159–167. https://doi.org/10.1016/j.buildenv.2014.07.018

Wadzuk BM, Hickman JM Jr, Traver RG (2015) Understanding the role of evapotranspiration in bioretention: Mesocosm study. J Sustain Water Built Environ 1:04014002. https://ascelibrary.org/doi/abs/10.1061/JSWBAY.0000794

Wallace J, McJanet D (2006) On interception modelling of a lowland coastal rainforest in northern Queensland, Australia. J Hydrol 329:477–488. https://doi.org/10.1016/j.jhydrol.2006.03.003
Walsh CJ, Roy AH, Feminella JW et al (2005) The urban stream syndrome: current knowledge and the search for a cure. J North Am Benthol Soc 24:706–723. https://doi.org/10.1899/04-028.1
Wang J, Endreny TA, Nowak DJ (2008) Mechanistic simulation of tree effects in an urban water balance model. J Am Water Resour Assoc 44:75–85. https://doi.org/10.1111/j.1752-1688.2007.00139.x
Wei X, Liu S, Zhou G et al (2005) Hydrologic processes in major types of Chinese forest. Hydrol Process 19:63–75. https://doi.org/10.1002/hyp.5777
Winston RJ, Dorsey JD, Hunt WF (2016) Quantifying volume reduction and peak flow mitigation for three bioretention cells in clay soils in northeast Ohio. Sci Total Environ 553:83–95. https://doi.org/10.1016/j.scitotenv.2016.02.081
Xiao Q, McPherson EG (2002) Rainfall interception by Santa Monica's municipal urban forest. Urban Ecosyst 6:291–302. https://doi.org/10.1023/B:UECO.0000004828.05143.67
Xiao Q, McPherson EG (2011a) Rainfall interception of three trees in Oakland, California. Urban Ecosyst 14:755–769. https://doi.org/10.1007/s11252-011-0192-5
Xiao Q, McPherson EG (2011b) Performance of engineered soil and trees in a parking lot bioswale. Urban Water J 8:241–253. https://doi.org/10.1080/1573062X.2011.596213
Xiao Q, McPherson EG (2016) Surface water storage capacity of twenty tree species in Davis, California. J Environ Qual 45:188–198. https://doi.org/10.2134/jeq2015.02.0092
Xiao Q, McPherson EG, Simpson JR et al (1998) Rainfall interception by Sacramento's urban forest. J Arboric 24:235–244
Xiao Q, McPherson EG, Forest U et al (2000a) A new approach to modeling tree rainfall interception. J Geophys Res 105:173–129. https://doi.org/10.1029/2000JD900343
Xiao Q, McPherson EG, Ustin SL et al (2000b) Winter rainfall interception by two mature open-grown trees in Davis, California. Hydrol Process 14:763–784. https://doi.org/10.1002/(SICI)1099-1085(200003)14:4<763::AID-HYP971>3.0.CO;2-7
Yuan J, Dunnett N, Stovin V (2017) The influence of vegetation on rain garden hydrological performance. Urban Water J 14:1083–1089. https://doi.org/10.1080/1573062X.2017.1363251
Zabret K, Rakovec J, Šraj M (2018) Influence of meteorological variables on rainfall partitioning for deciduous and coniferous tree species in urban area. J Hydrol 558:29–41. https://doi.org/10.1016/j.jhydrol.2018.01.025
Zeng N, Shuttleworth JW, Gash JHC (2000) Influence of temporal variability of rainfall on interception loss. Part I. Point analysis. J Hydrol 228:228–241. https://doi.org/10.1016/S0022-1694(00)00140-2
Zhang L, Dawes WR, Walker GR (2001) Response of mean annual evapotranspiration to vegetation changes at catchment scale. Water Resour Res 37:701–708. https://doi.org/10.1029/2000WR900325
Zipper SC, Schatz J, Kucharik CJ et al (2017) Urban heat island-induced increases in evapotranspirative demand. Geophys Res Lett 44:873–881. https://doi.org/10.1002/2016GL072190
Zölch T, Maderspacher J, Wamsler C et al (2016) Using green infrastructure for urban climate-proofing: an evaluation of heat mitigation measures at the micro-scale. Urban For Urban Green 20:305–316. https://doi.org/10.1016/j.ufug.2016.09.011
Zölch T, Henze L, Keilholz P (2017) Regulating urban surface runoff through nature-based solutions – an assessment at the micro-scale. Environ Res 157:135–144. https://doi.org/10.1016/j.envres.2017.05.023

Chapter 18
Urban Tree Canopy Effects on Water Quality via Inputs to the Urban Ground Surface

S. M. Decina, A. G. Ponette-González, and J. E. Rindy

18.1 Introduction: The Urban Tree Canopy

The word "forest" invokes images of wide swaths of land covered by an unbroken sea of timber and leaves. However, in the world's cities, many of which were built within forested biomes, a forest is often much different. Although many ecosystem models and forest inventories represent urban areas as biologically barren, quite the opposite is true (Hardiman et al. 2017). Around 2014, an estimated 39% of urban land across the USA was covered by trees, with variations in extent and composition across cities (Nowak and Greenfield 2018). The wetter northeastern, southeastern, and Pacific Northwest regions of the USA, for example, have a more extensive tree canopy in developed land compared to the drier Great Plains and western US regions (Fig. 18.1). In recent decades, the potential for this tree canopy to ameliorate urban environmental problems, including stormwater runoff and pollutant loading of receiving waterbodies, has garnered considerable attention (Bolund and Hunhammar 1999; Pataki et al. 2011; Elmqvist et al. 2015; Endreny 2018). With urban population and land area increasing at a rapid pace around the globe (Seto et al. 2012; UNDESAPD 2015), understanding the influence of urban trees and their canopies on the quantity and quality of water entering urban ecosystems and watersheds is now a major area of scientific research and public interest (Livesley et al. 2016; McGrane 2016; Berland et al. 2017; Decina et al. 2018).

But what is an urban tree, and what exactly comprises the urban tree canopy? Urban trees come in various sizes, species, and forms. In many cities, trees line both streets and streams; aerial images allow the eye to map sections of the urban street grid using the canopy itself. There are also trees, individually and in groups (e.g.,

S. M. Decina (✉)
American Association for the Advancement of Science (AAAS) Science and Technology Policy Fellow, Hosted by the U.S. Environmental Protection Agency, Washington, DC, USA
e-mail: sdecina@bu.edu

A. G. Ponette-González · J. E. Rindy
Department of Geography and the Environment, University of North Texas, Denton, TX, USA

D. F. Levia et al. (eds.), *Forest-Water Interactions*, Ecological Studies 240,
https://doi.org/10.1007/978-3-030-26086-6_18

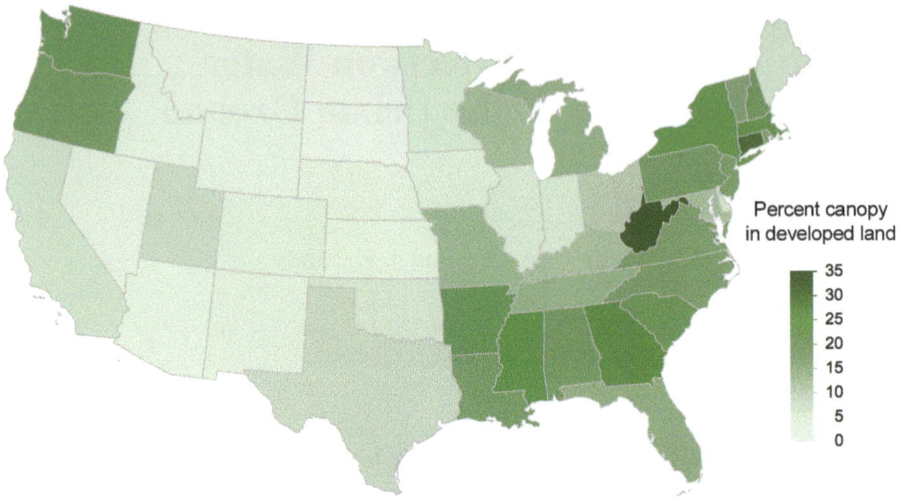

Fig. 18.1 Percent canopy cover in developed land (canopy in developed land/total developed land * 100) by state in the conterminous USA Developed land includes National Land Cover Database 2001 classes 21–24 (developed, open space; developed, low intensity; developed, medium intensity; and developed, high intensity). (Data from US Forest Service https://www.nrs.fs.fed.us/data/urban/)

pocket parks), on residential, municipal, and commercial properties. Within and just outside of city limits, there are frequently larger tracts of trees or remnant forest, resembling more of what is classically considered a forest. These tracts can exist on private land, in arboreta, on university campuses, and on public conservation land. With respect to size and species composition, urban trees can be broadleaf or needleleaf, native or exotic, and small or large, varying in height from <5 to well over 50 m in conifer-dominated regions. In the USA, many urban trees are species which are now mainly found in urban areas, such as the fast-growing, hardy, non-native tree of heaven (*Ailanthus altissima*) and Norway maple (*Acer platanoides*). The *urban tree canopy* (the layer of tree leaves, branches, and stems that provides tree coverage of the ground when viewed from above; USFS 2018) can thus be quite diverse and distinctive from its rural counterpart. Not only is the character (i.e., structure, species composition) of the urban tree canopy different from that of rural forest, but the overall form of the urban tree canopy – its configuration (e.g., spacing, arrangement) and spatial extent – differs as well (Nowak and Greenfield 2018).

In this chapter, we review how the urban tree canopy interacts with the urban atmosphere and precipitation to influence the quality of water delivered to the ground surface. We first describe the main processes by which tree canopies influence water quality and examine the major controls on atmospheric deposition to trees in urban systems. We then briefly review the sources of major elements in urban areas, their interaction with urban tree canopies, and recent measurements of particulate matter deposition to canopies and inputs to the ground. We conclude with a synthesis of the implications of urban tree canopy effects on water quality and biogeochemical cycling and outline opportunities for future research.

18.2 Processes by Which Tree Canopies Influence Water Quality

All plant canopies, including urban tree canopies, modify the quality of water transported to the ground in three ways: (1) through interactions with ions, particles, and gases deposited to canopy surfaces from the atmosphere (atmospheric deposition); (2) through the canopy processes of uptake and leaching; and (3) through the production of materials generated within the canopy, such as pollen (Fig. 18.2).

The urban atmosphere represents an abundant source of materials (e.g., nutrients, pollutants, microbes) for which urban tree canopies serve as a physical receptor surface. The process by which these materials are deposited to plant canopies (including leaves, stems, trunks, and epiphytes) is termed *atmospheric deposition* (Weathers and Ponette-González 2011). Materials incorporated into rain, snow, sleet, and fog droplets, including dissolved ions and insoluble particles, are deposited to canopies via *wet* and *fog deposition* (Fig. 18.2). Particles and gases also impact directly onto canopy surfaces via *dry deposition*. Particles and gases can deposit to external surfaces and dissolve upon interaction with water. Additionally, gases (e.g., nitrogen dioxide (NO_2)) can diffuse through leaf stomata and into the leaf mesophyll (Lovett 1994). Thus, the controls on the physical processes of atmospheric deposition can range from aerodynamic resistance and boundary layer physics to plant physiological factors controlling stomatal uptake (Lovett 1994).

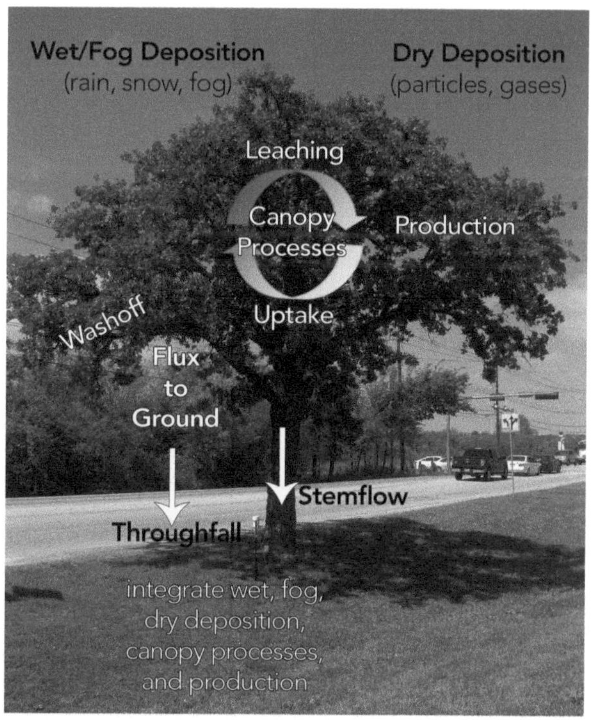

Fig. 18.2 Materials are deposited from the atmosphere to tree canopies by wet, fog, and dry deposition. Once deposited to the canopy, materials can be taken up, leached into solution, or washed from the canopy after relatively little interaction with canopy surfaces. New materials are also produced within urban tree canopies. Depending on the material, flux to the ground in stemflow and throughfall is not always the same as what is deposited to the canopy

Once deposited to the canopy, some materials may undergo biological and chemical transformations within the canopy. These *canopy processes* are broadly grouped into two categories: uptake and leaching (Fig. 18.2). *Uptake* is the process whereby materials are adsorbed, absorbed, or assimilated into the plant and can occur via abiotic and biotic mechanisms. Nitrogen (N), a critical limiting nutrient to productivity, for example, can be absorbed through leaf surfaces (Templer et al. 2015a) or adsorbed onto canopy organic matter (Lovett et al. 1985). *Leaching* is the process whereby materials are released from the canopy into external solution (Lovett et al. 1985). For example, some nutrients, such as potassium (K), that are strongly cycled by plants are easily leached from plant surfaces (Schlesinger and Bernhardt 2013). It is important to note that some materials, such as sulfate (SO_4^{2-}), chloride (Cl^-), and sodium (Na^+) ions, undergo little interaction with tree canopies (Ponette-González et al. 2016a). That is, both uptake and leaching are minimal relative to total (wet + fog + dry) deposition, such that most of what is deposited to the canopy is simply washed off with precipitation.

Urban tree canopies also serve as hotspots of biological activity (Stadler and Michalzik 1998; Osono et al. 2006; Kopacek et al. 2009). They contribute to the *production of materials*, including pollen, plant fragments, bird and insect excretions, and microorganisms, all of which can alter the quality of water as it passes through the canopy (Arango et al. In press). Some of this activity can be seasonal in nature, potentially influencing water chemistry beneath the canopy differently throughout the year (Draaijers et al. 1996; Decina et al. 2018).

As discussed by Carlyle-Moses et al. (Chap. 17, this volume), urban trees intercept and redistribute precipitation over the ground. The water that flows down tree stems (*stemflow*) and percolates through the canopy (*throughfall*) picks up materials from atmospheric deposition (wet, fog, dry) and the canopy itself and then transports these materials to the ground. As a result, the chemical composition of water reaching the ground below urban tree canopies is often different from that of rainfall. In the same way that streams and rivers act as integrators of watershed processes (Peterson et al. 2001), stemflow and throughfall act as integrators of the atmospheric and canopy processes described above (Fig. 18.2).

Because it is difficult to separate the importance of atmospheric deposition from that of canopy and production processes, comparison of the chemical or elemental composition of wet deposition (i.e., rainfall that has not interacted with plant canopies) with that of stemflow and throughfall, or throughfall alone, is often used to infer canopy processes (e.g., Ponette-González et al. 2014, 2017). In nutrient-rich urban areas, nutrient inputs measured beneath the canopy are often greater than those measured in rainfall. In this case, fog/dry deposition to the canopy and canopy leaching are assumed to be greater than any potential nutrient uptake by the canopy itself. Sometimes, however, nutrient inputs measured beneath the canopy are lower than those measured in rainfall (as often occurs in more nutrient-limited rural areas; Lajtha et al. 1995; Ponette-González et al. 2010; Templer et al. 2015a). In this case, the reverse is true, and nutrient uptake in the canopy is assumed to be greater than the sum of fog/dry deposition to the canopy and canopy leaching.

18.3 Controls on Atmospheric Deposition to Urban Tree Canopies

Atmospheric deposition is perhaps the most well studied of the processes influencing water quality below tree canopies. Although atmospheric deposition in urban areas is generally not measured by organized networks (but see, e.g., the East Asia Acid Deposition Monitoring Network (EANET)), field studies show that cities receive higher rates of wet and dry atmospheric deposition than nearby rural areas (Rao et al. 2014; Lewis et al. 2015; Khan et al. 2018). In many places, this urban enhancement is a factor of 2–4 times higher. However, in some cities, such as New Delhi, India, urban atmospheric deposition rates can be almost 20 times higher than in nearby rural areas (Tiwari et al. 2016). Urban-to-rural gradient studies also show that deposition rates often increase predictably along a transect from rural areas into the city (Lovett et al. 2000; Fang et al. 2011; Rao et al. 2014; Du et al. 2015; Lewis et al. 2015).

Although more limited than in rural areas, research conducted within urban landscapes indicates that the magnitude of deposition to urban trees and forests varies across space (King et al. 2014; Decina et al. 2017; Ponette-González et al. 2017), resulting in "hot" and "cold" depositional spots. In Boston, Massachusetts, deposition rates varied more than threefold over a distance of just a few kilometers, with some rates equivalent to those measured in nearby rural areas (Decina et al. 2017). This intra-urban variability is the result of several factors that influence the deposition process, including the number, strength, and location of urban emission sources, unique urban meteorology and transport patterns, urban form, and vegetation density and species composition (Lovett et al. 2000; Janhäll 2015; Decina et al. 2017). Thus, these factors represent important controls on the extent to which urban trees influence the composition of water delivered to the ground.

18.3.1 Urban Emission Sources of Gases and Particulate Matter

Urban areas contain a multitude of local atmospheric emission sources. The haze often found in large, polluted cities is a product of these emissions, which are largely the result of human activities. For example, combustion from power plants and factories emit N and sulfur (S) oxides directly into the atmosphere (hence, they are referred to as primary pollutants). In the USA, electricity generation and industrial and manufacturing processes are the primary source of sulfur dioxide (SO_2) to the atmosphere (Weathers et al. 2006). Vehicles also produce nitrogenous gases, such as N oxides (NO_x; Weathers et al. 2006). Urban sources of ammonia (NH_3) in cities include vehicles and wastewater treatment plants (Reche et al. 2015), as well as residential and municipal fertilizers (Decina et al. 2017), which can be applied at excessive rates on urban green spaces (Morton et al. 1988; Law et al. 2004), such as lawns, parks, and golf courses. After emission, these gases can deposit onto canopy surfaces in their gaseous form or as particulate matter (Lovett and Lindberg 1993).

Particulate matter (PM) is a growing problem in urban areas, and the sources of this PM are both abundant and numerous (Fuzzi et al. 2015). Primary particles are directly released into the atmosphere by wind or combustion processes and other human activities. For example, in industrialized countries, diesel-powered vehicles represent the dominant source of black carbon (BC) to the atmosphere (Bond et al. 2013). Phosphorus (P) is also present in soil particles, fertilizer, and pet waste (Michopoulos et al. 2007a; Berretta and Sansalone 2011; Hochmuth et al. 2012). Dust from roads, construction sites, landfills, and other urban industrial processes provide an enormous source of particles to the atmosphere, which contain P and other elements like calcium (Ca) and K (Wang et al. 2005). When lifted by wind, these particles can land on canopy surfaces and become incorporated into stemflow and throughfall.

In addition to these primary gaseous and particle emissions, gases can react with each other and other urban matter to create particles (secondary pollutants). Examples of secondary pollutants include acid rain (a form of wet deposition) that forms when SO_2 and nitrogen oxides (NO_x) react with water, oxygen, oxidants, and ozone (O_3) that forms when NO_x reacts with volatile organic carbons in the presence of sunlight.

18.3.2 Urban Form

Urban form – the spatial organization and arrangement of people, buildings, infrastructure, and green space (Bereitschaft and Debbage 2013) – affects atmospheric concentrations and deposition through its influence on urban emission sources, transport and dispersion, and vegetation exposure (Oke 1989; Cherin et al. 2015). As such, it is a key factor driving variability in atmospheric deposition rates within cities.

Urban form determines travel behavior (Ewing and Cervero 2010) and energy use (Ko and Radke 2014) and thus influences the location and strength of emission sources (McCarty and Kaza 2015). For example, Bereitschaft and Debbage (2013) found that on-road fine PM emissions were higher in low- compared to high-density residential areas. In cities, location relative to emission source is a particularly important factor affecting both atmospheric concentrations and deposition. Urban form also influences pollutant transport and dispersion. For example, urban street canyons – streets flanked by buildings on both sides – can reduce the dispersion of atmospheric pollutants (Boogaard et al. 2011). Here, deposition rates to vegetation may be enhanced as a result of the higher atmospheric concentrations and surface-to-volume ratio (Pugh et al. 2012). Finally, vegetation exposure to the atmosphere is in part the product of urban form. Due to the patchy nature of urban ecosystems, vegetation edges (here defined as structural discontinuities or transitions that expose vegetation to prevailing winds; after Weathers et al. 2000) are common features in cities. The abrupt change in vegetation between a closely spaced row of street trees and a wide road is one example. The outer canopy of a single open-grown tree in a parking lot or of a small cluster of trees in a park may also function as an edge. Thus, many urban trees (e.g., tree depicted in Fig. 18.2) act as "edge" trees – they experience higher than average exposure to the atmosphere (Weathers et al. 2001) and experience physical phenomena and influence biogeochemical cycling in different ways than trees not exposed to edges (Reinmann and Hutrya 2017; Smith et al. 2018).

18.3.3 Vegetation

Vegetation affects atmospheric deposition at landscape, stand/tree, and leaf scales. Wet deposition, particularly deposition in rain, is more influenced by the factors that control precipitation than by vegetation, whereas vegetation represents a major control on fog and dry deposition. Reviews, or partial reviews, on the influence of forest vegetation on fog and dry deposition have been published elsewhere (e.g., Lovett 1994; Wesely and Hicks 2000; De Schrijver et al. 2008; Ponette-González et al. 2016a). Therefore, here we limit our discussion of dry deposition to recent studies, especially concerning particulate deposition, to urban vegetation (see Janhäll 2015 for a recent review).

At the landscape scale, vegetation exposure strongly influences deposition patterns. As discussed above, vegetation edges (e.g., vegetation barriers) are prominent features in urban landscapes and should experience higher than average deposition rates. However, the complex arrangement of vegetation, buildings, and infrastructure may obscure these effects. For example, in Belgium, Hofman et al. (2013) compared PM deposition to edge and interior street canyon trees and found that the effect of tree position was dependent on azimuth (and by inference wind flow). This kind of interaction among vegetation, meteorology, and emission source explains why studies examining the effect of vegetation barriers on PM deposition produce varying results, with some studies showing the expected decrease with distance from a road (Mori et al. 2015) and others showing no such effect (Hofman et al. 2013).

At the stand or individual tree scale, vertical structure (e.g., tree height, vertical layering) and species composition affect deposition to trees and forests in cities. In an urban forest in Japan, Hara et al. (2014) found that the mass of black carbon (BC) deposited to tree leaves increased with canopy height. In contrast, PM deposition to leaf surfaces was found to decrease with canopy height in an urban street canyon. In the latter study, the authors suggested that high pollutant concentrations from street level (i.e., mobile) sources near the bottom of the canopy were more important than faster wind speeds at the top of the canopy in governing the relationship between PM deposition and canopy height. Although in theory dry deposition inputs should increase with canopy leaf area until a saturation point where canopies become too dense to efficiently filter particles (e.g., Lovett and Reiners 1986), it has been difficult to demonstrate this relationship empirically. Hofman et al. (2014) coupled measurements of leaf-scale PM with LiDAR to evaluate the effect of leaf density, among other variables, on PM deposition rates. They found that PM deposition decreased with increasing leaf density, but the relationship was dependent on azimuth. With respect to species composition, it is well established that dry deposition to coniferous species is higher than to broadleaved tree species (Cai et al. 2017;

Xu et al. 2018). For instance, in a study of 22 tree and 25 shrub species commonly planted in urban areas of Norway, some species, such as Scots pine (*Pinus sylvestris*), accumulated more coarse and fine particles compared to other species (Saebø et al. 2012). This is, in part, due to their tall stature and large leaf area index (LAI), two tree-scale characteristics that promote PM deposition (Pryor et al. 2014).

At the leaf scale, several leaf traits, including leaf micro-roughness (e.g., hairs, scales, glands, furrows, veins) and waxiness have been shown, or hypothesized, to affect particle deposition rates. In a well-known wind tunnel experiment, Beckett et al. (2000) demonstrated that conifer needles have higher particle capture efficiency than the flat leaves of broadleaved trees. This was attributed to the smaller boundary layer thickness of needles compared to flat leaves. Multiple studies also point to the importance of trichomes (hairlike projections; Fig. 18.3) in enhancing dry deposition (Dzierżanowski et al. 2011; Saebø et al. 2012; Cai et al. 2017; Liu et al. 2018). Additionally, the same studies show that waxy leaves capture more PM than less waxy leaves; the relative importance of wax quantity versus wax chemical composition in promoting capture, however, remains unclear. In sum, the unique characteristics of urban ecosystems work in concert to create highly variable spatial and temporal patterns of urban atmospheric deposition.

Fig. 18.3 Leaf trichomes on the abaxial (lower) surface of a *Quercus virginiana* (live oak) leaf in the City of Denton, Texas. Photo acquired with an FEI Quanta 200 ESEM at magnification 316x. This work was performed in part at the University of North Texas's Materials Research Facility: A shared research facility for multidimensional fabrication and characterization. (Photo credit: Jenna E. Rindy)

18.4 Movement of Major Elements To and Through Urban Tree Canopies

As reviewed earlier, materials coming through the urban tree canopy are influenced by atmospheric deposition, canopy processes, and production of new materials. Both deposited and canopy-derived materials interact with the hydrological cycle when they wash through tree canopies to the ground and enter soil water, groundwater, and/or runoff into urban waterways.

Below we review the sources of, and canopy processes associated with, major elements required by ecosystems for growth and development (nutrients) but that, in excess, can have negative effects on ecosystems (pollutants).

18.4.1 Nitrogen (N)

The urban canopy often augments N in precipitation by acting both as a surface for the deposition of gaseous and particulate N from the atmosphere and as a source of canopy-derived N (Lovett and Lindberg 1993). Deposition of nitrogenous gases, and the aerosol products that result from their secondary reactions, is a major source of N to the canopy and to stemflow and throughfall. Nitric acid (HNO_3) gas is produced in urban areas by the reaction of NO_x in vehicle exhaust with hydroxyl radicals in the urban atmosphere. This gas has a high deposition velocity (the speed at which a gas deposits to a surface), and thus it deposits rapidly on canopy surfaces (Lovett 1994). The other major gaseous component of urban N deposition is NH_3 gas, which also deposits quickly close to its emission source on the canopy surface or into leaf mesophyll through stomatal uptake (Wesely and Hicks 2000). While there are numerous sources of NH_3 in urban areas, a large, growing source of NH_3 comes from vehicle exhaust (Li et al. 2016; Sun et al. 2017). Over-reduction of NO_x in catalytic converters in gasoline vehicles, as well as the production of NH_3 from selective catalytic reduction (SCR) in diesel cars and trucks, has become a substantial source of atmospheric N (Li et al. 2016). Both HNO_3 and NH_3 can deposit onto canopy surfaces or interact with precipitation producing the soluble ions of nitrate (NO_3^-) and ammonium (NH_4^+), respectively (Fig. 18.4). The reaction of HNO_3 and NH_3 in the urban atmosphere also leads to the formation of particulate ammonium nitrate (NH_4NO_3), which is intercepted by canopy surfaces and, upon contact with water on leaf cuticles, separates into its soluble NO_3^- and NH_4^+ ionic components. Soluble NO_3^- can also be formed from the reaction of HNO_3 with Ca from urban dusts and soils present on leaf surfaces (Michopoulos et al. 2007a). Unlike HNO_3, NO_2 gas deposits mainly into leaf mesophyll through stomata and does not make up a large fraction of canopy-intercepted N (Lovett 1994).

Urban trees also represent sources of N. Dissolved organic N (DON) has been shown to make up roughly one third of N in throughfall measured in various cities (Araujo et al. 2015; de Souza et al. 2015; Decina et al. 2018). However, it remains

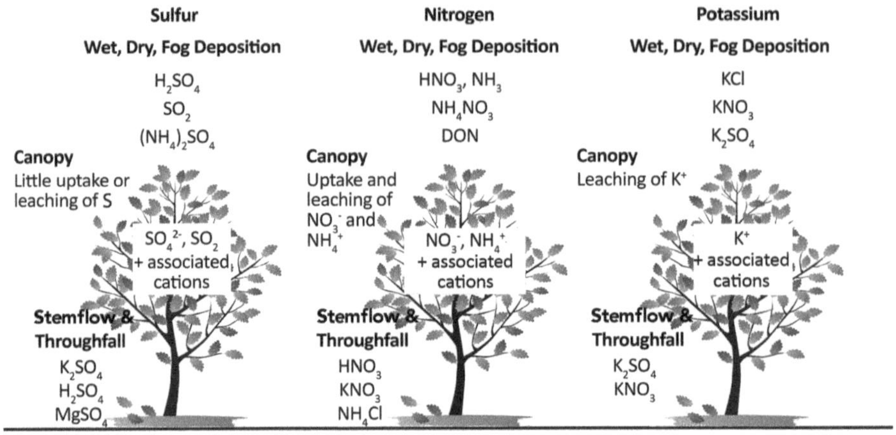

Fig. 18.4 Interactions of three elements with plant canopies: (a) sulfur (S), a conservative tracer of total (wet + fog + dry) atmospheric deposition; (b) nitrogen (N), an element subject to multiple canopy transformations; and (c) potassium (K), a mobile element in canopy throughfall. (From Ponette-González et al. 2016a)

unclear what proportion of organic N species is deposited to the canopy versus what proportion is produced by the tree canopy itself (Decina et al. 2018). The urban tree canopy also produces pollen and biological matter which contain N (Ferm 1993). Birds, insects, and other organisms are concentrated in urban trees and produce wastes (e.g., frass, honeydew, feces) rich in N. These materials can be directly deposited beneath tree canopies or can dissolve in water during precipitation events (Stadler and Michalzik 1998; Osono et al. 2006; Kopacek et al. 2009).

Comparisons of wet N deposition and throughfall N have shown that canopy uptake occurs routinely in rural trees (Lajtha et al. 1995; Ponette-González et al. 2010; Templer et al. 2015a), providing an additional source of N in N-limited environments. In contrast, in urban areas experiencing N pollution, numerous studies indicate that the amount of N in water collected beneath urban tree canopies is typically much higher compared to rainfall deposition measured in areas without canopy cover (Lovett et al. 2000; Bettez and Groffman 2013; Decina et al. 2018). In other words, in cities, atmospheric N deposition and canopy leaching outweigh canopy uptake. Therefore, urban trees often act as net *sources* of N to the ground below (Michopoulos et al. 2007a).

18.4.2 *Phosphorus (P)*

Unlike N, atmospheric P deposited to the tree canopy is almost completely particulate in nature (Newman 1995) as there are no common atmospheric gases containing P. In urban areas, P in deposition comes primarily from fertilizer, dust, pet waste, and biological materials, such as pollen. Additionally, fossil fuel combustion can produce particles containing P (Wang et al. 2014). Biomass burning, which can be an important source of energy in many countries, also produces P that can be dry-deposited to canopies and incorporated into throughfall (Ponette-González et al. 2016b).

In rural areas, wet atmospheric deposition of P is generally low (Ponette-González et al. 2018), whereas in urban areas, rates are comparatively higher (Eisenreich et al. 1977; Hou et al. 2012). Studies that have quantified P beneath tree canopies are, however, much more limited. In rural areas, some research shows that throughfall P can be much higher than P in rainfall (Das et al. 2011; Lequy et al. 2014), as much as 33 times greater (Ponette-González et al. 2016b). While P leaching from canopy surfaces is thought to be limited, there is some evidence to suggest that P can be taken up by microbes on leaf surfaces (Runyan et al. 2013). In urban areas, to our knowledge, only one study has quantified P in throughfall. That study, from Boston, Massachusetts, showed that P in throughfall was sevenfold higher compared to P in rainfall, indicating that the urban tree canopy is an important receptor surface for dry-deposited P and/or itself a source of P (Decina et al. 2018). As mentioned earlier, nutrient-rich materials, such as bird droppings, insect frass and honeydew, pollen, and other biological matter, are produced within urban tree canopies (specifically, pollen can be a substantial source of P; Draaijers et al. 1996; Arango et al. In press) High fluxes of urban P in throughfall measured in late spring may be due to a large seasonal input of pollen beneath the canopy (Draaijers et al. 1996; Decina et al. 2018). Phosphorus has also been quantified in stemflow. Although it varies by species, multiple studies show that P concentrations in stemflow can be up to six times higher than P concentrations in rainfall (Stevens et al. 1988; Neal et al. 2003).

18.4.3 Sulfur (S)

Fossil fuel combustion is the dominant source of S in urban areas, especially in places with a preponderance of coal burning (Quan et al. 2008). Emitted sulfur dioxide (SO_2) and sulfuric acid (H_2SO_4) react with urban dusts and NH_3 to form particulate calcium and ammonium sulfates ($CaSO_4$, $(NH_4)_2SO4$; Fig. 18.4), and SO_4^{2-} is emitted directly as well in gypsum production (Quan et al. 2008). The conversion of SO_2 to H_2SO_4 is catalyzed by the presence of metal cations, which are abundant in urban areas (Michopoulos et al. 2007a). Marine aerosols in coastal cities, like Los Angeles, can also be a source of S to the atmosphere (Fenn and Kiefer 1999), as can volcanic activity in cities near volcanoes (e.g., near Mexico City, Fenn et al. 1999). Canopy surfaces play a large part in S deposition in urban areas. Sulfur dioxide can deposit within leaf stomata under dry conditions, but its deposition velocity is enhanced when canopy surfaces are wet. The efficiency with which particles containing SO_4^{2-} deposit to surfaces is also dependent on their size, with fine particles depositing less efficiently (Lovett 1994).

In stemflow and/or throughfall studies, SO_4^{2-} is often used as a conservative chemical tracer to estimate rates of total (wet + fog + dry) atmospheric deposition, because interactions (i.e., uptake and leaching) between the SO_4^{2-} and plant canopies are considered minimal relative to total deposition (Lindberg and Lovett 1992; Fig. 18.4). Thus, together, measurements of wet deposition and total stemflow and throughfall inputs of SO_4^{2-} can be used to evaluate the importance of fog plus dry deposition to the canopy. In studies where S in urban (or urban-influenced) stemflow and/or throughfall has been measured, it is consistently higher than in rural stemflow and/or throughfall (Fenn and Kiefer 1999; Fenn et al. 1999; Chiwa et al. 2003; Aikawa et al. 2006; Juknys et al. 2007; Ponette-González et al. 2010) and wet deposition (Aikawa et al. 2006; Juknys et al. 2007; Michopoulos et al. 2007a; Quan et al. 2008; Ponette-González et al. 2010). A few studies (e.g., Fenn and Kiefer 1999) document foliar uptake of S in the tree canopy, as evidenced by lower S in throughfall than in precipitation and elevated foliar S content. This uptake may occur in conditions where SO_2 deposited through leaf stomata is not released into throughfall (Lindberg and Garten Jr. 1988).

18.4.4 Base Cations

While less abundant in the atmosphere than elements such as N, base cations, including Ca, Mg, and K, have numerous urban sources. Dusts from calcareous building materials and soils settling on canopy surfaces are a source of Ca (Liu et al. 2005). Soils also contain Mg, a nutrient which can also be of marine origin in coastal cities, produced by anthropogenic activity (Michopoulos et al. 2007b), and leached by the tree canopy itself (Michopoulos et al. 2007a; Michopoulos et al. 2007b). Dust, fire, and pollen represent sources of K to the atmosphere and plant canopies (Balestrini et al. 2007; Ponette-González et al. 2018), along with leaching from the canopy itself when acidic gases deposit on leaf surfaces and lower the pH of water in the tree canopy (Juknys et al. 2007; Michopoulos et al. 2007b).

In numerous studies, Ca, Mg, and K all display elevated concentrations in precipitation in urban areas as compared to rural areas and are elevated in stemflow and/or throughfall as compared to wet deposition (Takagi et al. 1997; Chiwa et al. 2003; Aikawa et al. 2006; Juknys et al. 2007; Michopoulos et al. 2007a; Michopoulos et al. 2007b; Quan et al. 2008, Schooling et al. 2017). In general, these studies have not suggested uptake of these base cations by the urban tree canopy, though Juknys et al. (2007) noted a reduction in Ca in throughfall as compared to wet deposition in a suburban forest. In addition to elevated deposition in urban areas and beneath trees, urban airborne Ca acts as an “urban scrubber,” reacting with acidic gases in the urban atmosphere to create particles which contribute to elevated urban PM pollution, increased deposition of S, and more alkaline conditions in urban areas (Lovett et al. 2000; Michopoulos et al. 2007b; Quan et al. 2008).

18.5 Movement of Particulate Matter (PM) To and Through Urban Tree Canopies

Over the last decade, the urban tree canopy has received considerable attention for its potential to remove PM from the atmosphere by dry deposition, in turn improving air quality (Nowak et al. 2013a; Nowak et al. 2013b; Popek et al. 2013; Saebø et al. 2012; Mori et al. 2015; Xu et al. 2018). A major component of the urban atmosphere, PM is generally divided into categories based on particle size: coarse, fine, and ultrafine. Coarse particles, known as PM_{10}, are less than 10 μm in diameter. The main sources of these particles are vehicular, industrial, and crustal (Viana et al. 2008). Fine PM, or $PM_{2.5}$, refers to particles less than 2.5 μm in diameter. These particles are largely composed of organic and elemental C and secondary aerosols such as sulfates, nitrates, and ammonium (Bell et al. 2007). The sources of $PM_{2.5}$ include fossil fuel combustion, agricultural emissions, solvents, and biomass burning (Monks et al. 2009). Ultrafine PM is less than 0.1 μm in diameter and often agglomerates upon emission into larger fine particles. Ultrafine PM is mostly derived from combustion and is found in high concentrations near busy highways (BAAQMD 2012).

As discussed previously, particles deposit to tree canopies through the mechanisms of wet deposition (precipitation) and dry deposition (direct particle deposition). Upon interaction with precipitation, some components of PM dissolve in water, thereby altering the chemical composition of water as it passes through the tree canopy, while other PM components remain in particulate form. Here we focus on studies that have examined the insoluble fraction of PM (i.e., operationally defined as the fraction that cannot pass through a 0.45-μm-diameter filter), which can affect biogeochemical cycling when it enters the soil or surface waters.

Particulate matter deposition from the atmosphere to urban tree canopies has been quantified using a variety of approaches: field, experimental, and modeling (Table 18.1). These studies indicate that urban trees can capture considerable amounts of PM and that, in general, PM deposition to urban vegetation is higher than to rural vegetation (Saebø et al. 2012; Przybysz et al. 2014). However, differences exist among urban areas. For example, in a 3-year study conducted at an urban site in Poland, an average of 17 $\mu g/m^2$ of PM accumulated on vegetation (Popek et al. 2013) compared to 62 $\mu g/m^2$ in Beijing, China (Mo et al. 2015). These studies also highlight species-specific differences in PM deposition within a single

Table 18.1 Recent publications on particulate matter deposition to urban vegetation

Urban area	Species	Method	References
Wind tunnel	*Cupressocyparis leylandii*, *Pinus nigra* var. *maritima*, *Sorbus intermedia*, *Acer campestre*, *Populus deltoides x trichocarpa*	Foliar water rinse	Beckett et al. (2000)
Warsaw, Poland City center	*Acer campestre*, *Fraxinus excelsior*, *Platanus x hispanica*, *Tilia cordata*, *Forsythia x intermedia*, *Hedera helix*, *Physocarpus opulifolius*, *Spiraea japonica*	Foliar water and chloroform rinse	Dzierżanowski et al. (2011)
Tokyo, Japan Urban forest	*Quercus serrata*, *Cryptomeria japonica*	Foliar water and chloroform rinse, modeling	Hara et al. (2014)
Antwerp, Belgium Street canyon	*Platanus xacerifolia*	Foliar water rinse	Hofman et al. (2014)
Wind tunnel	*Ilex cornuta*, *Quercus alba*, *Magnolia grandiflora*, *Lonicera fragrantissima*	Wind tunnel experiment	Huang et al. (2013)
Beijing, China	24 tree, 11 shrub species	Foliar water and chloroform rinse, SEM	Mo et al. (2015)
Stavanger, Norway Busy roadway	*Picea sitchensis*, *Pinus sylvestris*	Foliar water and chloroform rinse	Mori et al. (2015)
Pęchcin, Central Poland Near roadway	7 tree, 6 shrub species	Foliar water and chloroform rinse	Popek et al. (2013)
Stavanger, Norway, 2 cities, 1 rural area	*Taxus baccata*, *Pinus sylvestris*, *Hedera helix*	Foliar water and chloroform rinse	Przybysz et al. (2014)
Wind tunnel	*Quercus robur*	Wind tunnel	Reinap et al. (2009)
Stavanger, Norway, and urban Poland	22 tree, 25 shrub species	Foliar water and chloroform rinse	Saebø et al. (2012)
Terni, Italy	*Quercus ilex*	Foliar water and chloroform rinse	Sgrigna et al. (2015)

urban location. Mo et al. (2015) found that PM deposition to different plant species across Beijing Forestry University varied from less than 20 μg/m^2 to as much as 161 μg/m^2 of PM. Just as with dissolved nutrient deposition, PM deposition can vary across short distances within urban areas.

Estimates of PM deposition to urban trees and forests at the city scale have been obtained using various models (Yang et al. 2005; Nowak et al. 2006; Schaubroeck et al. 2014). A 3-year study conducted by McPherson et al. (1994) in Chicago's urban forest showed an ~234 tons of PM deposited to trees in the year 1991. In the city of Beijing, China, an urban forest model with multiple native trees estimated an annual deposition of PM_{10} totaling 772 tons (Yang et al. 2005). Nowak et al. (2006) conducted a meteorological modeling study to estimate the amount of gaseous and particulate air pollution deposition to urban forests across the conterminous USA. This study resulted in a total of approximately 711,000 metric tons of pollutants removed annually, with up to 3.5% of total emitted PM removed in individual cities. Together, these field, experimental, and modeling studies underscore the role that urban trees and forests play in serving as PM receptor sites at local to city scales.

After deposition to tree canopies, particles can reside on canopy surfaces or become embedded within the leaf epicuticular wax layer. The fraction retained within leaf waxes remains in the canopy until the wax layer decomposes or until leaf fall, and thus it represents a form of canopy uptake. An experimental study of 13 urban woody species in Poland revealed that, on average, ~60% of dry-deposited particulate matter was washable from leaves, while the remainder (40%) was immobilized in leaf waxes (Popek et al. 2013). This study suggests that in urban areas, temporary PM uptake (or retention) by the canopy may be quantitatively important and therefore worth consideration.

Particulate matter also moves to soils in stemflow and throughfall (Levia et al. 2013; Lequy et al. 2014). Although mostly conducted in forests, studies show that PM delivery from the canopy to the soil in throughfall can be considerable, amounting to hundreds of kilograms per hectare per year (Lequy et al. 2014). Other studies have quantified particulate inputs in throughfall in the context of insect infestations, showing their contribution to organic carbon and nitrogen fluxes to the forest floor (Michalzik and Stadler 2005; Le Mellec et al. 2011). Stemflow and throughfall PM inputs are influenced by myriad factors, including canopy cover and season. For example, a study by Levia et al. (2013) found that more particulates reach the forest floor during leafless as compared to leaf-on periods, but the average diameter of the particles was larger during the leaf-on period. Taken together, these studies highlight potential, but overlooked, effects of urban tree canopies on particulate fluxes to the soil, with impacts to groundwater and surface water both within and downstream of urban areas.

18.6 Future Opportunities and Directions

The urban tree canopy has a clear effect on atmospheric deposition and precipitation chemistry, altering the rate and composition of nutrient and pollutant inputs to urban ground surfaces and in turn impacting urban water quality. While researchers are beginning to study cities in and of themselves (rather than solely in comparison to rural areas), there is still much work to be done.

18.6.1 Trade-Offs of the Urban Tree Canopy to the Hydrologic Cycle

One area of consideration concerns quantifying the trade-offs of planting and maintaining urban trees, a topic debated both within and outside the scientific community (Vos et al. 2013; Endreny 2018). Apart from considerations outside the scope of this chapter, such as human health, the urban heat island, C sequestration, and emission of volatile organic compounds (e.g., isoprene), planting urban trees has both potential positive and negative effects on urban water quality. On the one hand, uptake of nutrients and pollutants by tree canopies and roots mitigates elevated flows of these materials into waterways (Ponette-González et al. 2017). However, concentration and augmentation of materials below tree canopies, including nutrients, pollutants, and leaf litter (Janke et al. 2017), can cause trees to enhance the flow of nutrients through the urban ecosystem (Hobbie et al. 2017). Though much of the urban tree litter (i.e., leaves and branches) is picked up and removed by municipal collection programs (Templer et al. 2015b; Hobbie et al. 2017), some fraction inevitably enters waterways, potentially influencing urban water quality. When considering that a significant amount of urban tree canopy overhangs non-permeable ground surfaces like streets and sidewalks, trees not only serve to augment below-canopy inputs to the ground but also to hasten its entry into urban waterways through runoff (Decina et al. 2018; Fig. 18.5). Further research should account for the optimal location, species, density, and substrate for urban tree planting to allow maximization of services and reduction of disservices to urban soils and waters.

18.6.2 Impacts on Urban Biogeochemical Cycling

While urban deposition studies have grown in prevalence in recent years, the coupling of these studies with biogeochemical effects in soil and water is still relatively uncommon. Of 174 urban N deposition studies, fewer than 20 have looked at urban biogeochemical cycling in tandem (Decina et al. in review). While some of these studies have noted biogeochemical effects related to deposition (Fang et al. 2011; Huang et al. 2012), others have not found changes in biogeochemical cycling when rates of deposition changed both along an urban-to-rural gradient and within the city itself (Rao et al. 2014; Decina et al. 2017). These results suggest that there are still many unknowns regarding the ability of urban trees and soils to process the excess nutrient inputs they receive, a factor which has direct impacts on water quality both within and downstream of urban areas.

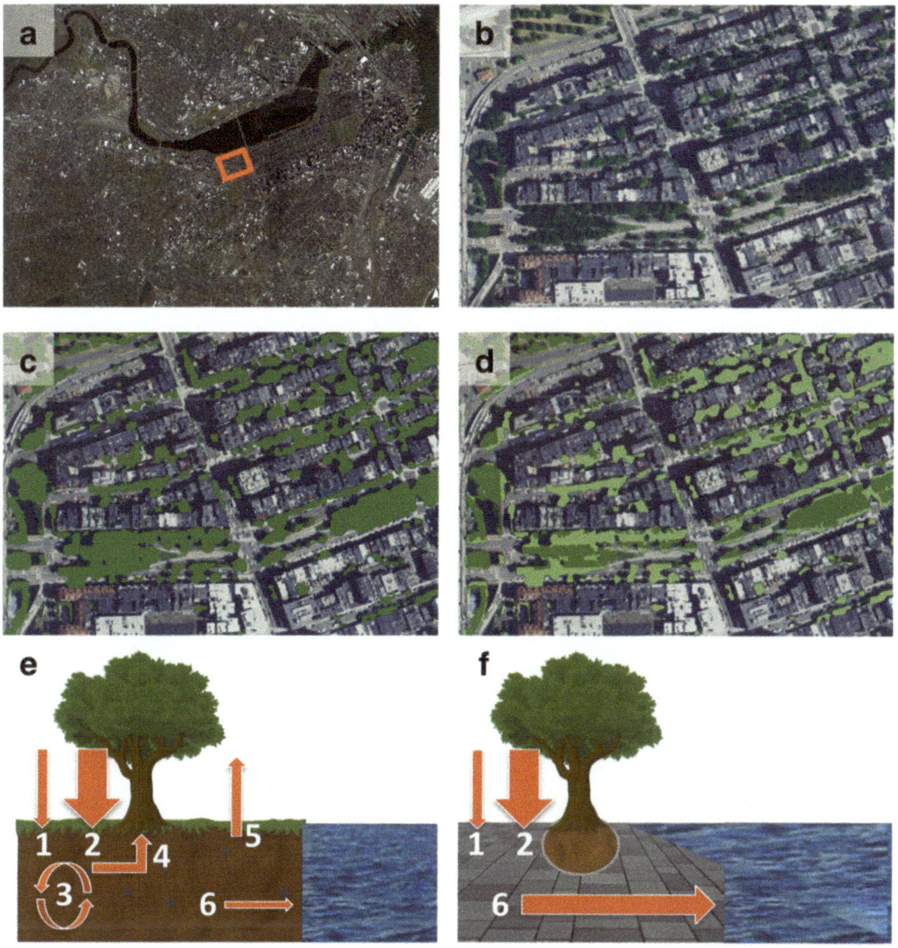

Fig. 18.5 Conceptual diagram hypothesizing the relationship between atmospheric deposition, the urban tree canopy, pavement, and waterways. (**a**) Aerial image with red box around area whose aerial image is shown in panels (**b**), (**c**), and (**d**). (**b**) Magnified aerial image. (**c**) Aerial image from panel (**b**) with all tree canopy colored dark green. (**d**) Aerial image from panel (**c**) with tree canopy above soil colored dark green and above paved surfaces colored light green. (**e**) The potential fate of atmospheric inputs above soil. (**f**) The potential fate of atmospheric inputs above pavement. Arrows represent the following fluxes: (1) wet deposition, (2) stemflow and throughfall, (3) soil nutrient cycling, (4) uptake by plant roots, (5) gas loss to the atmosphere (e.g., denitrification), and (6) flux to waterways through runoff. Inputs deposited to pavement have a higher potential to runoff into nearby waterways, while a portion of inputs deposited to soil are cycled by organisms, with a smaller fraction likely lost to waterways. (From Decina et al. 2018)

References

Aikawa M, Hiraki T, Tamaki M (2006) Comparative field study on precipitation, throughfall, stemflow, fog water, and atmospheric aerosol and gases at urban and rural sites in Japan. Sci Total Environ 366:275–285. https://doi.org/10.1016/j.scitotenv.2005.06.027

Arango C, Ponette-González AG, Neziri* I, Bailey J (In press) Western spruce budworm effects on throughfall C, N, P fluxes and soil nutrient status in the Pacific Northwest. Canadian Journal of Forest Research

Araujo TG, Souza MFL, De Mello WZ, Da Silva DML (2015) Bulk atmospheric deposition of major ions and dissolved organic nitrogen in the lower course of a tropical river basin, southern Bahia, Brazil. J Braz Chem Soc 26:1692–1701. https://doi.org/10.5935/0103-5053.20150143

Balestrini R, Arisci S, Cristina M, Mosello R, Rogora M, Tagliaferri A (2007) Dry deposition of particles and canopy exchange: comparison of wet, bulk and throughfall deposition at five forest sites in Italy. Atmos Environ 41:745–756. https://doi.org/10.1016/j.atmosenv.2006.09.002

Bay Area Air Quality Management District (BAAQMD) (2012) Summary of PM Report. Retrieved from http://www.baaqmd.gov/~/media/files/planning-and-research/plans/pm-planning/summary-of-pm-report_draft_aug-23.pdf

Beckett KP, Freer-Smith PH, Taylor G (2000) Particle pollution capture by urban trees: effect of species and windspeed. Glob Change Biol 6:995–1003. https://doi.org/10.1046/j.1365-2486.2000.00376.x

Bell ML, Dominici F, Ebisu K, Zeger SL, Samet JM (2007) Spatial and temporal variation in PM2.5 chemical composition in the United States for health effects studies. Environ Health Perspect 115:989–995. https://doi.org/10.1289/ehp.9621

Bereitschaft B, Debbage K (2013) Urban form, air pollution, and CO_2 emissions in large U.S. metropolitan areas. *Prof Geogr 65*:612–635. https://doi.org/10.1080/00330124.2013.799991

Berland A, Shiflett SA, Shuster WD, Garmestani AS, Goddard HC, Herrmann DL et al (2017) The role of trees in urban stormwater management. Landsc Urban Plan 162:167–177. https://doi.org/10.1016/j.landurbplan.2017.02.017

Berretta C, Sansalone J (2011) Speciation and transport of phosphorus in source area rainfall-runoff. Water Air Soil Pollut 222:351–365. https://doi.org/10.1007/s11270-011-0829-2

Bettez ND, Groffman PM (2013) Nitrogen deposition in and near an urban ecosystem. Environ Sci Technol 47:6047–6051. https://doi.org/10.1021/es400664b

Bolund P, Hunhammar S (1999) Ecosystem services in urban areas. Ecol Econ 29:293–301. https://doi.org/10.1016/S0921-8009(99)00013-0

Bond TC, Doherty SJ, Fahey DW, Forster PM, Berntsen T, DeAngelo BJ et al (2013) Bounding the role of black carbon in the climate system: A scientific assessment. J Geophys Res Atmos 118:5380–5552. https://doi.org/10.1002/jgrd.50171

Boogaard H, Kos GP, Weijers EP, Janssen NA, Fischer PH, van der Zee SC et al (2011) Contrast in air pollution components between major streets and background locations: particulate matter mass, black carbon, elemental composition, nitrogen oxide and ultrafine particle number. Atmos Environ 45:650–658. https://doi.org/10.1016/j.atmosenv.2010.10.033

Cai M, Xin Z, Yu X (2017) Spatio-temporal variations in PM leaf deposition: A meta-analysis. Environ Pollut 231:207–218. https://doi.org/10.1016/j.envpol.2017.07.105

Cherin N, Roustan Y, Musson-Genon L, Seigneur C (2015) Modelling atmospheric dry deposition in urban areas using an urban canopy approach. Geosci Model Dev 8:893–910. https://doi.org/10.5194/gmd-8-893-2015

Chiwa M, Oshiro N, Miyake T, Nakatani N, Kimura N, Yuhara T et al (2003) Dry deposition washoff and dew on the surfaces of pine foliage on the urban- and mountain-facing sides of Mt. Gokurakuji, western Japan. Atmos Environ 37:327–337. https://doi.org/10.1016/S1352-2310(02)00889-0

Das R, Lawrence D, Odorico PD, Delonge M (2011) Impact of land use change on atmospheric P inputs in a tropical dry forest. J Geophys Res 116:1–9. https://doi.org/10.1029/2010JG001403

De Schrijver A, Staelens J, Wuyts K, Van Hoydonck G, Janssen N, Mertens J et al (2008) Effect of vegetation type on throughfall deposition and seepage flux. Environ Pollut 153:295–303. https://doi.org/10.1016/j.envpol.2007.08.025
Decina SM, Templer PH, Hutyra LR, Gately CK, Rao P (2017) Variability, drivers, and effects of atmospheric nitrogen inputs across an urban area: emerging patterns among human activities, the atmosphere, and soils. Sci Total Environ 609:1524–1534. https://doi.org/10.1016/j.scitotenv.2017.07.166
Decina SM, Templer PH, Hutyra LR (2018) Atmospheric inputs of nitrogen, carbon, and phosphorus across an urban area: unaccounted fluxes and canopy influences. Earth's Future 6:134–148. https://doi.org/10.1002/2017EF000653
Decina SM, Templer PH, Hutyra LR (in review) Human alteration of the nitrogen cycle in urban ecosystems: a data synthesis
de Souza PA, Ponette-González AG, de Mello WZ, Weathers KC, Santos IA (2015) Atmospheric organic and inorganic nitrogen inputs to coastal urban and montane Atlantic Forest sites in southeastern Brazil. Atmos Res 160:126–137. https://doi.org/10.1016/j.atmosres.2015.03.011
Draaijers G, Erisman J, Spranger T, Wyers GP (1996) The application of throughfall measurements for atmospheric deposition monitoring. Atmos Environ 30:3349–3361. https://doi.org/10.1016/1352-2310(96)00030-1
Du E, de Vries W, Liu X, Fang J, Galloway JN, Jiang Y (2015) Spatial boundary of urban "acid islands" in southern China. Sci Rep-UK 5:12625. https://doi.org/10.1038/srep12625
Dzierżanowski K, Popek R, Gawrońska H, Saebø A, Gawroński SW (2011) Deposition of particulate matter of different size fractions on leaf surfaces and in waxes of urban forest species. Int J Phytoremediation 13:1037–1046. https://doi.org/10.1080/15226514.2011.552929
Eisenreich SJ, Emmling PJ, Beeton AM (1977) Atmospheric loading of phosphorus and other chemicals to Lake Michigan. J Great Lakes Res 3:291–304. https://doi.org/10.1016/S0380-1330(77)72261-0
Elmqvist T, Setala H, Handel SN, van der Ploeg S, Aronson J, Blignaut JN et al (2015) Benefits of restoring ecosystem services in urban areas. Curr Opin Env Sust 14:101–108. https://doi.org/10.1016/j.cosust.2015.05.001
Endreny TA (2018) Strategically growing the urban forest will improve our world. Nat Commun 9:1160. https://doi.org/10.1038/s41467-018-03622-0
Ewing R, Cervero R (2010) Travel and the built environment. J Am Plann Assoc 76:265–294. https://doi.org/10.1080/01944361003766766
Fang Y, Yoh M, Koba K, Zhu W, Takebayashi Y, Xiao Y et al (2011) Nitrogen deposition and forest nitrogen cycling along an urban-rural transect in southern China. Glob Change Biol 17:872–885. https://doi.org/10.1111/j.1365-2486.2010.02283.x
Fenn ME, De Bauer LI, Quevedo-Nolasco A, Rodriguez-Frausto C (1999) Nitrogen and sulfur deposition and forest nutrient status in the Valley of Mexico. Water Air Soil Pollut 113:155–174. https://doi.org/10.1023/A:1005033008277
Fenn ME, Kiefer JW (1999) Throughfall deposition of nitrogen and sulfur in a Jeffrey pine forest in the San Gabriel Mountains, southern California. Environ Pollut 104:179–187. https://doi.org/10.1016/S0269-7491(98)00195-X
Ferm M (1993) Throughfall measurements of nitrogen and sulphur compounds. Int J Anal Chem 50:29–43. https://doi.org/10.1080/03067319308027581
Fuzzi S, Baltensperger U, Carslaw K, Decesari S, Denier Van Der Gon H, Facchini MC et al (2015) Particulate matter, air quality and climate: lessons learned and future needs. Atmos Chem Phys 15:8217–8299. https://doi.org/10.5194/acp-15-8217-2015
Green ML, Ponette-González AG, McCullars J, Gough L (in preparation) Nitrogen deposition effects on native prairie grasses along an urban gradient in the Southern Great Plains
Hara H, Kashiwakura T, Kitayama K, Bellingrath-Kimura SD, Yoshida T, Takayanagi M et al (2014) Foliar rinse study of atmospheric black carbon deposition to leaves of konara oak (*Quercus serrata*) stands. Atmos Environ 97:511–518. https://doi.org/10.1016/j.atmosenv.2014.04.015

Hardiman BS, Wang JA, Hutyra LR, Gately CK, Getson JM, Friedl MA (2017) Accounting for urban biogenic fluxes in regional carbon budgets. Sci Total Environ 592:366–372. https://doi.org/10.1016/j.scitotenv.2017.03.028

Hobbie SE, Finlay JC, Janke BD, Nidzgorski DA, Millet DB, Baker LA (2017) Contrasting nitrogen and phosphorus budgets in urban watersheds and implications for managing urban water pollution. Proc Natl Acad Sci U.S.A. 114:4177–4182. https://doi.org/10.1073/pnas.1618536114

Hochmuth G, Nell T, Unruh JB, Trenholm L, Sartain J (2012) Potential unintended consequences associated with urban fertilizer bans in Florida – a scientific review. Horttechnology 22:600–616. https://doi.org/10.21273/HORTTECH.22.5.600

Hofman J, Bartholomeus H, Calders K, Van Wittenberghe S, Wuyts K, Samson R (2014) On the relation between tree crown morphology and particulate matter deposition on urban tree leaves: A ground-based LiDAR approach. Atmos Environ 99:130–139. https://doi.org/10.1016/j.atmosenv.2014.09.031

Hofman J, Stokkaer I, Snauwaert L, Samson R (2013) Spatial distribution assessment of particulate matter in an urban street canyon using biomagnetic leaf monitoring of tree crown deposited particles. Environ Pollut 183:123–132. https://doi.org/10.1016/j.envpol.2012.09.015

Hou P, Ren Y, Zhang Q, Lu F, Ouyang Z, Wang X (2012) Nitrogen and phosphorus in atmospheric deposition and roof runoff. Pol J Environ Stud 21:1621–1627

Huang C, Lin M, Khlystov A, Katul G (2013) The effects of leaf area density variation on the particle collection efficiency in the size range of ultrafine particles (UFP). Environ Sci Technol 47:11607–11615. https://doi.org/10.1002/2014JD022458

Huang L, Zhu W, Ren H, Chen H, Wang J (2012) Impact of atmospheric nitrogen deposition on soil properties and herb-layer diversity in remnant forests along an urban–rural gradient in Guangzhou, southern China. Plant Ecol 213:1187–1202. https://doi.org/10.1007/s11258-012-0080-y

Janhäll S (2015) Review on urban vegetation and particle air pollution–deposition and dispersion. Atmos Environ 105:130–137. https://doi.org/10.1016/j.atmosenv.2015.01.052

Janke BD, Finlay JC, Hobbie SE (2017) Trees and streets as drivers of urban stormwater nutrient pollution. Environ Sci Technol 51:9569–9579. https://doi.org/10.1021/acs.est.7b02225

Juknys R, Zaltauskaite J, Stakenas V (2007) Ion fluxes with bulk and throughfall deposition along an urban-suburban-rural gradient. Water Air Soil Pollut 178:363–372. https://doi.org/10.1007/s11270-006-9204-0

Khan FM, Maulud KNA, Latif MT, Chung JX, Amil N, Alias A et al (2018) Physicochemical factors and their potential sources inferred from long-term rainfall measurements at an urban and a remote rural site in tropical areas. Sci Total Environ 613–614:1401–1416. https://doi.org/10.1016/j.scitotenv.2017.08.025

King KL, Johnson S, Kheirbek I, Lu JW, Matte T (2014) Differences in magnitude and spatial distribution of urban forest pollution deposition rates, air pollution emissions, and ambient neighborhood air quality in New York City. Landsc Urban Plan 128:14–22. https://doi.org/10.1016/j.landurbplan.2014.04.009

Ko Y, Radke J (2014) The effect of urban forms on residential cooling energy use in Sacramento, California. Environ Plann B 41:573–593. https://doi.org/10.1068/b12038p

Kopacek J, Turek J, Hejzlar J, Santruckova H (2009) Canopy leaching of nutrients and metals in a mountain spruce forest. Atmos Environ 43:5443–5453. https://doi.org/10.1016/j.atmosenv.2009.07.031

Lajtha K, Seely B, Valiela I (1995) Retention and leaching losses of atmospherically-derived nitrogen in the aggrading coastal watershed of Waquoit Bay, MA. Biogeochemistry 28:33–54. https://doi.org/10.1007/BF02178060

Law N, Band L, Grove M (2004) Nitrogen input from residential lawn care practices in suburban watersheds in Baltimore county, MD. J Environ Plan Manag 47:737–755. https://doi.org/10.1080/0964056042000274452

Le Mellec A, Gerold G, Michalzik B (2011) Insect herbivory, organic matter deposition and effects on belowground organic matter fluxes in a central European oak forest. Plant Soil 342:393–403. https://doi.org/10.1007/s11104-010-0704-8

Lequy E, Calvaruso C, Conil S, Turpault MP (2014) Atmospheric particulate deposition in temperate deciduous forest ecosystems: Interactions with the canopy and nutrient inputs in two beech stands of Northeastern France. Sci Total Environ 487:206–215. https://doi.org/10.1016/j.scitotenv.2014.04.028

Levia DF, Michalzik B, Bischoff S, Näthe K, Legates DR, Gruselle MC et al (2013) Measurement and modeling of diameter distributions of particulate matter in terrestrial solutions. Geophys Res Lett 40:1317–1321. https://doi.org/10.1002/grl.50305

Lewis GP, Andersen CB, Moloney TP, Muthukrishnan S (2015) Relationships between dry deposition of ions and urban land cover in the South Carolina Piedmont. Water Air Soil Pollut 226:1–15. https://doi.org/10.1007/s11270-015-2525-0

Li Y, Schichtel BA, Walker JT, Schwede DB, Chen X, Lehmann CMB et al (2016) Increasing importance of deposition of reduced nitrogen in the United States. Proc Natl Acad Sci U.S.A. 113:5874–5879. https://doi.org/10.1073/pnas.1525736113

Liu J, Cao Z, Zou S, Liu H, Hai X, Wang S et al (2018) An investigation of the leaf retention capacity, efficiency and mechanism for atmospheric particulate matter of five greening tree species in Beijing, China. Sci Total Environ 616:417–426. https://doi.org/10.1016/j.scitotenv.2017.10.314

Liu X, Zhu J, Van Espen P, Adams F, Xiao R (2005) Single particle characterization of spring and summer aerosols in Beijing: formation of composite sulfate of calcium and potassium. Atmos Environ 39:6909–6918. https://doi.org/10.1016/j.atmosenv.2005.08.007

Lindberg SE, Garten CT Jr (1988) Sources of sulphur in forest canopy throughfall. Nature 336:148–151. https://doi.org/10.1038/336148a0

Lindberg S, Lovett G (1992) Deposition and forest canopy interactions of airborne sulfur: results from the integrated forest study. Atmos Environ 26A:1477–1492. https://doi.org/10.1016/0960-1686(92)90133-6

Livesley SJ, McPherson EG, Calfapietra C (2016) The urban forest and ecosystem services: impacts on urban water, heat, and pollution cycles at the tree, street, and city scale. J Environ Qual 124:119–124. https://doi.org/10.2134/jeq2015.11.0567

Lovett GM (1994) Atmospheric deposition of nutrients and pollutants in North America: an ecological perspective. Ecol Appl 4:629–650. https://doi.org/10.2307/1941997

Lovett GM, Lindberg SE (1993) Atmospheric deposition and canopy interactions of nitrogen in forests. Can J Forest Res 23:1603–1616. https://doi.org/10.1139/x93-200

Lovett GM, Reiners WA (1986) Canopy structure and cloud water deposition in subalpine coniferous forests. Tellus B 38:319–327. https://doi.org/10.3402/tellusb.v38i5.15140

Lovett GM, Lindberg SE, Richter DD, Johnson DW (1985) The effects of acidic deposition on cation leaching from three deciduous forest canopies. Can J Forest Res 15:1055–1060. https://doi.org/10.1139/x85-171

Lovett GM, Traynor MM, Pouyat RV, Carreiro MM, Zhu WX, Baxter JW (2000) Atmospheric deposition to oak forests along an urban-rural gradient. Environ Sci Technol 34:4294–4300. https://doi.org/10.1021/es001077q

McCarty J, Kaza N (2015) Urban form and air quality in the United States. Landsc Urban Plan 139:168–179. https://doi.org/10.1016/j.landurbplan.2015.03.008

McGrane SJ (2016) Impacts of urbanisation on hydrological and water quality dynamics, and urban water management: a review. Hydrolog Sci J 61:2295–2311. https://doi.org/10.1080/02626667.2015.1128084

McPherson GE, Nowak DJ, Rowntree RA (1994) Chicago's urban forest ecosystem: results of the Chicago Urban Forest Climate Project, Gen. Tech. Rep. NE-186. US Department of Agriculture, Forest Service, Northeastern Forest Experiment Station, Radnor, PA, 201 p

Michalzik B, Stadler B (2005) Importance of canopy herbivores to dissolved and particulate organic matter fluxes to the forest floor. Geoderma 127:227–236. https://doi.org/10.1016/j.geoderma.2004.12.006

Michopoulos P, Baloutsos G, Economou A, Samara C, Thomaidis NS, Grigoratos G (2007a) Nutrient cycling and foliar status in an urban pine forest in Athens, Greece. Plant Soil 294:31–39. https://doi.org/10.1007/s11104-007-9224-6

Michopoulos P, Baloutsos G, Economou A, Voulala M, Bourletsikas A (2007b) Bulk and throughfall deposition chemistry in three different forest ecosystems. Fresen Environ Bull 16:91–98

Monks PS, Granier C, Fuzzi S, Stohl A, Williams ML, Akimoto H et al (2009) Atmospheric composition change - global and regional air quality. *Atmos Environ 43*:5268–5350. https://doi.org/10.1016/j.atmosenv.2009.08.021

Mo L, Ma Z, Xu Y, Sun F, Lun X, Liu X, Jungang C, Yu X (2015) Assessing the capacity of plant species to accumulate particulate matter in Beijing, China. PLoS One 10:e0140664. https://doi.org/10.1371/journal.pone.0140664

Mori J, Hanslin HM, Burchi G, Sæbø A (2015) Particulate matter and element accumulation on coniferous trees at different distances from a highway. Urban For Urban Gree 14:170–177. https://doi.org/10.1016/j.ufug.2014.09.005

Morton TG, Gold AJ, Sullivan WM (1988) Influence of overwatering and fertilization on nitrogen losses from home lawns. J Environ Qual 17:124–130. https://doi.org/10.2134/jeq1988.00472425001700010019x

Neal C, Reynolds B, Neal M, Hughes S, Wickham H, Hill L et al (2003) Soluble reactive phosphorus levels in rainfall, cloud water, throughfall, stemflow, soil waters, stream waters and groundwaters for the Upper River Severn area, Plynlimon, mid Wales. Sci Total Environ 314:99–120. https://doi.org/10.1016/S0048-9697(03)00100-1

Newman EI (1995) Phosphorus inputs to terrestrial ecosystems. J Ecol 83:713–726. https://doi.org/10.2307/2261638

Nowak DJ, Greenfield EJ (2018) US urban forest statistics, values, and projections. J For 116:164–177. https://doi.org/10.1093/jofore/fvx004

Nowak DJ, Crane DE, Stevens JC (2006) Air pollution removal by urban trees and shrubs in the United States. Urban For Urban Gree 4:115–123. https://doi.org/10.1016/j.ufug.2006.01.007

Nowak DJ, Greenfield EJ, Hoehn RE, Lapoint E (2013a) Carbon storage and sequestration by trees in urban and community areas of the United States. Environ Pollut 178:229–236. https://doi.org/10.1016/j.envpol.2013.03.019

Nowak DJ, Hirabayashi S, Bodine A, Hoehn R (2013b) Modeled PM2.5 removal by trees in ten U.S. cities and associated health effects. Environ Pollut 178:395–402. https://doi.org/10.1016/j.envpol.2013.03.050

Oke TR (1989) The micrometeorology of the urban forest [and discussion]. Philos Trans Royal Soc B 324:335–349. https://doi.org/10.1098/rstb.1989.0051

Osono T, Hobara S, Koba K, Kameda K (2006) Reduction of fungal growth and lignin decomposition in needle litter by avian excreta. Soil Biol Biochem 38:1623–1630. https://doi.org/10.1016/j.soilbio.2005.12.001

Pataki DE, Carreiro MM, Cherrier J, Grulke NE, Jennings V, Pincetl S et al (2011) Coupling biogeochemical cycles in urban environments: Ecosystem services, green solutions, and misconceptions. Front Ecol Environ 9:27–36. https://doi.org/10.1890/090220

Peterson BJ, Wollheim WM, Mulholland PJ, Webster JR, Meyer JL, Tank JL et al (2001) Control of nitrogen export from watersheds by headwater streams. Science 292:86–90. https://doi.org/10.1126/science.1056874

Ponette-González AG, Weathers KC, Curran LM (2010) Tropical land-cover change alters biogeochemical inputs to ecosystems in a Mexican montane landscape. Ecol Appl 20:1820–1837. https://doi.org/10.1890/09-1125.1

Ponette-González AG, Marín-Spiotta E, Brauman KA, Farley KA, Weathers KC, Young KR (2014) Hydrologic connectivity in the high-elevation tropics: heterogeneous responses to land change. BioScience 64:92–104. https://doi.org/10.1093/biosci/bit013
Ponette-González AG, Ewing HA, Weathers KC (2016a) Interactions between precipitation and vegetation canopies. In: Johnson E, Martin Y (eds) A biogeoscience approach to ecosystems. Cambridge University Press, Cambridge, UK, pp 215–253
Ponette-González AG, Curran LM, Pittman AM, Carlson KM, Steele BG, Ratnasari D et al (2016b) Biomass burning drives atmospheric nutrient redistribution within forested peatlands in Borneo. Environ Res Lett:11. https://doi.org/10.1088/1748-9326/11/8/085003
Ponette-González AG, Perroni Y, Weathers KC, DeSouza P, Garcia-Oliva F, DeMello W (2017) Nitrogen cycling in tropical Atlantic Forest differing in exposure to urban atmospheric nitrogen deposition. Plant Soil 420:451–465. https://doi.org/10.1007/s11104-017-3421-8
Ponette-González AG, Collins JD, Manuel JE, Byers TA, Glass GA, Weathers KC et al (2018) Wet dust deposition during the 2012 US drought: An overlooked pathway for elemental flux to ecosystems. J Geophys Res Atmos:123. https://doi.org/10.1029/2018JD028806
Popek R, Gawrońska H, Wrochna M, Gawroński SW, Sæbø A (2013) Particulate matter on foliage of 13 woody species: deposition on surfaces and phytostabilisation in waxes - a 3-year study. Int J Phytoremediation 15:245–256. https://doi.org/10.1080/15226514.2012.694498
Pryor SC, Hornsby KE, Novick KA (2014) Forest canopy interactions with nucleation mode particles. Atmos Chem Phys 14:11985–11996. https://doi.org/10.5194/acp-14-11985-2014
Przybysz A, Sæbø A, Hanslin HM, Gawroński SW (2014) Accumulation of particulate matter and trace elements on vegetation as affected by pollution level, rainfall and the passage of time. Sci Total Environ 481:360–369. https://doi.org/10.1016/j.scitotenv.2014.02.072
Pugh TA, MacKenzie AR, Whyatt JD, Hewitt CN (2012) Effectiveness of green infrastructure for improvement of air quality in urban street canyons. Environ Sci Technol 46:7692–7699. https://doi.org/10.1021/es300826w
Quan J, Zhang X, Zhang Q, Guo J, Vogt RD (2008) Importance of sulfate emission to sulfur deposition at urban and rural sites in China. Atmos Res 89:283–288. https://doi.org/10.1016/j.atmosres.2008.02.015
Rao P, Hutyra LR, Raciti SM, Templer PH (2014) Atmospheric nitrogen inputs and losses along an urbanization gradient from Boston to Harvard Forest, MA. Biogeochemistry 121:229–245. https://doi.org/10.1007/s10533-013-9861-1
Reche C, Viana M, Karanasiou A, Cusack M, Alastuey A, Artiñano B et al (2015) Urban NH3 levels and sources in six major Spanish cities. Chemosphere 119:769–777. https://doi.org/10.1016/j.chemosphere.2014.07.097
Reinmann AB, Hutrya LR (2017) Edge effects enhance carbon uptake and its vulnerability to climate change in temperate broadleaf forests. Proc Natl Acad Sci U.S.A. 114:107–112. https://doi.org/10.1073/pnas.1612369114
Reinap A, Wiman BLB, Svenningsson B, Gunnarsson S (2009) Oak leaves as aerosol collectors: relationships with wind velocity and particle size distribution. Experimental results and their implications. Trees-Struct Funct 23:1263–1274. https://doi.org/10.1007/s00468-009-0366-4
Runyan CW, Odorico PD, Vandecar KL, Das R, Schmook B, Lawrence D (2013) Positive feedbacks between phosphorus deposition and forest canopy trapping, evidence from Southern Mexico. J Geophys Res Biogeosci 118:1521–1531. https://doi.org/10.1002/2013JG002384
Saebø A, Popek R, Nawrot B, Hanslin HM, Gawronska H, Gawronski HW (2012) Plant species differences in particulate matter accumulation on leaf surfaces. Sci Total Environ 427-428:347–354. https://doi.org/10.1016/j.scitotenv.2012.03.084
Schaubroeck T, Deckmyn G, Neirynck J, Staelens J, Adriaenssens S, Dewulf J et al (2014) Multilayered modeling of particulate matter removal by a growing forest over time, from plant surface deposition to washoff via rainfall. Environ Sci Technol 48:10785–10794. https://doi.org/10.1021/es5019724

Schlesinger WH, Bernhardt ES (2013) In: Schlesinger WH, Bernhardt ES (eds) Biogeochemistry, 3rd edn. Academic Press, Boston. https://doi.org/10.1016/B978-0-12-385874-0.00003-0
Schooling JT, Levia DF, Carlyle-Moses DE, Dowtin AL, Brewer SE, Donkor KK (2017) Stemflow chemistry in relation to tree size: a preliminary investigation of eleven urban park trees in British Columbia, Canada. Urban For Urban Green 21:129–133. https://doi.org/10.1016/j.ufug.2016.11.013
Seto KC, Güneralp B, Hutyra LR (2012) Global forecasts of urban expansion to 2030 and direct impacts on biodiversity and carbon pools. Proc Natl Acad Sci USA 109:16083–16088. https://doi.org/10.1073/pnas.1211658109
Sgrigna G, Sæbø A, Gawronski S, Popek R, Calfapietra C (2015) Particulate matter deposition on *Quercus ilex* leaves in an industrial city of Central Italy. Environ Pollut 197:187–194. https://doi.org/10.1016/j.envpol.2014.11.030
Smith IA, Hutrya LR, Reinmann AB, Marrs JK, Thompson JR (2018) Piecing together the fragments; piecing together edge effects on forest carbon dynamics. Front Ecol Environ 16:213–221. https://doi.org/10.1002/fee.1793
Stadler B, Michalzik B (1998) Linking aphid honeydew, throughfall, and forest floor solution chemistry of Norway spruce. Ecol Lett 1:13–16. https://doi.org/10.1046/j.1461-0248.1998.00006.x
Stevens PA, Adamson JK, Anderson MA, Hornung M (1988) Effects of clear felling on surface water quality and site nutrient status. In: Usher MB, Thompson DBA (eds) Ecological change in the uplands, Blackwell, British Ecological Society, Special Publ No. 7, pp 289–294
Sun K, Tao L, Miller DJ, Pan D, Golston LM, Zondlo MA et al (2017) Vehicle emissions as an important urban ammonia source in the United States and China. Environ Sci Technol 51:2472–2481. https://doi.org/10.1021/acs.est.6b02805
Takagi M, Shigeyuki S, Gyokusen K, Saito A (1997) Stemflow chemistry of urban street trees. Environ Pollut 96:107–109. https://doi.org/10.1016/S0269-7491(97)00005-5
Templer PH, Weathers KC, Lindsey A, Lenoir K, Scott L (2015a) Atmospheric inputs and nitrogen saturation status in and adjacent to Class I wilderness areas of the northeastern US. Oecologia 177:5–15. https://doi.org/10.1007/s00442-014-3121-5
Templer PH, Toll JW, Hutyra LR, Raciti SM (2015b) Nitrogen and carbon export from urban areas through removal and export of litterfall. Environ Pollut 197:256–261. https://doi.org/10.1016/j.envpol.2014.11.016
Tiwari R, Gupta GP, Kulshrestha UC (2016) Summer time dustfall fluxes of reactive nitrogen and other inorganic species over the tropical megacity of Indo-Gangetic Plains. Earth Interact 20. https://doi.org/10.1175/EI-D-15-0053.1
United Nations, Department of Economic and Social Affairs, Population Division (2015) World urbanization prospects: the 2014 revision, (ST/ESA/SER.A/366). United Nations, Department of Economic and Social Affairs, Population Division, New York
United States Forest Service. https://www.nrs.fs.fed.us/urban/utc/. Accessed 9 Sept 2018
Viana M, Kuhlbusch TAJ, Querol X, Alastuey A, Harrison RM, Hopke PK et al (2008) Source apportionment of particulate matter in Europe: a review of methods and results. J Aerosol Sci 39:827–849. https://doi.org/10.1016/j.jaerosci.2008.05.007
Vos PEJ, Maiheu B, Vankerkom J, Janssen S (2013) Improving local air quality in cities: to tree or not to tree? Environ Pollut 183:113–122. https://doi.org/10.1016/j.envpol.2012.10.021
Wang C, Chang C, Tsai S, Chiang H (2005) Characteristics of road dust from different sampling sites in northern Taiwan. J Waste Manag Assoc 55:1236–1244. https://doi.org/10.1080/10473289.2005.10464717
Wang R, Balkanski Y, Boucher O, Ciais P, Peñuelas J, Tao S (2014) Significant contribution of combustion-related emissions to the atmospheric phosphorus budget. Nature Geosci 8:48. Retrieved from https://doi.org/10.1038/ngeo2324
Weathers KC, Cadenasso ML, Pickett STA (2001) Forest edges as nutrient and pollutant concentrators: potential synergisms between fragmentation, forest canopies, and atmosphere. Conserv Biol 15(6):1506–1514

Weathers KC, Ponette-González AG (2011) Atmospheric deposition. In: Levia DF, Carlyle-Moses DE, Tanaka T (eds) Forest hydrology and biogeochemistry: synthesis of past research and future directions, Springer, Ecological Studies Series, vol 216, pp 335–370. https://doi.org/10.1007/978-94-007-1363-5_17

Weathers KC, Likens GE, Butler TJ (2006) Acid rain. In: Environmental and occupational medicine, 4th edn. Lippincott Williams & Wilkins, Philadelphia, pp 1507–1522

Weathers KC, Lovett GM, Likens GE, Lathrop R (2000) The effect of landscape features on deposition to Hunter Mountain, Catskill Mountains, New York. Ecol Appl 10:528–540. https://doi.org/10.1890/1051-0761(2000)010[0528:TEOLFO]2.0.CO;2

Wesely ML, Hicks BB (2000) A review of the current status of knowledge on dry deposition. Atmos Environ 34:2261–2282. https://doi.org/10.1016/S1352-2310(99)00467-7

Xu Y, Xu W, Mo L, Heal MR, Xu X, Yu X (2018) Quantifying particulate matter accumulated on leaves by 17 species of urban trees in Beijing, China. Environ Sci Pollut Res Int 25:12545–12556. https://doi.org/10.1007/s11356-018-1478-4

Yang J, McBride J, Zhou J, Sun Z (2005) The urban forest in Beijing and its role in air pollution reduction. Urban For Urban Gree 3:65–78. https://doi.org/10.1016/j.ufug.2004.09.001

Chapter 19
Modeling the Impact of Urban Trees on Hydrology

Robert Coville, Ted Endreny, and David J. Nowak

19.1 Introduction

Urban trees provide numerous benefits to society, as well as costs. These benefits include moderating climate and cooling the urban heat island; reducing building energy use and atmospheric carbon dioxide (CO_2); improving air and water quality; mitigating rainfall runoff and flooding; enhancing aesthetics, human health, and social well-being; and lowering noise impacts (Dwyer et al. 1992, Nowak and Dwyer 2007, Dobbs et al. 2017). Urban forests have various costs associated with tree planting and maintenance, along with other possible indirect costs through pollen production and allergic reactions, winter shade increasing building energy use, lowered wind speed and dispersion increasing pollutant concentrations, and invasive plants altering local biodiversity (Lyytimaki 2017; Long et al. 2018). Annually, urban trees in the United States produce a total of \$18.3 billion in value related to air pollution removal (\$5.4 billion), reduced building energy use (\$5.4 billion), carbon sequestration (\$4.8 billion), and avoided pollutant emissions (\$2.7 billion) (Nowak and Greenfield 2018b). As urban land in the United States is projected to increase by 384,000 km^2 between 2010 and 2060 (increasing from 275,000 km^2 (3.0% of US land) to 660, 000 km^2 (8.6%)) (Nowak and Greenfield 2018b), the value and importance of the urban forest resources will continue to rise in the coming years. One of the more important benefits of urban trees relates to their impact on surface stormwater runoff, stream flow, and water quality.

R. Coville
Davey Institute, 5 Moon Library, SUNY-ESF, Syracuse, NY, USA

T. Endreny
Department of Environmental Resources Engineering, 426 Baker Lab, SUNY-ESF, Syracuse, NY, USA

D. J. Nowak (✉)
USDA Forest Service, 5 Moon Library, SUNY-ESF, Syracuse, NY, USA
e-mail: dnowak@fs.fed.us

D. F. Levia et al. (eds.), *Forest-Water Interactions*, Ecological Studies 240,
https://doi.org/10.1007/978-3-030-26086-6_19

The removal of trees typically leads to increased stormwater runoff, potentially increasing localized and extensive flooding in urban areas. The economic impacts of flooding can be substantial (Nowak et al. 2007). The costs/impacts associated with urban flooding include wet structures with mold and potential increase in respiratory problems, increased insurance rates, lower property values, streambank erosion, degraded water quality, and reduced health of aquatic ecosystems. In Cook County, Illinois, total claims paid for urban flooding incidents over 5 years (2007–2011) were more than $773 million (CNT 2014). In addition to larger peak flows, increased stormwater can also lead to instability in drainage systems and reduced recharge of groundwater (Herricks 1995; Thorne 1998; FISRWG 1999). Instability in the drainage system can rapidly erode streambanks, damage streamside vegetation, and widen stream channels (Hammer 1972). Instability combined with reduced groundwater recharge results in lower water depths during non-storm periods, higher than normal water levels during wet weather periods, increased sediment loads, and higher water temperatures (Brookes 1988).

Trees can reduce stormwater runoff in many ways and help reduce the hydrologic impacts of urbanization (Kuehler et al. 2017). Trees affect stream flow rates primarily through three mechanisms: canopy rainfall interception, soil water infiltration, and evapotranspiration (US EPA 2016). These mechanisms result in the cumulative effect trees have on urban water volume. Trees also affect water quality, both by affecting water volume and thus pollutant transport and by affecting the concentration of pollutants in water.

According to US GAO (2001), when natural ground cover is present over the entire site, on average, 10% of precipitation runs off the land into nearby creeks, rivers, and lakes. In contrast, when a site is 75% impervious, and not all directly connected to receiving waters, on average, 55% of the precipitation runs off into receiving waters. Runoff from parking lots and other paved areas is estimated at 98% of storm event precipitation (USDA NRCS 1986). The impervious surfaces in a typical city block may generate nine times more runoff than a woodland area of the same size (US EPA 1996). Urban impervious cover in the conterminous United States averages 26.6% (Nowak and Greenfield 2018a). Runoff from urban land cover collects pollutants from the land surface and poses a threat to receiving waters (US GAO 2001).

Trees are capable of being dominant land cover elements in cities. In the conterminous United States, urban tree cover averages 39.3%, ranging from 10.1% in North Dakota to 61.6% in Connecticut (Nowak and Greenfield 2018b). To optimize land use planning and facilitate the inclusion of more tree canopy for their hydrologic ecosystem services, models can serve a useful role (Lin et al. 2007; Guswa et al. 2014). Due to the complexity of interactions among trees and the hydrologic and biogeochemical systems, models are often used to extend field observations and estimate the outcomes of these interactions in different scenarios. While models are simplifications of reality used to gain insight into select attributes of a particular system (US EPA 2009), hydrological models have proven useful tools in estimating the transport of water (Borah and Bera 2004), and biogeochemical models have proven useful in estimating the transformation of chemicals (Bouraoui and Grizzetti 2014). Chapters 17 (Carlyle-Moses et al. this volume) and 18 (Decina et al. this volume) describe urban impervious and urban tree impacts on hydrologic and biogeochemical processes, and this chapter summarizes and discusses various models used to quantify these urban tree hydrology impacts.

19.2 Modeling Urban Tree Effects on Water Volume and Quality

Computer models simulating how urban land cover impacts water quality and quantity are flexible, fast, and low-cost management options when compared with field or laboratory studies. Models can also be used to examine scientific hypotheses regarding the cause and effect feedback between land cover and water resources in urban systems. Given the importance of water quantity and quality in cities, and the range of components, systems, processes, and priorities involved, a wide variety of urban hydrology models have been developed. These models typically vary in their conceptual framework, input requirements, predictive goals, mathematical algorithms, applications, user support, and required user investment.

This chapter reviews a limited set of urban hydrology models, selected based on the following criteria that the model is (a) in the public domain and free to use; (b) considered useful to the broad range of urban hydrology management issues and not too specialized; and (c) capable of simulating tree or forest effects on hydrology. Readers interested in models that extend beyond these criteria can review a larger list of models compiled by the Minnesota Pollution Control Agency (MPCA 2018). A review of common modeling concepts is provided here to help compare and contrast between models, communicate with model developers, and appreciate how models are a simplified approximation of the actual system.

Spatial representation is a major model concept and can be explained using the following illustration. Consider a 1 km^2 area of an urban area, with pixelated maps for each watershed element, including soil types, land cover types, terrain elevation, precipitation, etc., with each pixel representing a 10 × 10 m sub-area. To continue with the illustration, consider a set of unique values for soil type (sandy loam, clay loam, silty clay) and land cover type (commercial, residential, forest), and ignore the variation in terrain elevation, precipitation, etc. Note that this illustration is using simple elements, and in most applications, there are more than three soil types or land cover types. Based on the 100 m^2 pixel size, there are 10,000 pixels in the 1 km^2 urban area, and map inventories determine the number of pixels, and hence percent of pixels, in each soil type (35% in sandy loam, 45% clay loam, 20% silty loam) and in each land cover type (15% in commercial, 60% in residential, 25% in forest).

In a spatially lumped model, the 1 km^2 area would be represented as a single area characterized by the percentages of soil type, land cover type, etc., and there would be no spatial relationship of adjacency within a map type (e.g., sandy loam adjacent to clay loam), nor spatial relationship of congruence between map types (e.g., forest congruent with sandy loam). Instead, the spatially lumped model could proceed with a water budget by treating each percent of cover independently and then averaging the response, or it could derive a set of effective parameters representing the composite land cover. In this illustration, if the spatially lumped model received a uniform 0.254 mm rainfall event, the area in forest might have 70% of this rain allocated to canopy interception and depression storage, and the area in commercial and residential might have 15% allocated to depression storage. This results in a lumped area average of 28.75% of the rain in storage and leaves 71.25% of rain available to proceed to infiltration in the three different soil types, each with a distinct infiltration rate. The spatially lumped model would likely ignore spatial variation in elevation and its influence on weather (e.g., cooler temperatures at higher elevations, stronger solar radiation on south-facing slopes).

In a spatially distributed model, the options include fully distributed, with each of the 10,000 pixels treated as a vertical stack with its unique precipitation, terrain elevation, land cover, soils, etc., and allow for lateral exchanges of water and constituents between pixels or, to partially simplify the system, often called semi-distributed or statistically distributed. One form of a semi-distributed model uses spatial congruence of watershed elements (e.g., forest above sandy loam, forest above silty loam, etc.) to identify systems that behave similarly in a water budget (e.g., same depth of interception and rate of infiltration for a given rainfall depth), which are referred to as hydrologic response units (HRUs). In the illustration of three soil types and three land cover types, the 1 km^2 urban area could have a maximum of 9 HRUs, which is computationally much faster than modeling 10,000 pixels and not much slower than modeling the lumped area. Another widely used form of the semi-distributed model uses the ratio of contributing area and slope for each pixel in the terrain elevation map to create a topographic index (TI) value, which represents wetness likelihood (Beven and Kirkby 1979), and then sorts the unique TI values for all pixel into a smaller set of bins, as in a histogram. In TI applications there are often 20–50 bins, and each bin might be assigned a lumped percent of land cover and soil types. The TI is then used in a function that laterally redistributes the precipitation that entered the soil as ground water at the end of each simulation time step, replacing a computationally expensive function to explicitly move water between 10,000 pixels.

Other ways to categorize models, along with the lumped vs. spatially distributed category above, include empirical vs. mechanistic modeling, stochastic vs. deterministic modeling, single-event vs. continuous modeling, event mean concentration (EMC) vs. buildup/washoff water quality routines, and object-oriented vs. function-based design. Empirical models represent the observed relationships in phenomena, such as rainfall partitioning to runoff, without representing the theory of cause and effect used in mechanistic models. As an example, if a field study measured the amount of runoff for several rainfall events in an area with residential land cover and silty loam soils, it could analyze the results and potentially derive a new parameter that estimated the amount of rainfall that becomes runoff for each event. Empirical models are limited to applications where the watershed conditions are similar to those of the study site, which is one constraint with regard to significant land or climate disturbance. With mechanistic models, there are varying levels of theoretical and mathematical sophistication, and first-order models tend to prefer greater simplicity over greater accuracy, parsimoniously keeping the governing terms in equations while removing higher-order terms.

Stochastic modeling can treat equation parameters as random variables with probability distributions to represent observed variability, while deterministic modeling assigns fixed values to parameters in a given model scenario. Single-event models simulate one precipitation event and typically ignore evapotranspiration, while continuous models simulate a period of time, at some time step, to represent precipitation and the evapotranspiration that follows. Length of time step is also an important defining factor in urban hydrology simulations, as there is a need for both shorter and longer time steps: sub-hourly time steps inform peak runoff rates green infrastructure may need to process and aid in proper sizing of drains, berms, and other flow conduits; and 24-hourly time steps inform total volumes that may need to be accommodated by the maximum capacity of green infrastructure. Shorter or longer time steps may be preferable depending on which type of resulting information is more pertinent and what kind of model inputs (temporal resolution of observed weather and/or discharge data) and computing resources (for data processing and storage) are available.

EMC-based estimates of pollutant loading, detailed in Sect. 19.2.1.1 of this chapter, are a parsimonious approach which can over- or underestimate pollutant concentration within a storm event, but this approach serves well to estimate total pollutant loading for an entire event. Buildup and washoff approaches to water quality estimates capture fluctuations in pollutant concentrations during an event and can more explicitly account for water quality BMPs (e.g., street sweeping), but these approaches tend to be difficult to parameterize and, in many cases, the important output is total pollutant loading, not a time series of pollutant washoff.

Object-oriented design organizes the model around watershed elements such as tree canopy and soil, which are called classes, and instances of the classes can handle data and functions acting on the data. Function-based design organizes the model around operations such as interception and infiltration, with algorithms linking these processes. The goal of object-oriented design is to make the model open for extension without modification, so that to the extent possible new ideas for the model only require new objects rather than modifying existing objects.

The models included in this chapter can be divided into three general categories of complexity, defined by number and availability of required inputs and parameters and the range of processes and systems simulated. The lower complexity models are (a) the Rational method approach and its application with the LMNO Rational Equation Calculator and the Minimal Impact Design Standards Calculator and (b) the Curve Number approach, and its application with WinTR-55 (NRCS 2018) and the Green Values National Stormwater Management Calculator. The moderate complexity models are i-Tree Hydro and the US EPA National Stormwater Calculator. The advanced complexity models are the (a) Storm Water Management Model (SWMM); (b) Hydrological Simulation Program-FORTRAN (HSPF); (c) Soil and Water Assessment Tool (SWAT); and (d) Regional Hydro-Ecological Simulation System (RHESSys). Models similar to the complexity of RHESSys used by research groups include the ecohydrologic (Ech2o) model (Maneta and Silverman 2013) and the Distributed Hydrology Soil Vegetation Model (DHSVM) (Wigmosta et al. 1994), while a fee-based decision support system by eWater of Australia is the Model for Urban Stormwater Improvement Conceptualization (MUSIC) (Wong et al. 2002). Models with a similar level of complexity can have substantially different methodologies in how they account for tree processes. Of these models, i-Tree Hydro (Wang et al. 2008) was explicitly designed to simulate tree effects in urban hydrological systems, and this model serves as a good starting point to introduce concepts and help the reader later compare and contrast simpler and more advanced approaches.

19.2.1 i-Tree Hydro

The i-Tree Hydro model, managed by the USDA Forest Service and Davey Tree Expert Company (Yang et al. 2011), is a spatially semi-distributed model that simulates runoff quantity and quality for watershed and non-watershed areas subjected to a single precipitation event or continuous weather. The i-Tree Hydro routines are divided into the hydrologic processes of the water balance, collectively with 100s of equations and descriptive parameters. The hydrologic processes include classifying precipitation as rain or snow based on air temperature, canopy interception, depression storage, impervious runoff, infiltration, soil moisture updating, pervious runoff, evaporation from surface water and canopy water, evapotranspiration from soils and leaves, subsurface runoff, lateral distribution of water, and estimates of water quality and quantity (Fig. 19.1).

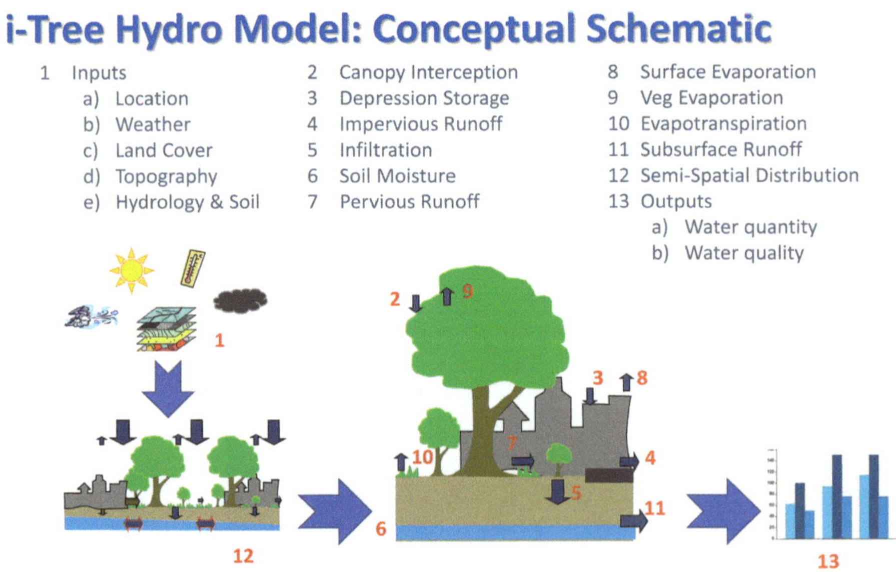

Fig. 19.1 Conceptual schematic of the i-Tree Hydro model

The model inputs require a time series of weather, terrain elevation data or a topographic index, and land cover with estimates of percent tree and impervious cover. The model maintains a water budget for snow and rain, implementing vertical redistribution of water between the vegetation canopy, depression storage, soil moisture, groundwater, and the atmosphere via evapotranspiration. A variable source area routine identifies saturation excess and infiltration excess surface runoff; the Green-Ampt model is used to determine infiltration, with a lateral redistribution of soil saturation implemented with the topographic index. Surface and subsurface runoff directed to the channel network is transformed to a hydrograph using a one- or two-parameter advection-diffusion routing model (Yang and Endreny 2013).

The user and input data define the time step (typically 1 h); land cover characteristics, including percent pervious or impervious cover beneath tree canopy and percent directly connected impervious area (DCIA); soil characteristics; and optionally other model parameters. Water quality constituents are represented with EMC values as described in the following section, and pollutant loads are determined by the volume of surface runoff. Storm sewer discharge is not simulated but approximated as the impervious runoff DCIA, which directly enters receiving waters. i-Tree Hydro is used to determine how changes in watershed management (e.g., percent tree cover, other land use, DCIA) or climate affect discharges based on preferred or regulated water quantity and water quality targets. A set of default soil parameters are provided based on prior model runs, and users can optionally use a custom implementation of the PEST tool (Doherty 2001a, b) for parameter calibration and evaluation when used in conjunction with a time series of observed discharge. This tool is based on the ObjTop and UFORE-Hydro models (Wang et al. 2005a, 2008), is actively updated, and is accessible on the www.itreetools.org website, with tutorials and technical support available. i-Tree Hydro is generally used as a scoping tool, estimating the effects of trees (Kirnbauer et al. 2013) and different land cover scenarios (Lefrançois 2015) on hydrology (Fig. 19.2).

19.2.1.1 Water Quality Modeling Using Event Mean Concentrations (EMCs)

The event mean concentration (EMC) method is considered a proven approach to quantify pollutant loading to receiving waters resulting from a runoff event (US EPA 2002). The EMC is not intended to predict the variation in concentration during an event, but rather to represent the total loading from an event. The EMC value itself (mg L^{-1}) is a statistical parameter representing the flow-proportional median concentration for a storm event and can be adjusted to represent other percentiles, such as the upper 90th or lower 10th (see Equations 4 and 5 in Stephan and Endreny 2016). Estimates of EMC are usually obtained from analysis of many flow-weighted composite samples taken during each storm, and not simply a time average of a single event. When an *EMC* from a look-up table is multiplied by the runoff volume,

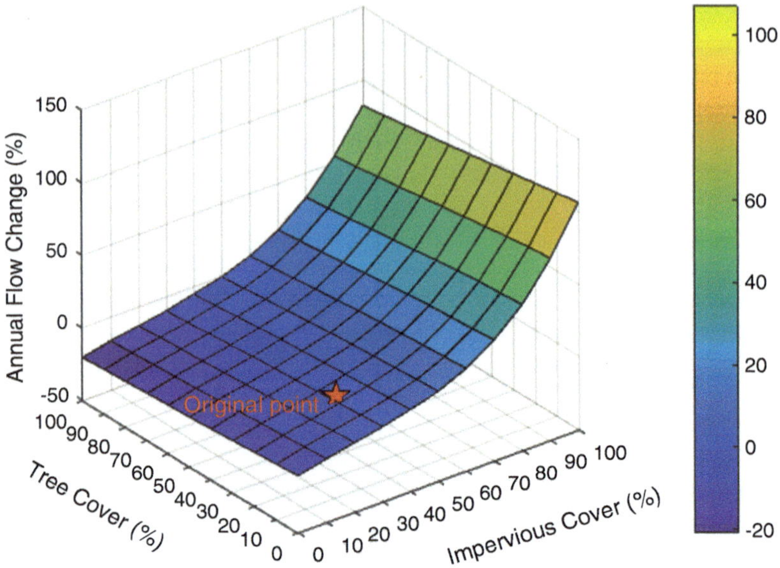

Fig. 19.2 i-Tree Hydro simulated effects of incremental changes to tree cover and impervious cover in 161 km^2 Rock Creek watershed near Washington, DC

V, for a new event, the resulting load, *L*, is the estimated total pollutant load for that event, with the equation given as:

$$L = EMC * V \quad (19.1)$$

To understand and control urban runoff pollution, the US Congress included the establishment of the Nationwide Urban Runoff Program (NURP) in the 1977 Amendments of the Clean Water Act (PL 95-217). In 1983, the US Environmental Protection Agency (US EPA 1983) published the results of the NURP, which nationally characterized urban runoff for 10 standard water quality pollutants in the United States, based on data from 2300 station-storms at 81 urban sites in 28 metropolitan areas. The USGS conducted a separate stormwater characterization from data measured through mid-1980s for more than 1100 stations at 97 urban sites located in 21 metropolitan areas (Driver et al. 1985). A third characterization of stormwater quality was compiled using data from stormwater discharge permits under the National Pollutant Discharge Elimination System (NPDES) for more than 30 cities, 800 station-storms, and 150 parameters (Smullen et al. 1999). The data from the three sources (NURP, USGS, and NPDES) were used to compute a pooled means with greater statistical confidence (Smullen et al. 1999), and the NURP and pooled mean EMCs for the ten constitutes are listed in Table 19.1. Pooled mean or NURP EMCs are based on field data collected from thousands of storm events and are representative of the United States nationwide rather than specific to any single site.

Table 19.1 National pooled EMCs and NURP EMCs

Constituent	Data Source[a]	EMCs (mg L^{-1})		No. of events
		Mean	Median	
Total suspended solids: TSS	Pooled	78.4	54.4	3047
	NURP	17.4	113	2000
Biochemical oxygen demand: BOD_5	Pooled[b]	14.1	11.5	1035
	NURP	10.4	8.39	474
Chemical oxygen demand: COD	Pooled	52.8	44.7	2639
	NURP	66.1	55	1538
Total phosphorus: TP	Pooled	0.315	0.259	3094
	NURP	0.337	0.266	1902
Soluble phosphorus: soluble P	Pooled[c]	0.129	0.103	1091
	NURP	0.1	0.078	767
Total Kjeldahl nitrogen: TKN	Pooled	1.73	1.47	2693
	NURP	1.67	1.41	1601
Nitrite and nitrate: NO_2 and NO_3	Pooled	0.658	0.533	2016
	NURP	0.837	0.666	1234
Copper: Cu	Pooled	0.0135	0.0111	1657
	NURP	0.0666	0.0548	849
Lead: Pb	Pooled	0.0675	0.0507	2713
	NURP	0.175	0.131	1579
Zinc: Zn	Pooled	0.162	0.129	2234
	NURP	0.176	0.140	1281

[a]Pooled data sources include NURP, USGS, and NPDES
[b]No BOD5 data available in the USGS dataset; pooled includes NURP+NPDES
[c]No TS data available in NPDES dataset; pooled includes NURP+USGS
Source: Smullen et al. (1999); reproduced with permission of IWA Publishing

19.2.1.2 Effect of Trees

There is explicit simulation of tree cover in i-Tree Hydro, as one of six land cover groups that also include shrub, herbaceous, water, impervious, and bare soil. Users identify land cover for a base case scenario and up to three alternative scenarios and give the percent of tree canopy over pervious cover and over impervious cover and the percent with evergreen canopy. The simulation includes a specific soil layer accessible to vegetation roots, using that soil water for transpiration, as well as a soil macropore fraction allowing precipitation falling on pervious soils to bypass the root zone layer (Aubertin et al. 1971). Canopy interception of liquid and snow precipitation are explicitly simulated at each time step. The model allows users to set the leaf on and off transition dates and transition duration, minimum and maximum leaf area index, bark area index, portion of canopy that is evergreen, and leaf storage depth available for water. Evaporation is explicitly modeled for canopy storage and surface (depression) storage and as evapotranspiration from soil pores and leaf stomata, which is used to reduce or reset the water storage in these layers. Additional details on these and related processes follow, to clarify how the model approximates a representation of the actual hydrologic cycle.

Interception is simulated in tree and short vegetation cover using methods based on the work of Rutter et al. (1971, 1975) rather than using empirical methods based on gross precipitation (Jackson 1975). Rutter interception theory was modified to account for throughfall in sparse vegetation (Gash et al. 1995; Valente et al. 1997) to represent the tree structure common in urban forests. Between-leaf throughfall is simulated as a function of canopy fraction, and when canopy storage is filled, canopy drip is simulated. Evaporation from the canopy reduces the storage. The interception depth is a function of weather dynamics of precipitation intensity and duration, wind, vapor pressure, and radiation values and tree characteristics of seasonally varying leaf area, storage capacity, and initial vegetation surface storage at each time step.

Depression storage is filled by precipitation reaching the ground; depressions include the water stored by leaf litter and other organic material as well as in potholes and impervious low spots. Impervious depression storage and pervious depression storage are modeled separately, and the maximum depth of storage is representative of a layer of water spread across the entire impervious or pervious area. Precipitation that exceeds the impervious depression storage depth is directly converted to surface runoff, with the DCIA fraction determining the amount going directly to the outlet as impervious runoff, while the remainder is passed to pervious areas. The amount of DCIA is determined by the user, with recommended values estimated based on work by Sutherland (2000).

Infiltration, ponding, and runoff are the partitions used to allocate precipitation over pervious areas and impervious runoff that exceeds pervious area storage. This partitioning uses the variable source area concept, where infiltration excess runoff occurs if precipitation rates are greater than the Green-Ampt infiltration rate and saturation excess runoff occurs if precipitation falls on saturated soils; otherwise precipitation will infiltrate or pond in queue for infiltration. Soil saturation can occur from a rising water table during the redistribution of subsurface water with the topographic index. The topographic index is modified for impervious surfaces, using the theory of TOPURBAN by Valeo and Moin (2000). Infiltration rates are a function of cumulative infiltration and soil properties, with hydraulic conductivity decaying with soil depth based on either an exponential function or power function (Wang et al. 2006).

Evaporation and evapotranspiration rates are calculated at each time step based on some fraction of the potential rate. The fraction is the ratio of actual water storage to maximum storage capacity based on work of Deardorff (1978) and Noilhan and Planton (1989), to represent the increasing resistance of a thinner layer of water. The maximum potential evaporation (*pE*) and evapotranspiration (*pET*) for various water storage zones are determined by preprocessing weather data (Hirabayashi and Endreny 2016). *pE* and *pET* are calculated in three distinct terms: *pE* from free water in the tree canopy; *pE* from free water in the short vegetation (shrub and herbaceous) and depression storages (pervious and impervious); and *pET* of soil water volumes through direct evaporation and vegetation transpiration. This distinction is used to allow for higher potential values above the turbulent tree canopy, lower values above the low-lying shrub and surface stores, and further constrained values in soils because of resistances. *pE* from vegetation is calculated using a modified Penman-Monteith equation (Shuttleworth 1993) with distinct values for canopy resistance. *pET* of soil water through vegetation is also based on Penman-Monteith formulations, where potential values are modified downward based on soil and leaf canopy moisture resistances. Direct evaporation from soil surfaces is based on the same soil moisture resistances. In allocating water to evaporation and evapotranspiration, an energy and water balance are maintained, guarding against taking more water than is physically possible.

Runoff in channels is the sum of precipitation falling directly on land cover types identified as water, subsurface flow, overland pervious runoff, and DCIA impervious area runoff. The water budget is a vertical balance in one dimension, recorded as depths, which can be integrated across the landscape area to generate volumes. The runoff hydrograph is presented as a volume per time, and the quantity of water in runoff can be reported with time step options of hourly, weekly, monthly, or yearly. The water quantity in runoff is reported in its component parts, showing the subsurface flow, the pervious runoff, and the impervious runoff. Pollutographs are also generated for any of the EMC constituents listed in Table 19.1. Both water quantity and quality for the base case and the alternative scenarios are plotted on a shared graph or table for quick comparison.

i-Tree Hydro's model architecture is designed for modularity and extendibility. This was originally achieved using object-oriented design (OOD), which is an approach to C++ software development which guides programmers to make programs, to the extent possible, open for extension and closed for modification. Using OOD, when simulation requirements change in the future, the model can be extended mostly by adding new code, not by changing old code that already works (Wang et al. 2005b). While i-Tree Hydro's model architecture has changed over time, adapting to changes in development constraints and goals, the principles of modularity and extendibility remain useful. This is true in the user-oriented side of model design as well: the program is designed to be flexible, allowing users to set up a simple project or extend it with additional parameterization and complexity as needed to meet a user's modeling goals.

19.2.2 Rational Method

The Rational method equation, used by engineers since the 1800s (Chin 2013), is a spatially lumped model that simulates peak runoff rates for the single outlet of a land parcel subjected to a single precipitation event. The Rational method water budget uses inputs of constant rainfall intensity for any duration equal to the time of concentration and represents the fraction of rainfall released as runoff. The model operates at one time step. The Rational peak runoff rate equation is:

$$Q_p = 0.00278\, C\, i\, A \tag{19.2}$$

where Q_p is peak runoff rate ($m^3\ s^{-1}$), C is runoff coefficient for a given frequency event (Table 19.2), i is rainfall intensity ($mm\ h^{-1}$), and A is drainage area (hectares),

Table 19.2 Simplified table of Rational method runoff coefficients

Ground cover type	Runoff coefficient, c
Lawns	0.05–0.35
Forest	0.05–0.25
Cultivated land	0.08–0.41
Meadow	0.1–0.5
Parks, cemeteries	0.1–0.25
Unimproved areas	0.1–0.3
Pasture	0.12–0.62
Residential areas	0.3–0.75
Business areas	0.5–0.95
Industrial areas	0.5–0.9
Asphalt streets	0.7–0.95
Brick streets	0.7–0.85
Roofs	0.75–0.95
Concrete streets	0.7–0.95

Source: LMNO 2018, reused with permission of LMNO Engineering, Research, and Software, Ltd.; https://www.LMNOeng.com/Hydrology/rational.php

and 0.00278 converts hectare millimeters per hour to $m^3 s^{-1}$. The Rational method is often used to design flow capacity for a structure, getting rainfall from intensity-duration-frequency (IDF) curves for the region of interest, with duration equal to the time of concentration and frequency set by local authorities (e.g., a 10-year storm frequency). The Rational method is predominantly used to design stormwater infrastructure for the peak runoff rate, including for low-impact development and green infrastructure designs (Montalto et al. 2007; Soulis et al. 2017), with a history of fusion into other design software, including geographical information systems (Djokic and Maidment 1991).

19.2.2.1 Effect of Trees

The impact of trees on peak flow or runoff reduction and water quality is not explicitly represented in this approach. The Rational method is not designed for simulating changes in tree cover in a mixed land cover scenario. While users could create a composite runoff coefficient to represent a parcel with sub-areas in tree and non-tree cover, the Rational method does not implicitly represent the drainage flow paths of sub-areas. Instead, users should apply the Rational method for each sub-area within the mixed area and route runoff from each sub-area to the outlet. Using this approach, users can adjust the sub-area in tree cover, based on the forest runoff coefficient value. This process has the advantage of being relatively simple in terms of calculations but relies on coefficients that do not necessarily capture the local tree processes and conditions that impact local stream flow and water quality.

19.2.3 Curve Number Approaches

The Curve Number (CN) model, developed by the USDA Natural Resources Conservation Service (Chin 2013), is a spatially lumped model that simulates runoff depth for a land parcel subjected to a single precipitation event. It has been combined with hydrograph models to estimate peak discharge (TR-55) and has been inserted into continuous weather water budget models (e.g., SWMM and SWAT; see below). The CN model water budget uses input of a 24-hour duration rainfall depth and represents the initial abstractions of rainfall prior to runoff. The model typically operates at one time step of 1 day. The CN runoff depth equation is:

$$Q = (P - (I_a S))^2/(P + ((1 - I_a)S)) \tag{19.3}$$

where Q is the runoff (cm), P rainfall (cm), S potential maximum storage (cm), and I_a initial abstraction (cm). The I_a is typically 0.2 and represents surface depression storage, canopy interception, and infiltration. The S is related to the CN value, with a maximum of 100 representing no storage:

$$S = (2540/\mathrm{CN}) - 25.4 \tag{19.4}$$

Factors that determine CN include the parcel land cover, soil hydrologic soil group, hydrologic condition, and antecedent moisture condition. The CN method is often used to determine how land use development changes CN values and affects runoff depths for design events taken from an IDF curve, with durations of 24 h and frequency set by local authorities (e.g., a 10-year storm frequency). The Curve Number model is predominantly used for estimating changes in stormwater runoff with changes to precipitation, land cover, and CN values by practitioners and academics (Chin 2017; Li et al., 2018), and the model is regularly being updated to increase flexibility and accuracy, including changes to antecedent moisture settings (Sahu et al. 2010) and initial abstractions (Jain et al., 2006).

Innovative uses of the Curve Number model for water quality analysis include the US-based National Green Values Calculator (CNT 2018) and the Center for Watershed Protection (CWP)'s Making Urban Trees Count tree crediting framework (CWP 2017). The Green Values Calculator compares the performance, costs, and benefits of BMPs to conventional stormwater practices. This calculator determines the runoff volume capture capacity of the BMPs and the total runoff volume produced by the pre-development, conventional development, and green development scenarios using CN values. The tool is meant for a single site or a campus of buildings contained on a single site. The CWP tree crediting framework uses a range of CN inputs to establish pollutant load reduction credits and stormwater performance-based credits for trees around the United States. This provides a standard method in the United States for trees to be accounted for in water quality regulation compliance and in site-based stormwater management requirements.

19.2.3.1 Effect of Trees

The impacts of trees on runoff reduction and water quality are not explicitly represented in the CN equations. It is common, however, to create a composite CN for mixed land cover when there is no internal drainage from one sub-area to another. An average tree cover is assumed with the urban type land cover (e.g., residential), but the amount of tree cover is unknown. As tree cover varies in urban areas across the United States, this average tree cover approach is a limitation in estimating the effect of urban trees on runoff across the United States. Trees are accounted for under land cover including woods-grass combination (orchard or tree farm) and woods classes under other agricultural land types.

The National Green Values Calculator also adds trees as a specific BMP. Trees only affect progress toward the runoff reduction goal by decreasing the site impervious area that is used to determine the volume of precipitation/runoff that must be captured on site. In addition, tree box filters can provide a volume capacity benefit when the user defines the area and depth of the filter box installed around the tree (CNT 2018).

The Making Urban Trees Count framework estimates stormwater credits for trees based on how trees influence precipitation, runoff generation, and soil water holding capacity over the course of a year in a range of locations, soil types, and tree types (small, medium, and large deciduous trees; and small and large coniferous trees). Precipitation is reduced as a function of leaf area index. Runoff generation is a function of CN values, with CN adjusted based on the temporal effects of stemflow, the improved physical structure of soils by tree roots, and the soil water holding capacity at a given time step. Soil water holding capacity is influenced by evapotranspiration, which in turn is a function of maximum potential evapotranspiration and soil water storage at a given time step (Hynicka and Caraco 2017). This framework extends the utility of the Curve Number approach, adding parsimonious accounting of tree effects to enable stormwater credits for trees in the United States.

19.2.4 EPA National Stormwater Calculator

The US Environmental Protection Agency (US EPA)'s National Stormwater Calculator (SWC), managed by the EPA (Rossman and Bernagros 2018), is a spatially semi-distributed model that simulates the amount of rainwater and frequency of runoff from a specific site, for either a continuous simulation period or event-based storm. Model parameters include local soil conditions, land cover, and historic rainfall records. The user interface accesses several US databases that provide soil, topography, rainfall, and evaporation information for a chosen site. The user supplies information about the site's land cover and selects low-impact development (LID) controls they would like to use. Land cover in the project area is described as percentages of forest, meadow, lawn, desert, and impervious cover. Different types of pervious cover capture different amounts of rainfall on vegetation or in natural depressions and have different surface roughness. Surface roughness affects the velocity of runoff and in turn the potential for infiltration. All non-pervious area is presumed to be DCIA, and there is an LID option for disconnecting some of the impervious area. The LID controls include seven green infrastructure practices: disconnection, rain harvesting, rain gardens, green roofs, street planters, infiltration basins, and permeable pavement. An LID cost estimation module is available to assist in LID project planning. Cost estimates are based on user-defined size configuration of the LID control along with other project and site-specific parameters. The SWC uses the EPA's Storm Water Management Model (SWMM; Rossman and Huber 2016) as its underlying computational engine, sending user inputs to SWMM and post-processing its outputs in the background without requiring user involvement. The SWC is intended to provide screening-level analysis of small footprint sites, offering an accessible tool for a broader audience than SWMM in most cases. The EPA National SWC is predominantly used by practitioners and developers to examine reduction in stormwater runoff with implementation of LID and other management measures, in locations supported by the preloaded data (Kertesz et al. 2014; Schifman and Shuster 2019).

19.2.4.1 Effect of Trees

There is no explicit simulation of tree cover in the SWC, but as in the SWMM model described below, there are features intended to represent the influence of trees on the water cycle. Forest land cover and certain LID controls can approximate tree canopy interception by adjusting depression storage and can approximate transpiration by adjusting evaporation.

19.2.5 SWMM

The Storm Water Management Model (SWMM), managed by the EPA (Rossman 2017; Rossman and Huber 2016), is a spatially semi-distributed model that simulates runoff quantity and quality for storm sewers and their contributing areas subjected to a single precipitation event or continuous weather. The SWMM routines are divided into hydrologic and hydraulic components, each using 100s of equations and descriptive parameters. The hydrologic component maintains a water budget (i.e., conservation of mass) for snow and rain for the area surrounding the storm sewer using weather inputs of precipitation, temperature, evaporation, and wind speed and then simulating surface runoff to storm sewer inlets, infiltration to soils, groundwater, snowmelt, and inflow and infiltration to sewers. The hydraulic component maintains conservation of energy, momentum, and mass equations within the storm sewer, using inputs of inflows from the hydrologic component, then simulating the storm sewer network conveyance of flow with dynamic wave or kinematic wave hydraulic methods, and considering the impact of pumps and flow regulators (e.g., orifices, weirs, outlets). The user and input data define the time step (typically

1 min to 1 h with the model able to interpolate evaporation from monthly data), storm sewer network characteristics, land use types, water quality constituents and sources, and buildup and washoff rates by land use type.

SWMM is often used to determine how changes in storm water management (e.g., installation of green infrastructure, street sweeping schedules, and sewer networks) affect discharges based on preferred or regulated water quantity and water quality targets. The analysis can use any duration or frequency design storms from an IDF curve or continuous weather from historical or predicted periods. SWMM is actively maintained and provides a user guide with a quick start tutorial and tables of common parameter values. SWMM is predominantly used by practitioners and academics to understand the hydraulics of runoff within the storm sewer, including the timing, frequency, and volume of discharges with changes in land cover or precipitation, with research into improved calibration (Barco et al. 2008), low-impact development (Palla and Gnecco 2015; Aad et al. 2010), and accounting for uncertainty (Muleta et al. 2013).

19.2.5.1 Effect of Trees

There is no explicit simulation of tree cover in SWMM, but there are features intended to represent the influence of trees on the water cycle. The SWMM low-impact development (LID) controls allow simulation of runoff capture and treatment prior to entry into the storm sewer, and these LID features include permeable pavement, rain gardens, green roofs, street planters, rain barrels, infiltration trenches, and vegetative swales. SWMM envisions trees and other vegetation would be in a certain LID, such as rain gardens and vegetative swales. For tree or other vegetation canopy interception, the SWMM manual recommends this depth of storage be included in the model depression storage property and that it will be reset when evaporation removes water from depression storage. For evapotranspiration by trees and other vegetation, SWMM allows for the user to define the soil depth to which evapotranspiration can occur and the fraction of total evaporation available to evapotranspiration. SWMM can also simulate the storage area within the LID that is occupied by vegetation, so water storage volumes can be adjusted. SWMM will simulate user-defined land uses, and their effect on water quality, including the removal efficiency of best management practices (BMPs) located with that land use. These BMPs are not specifically intended to represent the role of trees but could be parameterized to represent such effects.

19.2.6 HSPF

The Hydrological Simulation Program-FORTRAN (HSPF), managed by the EPA (Bicknell et al. 1993), is a spatially semi-distributed model that simulates runoff quantity and quality for watersheds, sub-watersheds, and interior land segment units with homogeneous hydrologic response, when subjected to continuous weather. The HSPF routines are divided into application and utility modules, most using 100s of equations and descriptive parameters. The application modules include pervious land, impervious land, reaches and reservoirs, and best management practices, while the utility modules include management and analysis of time series of weather input and water flux, storage, and quality output. The land segments maintain a water budget for snow and rain, considering elevation impacts on air temperature and implementing vertical redistribution of water between the vegetation canopy, depression storage, soil moisture, groundwater, and the atmosphere via evapotranspiration.

A lateral redistribution of surface and subsurface water is implemented to move excess water to the channel network, and the Green-Ampt model is used to determine infiltration and surface runoff. The reach and reservoir module simulates runoff hydrographs, transport and retention of pollutants from land segments, as well as water temperature, dissolved oxygen, carbon, and plankton. The user and input data define the time step (typically 1 min to 1 h, with a maximum of 1 day), reach and reservoir characteristics, depth to discharge relationships to determine hydrographs, land use types, water quality constituents and sources, and buildup and washoff rates by land use type. Water quality constituents can be in gaseous, soluble, or sediment-sorbed forms, and routines are available for sediment production and erosion and the fate of pesticides and nitrogen and phosphorus.

HSPF is used to determine how changes in watershed management (e.g., land use, reservoir operations, flow diversions, in-stream aeration, and best management practices including street sweeping) or climate affect discharges based on preferred or regulated water quantity and water quality targets. HSPF provides a database of parameters from prior model runs and supports use of the PEST tool for parameter calibration and evaluation. HSPF derives from the Stanford Watershed Model, is not actively updated, but is accessible in the actively maintained EPA BASINS, with tutorials. HSPF is predominantly used for analysis of changes to runoff quality and quantity to changes in weather, land cover, and configuration of routing or best management practices, often with regulatory applications such as total maximum daily load planning (Benham et al. 2006; Mohamoud and Zhang 2019; Lee et al. 2018)

19.2.6.1 Effect of Trees

There is no explicit simulation of tree cover in HSPF, but there are features intended to represent the influence of trees on the water cycle. The HSPF best management practice (BMP) module allows modification of the linkage between land and water segments through specification of generic functions, which could represent tree effects on runoff and pollutant reductions. HSPF expects trees and other vegetation would be in certain land uses, such as forest, agro-forestry, or urban green infrastructure, and simulates vegetation canopy interception, as well as the influence of vegetation on the nitrogen and phosphorus cycles. Evapotranspiration by trees and other vegetation will reset interception and reduce the water storage in both soil layers and subsurface discharge. The HSPF water temperature routines provide a parameter to reduce incoming radiation due to shading by trees and other structures.

19.2.7 SWAT

The Soil and Water Assessment Tool (SWAT) model, managed by the Texas Water Resources Institute (Neitsch et al. 2011), is a spatially semi-distributed model that simulates runoff quantity and quality for watersheds, sub-watersheds, and interior land segment units with homogeneous hydrologic response, when subjected to continuous weather. The SWAT routines can be divided into hydrologic and biogeochemical, including soil, vegetation and atmospheric features, and water quality components, each using 100s of equations and descriptive parameters. The hydrologic response units maintain a water budget for snow and rain, requiring weather inputs and considering elevation impacts on air temperature and precipitation, implementing vertical redistribution of water between the vegetation canopy, depression storage, soil moisture, groundwater, and the atmosphere via evapotranspiration. SWAT is used for a variety of applications around the world including assessments of hydrology, pollutants, conservation agriculture (Ullrich and Volk 2009), and climate change impacts, as reviewed by Gassman et al. (2007).

SWAT simulates crop growth and scheduled management operations (e.g., fertilization, grazing, tillage) to estimate their impact on water quantity and quality. A lateral redistribution of surface and subsurface water is implemented to move excess water to the channel network, with the CN or the Green-Ampt model used to determine infiltration and surface runoff. SWAT uses a set of routing routines to transport water, its temperature, sediment, nutrients, and other constituents through its channels and reservoirs and a version of the Rational method combined with estimates of velocity, flow distance, and lag time to determine hydrograph timing and peak flow.

The user and input data define the time step (typically 1 h to 1 day, with the model able to interpolate from monthly data), reach and reservoir characteristics, land use types, water quality constituents and sources, and buildup and washoff rates by land use type. Water quality constituents can be in gaseous, soluble, or sediment-sorbed forms, and routines are available for sediment erosion and the fate of pesticides, nitrogen and phosphorus, carbon, and bacteria. SWAT is typically used to determine how long-term (> 1-year) changes in watershed management (e.g., land use, reservoir operations, flow diversions, in-stream aeration, and best management practices including street sweeping) or climate affect discharges based on preferred or regulated water quantity and water quality targets. SWAT provides several model input parameter databases, including plant growth, and a weather generator and example watershed configurations. SWAT derives from water quantity and quality models known as CREAMS, GLEAMS, EPIC, SWRRB, CFARM, ROTO, and QUAL2E and is actively updated (SWAT+ is in development) as of early 2019.

19.2.7.1 Effect of Trees

There is explicit simulation of tree cover in SWAT, as one of seven agricultural land cover groups that also include perennial and warm and cold season annual crops. SWAT simulates the rooting depth as the maximum allowed for the tree and soil, partitions new growth between leaves and woody growth, and converts a fraction of biomass to residue at the end of each growth season. SWAT uses minimum and maximum leaf area index with annual phenology to grow out the canopy area. SWAT explicitly simulates canopy interception of precipitation when the Green-Ampt infiltration method is used, but this is implicitly represented when the CN method is used to estimate initial abstractions. SWAT simulates the influence of vegetation on the nitrogen and phosphorus cycles. Evapotranspiration by trees and other vegetation will reset interception and reduce the water storage in both soil layers and subsurface discharge. The SWAT best management practice (BMP) modules allow specification of generic functions and include a specific vegetative filter strip module to represent tree effects on runoff and pollutant reductions.

19.2.8 RHESSys

The Regional Hydro-Ecological Simulation System (RHESSys) model (Tague and Band 2004) is a spatially semi-distributed model that simulates runoff quantity and quality for small- to mid-sized river basins, interior hillslopes, and smaller patches defined as hydrologic response units, when subjected to continuous weather. The RHESSys routines can be divided into hydrologic and biogeochemical, including soil, vegetation and atmospheric features, and carbon and nitrogen components, using 100s of equations and descriptive parameters. The patches, representing pervious land, roads, or channels, maintain a water budget for snow and rain, requiring weather inputs specific to microclimate zones. It implements vertical redistribution of water between the vegetation canopy, leaf litter, depression storage, soil moisture, groundwater, and the atmosphere via evapotranspiration.

RHESSys simulates plant and tree growth and can schedule management operations (e.g., fire, clear-cutting) to estimate their impact on water quantity and quality. A lateral redistribution of surface and subsurface water can be explicitly routed between patches or approximated with the topographic index, and hillslope water is sent directly to the basin outlet, with the Phillip and Green-Ampt models used to determine infiltration and surface runoff. RHESSys uses detailed carbon and nitrogen cycle routines. The user and input data define the time step (typically 1 day), parameters regulating carbon and nitrogen fluxes and storage, and a geographic information system is recommended to pre-process elevation, soil, and land cover data. RHESSys is typically used to determine how changes in watershed management (e.g., land use and road networks) or climate affect the storage and flux of water, carbon, and nitrogen based on water quantity and water quality targets.

RHESSys input pre-processing is supported with GRASS GIS routines and documentation. RHESSys derives from water quantity and quality models known as TOPMODEL (Beven et al. 1995), DHSVM, MTN-CLIM, BIOME-BGC, and $CENTURY_{NGAS}$ and is actively updated. The RHESSys model is predominantly used to analyze hydro-ecological processes and test hypothesis useful for guiding management, and uses have varied from analysis of snow-dominated areas (Christensen et al. 2008) to regions at risk of fires (Tague et al. 2004).

19.2.8.1 Effect of Trees

There is explicit simulation of tree cover in RHESSys. Trees are within a landscape patch that can contain a mixture of other land cover types. RHESSys simulates tree and other vegetation leaves, stems, and roots and the storage and flux of water, carbon, and nitrogen in these plant components when they are living and as dead organic residue. The model estimates the rooting zone temperature and moisture content, partitions new growth between leaves and woody growth, and converts a fraction of biomass to residue at the end of each growth season. RHESSys uses minimum and maximum leaf area index with annual phenology to grow out the canopy area and maintains an upper and lower canopy layer. RHESSys explicitly simulates canopy interception of precipitation and the extinction of solar radiation used in photosynthesis (and respiration) routines, considering shaded and sunlit leaves separately. RHESSys simulates the influence of feedback between vegetation growth and the carbon and nitrogen cycles and simulates the nitrogen cycle in the soil, as well as its flushing from patches to the receiving water. Evaporation resets canopy interception and reduces the water content in depression storage, and RHESSys simulates stomatal conductance to update carbon storage in the vegetation and soil moisture storage.

19.3 Summary of Modeling Urban Forest Hydrology

There are numerous models used to estimate tree effects on hydrology. These models have varying methods and levels of simplifying assumptions. Some models use simplified approaches to facilitate ease of estimation by managers. Other models use more detail and sophisticated estimation procedures but, in doing so, can limit the ease of use by managers. By understanding the differences in the models discussed in this chapter, users can choose the best models for their application and expertise. The models described in this chapter are summarized to aid in understanding their key attributes related to trees and hydrologic functions (Table 19.3). The models described in this chapter were selected to meet the following criteria:

Table 19.3 List of models described in this chapter and their key attributes

Model	Feature			
	Input complexity	Runoff (R), hydrology (F), or hydraulic (P)	Event-based (E) or continuous (C)	Water quality methods
i-Tree Hydro	Moderate	R, F	E and C	EMC
Rational method	Low	R	E	N/A
Curve Number	Low	R	E	N/A
EPA SWC	Low	R	E and C	N/A
EPA SWMM	Moderate	R, F, P	E and C	Buildup/washoff, urban BMPs
HSPF	High	R, F	E	Chemical applications, Buildup/washoff, sediment, pesticide, nutrient cycles, Ag and urban BMPs
SWAT	High	R, F	C	Chemical applications, Buildup/washoff, sediment, pesticide, carbon, bacteria, nutrient cycles, Ag and urban BMPs
RHESSys	High	F	C	Carbon and nitrogen biogeochemistry

R – runoff: This modeling approach partitions rainfall to runoff, without necessarily representing the theory and processes of cause and effect used in mechanistic models

F – flow, hydrology: This modeling approach maintains conservation of mass (i.e., a water budget) using weather inputs. Model processes can include surface runoff to storm sewer inlets, infiltration to soils, groundwater, snowmelt, and inflow and infiltration to sewers

P – pressure, hydraulic: This modeling approach maintains conservation of energy, momentum, and mass equations using inputs of inflows from the hydrologic component. Model processes including storm sewer network conveyance of flow with dynamic wave or kinematic wave hydraulic methods and considering the impact of pumps and flow regulators (e.g., orifices, weirs, outlets)

E – event-based: This modeling approach simulates one precipitation event to estimate runoff response, typically ignoring evapotranspiration

C – continuous: This modeling approach simulates a period of time to represent any number of precipitation events and the evapotranspiration that follows

- Public domain/free to use
- Useful to the broad stormwater management community/not too specialized
- Capable of simulating tree or forest effects on hydrology

This list may not be comprehensive, as some models may have been omitted that meet these criteria.

Each of these models has some form of a water balance to partition precipitation into runoff, and the moderate- to high-complexity models typically account for other water storage and fluxes, such as soil moisture and evapotranspiration.

Biogeochemical processes are typically regulated by these water storages and fluxes. In general, hydrological processes drive the transport of pollutants, as well as affect their chemical fate by regulating temperature, residence time, chemical interactions, and exposure to reactions, uptake, or assimilation. Some models have algorithms to simulate specific pollutant cycles, including those of sediment, pesticide, carbon, bacteria, nitrogen, and phosphorus. These cycles are generally more complex to model than hydrological cycles; thus many models use a simplified approach for estimating water quality or do not estimate water quality. All of the high-complexity models covered in this chapter allow the user to ignore complexity and simulation of pollutants and focus instead on parameterization and simulation of hydrologic processes. i-Tree Hydro, a moderate-complexity model, uses a fixed estimate of pollutant concentration (EMC) that is independent of hydrology and simply multiplies surface runoff by the EMC of a pollutant to estimate pollutant loading. The models which do explicitly account for pollutant cycles and biogeochemical processes (e.g., SWAT, RHESSys) are advanced tools that have more extensive and uncertain input requirements. The level of expertise required to use high-complexity pollutant cycle models often limits use to specialized audiences with advanced training. Given the increasing challenges in managing water quality in landscape scales, advances in accessibility of modeling biogeochemical processes related to vegetation will be an important next step for modeling the hydrologic effects of trees.

The degree to which the trees affect water quality is a function of many factors, including the hydrology, soils, and vegetation. Trees will have fewer beneficial impacts on water quality when the runoff is decoupled from natural processes related to physical filtering, chemical transformation, or biological uptake. This decoupling occurs when concentrated surface flow bypasses filtering, such as street runoff in a gutter, water transport through pipes, or pervious runoff in a rill. It also happens when deeper subsurface flows pass below tree root zones, which can occur in urban areas when flood control measures lower the water table. Actions to help minimize water pollution include (1) reducing surface runoff (e.g., reducing impervious area or increasing infiltration), (2) increasing water contact with soils and vegetation (e.g., increasing infiltration and water retention), and (3) reducing surface pollutants (e.g., increasing street sweeping or reducing deposition of pollutants). Urban trees and natural surfaces help break up the continuity of impervious cover, which can substantially reduce runoff. In addition, trees can provide numerous other benefits to society (Nowak 2018).

19.3.1 Future Directions

There are opportunities and a need for collaboration and cross-pollination among the diversity of hydrology models. Currently, some strengths of hydrology models are limited to only one tool or specialized tools, for example, explicit simulation of trees (i.e., i-Tree Hydro), whole-system biogeochemical modeling (i.e., RHESSys), and explicit documentation of a tool's modeling philosophy and architecture (i.e., HSPF). Technology can facilitate bridging gaps among models and users. Increased connectivity and transparency are exemplified in OpenSWMM, a free knowledge base for the user community of EPA's SWMM, hosted by CHI (OpenSWMM 2018). OpenSWMM includes an email-based forum for users to share ideas and ask questions; a web-based source code viewer with an interface for commenting or suggesting changes to SWMM code; and a platform for sharing SWMM-based research projects, enabling the project itself to receive expert attention and feedback from the community while also providing examples of model use for the community. Leveraging advances in the technology can empower the next generation of models to be more accessible and provide better support for their users.

The future of eco-hydrology modeling is likely to become more advanced in accuracy and accessibility as advances in computer science are applied to the field. Broader and richer data sources (from advances in remote sensing, instrumentation, and open data sharing) along with more powerful means of analysis (with advances in machine learning and hardware for data storage and computations) could revolutionize how hydrology models are developed, maintained, parameterized, and used. Distributed computing can enable basic devices to access robust, computationally intensive models. There are already examples of innovations in computer science diffusing into hydrology modeling, as is the case with Google's AI-enabled flood forecasting (Matias 2018).

The overarching needs that users have for models of urban tree effects on hydrology will continue to shape future model development. Those needs include crediting of trees to meet requirements of stormwater regulations, decision support for optimization of hydrologic benefits from urban forests, and valuation of tree effects on water resources. Crediting trees for their role in meeting stormwater management requirements has been the focus of recent investigations (Kuehler et al. 2017; CWP 2017; CWP 2018) and is being implemented in some US cities (MPCA 2017). Decision support is becoming a focus of more recent models, including i-Tree Landscape and the EPA's SWC. Valuation of runoff reduction and water quality improvement is complex, but approaches do exist (e.g., McPherson et al. 2007). Each of these modeling needs presents opportunities for significant improvement. As hydrology models and field studies advance in estimation of tree effects, the capacity to meet these modeling needs will increase. While more research is needed regarding urban tree effects on hydrology, models are currently being used to assess the hydrologic benefits that trees provide. Water resource managers can refer to these tools today to better account for the hydrologic effects of trees and hopefully improve urban water management and quality through better urban forest management and designs.

19.4 Disclaimer

The use of trade, firm, or corporation names in this article is for the information and convenience of the reader. Such does not constitute an official endorsement or approval by the US Department of Agriculture Forest Service, Davey Tree, or the SUNY College of Environmental Science and Forestry of any product or service to the exclusion of others that may be suitable.

References

Aad MPA, Suidan MT, Shuster WD (2010) Modeling techniques of best management practices: rain barrels and rain gardens using EPA SWMM-5. J Hydrol Eng 15:434–443. https://doi.org/10.1061/(asce)he.1943-5584.0000136

Aubertin GM (1971) Nature and extent of macropores in forest soils and their influence on subsurface water movement. Northeastern Forest Experiment Station, Research Paper NE-192. Upper Darby, PA. 33 p

Barco J, Wong KM, Stenstrom MK (2008) Automatic calibration of the US EPA SWMM model for a large urban catchment. J Hydraul Eng 134:466–474. https://doi.org/10.1061/(asce)0733-9429(2008)134:4(466)

Benham BL, Baffaut C, Zeckoski RW, Mankin KR, Pachepsky YA, Sadeghi AA et al (2006) Modeling bacteria fate and transport in watersheds to support TMDLs. T ASABE 49:987–1002. https://doi.org/10.13031/2013.21739

Beven KJ, Kirkby MJ (1979) A physically based, variable contributing area model of basin hydrology. Hydrol Sci Bull 24:43–69. https://doi.org/10.1080/02626667909491834

Beven KJ, Lamb R, Quinn P, Romanowics R, Freer J (1995) Topmodel. In: Singh VP (ed) Computer models of watershed hydrology. Water Resources Publications, Colorado, pp 627–688

Bicknell BR, Imhoff JC, Kittle JL, Donigian AS, Johanson RC (1993) Hydrological simulation program - FORTRAN (HSPF): user's manual release 10. US Environmental Protection Agency, Athens, GA

Borah DK, Bera M (2004) Watershed-scale hydrologic and nonpoint-source pollution models: review of applications. T ASABE 47:789–803. https://doi.org/10.13031/2013.16110

Bouraoui F, Grizzetti B (2014) Modelling mitigation options to reduce diffuse nitrogen water pollution from agriculture. Sci Total Environ 468-469:1267–1277. https://doi.org/10.1016/j.scitotenv.2013.07.066

Brookes A (1988) Channelized rivers: perspectives for environmental management. Wiley, New York

Carlyle-Moses D, Livesley S, Baptista M, Thom J, Szota C (this volume) Urban trees as green infrastructure for stormwater mitigation and use. In: Levia DF, Carlyle-Moses DE, Iida S, Michalzik B, Nanko K, Tischer A (eds) Forest-water interactions, Ecological Studies. Springer, Cham

Center for Neighborhood Technology (CNT) (2014) The prevalence and cost of urban flooding: a case study of cook County, IL. Center for Neighborhood Technology. 26 p. https://www.cnt.org/sites/default/files/publications/CNT_PrevalenceAndCostOfUrbanFlooding2014.pdf (September 2018)

Center for Neighborhood Technology (CNT) (2018) Green values national stormwater management calculator. http://greenvalues.cnt.org/national/calculator.php (September 2018)

Center for Watershed Protection (CWP) (2017) Making urban trees count. CWP, Ellicott City, MD. https://www.cwp.org/making-urban-trees-count/ (January 2019)

Center for Watershed Protection (CWP) (2018) Accounting for trees in stormwater models. CWP, Ellicott City, MD. https://owl.cwp.org/mdocs-posts/accounting-for-trees-in-stormwater-models/ (October 2018)

Chin DA (2013) Water resources engineering, 3rd edn. Pearson, Upper Saddle River, NJ

Chin DA (2017) Estimating the parameters of the curve number model. J Hydrol Eng 22. https://doi.org/10.1061/(asce)he.1943-5584.0001495

Christensen L, Tague CL, Baron JS (2008) Spatial patterns of simulated transpiration response to climate variability in a snow dominated mountain ecosystem. Hydrol Process 22:3576–3588. https://doi.org/10.1002/hyp.6961

Deardorff JW (1978) Efficient prediction of ground surface temperature and moisture with inclusion of a layer of vegetation. J Geophys Res 83:1889–1903. https://doi.org/10.1029/JC083iC04p01889

Decina SM, Ponette-González AG, Rindy JE (this volume) Urban tree canopy effects on water quality via inputs to the urban ground surface. In: Levia DF, Carlyle-Moses DE, Iida S, Michalzik B, Nanko K, Tischer A (eds) Forest-water interactions, Ecological studies. Springer, Cham

Djokic D, Maidment DR (1991) Terrain analysis for urban stormwater modelling. Hydrol Process 5:115–124. https://doi.org/10.1002/hyp.3360050109

Dobbs C, Martinez-Harms MJ, Kendal D (2017) Ecosystem services. In: Ferrini F, Konijnendijk CC, Fini A (eds) Routledge handbook of urban forestry. Routledge, New York, pp 51–64

Doherty J (2001a) PEST-ASP user's manual. Watermark Numerical Computing, Brisbane, Australia

Doherty J (2001b) PEST surface water utilities user's manual. Watermark Numerical Computing, Brisbane, Australia

Driver NE, Mustard MH, Rhinesmith RB, Middelburg RF (1985) U.S. geological survey urban-stormwater data base for 22 metropolitan areas throughout the United States. United States Geological Survey, Open-File Report 85-337, Lakewood, CO.

Dwyer JF, McPherson EG, Schroeder HW, Rowntree RA (1992) Assessing the benefits and costs of the urban forest. J Arboric 18:227–234

Federal Interagency Stream Restoration Working Group (FISRWG) (1999) Stream corridor restoration – principles, processes, and practices, Federal Interagency Stream Restoration Working Group, Washington, DC. NTIS: PB98-158348INQ.

Gash JHC, Lloyd CR, Lachaud G (1995) Estimating sparse forest rainfall interception with an analytical model. J Hydrol 170:79–86. https://doi.org/10.1016/0022-1694(95)02697-N

Gassman PW, Reyes MR, Green CH, Arnold JG (2007) The soil and water assessment tool: historical development, applications, and future research directions. T Asabe 50:1211–1250. https://doi.org/10.13031/2013.23637

Guswa AJ, Brauman KA, Brown C, Hamel P, Keeler BL, Sayre SS (2014) Ecosystem services: challenges and opportunities for hydrologic modeling to support decision making. Water Resour Res 50:4535–4544. https://doi.org/10.1002/2014WR015497

Hammer TR (1972) Stream channel enlargement due to urbanization. Water Resour Res 8:1530–1540. https://doi.org/10.1029/WR008i006p01530

Herricks EE (1995) Stormwater Runoff and Receiving Systems: Impact, Monitoring, and Assessment CRC Lewis Publishers. York, New

Hirabayashi S, Endreny TA (2016) Surface and upper weather pre-processor for i-Tree Eco and Hydro. https://www.itreetools.org/eco/resources/Surface_weather_and_upper_air_preprocessor_description.pdf (October 2018).

Hynicka J, Caraco D (2017) Relative and absolute reductions in annual water yield and non-point source pollutant loads of urban trees. Crediting framework product #2 for the project making urban trees count: a project to demonstrate the role of urban trees in achieving regulatory compliance for clean water. Center for Watershed Protection, Ellicott City, MD.

Jackson IJ (1975) Relationships between rainfall parameters and interception by tropical forest. J Hydrol 24:215–238. https://doi.org/10.1016/0022-1694(75)90082-7

Jain MK, Mishra SK, Babu PS, Venugopal K, Singh VP (2006) Enhanced runoff curve number model incorporating storm duration and a nonlinear Ia-S relation. J Hydrol Eng 11:631–635. https://doi.org/10.1061/(asce)1084-0699(2006)11:6(631)

Kertesz R, Green OO, Shuster WD (2014) Modeling the hydrologic and economic efficacy of stormwater utility credit programs for US single family residences. Water Sci Technol 70:1746–1754. https://doi.org/10.2166/wst.2014.255

Kirnbauer MC, Baetz BW, Kenney WA (2013) Estimating the stormwater attenuation benefits derived from planting four monoculture species of deciduous trees on vacant and underutilized urban land parcels. Urban For Urban Green 12:401–407. https://doi.org/10.1016/j.ufug.2013.03.003

Kuehler E, Hathaway J, Tirpak A (2017) Quantifying the benefits of urban forest systems as a component of the green infrastructure stormwater treatment network. Ecohydrology:e1813. https://doi.org/10.1002/eco.1813

Lee EJ, Kim TG, Choi KS (2018) A study of the load allocation using watershed model and load duration curve in TMDL. KSCE J Civ Eng 22:3222–3232. https://doi.org/10.1007/s12205-018-0910-0

Lefrançois CB (2015) Designing effective stormwater management policies: the role of the urban forest and impervious cover in Vancouver, B.C. G. SCARP Graduating Projects. November 30. https://doi.org/10.14288/1.0300042

Li CL, Liu M, Hu YM, Shi T, Qu XQ, Walter MT (2018) Effects of urbanization on direct runoff characteristics in urban functional zones. Sci Total Environ 643:301–311. https://doi.org/10.1016/j.scitotenv.2018.06.211

Lin YP, Hong NM, Wu PJ, Lin CJ (2007) Modeling and assessing land-use and hydrological processes to future land-use and climate change scenarios in watershed land-use planning. Environ Geol 53:623–634. https://doi.org/10.1007/s00254-007-0677-y

LMNO Engineering, Research, and Software, Ltd (2018) Rational equation calculator. https://www.lmnoeng.com/Hydrology/rational.php (September 2018)

Long X, Bei N, Wu J, Li X, Feng T, Xing L et al (2018) Does afforestation deteriorate haze pollution in Beijing–Tianjin–Hebei (BTH), China? Atmos Chem Phys 18:10869–10879. https://doi.org/10.5194/acp-18-10869-2018

Lyytimaki J (2017) Disservices of urban trees. In: Ferrini F, Konijnendijk CC, Fini A (eds) Routledge handbook of urban forestry. Routledge, New York, pp 164–176

Maneta MP, Silverman NL (2013) A spatially distributed model to simulate water, energy, and vegetation dynamics using information from regional climate models. Earth Interact 17:1–44. https://doi.org/10.1175/2012ei000472.1

Matias Y (2018) Keeping people safe with AI-enabled flood forecasting. The Keyword. https://www.blog.google/products/search/helping-keep-people-safe-ai-enabled-flood-forecasting/ (September 2018).

McPherson EG, Simpson JR, Peper PJ, Gardner SL, Vargas KE, Xiao Q (2007) Northeast community tree guide: benefits, costs, and strategic planting. Gen. Tech. Rep. PSW-GTR-202. Albany, CA: U.S. Department of Agriculture, Forest Service, Pacific Southwest Research Station, 106 p

Minnesota Pollution Control Agency (MPCA) (2017) Examples of stormwater credits for urban trees. https://stormwater.pca.state.mn.us/index.php?title=Examples_of_stormwater_credits_for_urban_trees (January 2019).

Minnesota Pollution Control Agency (MPCA) (2018) Minnesota Stormwater Manual: Available stormwater models and selecting a model. https://stormwater.pca.state.mn.us/index.php?title=Available_stormwater_models_and_selecting_a_model (September 2018).

Mohamoud Y, Zhang H (2019) Applications of linked and nonlinked complex models for TMDL development: approaches and challenges. J Hydrol Engin 24. https://doi.org/10.1061/(asce)he.1943-5584.0001721

Montalto F, Behr C, Alfredo K, Wolf M, Arye M, Walsh M (2007) Rapid assessment of the cost-effectiveness of low impact development for CSO control. Landsc Urban Plan 82:117–131. https://doi.org/10.1016/j.landurbplan.2007.02.004

Muleta MK, McMillan J, Amenu GG, Burian SJ (2013) Bayesian approach for uncertainty analysis of an urban storm water model and its application to a heavily urbanized watershed. J Hydrol Eng 18:1360–1371. https://doi.org/10.1061/(asce)he.1943-5584.0000705

National Resources Conservation Service (NRCS) (2018) WinTR-55 watershed hydrology. https://www.nrcs.usda.gov/wps/portal/nrcs/detailfull/national/water/?cid=stelprdb1042901 (September 2018).

Neitsch SL, Arnold JG, Kiniry JR, Williams JR (2011) Soil and water assessment tool theoretical documentation version 2009, p 618, Texas Water Resources Institute, College Station, TX

Noilhan J, Planton S (1989) A simple parameterization of land surface processes for meteorological models. Mon Weather Rev 117:536–549. https://doi.org/10.1175/1520-0493(1989)117<0536:ASPOLS>2.0.CO;2

Nowak DJ (2018) Quantifying and valuing the role of trees and forests on environmental quality and human health. In: van den Bosch M, Bird W (eds) Nature and public health. Oxford textbook of nature and public health. Oxford University Press, Oxford, UK, pp 312–316

Nowak DJ, Dwyer JF (2007) Understanding the benefits and costs of urban forest ecosystems. In: Kuser J (ed) Urban and community forestry in the Northeast. Springer, New York, pp 25–46

Nowak DJ, Greenfield EJ (2018a) Declining urban and community tree cover in the United States. Urban For Urban Green 32:32–55. https://doi.org/10.1016/j.ufug.2018.03.006

Nowak DJ, Greenfield EJ (2018b) U.S. urban forest statistics, values and projections. J For 116:164–177. https://doi.org/10.1093/jofore/fvx004

Nowak DJ, Wang J, Endreny T (2007) Environmental and economic benefits of preserving forests within urban areas: air and water quality. In: de Brun CTF (ed) The economic benefits of land conservation. The Trust for Public Land, San Francisco, CA, pp 28–47

OpenSWMM (2018) About OpenSWMM – mission and intent. Retrieved from https://www.openswmm.org/Forum/About

Palla A, Gnecco I (2015) Hydrologic modeling of low impact development systems at the urban catchment scale. J Hydrol 528:361–368. https://doi.org/10.1016/j.jhydrol.2015.06.050

Rossman LA (2017) Storm water management model reference model volume II – hydraulics, p. 189, US Environmental Protection Agency, Cincinnati, OH

Rossman LA, Bernagros JT (2018) National stormwater calculator user's guide – version 1.2.0.1 (Revised), US Environmental Protection Agency, Cincinnati, OH

Rossman LA, Huber WC (2016) Storm water management model reference model volume I – hydrology (Revised), p. 231, US Environmental Protection Agency, Cincinnati, OH

Rutter AJ, Morton AJ, Robins PC (1971) A predictive model of rainfall interception in forests. I. derivation of the model from observations in a plantation of Corsican pine. Agric Meteorol 9:367–384. https://doi.org/10.1016/0002-1571(71)90034-3

Rutter AJ, Morton AJ, Robins PC (1975) A predictive model of rainfall interception in forests. II. generalization of the model and comparison with observations in some coniferous and hardwood stands. J Appl Ecol 12:367–380. https://doi.org/10.2307/2401739

Sahu RK, Mishra SK, Eldho TI (2010) An improved AMC-coupled runoff curve number model. Hydrol Process 24:2834–2839. https://doi.org/10.1002/hyp.7695

Schifman LA, Shuster WD (2019) Comparison of measured and simulated urban soil hydrologic properties. J Hydrol Eng 24. https://doi.org/10.1061/(asce)he.1943-5584.0001684

Shuttleworth JW (1993) Evaporation. In: Maidment DR (ed) Handbook of hydrology. McGraw-Hill, New York, pp 5.1–5.51

Smullen JT, Shallcross AL, Cave KA (1999) Updating the U.S. nationwide urban runoff quality database. Water Sci Technol 39:9–16. https://doi.org/10.1016/S0273-1223(99)00312-1

Soulis KX, Ntoulas N, Nektarios PA, Kargas G (2017) Runoff reduction from extensive green roofs having different substrate depth and plant cover. Ecol Eng 102:80–89. https://doi.org/10.1016/j.ecoleng.2017.01.031

Stephan E, Endreny T (2016) Weighting nitrogen and phosphorus pixel pollutant loads to represent runoff and buffering likelihoods. J Am Water Resour Assoc 52. https://doi.org/10.1111/1752-1688.12390

Sutherland RC (2000) Methods for Estimating the Effective Impervious Area of Urban Watersheds. Pract Water Protect 32:193–195

Tague CL, Band LE (2004) RHESSys: regional hydro-ecologic simulation system—an objectoriented approach to spatially distributed modeling of carbon, water, and nutrient cycling. Earth Interact 8:1–42. https://doi.org/10.1175/1087-3562(2004)8<1:RRHSSO>2.0.CO;2

Tague C, McMichael C, Hope A, Choate J, Clark R (2004) Application of the RHESSys model to a California semiarid shrubland watershed. J Am Water Resour Assoc 40:575–589. https://doi.org/10.1111/j.1752-1688.2004.tb04444.x

Thorne CR (1998) River width adjustment. I: processes and mechanisms. J Hydraul Eng 124:881–902. https://doi.org/10.1061/(ASCE)0733-9429(1998)124:9(881)

U.S. Department of Agriculture, Natural Resources Conservation Service (1986) Urban hydrology for small watersheds. Technical Release 55. 164 p

U.S. Environmental Protection Agency (US EPA) (1983) Results of the nationwide urban runoff program: Volume I – final report. U.S. Environmental Protection Agency, PB84-185552, Washington, DC

U.S. Environmental Protection Agency (US EPA) (1996) Managing urban runoff, EPA841-F-96-004G

U.S. Environmental Protection Agency (US EPA) (2002) Urban stormwater BMP performance monitoring, a guidance manual for meeting the national stormwater BMP database requirements

U.S. Environmental Protection Agency (US EPA) (2009) Guidance on the development, evaluation, and application of environmental models. Council for Regulatory Environmental Modeling U.S. Environmental Protection Agency Washington, DC. EPA/100/K-09/003, 90 p https://www.epa.gov/sites/production/files/2015-04/documents/cred_guidance_0309.pdf (September 2018)

U.S. Environmental Protection Agency (US EPA) (2016) Stormwater tees: technical memorandum. https://www.epa.gov/sites/production/files/2016-11/documents/final_stormwater_trees_technical_memo_508.pdf (January 2019)

U.S. General Accounting Office (GAO) (2001) Water quality: better data and evaluation of urban runoff programs needed to assess effectiveness. GAO-01-679

Ullrich A, Volk M (2009) Application of the Soil and Water Assessment Tool (SWAT) to predict the impact of alternative management practices on water quality and quantity. Agric Water Manag 96:1207–1217. https://doi.org/10.1016/j.agwat.2009.03.010

Valente F, David JS, Gash JHC (1997) Modeling interception loss for two sparse eucalypt and pine forests in central Portugal using reformulated Rutter and Gash analytical models. J Hydrol 190:141–162. https://doi.org/10.1016/S0022-1694(96)03066-1

Valeo C, Moin SMA (2000) Variable source area modelling in urbanizing watersheds. J Hydrol 228:68–81. https://doi.org/10.1016/S0022-1694(00)00153-0

Wang J, Endreny TA, Hassett JM (2005a) Flexible modeling package for topographically based watershed hydrology. J Hydrol 314:78–91. https://doi.org/10.1016/j.jhydrol.2005.03.030

Wang J, Endreny TA, Hassett JM (2005b) An object oriented approach to the description and simulation of watershed scale hydrologic processes. Comp Geosci 31:425–435. https://doi.org/10.1016/j.cageo.2004.09.025

Wang J, Endreny TA, Hassett JM (2006) Power function decay of hydraulic conductivity for a TOPMODEL-based infiltration routine. Hydrol Process 20:3825–3834. https://doi.org/10.1002/hyp.6159

Wang J, Endreny TA, Nowak DJ (2008) Mechanistic simulation of urban tree effects in an urban water balance model. J Am Water Resour Assoc 44:75–85. https://doi.org/10.1111/j.1752-1688.2007.00139.x

Wigmosta MS, Vail LW, Lettenmaier DP (1994) A distributed hydrology-vegetation model for complex terrain. Water Resour Res 30:1665–1679. https://doi.org/10.1029/94WR00436

Wong THF, Fletcher TD, Duncan HP, Coleman JR, Jenkins GA (2002) A model for urban stormwater improvement: conceptualization. In: Global solutions for urban drainage, ninth international conference on urban drainage (9ICUD), edited, American Society of Civil Engineers. doi:https://doi.org/10.1061/40644(2002)115

Yang Y, Endreny TA (2013) Watershed hydrograph model based on surface flow diffusion. Water Resour Res 49:507–516. https://doi.org/10.1029/2012WR012186

Yang Y, Endreny TA, Nowak DJ (2011) iTree-Hydro: snow budget and stormwater pollutant updates for the urban forest hydrology model. J Am Water Resour Assoc 47:1211–1218. https://doi.org/10.1111/j.1752-1688.2011.00564.x

Chapter 20
Using Community Planning to Conserve Green Infrastructure and Water Quality

William F. Elmendorf

20.1 Introduction

20.1.1 Green Infrastructure: Providing Ecosystem Processes and Services

Although valued differently by organizations and agencies, green infrastructure is used as a synonym for tree canopies, open space, and other resources found within the urban forest (Pearlmutter et al. 2017). Contrasted to the well-funded and planned gray infrastructure of pipes, roads, and utilities, green infrastructure has been defined as a large-scale planning philosophy to guide the conservation and funding of tree canopies, open space, water, and other resources in community development (Benedict and McMahon 2006). It has also been defined as the engineering of systems (bioretention basins, rain gardens) to provide for and protect water quality, including stormwater management. In this definition, green infrastructure is a cost-effective, resilient approach to managing wet weather impacts that provides many community benefits (US EPA 2017).

Although there are continued concerns about equitable access and provision (Donovan and Mills 2014; Nesbitt et al. 2019) and different ethnic and cultural groups have different landscape preferences (Elmendorf et al. 2005), the importance of this infrastructure to people and the places they live has a long history. Seemingly ahead of their time, in the Brooklyn Report of 1868, Frederick Law Olmsted and Clarence Vaux wrote that landscape beauty paid enterprising cities directly by raising property values and swelling tax revenues (Wilson 1989). They also wrote about the positive effects on human and societal health. Today, conserving green infrastructure in growth and development can help assure that the many economic,

W. F. Elmendorf (✉)
Department of Ecosystem Science and Management, Pennsylvania State University, University Park, PA, USA
e-mail: wfe1@psu.edu

D. F. Levia et al. (eds.), *Forest-Water Interactions*, Ecological Studies 240,
https://doi.org/10.1007/978-3-030-26086-6_20

environmental, and social ecological services, well documented and valued by many authors, are provided (e.g., National Park Service 1992; Irvin 2002; Wolf et al. 2003; Chiesura 2004; McPherson et al. 2005; Hein 2006; Maller et al. 2006; Nowak and Dwyer 2006; Elmendorf 2008; Wachter and Wong 2008; Geis 2009; Miller et al. 2015; Livesley 2016).

The negative ecological effects of spreading or sprawling development have been described by many authors (e.g., Falk et al. 1992; McHarg 1992; Johnson 2001; Nowak and Walton 2005; Benedict and McMahon 2006; Lichtenberg and Hardie 2007; Miller et al. 2015; Martel et al. 2017). Such adverse effects include fragmentation of forests, wildlife habitat, and recreational opportunities and destruction of and damage to productive forest and agricultural areas, loss of biodiversity, undocumented wildfire, erosion of soil and soil nutrients, and damage to and contamination of waterways and aquifers.

Given the negative effects of fast-moving and large-scale land conversion and development in many places, conserving green infrastructure, both at large, nonstructural system (e.g., open space and greenways) and smaller, structural system (landscape trees and bioretention basins) scales, is increasingly important in the provision of water and the management of stormwater (Pennsylvania DEP 2006; National Research Council 2009). However, the ecosystem processes and service conserved in and provided by green infrastructure are not mutually exclusive. Whether you view and assess water quality as a final service, or a contributor to many services, from recreation to human health (Keeler et al. 2012), conserving and providing water will be more successful if a comprehensive view that considers conserving green infrastructure at both large and small scales is provided through community planning and regulatory policy. Providing for water quality and other ecosystem processes and services is complicated by the fact that many elected officials, and the citizens they represent, simply do not understand the values of these services, especially those surrounding short- and long-term human health.

20.1.2 Chapter Overview and Orientation

This chapter supports the argument that ecological processes and services are fundamental to development of community and that water quality is enhanced using community planning and regulatory policy that conserves and protects a wide variety of green infrastructure including soil, trees and forests, riparian areas, steep slopes, and open space. The chapter provides a summary of community planning and policy evaluation. The use of regulatory and nonregulatory community planning principals is introduced to conserve large-scale, nonstructural systems of green infrastructure. In the last section, thoughts about the future of community planning and water quality are provided.

A fundamental proposition of this chapter is that the conservation and provision of ecological services found in green infrastructure are facilitated when complex ecological/social systems are correctly understood and valued. And, although much consideration continues to be given to economic values, many authors have discussed the importance of broader perspectives in understanding human values and the strong emotional values that can be held by people toward nature, especially in growing and changing places, that are not always predicted by utility value alone (Dwyer et al. 1991; O'Brien 2003).

20.2 Community and Green Infrastructure: A Pretext

20.2.1 Community, Community Development, and Green Infrastructure

Although the process, scale, location, basic definition, and measurement of community is debated (Harrington et al. 2008) from an interactional viewpoint (Hillery 1955; Warren 1972; Wilkinson 1991), community is the social field that emerges where people frequently act together in the common concern of life (Kaufman 1959). Community is not a place; rather, it is a place-oriented process. It is an emerging process that develops a synergistic, nonlinear social field, a field of fields. This field is place oriented and is where actions expressing a broad range of views are coordinated, often through conflict. Community is not necessarily a heterogeneous phenomenon, but at some point, actions, actors, and associations are oriented toward the general needs and concerns of the community.

Community is a process of purposive and interrelated action through which residents express their shared interests in local society through both horizontal interactions, including institutions of family, friends, church, schools, and health care, and vertical interactions with corporations, media, and government (Wilkinson 1991). People have and will continue to live in interaction with each other and with broader social fields and territories. The vertical interaction of community with outside institutions and markets, the idea of functional local societies, and the importance of ecological well-being are important in community. Wilkinson (1991: 75) wrote an opinion shared by other authors (Rolston 2000; Elmendorf 2008): "Societal and individual well-being cannot be achieved except in ways that also promote ecological well-being. Ecological well-being, which in a literal sense means the well-being of the 'house' of civilization refers explicitly to natural and other conditions that support and sustain human life. It is not accurate or appropriate to treat the environment as though it was somehow separate from the social life it supports. An active interdependency characterizes the relationship between social life and its surroundings."

From an interactional perspective, there is tension between those whose primary goal is development *in* community and those who prefer development *of* community. Those interested in the *development of* perspective view community as the quality of relationships among residents of a locality that supports community's well-being (Wilkinson 1979; Christenson and Robinson 1989). Development of community requires attention to the structure and fabric of social life (Kaufman 1959; Wilkinson 1991) including economic, social, and ecological spheres. In this viewpoint there is again a fundamental connection between healthy social systems and the ecological systems that support them. It also advocates increasing the capacity, or agency, of people to work together to address their common problems. In contrast, development in community views the community as a setting in which social, political, and economic investment and development activities occur. The locality is treated like a business firm and attention is given to the efficient use and maintenance of productive resources.

Economic development is often an overriding objective in community development (Wilkinson 1991), but economic development without development of community can be divisive, exploitative, and unsustainable (Krannich and Luloff 1991). A developed community is improved through economic, social, and ecological elements and empowered in decision-making and engagement.

20.2.2 Development of Community and Sustainability: A Connection

The 1987 Brundtland Report *Our Common Future* is credited with providing one of the first discussion and principles surrounding the concept of sustainability; economic, environment, and equity. Since 1987, these initial principles of sustainability had been further broadened into sustainability science (Wu 2013) and additional principles have been discussed including generalizing leadership, governance, and politics/policies. All the principles discussed in sustainability are important considerations in the development of community; there is a basic relationship between development of community and that of a sustainable system. This includes using the concept of diversity to decrease vulnerability and increase resilience of communities to environmental and economic changes. The concept of sustainability has been used as a key theme to generate discussions and directions when planning for community growth and development including the conservation of the natural environment in green infrastructure (Hana 2005).

20.2.3 Change and Community Planning

As places experience population growth and development, fueled by both local and ex-local forces, change is inevitable and often expressed in the social, economic, and ecological realities of a place. As a result, planning and managing growth, and in some places decline, can be some of the most baffling, toughest, troubling, and conflicting decisions a community can face. Some communities are better able to deal with change than others: that is, the competent community (Cottrell 1983). In competent communities, people collaborate effectively in identifying the needs of the community, achieve a working consensus of goals and priorities, agree on ways and means to implement agreed-upon goals, and collaborate effectively on required action. The processes found in competent communities can be guided by, or result in, community planning and regulatory policy.

Planners have been guiding growth and change and intervening in community development since the 1893 Chicago's World Fair and the Colombian Exposition gave impetus to planning in America. In theory, community planning attempts to deal with change by adapting individuals and organizations investing in a place to the values expected by that place, creating the desirable, avoiding undesirable environmental and social costs, and assisting private investment and development projects to meet community expectations. A comprehensive planning process is required to actively shape the internal and external forces of change, helping a community to remain vibrant and healthy. It is a cornerstone in both the development of community and in the integration and conservation of green infrastructure into changing landscapes.

20.3 Community Planning: Enabled by Federal and State Governments

In the United States, states, counties, and municipalities have the authority to pass and enforce laws to manage and regulate land use including those effecting water quality. This authority, established and expanded through federal and state judicial review and legislation, is based on the capacity provided by police power. Police power, as a basic authority of democratic government, authorizes government to pass laws to promote and protect public health, safety, general welfare, and morals. The greatest challenge to any land use regulation restricting private property rights is that it must effectively demonstrate that the regulation does promote or protect one of these components.

A court case (Village of Euclid v. Ambler Reality Company 272, U.S. 365 1926) helped confirm the constitutional foundation of police power authority in land use regulation (e.g., zoning and subdivision and land development) as an instrument by which people could attempt to order the destiny of their cities. In this decision, Associate Justice George Sutherland of the US Supreme Court wrote: "Until recently, life was comparatively simple; but with the increase and concentration of population, problems have developed, and constantly are developing, which require additional restrictions in respect to the use and occupation of private lands in communities" (Eisner et al. 1993: 113).

The authority of government to regulate the use of private property is associated with the "taking" issue (McElfish 2004). A "taking" is a problem of government overregulation that goes so far as to result in confiscation of private property without payment of fair compensation or due process. Considering regulation, United States

Supreme Court Justice Oliver Wendell Holmes (Pennsylvania Coal Company v. Mahon 260, U.S. 393, 415 1922) wrote: "The general rule at least is, that while property may be regulated to a certain extent, if regulation goes too far it will be recognized as a taking" (Eisner et al. 1993: 113). The wonder of this ruling is the ambiguity surrounding the term "too far." Because of the Supreme Court's unwillingness to become the "super legislature," and the fact that different places have different realties, "too far" continues to be defined on a case-by-case basis. Although state and federal legislation such as the 1972 Clean Water Act can preempt local municipal authority, very often in the USA, land use regulation is established by municipal government through a local democratic process, including community planning.

Municipalities are the creatures of the state, which has sovereign power over them. States pass the authority and responsibilities for community planning and land use regulation to local government through state constitutions and enabling legislation. There are different state approaches to planning and land use regulations, with some states providing more (e.g., Pennsylvania) authority for land use planning and regulation at the municipal level and some less (e.g., Maryland). In most cases, state-enabling legislation, at a minimum, empowers municipal government to plan for and govern development through zoning, subdivision, and other ordinances, official maps, and reservation and acquisition of land for public purposes; establish planning departments, commissions, committees, and zoning hearing boards; establish transferable development rights, conservation subdivisions, and other innovative processes; and provide for mediation and appeals to courts and penalties.

The protection of private property rights and encouragement of private investment continue to be consistent themes in land use planning and regulatory policy (Jacobson and Paulson 2009). Planning and land use regulation must balance the investment in and development of private property with reasonable and conforming land use and the health, safety, and welfare of a community, including economic, social, and ecological realms.

20.4 Community Planning: A Comprehensive Approach

Although there have been major planning movements including the reform policies of the City Beautiful (1890–1910), promotion of commerce in the City Practical (1920s), and protection of the environment in the environmental (1960s–1970s) (Daniels 2009), the basic theories and tools of planning, including environmental planning, have been well discussed over the last few decades in several publications (Greider and Garkovich 1994; Hawley 1950; Mumford 1961; Davidoff and Reiner 1962; Christopher et al. 1977; Kunstler 1994; Eisner et al. 1993; Smith 1972; Kaiser et al. 1995; McHarg 1992; Porterfield and Hall 1995; Hylton 1998; Selman 2000; Young 2011; USDA Forest Service 2014; Sanyal 2018). Along with the author's experiences and perceptions, the theory and tools of community planning are summarized in the following sections.

Community planning is defined by the American Planning Association (Eisner et al. 1993: 201) as: "a comprehensive, coordinated, and continuing process, the purpose of which is to help public and private decision makers arrive at decisions that promote the common good of society." Simply stated, planning is an organized approach to solving current and future community problems.

Planning can be used to identify strategic incremental solutions or comprehensive utopian plans. Any possibility of effective planning relies on the assumption that people have some control over their own destiny and an understanding of existing conditions including the velocity and direction of economic, social, and ecological change, desired or not. Where efficient means reduced cost, waste, and destruction, planning facilitates the conservation of resources through efficient and rational decision-making and action. It assists social choices about resource allocation by answering questions regarding how, when, to whom, to what purpose, and in what combination. But, in complicated political/power systems with varied attitudes, abilities, priorities, and agendas, "efficient" and "desired change" can be confused with that which satisfies more powerful actors.

While other authors are more optimistic about planning, Warren (1977) believed that most change that takes place in communities is not the result of purposive planning, but rather is hard-won by some group over hard opposition of another group. Furthermore, he discussed several flaws of planning, including planning that protects the status quo, can discourage major intervention and structural change; planning lulls people into thinking that positive change is being made; planning lulls people into thinking there will be fewer negative impacts as a result of growth; and many times planning itself can constitute a biased growth agency.

The orientation of community planning is to future costs; that is, taking appropriate actions today that will avoid costs in the future. There are four "costs" to municipalities that may result from not planning for community growth and development: direct mistakes that take municipal revenues to correct (e.g., cost of actual landslides, stormwater, and flooding), indirect mistakes that cause added municipal expense (e.g., cost of stormwater management and engineered flood control), loss of value mistakes that cause public and private property to depreciate (e.g., improper alignment of development), and failure to secure private investment that would have been made if appropriate development actions had been taken (e.g., not providing an attractive, functional place for people and business).

A sound planning process, also referred to as a model for rational decision-making, includes (Eisner et al. 1993): (a) putting together the planning group by including all those individuals interested, affected, and important; (b) mutually agreed-upon identification of problems and issues; (c) research and analysis to provide definitive understanding of problems and issues; (d) formulation of goals and objectives to be attained in alleviating problems or resolving issues; (e) development and evaluation of alternative method strategies to attain agreed-upon goals and objectives; (f) recommendations of appropriate courses of actions from alternatives; (g) assistance in implementation of approved plans and programs; (h) evaluation of actions taken to implement approved plans and programs in terms of progress toward agreed-upon goals and objectives; and (i) a continuing process of adjusting plans and programs in light of results of such evaluations or to consider changing circumstances. In steps eight and nine the idea of adaptive management is included as a fundamental concept of planning.

Planning guides growth and change through a number of different functions: development of a comprehensive or general plan to guide future development and land use; collection, analysis, and provision of information to decision-making; identification and consideration of different community alternatives; coordinating local development activities; review and modification of proposed land development; providing information and opportunities for public involvement (Burby 2003; Lurie and Hibbard 2008); and the formulation and update of ordinances and other policy (Arendt 1999; McElfish 2004).

20.4.1 Community Policy: Evaluation, Comparison, Monitoring, and Conflict

Policy can be defined as to display, make known. Policy is a principle, plan, or course of action, as pursued by a government, organization, or individual. Public policy can be formal, legal, and binding – as in the case of a municipal zoning ordinance that protects native trees from being removed during development. It can also be subtler, such as the will of a certain city council and how its members vote on certain issues.

When using community planning to develop regulatory and nonregulatory policy for the conservation of green infrastructure, a municipality should consider the following broad objectives: provide equity for all; be efficient by providing the most benefit for cost; strive to achieve a legitimate purpose and demonstrate positive impact on public safety, welfare, and health; and impede personal liberty and private property rights as little as possible by assuring property owners reasonable and conforming land use. Most conflict and debate over ordinances and other land management policies boil down to concerns and debates over one of these objectives (Stone 1988).

Many times, capital investment, ordinances, and other types of policy are formally evaluated for technical, economic, and administrative success (Fig. 20.1) and are monitored after enactment (Patton et al. 2013). Policy evaluation can also include the ranking and comparison of various policy alternatives to fulfill similar desires. To enable evaluation for both comparison and success, goals, objectives, criteria, and measures are identified and determined during the policy planning process (Fig. 20.1).

Goals are formal, broadly worded statements that identify what the policy is intended to achieve in the long run. They are often unachievable but serve to provide direction and incentive. Objectives are more focused, timely, and concrete statements that are more readily achieved. These targets and milestones work toward achieving desired goals and are often short term, perhaps one year in scope.

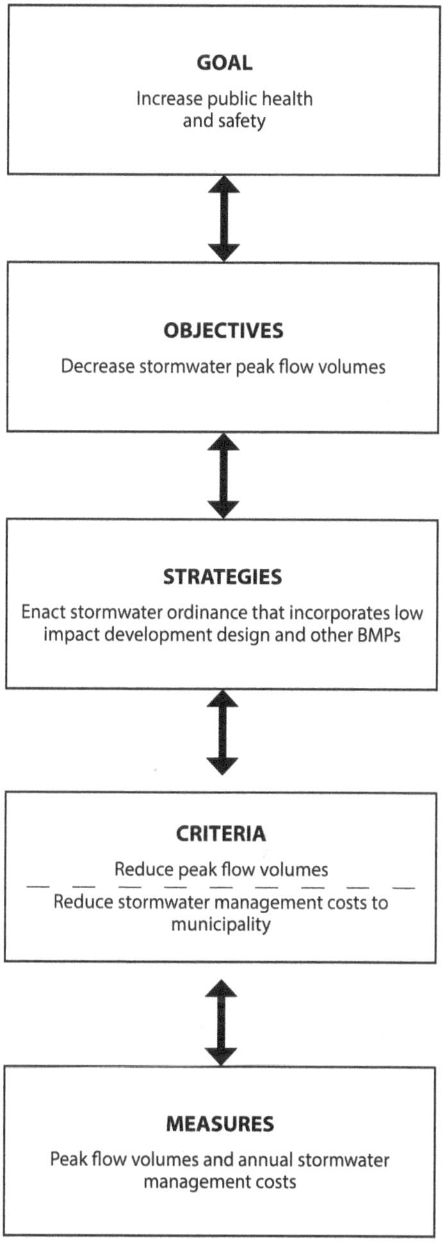

Fig. 20.1 Example of the policy evaluation process

Strategies are the day-to-day and month-to-month specific actions that will be taken to accomplish objectives. Criteria are the specific factors used to evaluate and monitor objectives and strategies related to a specific policy. Possible dimensions include costs, benefits, effectiveness, risk, liability, political viability, administrative ease, equity, and public sentiment. In general, accepted criteria address:

Technical feasibility – Does the ordinance achieve its purpose in a technical sense; that is, will it work?
Economic and financial possibility – Often evaluated through net present value or internal rate of return, what are the benefits and at what cost? Do the benefits outweigh the costs?
Political viability – Is the policy acceptable to decision-makers, public officials, businesses, citizens, and other influential groups?
Administrative operability – Is there authority to implement the policy? Does staff have the commitment and capability to administer and enforce the policy?

Measures are tangible or quantitative definitions of the criteria statements. Each of the criteria can have multiple measures.

It seems many planning processes assume that participants will sit together singing Kumbaya with no inherent conflict, but conflict is often at the heart of change and policy approaches used to manage it. A collaborative approach to conflict resolution (Gray 1989; Fisher and Ury 1991; Thompson et al. 2005) brings people together in a facilitated process to listen, define a problem, understand facts, share and understand different values, and develop mutually acceptable solutions rather than just deciding the winner. People's attitudes, including disagreement about facts, may be resolved by information from an impartial technical expert, but disagreement about deeply held values and beliefs can be difficult to resolve. Personal and direct explanation (educating/sharing) and acknowledgment (listening/understanding) may help to change deeply ingrained values. A collaborative process for conflict resolution includes: good communications where the importance of listening is emphasized; an inclusive process where diverse viewpoints are brought together and shared; mutual respect for all people at the table; a focus on interest (why people want), not position (what they want); and an identification of shared interests and the establishment of negotiations and solutions based on interests. In this process, positions are moved aside so that mutual interests can be identified. These mutual interests become the criteria for collaboration. Surrounding the collaborative process in communities is the idea of generalizing leadership where leadership as described by Rost (1991: 8) is "an influence relationship among leaders and followers who intend real changes that reflect their mutual purposes." Generalizing leadership that promotes collective action though lack of suppression, potential for acclamation, and engagement is crucial.

20.5 Regulatory and Nonregulatory Tools for Conserving Green Infrastructure

When considering the conservation of green infrastructure for water quality, broad objectives for planning that manages developing landscapes can include preserving the best agricultural lands for agriculture and the best forest lands for forestry and recreation, preserving connected natural areas of open space, preserving landscape and biodiversity, restricting intrusion into sensitive areas, considering the effect of fire and flood on residential and other development, protecting scenic views and historic/cultural resources, and preserving open space, greenbelts, riparian areas, steep slopes, woodlots, and other special landscapes such as wildlife habitat and watersheds. There are different regulatory and nonregulatory tools, many described in the philosophies of New Urbanism and Smart Growth (Nolen 2001; Freilich et al. 2010), that can be used in the community planning process (Fig. 20.2) to achieve the objectives of incorporating green infrastructure into community development and increasing water quality (Falk et al. 1992; Arendt 1994; Dwyer et al. 2003; Benedict and McMahon 2006; U.S. Forest Service 2014; Martel et al. 2017).

20.5.1 Nonregulatory Tools

Increased State Planning. State legislation like the and California Environmental Quality Act requires local agencies to follow a protocol of analysis and public disclosure of environmental impacts of proposed development, such as increased stormwater or water usage, and adopt all feasible measures to mitigate impacts. Municipalities are required to review development proposals in a comprehensive fashion including the completion of an environmental impact assessment for larger subdivisions and other proposed developments. The state takes a more active role in local planning processes to provide comprehensive and coordinated efforts as well as monitoring growth and change.

Comprehensive or General Plans. Using the planning process outlined in Sect. 20.4, a municipal or county comprehensive plan provides for community engagement that incorporates community input (Elmendorf and Luloff 2001) and background information (social and ecological assessments) to provide goals, objectives, and strategies to guide/manage future development and growth. Comprehensive plans also provide a nexus for the use of police power authority in land use ordinances, a land use map as a basis for zoning, and maps to guide future placement of roads, utilities, and other public services. The comprehensive plan essentially provides policies that set forth mandatory, optional, and prohibited courses of action in the development of a community or region.

Numerous authors, and state-enabling legislations (Kelly 2010; Pennsylvania Governor's Center 2015; Thousand Oaks n.d.), have discussed important sections of comprehensive plans including: (1) a background information section that

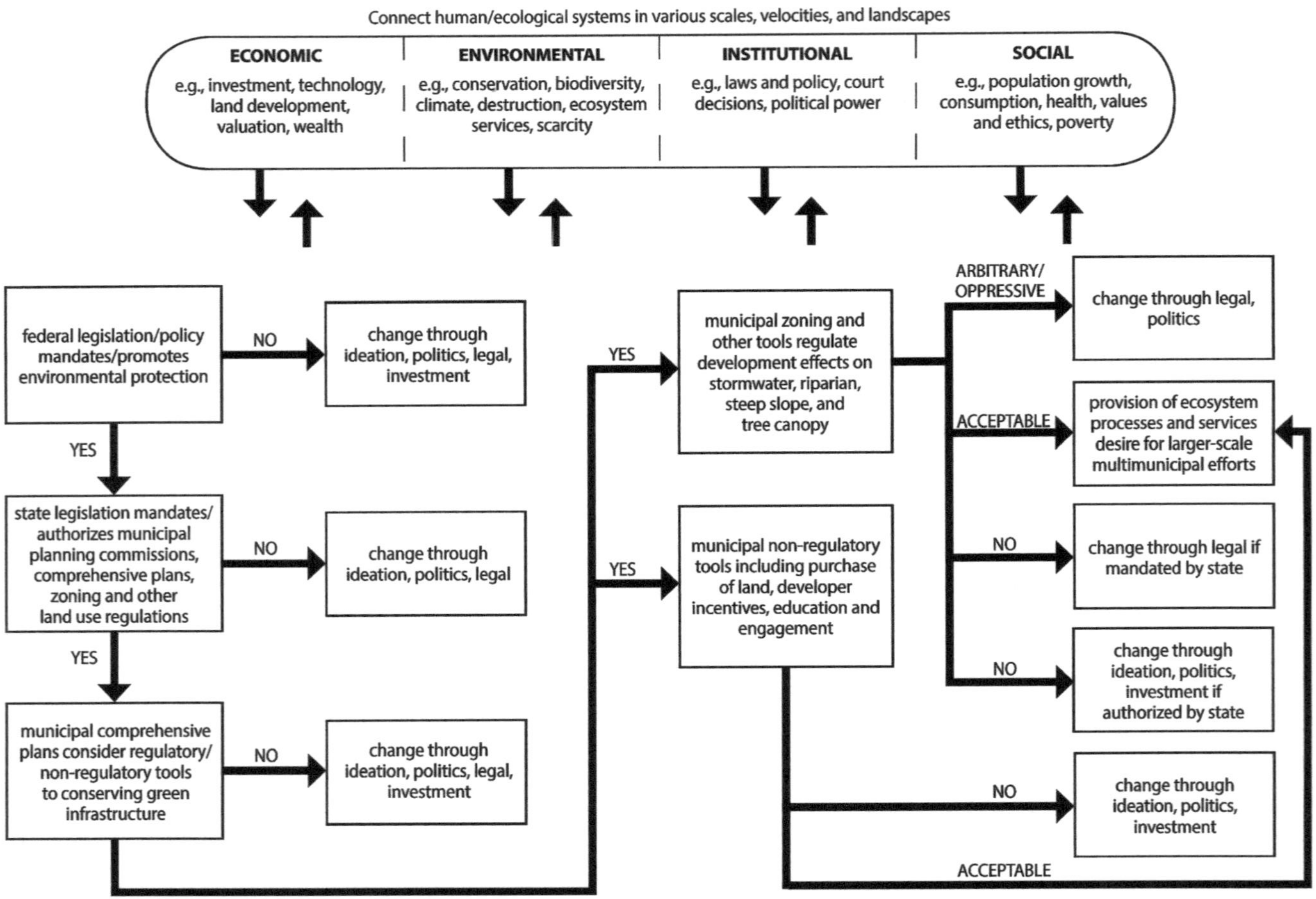

Fig. 20.2 Using community planning to conserve green infrastructure (no pathways support weak sustainability)

describes history, ecological features, population characteristics, community facilities, and other community characteristics; (2) a resource assessment that describes in detail the economic, social, ecological, and cultural resources existing in a community including a map of important locations, connections, and changes; and (3) statements of vision, goals, objectives, and strategies that relate the economic, social, and ecological resources of a community to change and development. A vision statement can include sentiments such as well-being, beauty, community image, attractiveness, sustainability, future generations, economic prosperity, ecological quality, and quality of growth and development. Goals can include increasing the enjoyment and quality of community life, providing a safe and healthy community environment, and increasing community commerce and business. Objectives describe specific achievements and requirements necessary to accomplish goals and often become the elements or chapters of a comprehensive plan.

The fourth section of a plan are the individual elements or chapters where strategies are organized and described including regulations and other policies, funding, and partnerships. The following are elements commonly found in comprehensive plans: (a) the land use element, which describes and maps the basic current and expected patterns of land use (residential, commercial, agricultural, rural residential, industrial, institutional, and open and natural space). This element becomes the logical basis for zoning map and regulations. (b) The transportation element uses maps to show the desired end states of road and other transportation systems. Transportation planning is not limited to roadways, but also includes parking, facilities, airports, public transit, bicycle paths, and pedestrian ways. (c) The community facilities and utilities element discusses the public, private, and institutional activities that serve the community. Development is encouraged and often follows the provision of sewer and water, as well as roads and other supporting infrastructure. This element identifies the provision of water, sewer, utilities, and other infrastructure to support and promote, or slow and exclude, future development. It also considers parks and recreation areas and facilities, schools, hospitals, churches, police and fire stations, civic buildings, cable TV, recycling centers, libraries, and the like. (d) The conservation, natural resources, environmental, or green infrastructure element describes how ecological processes and services will be conserved in development. This element may consider conservation of such things as historic landscapes, forests and woodlots, open space and greenways, sensitive environmental areas, and riparian and groundwater resources. Arendt (1994, 1999) described several strategies that can be considered in this element: both an understanding and documentation of a community's green infrastructure including how connected and the places best suited for development and those for conservation (also see McHarg 1992); reasonable objectives for both conservation of green infrastructure and economic development; a progressive land use plan and zoning framework that allows conservation subdivisions and other innovative developments; good working relationships between municipal officials, landowners, and developers; mechanisms for assessing, prioritizing, and purchasing and managing important landscapes; a process for multijurisdictional cooperation in planning and funding; and building strong leadership and citizen support though education and engagement. If green infrastructure is not considered in a comprehensive plan, it will not be adequately or correctly addressed in zoning, subdivision and land development, and other ordinances or in capital budgeting used to implement policies expressed within the plan. Furthermore, it will not be adequately considered in the permit review process for individual developments.

Finally, a capital budget and plan implementation element consider how public facilities (such as the purchase of open space) will be funded through bond issue, general taxes, municipal revenues, authorities, grants, and partnerships. It also describes how objectives and policies will be implemented including using municipal and private sector resources, developer incentives, and other approaches.

Comprehensive Plans: Water Supply and Stormwater. Considering increased growth and development and state and federal MS4 (municipal separate storm sewer system) permit requirements, elements for the reliable supply of water and stormwater management have received more attention in municipal and county comprehensive plans (Pennsylvania DEP 2006; Bucks County 2011). Plans for the reliable supply of water typically encourage integrated water resources planning through use of land development, water supply, stormwater, and wastewater techniques that protect and maintain natural functions of the hydrologic cycle. Although different from place-to-place water supply plans consider several strategies including: (a) establishment, review, and update, as necessary, of municipal ordinances related to water resource protection; (b) protecting water resources from overwithdrawal and contamination caused by incompatible uses or development including the protection of wellheads, source water, and riparian areas; (c) providing municipal water, and sewer, to either accelerate (growth areas) or slow growth in certain areas; (d) allowing growth placement, densities, and velocities and corresponding provision of water services that are compatible with underlying aquifers or other sources; (d) providing water for emergency periods; (e) coordinating efforts of government, private, and nonprofit organizations responsible for the planning and management of water resources; and (f) cooperation with federal, state, county, and other municipalities to implement requirements and recommendations of water quality legislation.

Stormwater management plans typically discuss the use of low impact development design (Pennsylvania DEP 2006; Cahill 2012; Lancaster 2015) to infiltrate, evapotranspirate, or capture and reuse as much stormwater runoff on-site as reasonably possible. They can consider several strategies to provide for human and ecological health and fulfill MS4 permit requirements including (a) adopting and enforcing stormwater ordinances that require new development to use structural and nonstructural systems to achieve on-site predevelopment hydrologic conditions that protect water quality, promote groundwater recharge, protect streambank erosion, and reduce pollutant discharge; (b) providing municipal sewer to either accelerate (growth areas) or slow growth in certain areas; (c) requiring redevelopment properties to address previously unaddressed stormwater control issues; (d) promoting the maximum use of stormwater runoff as a water resource; (e) identifying and eliminating unnecessary requirements for impervious coverage in subdivision and land development regulations (i.e., oversized parking facilities); (f) identifying and encouraging the removal of needless impervious coverage and replacing with landscaping or other pervious surfaces; (g) upgrading, including operational and maintenance plans, treatment capacity and processes of municipal treatment plants that are experiencing hydraulic and organic overloads or future reserve capacity issues; (h) providing funding options to adequately fund municipal stormwater management programs, fix existing stormwater problems, and to retrofit aging facilities; (g) and discussing implementation of a stormwater utility that taxes impervious surfaces and provides tax credits for their illumination. Stormwater management plans can also provide a comprehensive strategy for the removal/mitigation of nitrogen, phosphorus, and sediment including the use of agricultural and urban stream buffers, the conversion of turf areas to trees, the planting of trees in and around impervious areas, and the protection of riparian areas, steep slopes, and open space. For information on the role of green infrastructure in stormwater management, see (Dhakal and Chevalier 2017; Kuehler et al. 2017).

Planning Commissions. A distinctive feature of planning in America is the tradition of citizen or volunteer planning commissions. These citizen groups are a by-product of the nineteenth-century civic improvement associations, which were formed to increase the safety, beauty, and function of post-Industrial Revolution American cities. Typical duties and powers of a planning commission include: (a) preparing a comprehensive plan, (b) preparing an official map, (c) preparing zoning ordinances and maps, (d) preparing and administrating subdivision and other land development regulations, (e) preparing environmental studies, (f) preparing water surveys, (g) promoting of public interest and action in understanding the comprehensive plan and development issues, (h) holding public hearings and testimony, and (i) reviewing and amending zoning, subdivision, and other ordinances and the municipality's comprehensive plan.

Multi-municipal Planning and Zoning. Strong regional coordination is crucial for preserving viable systems of green infrastructure. The implementation of a multi-municipal comprehensive plan and zoning ordinance can provide an understanding of how and where green infrastructure will be preserved. It also can provide uniformity and consistency in planning and regulatory policy by providing larger scale goals, objectives, and strategies. Although states like Pennsylvania authorize multi-municipal planning and zoning, it is not often implemented, although multi-municipal authorities for water, police, fire, and recreation are.

Natural Resource Assessment. Identifying landscapes with important resources and placing this information in a Geographical Information System (GIS) allows planners to understand existing natural systems as well as areas of special concern for both conservation and development. In defining the natural environment, the following resource categories can be used: significant wildlife habitat, scenic areas, river and riparian corridors, recreational resources, productive agricultural resources, productive forest resources, woodlands and natural areas, open space, special or unique landscapes, riparian areas and wetlands and their buffers, floodways and floodplains, historical and cultural landscapes, mineral and oil resources, contiguous block, vulnerable landscapes, moderate and steep slopes, viewsheds, and greenways and corridors for the passage of people and wildlife and other natural elements.

Less-than-Fee-Simple Land Acquisition. Land is viewed as a bundle of private property rights including development, water, mineral, view, and easement. Less-than-fee-simple is ownership of less than all the legal rights associated with a property. Green infrastructure would be protected by the purchase of certain rights (development, recreation, river corridor, view) where the easement buyers pay only for land rights that needs to be protected. This purchasing concept is behind conservation easements where property owners are paid for their development rights, and the land is preserved and managed as open space. Depending on location, development rights can be 40–90 of full market value of property depending on location.

Full Fee Acquisition. This is the highest form of land ownership where all the legal rights associated with a property are purchased. This method is good for preserving high-quality lands with ecological and public values.

Tax Benefits. Tax incentives include no capital gains tax for landowners who sell land rights to governments and conservation groups for conservation purposes. Tax benefits, such as lower property taxes, also can be granted for individuals who keep their land in agriculture or forest for a period under contract with state government.

Developer Incentives. Many incentives such as increased development density, decreased development fees, preferential tax assessments, relaxation of parking and other requirements, and preferential permit processing can be used to help motivate developers and property owners to preserve green infrastructure.

Private-Public Partnerships. Nonprofit organizations such as the Nature Conservancy can partner with government agencies in the acquisition of green infrastructure. Often a conservancy has more flexibility to purchase lands in a timely fashion, then selling to a state or other agency later.

Public Empowerment. Through public forums, committees, surveys, education, training, and volunteerism, providing citizens abilities and opportunities for participation and engagement.

20.5.2 Regulatory Tools

Zoning Ordinances. In 1916, Edward Bassett, a New York City attorney, defined zoning as "The regulation by districts under the police power of the height, bulk, and use of buildings, and the use of land, and the density of population" (Eisner et al.: 113). Under this principle, zoning has survived numerous court challenges. Zoning provides standards for the height of buildings, the percentages of a parcel that may be developed, the density of development, and the districting of land use (e.g., residential, commercial, industrial, institutional, rural resource, agricultural, forest, open space). Zoning is not planning but is one implementation tool of planning. Zoning is accomplished through the development of a comprehensive plan and enacted by ordinance and map through public hearing and the vote of elected officials. One phrase found in many state-enabling statues is that zoning ordinances shall be prepared in accordance with a comprehensive plan.

Zoning Options. Zoning options to preserve green infrastructure include: (1) existing use zoning, which is the legal use of a parcel based on the use which the parcel is now in. This protects both traditional land uses and the prevailing character of a community and surrounding where the use could be forest or agricultural management areas. (2) Overlay zoning is often used to address areas with unique or important environmental benefits. Additional conservation regulations (e.g., standards for riparian buffers) are "overlaid" on underlying zoning. These overlays restrict or prohibit development of sensitive areas such as impervious surfaces, wetlands, important habitat, hillsides, riparian areas, fragile shorelines, fire-adapted landscapes, and worthy open space (University of Wisconsin 2005). (3) Conservation zoning sets the groundwork for conservation subdivisions by providing for smaller lot sizes, clustered development, and open space protection. A certain percentage of the parcel is conserved in open space and lots are clustered together on the parcel allowing for the protection of open space and other natural and cultural resources on the remaining parcel. (4) Sensitive area ordinances, zoning ordinances are used to restrict and regulate the removal of vegetation and amount of grading and development along riparian areas, on steep slopes, and other areas. As an example, depending on slope and stream quality, restrictive and prohibited setbacks are placed to protect riparian vegetation and banks. Restrictive areas would allow some authorized use, while prohibitive areas would only allow trails and other compatible uses. These ordinances are used to ensure that sensitive design and grading is accomplished, and a reasonable percentage of vegetative cover and natural land forms remain on a property after development. (6) Timber harvest ordinances provide standards for the completion of timber harvesting plans, types of harvesting methods used, protection of riparian and other areas, and construction of roads and landings. (7) Tree preservation ordinances provide standards for the preservation of trees and woods in development including the completion of tree inventories and tree preservation plans, the use of retaining walls and other engineered devices, and the use of proper arboricultural methods. (8) Official maps show proposed public parks, playgrounds, greenways, open space, pedestrian ways, and other reservations. If the owner of the land on an official map applies for a building permit or submits written notice of intention to subdivide or develop, the municipality has one year to acquire the property shown on the official map.

Stormwater Ordinances. Responding to both federal and state mandates concerning combined sewer systems and MS4 (municipal separate storm sewer systems) permitting requirements, many municipalities in the USA have enacted stormwater management ordinances (Lancaster 2012; Pennsylvania DEP 2019). The provisions of these ordinances are placed, when restricting the use of private property, in zoning (e.g., impervious surface limitations, which place limitations on the percentage of ground that can be covered with asphalt or other impervious surfaces) and, when providing standards for design, construction, and facilities, in subdivision and land development ordinances. Typical sections of these ordinances mandate: (a) based on size of development and amount of ground disturbance, developing a stormwater management plan that incorporates structural and nonstructural BMPs for water quality and rate controls so that post-peak flows are equal to predevelopment peak flow levels; (b) reducing vegetation removal and impervious surfaces and earth disturbances to increase evapotranspiration, infiltration, filtration, and groundwater recharge; (c) providing a soil erosion and sedimentation plan; (d) providing onsite systems for the filtering/treatment of water; (e) mandating the preservation of trees, woodlots, riparian areas, wetlands, steep slopes, and meadows; (f) mandating permit application information on soils, hydrology (e.g., pre- and post- development stormwater runoff computations), and natural features; (g) considering innovate and site-specific planning and design including different types of housing footprints, road widths, and curbing; (h) and designing tree and shrub landscapes not dominated by turf – a different type of landscaping – a new conception of beauty. These ordinances are comprehensive and use both small, structural (e.g., rain gardens and bioretention basins) and large, nonstructural (e.g., preservation of riparian areas and open space) systems to decrease to stormwater flow and velocity. Some municipalities, Philadelphia, Pennsylvania, and Washington, DC, have enacted stormwater utilities that tax property owners on the square footage of impervious surfaces and provide tax breaks when these areas are reduced.

Subdivision and Land Development Ordinances (SALDO). As with traditional zoning ordinances that simply define the minimum number of certain sized lots on a parcel, conventional SALDO ordinances often lead to cookie-cutter development of strips of lots and homes along streets and fitting in as many lots as possible on a piece of property, without regard for ecological processes and services. Some SALDO ordinances do not consider green infrastructure, but rather only provide standards for subdivision review process, design, and procedure such as parceling of land into building sites; a sequence for development permit process; required improvements such as recreational area, utilities, streets, and sewage; and bonding for required improvements. To conserve green infrastructure, SALDO ordinances must address the internal relationship of riparian areas, woodlots, steep slopes, and open space with the placement of house lots, streets, and utilities. SALDO ordinances can mandate completion of natural resource assessments and maps and, with zoning, allow for the design and placement of clustered development density that preserves green infrastructure (such as conservation subdivisions or lot averaging). They can provide for the mandatory dedication of open space in certain sized developments and, in some states, provide for in-lieu fees to be paid to acquire off-site open space. And, they often provide standards for the landscaping of parking lots, streetscapes, and other areas.

Cluster, Open Space, or Conservation Subdivisions. This subdivision design allows for development and community commerce while maintaining some integrity of worthy farmlands, forests, open space, and habitat. Objectives of conservation subdivision design include understanding the natural features of the site, conserving sensitive natural areas such as riparian areas and wetlands, and keeping fragmentation of resources to a minimum. As an example, the municipal SALDO ordinances must work with zoning ordinances. A zoning ordinance would require that 50% of a subdivision over 10 acres be conserved in open space, that a 100-ft buffer be provided for riparian areas, and that allowable building density be clustered on smaller lots. The SALDO ordinance would provide distinct standards for the management of stormwater and outline a design process for green infrastructure conservation: completion of a natural resources inventory and map to identify primary (e.g., riparian areas, hydric soils, woodlots, meadows) and secondary (e.g., views, worthy trees, geological features) conservation features and placing these on a site feature map, locating the best places for houses and streets, placement of clustered streets and building parcels, and placement of lot lines (Arendt 1994, 1999). The fragmentation and isolating of green infrastructure can be further reduced by increasing the scale of planning and designing and incorporating individual subdivisions as part of a larger continuous integrated system by using a context map that identifies the subject property in relationship with surrounding landscapes including those in other municipalities.

Lot Averaging. This approach allows flexibility in subdivision design and design by requiring the average size of all lots in a subdivision are equal to a specified minimum size rather than requiring that each individual lot meet the minimum size threshold.

Development Dedications. Dedications from developers, including impact fees for additional schools, public safety, and other services, are authorized in state-enabling legislation and placed in municipal zoning, SALDO, and other ordinances. Dedications to jurisdiction(s), such as land for greenways, open space, and parks, are mandated through conditions placed in the approval of subdivision and other development permits.

Growth Areas, Growth Boundaries, or Service Limitation Boundaries. Rural resource, agriculture, or other zoning is used to provide a horizontal limit, or boundary line, to development of certain types including in most cases large-scale subdivision, commercial, or industrial. Land use in the growth line is characterized by a firm, long-term commitment to increased density and compact development and low-density rural preservation outside the boundary line. Municipalities must commit themselves to infill housing and cluster development within the growth area. Furthermore, to limit development outside the growth area, the municipal provision of roads, public water, and public sewer is restricted.

Transferable Development Rights. Areas appropriate for conservation, such as valuable forest, agricultural, or recreational lands, are classified as sending districts and areas appropriate for development (growth areas) are identified and designated as receiving districts. Development rights in the sending district are purchased from the owner and transferred to the receiving district, restricting development in a sending district. This tool can be used to transfer development to growth areas where increased development densities and multiple use are allowed.

Planned Residential Development (PRD). This flexible regulatory technique combines aspects of both zoning and SALDO regulations. It encourages the mixing in the same area of a variety of different and compatible land uses. Development is not limited to residential uses, but mixes residences of all types, shopping, offices, and places of entertainment in the same development. It also encourages the incorporation and preservation of natural resources within the development. A key to this development is the preservation of large proportions of the site in open space. Perhaps 40, 50, or even 60% of the development would be preserved in open space.

Traditional Neighborhood Development (TND). Provisions adopted by the municipality for traditional neighborhood development consider the following: the amount, location, and proposed use of common open space to be distributed throughout the development; the location and physical characteristics of housing and other development to provide for preserving or enhancing of natural features such as wetlands, ponds, waterways, trees of high quality, wooded areas, and other significant natural features; and the encouragement of pedestrian traffic.

In summary, there are five major regulatory and nonregulatory strategies that can be used to help integrate green infrastructure into community development: subsidies and incentives, taxation, regulation, public investment, and public education and engagement. Subsidies and incentives include increased development density, allowing for alternative development design and placement, reduction of permit and development fees, and relaxation of development requirements. Taxation includes reduced property taxes for agriculture and forest lands held in these uses. Regulations include zoning, SALDO, and other land use regulations. Public investment includes fee and fee-simple purchase funded through general tax revenues, municipal revenues, and bond issues. Public education and engagement include planning commissions and ad hoc committees, public hearings, an unbiased media, working with youth and schools, and special celebrations and activities. Another way these strategies can be expressed is that concerns about conservation of green infrastructure can be directed by changing property ownership regime (e.g., private to state); reducing damage and extraction; using the coercion of rules and regulations; using market-based tools such as increased cost of residential water; increasing people's direct engagement; and using propaganda and education.

20.6 Community Planning, Green Infrastructure, and Water Quality: Today and Tomorrow

Considering the projected population growth (392 million people in the United States), distribution, and change between now and 2050, the conservation of green infrastructure and the water quality it provides in places experiencing growth and development will become increasingly important and most likely difficult. Continuing challenges to using community planning to conserve green infrastructure include tension over restricting private property to provide societal good, continued suburban and rural sprawl, small-scale planning and the ability of planning to work across a historic system of fragmented municipal governments, and – especially in rural places facing development – outdated zoning and other development policies that do not provide adequate standards or allow for flexibility and ingenuity. Some of these challenges are elaborated below.

20.6.1 Impediments to Conserving Green Infrastructure in Development

The long history of government reservation of property for public water supply includes Philadelphia's Fairmount Park (1876) and the Los Padres (1936) and other Western National Forests. More recently, using multiple partners and both regulatory and nonregulatory tools, the conservation of large areas of green infrastructure in open space continues, including the Santa Monica National Recreation Area and Conejo Valley Open Space Conservation Area in Southern California; the Boulder, Colorado, Open Space and Mountains Parks; and the Montgomery County, Pennsylvania, Open Space Plan and Program. Planners have expressed that from a human well-being standpoint that does not consider larger requirements for ecosystem health, 30 acres of open space should be provided per 1000 people, but in many places larger conservation efforts, and the water quality and other ecosystem service they provide, have not kept pace with growing development and populations (Center for the Study of Law and Politics 1991; Ehrlich 2002). Although land use planners continue to promote the necessity of green infrastructure in community development, conservation may be compromised in some places because of lack of technical assistance from the state, inadequate zoning and other land use policy, concerns for private property rights, the velocity of development, high land costs, speculative or short-term priorities, the profitable nature of land development, and a competitive land use arena (Barnett 1999; Miller et al. 2015). And, as time and land development continue, opportunities for green infrastructure conservation are reduced or eliminated by land conversion, fragmentation, and increasing land costs in a competitive real estate arena; land can become scarce and conservation is timely. In the future, stormwater, water quality and availability, and flooding will likely become major drivers of community planning that encourages the use of clustered development and other Smart Growth, low impact development techniques.

20.6.2 A Need for an Ecosystem Approach Including Watershed Scales

It has been argued that community planning to conserve water quality and other ecosystem services has failed because it is implemented by individual municipalities using zoning as the primary tool (Bosselman and Callies 1971). Most community planning policy is based on municipal boundaries, while ecosystems and landscapes are based on natural boundaries such as watersheds, and the conservation of high-quality ecosystem services, such as those provided by riparian areas, requires larger planning scales. This often means intermunicipal partnerships and cooperation, which can be difficult because of feuds, power struggles, and competition for tax dollars. The success of past attempts at watershed and larger-scale planning to break through structural barriers, such as autonomy and self-sufficiency of local governments long held central to the American individualist paradigm, to bring municipalities together has been questioned. The lack of cooperation and coordination between municipal jurisdictions is an obstacle in conserving viable large-scale green infrastructure when considering planning, administrative, and funding realities.

The importance of a holistic approach to planning and managing natural environments that does not concentrate on economic value alone, but also considers ecological, social, economic, and political values, has been described by several authors (Freudenburg and Keating 1985; Bengston 1994). Ecosystem-based management (Grumbine 1994) is a planning and decision-making process that should be considered to facilitate the integration of natural and social systems into community development. It is intended for larger-scale planning, such as at the watershed scale, and several themes revolve around this process: connected social/ecological systems, ecological boundaries, ecological integrity, data collection, monitoring, interagency cooperation, generalizing leadership, organizational change and adaptive management, social values, and humanity and ethics in nature.

20.6.3 Concerns with a Positivist Approach to Environmental Valuation

Historically, the main approach to valuing natural resources, and the development projects that affect them, has been based on the concepts of neoclassical, market-based economics including net present value and other cost/benefit modeling. Reliance purely on cost/benefit valuation has been criticized as using a single value dimension and narrow methodology; undervaluing cultural, historic, and other principled values; ignoring wealth and power effects; and lacking full costing of externalities, unknowns, and irreversibilities. And a question has been asked whether cost/benefit valuation alone adequately considers ethical dilemmas and debates (O'Brien 2003). To help understand both public value and environmental decision-making, authors have suggested the integration of interpretivism and other approaches that explore a wide range of people's values, and the reasons and judgments they give for having these, rather than focusing solely on economic realms.

20.6.4 Private Property

In the early twentieth century Frederick Law Olmsted believed that progress in America relied primarily upon individual initiative under the stimulus of private property (Olmstead 1916). Private property continues to have a special cultural significance in the United States and a historic role in the formation of environmental policy, including community planning and regulatory policy to conserve green infrastructure (Bromley 1991; Jacobson and Paulson 2009). Private property can be defined as a bundle of both rights and duties, including the owner's rights to future benefit streams, duty of others to respect rights of owner, and duties of owner to act in a socially desirable manner. *Socially desirable* is when "things" are being produced that have some value to society and the management values of the owner are somewhat in line with those of society. Private property is a triadic social arrangement between an owner, nonowner, and the state where both rights and duties are often defined and protected by the state. The allocation, alteration, and reallocation of private property by the state to provide for a greater public good, including beauty, order, justice, and economy, remains a consistent theme and source of conflict and tension in both community planning and environmental policy.

Although the concept of private property was modified during the 1970s environmental movement to consider land as a community resource rather than a purely private commodity (Bosselman and Callies 1971), a growing political ideology that supports privatization and deregulation has supported either compensating the property owner for restrictive regulation or revocation of zoning and other ordinances. The tension over governments' management and regulation of private property in growth and development will continue, increasing conflict and difficulty in using community planning to conserve green infrastructure, water quality, and other socially optimal results.

Although Hardin (1968) advocated private property as a tool to assure rational behavior to obtain socially desirable results (e.g., conservation of water quality and other ecosystem processes and services), a question remains: "Does the ownership of private property insure proper resource stewardship and management?" With increasing populations and resource scarcity, including water, authors suggest (Bromley 1991) that private property ownership alone will not provide for conservation and that collective action be taken to provide standards and rules that would provide more guidance to market-based process for implementation. The problem for water quality and provision arises when rules are too intrusive/oppressive and market-based decisions (e.g., consumption, cost, investment) are insufficient to influence policy or conservation.

20.6.5 An Environmental Esthetic and Ethic

Concentrating on people's regard for forestry and other extractive practices, Aldo Leopold's (1948: 224) land ethic stated: "Examine each question in terms of what is ethically and esthetically right as well is economically expedient. A thing is right when it tends to preserve the integrity, stability, and beauty of the biotic community. It is wrong when it tends otherwise." Callicott (1987) and other authors have built on Leopold's work by discussing a need for a land esthetic where, as a precursor to the burden of ethics, the beauties of natural landscapes are something to love, admire, and cherish. Perhaps today, discussions about environmental ethics should be broadened to not only consider natural resources on public and private forest lands but to also embrace the effects of subdivision and other land conversion and development on ecological systems. In doing this, an ethic implies an individual and/or social value system and, as something that touches the lives of people daily, a concentration on the significance of water for both social and ecological health may be one avenue to help evolve a stronger environmental value and corresponding ethic.

The ever-growing recognitions of the profound effects of land conversion and development, and other human actions, on global aquatic ecosystems including rising ocean levels, ocean acidification, salinization of freshwater, drought, stormwater, and flooding may also drive this cultural evolution at much larger scales. As demonstrated in the chapter, comprehensive planning is necessary to better utilize green infrastructure to maintain and enhance water quality and promote better stormwater management plans.

References

Arendt R (1994) Rural by design. American Planning Association, Chicago

Arendt R (1999) Growing greener: putting conservation into local plans and ordinances. Island Press, Washington, DC

Barnett H (1999) The land ethic and environmental crime. Crim Justice Policy Rev 10:161–191. https://doi.org/10.1177/088740349901000203

Benedict M, McMahon E (2006) Green infrastructure: linking landscapes and communities. Island Press, Washington, DC

Bengston D (1994) Changing forest values and ecosystem management. Soc Nat Res 7:515–533. https://doi.org/10.1080/08941929409380885

Bosselman F, Callies D (1971) The quiet revolution in land use control. U.S. Government Printing Office, Washington, DC

Bromley (1991) Property rights and property regimes in natural resource policy. In: Bromley, D (ed) Environment and economy: property rights and public policy. Blackwell, Oxford, UK, pp 14-40

Bucks County (2011) Bucks County comprehensive plan. Doylestown. Retrieved at http://www.buckscounty.org/docs/default-source/government-documents/bccp2011final.pdf?sfvrsn=

Burby R (2003) Making plans that matter: citizen involvement and government action. J Am Plann Assoc 69:33–49. https://doi.org/10.1080/01944360308976292

Cahill T (2012) Low impact development and sustainable stormwater management. Wiley, Hoboken

Callicott J (1987) The conceptual foundation of the land ethic. In: Callicott J (ed) A companion to a sand county almanac. University of Wisconsin Press, Madison, pp 157–171

Center for the Study of Law and Politics (1991) Building sustainable communities: open space preservation and acquisition. In: San Francisco

Chiesura A (2004) The role of urban parks for the sustainable city. Landsc Urban Plan 68:129–138. https://doi.org/10.1016/j.landurbplan.2003.08.003

Christenson J, Robinson J (1989) Community development in perspective. Iowa State University Press, Ames

Christopher S, Ishikawa S, Silverstein M (1977) A pattern language: towns, buildings, construction. Oxford University Press, New York

Cottrell L (1983) The competent community. In: Warren R, Lyon L (eds) New perspectives on the American community. Dorsey Press, Homewood, pp 401–412

Daniels D (2009) A trail across time: American environmental planning from city beautiful to sustainability. J Am Plann Assoc 25:178–208. https://doi.org/10.1080/01944360902748206

Davidoff P, Reiner T (1962) Choice theory of planning. J Am Inst Plann 28:103–115. https://doi.org/10.1080/01944366208979427

Dhakal K, Chevalier L (2017) Managing urban stormwater for urban sustainability: barriers and policy solutions for green infrastructure application. J Environ Manage 203:171–181. https://doi.org/10.1016/j.jenvman.2017.07.065

Donovan G, Mills J (2014) Environmental justice and factors that influence participation in tree planting in Portland, Oregon. U.S. Arboric Urban For 40:70–77

Dwyer J, Schroeder H, Gobster P (1991) The significance of urban trees and forests: a deeper understanding of values. J Arboric 17:276–284

Dwyer J, Nowak D, Heather M (2003) Sustaining urban forests. J Arboric 29:49–55

Ehrlich P (2002) Human natures, nature conservation, and environmental ethics. Bioscience 52:31–43. https://doi.org/10.1641/0006-3568(2002)052[0031:HNNCAE]2.0.CO;2

Eisner S, Gallon B, Eisner S (1993) The urban pattern: city planning and design. Van Nostrand Reinhold, New York

Elmendorf W (2008) The importance of trees and nature in community: a review of the relative literature. Arboric Urban For 34:152–256

Elmendorf W, Luloff A (2001) Using qualitative data collection methods when planning for ther community forest. J Arboric 27(3):139–151

Elmendorf W, Willits F, Sasidharan V (2005) Urban park and forest participation and landscape preference: a review of the relevant literature. Arboric Urban For 31:311–317

Falk L, Harper S, Rankin E (1992) The northern forest lands study of New England and New York. USDA Forest Service, Rutland

Fisher R, Ury W (1991) Getting to yes. Penguin Books, New York

Freilich R, Sitkowski S, Mennillo S (2010) From sprawl to sustainability: smart growth, new urbanism, green development and renewable energy. American Bar Association, Chicago

Freudenburg W, Keating K (1985) Applying sociology to policy: social science and the environmental impact statement. Rural Sociol 50:578–605

Geis E (2009) Conservation: an investment that pays. The Trust for Public Land, San Francisco

Gray B (1989) Collaborating: finding common ground for multiparty problems. Jossey-Bass, San Francisco

Greider T, Garkovich E (1994) Landscapes: the social construction of nature and environment. Rural Sociol 59:578–605. https://doi.org/10.1111/j.1549-0831.1994.tb00519.x

Grumbine E (1994) What is ecosystem management. Conser Biol 8:27–38. https://doi.org/10.1046/j.1523-1739.1994.08010027.x

Hana K (2005) Planning for sustainability: experiences in two contrasting communities. J Am Plann Assoc 75:27–40. https://doi.org/10.1080/01944360508976403

Hardin G (1968) The tragedy of the commons. Science 162:1243–1248. https://doi.org/10.1126/science.162.3859.1243

Harrington C, Curtis A, Black R (2008) Locating communities in natural resource management. J Environ Res Plan 10:199–215. https://doi.org/10.1080/15239080801928469

Hawley A (1950) The changing shape of metropolitan America: decentralization since 1920. Free Press, Glenco

Hein L, de Groot R, van Koppen C, van Ierland E (2006) Spatial scales, stakeholders, and valuation of ecosystem services. Ecol Econ 57:209–228. https://doi.org/10.1016/j.ecolecon.2005.04.005

Hillery G (1955) Definitions of community: areas of agreement. Rur Soc 20:111–125

Hylton T (1998) Save our lands, save our towns: a plan for Pennsylvania. RB Books, Harrisburg, PA

Irvin E (2002) The effects of open space on residential property values. Land Econ 78(4):465–480. https://doi.org/10.2307/3146847

Jacobson H, Paulson K (2009) Property rights: a neglected theme of 20th century planning. J Am Plann Assoc 75:134–143. https://doi.org/10.1080/01944360802619721

Johnson M (2001) Environmental impacts of urban sprawl: a survey of the literature and proposed research agenda. Env Plan 3:717–735. https://doi.org/10.1068/a3327

Kaiser E, Godschalk D, Chapin F (1995) Urban land use planning. University of Illinois Press, Urbana

Kaufman H (1959) Toward an interactional conception of community. Soc Forces 38:8–17. https://doi.org/10.2307/2574010

Keeler B, Polasky S, Brauman K, Johnson K, Finlay J, O'Neill A et al (2012) Linking water quality and wellbeing for improved assessment. Proc Natl Acad Sci 45:18619–18624. https://doi.org/10.1073/pnas.1215991109

Kelly E (2010) Community planning: an introduction to the comprehensive plan. Island Press, Washington, DC

Krannich K, Luloff A (1991) Problems of resource dependency in U.S. rural communities. Belhaven Press, London

Kuehler E, Hathaway J, Tirpak A (2017) Quantifying the benefits of urban forest systems as a component of the green infrastructure stormwater treatment network. Ecohydrology 10:1–10. https://doi.org/10.1002/eco.1813

Kunstler J (1994) The geography of nowhere. Simon and Shuster, New York

Lancaster (2012) City of Lancaster stormwater management ordinance. Lancaster Retrieved at https://ecode360.com/8119841

Lancaster (2015) City of Lancaster stormwater management plan. Retrieved at http://cityoflancasterpa.com/resident/stormwater-management

Leopold A (1948) A sand county almanac. Oxford University Press, New York

Lichtenberg E, Hardie I (2007) Open space, forest conservation and urban sprawl in Maryland suburban subdivisions. Am J Ag Econ 89:1198–1204. https://doi.org/10.1111/j.1467-8276.2007.01084.x

Livesley S, McPherson G, Calfapietra C (2016) The urban forest and ecosystem services: impacts of urban water, heat, and pollution cycles at the tree, street, and city scale. J Environ Qual 45:119–124. https://doi.org/10.2134/jeq2015.11.0567

Lurie S, Hibbard M (2008) Community-based natural resource management: ideas and realities of Oregon Watershed councils. Soc Nat Res 21(6):526–537. https://doi.org/10.1080/08941920801898085

Maller C, Townsend M, Pryor A, Brown P, St Leger L (2006) Healthy nature healthy people: contact with nature as an upstream health promotion intervention for populations. Health Promot Int 21:45–54. https://doi.org/10.1093/heapro/dai032

Martel M, Harper S, Propst L (2017) Creating successful communities: a guidebook to growth management strategies. Island Press, Washington, DC

McElfish J (2004) Nature friendly ordinances. Environmental Law Institute, Washington, DC

McHarg I (1992) Design with nature. Wiley, New York

McPherson G, Simpson J, Peper P, Maco S, Xiao Q (2005) Municipal forest benefits and costs in five U.S. cities. J For 103:411–416. https://doi.org/10.1093/jof/103.8.411

Miller R, Hauer R, Werner L (2015) Urban forestry: planning and managing urban greenspaces. Waveland Press, Long Grove, IL

Mumford L (1961) The city in history: its origins, its transformations, and its prospects. Harcourt, New York

National Park Service (1992) Economic impacts of protecting rivers, trails, and greenway corridors. River trails and conservation assistance program, National Park Service. U.S. Department of Interior, Washington, DC

National Research Council (2009) Urban stormwater management in the United States. The National Academies Press, Washington, DC

Nesbitt L, Meitner M, Girling C, Sheppard S (2019) Who has access to urban vegetation: a spatial analysis of distributional green infrastructure in U.S. cities. Landscape Urban Plan 181:51–79. https://doi.org/10.1016/j.landurbplan.2018.08.007
Nolen J (2001) Well grounded: using local land use authority to achieve smart growth. Environmental Law Institute, Washington, DC
Nowak D, Dwyer J (2006) Understanding the benefits and costs of urban forest ecosystems. In: Kuser J (ed) Urban forestry in the northeast. Springer, New York, pp 25–41. https://doi.org/10.1007/978-1-4020-4289-8_2
Nowak D, Walton J (2005) Projected urban growth (2000-2050) and its estimated impact on U.S. forest resources. J For 103:383–389. https://doi.org/10.1093/jof/103.8.383
O'Brien EA (2003) Human values and their importance to the development of forest policy in Britain: a literature review. Forestry 76:1–16. https://doi.org/10.1093/forestry/76.1.3
Olmstead FL (1916) Introduction. In: Nolen J (ed) City planning: a series of papers presenting the essential elements of a city plan. D. Appleton and Company, New York, pp 1–18
Patton C, Sawicki D, Clark J (2013) Basic methods of policy analysis and planning. Rutledge, New York
Pearlmutter D, Calfapietra C, Samson R, O'Brien L, Krajter Ostoić S, Sanesi G (2017) The urban forest: cultivating green infrastructure for people and the environment. Springer, New York. https://doi.org/10.1007/978-3-319-50280-9
Pennsylvania Department of Environmental Protection (2006) Pennsylvania stormwater best management practices manual. Harrisburg, PA. Retrieved at http://www.stormwaterpa.org/from-the-foreword.html
Pennsylvania Department of Environmental Protection (2019) Pennsylvania stormwater management act: act 167. Harrisburg. Retrieved at https://www.dep.pa.gov/Business/Water/CleanWater/StormwaterMgmt/Pages/Act-167.aspx
Pennsylvania Governor's Center for Local Government Services (2015) Pennsylvania municipalities planning code. Harrisburg. Retrieved at https://dced.pa.gov/download/pennsylvania-municipalities-planning-code-act-247-of-1968/?wpdmdl=56205
Philadelphia (n.d.) Stormwater. Philadelphia. Retrieved at https://www.phila.gov/water/wu/stormwater/Pages/default.aspx
Porterfield G, Hall K (1995) A concise guide to community planning. McGraw-Hill, New York
Rolston H (2000) The land ethic at the turn of the millennium. Biodivers Conserv 9:1045–1058. https://doi.org/10.1023/A:1008918517655
Rost J (1991) Leadership for the Twenty-first century. Praeger, New York
Sanyal B (2018) A planners' planner: John Friedmann's quest for a general theory of planning. J Am Plann Assoc 84:179–191. https://doi.org/10.1080/01944363.2018.1427616
Selman P (2000) Environmental planning: the conservation and development of biophysical resources. Sage, New York
Smith H (1972) The citizens guide to planning. Planners Press, Chicago
Stone D (1988) Policy paradox and political reason. Scott, Foresman and Company, Boston
Thompson J, Elmendorf W, McDonough M, Burban L (2005) Participation and conflict: lessons learned from community forestry. J For 103:174–178. https://doi.org/10.1093/jof/103.4.174
Thousand Oaks (n.d.) City of Thousand Oaks general plan. Thousand Oaks. Retrieved at https://www.toaks.org/departments/community-development/planning/general-plan
University of Wisconsin Center for Land Use Education (2005) Planning implementation tools: overlay zoning. University of Wisconsin Center for Land Use Education, Madison
US Environmental Protection Agency (2017) What is green infrastructure? Retrieved from: https://www.epa.gov/green-infrastructure/what-green-infrastructure
US Forest Service (2014) Changing landscapes. In: Clark S, Foster D (eds) Forest service, NE state and private forestry. US Forest Service, Newtown Square, PA
Wachter S, Wong G (2008) What is a tree worth: green-city strategies, signaling, and housing price. Re Econ 36:213–239. https://doi.org/10.1111/j.1540-6229.2008.00212.x
Warren R (1972) The Community in America. Rand McNally, Chicago

Warren R (1977) New perspectives on the American community: a book of readings. Rand McNally, Chicago
Wilkinson K (1979) Social well-being and community. J Comm Dev Soc 10:5–16. https://doi.org/10.1080/15575330.1979.9987075
Wilkinson K (1991) The community in rural America. Greenwood Press, New York
Wilson W (1989) The city beautiful movement. Johns Hopkins University Press, Baltimore
Wolf L, Westphal L, Kuo F (2003) Social aspects of urban forestry. J Arboric 29:117–155
Wu J (2013) Landscape sustainability science: ecosystem services and human well-being in changing landscapes. Landsc Ecol 28:999–1023. https://doi.org/10.1007/s10980-013-9894-9
Young R (2011) Planting the living city: best practices for planning green infrastructure for major U.S. cities. J Am Plann Assoc 77:368–381. https://doi.org/10.1080/01944363.2011.616996

Part V
Forest-Water Interactions over the Long-Term: Analysis and Synthesis from Several Experimental Watersheds

Chapter 21
Forest Influences on Streamflow: Case Studies from the Tatsunokuchi-Yama Experimental Watershed, Japan, and the Leading Ridge Experimental Watershed, USA

Koji Tamai, Elizabeth W. Boyer, Shin'ichi Iida, Darryl E. Carlyle-Moses, and Delphis F. Levia

21.1 Introduction

Forests are vital for rural and urban populations all over the world, providing multiple ecosystem benefits to society and playing a key role in the supply of fresh water and the regulation of climate. The influence of forest vegetation and soils on the generation of streamflow has long been recognized. Incoming precipitation to forested watersheds is subject to evapotranspiration by vegetation, as well as runoff delay via infiltration, storage, flow paths, and residence times through forest soils. Changes in vegetation or soil characteristics in turn affect the characteristics of discharge from forested watersheds. Changes in the landscape raise concerns about the volume of water flowing to streams, timing of those flows, water quality, and human well-being. Paired watershed studies have been widely

K. Tamai (✉)
Forestry and Forest Products Research Institute, Tsukuba, Ibaraki, Japan
e-mail: a123@affrc.go.jp

E. W. Boyer
Department of Ecosystem Science and Management, Pennsylvania State University, University Park, PA, USA

S. Iida
Department of Disaster Prevention, Meteorology and Hydrology, Forestry and Forest Products Research Institute, Tsukuba, Ibaraki, Japan

D. E. Carlyle-Moses
Department of Geography and Environmental Studies, Thompson Rivers University, Kamloops, BC, Canada

D. F. Levia
Departments of Geography and Plant & Soil Sciences, University of Delaware, Newark, DE, USA

D. F. Levia et al. (eds.), *Forest-Water Interactions*, Ecological Studies 240,
https://doi.org/10.1007/978-3-030-26086-6_21

used to quantitatively evaluate such changes (Hewlett 1982). With this method, two or more forested watersheds with similar hydrological conditions (such as lithology, geographical features, soils, weather, and vegetation) are monitored along with precipitation and discharge for a calibration period spanning a range of climatic variability. Next, the state of forests in one basin is changed (treatment watershed) while the other watershed is untouched (control watershed), and hydrological monitoring continues during this treatment period. The relationship between the state of a forest and the water balance is elucidated by comparing the relative precipitation-discharge relation at the control catchment and the treatment catchment between the control period and the treatment period.

In this chapter, we review two classic paired watershed studies carried out in temperate forests in contrasting world regions. We present results from watershed manipulation experiments at the Tatsunokuchi-Yama Experimental Watershed in Japan (Tamai et al. 2004; Tamai 2005 and 2008) and at the Leading Ridge Experimental Watershed in the United States (Lynch et al. 1985; Lynch and Corbett 1990; Hornbeck et al. 1993).

21.2 Overview of the Experimental Watersheds and Their Vegetation Manipulation

The Tatsunokuchi-Yama Experimental Watershed (34.7°N, 133.97°E) is located in a hilly area in Okayama Prefecture of Japan, consisting of two adjoining catchments of Minami-dani (MN, 22.6 ha) and Kita-dani (KT, 17.3 ha). Its lithology is classified as Paleozoic, consisting of hard sand stones and clay stones, except that igneous rocks, including quartz porphyry, occupy about one-third of the KT catchment, while its soil is a little cohesive unripe clay loam. The mean annual temperature is 14.3 °C, and the mean annual precipitation is approximately 1200 mm year^{-1}. Due to the rainy season and the occurrence of tropical cyclonic storms (typhoons), June, July, and September are the wettest months. Winter is the driest season and snow coverage is rare.

The vegetation history and forest management history of the Tatsunokuchi-Yama Experimental Watershed is described by Tamai (2010) and is summarized in Table 21.1. Both the KT and MN catchments were natural forests of *Pinus densiflora* Sieb & Zucc. (Japanese red pine) at the beginning of the observation period in 1937. However, all of the *Pinus densiflora* trees were cut between 1944 and 1947 because of die-off by a pine wilt disease (Mamiya 1988) that led to enormous timber loss throughout northern Japan. The change of streamflow discharge was investigated at the KT and MN catchments individually utilizing a single-catchment experiment methodology. The catchments were at that time covered in broadleaf forest, though disturbances of various scales and forms have occurred in the MN treatment catchment. The major events are included in cases when the brush within 19.5 ha of broadleaf forests was cut down in 1954–1957. Subsequently, *Chamaecyparis obtusa* Sieb. & Zucc. (Japanese cypress, hinoki) and *Pinus thunbergii* Parl. (black pine) were planted. However most of the planted trees died (Kishioka et al. 1981).

Table 21.1 Major events of vegetation change and management in the Tatsunokuchi-Yama Experimental Watershed (Japan) and the Leading Ridge Experimental Watersheds (United States)

Year	Tatsunokuchi-Yama Experimental Watershed, Japan
1937	KT and MN observations of hydrological balance began
1947	KT and MN vegetation removal via clearcut
1954–1957	MN vegetation removal (partial) via brush cutting (Case 1, Fig. 21.4)
1959	MN vegetation removal of completion via forest fire (Case 2, Fig. 21.4)
1962	MN vegetation removal (selected riparian locations) via clearcut
1967	MN vegetation removal via forest fire
1977	MN vegetation removal (selected locations) via clearcut
1978–1980	MN vegetation removal via pine wilt disease (Case 3, Fig. 21.4)
1998	MN vegetation change via thinning

Year	Leading Ridge Experimental Watershed, United States
1957	LR1 LR2, LR3 observations of hydrological balance began
1967	LR2 vegetation removal (phase 1, lower slope) via clearcut
1971–1972	LR2 vegetation removal (phase 2, middle slope) via clearcut
1974	LR2 vegetation removal (lower & mid slope) via herbicide
1975–1976	LR2 vegetation removal (phase 3, upper slope) via clearcut
1976	LR2 vegetation removal (whole basin) via clearcut
1976	LR3 vegetation removal (selected locations) via cutting with best practices
1977	LR2 vegetation removal (whole basin) via herbicide

In September 1959, a fire destroyed the vegetation and litter layer over most of the MN catchment. *Pinus thunbergii* was subsequently planted over the entire MN catchment in March 1960 (Kishioka et al. 1981). The *Pinus thunbergii* forest, however, was destroyed by pine wilt decease during 1978–1980 (Abe et al. 1983). A natural broadleaf deciduous forest has since prevailed.

The Leading Ridge Experimental Watershed (40.66°N, 77.93°W) is located in the Appalachian Mountain region of the eastern United States. The site was selected to be representative of approximately four million ha of forest land within the Ridge and Valley Province of central Pennsylvania. There are three adjacent catchments of varying sizes, including Leading Ridge catchment 1 (LR1, 123 ha), catchment 2 (LR2, 42 ha), and catchment 3 (LR3, 104 ha). The catchments have a southeastern aspect and range in elevation from 274 to 442 m. They are underlain by the Rose Hill shale formation (over 200 m thick, underlying the entire study area), Castanea sandstone (over 150 m thick), and Tuscarora quartzite. The forests covering the catchments are even-age coppice forests. *Quercus montana* Willd. (chestnut oak) is currently the most dominant tree species in the watershed in both tree density and basal area, followed by *Q. rubra* L. (northern red oak), *Q. alba* L. (white oak), and *Acer rubrum* L. (red maple). Other tree species occurring in the watershed include *Tsuga canadensis* (L.) Carrière (eastern hemlock), *Pinus strobus* L. (eastern white pine), and *Q. velutina* Lam. (black oak) (Brubaker 2011). The oak-hickory cover type dominates the watershed while an oak-hemlock community exists in the moist valley floor areas.

The vegetation and forest management history of the Leading Ridge Experimental Watershed is described by Lynch and Corbett (1990) and Hornbeck et al. (1993, 1995) and is summarized in Table 21.1. Catchment LR1 is the control watershed and has not been actively managed. The forest on the control watershed was mature and in reasonably steady state considering biomass accumulation and annual evapotranspiration. However, both the control and the treatment catchments were subject to occasional periods of defoliation by gypsy moths over the period of record (Hornbeck et al. 1993, 1995). As part of the vegetation manipulation experiments, catchments LR2 and LR3 received various treatments between 1967 and 1977. Catchment LR2 had 100% of its vegetation removed in a three-phase clear-cutting and herbicide experiment, after a 7-year calibration period from 1959 to 1966. Phase 1 in 1967 was a complete riparian clear-cut of 21 acres (8.5 ha), with subsequent spraying of 2,4,5-T and 2,4-D herbicides to remove any remaining vegetation. Phase 2 on LR2 in 1971–1972 was a middle slope clear-cut of 27 acres (10.9 ha). Herbicide was applied to both the lower slope (phase 1) and middle slope (phase 2) areas of watershed 2 in 1974. Phase 3 on LR2 in 1975–1976 was an upper slope and ridge top clear-cut of 42 acres. Herbicide was applied to the entire 90-acre (36.4 ha) clear-cut on watershed 2 to remove all vegetation throughout the basin in 1977. Catchment LR3 had about 43% of its vegetation removed, in an experiment aimed at evaluating the use of best forest management practices during forest management operations (Lynch et al. 1985; Lynch and Corbett 1990). A commercial clear-cut of a portion of LR3 following best practices occurred during 1976–1977.

Hydrological fluxes enabling water balance calculations have been made at both experimental watersheds, with observations at the Tatsunokuchi-Yama Experimental Watershed beginning in 1937 and at the Leading Ridge Experimental Watershed in 1957. Here, we present results from monitoring of these watersheds from 1937 to 2002 in Japan and from 1957 to 2007 in the United States. The annual hydrological fluxes from the watersheds during both the control and treatment periods over the years of observation are shown in Fig. 21.1. In the Tatsunokuchi-Yama sub-catchments (KT and MN), annual precipitation ranged from approximately 600 to 1730 mm with an average of 1220 mm year^{-1}. In the Leading Ridge sub-catchments (LR1, LR2, LR3), annual precipitation ranged from about 470 to 1470 mm with an average of 1060 mm year^{-1}. Using the average values of precipitation (P), evapotranspiration (ET), and discharge (Q) measured at the watersheds, about 33% of the average annual precipitation is delivered to streamflow in the KT and MN watersheds, while about 45% of the average annual precipitation is delivered to streamflow in the LR1, LR2, and LR3 watersheds. Results from experimental watersheds in Japan and the United States show that increases in water yield occur on small watersheds in response to removal of vegetation and that increases in streamflow diminish as vegetation is replanted or naturally recovers.

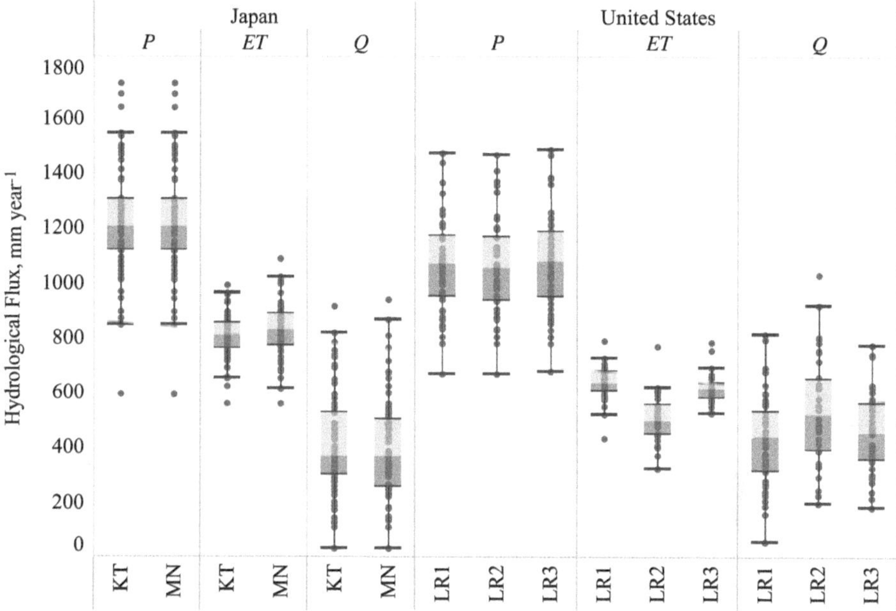

Fig. 21.1 Annual fluxes of precipitation (*P*), evapotranspiration (*ET*), and discharge (*Q*) observed over the period of long-term monitoring at the Tatsunokuchi-Yama Experimental Watershed, Japan (1937–2003, KT and MN catchments), and at the Leading Ridge Experimental Watershed, United States (1957–2007, LR1, LR2, and LR3 catchments). In this box and whisker plot, the boxes indicate the middle 50% (the middle two quartiles) of the data distribution. The line in the middle of each box represents the median value. The upper and lower whiskers are 50% higher and lower than the inner quartile range. Observations outside of the whiskers represent extreme events

21.3 Results and Discussion from the Tatsunokuchi-Yama Experimental Watershed, Japan

21.3.1 Determination of Periods in the Paired Catchment Experiments

The relative change in discharge under the influence of forest fires and pine wilt disease was investigated with a paired catchment experiment in the Tatsunokuchi-Yama experimental watershed (Abe and Tani 1985; Hattori 1994). Because it was in the MN catchment that forests were disturbed (Table 21.1), the MN catchment was designated as a treatment catchment and the KT catchment as a control catchment in these previous studies. ΔQ, the difference of annual discharge of the KT catchment from that of the MN catchment (Eq. 21.1), is positive for most years, but is negative for 1960–1966 and 1981–1994, presumably due to decreased transpiration as a consequence of forests in the MN catchment was recovering from forest fire damage in 1959 and pine wilt disease in 1978–1980, respectively. Discharge from the MN catchment increased in a relative sense (Hattori, 1994).

$$\Delta Q = Q_{\text{kobs}} - Q_{\text{mobs}} \tag{21.1}$$

where Q_{kobs} and Q_{mobs} represent the observed annual discharge from KT and MN catchments, respectively.

The paired catchment experiment analyses were conducted from 1960 to 1966 and 1981 to 1994 when ΔQ was negative, in 1959 and 1978–1980 when deforestation took place due to forest fire and pine wilt disease (disaster periods), while all other years were considered control periods. Years 1967, 1974, 1991, 1993, and 1995–1997 included days when discharge was not measured (due to observation system maintenance, etc.) and were excluded from analyses. Accordingly, a treatment period of 19 years (1960–1966, 1981–1994 except for 1991 and 1993), a disaster period of four years (1959, 1978–1980), and a control period of 37 years (1937–1958, 1967–1977, 1995–2003 except for 1967, 1974, 1995–1997) are defined.

21.3.2 Discharge Characteristics of the Control and Treatment Periods

A discharge duration curve sorts the 365 daily discharge values for a year in descending order. The larger discharge side of the curve represents the daily discharge at the time of a flood, whereas the smaller discharge side represents that at low discharge. The 95th most of the 365 daily discharges is designated as "plentiful water discharge," the 185th is "ordinary water discharge," the 275th is "low water discharge," and the 355th is "drought water discharge." These discharges from both catchments of Tatsunokuchi-Yama Experimental Watershed were compared. Figure 21.2 shows discharge from the MN watershed as a treatment watershed on the vertical axis and that from the KT watershed as a control watershed on the horizontal axis. Each figure has 19 open squares (□) representing the treatment periods, 36 closed diamonds (◆) representing the control periods, and 4 asterisks (∗) representing the disaster periods. Discharge values from the two catchments were highly linearly correlated for the treatment periods and for the control periods (Tamai et al. 2004):

$$Q_{\text{mobs}}(\text{year}, i) = a_{p,i} Q_{\text{kobs}}(\text{year}, i) + b_{p,i} \tag{21.2}$$

where $Q(\text{year}, i)$ represents the daily discharge (mm day^{-1}) on day i of the discharge duration curve of the year and $a_{p,i}$ and $b_{p,i}$, respectively, express the slope and y-intercept of the linear regression equation. Subscript p denotes the period of the paired catchment experiment, subscript t represents the treatment period, and subscript c denotes the control period.

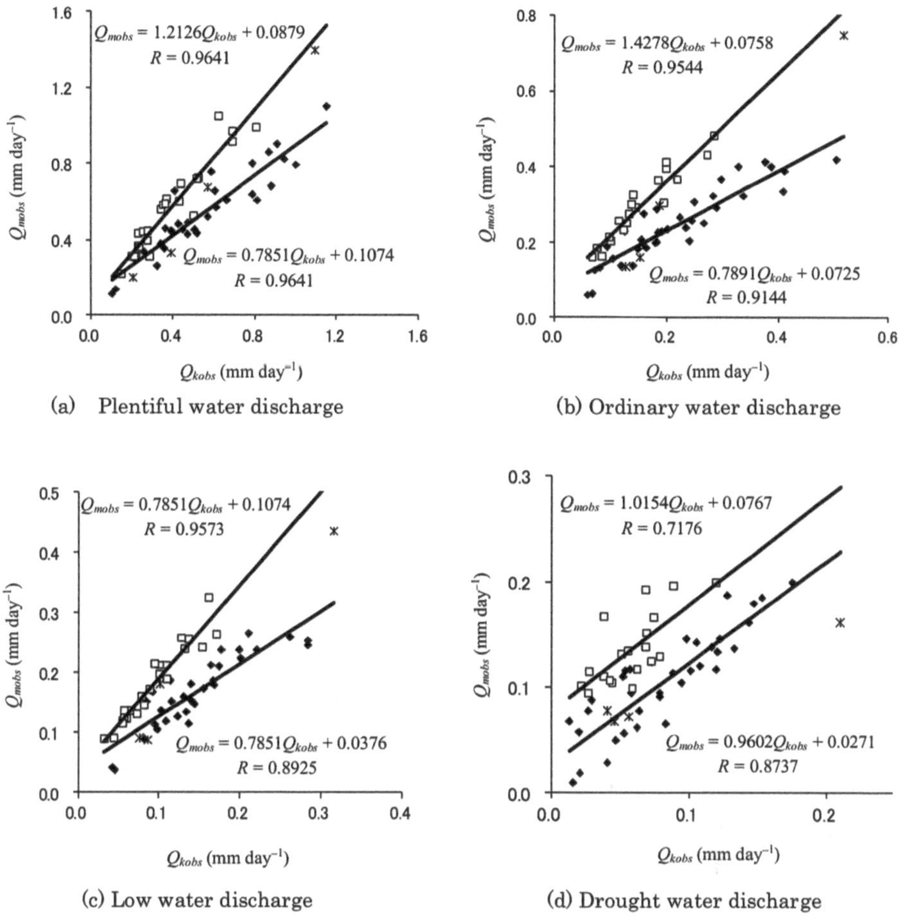

Fig. 21.2 Relationships between daily discharge in discharge duration curves for the Kita-dani and Minami-dani catchments, Tatsunokuchi-Yama Experimental Watershed (from Tamai et al. 2004). □: Treatment period (1960–1966, 1981–1994 except for 1991 and 1993). ◆: Control period (1937–1958, 1967–1977, 1995–2003 except for 1967, 1974, 1995–1997). ∗: Disaster period (1959, 1978–1980)

Equation 21.3 expresses $Q_{\mathrm{mcal}}(\text{year}, i)$ as computed by substituting the observed discharge $Q_{\mathrm{kobs}}(\text{year}, i)$ from the KT control catchment during the treatment period for Eq. (21.2). $Q_{\mathrm{mcal}}(\text{year}, i)$ is regarded as the estimated discharge from the MN catchment expected when the forest in that catchment is not disturbed (or stressed).

$$Q_{\mathrm{mcal}}(\text{year}, i) = a_{c,i} Q_{\mathrm{kobs}}(\text{year}, i) + b_{c,i} \tag{21.3}$$

As computed by Eq. (21.4), $\Delta Q_m(\text{year}, i)$ represents the relative change in discharge from the MN catchment with forest disturbance compared with the control KT catchment as indicated below, where Q_{mobs}(year, i) is the observed discharged from MN catchment during the treatment period. Thus, the difference between observed runoff of the MN catchment (Q_{mobs}) and estimated runoff of MN (Q_{mcal}) is assumed to be equal to the change in runoff caused by disturbance.

$$\Delta Q_m(\mathrm{year}, i) = Q_{\mathrm{mobs}}(\mathrm{year}, i) - Q_{\mathrm{mcal}}(\mathrm{year}, i) \qquad (21.4)$$

21.3.2.1 Indices for Evaluating the Influence of Forestry Operations

A regression line between daily discharge from the MN and KT catchments with high correlation was obtained for each of the control periods and treatment periods, respectively (Fig. 21.2). The forests within both catchments, especially within MN, have experienced disturbances of various scales in addition to the complete loss of forest by fire in 1959 and by pine wilt disease in 1978–1980. Brush cutting of 7.5 ha, 7.2 ha, and 4.8 ha was performed in 1954, 1955, and 1957, respectively. A 0.4 ha portion of forest along a mountain stream was clear-cut in 1962 and 1964, respectively, and left to natural recovery (Forest Influence Unit and Okayama Experimental Site, Kansai Branch Station, 1979). Forests of 0.4 ha were lost to fire in 1974 and were afforested with *Chamaecyparis obtusa* in 1976 (Tani and Abe 1987). Additionally, 0.4 ha of forest was clear-cut in 1977 and a further 2.3 ha was cut in 1982 and afforested with *Chamaecyparis obtusa*. Stands of *Chamaecyparis obtusa* having areas of 2.3 ha and 2.5 ha underwent thinning in 1998 and 2001, respectively (Goto et al. 2006). Consequently, the plots of daily discharge from both catchments are anticipated to fluctuate around two regression lines for the control and treatment periods presented in Fig. 21.2 according to recovery from disturbances of various scales and removal.

21.3.2.2 Definition of a Discharge Index at Plentiful, Ordinary, and Low Water Discharge Values

We define an index that expresses whether a particular annual Q value is closer to the control or treatment (i.e., disturbance) regression. This enables an investigation into the status and years of the influence of forest disturbances on discharge. The slopes of two regression lines ($a_{p,\ i}$) differ greatly at plentiful water discharge, ordinary water discharge, and low water discharge (Figs. 21.2a–c). Index I(year, i) is defined in Eq. (21.5) and Fig. 21.3a. Lines OT and OC represent the regression lines for treatment and control periods, respectively (Fig. 21.3).

$$I(\mathrm{year}, i) = 2 \cdot (\theta_{QOH} - \theta_{COH})/(\theta_{TOH} - \theta_{COH}) - 1 \qquad (21.5)$$

Therein, I(year, i) is an index of discharge on i-th day of the year on a discharge-duration curve, θ is an angle designated by a subscript, and OH is a horizontal

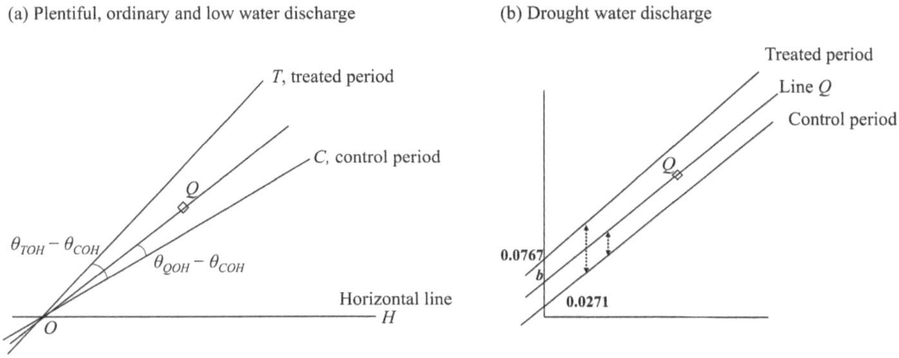

Fig. 21.3 Definition of a discharge index *I* (*year, i*) for the Tatsunokuchi-Yama Experimental Watershed. (Adapted from Tamai 2005)

straight line parallel to the x-axis passing through intersection O. When $I(\text{year}, i)$ is -1 or 1, point Q is located on a regression line for the control period or the treated period, respectively. Otherwise, when $I(\text{year}, i)$ is 0, Q is located at the midpoint of two regression lines.

21.3.2.3 Definition of a Discharge Index at Drought Water Discharge Values

The slopes of regression lines for the control and treatment periods at drought water discharge (the 355th water discharge in duration curve) are almost equal (see Fig. 21.2d). Accordingly, an index is computed using Eqs. 21.6 and 21.7 (Fig. 21.3b).

$$b(\text{year}) = Q_{\text{mobs}} \cdot (\text{year}, 355) - 0.9878 \cdot Q_{\text{kobs}}(\text{year}, 355) \tag{21.6}$$

$$I(\text{year}, 355) = 2 \cdot \frac{b(\text{year}) - 0.0271}{0.0767 - 0.0271} - 1 \tag{21.7}$$

The slope value 0.9878 of Eq. 21.6 represents the average values of slopes $a_{t,\ 355}$ (1.0154) and $a_{c,\ 355}$ (0.9602) given by Eq. (21.2) for the treatment and control periods at drought water discharge, respectively. $b(year)$ is the y-intercept of a straight line of slope 0.9878 that passes along point Q representing discharge in any year. Additionally, the values 0.0767 and 0.0271 in Eq. 21.7 are the respective values of $b_{t,\ 355}$ and $b_{c,\ 355}$, the y-intercepts given by Eq. 21.2 in the treatment and control periods at drought water discharge. When $I(year, i)$ is either -1 or 1, points Q are located on the regression lines for the control or the treatment periods, respectively; otherwise when $I(year, i)$ is 0, Q is located at the midpoint of the two regression lines.

21.3.2.4 Overview of Discharge Index Fluctuation

Figure 21.4 depicts the change of the discharge index *I*(year, *i*) over the years 1937–2003. When values of *I*(year, *i*) are positive and negative, open squares (□) and closed diamonds (◆), representing discharges from both the KT and MN catchments in Fig. 21.2, are located near the regression lines for the treatment period and the control period, respectively. The threshold of *I*(year, *i*) for estimating the influence of forest treatment is defined as 0, which implies that *I*(year, *i*) is located at the midpoint of the two regression lines for the control and treatment periods. Almost all values of *I*(year, *i*) are positive for the periods of 1960–1966 and 1981–1994, defined as treatment periods in Sect. 21.3.1, so that discharge from the MN catchment is regarded as having been influenced by forest removal. However, negative values were observed in 1981, the first year of the treatment period with pine wilt disease, demonstrating the effects of this stressor on discharge. This may be a consequence of defining a water year for the computation of the discharge-duration curves as spanning January 1 to December 31. Two dry seasons, one from January–April and another in December, are contained in one discharge-duration curve (Inaba et al. 2007). Only drought water discharge has the possibility to undergo the influence of precipitation events in the year prior to the year of analysis. Years when only the change of drought water discharge is delayed one year compared with that of other amounts of discharge and years with no delay might be mixed.

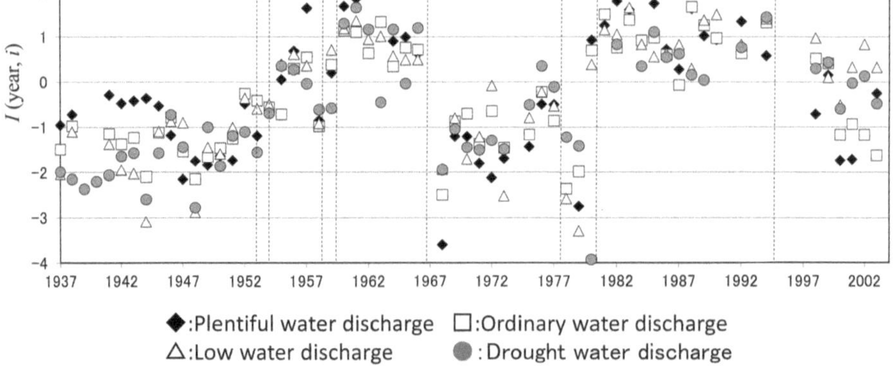

Fig 21.4 Long-term fluctuation of discharge index *I* (year, *i*) at the Tatsunokuchi-Yama Experimental Watershed from 1937 to 2003. (Adapted from Tamai 2005, 2008)

21.3.2.5 Influence of Disturbance and Forestry Operations on Discharge

The Minami-dani catchment has experienced various disturbances and forestry operations. Cases exist in which the increment in water discharge was or was not observed as a consequence of these events. Accordingly, incremental discharges that occurred in the following three cases were compared (Fig. 21.5). Two cases among three are forest fire in 1959 with 22.3 ha damaged area (Case 2) and pine wilt decease in 1978–1980 with 18.8 ha damaged area (Case 3). Their treatment periods are 1960–1966 and 1981–1994, respectively. The remaining Case 1 was from brush cutting in 1954–1957. However, the influenced years of this operation is not clear. Considering the continuity of the operated years of the brush cut and followed by Case 2 (i.e., forest fire), the disturbed period and the treatment period in Case 1 are defined to be 1954–1955 and 1956, respectively. The operated area in 1954–1955 was 14.7 ha. The brush cut in 1957 with 4.8 ha operated area is not evaluated here.

In Fig. 21.5, the ratio Q_{mobs}/Q_{mcal} in cases 2 and 3 is approximately equal in the range of $i < 270$, while the result in Case 1 is much smaller. This is attributed to the fact that the damaged area and expected decreased vegetation volume in cases 2 and 3 are wider and larger than those in Case 1. In comparison with the forest fire in 1959 where the vegetation was burned out, and the pine wilt disease in 1978–1980 when almost all trees died, the vegetation volume reduction by brush cut in 1954–1955 is considered to be small. This suggests that the decline of the amount of leaves reduced transpiration and consequently enhanced the incremental discharge. Further, Q_{mobs}/Q_{mcal} is greater for larger i in all cases. This also reflects the influence of transpiration by forest vegetation. Runoff at low discharge is constituted mainly by

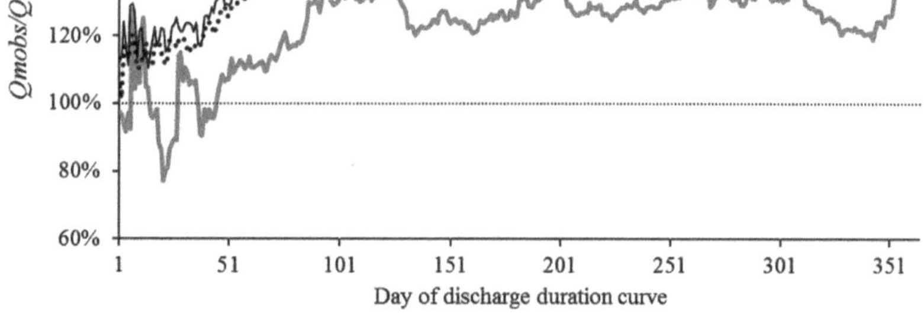

Fig. 21.5 Increased discharge ratio after forest disturbance or forestry operation in the Minami-dani catchment. Case 1: Treatment period is 1957 after brush cut in 1954–1955. Case 2: Treatment period is 1960–1966 after forest fire in 1959. Case 3: Treatment period is 1981–1994 after pine wilt decease in 1978–1980

base flow originating from groundwater. Rainwater that falls in a forest permeates the soil over a long time from the forest floor to a water table to turn into groundwater, under the continuous influence of transpiration. However, runoff during a flood includes a direct runoff component by which rainwater flows down promptly at the shallow part of forest soil, in addition to the base flow. The residence time of direct runoff in a catchment is so short that the influence of transpiration is small. Accordingly, it is inferred that Q_{mobs}/Q_{mcal} is larger when i in a discharge duration curve is larger.

Tani et al. (2012) optimized the parameters of a process-based model to simulate the water movements in the catchments, so that runoff data can be reproduced properly from each of two adjacent catchments (34.92°–34.93°N, 135.97–135.98°E) in Shiga Prefecture. One catchment has bare ground catchment (named Rachidani) and the forested catchment (named F2) protected as the temple grove of a temple founded in the ninth century. The difference between the simulated discharge duration curves shows the effects of the forest vegetation and soils. The flow duration curves of both catchments were estimated from discharge data computed by the process-based model given identical meteorological data and evapotranspiration rates. Thus the difference between the simulated discharge duration curves primarily reflects the delay of runoff by forest soil. The simulated discharge duration curve from the bare ground catchment (Rachidani) indicated more discharge at a flood and less at a drought compared with that from the forest catchment (F2) suggesting that runoff delay by forest soil has affected discharge. A simulated discharge duration curve was similarly obtained with the actual water balance difference between precipitation and discharge given as the amount of evapotranspiration. In this comparison, the given evapotranspiration rate in the bare ground catchment (Rachidani) was smaller than in the forest catchment (F2). So, the difference between the simulated discharge duration curves shows the effects of runoff delay by forest soil and of evapotranspiration by vegetation. Then discharge from the bare ground catchment (Rachidani) was greater than that from the forest catchment (F2) throughout the simulated discharge duration curve including at the low discharge. The smaller evapotranspiration in the bare ground catchment (Rachidani) led to the larger low water discharge than the forest catchment (F2). This implies that evapotranspiration by vegetation has more effects on runoff than runoff delay by forest soil.

Consequently, in the Japanese watersheds, the influences of forest extinction and forestry operations can have similar explanations (after Tamai, 2008). The number of years it took for discharge to return to the level of the control period was shorter in the case of the forest fire in 1959 when *P. thunbergii* was planted than in the case of pine wilt in 1978–1980 when the vegetation was left to natural recovery. Years taken for the discharge to return to the level of the control period were shorter at plentiful water discharge and ordinary water discharge than at low water discharge and drought water discharge in the case of the pine wilt decease in 1978–1980. Plentiful water discharge, ordinary water discharge, low water discharge, and drought water discharge were incremented in about 22.6 ha of the MN catchment during 1955–1956, caused by the forestry operations such as about 14.7 ha brush cutting and planting. No distinct influence on discharge was observed for periods when treatments were performed in an area smaller than 3.5 ha, such as a small forest fire in 1974, small-scale clear-cutting in 1977, brush cutting and *C. obtusa* planting in 1982, and *C. obtusa* thinning in 1998 and 2001. Cases with broader disturbed forests or greater tree volumes regarded as becoming extinct showed greater incremental discharge. Moreover, the rate of increase of discharge was greater on days with less discharge than days with more discharge in a discharge duration curve. This result highlights the markedly higher influence of evapotranspiration by forest vegetation than runoff delay by forest soil on streamflow.

21.4 Results and Discussion from the Leading Ridge Watershed, United States

A similar methodology using the paired watershed approach as presented for the experimental catchments in Japan was used at the Leading Ridge watersheds in the United States. Regression equations were used to develop the relationship between annual water yield from the control catchment (LR1) and the treatment catchments (LR2 and LR3) and to elucidate the effects of the vegetation removal treatments.

The manipulation experiment at watershed LR2 was a complete (100%) removal of the forest and other vegetation on the catchment (Table 21.2), in order to evaluate questions about the magnitude of change in streamflow that could result from removing the forest. Complete removal of vegetation from the LR2 catchment occurred via clear-cutting in three phases, along with the application of herbicides to eliminate vegetation regrowth. The results of the vegetation removal treatments on streamflow water yield are shown in Fig. 21.6, using data adapted from Hornbeck et al. (1995). The timeline shown begins at 1967, the time of the initial treatment on the catchment. Vegetation from the catchment was removed in three phases, with phase 1 clear-cut in the lower third (1967), phase 2 clear-cut in the middle third (1971–1972), and phase 3 clear-cut in the upper third (1975–1976). In 1977 herbicide was applied to the entire clear-cut across the watershed to remove all vegetation. A photo of the LR2 catchment after the second phase is shown in Fig. 21.7.

The vegetation removal treatments in LR2 resulted in substantial streamflow increases of up to 35%, with the maximum increases occurring (not surprisingly)

Table 21.2 Relative to the control watershed, the seasonal and annual discharge increased in the clear-cut basin, with the largest increases occurring once vegetation was fully removed

Forest vegetation removal in the treatment watershed	Water years	Growing season (%)	Annual year (%)
Lower third (phase 1)	1967-71	6.7	5.2
Middle third (phase 2)	1972-73	5.9	7.2
Upper third (phase 3)	1974-75	14.4	8.6
Entire basin (final phase)	1976-78	24.1	15.5

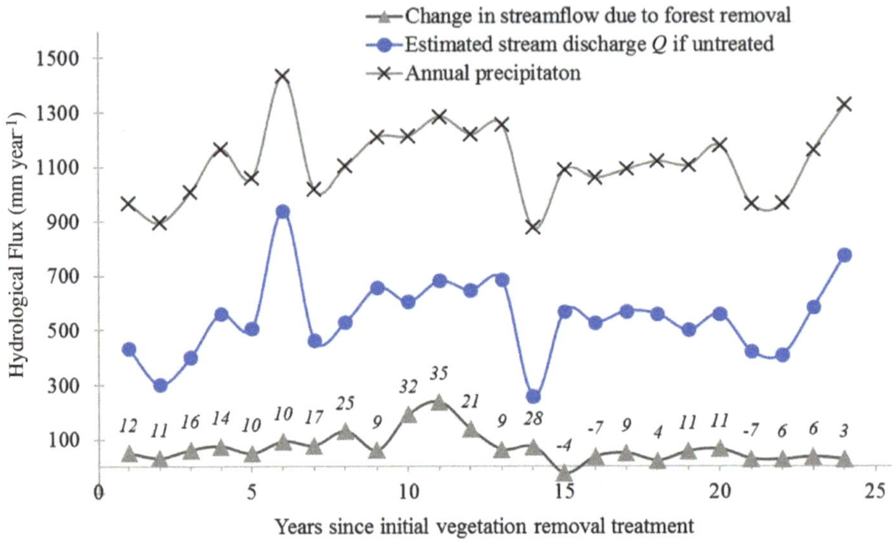

Fig. 21.6 Changes in streamflow water yield at the Leading Ridge Experimental Watershed LR2 in response to forest vegetation removal. Results are shown from 1967, the time of the initial treatment. Vegetation was removed in three phases, with phase 1 clear-cut in the lower third of the catchment (1967), phase 2 clear-cut in the middle third (1971–1972), and phase 3 clear-cut in the upper third (1975–1976). In 1977, herbicide was applied to the entire clear-cut across the catchment to remove all vegetation. Changes in streamflow from the vegetation removal treatments are shown with the triangles (in units of mm year^{-1}) and are labeled with the percentage change (in italics) attributed to vegetation removal

Fig. 21.7 Aerial view of Leading Ridge catchment LR2, after the second phase clear-cut of the middle third of the forest in 1972–1973

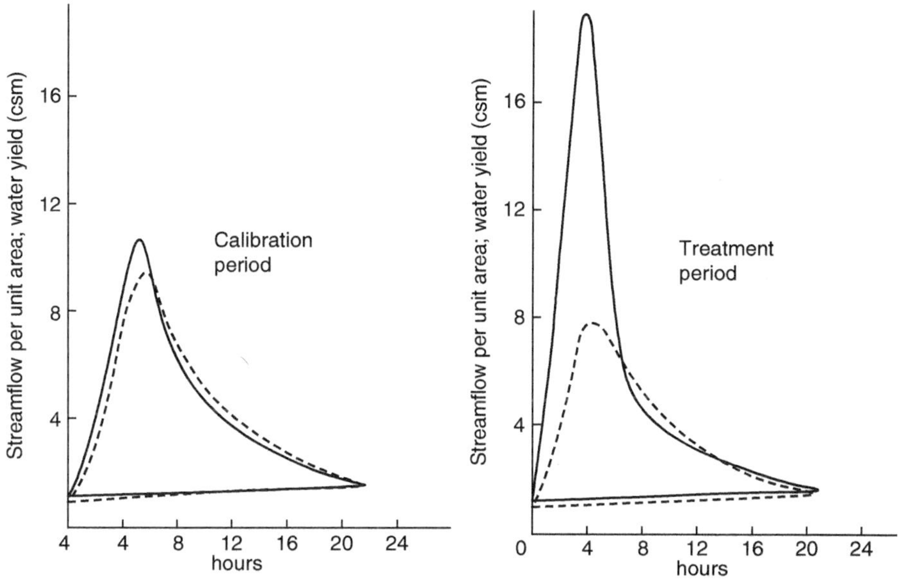

Fig. 21.8 Average hydrographs produced by Leading Ridge catchment LR1 (control, dotted line) and LR2 (clear-cut treatment, solid line), during the calibration period before vegetation removal (left) and during the treatment period after vegetation removal (right). The water yield is reported in cubic feet per second per square mile (csm)

after all of the vegetation was removed from the catchment. Water yield increases during the first 1–3 years after clear-cutting occurred during both the growing and dormant season, but mostly during the growing season as augmentation of base flow (Lynch et al. 1972). Mean peak flow during the growing season increased by more than 150%. Overall, the increases in water yield due to removing the forest vegetation declined quickly as natural regrowth of vegetation occurred (Hornbeck et al. 1993, 1995). Considering changes to individual precipitation and stormflow events, the clear-cutting greatly increased peak flows, as evidenced by individual hydrographs considered during the calibration period compared to during the treatment period after vegetation removal (Fig. 21.8). Results showed that substantial increases in peak flows on small watersheds occurred in response to clear-cutting, especially for small discharge events.

The manipulation experiment at catchment LR3 evaluated the question of how streamflow and water quality are affected by the use of best management practices during forest harvesting, with careful attention to protection of hydrologically sensitive areas. In this experiment, a timber sale was advertised and a commercial clear-cut operation took place on LR3 while following a comprehensive suite of forest best management practices (Lynch et al. 1985). During the vegetation removal operations, a protective buffer strip 100 ft (30.5 m) wide was left on each side of all streams. Skidding over perennial streams, except over approved culverts, was prohibited. Slash was not permitted within 25 ft (7.6 m) of any perennial or intermittent stream. Main skid trails, logging roads, and log landings were carefully laid out by a forester. After the operations, culverts were removed and water bars and other drainage devices installed on all logging roads and major skid trails, and roads were gated and grated to prelogging conditions. Lastly, filtration strips between road surfaces and stream channels were used.

In addition to the ongoing measurements of the water balance and streamflow, stream hydrochemistry data were collected on both the control catchment (LR1) and the treatment catchment from 1973 to 1979, with over three years of monitoring in the calibration phase before the vegetation removal (Lynch et al. 1985; Lynch and Corbett 1990). Comparing the control (LR1) to the catchment that was clear-cut along with using best practices (LR3) revealed only minor changes in stream turbidity and sedimentation. In contrast, turbidity and erosion were significantly impaired following vegetation removal in the LR3 catchment, where the forest was removed (via clear-cutting and application of herbicides) without considerations of protection of hydrologically sensitive areas or efforts to reduce erosional losses. Results showed that the use of forestry best management practices were effective at controlling nonpoint source pollution during and following the logging harvest operations (Lynch et al. 1985; Lynch and Corbett 1990). The results of this study were influential in establishing recommendations for use of forestry best management practices in operations throughout the eastern United States.

21.5 Future Directions

Paired watershed studies from around the world collectively have provided understanding of how land use changes affect the water balance and streamflow and how the responses vary across heterogeneous environmental settings. Results from the watersheds discussed here as well as other experimental watersheds highlight that the increases in water yield that accompany forest vegetation removal observed at individual watershed sites are highly variable (e.g., Bosch and Hewlett 1982). Best et al. (2003) and Brown et al. (2005) provide a critical review summarizing the limitations of the paired watershed approach, noting the results gleaned from the research across experimental watersheds are hard to generalize and that results from individual watershed studies cannot be easily translated to larger regions or to catchments with different conditions. Nonetheless, information from experimental watersheds have been key to informing strategies for watershed management and forest restoration, toward conserving and protecting water resources. Comprehensive monitoring data from watersheds are invaluable and foundational resources for understanding the effects of long-term changes in climate and landscapes on nutrient cycles, ecosystems, and water resources – where external effects shaping these processes can be subtle and long duration. Long-term watershed monitoring data are becoming increasingly useful to link the impacts of changing climate, vegetation, soil, and water as the data are synthesized and shared publicly, allowing application to new research questions and hypotheses (Tetzlaff et al. 2017).

References

Abe T, Tani M, Kishioka T, Kobayashi C (1983) Changes of direct runoff after killing of pine trees by the pine-wood nematode. 34th transactions of Kansai branch of the Japanese Forestry Society, 337–340 (in Japanese)
Abe T, Tani M (1985) Streamflow changes after killing of pine trees by the pine-wood nematode. J Jpn For Soc 67:261–270. https://doi.org/10.11519/jjfs1953.67.7_261. (in Japanese)
Best A, Zhang L, McMahon T, Western A, Vertessey R (2003) A critical review of paired catchment studies with reference to seasonal flows and climatic variability. CSIRO Land and Water Technical Report No 25/03, 56 p
Bosch JM, Hewlett JD (1982) A review of catchment experiments to determine the effect of vegetation changes on water yield and evapotranspiration. J Hydrol 55:3–23. https://doi.org/10.1016/0022-1694(82)90117-2
Brown AE, Zhang L, McMahon TA, Western AW, Vertessey RA (2005) A review of paired catchment studies for determining changes in water yield resulting from alterations in vegetation. J Hydrol 310:28–61. https://doi.org/10.1016/j.jhydrol.2004.12.010
Brubaker K (2011) Multi-scale lidar-based approaches to characterizing stream networks, surface roughness, and landforms of forest watersheds. PhD dissertation, The Pennsylvania State University, 162 p
Forest Influence Unit and Okayama Experimental Site, Kansai Branch Station (1979) Statistical report of hydrological observation at Tatsunokuchi-yama Experimental Watershed 1959–1977. Bull Forestry Forest Prod Res Inst 308:133–195. (in Japanese)
Goto Y, Tamai K, Miyama T, Kominami Y, Hosoda I (2006) Effects of disturbance on vertical stratification of broad-leaved secondary forests in Tatsunokuchi-yama Experimental Forest. Bull Forestry Forest Prod Res Inst 400:215–225. (in Japanese with English abstract
Hattori S (1994) Water balance in forest in small rainfall region. J Jpn Soc Reveg Technol 19:296–302. https://doi.org/10.7211/jjsrt.19.276. (in Japanese)
Hewlett JD (1982) Principles of forest hydrology. University of Georgia Press, Athens. 183 p
Hornbeck JW, Adams MB, Corbett ES, Verry ES, Lynch JA (1995) A summary of water yield experiments on hardwood forested watersheds in the northeastern United States. In: Gottschalk KW, Fosbroke Sandra LC (eds) United States Forest Service, Proceedings of the central hardwood forest conference, March 1995, Morgantown, WV. Gen Tech Rep NE-197, US Department of Agriculture, Forest Service, Northeastern Forest Experiment Station, pp 282–295
Hornbeck JW, Adams MB, Corbett ES, Verry ES, Lynch JA (1993) Long-term impacts of forest treatments on water yield: a summary for northeastern USA. J Hydrol 150:323–344. https://doi.org/10.1016/0022-1694(93)90115-P
Inaba N, Kondo K, Numamoto S, Hayashi S (2007) Influence of the definition of water-year period on discharge-duration analysis focused on low flow: in the case of the Tatsunokuchi-yama experimental watershed. J Jpn For Soc 89:412–415. https://doi.org/10.4005/jjfs.89.412. in Japanese with English summary
Kishioka T, Abe T, Tani M (1981) The influences on peak flow by forest fire and Hinoki plantation in Minamitani catchment, Tatsunokuchiyama experimental watershed. Annual report of Kansai Branch Station. Forestry Forest Prod Res Inst 23:55–58
Lynch JA, Corbett ES, Mussallem K (1985) Best management practices for controlling nonpoint-source pollution on forested watersheds. J Soil Water Conserva 40:164–167. https://doi.org/10.2489/jswc.67.4.300
Lynch JA, Sopper WE, Partridge DB (1972) Changes in streamflow following partial clearcutting on a forested watershed. In: Csallany SC, McLaughlin TB, Striffler WD (eds) National symposium on watersheds in transition proceedings. American Water Resources Association, Urbana, IL, pp 313–320
Lynch JA, Corbett ES (1990) Evaluation of best management practices for controlling nonpoint pollution from silvicultural operations. J Am Water Resour Assoc 26:41–52. https://doi.org/10.1111/j.1752-1688.1990.tb01349.x

Mamiya Y (1988) History of pine wilt disease in Japan. J Nematol 20:219–226
Tamai K (2005) A paired-catchment experiment in the Tatsunokuchi-yama experimental Forest, Japan: the influence of forest disturbance on water discharge. WIT Trans Ecol Environ 83:173–181. https://doi.org/10.2495/RM050171
Tamai K (2008) The things to be understood with water balance monitoring and expectation to the future: an example in Taktsunokuchiyama experimental watershed. Water Sci 61:1–21. https://doi.org/10.20820/suirikagaku.52.3_34. (in Japanese)
Tamai K (2010) Comparison of discharge duration curves from two adjacent forested catchments - effect of forest age and dominant tree species. J Water Resour Protect 2:742–750. https://doi.org/10.4236/jwarp.2010.28086
Tamai K, Goto Y, Miyama T, Kominami Y (2004) Influence of forest decline by forest fire and pine wilt disease on discharge and discharge duration curve, -in case of Tatsunokuchi-yama experimental forest. J Jpn For Soc 86:375–379. (in Japanese with English summary. https://doi.org/10.11519/jjfs1953.86.4_375
Tani M, Abe T (1987) Analysis of stormflow and its source area expansion through a simple kinematic wave equation. In: Forest Hydrology and Watershed Management, Proceedings of the Vancouver Symposium. IAHS Publ No 167, 609–615
Tani M, Fujimoto M, Katsuyama M, Kojima N, Hosoda I, Kosugi K, Kosugi Y, Nakamura S (2012) Predicting the dependencies of rainfall-runoff responses on human forest disturbances with soil loss based on the runoff mechanisms in granite and sedimentary rock mountains. Hydrol Process 26:809–826. https://doi.org/10.1002/hyp.8295
Tetzlaff D, Carey SK, McNamara JP, Laudon H, Soulsby C (2017) The essential value of long-term experimental data for hydrology and water management. Water Resour Res 53:2598–2604. https://doi.org/10.1002/2017WR020838

Chapter 22
The Biogeochemical Response of Nitrate and Potassium to Landscape Disturbance in Watersheds of the Hubbard Brook Experimental Forest, New Hampshire, USA

Habibollah Fakhraei, Timothy J. Fahey, and Charles T. Driscoll

22.1 Introduction

The Hubbard Brook Ecosystem Study (HBES) was initiated in the early 1960s with the goal of improving basic understanding of the ecosystem dynamics of forested landscapes. Bormann and Likens (1967) applied the small watershed budget approach, originally formulated to quantify the hydrologic cycle (Hewlett and Hibbert 1967) to biogeochemical cycles (Likens et al. 1967). By measuring the chemistry of hydrologic inputs (precipitation) and outputs (stream discharge) in a catchment with watertight bedrock, a mass balance for elements can be quantified precisely. After several years of pre-treatment monitoring, Likens et al. (1970) applied this approach to evaluate ecosystem responses to the deforestation of a small watershed (W2; 15.6 ha) at the Hubbard Brook Experimental Forest (HBEF). Since that beginning continued monitoring and experimental treatments in the Hubbard Brook watersheds have further elaborated the mechanisms underlying long-term changes in the biogeochemistry of northern hardwood ecosystems and responses to natural and anthropogenic disturbances (Likens 2013). The present chapter is a further contribution to this long legacy of discovery in biogeochemistry from the HBES.

In forest ecosystems, disturbances can occur as relatively discrete events that result in a drastic change in environmental conditions and resource availability, usually associated with mortality of overstory trees. In northeastern forests the natural disturbance cycle most commonly reflects the death of one or a few trees in small patches caused by windthrow (Bormann and Likens 1979), but occasionally extensive disturbances occur as a result of severe windstorms (microbursts,

H. Fakhraei (✉) · C. T. Driscoll
Department of Civil and Environmental Engineering, Syracuse University, Syracuse, NY, USA
e-mail: hfakhrae@syr.edu

T. J. Fahey
Department of Natural Resources, Cornell University, Ithaca, NY, USA

D. F. Levia et al. (eds.), *Forest-Water Interactions*, Ecological Studies 240,
https://doi.org/10.1007/978-3-030-26086-6_22

tornadoes, and hurricanes), wildfire, ice storms, insects, and pathogen irruptions. In the HBEF three severe storms have impacted the experimental watersheds and adjacent forests during the past 80 years: the hurricane of 1938 (Merrens and Peart 1992), a severe ice storm in 1998 (Rhoads et al. 2002), and a large microburst windstorm in 2013 that knocked down about 25 ha of forest immediately west of the gauged watersheds (Battles et al. 2017). Biogeochemical responses to such disturbance events are distinct from those creating small canopy gaps (Houlton et al. 2003). Thus, the northeastern forest landscape was originally characterized as a shifting mosaic of forest stands of different sizes, shapes, and ages resulting from this combination of disturbances (Bormann and Likens 1979). However, with the exception of a few virgin tracts (Runkle 1982), the modern northern forest landscape has been pervasively changed by extensive forest harvest, beginning in the mid-nineteenth century. Although forestry practices have varied through time and space, forest harvests then and now comprise large-scale disturbances that in many ways are comparable to the natural disturbances of northern forests, especially from the mortality of overstory trees. Thus, a better understanding of both natural forest ecosystem dynamics and those of the fully managed forest can be gleaned by studying watershed-scale responses to experimental treatments.

The early watershed-scale experiments at Hubbard Brook, together with other seminal studies of the biogeochemistry of forest soils (e.g., McColl and Cole 1968; Johnson and Cole 1980), have provided key insights into the mechanisms whereby forest disturbances could result in increased leaching of nutrients from soil and transport to surface water. Mortality of canopy trees greatly decreases root uptake of mineral nutrients, while microbial mineralization of soil organic matter and plant detritus continues to generate soluble cations and anions. Cations (K^+, Ca^{2+}, Mg^{2+}) can be retained by the soil cation exchange complex, whereas anions are typically more mobile and more readily transported through soil. Thus, the supply of mobile anions to soil solution can influence leaching of base cations (Johnson and Cole 1980). The relative mobility of nutrient cations differs, reflecting the lytropic or Hofmeister series, $NH_4^+ \sim K^+ > Mg^{2+} > Ca^{2+}$, so that cation exchange reactions can attenuate leaching responses. Moreover, decreases in pH associated with nitrification can decrease the mobility of the major anion, SO_4^{2-}, due to enhanced retention associated with the protonation of soil surfaces (Nodvin et al. 1986a). Finally, the supply and mobility of NO_3^- is altered by microbial processes, including N mineralization, immobilization, nitrification, and denitrification (Vitousek et al. 1997), and by atmospheric inputs (Galloway et al. 2004).

Largely because of their importance as nutrients, ionic properties, and high solubility, the most mobile solutes in forest soils following large-scale disturbance are usually K^+ and NO_3^-. Both are essential plant nutrients for which demand is greatly reduced by canopy tree mortality, and leaching of these nutrients might be expected to increase roughly in parallel after disturbance and to return to pre-disturbance values as vegetation recovers. In fact, Likens et al. (1978) observed that K^+ and NO_3^- exhibited by far the largest response of all nutrients measured to experimental deforestation of W2 at the HBEF, and their recovery roughly paralleled vegetation recovery. However, the sources and mobility of K^+ and NO_3^- differ in several important ways. The high mobility of K^+ results in part from its nonstructural form in plant tissues, so that it is rapidly leached from detritus following the death of plant tissues (Attiwill 1968). The mobility of K^+ in soil is consistent with its place in the lyotropic series, but unlike NO_3^-, K^+ is retained by cation exchange sites. Moreover, K^+ input is primarily associated with weathering of primary minerals. In contrast, NO_3^- is not present in significant amounts in plant tissues or minerals and is generated from decay of detritus only following mineralization of organic matter and subsequent microbial nitrification. Moreover, NO_3^- is supplied to north-eastern forests from the atmosphere over recent decades mostly by atmospheric deposition as a result of elevated emissions of nitrogen oxides (Aber et al. 1989; Driscoll et al. 2003). Finally, NO_3^- can be immobilized by microbial uptake to a much greater extent than for K^+, because of the relatively high requirement for N by microbial biomass. In sum, these contrasts between K^+ and NO_3^- likely shape differences in their behavior in response to large-scale forest disturbances.

The goal of this chapter is to contribute to a better understanding of the mechanisms of forest biogeochemical response to large-scale disturbances. Our objective is to compare and contrast the mobilization and subsequent recovery of K^+ and NO_3^- following natural and experimental disturbances of the Hubbard Brook watersheds. We hypothesize that the difference in sources and mobility of K^+ and NO_3^- should result in nonparallel behavior of these solutes. Moreover, long-term studies at the HBEF have documented unexpected and unexplained patterns of NO_3^- loss from the experimental watersheds; in particular, decreasing NO_3^- losses have occurred since the 1970s, despite maturing forest which assimilates less nitrogen and relatively small changes in atmospheric nitrogen inputs during 1970s–1990s (Yanai et al. 2013; Groffman et al. 2018). We hoped that coincident patterns of K^+ flux in soils might provide new insights into the mechanism underlying these patterns.

22.2 Methods and Approaches

The HBEF is a 3160 ha forest experiment station located in the White Mountain National Forest in central New Hampshire (43° 56′N, 71°45′W). The climate is cool temperate, humid continental, with mean July and January temperatures of 19 and −9 °C, respectively (at 450 m elevation). Annual precipitation averages about 1400 mm and is distributed nearly evenly throughout the year. Approximately 30% of annual precipitation occurs as snow (Federer et al. 1989).

The HBEF encompasses the entire catchment of the 4th order stream Hubbard Brook. The bedrock geology underlying the gauged watersheds that are the subject of this chapter is in the Littleton Formation, a metamorphic schist from the Devonian period. Detailed studies indicated that this bedrock is relatively watertight so that deep seepage of water is minimal and essentially all outflow is captured by V-notch weirs at the base of the six small gauged watersheds (W1–W6). The watersheds are situated on a moderately steep, south-facing slope and are drained by first- and second-order streams. Soils are derived from glacial till of varying thickness and are mostly well-drained, acid Spodosols (Bailey et al. 2014). Recent studies have emphasized spatial variation in the soils within the HBEF, including some lateral development near bedrock outcrops and the stream channel and more vertical development away from these hydrologic constraints (Bailey et al. 2014; Gannon et al. 2014). On ridgetops and upper slopes soils are thin and often bedrock limited whereas deeper glacial till is found on lower slopes.

The forest vegetation of the HBEF is mostly northern hardwood forest dominated by sugar maple (*Acer saccharum* Marsh.), American beech (*Fagus grandifolia* Ehrh.), and yellow birch (*Betula alleghaniensis* Britt.) with increasing amounts of the conifers red spruce (*Picea rubens* Sarg.) and balsam fir (*Abies balsamea* (L.) Mill) at the highest elevations. Forest on the reference watersheds (W1, W6) is mostly about 100 years old, recovering from intensive harvest in the early twentieth century, as well as from some earlier cutting (1870s–1880s) and damage by the 1938 hurricane.

Among the six gauged watersheds on the south-facing slope of the HBEF, two are reserved as references (W3, hydrologic reference, and W6, biogeochemical reference). The earliest experiment at the HBEF was the devegetation of W2 in which all trees were cut and left in place in 1965 and regrowth was prevented for three years using herbicides (Likens et al. 1970). W4 was logged by clear-cutting of 25-m-wide, horizontally oriented strips. One-third of the strips were logged in each of three years (1970, 1972, and 1974) and a buffer strip was retained along the stream channel in the lower watershed; this experiment was successful in minimizing nutrient outflow in stream water during and after logging (Hornbeck et al. 1986). In winter 1983–1984 W5 was clear-cut as a whole-tree harvest, in which all aboveground biomass of trees >2.5 cm DBH was removed from the watershed (Dahlgren and Driscoll 1994).

A "natural" large-scale disturbance experiment was provided across all the gauged watershed by the severe 1998 ice storm (Irland 1998). The upper two-thirds of the watersheds were severely damaged by the 1.2–1.4 cm of ice that accumulated on the tree crowns. Detailed measurements of damage indicated 28–30% crown damage watershed wide in the mature forests on W1 and W6; the most severe damage was measured on the 24–28-year-old W4, whereas minimal canopy damage occurred in the 14-year-old W5 (Rhoads et al. 2002).

Stream discharge has been measured continuously on the gauged watersheds since the early 1960s using V-notch weirs. Precipitation is measured on a network of rain gauges, and the annual water balance is calculated on a water year basis from June 1 to May 31 (Likens 2013). Stream water samples for chemical analysis have been collected weekly since 1963 immediately above the weir in each watershed. In addition, beginning in the 1980s stream water samples for chemical analysis have been collected at a series of locations arranged longitudinally upstream from the weir in W1, W5, and W6 (Fuss et al. 2015). Soil solution samples have been collected monthly from three soil horizons (Oa, Bh, Bs) at three elevations, in the lower and higher elevations of the hardwood zone and within the spruce-fir zone, of W1, W5, and a reference watershed immediately west of W6 using zero-tension lysimeters (Fuss et al. 2015; Fig. 22.1).

In this chapter we report on concentrations of K^+ and NO_3^- in the water samples described above. The analytical chemistry approaches are detailed in Buso et al.

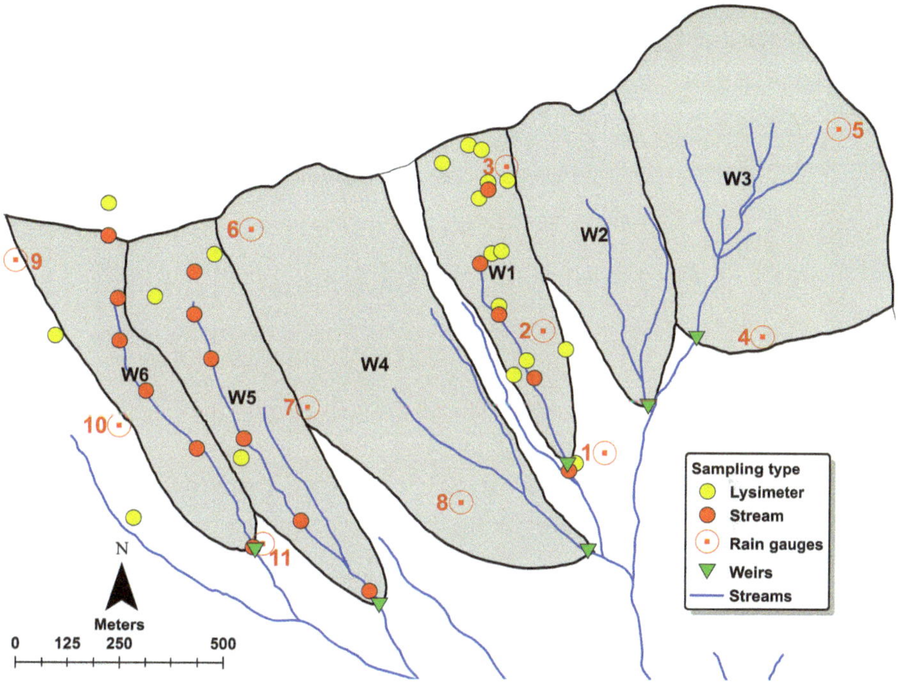

Fig. 22.1 Location of the experimental watersheds at the Hubbard Brook Experimental Forest, NH. Location of rain gauges, stream weirs, lysimeter, and stream chemistry sampling sites is shown

(2000). For some purposes in this chapter, we report volume-weighted mean concentrations. Appling a linear interpolation, we calculated daily concentration for days between sampling dates (Buso et al. 2000). The daily flux was estimated by multiplying the daily concentration by the daily streamflow. We calculated period (monthly and annual) volume-weighted mean concentrations by dividing the summation of daily fluxes to the summation of streamflows for the corresponding period. In describing overall disturbance effects on solute concentrations, we calculated average values for the interval during which concentrations were elevated above pre-disturbance background levels (Table 22.1). To test if post-disturbance concentrations of K^+ and NO_3^- in stream and soil solutions are significantly different from pre-disturbance concentrations, we used a two-sample t-test. In the *t*-test, the monthly observations during the elevated concentration interval (post-disturbance interval) were compared to the monthly observations during the pre-disturbance interval. The length of pre-disturbance period was determined based on availability of concentration measurements and to ensure a period that minimizes the seasonality in measurements. We also investigated long-term trends in concentrations of K^+ and NO_3^- in stream and soil solutions using the nonparametric seasonal Kendall's tau test (Hirsch et al. 1982). Seasonal trend analysis was conducted on monthly data. The slope of trends was estimated using the Sen Estimator (Sen 1968).

Table 22.1 Average concentrations of NO_3^- and K^+ in streamflow and soil solution for pre- and post- disturbance periods (in μmol/L). Relative changes in average concentration due to the disturbances are calculated (in μmol/L and %)

						NO_3^-					K^+				
			Pre-disturb. Interval[a]	Onset of response[b]	Post-disturb. Interval[c]	Pre-disturb. conc.	Post-disturb. conc.	Δ conc.	Δ conc.	P-value[d]	Pre-disturb. conc.	Post-disturb. conc.	Δ conc.	Δ conc.	P-value[d]
W	Disturbance	Sampling site	(year)	(month-year)	(year)	(μmol/l)	(μmol/l)	(μmol/l)	(%)		(μmol/l)	(μmol/l)	(μmol/l)	(%)	
1	Ice storm 1998	Weir 1	3	Sep-98	3	3.9	19.4	15.4	392	< 0.001	3.2	3.5	0.3	10	0.435
2	Devegetated 1965–67	Weir 2	2	Jun-66	6	10.2	560.9	550.8	5422	< 0.001	4.0	55.8	51.8	1280	< 0.001
3	Ice storm 1998	Weir 3	3	Sep-98	3	2.3	11.7	9.4	400	< 0.001	3.4	3.3	−0.1	−4	0.663
4	Strip cut 1970, 72, 74	Weir 4	5	Aug-70	6	16.5	55.7	39.2	238	< 0.001	5.5	9.1	3.5	64	< 0.001
5	Clear cut 1983–84	Weir 5	3	May-84	4	6.4	142.5	136.1	2135	< 0.001	5.2	20.0	14.8	285	< 0.001
6	Ice Storm 1998	Weir 6	3	Sep-98	3	1.8	8.2	6.4	354	< 0.001	3.9	4.2	0.4	9	0.379
1	Ice storm 1998	Oa	2	Jun-98	3	30.7	81.5	50.8	165	< 0.001	10.7	25.3	14.6	136	< 0.001
		Bh				36.5	68.1	31.6	87	0.044	8.1	16.9	8.8	108	0.023
		Bs				12.4	23.8	11.4	92	0.031	4.7	5.8	1.0	22	0.389
		Stream				4.0	28.7	24.7	617	< 0.001	3.2	4.9	1.8	56	0.014
5	Clear cut 1983–84	Oa	1	May-84	4	37.2	195.7	158.5	426	0.045	63.9	49.1	−14.9	−23	0.274
		Bh				6.4	212.8	206.4	3250	0.050	10.7	44.0	33.3	312	0.078
		Bs				9.6	290.0	280.5	2928	0.062	4.5	33.4	28.9	639	0.003
		Stream				5.7	175.3	169.5	2962	0.002	7.0	27.2	20.1	286	< 0.001
6	Ice storm 1998	Oa	2	Jun-98	2	31.6	35.7	4.1	13	0.558	33.6	15.8	−17.8	−53	0.046
		Bh				19.9	22.8	2.9	15	0.580	15.6	7.0	−8.7	−55	0.252
		Bs				4.3	6.5	2.2	51	0.236	3.1	4.5	1.4	45	0.257
		Stream				1.7	19.2	17.5	1060	< 0.001	3.3	5.2	1.9	58	0.004

[a]Number of years before disturbance that either the data are available or the concentration time series is relatively stable and represents the pre-disturbance condition.
[b]The time that initial response in concentration of nitrate to the disturbance was observed.
[c]Number of years after disturbance that concentration return to the pre-disturbance concentration.
[d]P-value of two-sample t-test for null hypothesis of the disturbance is not changing the concentrations.

22.3 Results

22.3.1 Reference Watershed: Patterns of NO_3^- and K^+

Concentrations of NO_3^- in the stream draining the biogeochemical reference watershed (W6) have significantly declined since the 1960s and 1970s (−0.16 μmol L^{-1} $year^{-1}$; Fig. 22.2; Yanai et al. 2013). This decline has preceded a marked decrease in atmospheric NO_3^- deposition since the early 2000s (−1 μmol L^{-1} $year^{-1}$) which was the result of controls on atmospheric nitrogen oxide emissions. The mechanisms driving this unexpected tightening of the nitrogen cycle, a so-called nitrogen oligotrophication (Groffman et al. 2018), remain unclear although the possible role of increased carbon supply associated with carbon dioxide fertilization to the forest, increases in the duration of the growing season, or increased mycorrhiza activity due to recovery from acid deposition has been posited. During most years since the early 1990s, stream NO_3^- concentrations have been at or near analytical detection limits. However, occasional large excursions from this strong N retention have been detected in 1981, 1991, 2000–2001, and 2015 (Fig. 22.2). The episode in 2000–2001 was clearly tied to canopy damage from the severe ice storm in 1998 (Rhoads et al. 2002) and consequent decrease in tree N uptake (Houlton et al. 2003). Earlier NO_3^- loss episodes were tentatively attributed

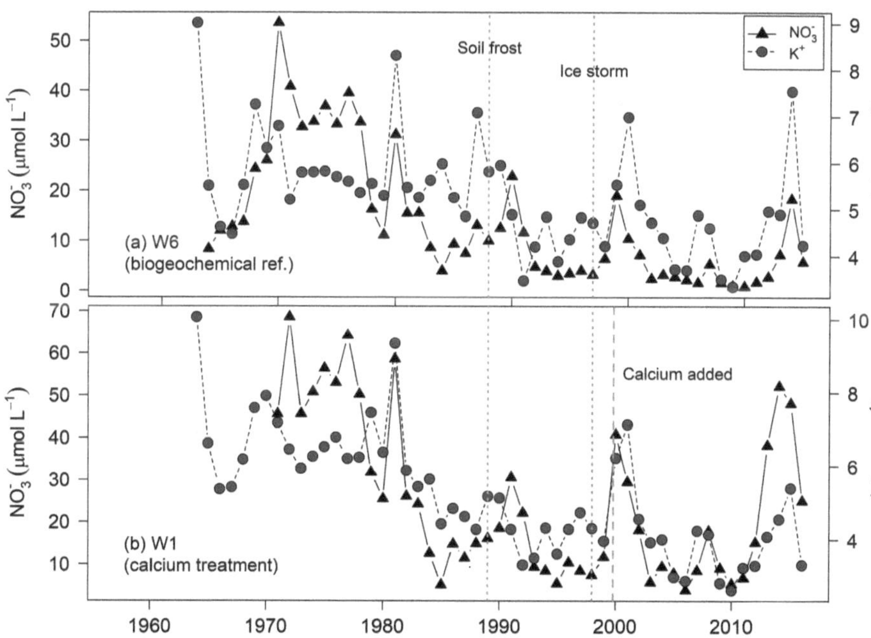

Fig. 22.2 Long-term annual volume-weighted concentrations of NO_3^- and K^+ in stream water in W6 (**a**) and W1 (**b**) at the HBEF the occurrence of the soil freezing and ice storm events are shown by vertical lines. Watershed 6 is the biogeochemical reference watershed. Watershed 1 was treated with wollastonite ($CaSiO_3$) in 1999

to effects of severe soil frost (Mitchell et al. 1996) and consequent damage to root systems depressing plant uptake (Tierney et al. 2001; Cleavitt et al. 2008). However, neither canopy damage nor severe soil frost coincided with the 2015 peak. It is also notable that the latter peak spanned three years, 2014–2016, in the adjacent W1 (Fig. 22.2) which was treated with a calcium silicate (wollastonite) to increase pH and soil exchangeable Ca^{2+} to levels that prevailed prior to the peak acid deposition era (1950–2000; Peters et al. 2004).

Concentrations of K^+ in stream water in the reference watershed (W6) have fluctuated to a similar degree over the years; however, the significant long-term temporal decline (-0.03 μmol L^{-1} $year^{-1}$) is less marked than for NO_3^- (-0.16 μmol L^{-1} $year^{-1}$). Some indications of the coupling between K^+ and NO_3^- losses are apparent from coincident peaks in annual volume-weighted mean concentrations of both solutes in a few years (1971, 1981, 2000–2001 and 2015; Fig. 22.2), but in other years high fluxes are decoupled, so that linear correlations between these solutes are moderate (annual volume-weighted concentrations, slope 0.18 mol K^+ (mol NO_3^-)$^{-1}$; $r^2 = 0.31$). Thus, while some common mechanism (e.g., reduced plant uptake) may contribute to leaching losses of both solutes, clearly other driving factors are involved.

The latter conclusion is reinforced by considering the seasonal patterns of stream NO_3^- and K^+ concentrations in the reference watershed (W6; Fig. 22.3). The long-term weekly average NO_3^- concentrations exhibit a clear increase beginning in early November and peaking in April prior to leaf out, followed by a marked decline to trace levels in early summer, presumably reflecting primarily high root uptake to supply canopy foliage expansion (Fig. 22.3). In contrast, K^+ concentrations increase and peak several weeks earlier than NO_3^- in fall, coinciding with leaching from senesced foliage, perhaps especially litter falling in and near stream channels. The decline in K^+ concentrations associated with leaf emergence in spring is modest and much delayed in comparison with that of NO_3^-, possibly due to interactions of K^+ with the soil exchanger (Fig. 22.3). Finally, the magnitude of the difference in stream K^+ concentration between the growing and dormant season is much less (ca. twofold) in comparison with that of NO_3^- (ca. tenfold).

During the unexplained flush of NO_3^- and K^+ from W6 in 2015 (Fig. 22.2a), the high NO_3^- flux was confined to the late winter peak as concentrations remained at detection limits (0.3 μmol L^{-1}; Buso et al. 2000) throughout the growing season and into early winter (Fig. 22.3). The increased K^+ concentrations in the W6 stream reflected primarily higher concentrations in early summer and early autumn and was clearly decoupled from the NO_3^- flush. The seasonal pattern of NO_3^- loss in 2015 is similar to that following the 1980 soil freezing event (Mitchell et al. 1996) although in 1981 increases in stream NO_3^- were evident in late fall (starting in November).

Concentrations of NO_3^- and K^+ in soil solutions (collected using zero-tension lysimeters; Fuss et al. 2015) decline in parallel from surface soil to deep soil horizons, with highest levels draining the forest floor organic horizons (Fig. 22.4). Presumably these depth patterns reflect the effects of differences between the supply of K^+ (leaching from detritus, mineral weathering) and NO_3^- (deposition, nitrification) and biological uptake (root, microbial) and, for K^+, exchange on soil surfaces. In upper soil horizons (Oa and Bh) concentrations of K^+ and NO_3^- in soil solutions

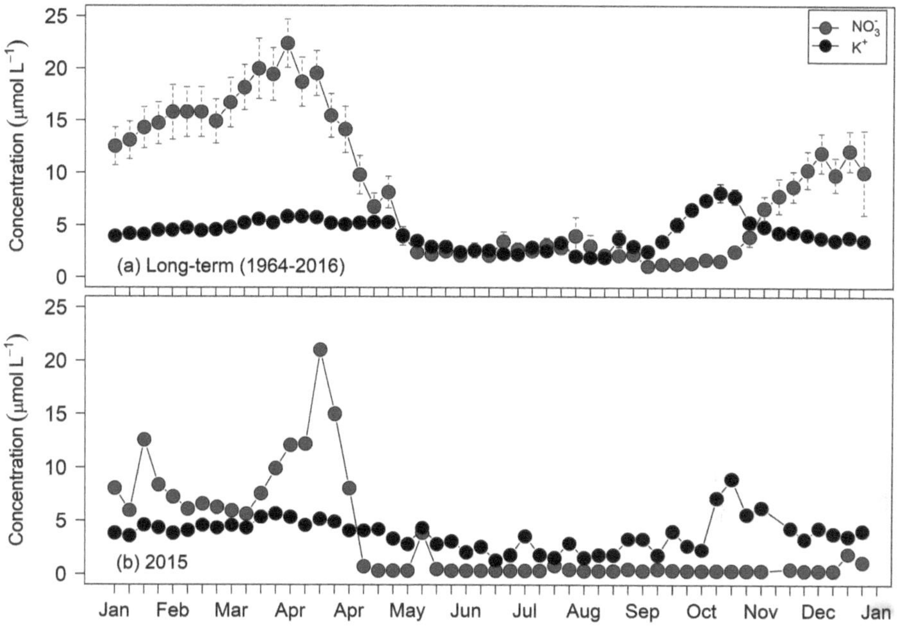

Fig. 22.3 Seasonal variations in NO_3^- and K^+ concentrations in stream water in W6: (**a**) long-term mean values (1964–2016) and (**b**) for 2015. The error bars indicate standard errors. In plot (**a**), because of limited observations and substantial variability in NO_3^- concentrations in the first week of observations, the related data point was not shown

decrease with increasing elevations in the watershed (Fig. 22.4). In the case of K^+, this may be due to the shallow mineral soil resulting in low K^+ supply. Dittman et al. (2007) suggested that low NO_3^- concentrations in the high elevation spruce-fir zone were due to N immobilization associated with high litter and soil C/N. One notable difference between the two solutes is that NO_3^- concentrations in streams are consistently lower than in soil solutions (including subsoil), whereas K^+ concentration is higher in streams than subsoil. The latter pattern is probably influenced by leaching from mineral soil or glacial till or from fresh detritus in streams, although shallow and rapid flow paths that short-circuit soil micropores also could contribute as detailed in the discussion. Decreases in NO_3^- concentrations are likely associated with immobilization in the riparian zone between the soil and the stream channel (Dittman et al. 2007).

22.3.2 Disturbance Effects: General

The response of soil and stream water chemistry to large-scale disturbance has been evaluated in a series of experiments, both natural (i.e., 1998 ice storm) and anthropogenic (harvesting) (W2, W4, W5). In Table 22.1 we compare and contrast the

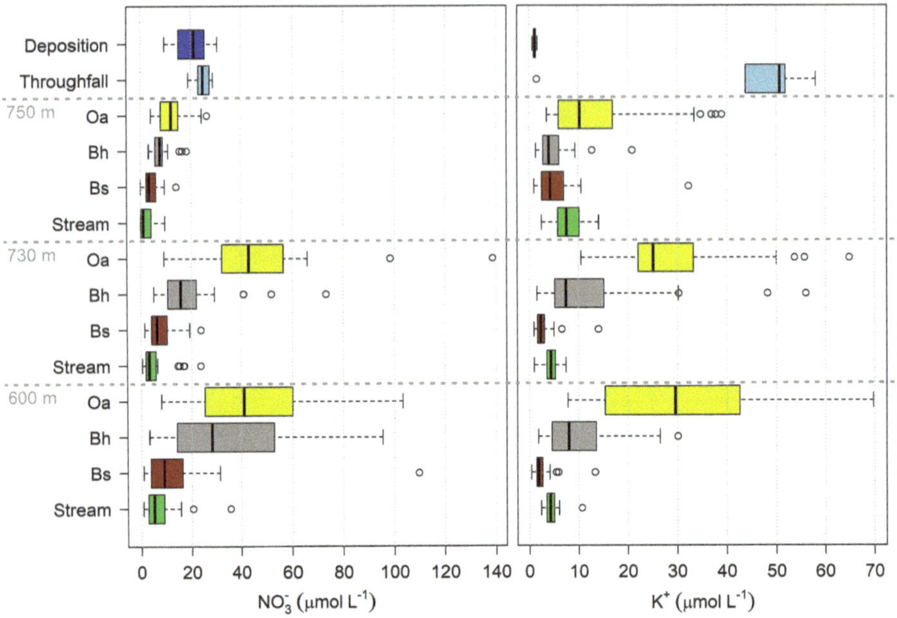

Fig. 22.4 Long-term mean NO_3^- and K^+ concentrations in bulk deposition, throughfall, soil solutions by horizon and stream water for three elevation zones: spruce-fir (750 m), high elevation hardwood (730 m), and low elevation hardwood (600 m) in reference sites west of W6 (1984–2016). Throughfall concentrations are average of measurements during the growing seasons of 1989–1992 in five sites at different elevation zones (Lovett et al. 1996)

magnitude of response of NO_3^- and K^+ concentrations in stream water during the three-to-six year interval following the disturbances during which concentrations were elevated in comparison with pre-disturbance average values. Clearly, NO_3^- is the most mobile solute in these northern hardwood forests following vegetation disturbance. Devegetation of W2 resulted in over a 50-fold increase in NO_3^- concentration. The elimination of nearly all plant uptake for three years caused marked accumulation of NH_4^+ on soil exchange sites, the product of organic matter mineralization, and subsequent microbial nitrification. By comparison, K^+ concentration increased 13-fold, partly from leaching of felled detritus, but especially by mobilization from cation exchange sites associated with acidity derived from nitrification (Van Miegroet and Cole 1984). The stoichiometric response of K^+ to NO_3^- associated with the W2 disturbance was 0.094 mol K^+ (mol $NO_3^-)^{-1}$ (Table 22.1).

Under less extreme forest harvest disturbances (W4, strip cut; Hornbeck et al. 1986), NO_3^- concentrations in streams increased by only two- to threefold as root uptake by both residual and recovering vegetation retained mineralized N. Similarly, K^+ losses were markedly lower, only 60% higher than pre-treatment, but the stoichiometric response of K^+ to NO_3^- associated with the strip cut was 0.090 mol K^+ (mol $NO_3^-)^{-1}$, similar to that of W2. The whole-tree harvest of W5 resulted in much higher concentrations of both NO_3^- (21-fold) and K^+ (nearly 3-fold) compared to the pre-treatment concentrations and intermediate between the W2 and W4 experiments. A slightly higher ratio of K^+ to NO_3^- leaching in the W5 experiment (0.109 mol K^+ (mol NO_3^-)$^{-1}$; Table 22.1) was evident compared to other treated watersheds (W2 and W4) despite the removal of aboveground biomass in W5 and hence a much lower supply of K^+ from leaching of dead organic matter (Likens et al. 1994).

The response of stream NO_3^- concentrations to the severe ice storm of 1998 was more modest than for the forest harvest experiments (Table 22.1); nevertheless, NO_3^- loss increased significantly ($p < 0.001$) for about three years after the ice storm in the mature forest watersheds: W1, W3, and W6 (Table 22.1; Houlton et al. 2003). The increase in NO_3^- concentration was similar across all three watersheds, 3.5- to 4-fold. Ice storm damage to the canopy in these watersheds was similar averaging 29% canopy loss with greater damage in mid to upper elevations (Rhoads et al. 2002). The response of stream NO_3^- to ice damage in the 24–28-year-old W4 was greater than for the mature forests, approaching values observed in response to forest harvest (Fig. 22.5a). This higher NO_3^- leaching in W4 reflects greater canopy damage in that younger forest owing to high susceptibility of the initially regrowing pin cherry (*Prunus pensylvanica* Sarg.) at this age (Rhoads et al. 2002). In striking contrast, no NO_3^- leaching pulse was observed in the adjacent, 14-year-old W5 (Fig. 22.5b) where canopy damage was minimal, despite similar ice loading, owing to flexible stems of younger trees.

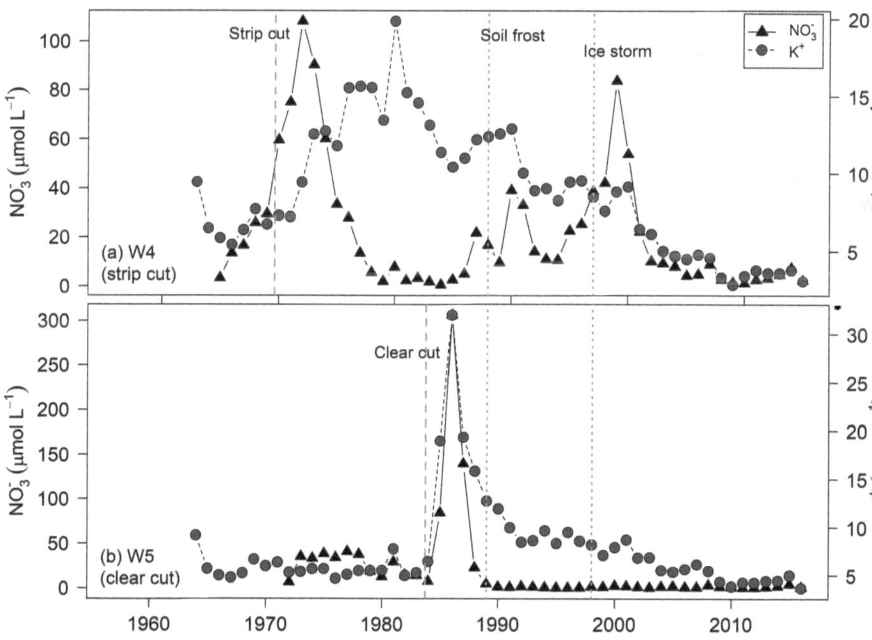

Fig. 22.5 Long-term annual volume-weighted concentrations of NO_3^- and K^+ in stream water in (**a**) W4 and (**b**) W5 at the HBEF the occurrence of the soil freezing and ice storm events are shown by vertical lines

The response of concentrations of K^+ to the ice storm disturbance was more muted than of NO_3^- in W1 and W6, but the stoichiometric response was similar to the harvesting treatments (W1 0.072 mol K^+ (mol NO_3^-)$^{-1}$; W6 0.109 mol K^+ (mol NO_3^-)$^{-1}$; Table 22.1). However, in W4 where the NO_3^- pulse was greatest, ice storm damage did not cause a significant increase in stream K^+ concentrations, further evidence of mechanistic decoupling of the mobilization of these two ions.

22.3.3 Seasonality of Response to Disturbances

Both NO_3^- and K^+ responded to the whole-tree harvest of W5 with a marked increase in stream water concentrations, with a distinct peak in 1986 two years after the 1983–1984 harvest. However, the seasonal pattern of these responses was distinctly different for NO_3^- and K^+ (Fig. 22.6) in this peak year of leaching loss. In particular, the NO_3^- concentration increase was most apparent during the dormant season, January to April, while concentrations declined dramatically in May and June, coincident with leaf expansion. In contrast, the seasonal pattern of K^+ concentration was muted, as the increase persisted through the year with only a minor decline in summer. Thus, in comparison with the long-term average behavior of W6 (Fig. 22.3) and the seasonal pattern during the 2015 pulse of loss, seasonal variation

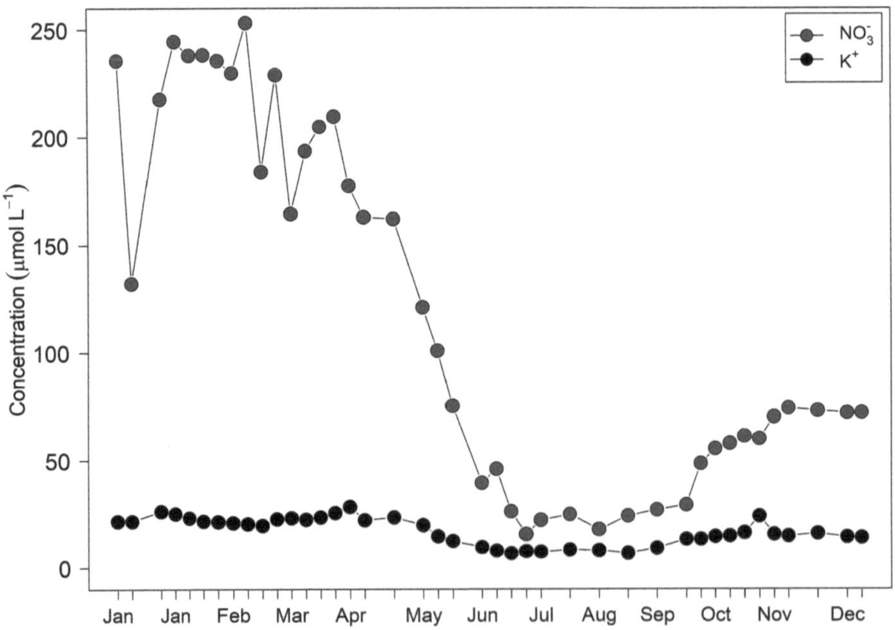

Fig. 22.6 Seasonal variation in concentrations of NO_3^- and K^+ in stream water in W5 for 1986 the peak year of solute loss following the whole-tree harvest

in K^+ during the postharvest of W5 was more muted, whereas NO_3^- leaching was similar.

22.3.4 Annual Patterns Following Disturbance

The timing and duration of the response of stream water solute concentrations to the watershed disturbances differed both between solutes and among the disturbances. In general, after forest harvest NO_3^- concentrations increased more quickly than those of K^+ and the subsequent decline was also more rapid for NO_3^-, as K^+ concentrations remained elevated for over a decade after harvest whereas NO_3^- concentrations returned to baseline levels within about 3 years (Fig. 22.5b). Not surprisingly the progressive strip harvest of W4 (1970–1974) resulted in a delayed and broader peak of stream concentrations than the more discrete whole-tree harvest (1983–1984) of W5 (Fig. 22.5a). Notably, the peak in K^+ concentrations in W4 occurred six years following the completion of harvest on W4 (1980) compared with two years on W5. In contrast, the timing and duration of high NO_3^- leaching was more similar between W4 and W5 than for K^+. Finally, following ice storm damage, the timing and duration of the NO_3^- response in the reference watersheds (Fig. 22.2) was quite similar to that following the harvest of W5 (Fig. 22.5b) and the peak of K^+ while smaller followed a similar rapid recovery pattern as NO_3^- (Fig. 22.2).

22.3.5 Response of Soils and Soil Solutions to Disturbance

In parallel with stream water chemistry, soil solution NO_3^- and K^+ concentrations increased markedly following large-scale disturbance (e.g., whole-tree harvest on W5; Fig. 22.7). Concentrations of NO_3^- in forest floor Oa horizon samples increased rapidly during the spring and summer for two years after the harvest, probably as a result of decreased root uptake and increased nitrification in the absence of mature forest vegetation. Increased K^+ concentration in Oa soil solution was roughly parallel with that of NO_3^- though the magnitude of increase was less. Following this early peak, concentrations of both solutes in Oa solution gradually declined during the next three years, probably in part as a result of increasing root uptake by recovering vegetation (Mou et al. 1993); for K^+, a decrease in the supply provided by decaying root systems certainly also contributed (Fahey et al. 1998). Thereafter, a marked increase in Oa solution K^+ concentrations was observed from years 3–8 following the harvest, reaching values similar to the initial peak immediately following harvest (Fig. 22.7). This interval was also marked by substantial season variation, with peak concentrations occurring in late winter. Notably, NO_3^- concentrations in Oa solutions during years 3–8 also were elevated compared to the baseline and followed similar seasonal cycling for K^+ (Fig. 22.8).

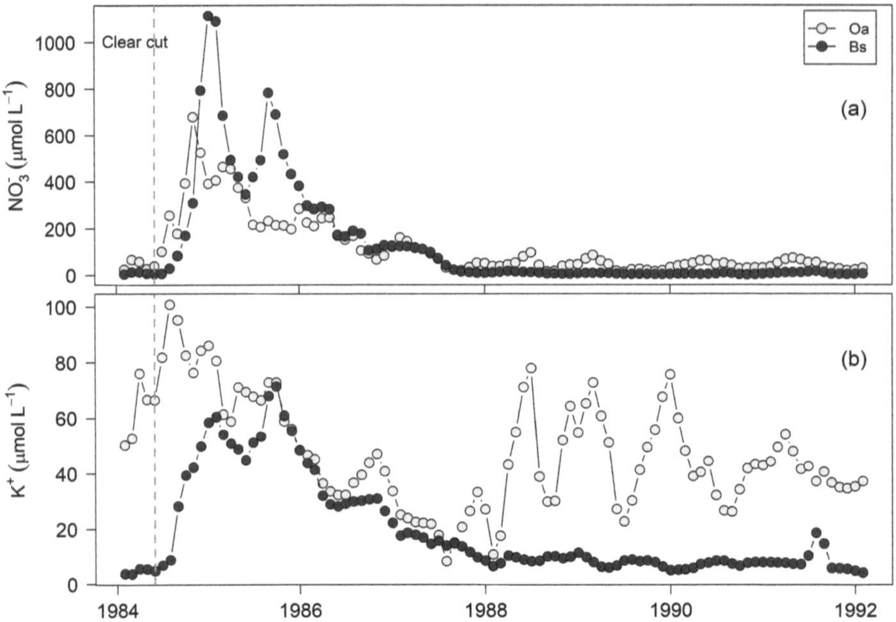

Fig. 22.7 Long-term concentrations of NO_3^- (**a**) and K^+ (**b**) in lysimeters at Oa and Bs depths in W5

Soil exchangeable K^+ was measured in the three elevation zones before the whole-tree harvest on W5 (1983) and at intervals during forest recovery (1986, 1991, 1998) (Romanowicz et al. 1996). Exchangeable K^+ in mineral soil horizons (Bs) increased markedly by the second summer after harvest. Thereafter, exchangeable K^+ in the Bs declined significantly. Although a small increase in exchangeable K^+ also was observed in the forest floor horizons in year 2 in the highest elevation zone of W5, on a watershed-wide basis this pool did not change significantly over time.

22.4 Discussion

Observations of the response of concentrations of NO_3^- and K^+, two relatively soluble nutrients in temperate forest soils, in ecosystem solutions draining northern hardwood forest watersheds following large-scale disturbances indicate considerable variation in the magnitude, timing, and duration of nutrient losses. Some of this variation can be attributed to specific mechanisms that influence the supply and solubility of NO_3^- and K^+ and provide insights into factors that may be responsible for long-term patterns of nutrient losses from the HBEF watersheds. In particular, we highlight the likely roles of (1) NO_3^- supply from atmospheric deposition and canopy leaching of K^+, (2) rapid leaching of K^+ from fresh plant detritus, (3) hydrologic flow paths and events, (4) root

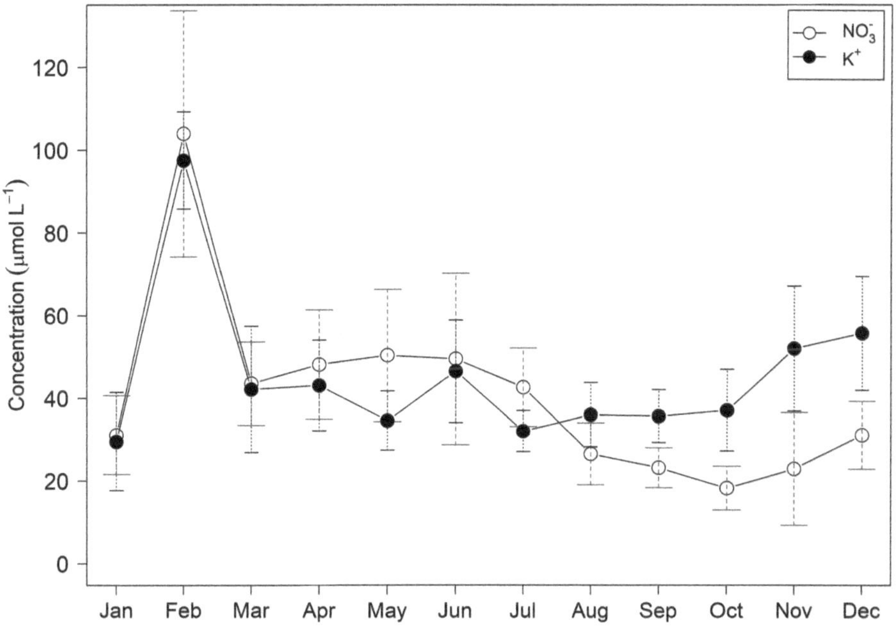

Fig. 22.8 Seasonal variation in concentrations of NO_3^- and K^+ in Oa lysimeters for 6 years (1988–1992) in W5. The error bars indicate standard errors

uptake by forest vegetation, (5) key microbial processes in the N cycle, and (6) soil cation exchange reactions. We evaluate the effect of each of these factors in accounting for differences in NO_3^- and K^+ behavior and their linkages and conclude with a synthesis some implications for biogeochemical dynamics in northern hardwood forest ecosystems and suggestions for future research.

22.4.1 Atmospheric Deposition and Canopy Leaching

A significant amount of NO_3^- is supplied to the HBEF in the form of atmospheric deposition which remained elevated through most of the duration of the HBES (1963~2000) until recent declines have been evident due to nitrogen oxide emission controls (Lloret and Valiela 2016); in fact, NO_3^- concentrations in precipitation are similar to levels in soil solutions in the mature forest (Fig. 22.4). However, tracer studies (e.g., Kendall et al. 1995; Campbell et al. 2007) indicate that most of the NO_3^- in soil solutions and streams is the product of internal N transformations involving microbial activity, rather than derived directly from precipitation NO_3^- which is rapidly immobilized in soil by biotic processes (see below). Thus, the direct effect of atmospheric NO_3^- on spatial and temporal patterns of NO_3^- concentrations in soil and stream water is likely minimal.

The input of K^+ from precipitation (long-term mean 16.65 mol ha^{-1} year^{-1}; 1.14 μmol L^{-1}; Fig. 22.4) is relatively minor compared with supply by mineral weathering (Likens et al. 1994). However, K^+ is readily supplied to the soil surface in large amounts by canopy leaching, owing to its soluble form in leaf cells (Parker 1983; Fig. 22.4). Thus, seasonal variation in K^+ supply to soils by canopy leaching results from leaf shedding in the deciduous canopy (see Lovett et al. 1996). In winter and spring we might expect lower K^+ inputs to soils associated solely with direct precipitation. On the other hand, however, fresh leaf litter is also subject to rapid leaching of K^+.

22.4.2 Leaching from Fresh Detritus

One of the principal mechanisms expected to cause decoupling between K^+ and NO_3^- concentrations in ecosystem solutions is the leaching of nutrients from fresh detritus, a large source of K^+ but not NO_3^-. In the reference watershed this effect is evident most clearly during and immediately after leaf litter fall in October–December: K^+ flux increases to a much greater degree than NO_3^- flux (Fig. 22.3).

The supply of K^+ from leaching of fresh detritus differed markedly among the watershed disturbances, which should be reflected in the relative increase of K^+ concentration compared to NO_3^- concentration in soil and streams. In particular, K^+ input in detritus followed the order W2 > W4 > W5 > ice storm. Although severe canopy damage occurred in the ice storms, tree mortality above the background rate only doubled (Battles et al. 2014); hence, mortality of fine roots, a principal source of K^+ in the harvested watersheds (Fahey et al. 1988), was minimal. This factor could help to explain the small increase in K^+ losses in soil solutions following the ice storm disturbance; however, K^+/NO_3^- stream concentration ratio was comparable to harvested watersheds (0.109 and 0.072 mol K^+ (mol NO_3^-)$^{-1}$ for W6 and W1, respectively; Table 22.1; see Synthesis below). At the other extreme, in W2 all aboveground biomass was left in place and as expected K^+ losses were particularly high in the forest 2–3 years after harvest (Likens et al. 1970), and these high K^+ losses were also undoubtedly facilitated by the elevated release of NO_3^- which enhanced the transport of K^+ due to its role as a mobile anion; however, again the K^+ released to stream water relative to NO_3^- was comparable to other disturbances (0.094 mol K^+ (mol NO_3^-)$^{-1}$).

In the case of the whole-tree harvest of W5, belowground detritus was a substantial contributor to K^+ losses (Fahey et al. 1988; Likens et al. 1994). The timing of the initial spike of K^+ concentration in streams was roughly coincident with that of NO_3^- (Fig. 22.5b) and probably resulted from rapid mortality of fine roots during the first summer postharvest (0.109 mol K^+ (mol NO_3^-)$^{-1}$). Thereafter, gradual mortality and leaching from the remaining dead root system could have supplied prolonged loss of K^+ for 3–4 years, contributing to the decoupling of K^+ and NO_3^- loss in 1987–1991 (Fig 22.5b). This mechanism also might help to account for some of the unexpected resurgence of K^+ leaching from surface soil horizons in

W5 in 1989–1991 (Fig. 22.7), although it is unclear why this pattern exhibited such a pronounced late winter peak (Fig. 22.8).

22.4.3 Hydrologic Effects

During low flow periods in summer, water generally percolates through the highly porous surface soil layers that predominate in the HBEF landscape, and stream discharge is mostly derived from subsurface flow; hence, stream water concentrations of K^+ and NO_3^- are similar to soil solutions in subsoil Bs horizon (Fig. 22.4). However, when evapotranspiration rates are low during leafless periods, water tables can migrate upward in the soil profile, especially in certain landscape positions (Bailey et al. 2014). At these times percolating water moves laterally in upper soil horizons, short-circuiting deep flow paths. Romanowicz et al. (1996) noted that during these high flow periods in W5, K^+ concentrations in stream water were similar to those in surface soil horizons, and much higher than in subsoil solutions. In contrast, in summer surface soil solution K^+ concentration was elevated (by leaching of detritus) and stream water levels were low. These observations are consistent with the hypothesized changing flow path mechanism outlined above.

A role of hydrologic factors in regulating soil solutions and stream water NO_3^- concentrations also seems likely. Notably, the high annual NO_3^- fluxes that occurred several times since the 1970s (Fig. 22.2) were associated primarily with high spring concentrations. Some of these unexplained NO_3^- flux events in the reference watersheds coincide with rain-on-snow events in late winter. These events could cause flushing of soil micropores or rapid flow of NO_3^--rich surface drainage to streams.

22.4.4 Plant Uptake of NO_3^- and K^+

Both NO_3^- and K^+ are macronutrients and are absorbed in large amounts by plant roots in the mature forest. Plant uptake undoubtedly contributes to declining concentrations of both solutes with increasing depth in soil in the mature forests (Fig. 22.4) and to seasonal differences between the dormant and growing season (Fig. 22.3).

Although plant demand for N exceeds that for K^+, because more N is resorbed from foliage than K^+ (Ryan and Bormann 1982), the difference in root uptake is smaller. Moreover, much of N uptake is in the form of NH_4^+ (Campbell et al. 2014); however, any decrease in root uptake of NH_4^+ can translate to increased nitrification and NO_3^- supply to leaching water. The large-scale disturbances of the HBEF watersheds reduced plant uptake, and the degree of reduction followed the order: W2 > W5 > W4 > ice storm. This differential reduction in root uptake played a dominant role in regulating the differences among the watersheds in NO_3^- and K^+ leaching (W2 > W5 > W4 > ice storms; Table 22.1). This reduction in root uptake would be expected to influence leaching losses during the growing season to a greater degree than during the dormant season; hence, we would expect the response to disturbance to be greater in the growing season. However, stream water concentrations of NO_3^- actually increased to a greater degree during the dormant season than the growing season in the peak flush year in W5 (1986; Fig. 22.6). Perhaps demand by the residual vegetation, the growth of new understory shrubs and herbs, and microbial biomass following whole-tree harvest were sufficient to buffer losses.

The influence of root uptake in directly regulating seasonal variations in K^+ losses is less clear. In W5 soil solutions, K^+ concentrations were elevated in comparison with the reference watershed in soil solutions from all depths and in stream water throughout the year (Romanowicz et al. 1996); thus, assimilation by roots apparently does not greatly reduce K^+ concentrations, at least in the mobile soil water collected by the zero-tension lysimeters. However, root uptake may be derived primarily from micropore water that is underrepresented in our soil solution collections. Moreover, as noted by Romanowicz et al. (1996), roots permeate near-stream saturated zones and may thereby directly influence stream water concentrations during the growing season.

22.4.5 Microbial Activity

The effect of microbial activity on NO_3^- dynamics is much more complex than for K^+ and certainly contributes to decoupling between the fluxes of these solutes. Although microbial decomposition of plant detritus contributes to K^+ leaching, this effect only slightly accelerates the leaching process. In contrast, NO_3^- dynamics are influenced by microbial immobilization, microbial mineralization of organic matter leading to ammonification and subsequent nitrification, and gaseous loss of N during microbial nitrification and denitrification. Researchers at the HBEF have long puzzled about the causes of long-term decrease in NO_3^- losses from the reference watershed despite continuing N deposition and maturing of the forest (thereby reducing net accumulation in biomass; Yanai et al. 2013). Some limited evidence suggests possible decreases in soil N mineralization and so-called N oligotrophication (Groffman et al. 2018) perhaps driven by increased C inputs and consequent microbial immobilization. Isotopic tracer studies indicate rapid microbial immobilization of precipitation NO_3^- in organic matter-rich surface soil horizons, even during the spring snowmelt period (Campbell et al. 2007). Another possible explanation is increasing denitrification losses, as suggested by some evidence for periodically high rates of denitrification (Morse et al. 2015) even in the mostly well-drained soils of the HBEF. Finally, a recent hypothesis (N-bank; Lovett et al. 2018) suggests a long-term shift in the mineral soil as a N sink vs source depending upon changing forest demand.

The contribution of these microbially mediated processes to the decoupling between NO_3^- and K^+ leaching following large-scale disturbance is not clear. Certainly the relatively high NO_3^- and K^+ leaching from W2 immediately after forest cutting and herbicide treatment resulted from the large supply of NO_3^- via nitrification owing to elimination of NH_4^+ uptake by plants and its accumulation in soils. In contrast, measurements of in situ nitrification following the ice storm disturbance did not detect a significant increase, and the direct role of reduced root uptake of NO_3^- was suggested (Houlton et al. 2003).

Following forest harvest, changes in conditions would seem to favor increased denitrification, including increased soil wetness (due to lower evapotranspiration), increased supply of NO_3^-, and increased dead organic matter. Unfortunately, measurements of denitrification response to forest cutting have been limited to observations of increases in nitrous oxide production after the whole-tree harvest of W5 which is suggestive of increases in rates of denitrification (Bowden and Bormann 1986).

22.4.6 Soil Cation Exchange

A key contrast between NO_3^- and K^+ is the ionic charge: K^+ is retained (albeit at low selectivity) on the cation exchange complex whereas NO_3^- is not. Although some soils with elevated pools of iron and aluminum exhibit anion adsorption (Johnson and Todd 1983), this is not the case for the Spodosols at the HBEF (Nodvin et al. 1986b). Moreover, NO_3^- is not a strongly adsorbing anion (Matson et al. 1987). The mineral soil horizons at the HBEF are coarse textured (loamy sands, low in clay), and most of the cation exchange capacity is associated with soil organic matter, especially in surface horizons (Johnson 2002); thus, K^+ is retained by cation exchange especially in the upper soil. This mechanism of K^+ retention would be expected to buffer K^+ losses in comparison to NO_3^- and thereby contribute to decoupling following large-scale disturbance. This pattern is clearly illustrated by the lower peak and longer tail on the annual NO_3^- compared to K^+ concentrations in stream water in W5 and W4 (Fig. 22.5). However, this pattern also is influenced by the dynamics of detrital sources (see above). Qualitative separation of these complementary effects is indicated by changes in soil exchangeable K^+ pools in W5 which peaked two years after harvest and declined to pre-treatment levels between 1991 and 1998. Notably, elevated K^+ concentrations in stream water persisted thereafter (Fig. 22.5b) suggesting continued increased supply from detritus. Nevertheless, the decoupling between NO_3^- and K^+ concentrations during the first several years after disturbance (Fig. 22.5b) is certainly explained in large part by K^+ buffering by the cation exchange complex (Romanowicz et al. 1996).

22.5 Synthesis

Although there is considerable variation in the patterns of loss and the source-sink processes of K^+ and NO_3^- (discussed above) with the greatest losses occurring in W2, followed by W5 and the ice storm and W4, at a coarse scale across disturbance type and time scale, there is remarkable consistency in their stoichiometric behavior in solutions with an approximately 0.1 molar ratio of K^+/NO_3^- (Fig. 22.9). At a first approximation, the processes driving NO_3^- loss control K^+ in an approximately stoichiometric manner. Stream sites show this behavior most consistently, particularly for those stream sites near the gauging stations lower in the watersheds. There is some deviation in steam sites at the highest elevations (732 and 701 m) draining shallow soils which emerge from bedrock in W5 (so called E podzols; Bailey et al. 2014). These sites are enriched in K^+ relative to NO_3^- compared to lower elevation stream sites. The stoichiometric deviation at the high elevation stream sites in W5 is consistent with the pattern showing higher K^+ to NO_3^- in Oa solutions with decreasing values in the Bh and Bs solutions. Bs solutions in W5 are similar to stream values (although the concentrations of both solutes are markedly lower in stream water). There is also some deviation in soil solutions from the ice storm, which in contrast are enriched in NO_3^- relative to K^+. This deviation probably reflects lower mobilization of K^+ due to limited release from dead roots given the much lower tree mortality in the ice storm in comparison to cutting treatments.

Despite the remarkable consistency in the stoichiometric behavior of K^+ and NO_3^- (Fig. 22.9), a highly variable degree of coupling between NO_3^- and K^+ concentrations in the stream water draining the mature forest watersheds, W1 and W6, at the HBEF suggests that different mechanisms contribute to the pulses of high NO_3^- leaching observed over the 45 years of record. An important consideration is

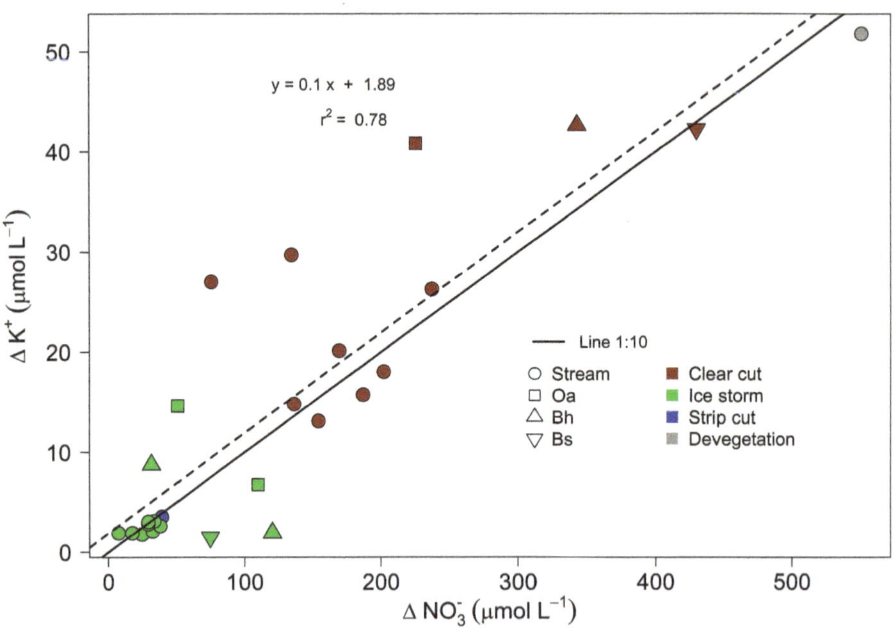

Fig. 22.9 Overall response in the response of the change in K^+ relative to the change in NO_3^- in solutions to disturbances of different watersheds at the HBEF

the distribution of anions and their changes over time. Cation leaching is closely coupled with anion loss due to constraints in solution electroneutrality (Cronan et al. 1978). Under most conditions sulfate is the dominant anion in HBEF drainage waters due to elevated inputs from atmospheric deposition (Likens et al. 2002) coupled with low mineral weathering rates and limited production of HCO_3^-. Note that atmospheric deposition and stream sulfate concentrations have been significantly decreasing (−0.79 and −0.49 μmol L^{-1} $year^{-1}$, respectively) throughout the biogeochemical record at Hubbard Brook, due to emission controls (Likens et al. 2002; Fuss et al. 2015). The long-term decline in stream sulfate and NO_3^- concentrations undoubtedly drive the long-term stream decline in K^+ concentration. However, superimposed on this pattern of chronic sulfate and nitrate leaching are periodic forest disturbance events (forest cutting, climatic disturbance) which add pulsed inputs of NO_3^-. Elevated anion leaching will facilitate cation leaching. The degree to which K^+ loss occurs under these conditions depends on the net supply of this nutrient controlled by mobilization and immobilization process (mineralization, plant uptake, cation adsorption/desorption). In most cases the behavior of W1 and W6 was largely concordant in terms of this coupling during years of high NO_3^- concentrations. In particular, the NO_3^- and K^+ concentrations were strongly coupled during four years of high stream NO_3^- concentrations (1981, 2000, 2008, 2015), whereas they were clearly decoupled in 1973 (W6 only) and in 1991 (Fig. 22.2). The flushes of NO_3^- from the Hubbard Brook watersheds have been attributed to reductions in plant uptake owing to disturbance by soil freezing (Mitchell et al. 1996) and ice storms (Houlton et al. 2003) and insect damage to the canopy (Aber et al. 2002). Clearly, a reduction in plant uptake of nitrogen has a much greater effect on stream NO_3^- flux than for K^+ due to the greater nutritional requirement coupled with the fact that K^+ leaching can be facilitated by elevated sulfate in addition to NO_3^-; see, for example, the relative responses of these solutes to the forest harvest experiments (Table 22.1) as well as the degree of reduction during the growing season (Fig. 22.3). Therefore, if reduced root uptake is the principal cause of elevated NO_3^- flux, we might not expect a very tight coupling between these solutes in high NO_3^- concentration years; other mechanisms (sulfate leaching, cation exchange) may be equally or more important.

The coupling between NO_3^- and K^+ concentrations following large-scale disturbance experiments in W2, W4, and W5 provides further insights into mechanisms contributing to leaching fluxes. In W5 during the second summer after harvest (1986), concentrations of K^+ peaked several weeks earlier than NO_3^- in Oa horizon soil solutions presumably due to a delay in peak rates of nitrification compared with rapid K^+ leaching from dead fine roots. In contrast, K^+ and NO_3^- concentrations in Bs soil solution were strongly coupled over this interval, with both delayed relative to Oa solutions (Fig. 22.7); this pattern could reflect either or both delayed leaching of K^+ from dead mineral soil roots or retention on the cation exchange complex. Notably exchangeable K^+ nearly doubled in 1986 in comparison with pre-treatment levels (Romanowicz et al. 1996). However, the seasonal pattern of stream NO_3^- and K^+ during this peak year of leaching (1986, Fig. 22.6) exhibited weak coupling, probably reflecting the lack of NO_3^- retention due to both minimal root uptake and lack of anion exchange in soil. In W4, where the forest was progressively cut over a four-year period, the flush of K^+ was strongly delayed in comparison to NO_3^-; for example, NO_3^- concentrations returned to background levels even before the peak in K^+ concentration was observed in 1980–1981 (Fig. 22.5a). The causes of this extended delay in K^+ flux are not clear, nor is the extended interval of elevated stream K^+ until the early 2000s. Deeper soils and the larger watershed area on W4 may contribute. It is notable that following the 1999 ice storm, which caused considerable canopy damage on W4, a coupled increase in NO_3^- and K^+ concentrations occurred in 2000–2001 but the magnitude of the NO_3^- flush greatly exceeded that for K^+ (0.013 mol K^+ (mol NO_3^-)$^{-1}$; Fig. 22.5a).

There is a curious seasonal coupling of increased NO_3^- and K^+ concentrations in Oa soil solutions on W5 during a five-year interval of high K^+ leaching from those surface horizons (Figs. 22.7 and 22.8). This appears to be controlled by NO_3^- ($r^2 = 0.58$) leaching in the nongrowing season which drives the release of K^+ presumably from the cation exchange complex.

The most recent flush of NO_3^- from the uncut watersheds at the HBEF (W1, W6) was larger and of longer duration in W1 than W6 (Fig. 22.2). The accompanying flush of K^+ also differed between these two catchments. In 1999 W1 was treated with a calcium silicate (wollastonite) which increased pH in soils, soil solutions, and surface waters (Cho et al. 2012), causing a marked recovery of forest health and productivity (Battles et al. 2014). The cause of the NO_3^- pulse in 2015 has not been determined; neither extensive soil freezing nor canopy damage was observed at this time. However, a reduction in forest floor mass in W1 (Johnson et al. 2014; Rosi-Marshall et al. 2016) may have contributed to this NO_3^- flush via enhanced mineralization and nitrification. The 3-year peak in NO_3^- concentration (2013, 2014, 2015) on W1 was much greater than the coincident K^+ flush (Fig. 22.2), perhaps supporting this mechanism. However, the high K^+ flush that accompanied the NO_3^- peak in W6 (Fig. 22.2) may suggest otherwise.

22.6 Suggestions for Future Research

In this study, we evaluated the mobilization of potassium and nitrate in response to disturbances to forest ecosystems (ice storm and harvest) at the Hubbard Brook Experimental Forest, New Hampshire. We envision that the mobility of other nutrient cations such as calcium and magnesium could also be influenced by various discrete natural and anthropogenic disturbance events including windstorms, wildfire, ice storms, insects, and pathogen irruptions and harvest. Calcium is known to be in short supply in these acid soils due to inherently low weathering rates coupled with legacy effects of acidic deposition and might be considered as a limiting nutrient due to its control of the acid base status of the ecosystems (Likens et al. 1998). Understanding the consequences of various disturbances on calcium and magnesium availability is therefore research interests that should be pursued in the future.

Here, we examined the influence of disturbances on nutrient mobility over both immediate and longer periods. Most atmospheric inputs to the study ecosystem have been relatively constant over the long term; however, inputs and leaching of both nitrate and sulfate have substantially decreased in recent years due to air pollution controls. It would be useful to examine the interactive effects of changes in strong anion inputs and land disturbances on the ecosystem behavior. For example, we suspect that aluminum mobilization following disturbance was greater during the historical period when acidic deposition was more severe and mobile anion leaching was greater.

As we discussed in this chapter, Groffman et al. (2018) hypothesized that climate change is driving an "oligotrophication" of these forested watersheds. Our unpublished observations suggest that nitrate leaching following natural disturbances like ice storms and windstorms is lower now than just two decades ago. The mechanisms contributing to this hypothesized oligotrophication could be elucidated by conducting a forest harvest experiment today in the same manner as the earlier experiments to examine differences in the stoichiometry of leaching.

At Hubbard Brook Experimental Forest, long-term monitoring of gauged watersheds has provided a quantitative basis for evaluating disturbance effects on stream water chemistry. Other sites with long-term records and whole-catchment disturbance experiments, for example, Coweeta Hydrologic Laboratory and HJ Andrews Experimental Forest, would be informative choices to conduct similar analysis. We envision different responses to disturbances given differences in soils, vegetation, climate, and atmospheric inputs. For example, in soils with elevated pools of amorphous iron and aluminum oxides and high sulfate adsorption capacity, such as the non-glaciated regions of the southeastern USA (Coweeta), historical atmospheric sulfate deposition has been retained in soil to a greater extent than at HBEF. We hypothesize that in response to discrete events such as forest harvest, abrupt changes in soil acidity could release sulfate previously retained in soil to surface waters and cause more severe episodic acidification. In contrast, in the relatively pristine Pacific Northwest, lower mobility of nitrate might be expected to result in different stoichiometry of K and N following disturbance at HJ Andrews. We believe that a better understanding of the complex controls on the stoichiometry of nutrient mobility would be forthcoming with a broader evaluation of disturbance responses across a global array of forested watersheds.

A major driver of changes in soil, soil solution, and stream chemistry at the HBEF has been long-term decreases in leaching by sulfate as a result of decreases in acid deposition and the apparent and unexplained "tightening" or oligotrophication of the nitrogen cycle (Groffman et al. 2018). A critical research question is what is driving the long-term decrease in NO_3^- leaching at Hubbard Brook and other forested watersheds in the Northeast United States (e.g., Driscoll et al. 2016). What is the role of decreases in NO_3^- concentration and acid deposition in this phenomenon, or do other factors such as climate change contribute? Do these long-term changes in the biogeochemistry at Hubbard Brook affect the pulsed response of watersheds to climatic or harvesting disturbance? Related questions are: To what extent do decreases in sulfate and NO_3^- leaching affect cation loss and the status of nutrient cations in soil? And do these changes alter the species composition and vigor of the forest community and soil development processes? If so, what are the implications for long-term ecosystem function?

Acknowledgments This is a contribution of the Hubbard Brook Ecosystem Study. Support for this work was largely provided by a number of grants from the National Science Foundation, particularly through the Long-Term Ecological Research program. Support is also provided by the US Forest Service. The Hubbard Brook Experimental Forest is administered by the US Forest Service. We thank Gene Likens and John Campbell for providing us chemistry of streamwater and streamflow data measured in the Hubbard Brook watersheds.

References

Aber JD, Nadelhoffer KJ, Steudler P, Melillo JM (1989) Nitrogen saturation in northern forest ecosystems. BioScience 39:378–286. https://doi.org/10.2307/1311067

Aber JD, Ollinger SV, Driscoll CT, Likens GE, Holmes RT, Freuder RJ et al (2002) Inorganic nitrogen losses from a forested ecosystem in responseto physical, chemical, biotic, and climatic perturbations. Ecosystems 5:0648–0658. https://doi.org/10.1007/s10021-002-0203-8

Attiwill PM (1968) The loss of elements from decomposing litter. Ecology 49:142–145. https://doi.org/10.2307/1933568

Bailey SW, Brousseau PA, McGuire KJ, Ross DS (2014) Influence of landscape position and transient water table on soil development and carbon distribution in a steep, headwater catchment. Geoderma 226:279–289. https://doi.org/10.1016/j.geoderma.2014.02.017

Battles JJ, Cleavitt NL, Saah DS, Poling BT, Fahey TJ (2017) Ecological impact of a microburst windstorm in a northern hardwood forest. Can J For Res 47:1695–1701. https://doi.org/10.1139/cjfr-2017-0206

Battles JJ, Fahey TJ, Driscoll CT Jr, Blum JD, Johnson CE (2014) Restoring soil calcium reverses forest decline. Environ Sci Technol Lett 1:15–19. https://doi.org/10.1021/ez400033d

Bormann FH, Likens GE (1967) Nutrient cycling. Science 155:424–429. https://doi.org/10.1126/science.155.3761.424

Bormann FH, Likens GE (1979) Catastrophic disturbance and the steady state in northern hardwood forests: A new look at the role of disturbance in the development of forest ecosystems suggests important implications for land-use policies. Am Sci 67:660–669

Bowden WB, Bormann FH (1986) Transport and loss of nitrous oxide in soil water after forest clear-cutting. Science 233:867–869. https://doi.org/10.1126/science.233.4766.867

Buso DC, Likens GE, Eaton JS (2000) Chemistry of precipitation, streamwater, and lakewater from the Hubbard Brook Ecosystem Study: a record of sampling protocols and analytical procedures. Gen Tech Rep NE-275 Newtown Sq PA US Dep Agric For Serv Northeast Res Stn 52 P 275

Campbell JL, Mitchell MJ, Mayer B, Groffman PM, Christenson LM (2007) Mobility of nitrogen-15-labeled nitrate and sulfur-34-labeled sulfate during snowmelt. Soil Sci Soc Am J 71:1934–1944. https://doi.org/10.2136/sssaj2006.0283

Campbell JL, Socci AM, Templer PH (2014) Increased nitrogen leaching following soil freezing is due to decreased root uptake in a northern hardwood forest. Glob Change Biol 20:2663–2673. https://doi.org/10.1111/gcb.12532

Cho Y, Driscoll CT, Johnson CE, Blum JD, Fahey TJ (2012) Watershed-level responses to calcium silicate treatment in a northern hardwood forest. Ecosystems 15:416–434. https://doi.org/10.1007/s10021-012-9518-2

Cleavitt NL, Fahey TJ, Groffman PM, Hardy JP, Henry KS, Driscoll CT (2008) Effects of soil freezing on fine roots in a northern hardwood forest. Can J For Res 38:82–91. https://doi.org/10.1139/X07-133

Cronan CS, Reiners WA, Reynolds RC, Lang GE (1978) Forest floor leaching: contributions from mineral, organic, and carbonic acids in New Hampshire subalpine forests. Science 200:309–311. https://doi.org/10.1126/science.200.4339.309-a
Dahlgren RA, Driscoll CT (1994) The effects of whole-tree clear-cutting on soil processes at the Hubbard Brook Experimental Forest, New Hampshire, USA. Plant Soil 158:239–262. https://doi.org/10.1007/BF00009499
Dittman JA, Driscoll CT, Groffman PM, Fahey TJ (2007) Dynamics of nitrogen and dissolved organic carbon at the Hubbard Brook Experimental Forest. Ecology 88:1153–1166. https://doi.org/10.1890/06-0834
Driscoll C, Whitall D, Aber J, Boyer E, Castro M, Cronan C et al (2003) Nitrogen pollution: Sources and consequences in the US northeast. Environ Sci Policy Sustain Dev 45:8–22. https://doi.org/10.1080/00139150309604553
Driscoll CT, Driscoll KM, Fakhraei H, Civerolo K (2016) Long-term temporal trends and spatial patterns in the acid-base chemistry of lakes in the Adirondack region of New York in response to decreases in acidic deposition. Atmos Environ 146:5–14. https://doi.org/10.1016/j.atmosenv.2016.08.034
Fahey TJ, Battles JJ, Wilson GF (1998) Responses of early successional northern hardwood forests to changes in nutrient availability. Ecol Monogr 68:183–212. https://doi.org/10.2307/2657200
Fahey TJ, Hughes JW, Pu M, Arthur MA (1988) Root decomposition and nutrient flux following whole-tree harvest of northern hardwood forest. For Sci 34:744–768. https://doi.org/10.1093/forestscience/34.3.744
Federer CA, Hornbeck JW, Tritton LM, Martin CW, Pierce RS, Smith CT (1989) Long-term depletion of calcium and other nutrients in eastern US forests. Environ Manage 13:593–601. https://doi.org/10.1007/BF01874965
Fuss CB, Driscoll CT, Campbell JL (2015) Recovery from chronic and snowmelt acidification: Long-term trends in stream and soil water chemistry at the Hubbard Brook Experimental Forest, New Hampshire, USA. J Geophys Res Biogeosciences 120:2360–2374. https://doi.org/10.1002/2015JG003063
Galloway JN, Dentener FJ, Capone DG, Boyer EW, Howarth RW, Seitzinger SP et al (2004) Nitrogen cycles: past, present, and future. Biogeochemistry 70:153–226. https://doi.org/10.1007/s10533-004-0370-0
Gannon JP, Bailey SW, McGuire KJ (2014) Organizing groundwater regimes and response thresholds by soils: A framework for understanding runoff generation in a headwater catchment. Water Resour Res 50:8403–8419. https://doi.org/10.1002/2014WR015498
Groffman PM, Driscoll CT, Durán J, Campbell JL, Christenson LM, Fahey TJ et al (2018) Nitrogen oligotrophication in northern hardwood forests. Biogeochemistry 141:523–539. https://doi.org/10.1007/s10533-018-0445-y
Hewlett JD, Hibbert AR (1967) Factors affecting the response of small watersheds to precipitation in humid areas. In: Sopper WE, Lull HW (eds) Forest hydrology. Pergamon Press, New York, pp 275–290
Hirsch RM, Slack JR, Smith RA (1982) Techniques of trend analysis for monthly water quality data. Water Resour Res 18:107–121. https://doi.org/10.1029/WR018i001p00107
Hornbeck JW, Martin CW, Pierce RS, Bormann FH, Likens GE, Eaton JS (1986) Clearcutting northern hardwoods: effects on hydrologic and nutrient ion budgets. For Sci 32:667–686. https://doi.org/10.1093/forestscience/32.3.667
Houlton BZ, Driscoll CT, Fahey TJ, Likens GE, Groffman PM, Bernhardt ES et al (2003) Nitrogen dynamics in ice storm-damaged forest ecosystems: implications for nitrogen limitation theory. Ecosystems 6:431–443. https://doi.org/10.1007/s10021-002-0198-1
Irland LC (1998) Ice storm 1998 and the forests of the Northeast: A preliminary assessment. J For 96:32–40. https://doi.org/10.1093/jof/96.9.32
Johnson CE (2002) Cation exchange properties of acid forest soils of the northeastern USA. Eur J Soil Sci 53:271–282. https://doi.org/10.1046/j.1365-2389.2002.00441.x

Johnson CE, Driscoll CT, Blum JD, Fahey TJ, Battles JJ (2014) Soil chemical dynamics after calcium silicate addition to a northern hardwood forest. Soil Sci Soc Am J 78:1458–1468. https://doi.org/10.2136/sssaj2014.03.0114

Johnson DW, Cole DW (1980) Anion mobility in soils: relevance to nutrient transport from forest ecosystems. Environ Int 3:79–90. https://doi.org/10.1016/0160-4120(80)90040-9

Johnson DW, Todd DE (1983) Relationships among iron, aluminum, carbon, and sulfate in a variety of forest soils 1. Soil Sci Soc Am J 47:792–800. https://doi.org/10.2136/sssaj1983.03615995004700040035x

Kendall C, Campbell DH, Burns DA et al (1995) Tracing sources of nitrate in snowmelt runoff using the oxygen and nitrogen isotopic compositions of nitrate. In: Tonnessen KA, Williams MW, Tranter M (eds) Biogeochemistry of seasonally snow-covered catchments, IAHS Publ No. 228. IAHS, Wallingford, UK, pp 339–348

Likens GE (2013) Biogeochemistry of a forested ecosystem. Springer, New York

Likens GE, Bormann FH, Johnson NM, Fisher DW, Pierce RS (1970) Effects of forest cutting and herbicide treatment on nutrient budgets in the Hubbard Brook Watershed-Ecosystem. Ecol Monogr 40:23–47. https://doi.org/10.2307/1942440

Likens GE, Bormann FH, Johnson NM, Pierce RS (1967) The calcium, magnesium, potassium, and sodium budgets for a small forested ecosystem. Ecology 48:772–785. https://doi.org/10.2307/1933735

Likens GE, Bormann FH, Pierce RS, Reiners WA (1978) Recovery of a deforested ecosystem. Science 199:492–496. https://doi.org/10.1126/science.199.4328.492

Likens GE, Driscoll CT, Buso DC, Siccama TG, Johnson CE, Lovett GM et al (1998) The biogeochemistry of calcium at Hubbard Brook. Biogeochemistry 41:89–173. https://doi.org/10.1023/A:1005984620681

Likens GE, Driscoll CT, Buso DC, Mitchell MJ, Lovett GM, Bailey SW et al (2002) The biogeochemistry of sulfur at Hubbard Brook. Biogeochemistry 60:235–316. https://doi.org/10.1023/A:1020972100496

Likens GE, Driscoll CT, Buso DC, Siccama TG, Johnson CE, Lovett GM et al (1994) The biogeochemistry of potassium at Hubbard Brook. Biogeochemistry 25:61–125. https://doi.org/10.1007/BF00000881

Lloret J, Valiela I (2016) Unprecedented decrease in deposition of nitrogen oxides over North America: the relative effects of emission controls and prevailing air-mass trajectories. Biogeochemistry 129:165–180. https://doi.org/10.1007/s10533-016-0225-5

Lovett GM, Goodale CL, Ollinger SV, Fuss CB, Ouimette AP, Likens GE (2018) Nutrient retention during ecosystem succession: a revised conceptual model. Front Ecol Environ 16:532–538. https://doi.org/10.1002/fee.1949

Lovett GM, Nolan SS, Driscoll CT, Fahey TJ (1996) Factors regulating throughfall flux in a New Hampshire forested landscape. Can J For Res 26:2134–2144. https://doi.org/10.1139/x26-242

Matson PA, Vitousek PM, Ewel JJ, Mazzarino MJ, Robertson GP (1987) Nitrogen transformations following tropical forest felling and burning on a volcanic soil. Ecology 68:491–502. https://doi.org/10.2307/1938454

McColl JG, Cole DW (1968) A mechanism of cation transport in a forest soil. Northwest Sci 42:134–141

Merrens EJ, Peart DR (1992) Effects of hurricane damage on individual growth and stand structure in a hardwood forest in New Hampshire, USA. J Ecol 80:787–795. https://doi.org/10.2307/2260866

Mitchell MJ, Driscoll CT, Kahl JS, Murdoch PS, Pardo LH (1996) Climatic control of nitrate loss from forested watersheds in the Northeast United States. Environ Sci Technol 30:2609–2612. https://doi.org/10.1021/es9600237

Morse JL, Durán J, Groffman PM (2015) Soil denitrification fluxes in a northern hardwood forest: the importance of snowmelt and implications for ecosystem N budgets. Ecosystems 18:520–532. https://doi.org/10.1007/s10021-015-9844-2

Mou P, Fahey TJ, Hughes JW (1993) Effects of soil disturbance on vegetation recovery and nutrient accumulation following whole-tree harvest of a northern hardwood ecosystem. J Appl Ecol 30:661–675. https://doi.org/10.2307/2404245

Nodvin SC, Driscoll CT, Likens GE (1986a) The effect of pH on sulfate adsorption by a forest soil. Soil Sci 142:69–75

Nodvin SC, Driscoll CT, Likens GE (1986b) Simple partitioning of anions and dissolved organic carbon in a forest soil. Soil Sci 142:27–35

Parker GG (1983) Throughfall and stemflow in the forest nutrient cycle. Adv Ecol Research 13:57–133. https://doi.org/10.1016/S0065-2504(08)60108-7

Peters SC, Blum JD, Driscoll CT, Likens GE (2004) Dissolution of wollastonite during the experimental manipulation of Hubbard Brook Watershed 1. Biogeochemistry 67:309–329. https://doi.org/10.1023/B:BIOG.0000015787.44175.3f

Rhoads AG, Hamburg SP, Fahey TJ, Siccama TG, Hane EN, Battles J et al (2002) Effects of an intense ice storm on the structure of a northern hardwood forest. Can J For Res 32:1763–1775. https://doi.org/10.1139/x02-089

Romanowicz RB, Driscoll CT, Johnson CE, Fahey TJ, Likens GE, Siccama TG (1996) Changes in the biogeochemistry of potassium following a whole-tree harvest. Soil Sci Soc Am J 60:1664–1674. https://doi.org/10.2136/sssaj1996.03615995006000060009x

Rosi-Marshall EJ, Bernhardt ES, Buso DC, Driscoll CT, Likens GE (2016) Acid rain mitigation experiment shifts a forested watershed from a net sink to a net source of nitrogen. Proc Natl Acad Sci 113:7580–7583. https://doi.org/10.1073/pnas.1607287113

Runkle JR (1982) Patterns of Disturbance in some old-growth mesic forests of eastern North America. Ecology 63:1533–1546. https://doi.org/10.2307/1938878

Ryan DF, Bormann FH (1982) Nutrient resorption in northern hardwood forests. BioScience 32:29–32. https://doi.org/10.2307/1308751

Sen PK (1968) On a class of aligned rank order tests in two-way layouts. Ann Math Stat 39:1115–1124. https://doi.org/10.1214/aoms/1177698236

Tierney GL, Fahey TJ, Groffman PM, Hardy JP, Fitzhugh RD, Driscoll CT (2001) Soil freezing alters fine root dynamics in a northern hardwood forest. Biogeochemistry 56:175–190. https://doi.org/10.1023/A:1013072519889

Van Miegroet H, Cole DW (1984) The impact of nitrification on soil acidification and cation leaching in a red alder ecosystem 1. J Environ Qual 13:586–590. https://doi.org/10.2134/jeq1984.00472425001300040015x

Vitousek PM, Aber JD, Howarth RW, Likens GE, Matson PA, Schindler PW et al (1997) Human alteration of the global nitrogen cycle: sources and consequences. Ecol Appl 7:737–750. https://doi.org/10.1890/1051-0761(1997)007[0737:HAOTGN]2.0.CO;2

Yanai RD, Vadeboncoeur MA, Hamburg SP, Arthur MA, Fuss CB, Groffman PM et al (2013) From missing source to missing sink: long-term changes in the nitrogen budget of a northern hardwood forest. Environ Sci Technol 47:11440–11448. https://doi.org/10.1021/es4025723

Chapter 23
Water and Nutrient Budgets of Organic Layers and Mineral Topsoils Under Tropical Montane Forest in Ecuador in Response to 15 Years of Environmental Change

W. Wilcke, A. Velescu, S. Leimer, and C. Valarezo

23.1 Introduction

The Andean tropical montane rain forest in the Ecuadorian Province of Zamora-Chinchipe on the western rim of the Amazon basin is shielded by the Andean cordillera from westerly winds during most of the time. As a consequence, westerly winds, potentially carrying polluted air from the coastal and inner Andean regions of Ecuador where most of the Ecuadorian population lives, rarely reach the east-exposed slope of the east Andean cordillera. However, westerly winds contributed to N deposition on the eastern slope of the east cordillera, where they blow during up to 40% of the time during October to December but <10% in the other months (Wilcke et al. 2013b). Westerly winds did not carry base metals (Wilcke et al. 2001; Boy and Wilcke 2008). Downwind of the dominating northeasterly trade winds to the east, the next big city is Manaus in Brazil at an aerial distance of ca. 2100 km. Although there is some local smallholder agriculture (Peters et al. 2013), it can be argued that the Ecuadorian montane rain forest on the eastern slope of the east Andean cordillera between ca. 1850 and 2200 m above sea level (a.s.l.) is still a remote ecosystem and therefore a suitable monitoring site for global change effects with minor local influence. Among the global change effects, changes in water-forest interactions and nutrient deposition with rainfall play a pivotal role for the response of the few remaining native Andean tropical montane forests.

In the north Andes, several simultaneously occurring environmental changes have been observed. For the eastern part of the north Andes, a warmer climate and

W. Wilcke (✉) · A. Velescu · S. Leimer
Institute of Geography and Geoecology, Karlsruhe Institute of Technology (KIT), Karlsruhe, Germany
e-mail: wolfgang.wilcke@kit.edu

C. Valarezo
National University of Loja, Loja, Ecuador

D. F. Levia et al. (eds.), *Forest-Water Interactions*, Ecological Studies 240,
https://doi.org/10.1007/978-3-030-26086-6_23

locally variable positive and negative changes of rainfall have been predicted by model projections (Vuille et al. 2003; Urrutia and Vuille 2009). The evaluation of the available climate data of the past four decades confirmed the predicted increasing temperatures. For the city of Loja in south Ecuador an increase rate of 0.13 °C per decade was estimated (Peters et al. 2013). Moreover, Wilcke et al. (2013b) reported for the period from 1998 to 2010 an increasingly dry microclimate for the Ecuadorian tropical montane forest on the Amazon-exposed slope of the Andes, because of changing temporal rainfall distribution, while total annual rainfall did not change. Changing rainfall intensity and temporal distribution might influence nutrient cycling in the Andean forests by their impact on nutrient leaching from the canopy and thus nutrient input into the soil and by their influence on soil water fluxes. Increasing dry spells result in drier soil conditions as reflected by decreasing times of waterlogging (Wilcke et al. 2013a).

Furthermore, N deposition increases strongly in the tropical montane forests of Ecuador (Wilcke et al. 2013b), possibly shifting the Andean forest from current N and P co-limitation to pure P limitation (Homeier et al. 2012; Velescu et al. 2016). Boy et al. (2008a) additionally observed seasonal acid deposition originating from Amazonian forest fires, which was associated with a significantly elevated N and Mn deposition. Besides the increasing N and the seasonal acid deposition, episodic alkaline dust deposition, likely related with long-range Sahara dust transport during strong La Niña conditions, reaches the tropical montane forests in Ecuador (Boy and Wilcke 2008; Wilcke et al. 2013a). Changing soil acidity is an important driver of nutrient availability, because in the acid soils of the Andean tropical montane forests, increasing pH improves living conditions for microorganisms (Pepper et al. 2015) and therefore might also accelerate microbial nutrient release including P from organic matter by mineralization.

The soils of the tropical Andes are little developed and therefore shallow and rich in stones (Frei 1958; Schrumpf et al. 2001). The mineralogy reflects their young age with moderate Fe and Al concentrations and the presence of primary minerals such as mica, illites, and three-layer clay minerals (Schrumpf et al. 2001). In the absence of carbonate minerals, the soils are usually acid and thus the effective cation-exchange capacity and base saturation are low (Wilcke et al. 2017). Consequently, the nutrient supply in forests on these acid young soils strongly depends on mineralization of the thick organic layers (Grubb 1995; Wilcke et al. 2002; Dietrich et al. 2017). Thick organic layers are typical for tropical montane forests and frequently host a large part of the plant roots (Tanner et al. 1998; Wilcke et al. 2002; Soethe et al. 2006). In the humid tropical montane forests, soil organic matter mineralization is limited by episodic waterlogging contributing to the accumulation of thick organic layers (Schuur and Matson 2001; Roman et al. 2010). The increasingly drier soil conditions in response to the increasing temperatures and changing rainfall might therefore further enhance organic matter mineralization.

Nitrogen and P co-limitation is common in the forests at the rim of the Amazon basin in Ecuador and Peru (Homeier et al. 2012; Fisher et al. 2013; Velescu et al. 2016). In a fertilizer experiment, Wullaert et al. (2010, 2013) reported that N, P, and Ca added to a tropical montane forest in Ecuador on the eastern slope of the Andes at 2000–2150 m a.s.l. remained in the ecosystem and were used by the aboveground organism community. In the same forest, Wilcke et al. (2017) found that the base metal cycling (K, Ca, Mg) was strongly biotically controlled with little losses from a small water catchment. Consequently, it can be expected that deposited nutrients remain to a large part in the ecosystem and change the nutrient supply of the biota. Changing nutrient supply, particularly of N and P, might have detrimental effects on the biodiversity of the humid Andean tropical montane forests in Ecuador (Sala et al. 2000; Wassen et al. 2005), which belong to the few mega hotspots of biodiversity in the world (Barthlott et al. 2007).

Because of their immediate response to environmental changes, the aqueous ecosystem fluxes serve as early indicator of a shifting ecosystem state (Bormann and Likens 1967; Bruijnzeel 1991; Likens 2013). Consequently, budgets of whole ecosystems or ecosystem compartments such as the organic layer and the uppermost part of the mineral soil of tropical montane rain forests, where almost all roots are located (Soethe et al. 2006) and where the microbial activity is elevated, might inform about ongoing changes.

Our objective was to detect possible trends of changes in the nutrient status of a humid tropical montane rain forest on the eastern slope of the east Andean cordillera at 1850–2200 m a.s.l. in response to forest-water interaction. Therefore, we quantified the budgets of the organic layer and the uppermost 30 cm of the mineral soil, by determining all aqueous fluxes (throughfall, stemflow, litter leachate, stream flow) of macronutrients (N, P, K, Ca, Mg, S) during 15 years of naturally observed nutrient deposition and climate warming. We furthermore calculated hypothetical accumulation and depletion times of the current nutrient storages neglecting any potential changes in litterfall quantity and quality and the rate of mixing of the organic layer and the mineral soil by faunal activity. We hypothesized that (i) the increasing N deposition results in the accumulation of N in the organic layer and the mineral topsoil via N uptake by the vegetation and recycling to the soil with litterfall, (ii) the low mobility of P in soils and increasing P retention by the biota community of the ecosystem in response to increasingly stronger P limitation result in P accumulation in the organic layer and the mineral soil, (iii) the seasonal acid deposition accelerates the leaching of base metals from the organic layer and mineral topsoil particularly of K, and (iv) the increasingly dry soil conditions favor organic matter mineralization and thus S is released and leached from the organic layer and also not retained in the mineral soil because of the low anion-exchange capacity.

23.2 Study Site and Methods

We studied a 30–50° steep ca. 9.1 ha large microcatchment under old-growth montane rain forest between 1850 and 2200 m a.s.l. (Wilcke et al. 2001; Boy et al. 2008b). Soils are mainly shallow Dystric Cambisols developed from metamorphic rock (phyllites and metasandstones) (IUSS Working Group WRB 2014). The studied forest can be classified as “bosque siempreverde montaño” (evergreen montane forest, Balslev and Øllgaard 2002) or as lower montane forest (Bruijnzeel and Hamilton 2000). More information on the composition of the forest can be found in the work of Homeier (2004). Mean annual temperature was 14.9 °C (Wilcke et al. 2013b, 2017).

We sampled the Oi, Oe, Oa, A, and B horizons of 29 soil profiles. The studied profiles comprise nine soils at midslope position and ten soils along each of two transects near the stream and near the ridge, respectively, covering the whole extension of the study catchment. Samples were taken from the walls of soil pits and air-dried. The organic layer samples were ground with a ball mill and the mineral soil was sieved to < 2 mm. Aliquots of the sieved mineral soil samples were also ground with a ball mill for elemental analysis and total digestions. Organic horizons were separated from the underlying mineral soil at the point where there was an abrupt increase in density from about 0.1 to 0.2 g cm^{-3} in the organic layer to > 1 g cm^{-3} in the mineral soil. All samples were stored in closed plastic bags at room temperature until analysis. All soil samples were collected in the year 1998 at the beginning of our observation period.

To determine the bulk density of the organic horizons, we extracted organic layer samples with a metal frame (0.2 m × 0.2 m). The organic layers were separated into Oi, Oe, and Oa horizons. The Oi horizon consists of fresh, little fragmented litter, the Oe horizon contains litter fragments and many visible fungi hyphens, and the Oa horizon consists of amorphous organic matter at an advanced stage of decomposition. The volumes of the horizons were calculated by multiplying the surface area of the metal frame by the thickness of the horizon. The samples were dried to constant mass at 40 °C in an oven. We determined the bulk density at ten randomly selected profiles and used the mean density of the Oi, Oe, and Oa horizons for the calculation of the element storages in the organic layer, because of the difficulty to obtain precise measurement values at individual sites as a consequence of disturbances by larger roots and stones. In the mineral soil, we determined the bulk density at each of the 29 study sites with the help of stainless steel cylinders with a volume of 100 cm^3, which were driven laterally into the horizons from the wall of the pits.

We used a frequency domain reflectometry (FDR) probe to measure soil water contents in the organic layer and data were recorded hourly with a data logger. Incident precipitation was collected weekly at 2–4 gauging stations, each consisting of five Hellmann-type collectors placed in clear-cut areas. To determine internal element cycling in the forest, we set up three ca. 20 m-long measurement transects (aligned downhill), covering about 10 m in elevation, on the lower to midslope (starting at 1900, 1950, and 2000 m a.s.l., respectively). This geomorphic form represented the largest part of the catchment. The elevational gradient covered was a compromise between reasonable accessibility and representativeness of the ca. 200 m gradient of the whole catchment. At each measurement transect, we collected throughfall with 5–20 fixed-positioned funnel collectors; we started with 5 collectors in 1998 and increased the number of collectors to 8 in 2000 and to 20 in 2002. Furthermore, we collected litter leachate (three zero-tension lysimeters made of plastic boxes with a polyethylene net as collecting surface area of 0.15 m × 0.15 m, installed below the organic layer) and fine litterfall (three litter traps, 0.3 m × 0.3 m from April 1998 to October 2005 and 0.5 m × 0.5 m thereafter). Stemflow was collected with polyurethane collars at five trees. Details of tree species and properties are summarized in Fleischbein et al. (2005). Stream water was sampled above a weir, which was used to gauge stream flow

(Fleischbein et al. 2006; Boy et al. 2008b). Water level at the weir was measured hourly with a pressure transducer in the first five years, although there were several and partly longer periods when the pressure transducer did not work properly (Fleischbein et al. 2006). After 2003, water levels were recorded weekly by manual measurement.

All ecosystem solution samples were collected in weekly resolution between April 1998 and March 2013. Rainfall, throughfall, and stemflow volumes were measured with a graduated cylinder in the field. After volume measurement, the samples of the replicate collectors of each ecosystem solution type at each measurement station/transect were bulked to result in one composite volume-weighted sample per collection date, ecosystem solution type, and measurement station/ transect. Litter leachates were directly bulked into one collection vessel per measurement transect to yield one individual sample per measurement site and collecting date, because zero-tension lysimeters do not collect water quantitatively (Jemison and Fox 1992). Litterfall was combined to one sample per measurement transect and sampling date and air-dried. Our hydrological year lasted from 01 April to 31 March of the following year because the experiment was established in March 1998.

The weekly litterfall samples were ground with a ball mill and composited to monthly samples. The ground litterfall and organic layer samples were digested in 8 mL concentrated HNO_3 and 2 mL H_2O_2 in a microwave oven. Ground mineral soil samples were digested with concentrated H_2O_2, HF, and HNO_3 in open Teflon vessels on a sand bath until dryness and redissolved in 5 M HNO_3.

We determined total N concentrations in ground litterfall, organic layer, and mineral soil samples with an elemental analyzer (EA). In the filtered 100 mL aliquots of ecosystem fluxes (pore size 4–7 μm, folded filter type 392), we determined total concentrations of N after oxidation by UV digestion with $K_2S_2O_8$ and P (in the last 10 years of the observation period) after digestion with H_2SO_4 with colorimetric methods in a continuous flow analyzer (CFA). Concentrations of Ca, K, and Mg in ecosystem solutions and litterfall, organic layer, and mineral soil digests were determined with a flame atomic absorption spectrometer (AAS) and those of S and P (in the digests and years 2–5 of the observation period) with an inductively coupled plasma-optical emission spectrometer (ICP-OES). Measurements of N and P were only conducted from the second year on (i.e., for the period 1999–2013) and those of S in years 1–8 (i.e., for the period 1998–2006).

For annual flux and budget calculations, weekly data of the aqueous nutrient fluxes were aggregated to monthly values to reduce the influence of outliers and the number of missing values. Monthly mean concentrations were calculated by arithmetically averaging weekly concentrations. Monthly fluxes were calculated by multiplying monthly mean concentrations with monthly water fluxes.

For the first five years (1998–2003), water fluxes through the organic layer (litter leachate) were calculated with a one-dimensional soil water budget model and taken from Boy et al. (2008b). For the later years (2004–2013), weekly litter leachate was calculated via a regression of throughfall on the modeled water fluxes of the first five observation years ($R^2 = 0.85$). In the first five years (1998–2003) we used a modeled streamflow taken from Fleischbein et al. (2006) and Boy et al. (2008b). For the later years (2004–2013) we used weekly measurements and assumed that the water level was constant during the measurement week.

The organic layer budget was calculated as the difference between the aqueous soil input (i.e., throughfall+stemflow) and the aqueous organic layer output (litter leachate) assuming that the organic layer was in steady state with respect to litterfall quantity and quality and mixing into the mineral soil. The budget of the uppermost 30 cm of the mineral soil was calculated as the difference between litter leachate and the stream flow, neglecting a possible retention of nutrients infiltrated with litter leachate by the deeper subsoil (below 30 cm) and the input of P, K, Ca, and Mg by weathering. The accumulation and depletion times were calculated by dividing the storages in the organic layer or the uppermost 30 cm of the mineral soils (element concentrations x volume x bulk density) by the mean annual budget.

The monthly time series data sets were tested for significant temporal trends with the seasonal Mann-Kendall test (Hirsch et al. 1982) as implemented in the function *SeasonalMannKendall()* from the package *Kendall* (McLeod 2011) in the statistical software R (R Core Team 2017). The seasonal Mann-Kendall test is a nonparametric trend test considering the most important autocorrelation (i.e., seasonal) which we detected. The test is not sensitive to a small number of outliers. The application of the test does not necessarily require a linear trend (Helsel and Hirsch 2002). To test, if the organic layer and mineral soil budgets of N, P, K, Ca, Mg, and S significantly differ from zero, we used a one-sample t-test as implemented in the R function *t.test ()* (R Core Team 2017). Temporal trends of the annual budget were tested with the function *zyp.trend.dataframe()* from the package *zyp()*, which uses prewhitening to account for serial correlation (Bronaugh and Werner 2015).

23.3 Results

From 1998 to 2013, the mean annual incident precipitation was 2320 mm $\pm$ 238 mm standard deviation of the 15 hydrological years, of which 1390 $\pm$ 199 mm reached the soil as throughfall and 30.8 $\pm$ 7.02 mm as stemflow. Of the 1420 $\pm$ 200 mm reaching the forest floor, 1160 $\pm$ 194 mm infiltrated the mineral soil as litter leachate, implying that 264 $\pm$ 114 mm were lost from the organic layer via evapotranspiration. Finally, 1060 $\pm$ 156 mm reached the stream resulting in an additional evapotranspiration loss of 100 $\pm$ 210 mm from the mineral soil (Wilcke et al. 2017). Mean annual litterfall from 1998 to 2013 was 1040 $\pm$ 115 g m^{-2}.

The organic layer had a mean thickness of 13.3 $\pm$ 6.85 cm ($n = 29$) and the mineral soil of 60.5 $\pm$ 18.8 cm (as far as it could be dug with a hand spade). The uppermost 30 cm of the mineral soil contained consistently more nutrients than the organic layer (Table 23.1). The nutrient storage of the organic layer accounted for 1.89 (K) to 44.4% (Ca) of the total nutrient storage in the organic layer and the uppermost 30 cm of the mineral soil.

To determine the total nutrient input to the soil via aqueous solutions, we summed the fluxes with throughfall and stemflow. The N and P fluxes in throughfall +stemflow increased significantly from 1999 to 2013 (Fig. 23.1a, b) and the S fluxes in throughfall+stemflow from 1998 to 2006 ($\tau = 0.207$, $p = 0.027$). In contrast, the

Table 23.1 Mean storage of macronutrients in the organic layer and the uppermost 30 cm of the mineral soil in a ca. 9-ha microcatchment under tropical montane forest in Ecuador (±standard deviation, $n = 29$ soils)

Nutrient	Organic layer	Contribution of organic layer	Uppermost 30 cm of the mineral soil
	g m^{-2}	%	g m^{-2}
N	478 ± 283	42.4 ± 19.4	726 ± 610
P	20.3 ± 10.7	14.4 ± 7.41	131 ± 50.9
K	75.7 ± 51.2	1.89 ± 1.30	4450 ± 1720
Ca	112 ± 93.0	44.4 ± 16.1	144 ± 189
Mg	36.0 ± 23.7	16.1 ± 8.93	197 ± 92.5
S	53.3 ± 31.7	38.9 ± 16.2	81.6 ± 37.5

Ca fluxes with throughfall+stemflow decreased significantly (Fig. 23.1c). The K and Mg fluxes with throughfall+stemflow did not change significantly from 1998 to 2013.

On average of the 14 and 15 budgeted years, respectively, P and K accumulated significantly in the organic layer, while all other macronutrients were on average significantly leached from the organic layer ($p \leq 0.017$; Fig. 23.2). In the uppermost 30 cm of the mineral soil, all elements accumulated significantly on average of the 8 (S), 14 (N, P), or 15 (all other elements) budgeted years ($p < 0.001$).

Except for N, all element budgets showed significant temporal trends. The P (14 years) and K (15 years) budgets of the organic layer increased significantly, while those of the uppermost 30 cm of the mineral soil decreased (Fig. 23.2, inlets). The same was true for S, which was only measured during 8 years (organic layer: $\tau = 0.714, p = 0.035$; mineral soil: $\tau = -0.714, p = 0.035$). The Ca and Mg budgets showed the reverse temporal trends, i.e., negative ones in the organic layer (Ca: $\tau = -0.670$, $p = 0.001$; Mg: $\tau = -0.670$, $p = 0.001$) and positive ones in the mineral soil (Ca: $\tau = 0.538$, $p = 0.009$; Mg: $\tau = 0.780$, $p < 0.001$).

Under the current conditions, the K and P inputs would double the organic layer P and K storages in 197 and 27 years, respectively (Fig. 23.3). In contrast, the loss of N, Ca, Mg, and S could cause a depletion of the organic layer storages in 38 (Mg) to 281 years (N). In the mineral soil, all macronutrients accumulated on average at a rate which would double their current storages in the 0–30 cm layer in 57 (Ca) to 601 years (P, Fig. 23.3).

23.4 Discussion

The hydrological budget of the studied forest is characterized by a high annual interception loss of 587–1190 mm representing 28–50% (mean 39%) of rainfall in the 15 study years, which is at the upper end of the range reported for tropical montane rain forests with little fog influence by Bruijnzeel et al. (2011). The stemflow only accounted for 1.3 ± 0.32% of the rainfall, which was at the lower

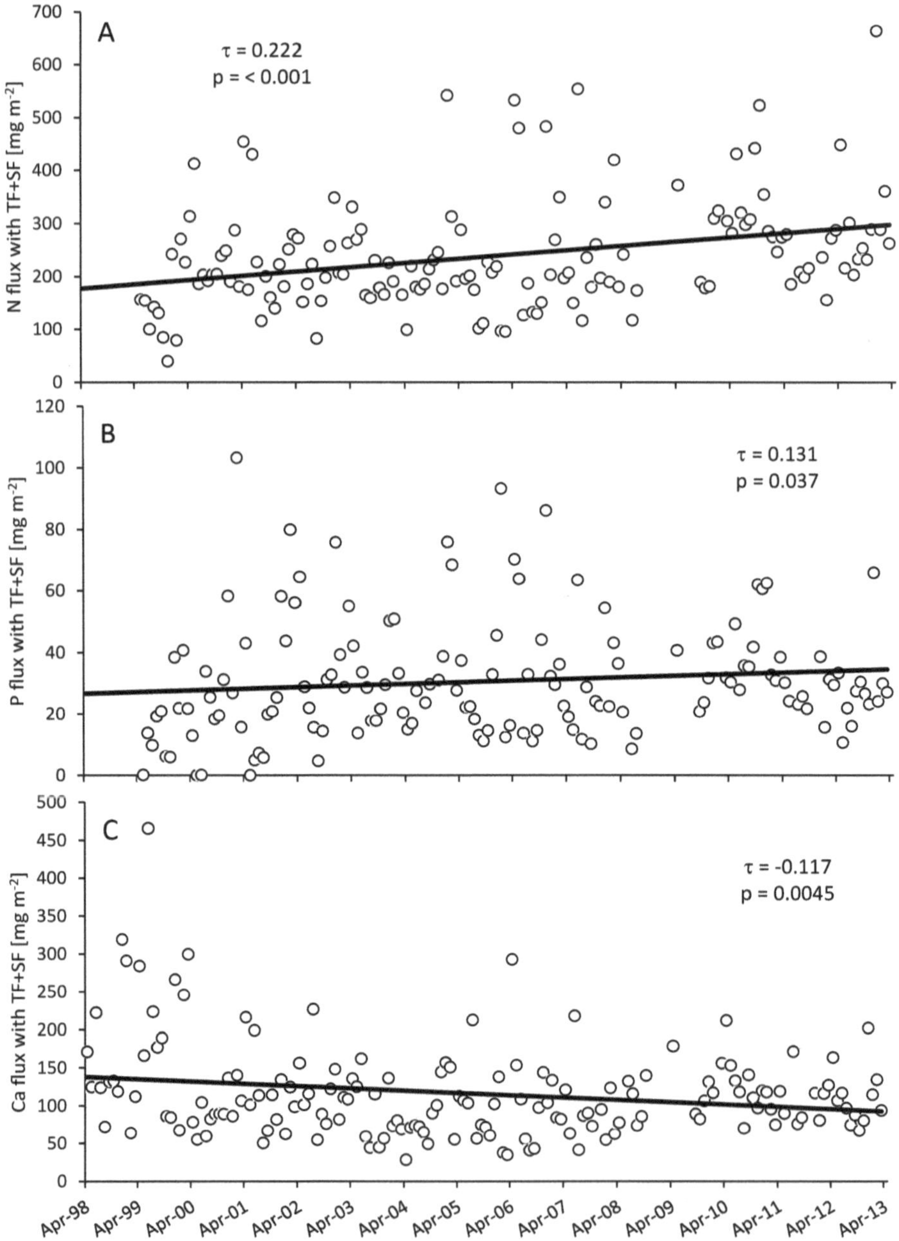

Fig. 23.1 Temporal course of the fluxes of (**a**) N, (**b**) P, and (**c**) Ca with throughfall (TF)+stemflow (SF) from April 1998 to March 2013. The measurement of N and P started only in April 1999. Temporal trends of monthly data were statistically evaluated with the seasonal Mann-Kendall test (Hirsch et al. 1982). Regression lines are shown to illustrate significant trends but do not necessarily imply that the trend is linear

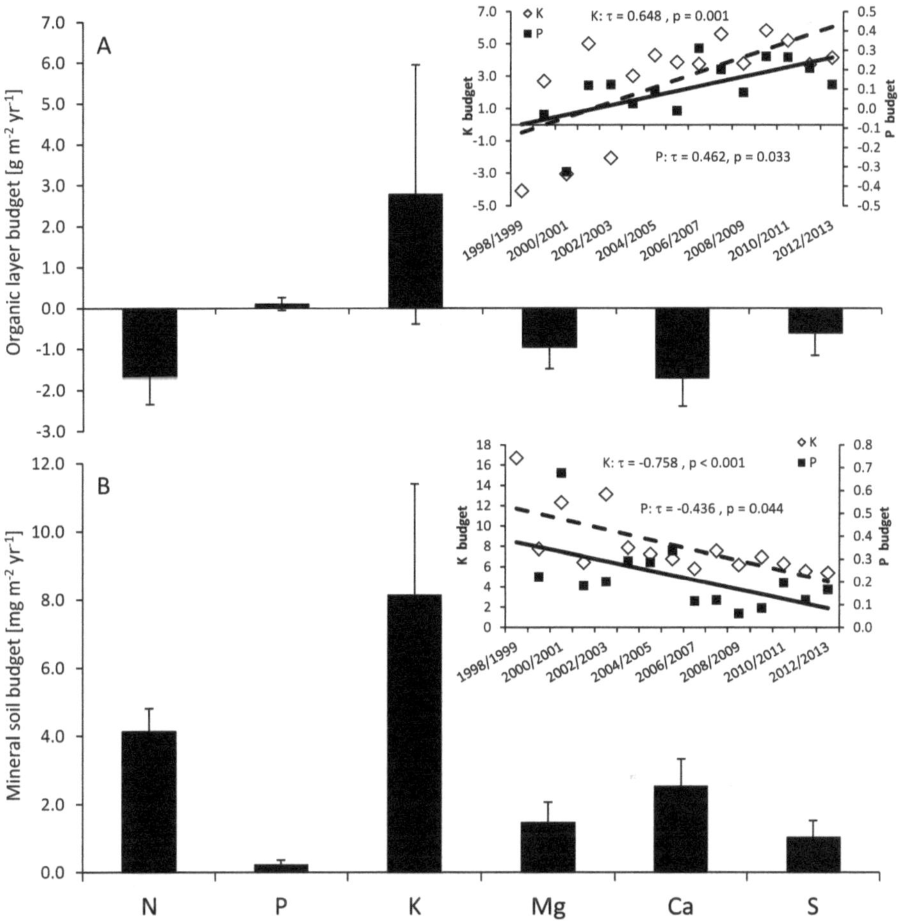

Fig. 23.2 Mean budgets of the aqueous nutrient fluxes through (**a**) the organic layer and (**b**) the uppermost 30 cm of the mineral soil from 1998 to 2013. Negative values indicate net losses, positive ones net accumulations. Whiskers indicate standard deviations ($n = 14$ years for N and P; $n = 15$ years for K, Mg, and Ca; and $n = 8$ years for S). Mean values were significantly different from zero according to one-sample t-tests. The inlet figures show the relationships between the sampling year and (**a**) the K and P budgets of the organic layer and (**b**) the K and P budgets of the uppermost 30 cm of the mineral soil from 1998 to 2013. The units of the *y*-axes in the inlet figures are the same as of the main figures. Temporal trends of monthly data were statistically evaluated with the Mann-Kendall test (Bronaugh and Werner 2015). Regression lines are shown to illustrate significant trends but do not necessarily imply that the trend is linear

end of the range of 0.1–8.8% in Bruijnzeel et al. (2011). The annual runoff coefficient (i.e., the ratio of streamflow to rainfall) varied from 0.38 to 0.58 resulting in an evapotranspiration of 42–62% of the rainfall, which is at the lower end of the range of lower montane rain forests with little influence of fog of 50–72% (mean 62% calculated from Table V in Bruijnzeel et al. 2011). The higher water flow below

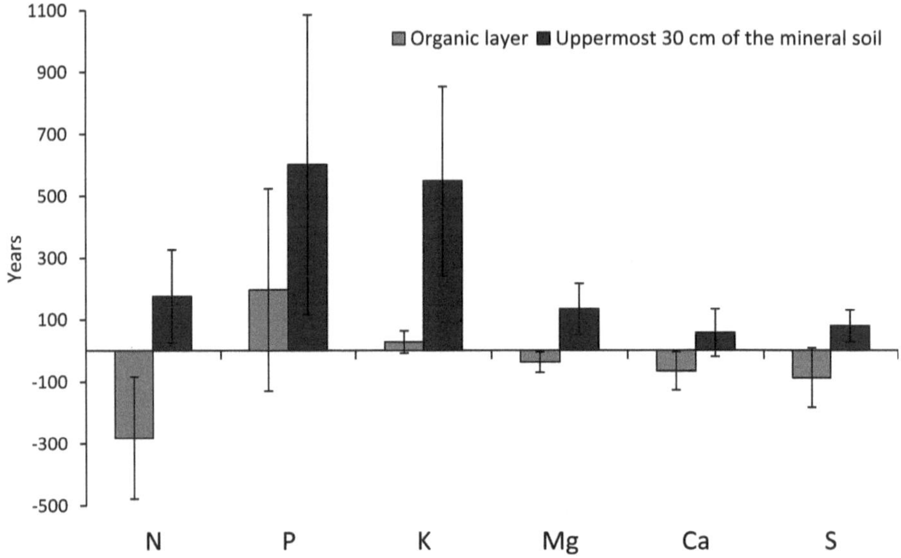

Fig. 23.3 Time needed to double (if positive) or entirely deplete (if negative) the nutrient storage in the organic layer or uppermost 30 cm of the mineral soil considering only the mean aqueous element fluxes from 1998 to 2013 (assuming that litterfall input, organic matter decomposition, and mixing of the organic layer with the mineral soil were and remain at steady state and that erosion and soil water fluxes do not change). The error bars show standard deviations calculated considering error propagation

the organic layer (litter leachate) than streamflow suggested that the vegetation took up some of its transpiration water from the mineral soil, while we considered direct evaporation from the soil as negligible (Wilcke et al. 2017). Because there were only few roots deeper than 30 cm in the mineral soil, we assume that further water loss by plant water uptake between the 0–30 cm layer and the stream is negligible and therefore used the streamflow as water flow out of the 0–30 cm layer.

Wilcke et al. (2013b) reported a significant decrease in throughfall from 1998 to 2010 in our study catchment. Because rainfall did not change significantly in the study of Wilcke et al. (2013b), interception loss (in % of rainfall) increased significantly. The same authors reported furthermore that the relative air humidity in our study forest significantly decreased during 1998–2010. Reduced humidity could be driven by the temperature increases predicted by Vuille et al. (2003) and Urrutia and Vuille (2009) which was confirmed for the greater study area with the help of a time series since 1964 (Peters et al. 2013). The generally increasing temperatures created an increasing evaporative demand, which was further enhanced by the fact that monthly hours of extreme temperatures >25 °C increased significantly during 1998–2010 (Wilcke et al. 2013b). Moreover, Peters et al. (2013) observed an increasing number of dry periods lasting for 4–14 days from 1998 to 2010, during which evapotranspiration was pronounced.

In response to the increasingly drier microclimatic conditions, Wilcke et al. (2013b) observed significantly decreasing water contents in the Oa horizon of the organic layer at 1900 m a.s.l. during 2000–2007 and during 2007–2010 at another measurement station at 2070 m a.s.l. However, the trend of the record from 2007–2010 was not significant, possibly because of its short duration. These findings are corroborated by the report of Wilcke et al. (2013a) that the frequency of waterlogged conditions decreased during 1998–2010 in the study catchment. Wilcke et al. (2013a) recorded soil matric potentials at the 0.15 and 0.30 m mineral soil depths at measurement stations in central position of each of the three measurement transects in the study catchment with manually read tensiometers. Reduced waterlogging could potentially increase the time period during which favorable conditions for mineralization of soil organic matter prevailed and thus increase the release of organically bound N, P, and S.

The annual litterfall was greater than that reported in two other studies on north Andean montane forests (Steinhardt 1979; Veneklaas 1991) and at the upper end of the range for various tropical montane forests globally (50–1100 g m^{-2} $year^{-1}$, Bruijnzeel and Proctor 1995). The forest produced as much biomass as found in tropical lowland forests (750–1330 g m^{-2} $year^{-1}$; Proctor 1987) and therefore belongs to the more productive tropical montane forests.

The organic layers in the study catchment stored substantially more macronutrients than those in temperate forests from North America and Europe in the "Integrated Forest Study" (Johnson and Lindberg 1992) where the nutrient storage was on average (with ranges) 92 g m^{-2} (2.8–264, N), 6.7 g m^{-2} (0.3–21.3, P), 7.8 g m^{-2} (0.4–19.8, K), 23 g m^{-2} (0.9–66.6, Ca), 4.0 g m^{-2} (0.1–8.8, Mg), and 8.5 g m^{-2} (0.3–29.2, S, compare with Table 23.1). The N and P storages of the organic layer at our study site surpassed those reported by Vogt et al. (1986) for the whole organic layer in evergreen tropical broadleaf forests presumably at tropical lowland locations of 32 ± 10 g m^{-2} (N) and 8.1 ± 2.3 g m^{-2} (P). This underlines the strong accumulation of organic matter in tropical montane forests, mainly because of the wet conditions, which hamper organic matter mineralization in spite of the high mean annual temperature of ca. 15 °C at 1950 m a.s.l. (Rollenbeck et al. 2015) and the lack of a frost period (Schuur and Matson 2001; Roman et al. 2010).

Nitrogen storages in the uppermost 30 cm of the mineral soils at our study site fell in the range of those reported for various soil types of the Amazon of 210–1370 g m^{-2} (Batjes and Dijkshoorn 1999) and of Central and Eastern Europe of 210–1520 g m^{-2} (Batjes 2002; Table 23.1). Storages of other nutrients in mineral soils have rarely been determined. The only studies from the tropical climate zone with which we could compare our results originated from the Brazilian Cerrado, a savanna ecosystem on strongly weathered Ferralsols. Lilienfein and Wilcke (2003) and Wilcke and Lilienfein (2004) reported for the same non-fertilized native Cerrado soils storages of 350 ± 11 g m^{-2} N, 63 ± 8.3 g m^{-2} P, 8.7 ± 1.5 g m^{-2} (Ca), 37 ± 5.5 g m^{-2} (K), 7.5 ± 1.4 g m^{-2} (Mg), and 53 ± 2.8 g m^{-2} (S). These storages are considerably lower than in our study soils, particularly with respect to the base metals (K, Ca, and Mg, compare with Table 23.1). The reason is the young age of the soils at our study site associated with a low degree of weathering.

Thus, the studied tropical montane forest harbors high nutrient storages in the organic layer and the mineral topsoil, which, however, are only to a limited degree bioavailable. Wilcke et al. (2002) reported turnover times of the organic layer ranging 15–18 years (N), 12–16 years (P), 6.6–20 years (K), 5.2–10 years (Ca), 4.5–7.8 years (Mg), and 15–20 years (S) for our study area. This illustrates that particularly the elements mainly released by organic matter mineralization, i.e., N, P, and S, become only bioavailable at the decadal scale. For P, Dietrich et al. (2016) reported that on average 13% of the total P concentrations in the mineral soil were extractable with the Bray method ($HCl+NH_4F$; Bray and Kurtz 1945) thought to represent bioavailable P at a study site next to our study catchment. In the mineral topsoil, the directly bioavailable exchangeable base metal concentrations account for only $0.37 \pm 0.24\%$ (K), $8.3 \pm 6.4\%$ (Ca), and $3.9 \pm 4.1\%$ (Mg) of the total storage of these elements (Wilcke et al. 2017). Thus, the by far largest part of the base metal storage is contained in minerals, which become bioaccessible in the long term only via weathering. Weathering rates of the studied macronutrients were generally low (Boy et al. 2008b; Wilcke et al. 2017, 2019). These studies reported weathering rates of 79 ± 63 g m^{-2} year^{-1} for P, 0.71 ± 0.77 g m^{-2} year^{-1} for K, 0.14 ± 0.24 g m^{-2} year^{-1} for Ca, 0.33 ± 0.22 g m^{-2} year^{-1} for Mg, and 0.04 g m^{-2} year^{-1} for S. The weathering rates of P, K, Ca, and Mg were estimated with three different methods and the standard deviation was calculated from the results of these methods (see the freely accessible supplementary materials of Wilcke et al. (2017) for a detailed description of the methods). The weathering rate of S was estimated as the sum of the S export with stream water during super dry and baseflow conditions as defined in Boy et al. (2008b).

Although the uppermost mineral soil stored 56–98% of the total nutrients in the root zone, which is more than in the organic layer (Table 23.1), the latter contained with 51% (at 1900 m a.s.l.) to 61% (at 2400 m a.s.l.) the higher root length density (Soethe et al. 2006). Bruijnzeel and Veneklaas (1998) speculated about a role of Al toxicity in the mineral soil in reducing root growth and contributing to the low biomass productivity of tropical montane forests with their increasingly stunted tree growth with increasing elevation. However, in spite of the acid character of the soil (mean pH of 4.3 in the organic layer and of 4.4 in the A horizons; Wilcke et al. 2017), there were hardly any indications of Al toxicity in our study forest, particularly because of the high degree of organo-complexation of Al (Rehmus et al. 2014, 2015, 2017). Therefore, there are no indications that the reason for the lower root length in the mineral soil than in the organic layer was Al toxicity. The root length density might instead be driven by the availability of the main limiting nutrients N and P, which is higher in the organic layer than in the mineral soil. Possibly, the temporally high soil moisture might also have contributed to reduce root penetration into the mineral soil.

The increasing N fluxes with throughfall+stemflow (Fig. 23.2a) parallel similar increases in N deposition with rainfall (Wilcke et al. 2013b). Boy et al. (2008a) identified Amazonian forest fires as a major source of N deposition to our study forest. Boy et al. (2008a) used two different indices of forest fire intensity in Amazonia, a fire pixel count along the wind pathways based on satellite imagery and an integrative satellite-based index of the NO concentrations above our study area to distinguish between "fire" and "no fire" periods. They found significantly increased N concentrations in rainfall during "fire" periods relative to "no fire" periods. The N concentrations in rainfall were not or only weakly correlated with rainfall volume, indicating that the variations in concentrations could not be explained by concentration/dilution effects. Because the "fire conditions" were drier (5-year mean weekly rainfall of 34–41 mm in the study) than the "no fire" conditions (5-year mean weekly rainfall of 62–63 mm), the N deposition with rainfall was not different between the "fire" and "no fire" periods in spite of the higher N concentration in rainfall during the former than the latter periods. However, the drier conditions during "fire" periods also resulted in significantly higher dry deposition of elements related with biomass burning. Consequently, total deposition of N was higher during the "fire" than the "no fire" periods. Total P deposition also tended to be higher during "fire" than "no fire" periods, although differences were not significant. In line with our findings, the release of N and P during biomass burning in Amazonia was reported in a range of publications (Williams et al. 1997; Artaxo et al. 2002; Mahowald et al. 2005; Hoffer et al. 2006). Thus, biomass burning in Amazonia is a major driver of element deposition even to distant montane forest at the outer rim of Amazonia, such as our study site. Biomass burning intensity is linked to the El Niño-Southern Oscillation (ENSO) (e.g., Alencar et al. 2006; van der Werf et al. 2006). Thus, a proposed frequency shift of future ENSO as a consequence of global warming (Timmermann et al. 1999) as well as increased biomass burning because of growing interest in soybean cropping in the Amazon basin (Arima et al. 2017) suggests rising deposition rates of aerosols related with biomass burning to our study site in the future.

The increasing P fluxes with throughfall+stemflow reflect the accumulation of deposited P in the biological part of the study catchment and indicate that the bioavailability of P increased (Fig. 23.2b; Wilcke et al. 2019). However, in the study of Wilcke et al. (2019), the annual increase in the P cycling with throughfall +stemflow was much lower than the annual total deposition of P, which was almost entirely retained in the study catchment. Because the studied forest has reached the mature old-growth stage, there should be no net annual P accretion in wood. This suggests that most of the deposited P retained in the catchment is stored in the soil (Wilcke et al. 2019). Because only a small part of the total soil input of P is leached into the mineral soil, the soil organic layer is likely the most important storage compartment of deposited P in the studied forest (Fig. 23.3; Wilcke et al. 2019). As a consequence, the studied N+P co-limited forest did not only receive a high N deposition but also accumulated P from 1999 to 2013. If the N deposition removes the N limitation in the near future and thus P becomes the only limiting nutrient, P accumulation will increase the risk of P eutrophication with possibly detrimental consequences for the biodiversity of the study forest (Wassen et al. 2005).

The decrease in the Ca input to the soil with throughfall+stemflow is attributable to a particularly strong deposition event of alkaline dust at the beginning of our study period (in the hydrological year 1999/2000 (Fig. 23.2c). Boy and Wilcke (2008) attributed the occasionally observed alkaline dust deposition events, which are rich in Ca and Mg, to Sahara dust transport all the way over the Atlantic Ocean and the

Amazon basin during strong La Niña events. Although the Amazon basin is even wetter during such events, there are dry spells that allow for the transport of Sahara dust without being washed out of the atmosphere before the dust reaches our study site. The 1999/2000 event was particularly strong. Thereafter, we observed only one other strong event in 2010/2011, which was, however, weaker than in 1999/2000.

The increasing S deposition with throughfall+stemflow from 1998 to 2006 was not related with the Amazonian forest fires, because neither rainfall nor dry deposition of S was higher during "fire" than "no fire" periods (Boy et al. 2008a). We speculate that the increasing S return with aqueous solutions from the canopy to the soil, reflects an enhanced S uptake by the vegetation following the improved N and P supply.

The mean accumulation of P in the organic layer can be attributed to the scarcity of this nutrient in the studied forest and indicates strong biological retention (Fig. 23.3; Wilcke et al. 2019). The retention of P in the organic layer increased with time as reflected by the significant positive trend of the P budget (Fig. 23.2a, inlet). This might indicate that P becomes increasingly scarcer relative to N, because of the much higher N than P deposition from the atmosphere.

The accumulation of the highly mobile K in the organic layer is surprising, because under the acid conditions of the soils, K should be strongly leached as a consequence of its low competitiveness for cation-exchange sites. One explanation could be that the plants take up K from the K-rich mineral soil and recycle it to the organic layer to a considerable extent (Jobbagy and Jackson 2001, 2004; Wilcke et al. 2017). Alternatively, K could be accumulated in the organic layer in response to the overall higher nutrient uptake by the biota of the forest ecosystem, because of the improved supply of the two co-limiting elements N and P, which generates an additional demand of all other nutrients, particularly of the third-most needed K (Marschner 2012). The latter assumption is supported by the fact that the K budgets of the organic layer showed a significantly increasing trend (Fig. 23.2a, inlet).

The depletion of N and S from the organic layer likely reflects a higher release by mineralization from the organic matter than is needed by the plants and soil organisms so that the two nutrients are leached (Fig. 23.3). High release rates by organic matter mineralization of 63.5 g m^{-2} $year^{-1}$ of N had already been reported by Wilcke et al. (2002) for our study area, which is much higher than the demand of N of the forest as approximated by the N flux with litterfall (18.7 ± 0.36 g m^{-2} $year^{-1}$). The N richness is also reflected by the strong nitrification in the organic layer identified by Schwarz et al. (2011) with the help of the natural abundance of the stable N isotope ratio in nitrate. Such strong nitrification only occurs if the system is near to N saturation (Aber et al. 1998). As Homeier et al. (2012) reported that there was a N+P limitation of several ecosystem processes, the loss of N from the organic layer is surprising. Perhaps, some part of this N is quickly leached out of the organic layer by stronger rain events and can therefore not be used by the biota community.

The S release of the organic layer was identical to the S flux with litterfall (as proxy of the S demand of the vegetation) of 2.2 ± 0.08 g m^{-2} $year^{-1}$ (Wilcke et al. 2002), making the S losses from the organic layer again unexpected. We speculate that the forest satisfies a part of its S demand directly from the atmospheric deposition so that some of the available S in the organic layer is not needed and can be leached or that some part of the released S in the organic layer is too quickly leached. The positive trend of the S budgets of the organic layer from 1998 to 2006 might indicate that the S supply of the vegetation is tightening so that S is increasingly accumulated in the organic layer.

The losses of Ca and Mg from the organic layer are fully expected because of the acid conditions, which favor the leaching of base metals and the smaller demand of the vegetation of these two elements than of N, P, and K. The negative trends of the Ca and Mg budgets of the organic layer might be attributable to the decreasing pH of the litter leachate from 1998 to 2013 ($\tau = 0.341$, $p < 0.001$).

The accumulation of N in the mineral soil could be attributed to its microbial immobilization, because of the sharp drop of the N concentrations from the litter leachate to the mineral soil solutions (Wilcke et al. 2013b). The accumulation of P in the mineral soil is likely mainly attributable to the sequestration of phosphate by iron (oxy)hydroxides and to the precipitation as Al-phosphates (Hinsinger 2001). The retention of the base metals (K, Ca, Mg) is likely attributable to their affiliation with the cation-exchange sites (Wilcke et al. 2017). For K, sequestration by illites, which occur in the study soils, might play an additional role (Schrumpf et al. 2001). The accumulation of S in the mineral soil can mainly be attributed to the strong retention of S-containing dissolved organic matter, because Goller et al. (2006) reported that at our study site, the contributions of dissolved organic S to the total dissolved S concentrations dropped markedly between the litter leachate (65 ± 17% of total dissolved S) and the mineral soil solution (at 30 cm depth, 18 ± 16%), while the total S concentrations did not change (0.42 ± 0.36 in the litter leachate vs. 0.43 ± 0.07 mg L^{-1} in the mineral soil solution at 30 cm depth). The reason for little changing total S concentrations in the soil solution of the mineral soil relative to the litter leachate and the even higher total S concentrations in the stream flow than in rainfall is the release of SO_4^{2-} by weathering in the subsoil (Goller et al. 2006; Boy et al. 2008b).

The temporal trends of the element budgets of the uppermost 30 cm of the mineral soils had generally the opposite sign than those of the organic layers, i.e., Ca and Mg were increasingly lost from the organic layer but at approximately the same rate increasingly accumulated in the mineral soil from 1998 to 2013, while for P, K, and S the contrary was true. We attribute the finding that P (1999–2013), K (1998–2013), and S (1998–2006) were increasingly less accumulated in the uppermost 30 cm of the mineral soil (Fig. 23.2b, inlet) to the increasing retention of the three nutrients in the organic layer, which resulted in reduced P, K, and S fluxes into the mineral soil. Furthermore, the increasing demand of P, K, and S in response to the high N deposition might also have stimulated the plant pump from the subsoil to the organic layer (Jobbagy and Jackson 2001, 2004). However, when discussing the temporal trend of the nutrient budgets in the organic layer and the mineral soil (Fig. 23.2a and b, inlets), it has to be borne in mind that the litter leachate and streamflow fluxes were differently determined in the early five than in the later ten years. It is therefore possible that at least a part of the apparent trend is attributable to the different determination methods. While there might be a stronger scatter in the first five years than later, we did not observe a sharp change of the slope of the regression lines or breakpoints (Fig. 23.2a and b).

Under the current conditions, the P and K inputs would double the organic layer P and K storages in 197 and 27 years, respectively, resulting in improving P and K bioavailability at the scale of a few decades (K) to a few centuries with unknown implications for the biodiversity of the studied forest, particularly with respect to P (Wassen et al. 2005). In contrast, the loss of N, Ca, Mg, and S could cause a depletion of the organic layer storages in 38 (Mg) to 281 years (N). However, the mean depletion rate of N with 1.7 g m^{-2} $year^{-1}$ from the organic layer is still low relative to the increase of the bulk deposition of N of roughly 5 g m^{-2} $year^{-1}$, which even does not include dry deposition (Wilcke et al. 2013b). The average losses of Ca (1.7 g m^{-2} $year^{-1}$) and of Mg (0.96 g m^{-2} $year^{-1}$) were similar to or even higher than the highest deposition of Ca (1.5 g m^{-2} $year^{-1}$) and Mg (0.66 g m^{-2} $year^{-1}$) with rainfall in the hydrological year 1999/2000. Therefore, it is unlikely that the Ca and Mg losses can be compensated by the occasional alkaline dust inputs. Consequently, these two base metals will become increasingly scarce. However, it has to be borne in mind that our organic layer budgets assume that the litterfall quantity and quality do not change significantly in the coming years, which might, however, not be the case.

The accumulation of all nutrients in the mineral soil would double their storages in the 0–30 cm layer in between 57 (Ca) and 601 years (P), again assuming no other changes such as increased leaching, erosion, or weathering. This illustrates that the nutrients are hardly lost from the whole ecosystem, which is in line with findings of Wullaert et al. (2010) for N and P, Wullaert et al. (2013) for Ca, and Wilcke et al. (2017) for the base metals (K, Ca, Mg). The shift of nutrient storages, particularly of the depleting Ca and Mg, and the already now scarce S from the organic layer into the mineral soil will, however, potentially force the plants to grow more roots into the mineral soil. Because there are indications that the currently low root density in the mineral soil is more driven by the frequent waterlogging than by other constraints including Al toxicity, the increasingly dry soils might remove this obstacle and allow for increased root growth into the mineral soil.

23.5 Conclusions

Contrary to our first hypothesis, N was lost from the organic layer and only accumulated in the mineral soil. This illustrated that enhanced mineralization because of the reduced soil moisture overwhelmed the accumulation of N from atmospheric deposition. Therefore, there was a shift of N storages from the organic layer into the less rooted mineral soil.

Our second hypothesis that P accumulates in the organic layer and the mineral soil was confirmed. This accumulation was particularly strong in the organic layer where it increased with time during 1999–2013. The strong retention of accumulated P in the organic layer even reduced the P input into the mineral soil during the observation period.

Surprisingly, K accumulated in the organic layer and the annual organic layer budgets even showed a positive trend in spite of the acidic soils, likely reflecting an increased K need of the biota in response to increasing N and P availability, which resulted in a tightened K cycling between the soil and the vegetation. As expected, Ca and Mg were lost from the organic layer and the budgets (i.e., losses) even increased with time during 1998–2013 in line with the observed acidification of the litter leachates. However, all three base metals accumulated in the acid mineral soil in contrast to Hypothesis iii.

The S budget of the organic layer was negative as hypothesized, although the S release by mineralization just matched the S demand of the vegetation. Thus, the vegetation must have made use of other S sources such as the deposition from the atmosphere or the mineral soil so that some of the released S in the organic layer could be leached.

Our results demonstrate that the current environmental changes enhance the internal N, P, and S cycling in the ecosystem as reflected by increasing fluxes of these nutrients with throughfall+stemflow. The enhanced N and P cycling improves the short-term availability of these two most limiting nutrients. However, the accumulation of N via deposition is stronger than that of P, possibly shifting the ecosystem from the current N+P co-limitation to pure P limitation.

Our results demonstrate that the soils of the studied tropical montane forest in Ecuador respond rapidly to the changing climate, hydrological cycle, and matter deposition, which in the long run might drive the studied remote native forest ecosystem to a new state. This new state might involve a reduced species richness, which is commonly expected in response to fertilization.

23.6 Future Directions

Our results illustrate how important long-term observations of the effects of water-forest interactions on the nutrient status of remote native forests are. Only at least decade-long observations can reveal trends in nutrient cycling in response to environmental change. Once such trends have been identified, their reasons can be explored by targeted experiments such as nutrient additions to simulate atmospheric deposition, litterfall manipulations to simulate changing biomass productivity, or temperature and moisture manipulations to account for climate warming. As native forest ecosystems are usually well buffered, such ecosystem manipulations must again be run for longer periods of at least a decade.

The currently observed biogeochemical changes in tropical montane forests might be detrimental for the high biodiversity existing in these forests. Therefore, in addition to the observation of water and nutrient cycles, the species composition, growth responses, and rejuvenation of individual species to the changing conditions need to be monitored and linked with specific changes by targeted manipulative experiments.

We observed indications of changing soil organic mineralization rates in response to increasingly reduced waterlogging of the soils. Increasing organic matter mineralization is also associated with the release of CO_2. An open question to be addressed in the future will therefore be whether the tropical montane forests with their particularly high C storages in the soil organic layer and mineral topsoil might become a future CO_2 source with negative feedback on climate change.

Finally, the current frequent replacement of the native tropical montane forest by pastures causes the complete loss of the living aboveground biomass and the soil organic layer, which is only partly compensated by higher C storages in the mineral topsoil under pastures than forests. This results not only in considerable nutrient losses but also in the additional release of CO_2 into the atmosphere. Future research should therefore address the landscape-scale element budgets including the remains of the native forests and their anthropogenic replacement systems. Besides effects on element cycles, the land-use change might also feed back on the water cycle, particularly via changed evapotranspiration, and thus on the local climate, which needs to be addressed to improve local predictions of global climate change.

Acknowledgments We thank E. Beck, K. Müller-Hohenstein, M. Richter, and W. Zech for co-initiating the long-term study; K. Fleischbein, R. Goller, M. Meyer-Grünefeldt, M. Sequeira, H. Wullaert, S. Yasin, and numerous undergraduate students for data acquisition during parts of the observation period; the Ecuadorian Environmental Ministry for the research permits; Naturaleza y Cultura Internacional (NCI) in Loja for providing the study area and the research station; and the Deutsche Forschungsgemeinschaft (DFG) for funding (FOR 402 and 816).

References

Aber J, McDowell W, Nadelhoffer K, Magill A, Berntson G, Kamakea M et al (1998) Nitrogen saturation in temperate forest ecosystems. BioSci 48:921–934. https://doi.org/10.2307/1313296

Alencar A, Nepstad D, Diaz MDV (2006) Forest understory fire in the Brazilian Amazon in ENSO and non-ENSO years: area burned and committed carbon emissions. Earth Interactions 10(6). https://doi.org/10.1175/EI150.1

Arima EY, Richards P, Walker RT (2017) Biofuel expansion and the spatial economy: Implications for the Amazon basin in the 21st century. In: Qin ZC, Mishra U, Hastings A (eds) Bioenergy and land use change: Impact on natural capital and ecosystem services. American Geophysical Union, Wiley, Hoboken, NJ, pp 53–62. https://doi.org/10.1002/9781119297376.ch4

Artaxo P, Martins JV, Yamasoe MA, Procopio AS, Pauliquevis TM, Andreae MO et al (2002) Physical and chemical properties of aerosols in the wet and dry seasons in Rondonia, Amazonia. J Geophys Res Atmos 107(D20). https://doi.org/10.1029/2001JD000666

Barthlott W, Hostert A, Kier G, Küper W, Kreft H, Mutke J et al (2007) Geographic patterns of vascular plant diversity at continental to global scales. Erdkunde 61:305–315. https://doi.org/10.3112/erdkunde.2007.04.01

Balslev H, Øllgaard B (2002) Mapa de vegetación del sur de Ecuador. In: Aguirre MZ, Madsen JE, Cotton E, Balslev H (eds) Botánica Austroecuatoriana. Estudios sobre los recursos vegetales en las provincias de El Oro, Loja y Zamora-Chinchipe. Ediciones Abya-Yala, Quito, pp 51–64

Batjes NH (2002) Carbon and nitrogen stocks of the soils of Central and Eastern Europe. Soil Use Manage 18:324–329. https://doi.org/10.1111/j.1475-2743.2002.tb00248.x

Batjes NH, Dijkshoorn JA (1999) Carbon and nitrogen stocks of the soils in the Amazon region. Geoderma 89:273–286. https://doi.org/10.1016/S0016-7061(98)00086-X

Bormann FH, Likens GE (1967) Nutrient cycling. Science 155:424–429. https://doi.org/10.1126/science.155.3761.424

Boy J, Wilcke W (2008) Tropical Andean forest derives calcium and magnesium from Saharan dust. Glob Biogeochem Cycle 22:GB1027. https://doi.org/10.1029/2007GB002960

Boy J, Rollenbeck R, Valarezo C, Wilcke W (2008a) Amazonian biomass burning-derived acid and nutrient deposition in the north Andean montane forest of Ecuador. Glob Biogeochem Cycle 22: GB4011. https://doi.org/10.1029/2007GB003158

Boy J, Valarezo C, Wilcke W (2008b) Water flow paths in soil control element exports in an Andean tropical montane forest. Eur J Soil Sci 59:1209–1227. https://doi.org/10.1111/j.1365-2389.2008.01063.x

Bray RH, Kurtz LT (1945) Determination of total organic and available forms of phosphorus in soils. Soil Sci 59:39–46. https://doi.org/10.1097/00010694-194501000-00006

Bronaugh D, Werner A (2015) Zhang + Yue-Pilon trends package. R package version 0.10-1, http://www.r-project.org

Bruijnzeel LA (1991) Nutrient input-output budgets of tropical forest ecosystems: A review. J Trop Ecol 7:1–24. https://doi.org/10.1017/S0266467400005010

Bruijnzeel LA, Hamilton LS (2000) Decision Time for Cloud Forests. IHP Humid Tropics Programme Series, 13. IHP-UNESCO, Paris

Bruijnzeel LA, Proctor J (1995) Hydrology and biogeochemistry of tropical montane cloud forests: what do we really know? In: Juvik JO, Scatena FN (eds) Tropical montane cloud forests, Ecol. Stud. 110. Springer, New York, pp 38–78. https://doi.org/10.1007/978-1-4612-2500-3_3

Bruijnzeel LA, Veneklaas EJ (1998) Climatic conditions and tropical montane forest productivity: The fog has not lifted yet. Ecology 79:3–9. https://doi.org/10.1890/0012-9658(1998)079[0003:CCATMF]2.0.CO;2

Bruijnzeel LA, Mulligan M, Scatena FM (2011) Hydrometeorology of tropical montane cloud forests: Emerging patterns. Hydrol Process 25:465–498. https://doi.org/10.1002/hyp.7974

Dietrich K, Spoeri E, Oelmann Y (2016) Nutrient addition modifies phosphatase activities along an altitudinal gradient in a tropical montane forest in southern Ecuador. Front Earth Sci 4:12. https://doi.org/10.3389/feart.2016.00012

Dietrich K, Spohn M, Villamagua M, Oelmann Y (2017) Nutrient addition affects net and gross mineralization of phosphorus in the organic layer of a tropical montane forest. Biogeochemistry 136:223–236. https://doi.org/10.1007/s10533-017-0392-z

Fisher JB, Malhi Y, Cuba Torres I, Metcalfe DB, van de Weg MJ, Meir P et al (2013) Nutrient limitation in rainforests and cloud forests along a 3000-m elevation gradient in the Peruvian Andes. Oecologia 172:889–902. https://doi.org/10.1007/s00442-012-2522-6

Fleischbein K, Wilcke W, Goller R, Valarezo C, Zech W, Knoblich K (2005) Rainfall interception in a lower montane forest in Ecuador: effects of canopy properties. Hydrol Proc 19:1355–1371. https://doi.org/10.1002/hyp.5562

Fleischbein K, Wilcke W, Valarezo C, Zech W, Knoblich K (2006) Water budget of three small catchments under montane forest in Ecuador. Hydrol Proc 20:2491–2507. https://doi.org/10.1002/hyp.6212

Frei E (1958) Eine Studie über den Zusammenhang zwischen Bodentyp, Klima und Vegetation in Ecuador. Plant Soil 9:215–236. https://doi.org/10.1007/BF01394152

Goller R, Wilcke W, Fleischbein K, Valarezo C, Zech W (2006) Dissolved inorganic nitrogen, phosphorus, and sulfur in the nutrient cycle of a montane forest in Ecuador. Biogeochemistry 77:57–89. https://doi.org/10.1007/s10533-005-1061-1

Grubb PJ (1995) Mineral nutrition and soil fertility in tropical rain forests. In: Lugo AE, Lowe C (eds) Tropical forests: management and ecology, Ecol Stud 112. Springer, New York, pp 308–330. https://doi.org/10.1007/978-1-4612-2498-3_12

Helsel DR, Hirsch RM (2002) Chapter A3: Statistical methods in water resources, in Techniques of water-resources investigations. In: United States Geological Survey (ed.) Book 4, Reston, VA, USA

Hinsinger P (2001) Bioavailability of soil inorganic P in the rhizosphere as affected by root-induced chemical changes: A review. Plant Soil 237:173–195. https://doi.org/10.1023/A:1013351617532

Hirsch RM, Slack JR, Smith RA (1982) Techniques for trend analysis for monthly water quality data. Water Resour Res 18:107–121. https://doi.org/10.1029/WR018i001p00107

Hoffer A, Gelencser A, Blazso M, Guyon P, Artaxo P, Andreae MO (2006) Diel and seasonal variations in the chemical composition of biomass burning aerosol. Atmos Chem Phys 6:3505–3515. https://doi.org/10.5194/acp-6-3505-2006

Homeier J (2004) Baumdiversität, Waldstruktur und Wachstumsdynamik zweier tropischer Bergregenwälder in Ecuador und Costa Rica. Dissertationes Botanicae 391. J Cramer, Berlin, Germany

Homeier J, Hertel D, Camenzind T, Cumbicus NL, Maraun M, Martinson GO et al (2012) Tropical Andean forests are highly susceptible to nutrient inputs – rapid effects of experimental N and P addition to an Ecuadorian montane forest. PLoS One 7:e47128. https://doi.org/10.1371/journal.pone.0047128

IUSS Working Group WRB (2014) World Reference Base for Soil Resources 2014. International soil classification system for naming soils and creating legends for soil maps. World Soil Resource Reports No. 106. FAO, Rome

Jemison JM, Fox RH (1992) Estimation of zero-tension pan lysimeter collection efficiency. Soil Sci. 154:85–94. https://doi.org/10.1097/00010694-199208000-00001

Jobbagy EG, Jackson RB (2001) The distribution of soil nutrients with depth: Global patterns and the imprint of plants. Biogeochemistry 53:51–77. https://doi.org/10.1023/A:1010760720215

Jobbagy EG, Jackson RB (2004) The uplift of soil nutrients by plants: Biogeochemical consequences across scales. Ecology 85:2380–2389. https://doi.org/10.1890/03-0245

Johnson W, Lindberg SE (1992) Appendix. In: Johnson W, Lindberg SE (eds) Atmospheric deposition and forest nutrient cycling – a synthesis of the integrated forest study, Ecol Stud 91. Springer, New York, pp 610–688

Likens GE (2013) Biogeochemistry of a forested ecosystem, 3rd edn. Springer, New York

Lilienfein J, Wilcke W (2003) Element storage in native, agri- and silvicultural ecosystems of the Brazilian savanna I. Biomass, carbon, nitrogen, phosphorus, and sulfur. Plant Soil 254:425–442. https://doi.org/10.1023/A:1025579932395

Mahowald NM, Artaxo P, Baker AR, Jickells TD, Okin GS, Randerson JT et al (2005) Impacts of biomass burning emissions and land use change on Amazonian atmospheric phosphorus cycling and deposition. Glob Biogeochem Cycle 19:GB4030. https://doi.org/10.1029/2005GB002541

McLeod AI (2011) Kendall: Kendall rank correlation and Mann-Kendall trend test. R package version 2.2, https://CRAN.R-project.org/package=Kendall

Marschner P (2012) Marschner's mineral nutrition of higher plants, 3rd edn. Elsevier, Amsterdam

Pepper IL, Gerba P, Gentry TJ (2015) Environmental microbiology, 3rd edn. Elsevier, Amsterdam

Peters T, Drobnik T, Meyer H, Rankl M, Richter M, Rollenbeck R et al (2013) Chapter 2: Environmental changes affecting the Andes of Ecuador. In: Bendix J, Beck E, Bräuning A, Makeschin F, Mosandl R, Scheu S, Wilcke W (eds) Ecosystem services, biodiversity and environmental change in a tropical mountain ecosystem of South Ecuador, Ecol Stud 221. Springer, Berlin, pp 19–29. https://doi.org/10.1007/978-3-642-38137-9_2

Proctor J (1987) Nutrient cycling in primary and old secondary rainforests. Appl Geogr 7:135–152. https://doi.org/10.1016/0143-6228(87)90046-4

R Core Team (2017) R: A language and environment for statistical computing. R Foundation for Statistical Computing, Vienna, Austria. URL: https://www.R-project.org/

Rehmus A, Bigalke M, Valarezo C, Castillo JM, Wilcke W (2014) Aluminum toxicity to tropical montane forest tree seedlings in southern Ecuador: Response of biomass and plant morphology to elevated Al concentrations. Plant Soil 382:301–315. https://doi.org/10.1007/s11104-014-2110-0

Rehmus A, Bigalke M, Valarezo C, Castillo JM, Wilcke W (2015) Aluminum toxicity to tropical montane forest tree seedlings in southern Ecuador: Response of nutrient status to elevated Al concentrations. Plant Soil 388:87–97. https://doi.org/10.1007/s11104-014-2110-0

Rehmus A, Bigalke M, Boy J, Valarezo C, Wilcke W (2017) Aluminum cycling in a tropical montane forest ecosystem in southern Ecuador. Geoderma 288:196–203. https://doi.org/10.1016/j.geoderma.2016.11.002

Rollenbeck R, Peters T, Emck P, Richter M (2015) ECSF_climate station data best estimate ver. 1. Available online from DFG-FOR816dw, http://www.tropicalmountainforest.org/data_pre.do?citid=1382

Roman L, Scatena FN, Bruijnzeel LA (2010) Chapter 6: Global and local variations in tropical montane cloud forest soils. In: Bruijnzeel LA, Scatena FN, Hamilton LS (eds) Tropical montane cloud forests, International Hydrology Series. Cambridge University Press, Cambridge, UK, pp 77–89. https://doi.org/10.1017/CBO9780511778384.008

Sala OE, Stuart Chapin IIIF, Armesto JJ, Berlow E, Bloomfield J, Dirzo R et al (2000) Global biodiversity scenarios for the Year 2100. Science 287:1770–1774. https://doi.org/10.1126/science.287.5459.1770

Schrumpf M, Guggenberger G, Schubert C, Valarezo C, Zech W (2001) Tropical montane rain forest soils: Development and nutrient status along an altitudinal gradient in the south Ecuadorian Andes. Die Erde 132:43–59

Schuur EAG, Matson PA (2001) Net primary productivity and nutrient cycling across a mesic to wet precipitation gradient in Hawaiian montane forest. Oecologia 128:431–442. https://doi.org/10.1007/s004420100671

Schwarz MT, Oelmann Y, Wilcke W (2011) Stable N isotope composition of nitrate reflects N transformations during the passage of water through a montane rain forest in Ecuador. Biogeochemistry 102:195–208. https://doi.org/10.1007/s10533-010-9434-5

Soethe N, Lehmann J, Engels C (2006) The vertical pattern of rooting and nutrient uptake at different altitudes of a south Ecuadorian montane forest. Plant Soil 286:287–299. https://doi.org/10.1007/s11104-006-9044-0

Steinhardt U (1979) Untersuchungen über den Wasser- und Nährstoffhaushalt eines andinen Wolkenwaldes in Venezuela. Göttinger Bodenkundliche Berichte 56: 1–146, University of Göttingen, Germany.

Tanner EVJ, Vitousek PM, Cuevas E (1998) Experimental investigations of nutrient limitations of forest growth on wet tropical mountains. Ecology 79:10–22. https://doi.org/10.1890/0012-9658(1998)079[0010:EIONLO]2.0.CO;2

Timmermann A, Oberhuber J, Bacher A, Esch M, Latif M, Roeckner E (1999) Increased El Nino frequency in a climate model forced by future greenhouse warming. Nature 398:694–697. https://doi.org/10.1038/19505

Urrutia R, Vuille M (2009) Climate change projections for the tropical Andes using a regional climate model: Temperature and precipitation simulations for the end of the 21st century. J Geophys Res 114:D02108. https://doi.org/10.1029/2008JD011021

van der Werf GR, Randerson JT, Giglio L, Collatz GJ, Kasibhatla PS, Arellano AF (2006) Interannual variability in global biomass burning emissions from 1997 to 2004. Atmos Chem Phys 6:3423–3441. https://doi.org/10.5194/acp-6-3423-2006

Velescu A, Valarezo C, Wilcke W (2016) Response of dissolved organic matter to moderate N, P, N +P and Ca amendments in a tropical montane forest of south Ecuador. Front Earth Sci 4:58. https://doi.org/10.3389/feart.2016.00058

Veneklaas EJ (1991) Litterfall and nutrient fluxes in two montane tropical rain forests, Colombia. J Trop Ecol 7:319–336. https://doi.org/10.1017/S0266467400005587

Vogt KA, Grier CC, Vogt DJ (1986) Production, turnover, and nutrient dynamics of above- and belowground detritus of world forests. Adv Ecol Res 15:303–377. https://doi.org/10.1016/S0065-2504(08)60122-1

Vuille M, Bradley RS, Werner M, Keimig F (2003) 20th century climate change in the tropical Andes: observations and model results. Clim Change 59:75–99. https://doi.org/10.1023/A:1024406427519

Wassen MJ, Venterink HO, Lapshina ED, Tanneberger F (2005) Endangered plants persist under phosphorus limitation. Nature 437:547–550. https://doi.org/10.1038/nature03950

Wilcke W, Lilienfein J (2004) Element storage in native, agri- and silvicultural ecosystems of the Brazilian savanna II. Metals. Plant Soil 258:31–41. https://doi.org/10.1023/B:PLSO.0000016503.59527.ea

Wilcke W, Yasin S, Valarezo C, Zech W (2001) Change in water quality during the passage through a tropical montane rain forest in Ecuador. Biogeochemistry 55:45–72. https://doi.org/10.1023/A:1010631407270

Wilcke W, Yasin S, Abramowski U, Valarezo C, Zech W (2002) Nutrient storage and turnover in organic layers under tropical montane rain forest in Ecuador. Eur J Soil Sci 53:15–27. https://doi.org/10.1046/j.1365-2389.2002.00411.x

Wilcke W, Boy J, Hamer U, Potthast K, Rollenbeck R, Valarezo C (2013a) Chapter 11. Current regulating and supporting services: nutrient cycles. In: Bendix J, Beck E, Bräuning A, Makeschin F, Scheu S, Wilcke W (eds) Ecosystem services, biodiversity and environmental change in a tropical mountain ecosystem of South Ecuador, Ecol Stud 221. Springer, Heidelberg, pp 141–151. https://doi.org/10.1007/978-3-642-38137-9_11

Wilcke W, Leimer S, Peters T, Emck P, Rollenbeck R, Trachte K et al (2013b) The nitrogen cycle of tropical montane forest in Ecuador turns inorganic under environmental change. Glob Biogeochem Cycle 27:1194–1204. https://doi.org/10.1002/2012GB004471

Wilcke W, Velescu A, Leimer S, Bigalke M, Boy J, Valarezo C (2017) Biological vs. geochemical control and environmental change drivers of the base metal budgets of a tropical montane forest in Ecuador during 15 years. Biogeochemistry 136:167–189. https://doi.org/10.1007/s10533-017-0386-x

Wilcke W, Velescu A, Leimer S, Bigalke M, Boy J, Valarezo C (2019) Temporal trends of phosphorus cycling in a tropical montane forest in Ecuador during 14 years. J Geophys Res-Biogeosci 124:1370–1386. https://doi.org/10.1029/2018JG004942

Williams MR, Fisher TR, Melack JM (1997) Chemical composition and deposition of rain in the central Amazon, Brazil. Atmos Environ 31:207–217. https://doi.org/10.1016/1352-2310(96)00166-5

Wullaert H, Homeier J, Valarezo C, Wilcke W (2010) Response of the N and P cycle of an old-growth montane forest in Ecuador to experimental low-level N and P amendments. For Ecol Manage 260:1434–1445. https://doi.org/10.1016/j.foreco.2010.07.021

Wullaert H, Bigalke M, Homeier J, Cumbicus NL, Valarezo C, Wilcke W (2013) Short-term response of the Ca cycle of a montane forest in Ecuador to low experimental $CaCl_2$ additions. J Plant Nutr Soil Sci 176:892–903. https://doi.org/10.1002/jpln.201300146

Part VI
Final Synthesis: Future Forest Hydrology

Chapter 24
Forest-Water Interactions Under Global Change

Julia A. Jones, Xiaohua Wei, Emma Archer, Kevin Bishop, Juan A. Blanco, David Ellison, Mark B. Gush, Steven G. McNulty, Meine van Noordwijk, and Irena F. Creed

24.1 Introduction

This chapter aims to characterize the effects of global changes on forest-water interactions and water availability to ecosystems and people. It synthesises current understanding of the implications of present and anticipated changes to forests and tree cover for local and global hydrology and provides an overview of contemporary global change processes and their interactions with forests and water. It focuses specifically on natural and human disturbances and their effects on biotic and abiotic properties of forests and their consequences for hydrological processes.

Forests are dynamic on time scales of years to centuries and beyond, as a result of natural and anthropogenic disturbances. Large infrequent disturbances, both natural and human, may significantly modify forest characteristics and post-disturbance forest succession on a seasonal timescale, to decades, and to millenia (Foster et al. 1998), with corresponding effects on associated hydrologic processes. Natural forest

---

This chapter was adapted from the International Union of Forest Research Organizations (IUFRO) report titled 'Forest and Water on a Changing Planet: Vulnerability, Adaptation and Governance Opportunities.'

J. A. Jones
College of Earth, Ocean, and Atmospheric Sciences (CEOAS), Oregon State University, Corvallis, OR, USA

X. Wei
Earth, Environmental and Geographical Sciences, University of British Columbia (UBC), Kelowna, Canada

E. Archer
Centre for Environmental Studies/Department of Geography, Geoinformatics and Meteorology, University of Pretoria, Hatfield, South Africa

K. Bishop
Department of Aquatic Sciences and Assessment, Swedish University of Agricultural Sciences, Uppsala, Sweden

D. F. Levia et al. (eds.), *Forest-Water Interactions*, Ecological Studies 240,
https://doi.org/10.1007/978-3-030-26086-6_24

disturbances can have significant economic and social costs, including those associated with water ecosystem services from forests. Climate change may alter the frequency, intensity, and timing of natural disturbances and forest characteristics (Overpeck et al. 1990; Dale et al. 2001), consequently affecting hydrological processes.

Human disturbances have further exacerbated forest disturbances by natural processes, and forest changes are increasingly evident in global-scale land cover change detection (Fig. 24.1). In the 1970s, there was considerable uncertainty in estimations of global forest cover (Allen and Barnes 1985), but global high-resolution satellite coverage now permits more accurate estimates of forest cover and the density of trees within forests (Hansen et al. 2013). As of 2000, the largest area of forest within climate domains (e.g., boreal zone, tropics, etc.) contained less than 25% tree cover (Fig. 24.1a; Hansen et al. 2013), attesting to the widespread prevalence of open forest types and the roles of human disturbance in reducing tree cover in forest areas.

There has been a net loss of forest cover in the past half-century (Fig. 24.1b). Over this period, more than 800,000 km^2 of forest area was gained, but 2.3 million km^2 of forest was lost (Hansen et al. 2013). The area of forest loss was highest in the tropics, and least in the temperate zone, whereas the area of forest gain was similar across all climate domains (Fig. 24.1b).

The locations and estimated rates of forest cover change differ depending on the type of forest and the time period, as well as improvements in the accuracy of forest cover estimates. In the 1970s, as a result of population growth and agricultural expansion, estimated forest cover loss (both area and rate of loss) was especially high in the native tropical rainforest in Africa (Ghana, Cote d'Ivoire, Cameroon, and present-day Democratic Republic of the Congo), Asia (Thailand, Indonesia, Malaysia), and South America (Brazil) (Allen and Barnes 1985). However, in the recent past (2000–2012), the most rapid rates of forest change have occurred as a result of the expansion of intensive plantation forestry, which produces short-term gains in

J. A. Blanco
Departamento de Ciencias, Universidad Publica de Navarra, Pamplona, Spain

D. Ellison
External Expert Consultant, Zurich, Switzerland

M. B. Gush
Science & Collections, Royal Horticultural Society, London, UK

S. G. McNulty
USDA Forest Service, Eastern Forest Environmental Threat Assessment Center, Raleigh, NC, USA

M. van Noordwijk
World Agroforestry Centre (ICRAF), Indonesia Wageningen University, Wageningen, The Netherlands

I. F. Creed (✉)
School of Environment and Sustainability (SENS), University of Saskatchewan, Saskatoon, Canada
e-mail: irena.creed@usask.ca

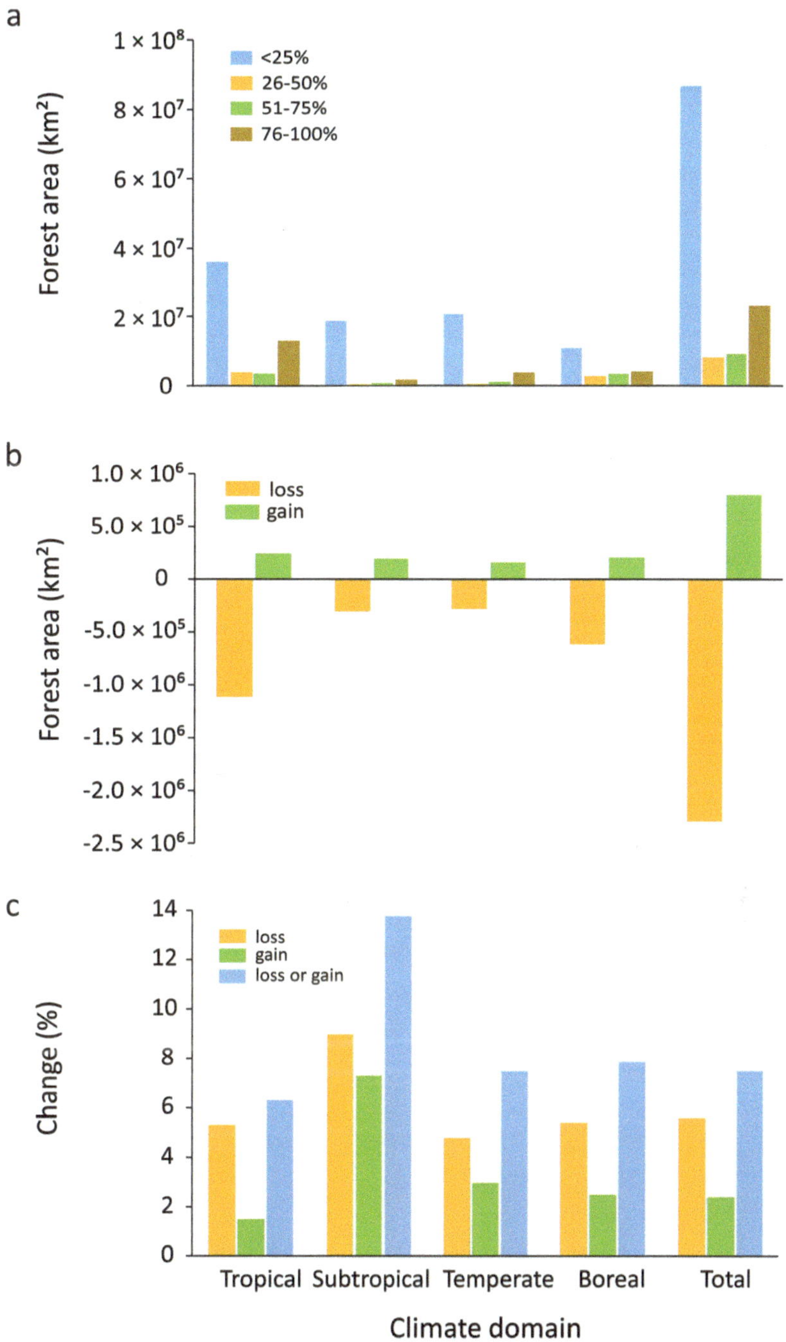

Fig. 24.1 Forest cover area and change from 2000 to 2012. (**a**) Estimated forest area in 2000 by tree density classes in different climate domains. (**b**) Total forest area loss and gain from 2000 to 2012 in different climate domains. (**c**) Change in area of forests with >50% tree cover from 2000 to 2012 (loss defined as loss >50%). (Source: Hansen et al. (2013) High-resolution global maps of twenty-first-century forest cover change. Science 342: 850–853. Reprinted with permission from AAAS)

particular types of forest cover, and associated clear-cut. This may in some cases produce forest loss when natural forest is replaced by plantation forests, e.g., in subhumid forests of North America (southeastern United States) and South America (Uruguay, southern Brazil) (Fig. 24.1c; Hansen et al. 2013). At the same time, as indicated earlier, the rates of forest loss in the tropics were two to three times higher than rates of forest gain in the temperate and boreal latitudes (Fig. 24.1c).

These patterns and trends demonstrate that forest change is dynamic and global, with an increasing impact from large-scale forestry plantations that are often monocultures. Forest change affects countries at all levels of economic development and is highly sensitive to global economic, social, and political factors. As a result of forest dynamics and net loss, water ecosystem services from forests are continually changing and being incrementally lost at the global scale.

24.2 Current Changes to Forest-Water Interactions

Drivers and determinants of change operating at a wide range of spatial and temporal scales modify a range of aspects of the climate-vegetation-soil-streamflow system. This section reviews examples of natural and human disturbances and their effects on components of forests (e.g., canopy and leaf area, litter and soil surface, rooting depth, and soil porosity) (Table 24.1) that in turn affect hydrological processes (e.g., precipitation, interception, infiltration, soil moisture storage, percolation, and evapotranspiration) (Table 24.2; Fig. 24.2). These characteristics of forest ecosystems strongly modulate the response of hydrology to forest disturbances (Jones and Post 2004; Peña-Arancibia et al. 2019; Zhang et al. 2019a). The section focuses on five natural disturbance processes (insects and pathogens, wildfire, drought, windthrow, and ice storms) and five human disturbances (silviculture including forest harvest, forest plantations, forestation, agroforestry, and urban and peri-urban forestry) (Fig. 24.2).

24.2.1 Hydrological Consequences of Natural Forest Disturbance

Natural disturbances have the potential to significantly alter a range of watershed processes (e.g., water quality, hydrology, channel morphology), as well as ecological functions in forested watersheds. Natural disturbance is intrinsic to forest dynamics (Attiwill 1994; Lertzman et al. 1997), producing spatial heterogeneity and ecosystem complexity and, under certain circumstances, supporting ecosystem

Table 24.1 Short-term effects of selected natural and anthropogenic disturbances discussed in this chapter on forest characteristics that influence water ecosystem services

	Canopy, leaf area	Litter, soil surface	Rooting depth	Soil porosity
Insects and pathogens	X		X	
Wildfire	X	X	X	X
Drought	X	X	X	
Ice storms	X			
Windthrow	X	X	X	
Silviculture, forest harvest	X	X	X	X
Plantations	X	X	X	X
Forestation	X	X	X	X
Agroforestry	X	X	X	X
Urban, peri-urban forestry	X	X	X	X

Table 24.2 Summary of influence of forest characteristics on hydrological processes

	Precipitation	Interception	Infiltration	Soil moisture storage	Percolation to groundwater	Evapo-transpiration
Canopy, leaf area	X	X		X	X	X
Litter, soil surface		X	X	X		X
Rooting depth			X	X		X
Soil porosity				X	X	X

resilience. Catastrophic forest disturbance (e.g., stand-replacing wildfire or large-scale insect pest outbreaks) may, however, cause undesired ecological and economic consequences. Wildfire, insects and pathogens are major natural disturbance agents that may affect global forests at a range of scales (van Lierop et al. 2015).

Disturbance events may be driven by natural or anthropogenic processes. The disturbance regime is a collection of events, with a characteristic frequency, severity, and extent. Natural and human disturbances in forests affect watershed processes. The hydrologic effects of some of the most widely studied natural forest disturbances are described below.

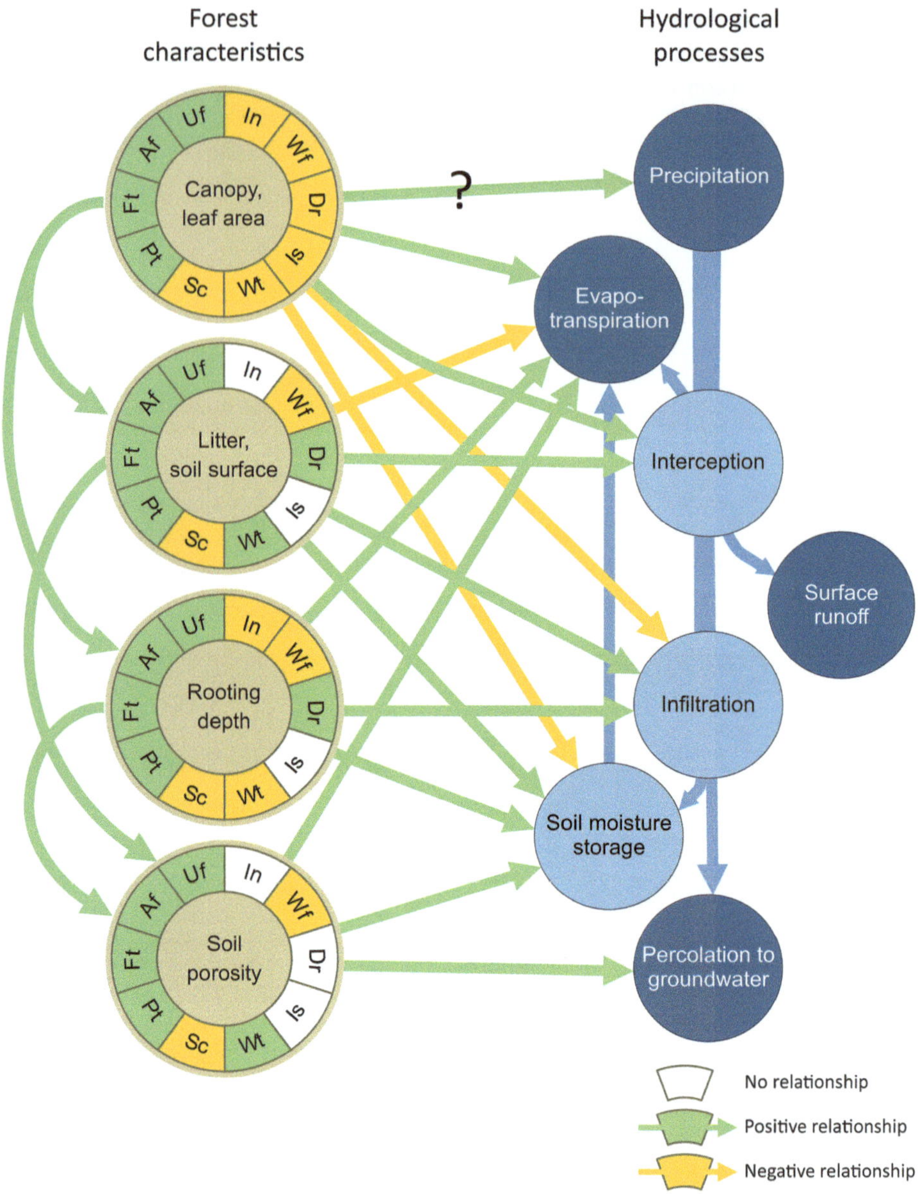

Fig. 24.2 Relationships between selected forest disturbances, forest characteristics, and hydrological processes that influence water ecosystem services. Disturbances are arranged around each forest characteristic; colours of slices indicate no (white), positive (green), or negative (yellow) short-term relationships with forest characteristics. Arrows between forest characteristics indicate relationships between forest characteristics. Arrows between forest characteristics and hydrological processes indicate positive (green) and negative (yellow) relationships with hydrological processes; no arrow indicates no relationship. The question mark (?) indicates that future research is needed to understand the degree and scale of the relationship. Arrows between hydrological processes indicate water transfers. Dark blue circles indicate hydrological processes that are included in water balance; light blue circles indicate hydrological processes that are intermediate. Disturbances: In (Insects), Wf (Wildfire), Dr (Drought), Is (Ice storms), Wt (Windthrow), Sc (Silviculture), Pt (Plantations), Ft (Forestation), Af (Agroforestry), Uf (Urban forestry)

24.2.1.1 Insects and Pathogens

Insects and pathogens are common disturbance agents affecting forests and can significantly influence hydrological processes. A characteristic example is the case of the mountain pine beetle (*Dendroctonus ponderosae* Hopkins) (MPB), a native insect to pine forests of western North America. Unusually hot, dry summers and mild winters, along with forests filled with mature pine, have led to an unprecedented epidemic of MPB in the Rocky Mountains of Colorado and much of British Columbia. Tree mortality from the MPB is caused by larval egg galleries and their symbiotic blue stain fungi in the inner bark of the trunk (Dhar et al. 2016a). In contrast with other major disturbances such as clear-cut harvesting, this disturbance may allow non-affected, non-target overstory and understory trees and shrubs to form new structurally diverse stands. There are three stages to an MPB attack. In the first summer, affected trees stop transpiring, but needle colour does not change, so this is termed a 'green attack'. In the following two to three years, the needles turn red and start to fall; this is the 'red attack' stage. In the third year, trees have died and lost their needles so the stage is termed 'grey attack'. Although changes in canopy colour increase albedo, reducing winter and early spring temperatures (O'Halloran et al. 2012; Vanderhoof et al. 2014), these changes are partially offset by a reduction in latent heat resulting from reduced evapotranspiration and associated increases in soil moisture, with the net effect that temperature increases in MPB-affected stands (Cooper et al. 2017). From the red attack stage onwards, surviving vegetation makes use of the increase in available resources (i.e., water, light, nutrients), often growing at an enhanced rate (Dhar et al. 2016a). The changes in energy and vegetation caused by an MPB infestation may thus drive hydrological responses dynamically, as the stand moves through the stages of attack and post-attack recovery.

An MPB outbreak serves as a clear example of how pest/pathogen infestation may affect all forest hydrologic processes. Tree mortality following MPB attack reduces foliage cover and density and consequently decreases canopy interception. The more open canopy after an MPB attack speeds snow ablation and advances spring melt by days to weeks compared to unaffected stands (Redding et al. 2008; Winkler et al. 2014). Tree transpiration is reduced following MPB attack, in magnitude proportional to the severity of mortality (Clark et al. 2014). The opening of the forest canopy concurrently increases sun exposure, however, which increases soil evaporation. These competing processes offset each other (Bearup et al. 2014; Biederman et al. 2014), to a degree that is not well quantified. Understory and surviving overstory trees and other vegetation also affect water dynamics after disturbance (Reed et al. 2014). Although evapotranspiration is reduced after insect attack (Dhar et al. 2016b), this may be short-lived as rapid growth of understory vegetation and regeneration increase evapotranspiration to the level prior to disturbance. This rapid growth of understory also suggests that the warming effects due to albedo change in the grey stage would be short-lived as the albedo change favours cooling. The effects of MPB outbreaks on streamflow are controversial (Biederman et al. 2015; Penn et al. 2016), but appear to depend on the extent and severity of tree mortality and remaining vegetation recovery (Weiler et al. 2009; Reed et al. 2014). Salvage logging following MPB infestation significantly increases high flows and advances their timing (Lin and Wei 2008; Zhang and Wei 2014), which can potentially increase flood risk.

Other pest and pathogen attacks on forests may have similar, mixed, and transient effects (Jones et al. 2009; Jones 2011; Adams et al. 2012). For example, water yield and peak flows of large events increased in watersheds where eastern hemlock was lost due to hemlock woolly adelgid infestation (Kim et al. 2017).

24.2.1.2 Wildfire

Wildfire is the most dominant natural disturbance in global forests, particularly in boreal and Mediterranean forests, although fire severity, intensity, and frequency vary according to forest type (Hansen et al. 2013; van Lierop et al. 2015; IUFRO 2018). Wildfires affect millions of square kilometers of land each year, and experience decreases or increases depending on fire suppression, fuel accumulation, and climate change (Westerling 2016; Andela et al. 2017).

In Canada alone, wildfire, on average, disturbs 1.6 Mha annually and accounts for 2.5 times more area disturbed than harvesting (White et al. 2017). Wildfire affects both the terrestrial environment and aquatic ecological processes. Severity of forest wildfire depends on meteorological conditions, vegetation type, stand fuel loading, and topographic properties (Oliveras et al. 2009). Implementation of particular types of fire suppression practices can lead to an accumulation of more fuels, which in turn may increase the chance of more catastrophic fires in the future (Collins et al. 2013).

In the immediate aftermath of wildfires, the burnt soil is bare and dark and highly susceptible to erosion and even mudflows. In the absence of all-consuming crown fires, subsequent tree mortality and litter fall can restore a protective litter layer, but water repellency of soils, caused by accumulation in the soil of organics volatilised during the fire, may cause high overland flow rates, especially on steep slopes (Neary et al. 2005), with enough energy to carry freshly fallen litter downhill.

The impact on forest water resources can be highly variable, for several reasons. Intensity, duration, and size all help determine wildfire impacts on tree mortality (Dunn and Bailey 2016; Iverson et al. 2017); it is not always straightforward to distinguish the effects of fire from the tree mortality caused by drought episodes that triggered the fire, but some studies have been able to quantify the relative contribution of both (e.g., van Nieuwstadt and Sheil 2005). The impacts of wildfire on forest water quantity and quality are thus also highly variable (Riggan et al. 1994; Vieira et al. 2015; Hallema et al. 2018). As a rule, streamflow response depends on the extent and severity of wildfire. Variability in climate and the heterogeneous nature of wildfire-induced forest loss can mask this relationship (Hallema et al. 2017). Wildfire impacts on water quality are even more complicated. On forests with little or no slope, wildfires will likely have minimal impact on forest water quality, if the areas are left to regenerate naturally (Hallema et al. 2017). Determinants of both wildfire and post-wildfire impacts (e.g., vegetation loss, soil infiltration change) need to be considered to better assess changes in forest water quantity and quality. Surface or ground fires can change the composition and porosity of soil. Forest fires tend to volatilise waxes and oils from litter, which may condense on soil particles, producing hydrophobic (water repellent) conditions in soils (Doerr et al. 1996) that in turn reduce infiltration and increase overland flow (Neary et al. 2005). Soil texture, state of aggregation, pH, mineral composition of the clay fraction, and microbial activity may also affect soil water repellency (Cesarano et al. 2016).

Crown fires or stand-replacing fires are more severe, not only affecting soils but also destroying canopy structures, potentially impacting all hydrological processes. Crown fires eliminate above-ground biomass which greatly reduces canopy interception and evapotranspiration (Bond-Lamberty et al. 2009; Montes-Helu et al. 2009) and increases soil evaporation, as soils with altered albedo become exposed to solar radiation. As a result of changes in the relative partitioning of water resources, soil hydrophobicity, and decreased evapotranspiration, crown fires increase annual runoff at the hillslope and catchment scales (Hallema et al. 2017; Kopp et al. 2017). In rain-dominated watersheds burned by crown fires, canopy removal and hydrophobic soils increase the kinetic energy of rainfall, limit soil infiltration capacity, and shorten flow paths. Consequently, the magnitude of peak flows is increased, and their timing is advanced (Liu et al. 2015). Unlike peak flow and annual runoff, the effect on base flow is uncertain, with great climatic, spatial and temporal variability. For example, base flow increased following fires during the dry season in many Mediterranean regions (Kinoshita and Hogue 2011; Bart and Tague 2017), while in northern Mongolia, baseflow declined in the dry season after wildfire partly due to the diminished water retention capacity of the organic surface layer (Kopp et al. 2017).

24.2.1.3 Drought

Extreme drought is associated with water stress and tree mortality (Bréda et al. 2006). Trees respond to drought by shifting the allocation of carbon from foliage to roots (Doughty et al. 2014). Drought also influences the hydrologic function of the soil (Gimbel 2016).

24.2.1.4 Windthrow

Windthrow, the process by which live trees are toppled by high winds, is a natural process in forests in boreal, temperate, and tropical regions (Bormann and Likens 1979). High winds during severe thunderstorms (Canham and Loucks 1984) and hurricanes (Boose et al. 1994) topple dominant trees in forest stands. In other contexts strong winds (as in typhoons) can lead to defoliation without windthrow, inducing strong hydrologic responses in the short term, but a more rapid recovery (Zhang et al. 2019b). Windthrow releases shade-tolerant tree species when dominant trees are toppled (e.g., Lorimer 1977; Foster 1988). Windthrow increases light to the forest understory and creates spatial heterogeneity of habitats in the form of root mounds and pits ('pit and mound topography') associated with the root wads of downed trees (Henry and Swan 1974; Carlton and Bazzaz 1998). Forests in certain locations, such as coastal areas, may be especially susceptible to natural windthrow as a disturbance agent (Kramer et al. 2001). Trees in areas exposed to high winds are especially susceptible to windthrow (Foster and Boose 1992), and by creating fresh edges, forest harvest may accelerate windthrow (Sinton et al. 2000).

24.2.1.5 Ice Storms

Ice storms are winter events characterised by freezing rain and are common in East Asia (Ding et al. 2008) and North America (Irland 2000). An ice storm forms along a narrow band on the cold side of a warm front, where surface temperatures are at, or just below, freezing. Under these conditions, rain becomes supercooled and freezes upon impact with cold surfaces (Irland 2000). Ice storms often have a large spatial extent and may catalyse other types of forest disturbance. For example, ice storm mortality and weakening of trees promote bark beetle population increases (de Groot et al. 2018). The relationship between damage severity, topography, and forest type was found to be significant at the watershed scale (Isaacs et al. 2014). Trees with narrow crown, coarse branching, strong branch attachments, or low surface area have greater resistance to damage from ice storms (Hauer et al. 1994). Research on the effects of ice storms on forest hydrology is very limited, but as ice storms cause stem and branch breakage and uprooting of trees, it is expected that their effects could be similar to a combination of effects of insects (i.e., dead trees) and windstorms (i.e., uprooted trees). A case study conducted in the Hubbard Brook Experimental Forest to monitor the impacts of the 1998 ice storm on hydrology and biogeochemistry found that stream discharge was not significantly altered (Houlton and Driscoll 2011). In 2011, the Hubbard Brook Experimental Forest ice storm experiment produced significant canopy damage and increased canopy openness (Rustad and Campbell 2012).

24.2.2 Hydrological Effects of Human-Driven Forest Changes

24.2.2.1 Silviculture and Forest Harvest

Specific growing conditions and silvicultural practices have an important bearing on the hydrology of forested watersheds. Managed and unmanaged forests vary with respect to stand density (stems per hectare), tree age distribution (rotation lengths), tree species, stand management practices (weeding, pruning, thinning, etc.), and tree health (du Toit et al. 2014).

Diversity of Species Opportunities to actively manage water use by forests derive from the fact that tree species vary in their use of water from different soil depths or at different times of year (Moore et al. 2011a; Kerhoulas et al. 2013). Forest stands of mixed-species display complementary water resource utilisation and may have higher water use efficiency in both temperate and tropical climates (Forrester 2015; Schwendenmann et al. 2015). Such complementary water resource utilisation suggests that mixed-species forests may be more resilient to drought. In dry regions, management actions that maintain or create low-density stands of large, deeply rooted trees increase tree access to water from winter precipitation stored in deep soil layers (Kerhoulas et al. 2013). The effects of species and leaf area on stand-level water use may be countered by differences in soil moisture and nutrient status among sites (Moore et al. 2011a).

Age Water use by individual trees in forest stands increases from the seedling stage to the closed canopy stage (Scott and Smith 1997; Dye and Bosch 2000), but stand-level transpiration appears to decline in old-growth native forests or mature plantations (Scott and Prinsloo 2008). Tree age has the greatest effect on differences in water use, with young forest stands using significantly more water than old-growth forest stands, followed by differences in basal area and, finally, species composition (Moore et al. 2004). Transpiration is more strongly coupled to streamflow when soils are wet, but transpiration may produce lagged, diel (24-hour cyclical) variations in streamflow during dry seasons (Moore et al. 2011b).

Very few studies have attempted to scale tree and forest stand water use to the watershed. At the watershed scale, native forest stands of old-growth trees use more water in the wet season, thus mitigating floods, while simultaneously using less water during dry periods, compared to closed canopy managed forests of native tree species (Jones 2000; Jones and Post 2004). Due to high water use by young, densely spaced trees, forest plantations of native tree species aged 25–45 years produce persistent dry season streamflow deficits exceeding 50% relative to native old-growth forests (Perry and Jones 2017). In addition, vegetation cover transition can greatly affect evapotranspiration and, consequently, long-term water balance responses at the watershed scale (Naranjo et al. 2011). Overall, the landscape scale effects of forest cover and management on hydrology depend upon the spatial arrangement of forest stands, which vary in age, density, and species composition.

Thinning Studies of how forest stand conditions (density, species, age) affect water use are typically conducted at the scale of individual trees or small forest stands. Stand-level transpiration is higher in stands with greater stem density (e.g., Whitehead et al. 1984). Thinning reduces interception and transpiration and consequently increases soil moisture and leaf water potential. Thinning also increases water availability, benefitting growth of dominant trees (Nnyamah and Black 1977; Bréda et al. 1995; Lechuga et al. 2017). However, thinning can increase soil evaporation due to more exposure of soil surface after thinning, which may partially offset water saving from thinning. In general, canopy conductance (stand-level transpiration) increases with leaf area when soil moisture is not limiting. But vapour pressure deficit and soil moisture deficits can limit transpiration (Granier et al. 2000). Thinning decreased water uptake for two tree species, especially during dry seasons and dry years, but effects differed between conifer and broadleaf species (Cardil et al. 2018).

Harvesting Timber harvesting removes trees and causes substantial changes in evapotranspiration, which in turn may alter water yield from a watershed (le Maitre et al. 2015). A range of literature reviews based on experimental studies of small paired watersheds have shown that harvesting operations reduce evapotranspiration and consequently increase annual streamflow (e.g., Bosch and Hewlett 1982; Andréassian 2004; Brown et al. 2005; Boggs et al. 2013), although it must be noted that there are large variations in changing magnitudes of streamflow. Several recent reviews based on large watersheds (Li et al. 2017) or both small and large watersheds (Zhang et al. 2017) reach similar conclusions.

Timber harvesting can significantly alter other components of streamflow (Zhang and Wei 2014; Li et al. 2018). For example, in northwest North America, forest harvesting increased large flood events, with effects persisting for multiple decades (Jones and Grant 1996; Jones 2000; Moore and Wondzell 2005). Forest roads shorten flow path lengths and advance peak flow timing in steep forest lands, permanently modifying streamflow response (La Marche and Lettenmaier 2001; Wemple and Jones 2003). Forest harvest affects snow accumulation and melt, which in turn increases the magnitude of extreme rain-on-snow floods (Harr 1986; Jones and Perkins 2010), and associated landslides, contributing sediment to streams (Wemple et al. 2001). On the other hand, as shown in the southeastern USA, sediment delivery from forest harvest may not significantly affect aquatic biota (Boggs et al. 2015).

24.2.2.2 Plantations

Plantation forests are becoming increasingly common and represent approximately 7% of the world's total forest area (FAO 2015; Payn et al. 2015). Highly managed conditions, which include stand fertilisation, thinning, regular tree spacing, genetically improved growing stock, controlled burning, and other practices, are designed to increase growth rate and wood quality (Fox et al. 2004). Management practices increase growth by maximising leaf area and growth efficiency (Waring 1982). Increased leaf area can increase water demand by trees (Scott et al. 2004). Fast-growing plantation forests may, under certain conditions, have greater water use efficiency (WUE) than unmanaged vegetation (Gyenge et al. 2008). Site condition also affects WUE in monoculture and mixed-species plantations of native tree species (Douglas-fir (*Pseudotsuga menziesii*) and red alder (*Alnus rubra*)) (Moore et al. 2011a, b).

Numerous studies, many in the form of paired catchment experiments, have shown conclusively that forest plantations of non-native fast-growing tree species, which are often densely planted, generally consume more water than native forests, grasslands, or shrublands and thus reduce water yield (streamflow) from reforested/afforested catchments (Bosch and Hewlett 1982; Farley et al. 2005; Jackson et al. 2005; Amazonas et al. 2017). Plantations of eucalypts in South Africa have been convincingly demonstrated to utilise more water than the vegetation they replaced

(Dye and Versfeld 2007; Scott and Prinsloo 2008). Reforestation improved soil infiltration in the Philippines (Zhang et al. 2019a). Evapotranspiration from grasslands was lower than from plantation forests along a rainfall gradient (Zhang et al. 1999, 2001). These findings have helped support the view that forest management involves a trade-off between 'blue' water (to streams) and 'green' water (to trees) (Calder et al. 2007; Cristiano et al. 2015). This literature, however, has largely remained silent about the downwind impacts of forest cover change (Ellison 2018).

Procedures have been developed to estimate the expected impact of plantation forestry relative to a baseline of natural vegetation in South Africa (Gush et al. 2002; Gush 2010), and similar calculations have been used to estimate effects of removal of invasive exotic species from riparian zones (Dzikiti et al. 2016), as well as expanding rubber plantations in SE China (Guardiola-Claramonte et al. 2010). To date, no procedures have been developed to estimate downwind impacts on rainfall and water availability.

Monoculture plantations also have less biodiversity compared to natural stands (Brockerhoff et al. 2008), which can increase the risk of episodic insect and disease outbreaks or fire that can threaten the health of the entire stand (Mitchell et al. 1983; McNulty et al. 2014). While complete stand or catchment mortality can significantly increase streamflows, tree mortality may also decrease water quality (Hibbert 1965; Swank et al. 2001).

24.2.2.3 Forestation

Forestation (used here as a generic term to reflect any increase in tree cover, regardless of methods applied or prior land use) may increase infiltration and soil moisture storage, reduce kinetic energy or precipitation and hence reduce erosion, and contribute to improvement of water quality and quantity. However, this depends on what it replaces, the species used and the approach taken. Forestation projects may involve plantation forestry or regeneration of native forests, which may have different effects on water yield. In a review of forest restoration effects on streamflow by Filoso et al. (2017), most studies reported decreases in water yields following forestation. Studies of forest restoration effects on local water yield are limited, especially for the humid tropics and subtropics.

Although forestation may reduce streamflow locally, larger-scale impacts of forestation, especially downwind from the forestation site and beyond the basin, and interactions with local climate, orography, and other features, may alter precipitation within and beyond the forested areas in ways that increase water availability (e.g., Van der Ent et al. 2010; Ellison et al. 2012, 2017; Keys et al. 2016; Creed and van Noordwijk 2018; Ellison 2018; Staal et al. 2018). Water yield responses to changes in forest cover also may occur over very long time periods (Naranjo et al. 2011).

In the northeastern United States, much of the Allegheny Mountain range was harvested in the early twentieth century (Cleland 1910). This loss of forest area drove both an increase in streamflow and a severe deterioration of water quality. Having recognised the forest area problem, much of the region was placed under strict protection to encourage restoration (natural regeneration) and prohibit cutting. A century later, the region is now again covered in mature forest and supplies New York City with some of the highest quality drinking water in the USA (NY EPA 2015). More recently, China implemented a 'Greening China' initiative. Over a decade, tens of millions of ha of forest were planted to stabilise soil and improve drinking water standards (Cao et al. 2011). An adverse side effect of this practice has been reductions in groundwater tables in areas of planted forest and competition between farmers and foresters for limited water resources (Cao et al. 2011). Effects of reforestation vary, depending on the choice of species, methods, and objectives (Mansourian et al. 2017).

24.2.2.4 Agroforestry

Like forestation, agroforestry – the interplanting of trees with crops – can significantly improve water infiltration, soil organic matter, and nutrient status; but, in contrast to forestation, the addition of trees to crop fields may also enhance local water availability to crops (Ong et al. 2014; Zomer et al. 2016). Tree litter enhances soil organic matter content, which in turn increases soil water holding capacity and may offset the water use of trees and crops (Mutegi et al. 2008). Tree roots compete for water with those of annual crops, but can also draw water from deeper layers of soil and, through hydraulic redistribution, replenish dry topsoil layers at night (Bayala et al. 2008). The lower density of trees in agroforestry systems, compared to forestation projects, in addition to improving infiltration, may also limit resulting reductions in groundwater (Ilstedt et al. 2016). Agroforestry systems can redistribute soil water belowground and along slopes (Wu et al. 2017).

Trees on farms can, under certain circumstances, help mitigate the effects of weather extremes on crops such as droughts, heat waves, and heavy rain, as well as provide shelter for livestock in extreme wind, high temperatures and extreme rainfall events. The tree roots in agroforestry systems are also able to take up nitrogen, phosphorus, and pesticide residues, as well as heavy metals, and therefore improve groundwater and downstream water quality (Pavlidis and Tsihrintzis 2018). The tree components of agroforestry systems stabilise soils against landslides, raise infiltration rates to limit surface flow during the rainy seasons, and increase groundwater release during the dry seasons, which can help crops to cope with drought and flood risks under future climate change (Ma et al. 2009; van Noordwijk et al. 2015, 2019). Further, appropriate agroforestry species can provide additional household food sources, fodder and shade for animals while providing organic fertilisers for the annual crops during the rainy season (Boffa 1999).

24.2.2.5 Urban and Peri-urban Forestry

The majority of the global population now lives in cities (54.5% in 2016 and around 80% in developed nations (United Nations 2016)). The global proportion is expected to reach 60% by 2030, with the large majority of that population growth occurring in rapidly expanding cities in Asia and Africa (United Nations 2016). This represents a dramatic demographic change from a population that was just 10% urban at the start of the twentieth century. While cities only represent approximately 3% of the terrestrial surface of the planet, they concentrate global environmental effects (e.g., carbon emissions) and place high demands for ecosystem services within cities (e.g., recreation), adjacent to them (e.g., water supply), and across the world (e.g., food, consumer goods) (Grimm et al. 2008; Millennium Ecosystem Assessment 2005). Perhaps the most substantive adverse effects of urbanisation on water quality and quantity are due to the increased amount of impervious surface cover in urban watersheds (Shuster et al. 2005).

The ecosystem services provided by urban and peri-urban forests have been the subject of a growing body of research (e.g., Vailshery et al. 2013; Duinker et al. 2015; Sanusi et al. 2017). Urban trees and forests provide surface cooling (Bounoua et al. 2015), increased infiltration, and attenuation of urban stormwater during precipitation events. Trees in cities mitigate stormwater runoff in at least three ways: by intercepting and storing water in their leaves and branches, by transpiring and thus reducing soil moisture (Chang and Li 2014), and by increasing litter fall and soil carbon, and thus the water holding capacity of soils, as well as increasing soil porosity through root expansion (Bartens et al. 2008). Trees also improve urban water quality by reducing sediments and particulate pollution (Sanders 1986).

Municipalities are increasingly looking to green infrastructure solutions that combine built environments with vegetation (Seitz and Escobedo 2011). For instance, bioretention installations, permeable pavements, and structural soil cells are increasingly using technologies for stormwater management that also provide sufficient soil volumes and irrigation for trees (Ow and Ghosh 2017; Scholz and Grabowiecki 2007). Such urban greening initiatives help to simultaneously provide necessary conditions to establish and grow trees in difficult urban settings while also mitigating the adverse effects of urbanisation on hydrological processes. Sustainably managed urban and peri-urban forests also represent green infrastructure that can play a central role in helping cities to adapt to the changing climate (Brandt et al. 2016).

24.3 Anticipated Changes to Forest-Water Interactions

Multiple future changes are expected to affect forest-water interactions in a range of ways. Climate change will and appears already to directly alter forest hydrological processes. At the same time, social and economic factors will directly alter forest management, through processes such as the expansion of intensive plantations, deforestation, forest degradation, selective logging, loss of riparian forest, and loss of urban trees, all of which will affect the hydrologic cycle. In addition, climate change is likely to alter the disturbance regimes of forests, indirectly influencing forest hydrology. Moreover, forest management will respond to climate change, which in turn will affect forest hydrological processes. These future changes are described below.

24.3.1 Climate Change and Future Forest Hydrology

Climate change will likely lead to an intensification of the hydrologic cycle in areas where vegetation water use is currently energy limited, but can elsewhere lead to a net drying effect (Huntington 2006; Cook et al. 2014; Burt et al. 2015; Li et al. 2017). More extreme precipitation regimes will imply a greater need for the flow-modulating effects of vegetation (Knapp et al. 2008), with flood risk increasing. Climate change is altering forest structure and species composition, and forest cover is expanding to higher latitudes and higher elevation, which may mitigate or exacerbate the direct effects of climate change (Hinzman et al. 2005; Lindner et al. 2010).

Rising atmospheric CO_2 concentrations can increase forest growth, while rising temperatures will promote forest expansion. Both processes may increase evapotranspiration (Wramneby et al. 2010), particularly in regions where there is no significant water or nutrient shortage (Holtum and Winter 2010). Rising atmospheric CO_2 concentrations can, however, also induce a partial closure of vegetation stomata and thus suppress evapotranspiration and increase runoff (Gedney et al. 2006). Rising atmospheric CO_2 concentrations have mixed and interacting effects on evapotranspiration, tree growth, and runoff (Hickler et al. 2008; Norby et al. 2010; Norby and Zak 2011; Silva and Anand 2013).

Forests, especially native and old-growth forests, influence future hydrology indirectly by storing and sequestering carbon and potentially limiting increases in air temperature and evapotranspiration. Many studies have shown that old-growth forests and old trees can continue to accumulate carbon in vegetation and in soils, while harvesting old-growth forests results in net carbon release (Harmon et al. 1990; Zhou et al. 2006; Luyssaert et al. 2008; Stephenson et al. 2014). At a timescale of decades, the use of forest products to substitute for fossil fuels or concrete in construction can lead to managed forests becoming net carbon sinks despite recurring harvests (Lundmark et al. 2018).

On a global scale, evapotranspiration increased between 1982 and 2008 (Jung et al. 2010), although changes in evapotranspiration are variable among regions. Evapotranspiration declined in parts of the world where soil moisture limits vegetation water use – including parts of Australia, East Africa, and South America, but evapotranspiration increased in regions such as China and southern India (Jung et al. 2010). Factors including soil moisture, stomatal closure resulting from rising CO_2 concentrations, land-use change, or declining wind speed all may cause evapotranspiration changes (Piao et al. 2007; McVicar et al. 2012; Rowland et al. 2015).

Forest die-off from drought and heat stress has, as mentioned previously, occurred around the world and is expected to increase with climate change (Anderegg et al. 2013). In northern and western Europe, where soil moisture may not be limiting, increased atmospheric CO_2 concentrations and warmer temperatures are expected to increase forest growth, whereas in southern and eastern Europe increasing drought and fire risks are expected to reduce forest productivity (Lindner et al. 2010). In southern European forests, progressive crown defoliation occurred from 1987 to 2007 apparently in response to increased water deficit (Carnicer et al. 2011). It has been argued that tall trees of old-growth forests are at the greatest risk of mortality due to moisture stress (McDowell and Allen 2015). However, in unmanaged old forests in the western USA, non-catastrophic mortality rates increased rapidly in recent decades, targeting small trees (van Mantgem et al. 2009). Forest mortality to drought may also increase in areas that currently or previously received high rates of atmospheric nitrogen deposition due to changes in tree morphology (McNulty et al. 2017).

Analyses of long-term records at 35 small watersheds (0.01–1 km^2) in the USA and Canada indicate that climate change effects on streamflow are not as clear as might be expected, arguably because of ecosystem processes and human influences (Jones et al. 2012). Although air temperature increased at 17 out of 19 sites with 20–60 yr records, climate trends were directly related to streamflow trends at only seven sites, and all of these involved changes in ice and snow. At other forest sites undergoing warming, other factors such as past forest disturbance and forest succession mimicked, exacerbated, counteracted, or masked the effects of climate change (Jones et al. 2012).

Interannual variability of climate significantly influences interannual variability of streamflow at forested headwater sites. For example, in the above-mentioned North American dataset (Jones et al. 2012), streamflow was significantly correlated with the El Niño-Southern Oscillation (ENSO), the Pacific Decadal Oscillation (PDO), and/or the Northern Atlantic Oscillation (NAO) at 26 of 30 forested headwater reference watersheds.

Forested sites differ in their sensitivity to interannual climate variability. An experimental analysis of long-term experimental watersheds in Canada and the USA was conducted over 5-year cool and warm periods to test whether changes in dryness were associated with consistent responses of water yield (Creed et al. 2014). Alpine sites, whose hydrology was dominated by water stored in snow and ice, showed the greatest sensitivity to warming, and any warming led to increased water yields.

Studies in small experimental catchments indicate that forest dynamics, including legacies of past disturbance and forest management, as well as forest succession and tree diversity, produce a wide range of forest hydrologic response to climate change at individual sites. For example, in the northern hardwood forest of Hubbard Brook (USA), climate change effects on ecosystem structure and function and hydrology appear to be modified by interactions with a spatially variable history of land use and a wide range of current human activities and concurrent environmental changes (Groffman et al. 2012). At Hubbard Brook, both air temperature and precipitation have increased, but winter precipitation has increased less. As a result of reduced snowpack accumulation, snowmelt-induced peak flows in spring have declined (Campbell et al. 2011) and have occurred earlier (Hamburg et al. 2013). In contrast, both winter and summer streamflows have increased. In winter, the increase is due to reduced storage of precipitation in the snowpack, whereas in summer (typically a low-flow season), streamflow has increased due to increasing precipitation and declining evapotranspiration (which has shown slight but significant declines since 1959). The cause of the decline in forest evapotranspiration is not known but may result from changes in vegetation composition, structure, or productivity, or forest response to increasing atmospheric carbon dioxide concentrations, or other factors (Groffman et al. 2012).

In mixed oak-hickory hardwood forests of the southeastern USA (Coweeta, North Carolina), forest succession has responded in unexpected ways to long-term changes in climate, perhaps reflecting long-term forest responses to burning, grazing, and logging more than one hundred years ago. At Coweeta, air temperature, drought severity, and precipitation extremes have increased since the late 1970s (Laseter et al. 2012). Annual water yield increased by as much as 55% from 1938 to the mid-1970s in some watersheds, which were undergoing forest succession after logging in the early 1900s (Caldwell et al. 2016). However, from the 1970s to 2013, water yield declined by 22%, associated with a shift in dominance from xerophytic oak and hickory tree species to mesophytic tree species including red maple (*Acer rubrum*) and tulip poplar (*Liriodendron tulipifera*) (Caldwell et al. 2016).

Forest vegetation succession provides strong negative feedbacks that, under certain circumstances, make permafrost resilient to large increases in air temperatures. However, as seen in the boreal forests of Alaska, climate warming is associated with reduced growth of dominant tree species, plant disease and insect outbreaks, warming and thawing of permafrost, drying of lakes, increased wildfire extent (and frequency), and increased post-fire recruitment of deciduous trees. These changes have reduced the effects of upland permafrost on regional hydrology (Chapin et al. 2010). Surface water, in contrast, provides positive feedbacks that make permafrost vulnerable to thawing even under cold temperatures (Jorgenson et al. 2010). In watersheds with low permafrost, base flow is higher, and annual water yield varies with summer temperature, whereas in watersheds with high permafrost, annual water yield varies with precipitation (Jones and Rinehart 2010). With climate warming and loss of permafrost, streamflows will become less responsive to precipitation and headwater streams may become ephemeral (Jones and Rinehart 2010).

Beyond the scale of individual catchments, it is, as yet, unclear how climate change will affect atmospheric moisture flows across short and long distances, and at seasonal and interannual time scales, and how these changes will interact with future forests to affect the global hydrologic cycle.

24.3.2 Forest Cover, Forest Management, and Future Forest Hydrology

Clearly, as shown thus far, anticipated future changes in forest cover and forest management are diverse. They may include expansion of intensive plantations, deforestation, selective logging, loss of riparian forest, and loss of urban trees, all of which will affect the hydrologic cycle at multiple temporal and spatial scales. Deforestation has been high, especially in the tropics, since records of global forest cover began (Allen and Barnes 1985). From 2000 to 2012, globally 2.3 million km^2 of forest were lost, and 0.8 million km^2 of new forest were gained (Hansen et al. 2013). Intensive forestry practised within subtropical forests resulted in the highest rates of forest change globally (Hansen et al. 2013; Jones et al. 2017). Boreal forest loss, due largely to fire and forestry, was second in absolute and proportional terms.

As shown earlier, deforestation tends to increase streamflow, and forestation decreases streamflow (Li et al. 2017) even though there are exceptions and cases where forest cover change is not reflected in the hydrological regime, as in a study of four watersheds from the Ethiopian Highlands (Gebrehiwot et al. 2013). But changes in forest cover also extend to other parts of the hydrologic cycle, including evapotranspiration and precipitation recycling. Tropical deforestation results in warmer, drier local conditions (Lawrence and Vandecar 2015; Lovejoy and Nobre 2018). Climate model simulations of Amazonia indicate that deforestation was associated with reduced rainfall (Spracklen and Garcia-Carreras 2015). Forest cover in Amazonia was not correlated with precipitation at the local scale (1–15 km) but was positively correlated with measured precipitation at the regional scale (30–50 km) (Debortoli et al. 2017). Thus, increases in forest cover may also contribute to higher precipitation locally or at broader scales (Ellison et al. 2012, 2017).

24.3.3 Climate Change Effects on Forest Disturbance, Succession, and Future Forest Hydrology

Anticipated future changes in forest disturbance regimes include, as shown earlier, more wildfire, more frequent and intense storms, and spatial changes in insect/pathogen outbreaks. Disturbances from wind, bark beetles, and wildfires have increased in Europe's forests throughout the twentieth century (Schelhaas et al. 2003). For example, the mountain pine beetle outbreak in British Columbia (Canada) produced changes in carbon cycling equivalent to approximately 75% of the average annual direct forest fire emissions from all of Canada during 1959–1999 (Kurz et al. 2008). Climate change is expected to interact with forest disturbance regimes (Dale et al. 2001). Models predict a lengthened fire season and significant increases in the area experiencing high to extreme fire danger in both Canada and Russia (Stocks et al. 1998).

Anticipated future forest succession processes will include changes in forest age, structure, and species composition that may increase or reduce water yield and water storage – some of the complexities of processes involved are shown earlier in this chapter. Hydrologic responses to drought can be either mitigated or exacerbated by forest vegetation depending upon vegetation water use and how forest population dynamics respond to drought (Vose et al. 2016). Tree species differ in canopy- and leaf-level stomatal conductance response to vapour pressure deficit, so ecophysiological differences, as well as structural differences among species, influence evapotranspiration (Ford et al. 2011). Although limited research has, thus far, been conducted on this topic, there is some evidence that species distribution models can usefully predict trends in average ecological conditions, but not species composition, of future forests (Scherrer et al. 2017).

24.3.4 Climate Change and Future Forest Management

Climate change implies an increased need for hydrological (and other) resilience in our landscapes (e.g., limiting floods and withstanding drought) (e.g., Hatcher and Jones 2013). Forest change may reduce floods and resist droughts, but natural disturbance and human disturbance, including ongoing deforestation, as well as ongoing climate change, continually alter forest hydrology. Interannual climate variation has a much greater impact on flow than forest management at the scale in which it normally practiced (Duan et al. 2016). Under such circumstances, forest management for the future should focus on managing in the face of uncertainty while managing for multiple benefits (Millar et al. 2007). Forest hydrology will continue to respond to multiple system drivers of change. Decisions should thus not be based on expected responses to single factors and be informed by what works under which conditions (Lindner et al. 2014). Forests and forest hydrology have experienced varied responses to change over the past 50–100 years; this underscores the importance of incorporating stochastic variability into projections of future forest conditions and forest hydrology (Daniel et al. 2017). Nevertheless, the demonstrated hydrologic resilience and adaptive capacities of native, unmanaged forests argues for the preservation and improved management of native forests. At the same time, in the face of future uncertainty, the most robust managed forests should have both structural and species diversity, for example, by retaining a large proportion of mature trees while using thinning to create spatial heterogeneity and enhance structural and compositional complexity (D'Amato et al. 2011).

24.4 Data Needs and Knowledge Gaps

Despite ongoing global environmental change, knowledge of forest responses to natural and anthropogenic disturbances and their effects on forest hydrology (as described above) is robust (and growing) and highly useful for predicting and managing the effects of current and future forest change. Ongoing and anticipated forest change will involve many of the same disturbance agents, including insects and pathogens, wildfire, wind, and ice, as well as forest harvest and tree planting, whose effects on forest hydrology have been thoroughly studied.

Nevertheless, the nature of forest changes and their effects on hydrology are poorly documented in many areas of the world, and novel combinations of processes and contexts may produce surprising outcomes. There is thus a clear need for more geographically extensive and long-term place-based studies of forest and water. Finally, new techniques and concepts, such as the tracing of evapotranspired water from forests and its recycling as precipitation, deserve greater attention.

Research topics include (but are not limited to) increased basic data on runoff and precipitation in currently ungauged watersheds, studies of the fate and transport of atmospheric water, and its impact on downwind rainfall and water availability, along with improved models and data products. Action is also needed to make forest hydrology data publicly accessible.

As mentioned above, there is a general lack of hydrological and meteorological monitoring, particularly in developing countries (McNulty et al. 2016). Paired watershed experimental studies became key to the development of forest hydrology as a science, a century ago. But these studies are resource intensive and limited in occurrence and scope. Data from paired watershed experiments have been mostly obtained in temperate moist climate zones (e.g., Hibbert 1967; Bosch and Hewlett 1982; Andreassian 2004; Jackson et al. 2005).

Increasingly, studies are based on remotely sensed data and process-based models. But these studies are weakened by a lack of reliable ground-truthed data (Meijninger and Jarmain 2014). However, intermodel comparisons and validation can be informative when databases are available for such studies (Caldwell et al. 2015). The intricate linkages between terrain, climate, forest conditions, disturbances (either natural or human-made), and hydrological processes often prevent adequate transferability of findings from well-studied watersheds to those with limited data. In many cases, empirical models of hydro-ecological interactions cannot be extended outside the site where they were created (Kimmins et al. 2010).

Improved sharing of data relevant to forests and water would greatly enhance research and related policy discussions (Emanuel et al. 2015). Repositories that archive and make freely available data on precipitation and runoff are rare. One valuable example is the Climate and Hydrology Database Projects (CLIMDB/HYDRODB), https://climhy.lternet.edu, supported jointly by the US Forest Service, US Geological Service, and US National Science Foundation's Long-Term Ecological Research Programme.

Uncertainty associated with the transport and fate of water evapotranspired from forests continues to pose a major challenge for science, land management and policy (van Noordwijk et al. 2014). The fraction of local precipitation that is derived from terrestrial sources varies from 0 to nearly 100%, depending on location and distance from upwind oceanic sources. Paired watershed studies, which are limited to usually small catchments, do not account for the water that is evapotranspired and leaves the confines of the basin; thus, these studies are unsuited to testing the hypothesis about the impact of forests on rainfall. Water isotopes are providing insights into the origins of atmospheric water, as well as water residence times in soil, vegetation, rivers, and the atmosphere (Wright et al. 2017; McDonnell et al. 2018). More work is needed to develop and test models of spatial transfers of atmospheric moisture (e.g., Weng et al. 2018). Although, as shown earlier, studies on precipitation recycling have been conducted in the Amazon (e.g., Bagley et al. 2014; Nobre et al. 2014, Spera et al. 2016; Wright et al. 2017), in Africa (Arnault et al. 2016; Dyer et al. 2017; Gebrehiwot et al. 2019), and in Asia (Hua et al. 2016; Kong and Pang 2016), additional studies are needed to clarify how forests, wetlands, and irrigated agriculture influence the fate and transport of evapotranspired water. Mechanistic studies also are needed to link vegetation, evapotranspiration and emission of biological compounds, and precipitation processes.

24.5 Conclusions

Changes in forest (due to both natural disturbance and human activities) affect how incoming precipitation is partitioned between evapotranspiration and streamflow. A recent global assessment found that forest changes explained, on average, 30% of annual streamflow variations (Wei et al. 2018). At a daily, or event, timescale forests with adequate infiltration maintain a high 'flow persistence' that reduces flood risks (van Noordwijk et al. 2017). This study and a range of other findings in forest hydrology reviewed in this chapter show that the relationship of forests and water is critical. It is, however, more difficult to generalise how forests and forest change will affect water in the future. The effects of forests on the hydrologic cycle occur at multiple spatial and temporal scales, and forest succession and disturbance alter the relationships of forests to water over time. Moreover, because water cannot be created or destroyed, local changes of forest-water relationships may be balanced by regional or global changes (Ellison 2018). As a result, the knowledge base to make specific predictions of the consequences of forest change at all relevant spatial and temporal scales is still evolving.

Natural disturbance such as wildfire, insect pests, diseases, and windthrow, among others, can, under certain circumstances, significantly alter a range of watershed processes and ecological functions in forested watersheds. Natural disturbance is part of the dynamics of forest ecosystems, but extreme events (e.g., stand-replacing wildfire or large-scale insect pest outbreaks) may, under certain circumstances, cause undesired ecological and economic consequences.

Human-driven changes to forests include deforestation, forest degradation, forestation, agroforestry, plantation forestry, and urban and peri-urban forestry. Native forests provide more sustained water yield compared to managed forest plantations. Re-establishment of forests may enhance sustained water yield, but effects vary depending on site conditions and may require years to decades to be detectable. The role of forests in hydrological regimes and associated watershed functions varies globally and needs to be well understood to best provide an evidence base for informed management.

Climate change is altering hydrological processes directly and is affecting forests, thereby altering hydrology indirectly. Climate change may also alter forest structure and species composition, which may mitigate or exacerbate direct effects of climate change. Forest dynamics, including legacies of past disturbance and forest management, as well as forest succession, produce a wide range of forest hydrologic response to climate change at individual sites.

Forest management for the future should factor in uncertainty as well as assume that uncertainty is likely to increase. Decisions should be based on expected responses to multiple factors, not single factors, as well as supporting resilience and multiple benefits. Evidence of how forests and forest hydrology have responded to change over the past 50–100 years highlights the need to incorporate stochastic variability into projections of future ecosystem condition and to utilise the precautionary principle in forest management.

Acknowledgements This book chapter is based on the collaborative work of scientific experts who authored the Global Forest Expert Panels (GFEP) assessment on Forests and Water that was published as a report titled 'Forest and Water on a Changing Planet: Vulnerability, Adaptation and Governance Opportunities'. We acknowledge the reviewers of this book chapter for their insightful comments that produced an improved product. We also acknowledge the financial and in-kind support provided by the Ministry for Foreign Affairs of Finland, the US Forest Service, The World Bank Group/PROFAR, and the Austrian Federal Ministry of Sustainability and Tourism.

References

Adams HD, Luce CH, Breshears DD, Allen CD, Weiler M, Hale VC et al (2012) Ecohydrological consequences of drought-and infestation-triggered tree die-off: insights and hypotheses. Ecohydrology 5:145–159. https://doi.org/10.1002/eco.233

Allen JC, Barnes DF (1985) The causes of deforestation in developing countries. Ann Assoc Am Geogr 75:163–184. https://doi.org/10.1111/j.1467-8306.1985.tb00079.x

Amazonas NT, Forrester DI, Oliveira RS, Brancalion PH (2017) Combining Eucalyptus wood production with the recovery of native tree diversity in mixed plantings: Implications for water use and availability. Forest Ecol Manag 418:34–40. https://doi.org/10.1016/j.foreco.2017.12.006

Andela N, Morton DC, Giglio L, Chen Y, van der Werf GR, Kasibhatla PS et al (2017) A human-driven decline in global burned area. Science 356:1356–1362. https://doi.org/10.1126/science.aal4108

Anderegg WR, Kane JM, Anderegg LD (2013) Consequences of widespread tree mortality triggered by drought and temperature stress. Nat Clim Change 3:30–36. https://doi.org/10.1038/nclimate1635

Andréassian V (2004) Waters and forests: from historical controversy to scientific debate. J Hydrol 291:1–27. https://doi.org/10.1016/j.jhydrol.2003.12.015

Arnault J, Knoche R, Wei J, Kunstmann H (2016) Evaporation tagging and atmospheric water budget analysis with WRF: a regional precipitation recycling study for West Africa. Water Resour Res 52:1544–1567. https://doi.org/10.1002/2015WR017704

Attiwill PM (1994) Ecological disturbance and the conservative management of eucalypt forests in Australia. Forest Ecol Manag 63:301–346. https://doi.org/10.1016/0378-1127(94)90115-5

Bagley JE, Desai AR, Harding KJ, Snyder PK, Foley JA (2014) Drought and deforestation: has land cover change influenced recent precipitation extremes in the Amazon? J Clim 27:345–361. https://doi.org/10.1175/JCLI-D-12-00369.1

Bart RR, Tague CL (2017) The impact of wildfire on baseflow recession rates in California. Hydrol Process 31:1662–1673. https://doi.org/10.1002/hyp.11141

Bartens J, Day SD, Harris JR, Dove JE, Wynn TM (2008) Can urban tree roots improve infiltration through compacted subsoils for stormwater management? J Environ Qual 37:2048–2057. https://doi.org/10.2134/jeq2008.0117

Bayala J, Heng LK, van Noordwijk M, Ouedraogo SJ (2008) Hydraulic redistribution study in two native tree species of agroforestry parklands of West African dry savanna. Acta Oecol 34:370–378. https://doi.org/10.1016/j.actao.2008.06.010

Bearup LA, Maxwell RM, Clow DW, McCray JE (2014) Hydrological effects of forest transpiration loss in bark beetle-impacted watersheds. Nat Clim Change 4:481–486. https://doi.org/10.1038/nclimate2198

Biederman JA, Harpold AA, Gochis DJ, Ewers BE, Reed DE, Papuga SA et al (2014) Increased evaporation following widespread tree mortality limits streamflow response. Water Resour Res 50:5395–5409. https://doi.org/10.1002/2013WR014994

Biederman JA, Somor AJ, Harpold AA, Gutmann ED, Breshears DD, Troch PA et al (2015) Recent tree die-off has little effect on streamflow in contrast to expected increases from historical studies. Water Resour Res 51:9775–9789. https://doi.org/10.1002/2015WR017401

Boffa JM (1999) Agroforestry Parklands in sub-Saharan Africa, FAO conservation guide 34. FAO, Rome

Boggs J, Sun G, Jones D, McNulty SG (2013) Effect of soils on water quantity and quality in Piedmont forested headwater watersheds of North Carolina. J Am Water Resour Assoc 49:132–150. https://doi.org/10.1111/jawr.12001

Boggs J, Sun G, McNulty S (2015) Effects of timber harvest on water quantity and quality in small watersheds in the Piedmont of North Carolina. J For 114:27–40. https://doi.org/10.5849/jof.14-102

Bond-Lamberty B, Peckham SD, Gower ST, Ewers BE (2009) Effects of fire on regional evapotranspiration in the central Canadian boreal forest. Global Change Biol 15:1242–1254. https://doi.org/10.1111/j.1365-2486.2008.01776.x

Boose ER, Foster DR, Fluet M (1994) Hurricane impacts to tropical and temperate forest landscapes. Ecol Monogr 64:369–400. https://doi.org/10.2307/2937142

Bormann FH, Likens GE (1979) Catastrophic disturbance and the steady state in northern hardwood forests: a new look at the role of disturbance in the development of forest ecosystems suggests important implications for land-use policies. Am Sci 67:660–669

Bosch JM, Hewlett JD (1982) A review of catchment experiments to determine the effect of vegetation changes on water yield and evapotranspiration. J Hydrol 55:3–23. https://doi.org/10.1016/0022-1694(82)90117-2

Bounoua L, Zhang P, Mostovoy G, Thome K, Masek J, Imhoff M et al (2015) Impact of urbanization on US surface climate. Environ Res Lett 10:084010. https://doi.org/10.1088/1748-9326/10/8/084010

Brandt L, Lewis AD, Fahey R, Scott L, Darling L, Swanston C (2016) A framework for adapting urban forests to climate change. Environ Sci Policy 66:393–402. https://doi.org/10.1016/j.envsci.2016.06.005

Bréda N, Granier A, Aussenac G (1995) Effects of thinning on soil and tree water relations, transpiration and growth in an oak forest (*Quercus petraea* (Matt.) Liebl.). Tree Physiol 15:295–306. https://doi.org/10.1093/treephys/15.5.295

Bréda N, Huc R, Granier A, Dreyer E (2006) Temperate forest trees and stands under severe drought: a review of ecophysiological responses, adaptation processes and long-term consequences. Ann For Sci 63:625–644. https://doi.org/10.1051/forest:2006042

Brockerhoff EG, Jactel H, Parrotta JA, Quine CP, Sayer J (2008) Plantation forests and biodiversity: oxymoron or opportunity? Biodivers Conserv 17:925–951. https://doi.org/10.1007/s10531-008-9380-x

Brown AE, Zhang L, McMahon TA, Western AW, Vertessy RA (2005) A review of paired catchment studies for determining changes in water yield resulting from alterations in vegetation. J Hydrol 310:28–61. https://doi.org/10.1016/j.jhydrol.2004.12.010

Burt TP, Howden NJK, McDonnell JJ, Jones JA, Hancock GR (2015) Seeing the climate through the trees: observing climate and forestry impacts on streamflow using a 60-year record. Hydrol Process 29:473–480. https://doi.org/10.1002/hyp.10406

Calder IR, Hofer T, Vermont S, Warren P (2007) Towards a new understanding of forests and water. Unasylva No 229, vol 58. Food and Agriculture Organization of the United. Nations, Rome, pp 3–10
Caldwell PV, Kennen JG, Sun G, Kiang JE, Butcher JB, Eddy MC et al (2015) A comparison of hydrologic models for ecological flows and water availability. Ecohydrology 8:1525–1546. https://doi.org/10.1002/eco.1602
Caldwell PV, Miniat CF, Elliott KJ, Swank WT, Brantley ST, Laseter SH (2016) Declining water yield from forested mountain watersheds in response to climate change and forest mesophication. Global Change Biol 22:2997–3012. https://doi.org/10.1111/gcb.13309
Campbell JL, Driscoll CT, Pourmokhtarian A, Hayhoe K (2011) Streamflow responses to past and projected future changes in climate at the Hubbard Brook Experimental Forest, New Hampshire, United States. Water Resour Res 47:W02514. https://doi.org/10.1029/2010WR009438
Canham CD, Loucks OL (1984) Catastrophic windthrow in the presettlement forests of Wisconsin. Ecology 65:803–809. https://doi.org/10.2307/1938053
Cao S, Sun G, Zhang Z, Chen L, Feng Q, Fu B et al (2011) Greening China naturally. Ambio 40:828–831. https://doi.org/10.1007/s13280-011-0150-8
Cardil A, Imbert JB, Camarero JJ, Primicia I, Castillo F (2018) Temporal interactions among throughfall, type of canopy and thinning drive radial growth in an Iberian mixed pine-beech forest. Agr Forest Meteorol 252:62–74. https://doi.org/10.1016/j.agrformet.2018.01.004
Carlton GC, Bazzaz FA (1998) Resource congruence and forest regeneration following an experimental hurricane blowdown. Ecology 79:1305–1319. https://doi.org/10.1890/0012-9658(1998)079[1305,RCAFRF]2.0.CO;2
Carnicer J, Coll M, Ninyerola M, Pons X, Sanchez G, Penuelas J (2011) Widespread crown condition decline, food web disruption, and amplified tree mortality with increased climate change-type drought. Proc Natl Acad Sci USA 108:1474–1478. https://doi.org/10.1073/pnas.1010070108
Cesarano G, Incerti G, Bonanomi G (2016) The influence of plant litter on soil water repellency: Insight from 13C NMR spectroscopy. PLoS One 11:e0152565. https://doi.org/10.1371/journal.pone.0152565
Chang CR, Li MH (2014) Effects of urban parks on the local urban thermal environment. Urban For Urban Green 13:672–681. https://doi.org/10.1016/j.ufug.2014.08.001
Chapin FS, McGuire AD, Ruess RW, Hollingsworth TN, Mack MC, Johnstone JF et al (2010) Resilience of Alaska's boreal forest to climatic change. Can J For Res 40:1360–1370. https://doi.org/10.1139/X10-074
Clark KL, Skowronski NS, Gallagher MR, Renninger H, Schäfer KVR (2014) Contrasting effects of invasive insects and fire on ecosystem water use efficiency. Biogeosciences 11:6509–6523. https://doi.org/10.5194/bg-11-6509-2014
Cleland HF (1910) The effects of deforestation in New England. Science 32:82–83. https://doi.org/10.1126/science.32.811.82
Collins R, de Neufville R, Claro J, Oliveira T, Pacheco A (2013) Forest fire management to avoid unintended consequences: a case study of Portugal using system dynamics. J Environ Manag 130:1–9. https://doi.org/10.1016/j.jenvman.2013.08.033
Cook BI, Smerdon JE, Seager R, Coats S (2014) Global warming and 21st century drying. Clim Dynam 43:2607–2627. https://doi.org/10.1007/s00382-014-2075-y
Cooper LA, Ballantyne AP, Holden ZA, Landguth EL (2017) Disturbance impacts on land surface temperature and gross primary productivity in the western United States. J Geophys Res-Biogeosci 122:930–946. https://doi.org/10.1002/2016JG003622
Creed IF, van Noordwijk M (2018) Forests, trees and water on a changing planet: a comtemporary scientific perspective. In: Creed IF, van Noordwijk M (eds) Forests and water on a changing planet: vulnerability, adaptation, and governance opportunities, IUFRO World Series, vol 38. International Union of Forest Research Organizations [IUFRO], Vienna. 192p
Creed IF, Spargo AT, Jones JA, Buttle JM, Adams MB, Beall FD et al (2014) Changing forest water yields in response to climate warming: Results from long-term experimental watershed sites across North America. Global Change Biol 20:3191–3208. https://doi.org/10.1111/gcb.12615

Cristiano PM, Campanello PI, Bucci SJ, Rodriguez SA, Lezcano OA, Scholz FG et al (2015) Evapotranspiration of subtropical forests and tree plantations: a comparative analysis at different temporal and spatial scales. Agr Forest Meteorol 203:96–106. https://doi.org/10.1016/j.agrformet.2015.01.007

D'Amato AW, Bradford JB, Fraver S, Palik BJ (2011) Forest management for mitigation and adaptation to climate change: insights from long-term silviculture experiments. Forest Ecol Manag 262:803–816. https://doi.org/10.1016/j.foreco.2011.05.014

Dale VH, Joyce LA, McNulty S, Neilson RP, Ayres MP, Flannigan MD et al (2001) Climate change and forest disturbances. BioScience 51:723–734. https://doi.org/10.1641/0006-3568(2001)051[0723,CCAFD]2.0.CO;2

Daniel CJ, Ter-Mikaelian MT, Wotton BM, Rayfield B, Fortin MJ (2017) Incorporating uncertainty into forest management planning: timber harvest, wildfire and climate change in the boreal forest. Forest Ecol Manag 400:542–554. https://doi.org/10.1016/j.foreco.2017.06.039

de Groot M, Ogris N, Kobler A (2018) The effects of a large-scale ice storm event on the drivers of bark beetle outbreaks and associated management practices. Forest Ecol Manag 408:195–201. https://doi.org/10.1016/j.foreco.2017.10.035

Debortoli NS, Dubreuil V, Hirota M, Lindoso DP, Nabucet J (2017) Detecting deforestation impacts in Southern Amazonia rainfall using rain gauges. Int J Climatol 37:2889–2900. https://doi.org/10.1002/joc.4886

Dhar A, Parrott L, Hawkins CD (2016a) Aftermath of mountain pine beetle outbreak in British Columbia: stand dynamics, management response and ecosystem resilience. Forests 7:171. https://doi.org/10.3390/f7080171

Dhar A, Parrott L, Heckbert S (2016b) Consequences of mountain pine beetle outbreak on forest ecosystem services in western Canada. Can J For Res 46:987–999. https://doi.org/10.1139/cjfr-2016-0137

Ding Y, Wang Z, Song Y, Zhang J (2008) Causes of the unprecedented freezing disaster in January 2008 and its possible association with the global warming (in Chinese). Acta Meteorol Sin 4: 14 p

Doerr SH, Shakesby RA, Walsh RP (1996) Soil hydrophobicity variations with depth and particle size fraction in burned and unburned *Eucalyptus globulus* and *Pinus pinaster* forest terrain in the Agueda Basin, Portugal. Catena 27:25–47. https://doi.org/10.1016/0341-8162(96)00007-0

Doughty CE, Malhi Y, Araujo-Murakami A, Metcalfe DB, Silva-Espejo JE, Arroyo L et al (2014) Allocation trade-offs dominate the response of tropical forest growth to seasonal and interannual drought. Ecology 95:2192–2201. https://doi.org/10.1890/13-1507.1

Du Toit B, Gush MB, Pryke JS, Samways MJ, Dovey SB (2014) Ecological impacts of biomass production at stand and landscape levels. In: Seifert T (ed) Bioenergy from wood: sustainable production in the tropics, Managing forest ecosystems, vol 26. Springer, Heidelberg, pp 211–236. https://doi.org/10.1007/978-94-007-7448-3_10

Duan K, Sun G, Sun S, Caldwell PV, Cohen EC, McNulty SG, Aldridge HD, Zhang Y (2016) Divergence of ecosystem services in U.S. National Forests and grasslands under a changing climate. Sci Rep-UK 6:24441

Duinker PN, Ordóñez C, Steenberg JWN, Miller KH, Toni SA, Nitoslawski SA (2015) Trees in Canadian cities: an indispensable life form for urban sustainability. Sustainability 7:7379–7396. https://doi.org/10.3390/su7067379

Dunn CJ, Bailey JD (2016) Tree mortality and structural change following mixed-severity fire in Pseudotsuga forests of Oregon's western Cascades, USA. Forest Ecol Manag 365:107–118. https://doi.org/10.1016/j.foreco.2016.01.031

Dye PJ, Bosch JM (2000) Sustained water yield in afforested catchments – the South African experience. In: von Gadow K, Pukkala T, Tomé M (eds) Sustainable forest management. Kluwer Academic Publishers, Dordrecht, pp 99–120. https://doi.org/10.1007/978-94-010-9819-9_3

Dye PJ, Versfeld D (2007) Managing the hydrological impacts of South African plantation forests: an overview. Forest Ecol Manag 251:121–158. https://doi.org/10.1016/j.foreco.2007.06.013

Dyer EL, Jones DB, Nusbaumer J, Li H, Collins O, Vettoretti G, Noone D (2017) Congo Basin precipitation: assessing seasonality, regional interactions, and sources of moisture. J Geophys Res- Atmos 122:6882–6898. https://doi.org/10.1002/2016JD026240

Dzikiti S, Gush MB, Le Maitre DC, Maherry A, Jovanovic NJ, Ramoelo A et al (2016) Quantifying potential water savings from clearing invasive alien *Eucalyptus camaldulensis* using in situ and high resolution remote sensing data in the Berg River Catchment, Western Cape, South Africa. Forest Ecol Manag 361:69–80. https://doi.org/10.1016/j.foreco.2015.11.009

Ellison D (2018) From myth to concept and beyond – the biogeophysical revolution and the forest-water paradigm, UNFF13 Background Analytical Study 2 on Forests and Water, report commissioned for the thirteenth session of the United Nations Forum on Forests, April

Ellison D, Futter MN, Bishop K (2012) On the forest cover–water yield debate: from demand-to supply-side thinking. Global Change Biol 18:806–820. https://doi.org/10.1111/j.1365-2486.2011.02589.x

Ellison D, Morris CE, Locatelli B, Sheil D, Cohen J, Murdiyarso D et al (2017) Trees, forests and water: cool insights for a hot world. Global Environ Change 43:51–61. https://doi.org/10.1016/j.gloenvcha.2017.01.002

Emanuel RE, Buckley JJ, Caldwell PV, McNulty SG, Sun G (2015) Influence of basin characteristics on the effectiveness and downstream reach of interbasin water transfers: displacing a problem. Environ Res Lett 10:124005. https://doi.org/10.1088/1748-9326/10/12/124005

FAO (2015) Global forest resources assessment 2015, FAO forestry paper no. 1. FAO, Rome

Farley KA, Jobbágy EG, Jackson RB (2005) Effects of afforestation on water yield: a global synthesis with implications for policy. Global Change Biol 11:1565–1576. https://doi.org/10.1111/j.1365-2486.2005.01011.x

Filoso S, Bezerra MO, Weiss KC, Palmer MA (2017) Impacts of forest restoration on water yield: a systematic review. PloS One 12:e0183210. https://doi.org/10.1371/journal.pone.0183210

Ford CR, Hubbard RM, Vose JM (2011) Quantifying structural and physiological controls on variation in canopy transpiration among planted pine and hardwood species in the southern Appalachians. Ecohydrology 4:183–195. https://doi.org/10.1002/eco.136

Forrester DI (2015) Transpiration and water-use efficiency in mixed-species forests versus monocultures: effects of tree size, stand density and season. Tree Physiol 35:289–304. https://doi.org/10.1093/treephys/tpv011

Foster DR (1988) Species and stand response to catastrophic wind in central New England, USA. J Ecol 76:135–151. https://doi.org/10.2307/2260458

Foster DR, Boose ER (1992) Patterns of forest damage resulting from catastrophic wind in central New England, USA. J Ecol 80:79–98. https://doi.org/10.2307/2261065

Foster DR, Knight DH, Franklin JF (1998) Landscape patterns and legacies resulting from large, infrequent forest disturbances. Ecosystems 1:497–510. https://doi.org/10.1007/s100219900046

Fox TR, Jokela EJ, Allen HL (2004) The evolution of pine plantation silviculture in the southern United States. In: Gen. Tech. Rep. SRS 75. US Department of Agriculture, Forest Service, Southern Research Station, Asheville

Gebrehiwot SG, Seibert J, Gardenas AI, Mellander P-E, Bishop K (2013) Hydrological change detection using modeling: Half a century of runoff from four rivers in the Blue Nile Basin. Water Resour Res 49:3842–3851. https://doi.org/10.1002/wrcr.20319

Gebrehiwot SG, Ellison D, Bewket W, Seleshi Y, Inogwabini BI, Bishop K (2019) The Nile Basin waters and the West African rainforest: rethinking the boundaries. WIRES Water 6:e1317. https://doi.org/10.1002/wat2.1317

Gedney N, Cox PM, Betts RA, Boucher O, Huntingford C, Stott PA (2006) Detection of a direct carbon dioxide effect in continental river runoff records. Nature 439:835–838. https://doi.org/10.1038/nature04504

Gimbel KF (2016) Does drought alter hydrological functions in forest soils? Hydro Earth Syst Sci 20:1301–1317. https://doi.org/10.5194/hess-20-1301-2016

Granier A, Loustau D, Bréda N (2000) A generic model of forest canopy conductance dependent on climate, soil water availability and leaf area index. Ann For Sci 57:755–765. https://doi.org/10.1051/forest:2000158

Grimm NB, Faeth SH, Golubiewski NE, Redman CL, Wu J, Bai X et al (2008) Global change and the ecology of cities. Science 319:756–760. https://doi.org/10.1126/science.1150195

Groffman PM, Rustad LE, Templer PH, Campbell JL, Christenson LM, Lany NK et al (2012) Long-term integrated studies show complex and surprising effects of climate change in the northern hardwood forest. BioScience 62:1056–1066. https://doi.org/10.1525/bio.2012.62.12.7

Guardiola-Claramonte M, Troch PA, Ziegler AD, Giambelluca TW, Durcik M, Vogler JB et al (2010) Hydrologic effects of the expansion of rubber (Hevea brasiliensis) in a tropical catchment. Ecohydrology 3:306–314. https://doi.org/10.1002/eco.110

Gush MB (2010) Assessing hydrological impacts of tree-based bioenergy feedstock. In: Amezaga JM, von Maltitz G, Boyes S (eds) Assessing the sustainability of bioenergy projects in developing countries: a framework for policy evaluation. Newcastle University Press, Newcastle, pp 37–52

Gush MB, Scott DF, Jewitt GPW, Schulze RE, Hallowes LA, Gorgens AHM (2002) A new approach to modelling streamflow reductions resulting from commercial afforestation in South Africa. S Afr For J 196:27–36. https://doi.org/10.1080/20702620.2002.10434615

Gyenge J, Fernández ME, Sarasola M, Schlichter T (2008) Testing a hypothesis of the relationship between productivity and water use efficiency in Patagonian forests with native and exotic species. Forest Ecol Manag 255:3281–3287. https://doi.org/10.1016/j.foreco.2008.01.078

Hallema DW, Sun G, Bladon KD, Norman SP, Caldwell PV, Liu Y et al (2017) Regional patterns of postwildfire streamflow response in the Western United States: the importance of scale-specific connectivity. Hydrol Process 31:2582–2598. https://doi.org/10.1002/hyp.11208

Hallema DW, Sun G, Caldwell PV, Norman SP, Cohen EC, Liu Y et al (2018) Burned forests impact water supplies. Nat Commun 9:1307. https://doi.org/10.1038/s41467-018-03735-6

Hamburg SP, Vadeboncoeur MA, Richardson AD, Bailey AS (2013) Climate change at the ecosystem scale: a 50-year record in New Hampshire. Clim Change 116:457–477. https://doi.org/10.1007/s10584-012-0517-2

Hansen MC, Potapov PV, Moore R, Hancher M, Turubanova S, Tyukavina A et al (2013) High-resolution global maps of 21st-century forest cover change. Science 342:850–853. https://doi.org/10.1126/science.1244693

Harmon ME, Ferrell WK, Franklin JF (1990) Effects on carbon storage of conversion of old-growth forests to young forests. Science 247:699–702. https://doi.org/10.1126/science.247.4943.699

Harr RD (1986) Effects of clearcutting on rain-on-snow runoff in Western Oregon: a new look at old studies. Water Resour Res 22:1095–1100. https://doi.org/10.1029/WR022i007p01095

Hatcher KL, Jones JA (2013) Climate and streamflow trends in the Columbia River Basin: evidence for ecological and engineering resilience to climate change. Atmos Ocean 51:436–455. https://doi.org/10.1080/07055900.2013.808167

Hauer RJ, Hruska MC, Dawson JO (1994) Trees and ice storms: The development of ice storm–resistant urban tree populations. Special Publication 94-1. Department of Forestry, University of Illinois at Urbana-Champaign, Urbana

Henry JD, Swan JMA (1974) Reconstructing forest history from live and dead plant material--an approach to the study of forest succession in southwest New Hampshire. Ecology 55:772–783. https://doi.org/10.2307/1934413

Hibbert AR (1965) Forest treatment effects on water yield. Coweeta Hydrologic Laboratory, Southeastern Forest Experiment Station, Asheville

Hibbert AR (1967) Forest treatment effects on water yield. In: Sopper WE, Lul HW (eds) International symposium of Forest hydrology. Pergamon Press, Oxford, pp 527–543

Hickler T, Smith B, Prentice IC, Mjöfors K, Miller P, Arneth A et al (2008) CO_2 fertilization in temperate FACE experiments not representative of boreal and tropical forests. Glob Change Biol 14:1531–1542. https://doi.org/10.1111/j.1365-2486.2008.01598.x

Hinzman LD, Bettez ND, Bolton WR, Chapin FS, Dyurgerov MB, Fastie CL et al (2005) Evidence and implications of recent climate change in northern Alaska and other arctic regions. Clim Change 72:251–298. https://doi.org/10.1007/s10584-005-5352-2

Holtum JA, Winter K (2010) Elevated [CO_2] and forest vegetation: more a water issue than a carbon issue? Funct Plant Biol 37:694–702. https://doi.org/10.1071/FP10001

Houlton BZ, Driscoll CT (2011) The effects of ice storms on the hydrology and biogeochemistry of forests. In: Levia DF, Carlyle-Moses D, Tanaka T (eds) Forest hydrology and biogeochemistry. Springer Science & Business Media, Heidelberg, pp 623–641. https://doi.org/10.1007/978-94-007-1363-5_31

Hua L, Zhong L, Ke Z (2016) Precipitation recycling and soil–precipitation interaction across the arid and semi-arid regions of China. Int J Climatol 36:3708–3722. https://doi.org/10.1002/joc.4586

Huntington TG (2006) Evidence for intensification of the global water cycle: review and synthesis. J Hydrol 319:83–95. https://doi.org/10.1016/j.jhydrol.2005.07.003

Ilstedt U, Tobella AB, Bazié HR, Bayala J, Verbeeten E, Nyberg G et al (2016) Intermediate tree cover can maximize groundwater recharge in the seasonally dry tropics. Sci Rep 6:21930. https://doi.org/10.1038/srep21930

Irland LC (2000) Ice storms and forest impacts. Sci Total Environ 262:231–242. https://doi.org/10.1016/S0048-9697(00)00525-8

Isaacs RE, Stueve KM, Lafon CW, Taylor AH (2014) Ice storms generate spatially heterogeneous damage patterns at the watershed scale in forested landscapes. Ecosphere 5:1–14. https://doi.org/10.1890/ES14-00234.1

IUFRO (2018) Global fire challenges in a warming world. In: Robinne F-N, Burns J, Kant P, de Groot B, Flannigan MD, Kleine M, Wotton DM (eds) Occasional paper no. 32. International Union of Forest Research Organizations (IUFRO), Vienna. https://www.iufro.org/news/article/2019/01/23/occasional-paper-32-global-fire-challenges-in-a-warming-world/

Iverson LR, Hutchinson TF, Peters MP, Yaussy DA (2017) Long-term response of oak-hickory regeneration to partial harvest and repeated fires: influence of light and moisture. Ecosphere 8: e01642. https://doi.org/10.1002/ecs2.1642

Jackson RB, Jobbágy EG, Avissar R, Roy SB, Barrett DJ, Cook CW et al (2005) Trading water for carbon with biological carbon sequestration. Science 310:1944–1947. https://doi.org/10.1126/science.1119282

Jones JA (2000) Hydrologic processes and peak discharge response to forest removal, regrowth, and roads in 10 small experimental basins, western Cascades, Oregon. Water Resour Res 36:2621–2642. https://doi.org/10.1029/2000WR900105

Jones JA (2011) Hydrologic responses to climate change: considering geographic context and alternative hypotheses. Hydrol Process 25:1996–2000. https://doi.org/10.1002/hyp.8004

Jones JA, Grant GE (1996) Peak flow responses to clear-cutting and roads in small and large basins, western Cascades, Oregon. Water Resour Res 32:959–974. https://doi.org/10.1029/95WR03493

Jones JA, Perkins RM (2010) Extreme flood sensitivity to snow and forest harvest, western Cascades, Oregon, United States. Water Resour Res 46:W12512. https://doi.org/10.1029/2009WR008632

Jones JA, Post DA (2004) Seasonal and successional streamflow response to forest cutting and regrowth in the northwest and eastern United States. Water Resour Res 40:W05203. https://doi.org/10.1029/2003WR002952

Jones JB, Rinehart AJ (2010) The long-term response of stream flow to climatic warming in headwater streams of interior Alaska. Can J For Res 40:1210–1218. https://doi.org/10.1139/X10-047

Jones JA, Achterman GL, Augustine LA, Creed IF, Ffolliott PF, MacDonald L et al (2009) Hydrologic effects of a changing forested landscape—challenges for the hydrological sciences. Hydrol Process 23:2699–2704. https://doi.org/10.1002/hyp.7404

Jones JA, Creed IF, Hatcher KL, Warren RJ, Adams MB, Benson MH et al (2012) Ecosystem processes and human influences regulate streamflow response to climate change at long-term ecological research sites. BioScience 62:390–404. https://doi.org/10.1525/bio.2012.62.4.10

Jones JA, Almeida A, Cisneros F, Iroumé A, Jobbágy E, Lara A et al (2017) Forests and water in South America. Hydrol Process 31:972–980. https://doi.org/10.1002/hyp.11035

Jorgenson MT, Romanovsky V, Harden J, Shur Y, O'Donnell J, Schuur EA et al (2010) Resilience and vulnerability of permafrost to climate change. Can J For Res 40:1219–1236. https://doi.org/10.1139/X10-060

Jung M, Reichstein M, Ciais P, Seneviratne SI, Sheffield J, Goulden ML et al (2010) Recent decline in the global land evapotranspiration trend due to limited moisture supply. Nature 467:951–954. https://doi.org/10.1038/nature09396

Kerhoulas LP, Kolb TE, Koch GW (2013) Tree size, stand density, and the source of water used across seasons by ponderosa pine in northern Arizona. Forest Ecol Manag 289:425–433. https://doi.org/10.1016/j.foreco.2012.10.036

Keys PW, Wang-Erlandsson L, Gordon LJ (2016) Revealing invisible water: moisture recycling as an ecosystem service. PloS One 11:e0151993. https://doi.org/10.1371/journal.pone.0151993

Kim J, Hwang T, Schaaf CL, Orwig DA, Boose E, Munger JW (2017) Increased water yield due to the hemlock woolly adelgid infestation in New England. Geophys Res Lett 44:2327–2335. https://doi.org/10.1002/2016GL072327

Kimmins JP, Blanco JA, Seely B, Welham C, Scoullar K (2010) Forecasting forest futures: a hybrid modelling approach to the assessment of sustainability of forest ecosystems and their values. Earthscan, London

Kinoshita AM, Hogue TS (2011) Spatial and temporal controls on post-fire hydrologic recovery in Southern California watersheds. Catena 87:240–252. https://doi.org/10.1016/j.catena.2011.06.005

Knapp AK, Beier C, Briske DD, Classen AT, Luo Y, Reichstein M et al (2008) Consequences of more extreme precipitation regimes for terrestrial ecosystems. BioScience 58:811–821. https://doi.org/10.1641/B580908

Kong Y, Pang Z (2016) A positive altitude gradient of isotopes in the precipitation over the Tianshan Mountains: effects of moisture recycling and sub-cloud evaporation. J Hydrol 542:222–230. https://doi.org/10.1016/j.jhydrol.2016.09.007

Kopp BJ, Lange J, Menzel L (2017) Effects of wildfire on runoff generating processes in northern Mongolia. Reg Environ Change 17:1951–1963. https://doi.org/10.1007/s10113-016-0962-y

Kramer MG, Hansen AJ, Taper ML, Kissinger EJ (2001) Abiotic controls on long-term windthrow disturbance and temperate rain forest dynamics in southeast Alaska. Ecology 82:2749–2768. https://doi.org/10.1890/0012-9658(2001)082[2749,ACOLTW]2.0.CO;2

Kurz WA, Dymond CC, Stinson G, Rampley GJ, Neilson ET, Carroll AL et al (2008) Mountain pine beetle and forest carbon feedback to climate change. Nature 452:987–990. https://doi.org/10.1038/nature06777

La Marche JL, Lettenmaier DP (2001) Effects of forest roads on flood flows in the Deschutes River, Washington. Earth Surf Process Land 26:115–134. https://doi.org/10.1002/1096-9837(200102)26:2<115::AID-ESP166>3.0.CO;2-O

Laseter SH, Ford CR, Vose JM, Swift LW (2012) Long-term temperature and precipitation trends at the Coweeta Hydrologic Laboratory, Otto, North Carolina, USA. Hydrol Res 43:890–901. https://doi.org/10.2166/nh.2012.067

Lawrence D, Vandecar K (2015) Effects of tropical deforestation on climate and agriculture. Nat Clim Change 5:27–36. https://doi.org/10.1038/nclimate2430

Le Maitre DC, Gush MB, Dzikiti S (2015) Impacts of invading alien plant species on water flows at stand and catchment scales. AoB Plants 7:plv043. https://doi.org/10.1093/aobpla/plv043

Lechuga V, Carraro V, Viñegla B, Carreira JA, Linares JC (2017) Managing drought-sensitive forests under global change. Low competition enhances long-term growth and water uptake in *Abies pinsapo*. Forest Ecol Manag 406:72–82. https://doi.org/10.1016/j.foreco.2017.10.017

Lertzman K, Spies T, Swanson F (1997) From ecosystem dynamics to ecosystem management. In: Schoonmaker PK, von Hagen B, Wolf EC (eds) The rain forests of home: profile of a North American bioregion. Island Press, Washington, DC

Li Q, Wei X, Zhang M, Liu W, Fan H, Zhou G et al (2017) Forest cover change and water yield in large forested watersheds: a global synthetic assessment. Ecohydrology 10:e1838. https://doi.org/10.1002/eco.1838

Li Q, Wei X, Zhang M, Liu W, Giles-Hansen K, Wang Y (2018) The cumulative effects of forest disturbance and climate variability on streamflow components in a large forest-dominated watershed. J Hydrol 557:448–459. https://doi.org/10.1016/j.jhydrol.2017.12.056

Lin Y, Wei X (2008) The impact of large-scale forest harvesting on hydrology in the Willow Watershed of Central British Columbia. J Hydrol 359:141–149. https://doi.org/10.1016/j.jhydrol.2008.06.023
Lindner M, Maroschek M, Netherer S, Kremer A, Barbati A, Garcia-Gonzalo J et al (2010) Climate change impacts, adaptive capacity, and vulnerability of European forest ecosystems. Forest Ecol Manag 259:698–709. https://doi.org/10.1016/j.foreco.2009.09.023
Lindner M, Fitzgerald JB, Zimmermann NE, Reyer C, Delzon S, van der Maaten E et al (2014) Climate change and European forests: what do we know, what are the uncertainties, and what are the implications for forest management? J Environ Manag 146:69–83. https://doi.org/10.1016/j.jenvman.2014.07.030
Liu W, Wei X, Liu S, Liu Y, Fan H, Zhang M et al (2015) How do climate and forest changes affect long-term streamflow dynamics? a case study in the upper reach of Poyang River basin. Ecohydrology 8:46–57. https://doi.org/10.1002/eco.1486
Lorimer CG (1977) The presettlement forest and natural disturbance cycle of northeastern Maine. Ecology 58:139–148. https://doi.org/10.2307/1935115
Lovejoy TE, Nobre C (2018) Amazon tipping point. Sci Adv 4:eaat2340. https://doi.org/10.1126/sciadv.aat2340
Lundmark T, Poudel BC, Stål G, Nordin A, Sonesson J (2018) Carbon balance in production forestry in relation to rotation length. Can J For Res 48:672–678. https://doi.org/10.1139/cjfr-2017-0410
Luyssaert S, Schulze ED, Börner A, Knohl A, Hessenmöller D, Law BE et al (2008) Old-growth forests as global carbon sinks. Nature 455:213–215. https://doi.org/10.1038/nature07276
Ma X, Xu J, Luo Y, Prasad Aggarwal S, Li J (2009) Response of hydrological processes to land-cover and climate changes in Kejie watershed, south-west China. Hydrol Process 23:1179–1191. https://doi.org/10.1002/hyp.7233
Mansourian S, Stanturf JA, Derkyi MAA, Engel VL (2017) Forest Landscape Restoration: increasing the positive impacts of forest restoration or simply the area under tree cover? Restor Ecol 25:178–183. https://doi.org/10.1111/rec.12489
McDonnell JJ, Evaristo J, Bladon KD, Buttle J, Creed IF, Dymond SF et al (2018) Water sustainability and watershed storage. Nat Sustain 1:378–379. https://doi.org/10.1038/s41893-018-0099-8
McDowell NG, Allen CD (2015) Darcy's law predicts widespread forest mortality under climate warming. Nat Clim Change 5:669–672. https://doi.org/10.1038/nclimate2641
McNulty SG, Boggs JL, Sun G (2014) The rise of the mediocre forest: why chronically stressed trees may better survive extreme episodic climate variability. New Forest 45:403–415. https://doi.org/10.1007/s11056-014-9410-3
McNulty S, Cohen E, Sun G, Caldwell P (2016) Hydrologic modeling for water resource assessment in a developing country: the Rwanda case study. In: Lachassagne P, Lafforgue M (eds) Forest and the water cycle: quantity, quality, management. Cambridge Scholars Publishing, Newcastle upon Tyne, pp 181–203
McNulty SG, Boggs JL, Aber JD, Rustad LE (2017) Spruce-fir forest changes during a 30-year nitrogen saturation experiment. Sci Total Environ 605–606:376–390. https://doi.org/10.1016/j.scitotenv.2017.06.147
McVicar TR, Roderick ML, Donohue RJ, Li LT, Van Niel TG, Thomas A et al (2012) Global review and synthesis of trends in observed terrestrial near-surface wind speeds: implications for evaporation. J Hydrol 416:182–205. https://doi.org/10.1016/j.jhydrol.2011.10.024
MEA [Millennium Ecosystem Assessment] (2005) Ecosystems and human wellbeing: framework for assessment. Island Press, Washington, DC
Meijninger WML, Jarmain C (2014) Satellite-based annual evaporation estimates of invasive alien plant species and native vegetation in South Africa. Water SA 40:95–107. https://doi.org/10.4314/wsa.v40i1.12
Millar CI, Stephenson NL, Stephens SL (2007) Climate change and forests of the future: managing in the face of uncertainty. Ecol Appl 17:2145–2151. https://doi.org/10.1890/06-1715.1

Mitchell RG, Waring RH, Pitman GB (1983) Thinning lodgepole pine increases tree vigor and resistance to mountain pine beetle. For Sci 29:204–211. https://doi.org/10.1093/forestscience/29.1.204

Montes-Helu MC, Kolb T, Dore S, Sullivan B, Hart SC, Koch G et al (2009) Persistent effects of fire-induced vegetation change on energy partitioning and evapotranspiration in ponderosa pine forests. Agric Forest Meteorol 149:491–500. https://doi.org/10.1016/j.agrformet.2008.09.011

Moore RD, Wondzell SM (2005) Physical hydrology and the effects of forest harvesting in the Pacific Northwest: a review. J Am Water Resour Assoc 41:763–784. https://doi.org/10.1111/j.1752-1688.2005.tb03770.x

Moore GW, Bond BJ, Jones JA, Phillips N, Meinzer FC (2004) Structural and compositional controls on transpiration in 40-and 450-year-old riparian forests in western Oregon, USA. Tree Physiol 24:481–491. https://doi.org/10.1093/treephys/24.5.481

Moore GW, Bond BJ, Jones JA (2011a) A comparison of annual transpiration and productivity in monoculture and mixed-species Douglas-fir and red alder stands. For Ecol Manag 262:2263–2270. https://doi.org/10.1016/j.foreco.2011.08.018

Moore GW, Jones JA, Bond BJ (2011b) How soil moisture mediates the influence of transpiration on streamflow at hourly to interannual scales in a forested catchment. Hydrol Process 25:3701–3710. https://doi.org/10.1002/hyp.8095

Mutegi JK, Mugendi DN, Verchot LV, Kung'u JB (2008) Combining napier grass with leguminous shrubs in contour hedgerows controls soil erosion without competing with crops. Agroforest Syst 74:37–49. https://doi.org/10.1007/s10457-008-9152-3

Naranjo JB, Weiler M, Stahl K (2011) Sensitivity of a data-driven soil water balance model to estimate summer evapotranspiration along a forest chronosequence. Hydrol Earth Syst Sci 15:3461–3472. https://doi.org/10.5194/hess-15-3461-2011

Neary DG, Ryan KC, DeBano LF (2005) Wildland fire in ecosystems: effects of fire on soils and water. Gen. Tech. Rep. RMRS-GTR-42-vol, 4, 250 p

Nnyamah JU, Black TA (1977) Rates and patterns of water uptake in a Douglas-fir forest. Soil Sci Soc Am J 41:972–979. https://doi.org/10.2136/sssaj1977.03615995004100050033x

Nobre AD, Oyama MD, Oliveira GS (2014) The future climate of Amazonia. Scientific Assessment Report. Articulación Regional Amazonica (ARA), São José dos Campos. Retrieved from: http://www.ccst.inpe.br/wpcontent/uploads/2014/11/The_Future_Climate_of_Amazonia_Report.pdf

Norby RJ, Zak DR (2011) Ecological lessons from free-air CO_2 enrichment (FACE) experiments. Annu Rev Ecol Evol Syst 42:181–203. https://doi.org/10.1146/annurev-ecolsys-102209-144647

Norby RJ, Warren JM, Iversen CM, Medlyn BE, McMurtrie RE (2010) CO_2 enhancement of forest productivity constrained by limited nitrogen availability. Proc Natl Acad Sci U S A 107:19368–19373. https://doi.org/10.1073/pnas.1006463107

NY EPA [New York Environmental Protection Agency] (2015) New York City drinking water supply and quality report. New York Environmental Protection Agency, New York

O'Halloran TL, Law BE, Goulden ML, Wang Z, Barr JG, Schaaf C et al (2012) Radiative forcing of natural forest disturbances. Glob Change Biol 18:555–565. https://doi.org/10.1111/j.1365-2486.2011.02577.x

Oliveras I, Gracia M, Moré G, Retana J (2009) Factors influencing the pattern of fire severities in a large wildfire under extreme meteorological conditions in the Mediterranean basin. Int J Wildland Fire 18:755–764. https://doi.org/10.1071/WF08070

Ong C, Black CR, Wilson J, Muthuri C, Bayala J, Jackson NA (2014) Agroforestry: hydrological impacts. In: Van Alfen NK (ed) Encyclopedia of agriculture and food systems, vol 1, 2nd edn. Academic Press, Amsterdam

Overpeck JT, Rind D, Goldberg R (1990) Climate-induced changes in forest disturbance and vegetation. Nature 343:51–53. https://doi.org/10.1038/343051a0

Ow LF, Ghosh S (2017) Urban tree growth and their dependency on infiltration rates in structural soil and structural cells. Urban For Urban Green 26:41–47. https://doi.org/10.1016/j.ufug.2017.06.005

Pavlidis G, Tsihrintzis VA (2018) Environmental benefits and control of pollution to surface water and groundwater by agroforestry systems: a review. Water Resour Manag 32:1–29. https://doi.org/10.1007/s11269-017-1805-4

Payn T, Carnus JM, Freer-Smith P, Kimberley M, Kollert W, Liu S et al (2015) Changes in planted forests and future global implications. For Ecol Manag 352:57–67. https://doi.org/10.1016/j.foreco.2015.06.021

Peña-Arancibia JL, Bruijnzeel LA, Mulligan M, van Dijk AI (2019) Forests as 'sponges' and 'pumps': assessing the impact of deforestation on dry-season flows across the tropics. J Hydrol. https://doi.org/10.1016/j.jhydrol.2019.04.064

Penn CA, Bearup LA, Maxwell RM, Clow DW (2016) Numerical experiments to explain multiscale hydrological responses to mountain pine beetle tree mortality in a headwater watershed. Water Resour Res 52:3143–3161. https://doi.org/10.1002/2015WR018300

Perry TD, Jones JA (2017) Summer streamflow deficits from regenerating Douglas-fir forest in the Pacific Northwest, USA. Ecohydrology 10:e1790. https://doi.org/10.1002/eco.1790

Piao S, Friedlingstein P, Ciais P, de Noblet-Ducoudré N, Labat D, Zaehle S (2007) Changes in climate and land use have a larger direct impact than rising CO_2 on global river runoff trends. Proc Natl Acad Sci U S A 104:15242–15247. https://doi.org/10.1073/pnas.0707213104

Redding T, Winkler R, Teti P, Spittlehouse D, Boon S, Rex J et al (2008) Mountain pine beetle and watershed hydrology. BC J Ecosyst Manag 9:33–50

Reed DE, Ewers BE, Pendall E (2014) Impact of mountain pine beetle induced mortality on forest carbon and water fluxes. Environ Res Lett 9:105004. https://doi.org/10.1088/1748-9326/9/10/105004

Riggan PJ, Lockwood RN, Jacks PM, Colver CG, Weirich F, DeBano LF et al (1994) Effects of fire severity on nitrate mobilization in watersheds subject to chronic atmospheric deposition. Environ Sci Technol 28:369–375. https://doi.org/10.1021/es00052a005

Rowland L, da Costa ACL, Galbraith DR, Oliveira RS, Binks OJ, Oliveira AAR et al (2015) Death from drought in tropical forests is triggered by hydraulics not carbon starvation. Nature 528:119–122. https://doi.org/10.1038/nature15539

Rustad LE, Campbell JL (2012) A novel ice storm manipulation experiment in a northern hardwood forest. Can J For Res 42:1810–1818. https://doi.org/10.1139/X2012-120

Sanders RA (1986) Urban vegetation impacts on the hydrology of Dayton, Ohio. Urban Ecol 9:361–376. https://doi.org/10.1016/0304-4009(86)90009-4

Sanusi R, Johnstone D, May P, Livesley SJ (2017) Microclimate benefits that different street tree species provide to sidewalk pedestrians relate to differences in Plant Area Index. Landsc Urban Plan 157:502–511. https://doi.org/10.1016/j.landurbplan.2016.08.010

Schelhaas MJ, Nabuurs GJ, Schuck A (2003) Natural disturbances in the European forests in the 19th and 20th centuries. Glob Change Biol 9:1620–1633. https://doi.org/10.1046/j.1365-2486.2003.00684.x

Scherrer D, Massy S, Meier S, Vittoz P, Guisan A (2017) Assessing and predicting shifts in mountain forest composition across 25 years of climate change. Divers Distrib 23:517–528. https://doi.org/10.1111/ddi.12548

Scholz M, Grabowiecki P (2007) Review of permeable pavement systems. Build Environ 42:3830–3836. https://doi.org/10.1016/j.buildenv.2006.11.016

Schwendenmann L, Pendall E, Sanchez-Bragado R, Kunert N, Hölscher D (2015) Tree water uptake in a tropical plantation varying in tree diversity: interspecific differences, seasonal shifts and complementarity. Ecohydrology 8:1–12. https://doi.org/10.1002/eco.1479

Scott DF, Prinsloo FW (2008) Longer-term effects of pine and eucalypt plantations on streamflow. Water Resour Res 44:W00A08. https://doi.org/10.1029/2007WR006781

Scott DF, Smith RE (1997) Preliminary empirical models to predict reductions in total and low-flows resulting from afforestation. Water SA 23:135–140

Scott DF, Bruijnzeel LA, Vertessy RA, Calder IR, Burley J, Evans J et al (2004) Impacts of forest plantations on streamflow. In: Encyclopedia of forest sciences, pp 367–377. https://doi.org/10.1016/B0-12-145160-7/00272-6

Seitz J, Escobedo F (2011) Urban forests in Florida: trees control stormwater runoff and improve water quality. City 393(6) University of Florida, IFAS Extension, Document No FOR184. http://edis.ifas.ufl.edu/
Shuster WD, Bonta J, Thurston H, Warnemuende E, Smith DR (2005) Impacts of impervious surface on watershed hydrology: a review. Urban Water J 2:263–275. https://doi.org/10.1080/15730620500386529
Silva LC, Anand M (2013) Probing for the influence of atmospheric CO_2 and climate change on forest ecosystems across biomes. Glob Ecol Biogeogr 22:83–92. https://doi.org/10.1111/j.1466-8238.2012.00783.x
Sinton DS, Jones JA, Ohmann JL, Swanson FJ (2000) Windthrow disturbance, forest composition, and structure in the Bull Run basin, Oregon. Ecology 81:2539–2556. https://doi.org/10.1890/0012-9658(2000)081[2539,WDFCAS]2.0.CO;2
Spera SA, Galford GL, Coe MT, Macedo MN, Mustard JF (2016) Land-use change affects water recycling in Brazil's last agricultural frontier. Glob Change Biol 22:3405–3413. https://doi.org/10.1111/gcb.13298
Spracklen DV, Garcia-Carreras L (2015) The impact of Amazonian deforestation on Amazon basin rainfall. Geophys Res Lett 42:9546–9552. https://doi.org/10.1002/2015GL066063
Staal A, Tuinenburg OA, Bosmans JH, Holmgren M, van Nes EH, Scheffer M et al (2018) Forest-rainfall cascades buffer against drought across the Amazon. Nat Clim Change 8:539–543. https://doi.org/10.1038/s41558-018-0177-y
Stephenson NL, Das AJ, Condit R, Russo SE, Baker PJ, Beckman NG et al (2014) Rate of tree carbon accumulation increases continuously with tree size. Nature 507:90–93. https://doi.org/10.1038/nature12914
Stocks BJ, Fosberg MA, Lynham TJ, Mearns L, Wotton BM, Yang Q et al (1998) Climate change and forest fire potential in Russian and Canadian boreal forests. Clim Change 38:1–13. https://doi.org/10.1023/A:1005306001055
Swank WT, Vose JM, Elliott KJ (2001) Long-term hydrologic and water quality responses following commercial clearcutting of mixed hardwoods on a southern Appalachian catchment. For Ecol Manag 143:163–178. https://doi.org/10.1016/S0378-1127(00)00515-6
United Nations (2016) The World's cities in 2016. United Nations, Department of Economic and Social Affairs, Population Division, New York
Vailshery LS, Jaganmohan M, Nagendra H (2013) Effect of street trees on microclimate and air pollution in a tropical city. Urban For Urban Green 12:408–415. https://doi.org/10.1016/j.ufug.2013.03.002
Van der Ent RJ, Savenije HH, Schaefli B, Steele-Dunne SC (2010) Origin and fate of atmospheric moisture over continents. Water Resour Res 46:W09525. https://doi.org/10.1029/2010WR009127
van Lierop P, Lindquist E, Sathyapala S, Franceschini G (2015) Global forest area disturbance from fire, insect pests, diseases and severe weather events. For Ecol Manag 352:78–88. https://doi.org/10.1016/j.foreco.2015.06.010
Van Mantgem PJ, Stephenson NL, Byrne JC, Daniels LD, Franklin JF, Fulé PZ et al (2009) Widespread increase of tree mortality rates in the western United States. Science 323:521–524. https://doi.org/10.1126/science.1165000
Van Nieuwstadt MG, Sheil D (2005) Drought, fire and tree survival in a Borneo rain forest, East Kalimantan, Indonesia. J Ecol 93(1):191–201. https://doi.org/10.1111/j.1365-2745.2004.00954.x
van Noordwijk M, Namirembe S, Catacutan D, Williamson D, Gebrekirstos A (2014) Pricing rainbow, green, blue and grey water: tree cover and geopolitics of climatic teleconnections. Curr Opin Environ Sust 6:41–47. https://doi.org/10.1016/j.cosust.2013.10.008
van Noordwijk M, Leimona B, Xing M, Tanika L, Namirembe S, Suprayogo D (2015) Water-focused landscape management. In: Minang PA, van Noordwijk M, Freeman OE, Mbow C, Leeuw JD, Catacutan D (eds) Climate-smart landscapes: multifunctionality in practice. World Agroforestry Centre (ICRAF), Nairobi

van Noordwijk M, Tanika L, Lusiana B (2017) Flood risk reduction and flow buffering as ecosystem services–part 1: theory on flow persistence, flashiness and base flow. Hydrol Earth Syst Sci 21(5):2321–2340

van Noordwijk M, Bargues-Tobella A, Muthuri CW, Gebrekirstos A, Maimbo M, Leimona B, Bayala J, Ma X, Lasco R, Xu J, Ong CK (2019) Agroforestry as part of nature-based water management. In: van Noordwijk M (ed) Sustainable development through trees on farms: agroforestry in its fifth decade. World agroforestry (ICRAF). Bogor, Indonesia, pp 261–287

Vanderhoof M, Williams CA, Shuai Y, Jarvis D, Kulakowski D, Masek J (2014) Albedo-induced radiative forcing from mountain pine beetle outbreaks in forests, south-central Rocky Mountains: magnitude, persistence, and relation to outbreak severity. Biogeosciences 11:563–575. https://doi.org/10.5194/bg-11-563-2014

Vieira DCS, Fernández C, Vega JA, Keizer JJ (2015) Does soil burn severity affect the post-fire runoff and interrill erosion response? a review based on meta-analysis of field rainfall simulation data. J Hydrol 523:452–464. https://doi.org/10.1016/j.jhydrol.2015.01.071

Vose JM, Miniat CF, Luce CH, Asbjornsen H, Caldwell PV, Campbell JL et al (2016) Ecohydrological implications of drought for forests in the United States. For Ecol Manag 380:335–345. https://doi.org/10.1016/j.foreco.2016.03.025

Waring RH (1982) Estimating forest growth and efficiency in relation to canopy leaf area. Adv Ecol Res 13:327–354. https://doi.org/10.1016/S0065-2504(08)60111-7

Wei X, Li Q, Zhang M, Giles-Hansen K, Liu W, Fan H, Wang Y et al (2018) Vegetation cover—another dominant factor in determining global water resources in forested regions. Glob Change Biol 24:786–795. https://doi.org/10.1111/gcb.13983

Weiler M, Scheffler C, Tautz A, Rosin K (2009) Development of a hydrologic process model for mountain pine beetle affected areas in British Columbia. Institut für Hydrologie, Universität Freiburg, Freiburg

Wemple BC, Jones JA (2003) Runoff production on forest roads in a steep, mountain catchment. Water Resour Res 39:1220. https://doi.org/10.1029/2002WR001744

Wemple BC, Swanson FJ, Jones JA (2001) Forest roads and geomorphic process interactions, Cascade Range, Oregon. Earth Surf Process Land 26:191–204. https://doi.org/10.1002/1096-9837(200102)26:2<191::AID-ESP175>3.0.CO;2-U

Weng W, Luedeke MK, Zemp DC, Lakes T, Kropp JP (2018) Aerial and surface rivers: downwind impacts on water availability from land use changes in Amazonia. Hydrol Earth Syst Sci 22:911–927. https://doi.org/10.5194/hess-22-911-2018

Westerling AL (2016) Increasing western US forest wildfire activity: sensitivity to changes in the timing of spring. Philos Trans Royal Soc B 371:20150178. https://doi.org/10.1098/rstb.2015.0178

White JC, Wulder MA, Hermosilla T, Coops NC, Hobart GW (2017) A nationwide annual characterization of 25 years of forest disturbance and recovery for Canada using Landsat time series. Remote Sens Environ 194:303–321. https://doi.org/10.1016/j.rse.2017.03.035

Whitehead D, Jarvis PG, Waring RH (1984) Stomatal conductance, transpiration, and resistance to water uptake in a *Pinus sylvestris* spacing experiment. Can J For Res 14:692–700. https://doi.org/10.1139/x84-124

Winkler R, Boon S, Zimonick B, Spittlehouse D (2014) Snow accumulation and ablation response to changes in forest structure and snow surface albedo after attack by mountain pine beetle. Hydrol Process 28:197–209. https://doi.org/10.1002/hyp.9574

Wramneby A, Smith B, Samuelsson P (2010) Hot spots of vegetation-climate feedbacks under future greenhouse forcing in Europe. J Geophys Res- Atmos 115:D21. https://doi.org/10.1029/2010JD014307

Wright JS, Fu R, Worden JR, Chakraborty S, Clinton NE, Risi C et al (2017) Rainforest-initiated wet season onset over the southern Amazon. Proc Natl Acad Sci USA 114:8481–8486. https://doi.org/10.1073/pnas.1621516114

Wu J, Liu W, Chen C (2017) How do plants share water sources in a rubber-tea agroforestry system during the pronounced dry season? Agric Ecosyst Environ 236:69–77. https://doi.org/10.1016/j.agee.2016.11.017

Zhang M, Wei X (2014) Alteration of flow regimes caused by large-scale forest disturbance: a case study from a large watershed in the interior of British Columbia, Canada. Ecohydrology 7:544–556. https://doi.org/10.1002/eco.1374

Zhang L, Dawes WR, Walker GR (1999) Predicting the effect of vegetation changes on catchment average water balance. CSIRO Technical Report 99/12, CRC for Catchment Hydrology. CSIRO, Canberra

Zhang L, Dawes WR, Walker GR (2001) Response of mean annual evapotranspiration to vegetation changes at catchment scale. Water Resour Res 37:701–708. https://doi.org/10.1029/2000WR900325

Zhang MF, Liu N, Harper R, Li Q, Liu K, Wei X et al (2017) A global review on hydrological responses to forest change across multiple spatial scales: importance of scale, climate, forest type and hydrological regime. J Hydrol 546:44–59. https://doi.org/10.1016/j.jhydrol.2016.12.040

Zhang J, Bruijnzeel LA, Quiñones CM, Tripoli R, Asio VB, van Meerveld HJ (2019a) Soil physical characteristics of a degraded tropical grassland and a 'reforest': implications for runoff generation. Geoderma 333:163–177. https://doi.org/10.1016/j.geoderma.2018.07.022

Zhang J, Bruijnzeel LA, Tripoli R, van Meerveld HJ (2019b) Water budget and run?off response of a tropical multispecies "reforest" and effects of typhoon disturbance. Ecohydrology 12(2): e2055. https://doi.org/10.1002/eco.2055

Zhou G, Liu S, Li Z, Zhang D, Tang X, Zhou C et al (2006) Old-growth forests can accumulate carbon in soils. Science 314:1417. https://doi.org/10.1126/science.1130168

Zomer RJ, Neufeldt H, Xu J, Ahrends A, Bossio D, Trabucco A et al (2016) Global tree cover and biomass carbon on agricultural land: The contribution of agroforestry to global and national carbon budgets. Sci Rep 6:29987. https://doi.org/10.1038/srep29987

Correction to: Cracking "Open" Technology in Ecohydrology

B. Turner, D. J. Hill, and K. Caton

Correction to:
Chapter 1 in: D. F. Levia et al. (eds.), *Forest-Water Interactions*, Ecological Studies 240, https://doi.org/10.1007/978-3-030-26086-6_1

The chapter "Cracking "Open" Technology in Ecohydrology" was previously published as non-open access. It has now been changed to open access under a CC BY-NC-ND 4.0 license and the copyright holder updated to "The Author(s)." The book has also been updated with these changes.

The updated online version of this chapter can be found at
https://doi.org/10.1007/978-3-030-26086-6_1

D. F. Levia et al. (eds.), *Forest-Water Interactions*, Ecological Studies 240,
https://doi.org/10.1007/978-3-030-26086-6_25

Index

D. F. Levia et al. (eds.), *Forest-Water Interactions*, Ecological Studies 240,
https://doi.org/10.1007/978-3-030-26086-6

FSC
www.fsc.org
MIX
Papier | Fördert gute Waldnutzung
FSC® C083411

Kolibri 360 GmbH
Ferdinand-Jühlke-Straße 7
99095 Erfurt, Deutschland
produktsicherheit@kolibri360.de